DNA–Protein Interactions

Methods in Molecular BIOLOGY™

John M. Walker, *Series Editor*

178. **Antibody Phage Display:** *Methods and Protocols*, edited by *Philippa M. O'Brien and Robert Aitken, 2001*
177. **Two-Hybrid Systems:** *Methods and Protocols*, edited by *Paul N. MacDonald, 2001*
176. **Steroid Receptor Methods:** *Protocols and Assays*, edited by *Benjamin A. Lieberman, 2001*
175. **Genomics Protocols**, edited by *Michael P. Starkey and Ramnath Elaswarapu, 2001*
174. **Epstein-Barr Virus Protocols**, edited by *Joanna B. Wilson and Gerhard H. W. May, 2001*
173. **Calcium-Binding Protein Protocols, Volume 2:** *Methods and Techniques*, edited by *Hans J. Vogel, 2001*
172. **Calcium-Binding Protein Protocols, Volume 1:** *Reviews and Case Histories*, edited by *Hans J. Vogel, 2001*
171. **Proteoglycan Protocols,** edited by *Renato V. Iozzo, 2001*
170. **DNA Arrays:** *Methods and Protocols*, edited by *Jang B. Rampal, 2001*
169. **Neurotrophin Protocols,** edited by *Robert A. Rush, 2001*
168. **Protein Structure, Stability, and Folding,** edited by *Kenneth P. Murphy, 2001*
167. **DNA Sequencing Protocols,** *Second Edition*, edited by *Colin A. Graham and Alison J. M. Hill, 2001*
166. **Immunotoxin Methods and Protocols**, edited by *Walter A. Hall, 2001*
165. **SV40 Protocols**, edited by *Leda Raptis, 2001*
164. **Kinesin Protocols,** edited by *Isabelle Vernos, 2001*
163. **Capillary Electrophoresis of Nucleic Acids, Volume 2:** *Practical Applications of Capillary Electrophoresis*, edited by *Keith R. Mitchelson and Jing Cheng, 2001*
162. **Capillary Electrophoresis of Nucleic Acids, Volume 1:** *Introduction to the Capillary Electrophoresis of Nucleic Acids*, edited by *Keith R. Mitchelson and Jing Cheng, 2001*
161. **Cytoskeleton Methods and Protocols**, edited by *Ray H. Gavin, 2001*
160. **Nuclease Methods and Protocols,** edited by *Catherine H. Schein, 2001*
159. **Amino Acid Analysis Protocols,** edited by *Catherine Cooper, Nicole Packer, and Keith Williams, 2001*
158. **Gene Knockoout Protocols,** edited by *Martin J. Tymms and Ismail Kola, 2001*
157. **Mycotoxin Protocols,** edited by *Mary W. Trucksess and Albert E. Pohland, 2001*
156. **Antigen Processing and Presentation Protocols,** edited by *Joyce C. Solheim, 2001*
155. **Adipose Tissue Protocols,** edited by *Gérard Ailhaud, 2000*
154. **Connexin Methods and Protocols,** edited by *Roberto Bruzzone and Christian Giaume, 2001*
153. **Neuropeptide Y Protocols**, edited by *Ambikaipakan Balasubramaniam, 2000*
152. **DNA Repair Protocols:** *Prokaryotic Systems*, edited by *Patrick Vaughan, 2000*
151. **Matrix Metalloproteinase Protocols,** edited by *Ian M. Clark, 2001*
150. **Complement Methods and Protocols,** edited by *B. Paul Morgan, 2000*
149. **The ELISA Guidebook**, edited by *John R. Crowther, 2000*
148. **DNA–Protein Interactions:** *Principles and Protocols* **(2nd ed.)**, edited by *Tom Moss, 2001*
147. **Affinity Chromatography:** *Methods and Protocols*, edited by *Pascal Bailon, George K. Ehrlich, Wen-Jian Fung, and Wolfgang Berthold, 2000*
146. **Mass Spectrometry of Proteins and Peptides,** edited by *John R. Chapman, 2000*
145. **Bacterial Toxins:** ***Methods and Protocols***, edited by *Otto Holst, 2000*
144. **Calpain Methods and Protocols**, edited by *John S. Elce, 2000*
143. **Protein Structure Prediction:** *Methods and Protocols*, edited by *David Webster, 2000*
142. **Transforming Growth Factor-Beta Protocols**, edited by *Philip H. Howe, 2000*
141. **Plant Hormone Protocols**, edited by *Gregory A. Tucker and Jeremy A. Roberts, 2000*
140. **Chaperonin Protocols**, edited by *Christine Schneider, 2000*
139. **Extracellular Matrix Protocols**, edited by *Charles Streuli and Michael Grant, 2000*
138. **Chemokine Protocols**, edited by *Amanda E. I. Proudfoot, Timothy N. C. Wells, and Christine Power, 2000*
137. **Developmental Biology Protocols, Volume III**, edited by *Rocky S. Tuan and Cecilia W. Lo, 2000*
136. **Developmental Biology Protocols, Volume II,** edited by *Rocky S. Tuan and Cecilia W. Lo, 2000*
135. **Developmental Biology Protocols, Volume I,** edited by *Rocky S. Tuan and Cecilia W. Lo, 2000*
134. **T Cell Protocols:** *Development and Activation*, edited by *Kelly P. Kearse, 2000*
133. **Gene Targeting Protocols,** edited by *Eric B. Kmiec, 2000*
132. **Bioinformatics Methods and Protocols,** edited by *Stephen Misener and Stephen A. Krawetz, 2000*
131. **Flavoprotein Protocols,** edited by *S. K. Chapman and G. A. Reid, 1999*
130. **Transcription Factor Protocols,** edited by *Martin J. Tymms, 2000*
129. **Integrin Protocols,** edited by *Anthony Howlett, 1999*
128. **NMDA Protocols,** edited by *Min Li, 1999*
127. **Molecular Methods in Developmental Biology:** Xenopus *and Zebrafish*, edited by *Matthew Guille, 1999*
126. **Adrenergic Receptor Protocols,** edited by *Curtis A. Machida, 2000*
125. **Glycoprotein Methods and Protocols:** *The Mucins,* edited by *Anthony P. Corfield, 2000*
124. **Protein Kinase Protocols,** edited by *Alastair D. Reith, 2001*
123. ***In Situ* Hybridization Protocols (2nd ed.),** edited by *Ian A. Darby, 2000*
122. **Confocal Microscopy Methods and Protocols,** edited by *Stephen W. Paddock, 1999*
121. **Natural Killer Cell Protocols:** *Cellular and Molecular Methods*, edited by *Kerry S. Campbell and Marco Colonna, 2000*
120. **Eicosanoid Protocols,** edited by *Elias A. Lianos, 1999*
119. **Chromatin Protocols,** edited by *Peter B. Becker, 1999*
118. **RNA–Protein Interaction Protocols,** edited by *Susan R. Haynes, 1999*
117. **Electron Microscopy Methods and Protocols,** edited by *M. A. Nasser Hajibagheri, 1999*
116. **Protein Lipidation Protocols,** edited by *Michael H. Gelb, 1999*
115. **Immunocytochemical Methods and Protocols (2nd ed.),** edited by *Lorette C. Javois, 1999*

Methods in Molecular BIOLOGY™

DNA–Protein Interactions

Principles and Protocols

Second Edition

Edited by

Tom Moss

*Centre de Recherche en Cancérologie de l'Université Laval,
Centre Hopital Universitaire de Québec et Départment
de biologie médicale, Université Laval,
Québec, QC, Canada*

Humana Press **Totowa, New Jersey**

Softcover reprint of the hardcover 2nd edition 2001
999 Riverview Drive, Suite 208
Totowa, New Jersey 07512

Cover design by Patricia F. Cleary

Cover Figure: A structural model for the RNA polymerase II open complex as determined by site-specific protein-DNA UV photo-cross-linking. Promoter DNA is wrappedaround RNA polymerase II (POL II), allowing contacts by the Xeroderma Pigmentosum Group B (XPB) helicase of transcription factor TFIIH to the template strand of the melted DNA duplex immediately upstream of the transcription initiation site. Transcription factors TBP, TFIIB, TFIIE and TFIIF, which are part of the complex, are not shown. For additional details, see Douziech et al. (2000) Mol. Cell. Biol. 20: 8168-8177.

Cover image kindly provided by Dr. Benoit Coulombe, Univerity of Sherbrooke, Quebec, Canada; Imaging: MOLECULAR IMAGE, University ofSherbrooke, Quebec, Canada.

Production Editor: Jason Runnion

10 9 8 7 6 5 4 3 2 1

Library of Congress Cataloging in Publication Data

DNA-protein interactions : principles and protocols / edited by Tom Moss.--2nd ed.
p. cm.--(Methods in molecular biology ; v. 148)
Includes bibliographical references and index.
ISBN 978-1-61737-125-7
1. DNA-protein interactions. I. Moss, Tom. II. Series.

QP624.75.P74 D57 2001
572.8'6--dc21 00-054100
CIP

Preface

DNA–protein interactions are fundamental to the existence of life forms, providing the key to the genetic plan as well as mechanisms for its maintenance and evolution. The study of these interactions is therefore fundamental to our understanding of growth, development, differentiation, evolution, and disease. The manipulation of DNA–protein interactions is also becoming increasingly important to the biotechnology industry, permitting among other things the reprogramming of gene expression. The success of the first edition of *DNA–Protein Interactions; Principles and Protocols* was the result of Dr. G. Geoff Kneale's efforts in bringing together a broad range of relevant techniques. In producing the second edition of this book, I have tried to further increase this diversity while presenting the reader with alternative approaches to obtaining the same information.

A major barrier to the study of interactions between biological macromolecules has always been detection and hence the need to obtain sufficient material. The development of molecular cloning and subsequently of protein overexpression systems has essentially breached this barrier. However, in the case of DNA–protein interactions, the problem of quantity and hence of detection is often offset by the high degree of selectivity and stability of DNA–protein interactions. DNA–protein binding reactions will often go to near completion at very low component concentrations even within crude protein extracts. Thus, although many techniques described in this volume were initially developed to study interactions between highly purified components, these same techniques are often just as applicable to the identification of novel DNA–protein interactions within systems as undefined as a whole cell extract. In general, these techniques use a DNA rather than a protein detection system because the former is more sensitive. Radiolabeled DNA fragments are easily produced by a range of techniques commonly available to molecular biologists.

DNA–protein complexes may be studied at three distinct levels—at the level of the DNA, of the protein, and of the complex. At the level of the DNA, the DNA binding site may be delimited and exact base sequence requirements defined. The DNA conformation can be studied and the exact bases contacted by the protein identified. At the protein level, the protein species binding a given DNA sequence can be identified. The amino acids contacting DNA and the protein surface facing the DNA may be defined and the amino acids essential to the recognition process can be identified. Furthermore, the protein's tertiary structure and its conformational changes on complex formation can be studied. Finally, global parameters of a DNA–protein complex such as stoichiometry, the kinetics of its formation and dissociation, its stability, and the energy of interaction can be measured.

Filter binding, electrophoretic mobility shift assay (EMSA/gel shift), DNaseI footprinting, and Southwestern blotting have been the most commonly used techniques to identify potentially interesting DNA target sites and to define the proteins that bind them. For example, gel shift or footprinting of a cloned gene regulation sequence by proteins in a crude cell extract may define binding activities for a given DNA sequence that correlates with gene expression or silencing. These techniques can be used as an assay during subsequent isolation of the protein(s) responsible. Interference assays, SELEX, and more refined footprinting techniques, such as hydroxy radical footprinting and DNA bending assays, can then be used to study the DNA component of the DNA–protein complex, whereas the protein binding surface can be probed by amino acid side chain modification, DNA–protein crosslinking, and of course by the production of protein mutants. Genetic approaches have also opened the way to engineer proteins recognizing chosen DNA targets.

DNA–protein crosslinking has in recent years become a very important approach to investigate the relative positions of proteins in multicomponent protein–DNA complexes such as the transcription initiation complex. Here, crosslinkable groups are incorporated at specific DNA sequences and these are used to map out the "positions" of different protein components along the DNA. Extension of this technique can also allow the mapping of the crosslink within the protein sequence. Similar data can be obtained by incorporating crosslinking groups at known sites within the protein and then identifying the nucleotides targeted.

Once the basic parameters of a DNA–protein interaction have been defined, it is inevitable that a deeper understanding of the driving forces behind the DNA–protein interaction and the biological consequences of its formation will require physical and physicochemical approaches. These can be either static or dynamic measurements, but most techniques have been developed to deal with steady-state situations. Equilibrium constants can be obtained by surface plasmon resonance, by spectroscopic assays that differentiate complexed and uncomplexed components, and, for more stable products, by footprinting and gel shift. Spectroscopy can also give specific answers about the conformation of proteins and any conformational changes they undergo on interacting with DNA as well as providing a rapid quantitative measure of complex formation. Microcalorimetry gives a global estimation of the forces stabilizing a given complex. Static pictures of protein–DNA interactions can be obtained by several techniques. At atomic resolution, X-ray crystallography, and nuclear magnetic resonance (NMR) studies require large amounts of highly homogeneous material. Lower resolution images can be obtained by electron and, more recently, by atomic force microscopies. Large multiprotein complexes are generally beyond the scope of NMR or even of X-ray crystallography. These are therefore more often studied using the electron microscope, either in a direct imaging mode or via the analysis of data obtained from 2D pseudocrystalline arrays.

Dynamic measurements of complex formation or dissociation can be obtained by biochemical techniques when the DNA–protein complexes have half-lives of several minutes to several hours. For footprinting and crosslinking, a general rule is that the complexes should be stable for a time well in excess of the proposed period of the enzymatic or chemical reaction. For gel shift, the complex half-life should at least approach that of the time of gel migration, although the cage effect may tend to stabilize the complex within the gel matrix, extending the applicability of this technique. More rapid assembly kinetics, multistep assembly processes, and short-lived DNA–protein complexes require much more rapid techniques such as UV laser-induced crosslinking, surface plasmon resonance, and spectroscopic assays. UV-laser induced DNA–protein crosslinking is a promising development because it potentially permits the kinetics of complex assembly to be followed both in vitro and in vivo.

When I decided to edit a second edition of the present volume, I was of course aware of the limitations of many of the more commonly used techniques. But as I read the various chapters I realized that each technique was at least as much limited by the conditions necessary for the probing reaction itself as by the type of information the probe could deliver. This is perhaps most evident for in vivo applications, which require agents that can easily enter cells, e.g., DMS and potassium permanganate are able to penetrate cells while DNaseI and DEPC are either too large or insufficiently water soluble to enter cells unaided. (Appendix II presents a summary of the activities and applications of the various DNA modification and cleavage reagents described in this book.) Gel shift assays are limited by the finite range of useable electrophoresis conditions. Because buffers must have low conductance, the KCl or NaCl solutions typically used for DNA–protein binding reactions are generally inappropriate. (Appendix I contains a list of the different gel shift conditions described in various chapters of this book.) Thus, it is often as important to choose a technique appropriate to the conditions under which one wishes to observe the DNA–protein interaction as it is to choose the appropriate probing activity.

The present volume attempts to bring together a broad range of techniques used to study DNA–protein interactions. Such a volume can never be complete nor definitive, but I hope this book will provide a useful source of technical advice for molecular biologists. Its preparation required the cooperation of many people. In particular I would like to thank all the authors for their very significant efforts. Thanks are also due to John Walker for his encouragement and to the previous editor Geoff Kneale and to Craig Adams of Humana Press for their help. I also thank Margrit and Peter Wittwer for providing space in the Pfarrhaus of the Predigerkirche, Zürich, where much of the chapter editing was done, and Bernadette for her patience, understanding, corrections, and advice.

Tom Moss

Contents

Contributors

SANKAR ADHYA • *Laboratory of Molecular Biology, National Institutes of Health, NCI, Bethesda, MD*
MARTIN ANGERS • *Division de Pathologie, Department de Biologie Médicale, Université Laval, et Unité de Recherche en Génétique Humaine et Moléculaire, Centre de Recherche, Pavilion Saint-Francois d'Assise, Québec, Canada*
BLAINE BARTHOLOMEW • *Department of Biochemistry and Molecular Biology, School of Medicine, Southern Illinois University, Carbondale, IL*
MIGUEL BEATO • *Insitute für Molekularbiologie und Tumorforshung, Philipps-Universität Marburg, Marburg, Germany*
RENE BEIGANG • *Fachbereich Physik, Universität Kaiserlautern, Germany*
ALAIN BÉLIVEAU • *Laboratory of Molecular Endocrinologie, Centre Hopitalier Universitaire de Québec, Université Laval, Québec, Canada*
NICOLAS BISCHLER • *Faculté de Médicine, IGBMC, Illkirch, France*
FREDERIC BOCCARD • *Centre de Génétique Moléculaire, CNRS, Yvette, France*
PHILIPPE BOUVET • *Laboratoire de Pharmacologie et de Biologie Structurale, CNRS, Toulouse, France*
RAYMOND S. BROWN • *Laboratory of Molecular Medicine, Howard Hughes Medical Institute, Children's Hospital, Boston, MA*
MALCOLM BUCKLE • *Unité Physicochimie des Macromolécules Biologiques, Institut Pasteur, Paris, France*
MICHAEL J. BUMBULIS • *Department of Molecular Biology and Microbiology, School of Medicine, Case Western Reserve University, Cleveland, and the Department of Biology, Baldwin-Wallace College, Berea, OH*
STEPHEN BUSBY • *School of Biochemistry, University of Birmingham, Birmingham, UK*
MARK L. CARPENTER • *University of Oxford, Oxford, UK*
DAVID R. CHAFIN • *Department of Biochemistry, University of Rochester, Rochester, NY*
YEN CHOO • *Laboratory of Molecular Biology, Medical Research Council, Cambridge, UK*
RACHEL M. CONLIN • *Laboratory of Molecular Medicine, Howard Hughes Medical Institute, Children's Hospital, Boston, MA*

BERNARD A. CONNOLLY • *Department of Biochemistry and Genetics, Medical School, University of Newcastle upon Tyne, Newcastle upon Tyne, UK*
BENOÎT COULOMBE • *Départment de Biologie, Centre de Recherche sur les Méchanismes d'Expression Génétique, Université de Sherbrooke, Sherbrooke, Québec, Canada*
LUDMILLA DENISSOVA • *Max Planck Institute of Biochemistry, Martinsried, Germany*
STEFAN I. DIMITROV • *Faculté de Médecine, Institut Albert Bonniot, Université Joseph Fourier Grenoble I, La Tronche, France*
QIANPING DONG • *Waksman Institute and Department of Chemistry, Howard Hughes Medical Institute, Rutgers University, Piscataway, NJ*
KLAUS DOSE • *Institut für Biochemie, Johannes Gutenberg-Universität, Mainz, Germany*
RÉGEN DROUIN • *Department de Biologie Médicale, Université Laval, et Unité de Recherche en Génétique Humaine et Moléculaire, Centre de Recherche, Pavilion Saint-Francois d'Assise, Québec, Canada*
RICHARD H. EBRIGHT • *Waksman Institute and Department of Chemistry, Howard Hughes Medical Institute, Rutgers University, Piscataway, NJ*
LISA E. ENGLER • *Department of Biological Sciences, University of Pittsburgh, Pittsburgh, PA*
ALEXANDER A. GALL • *Seattle Genetics, Bothell, WA*
JOHANNES GEISELMANN • *Plasticité et Expression des Génomes Microbiens, Université Joseph Fourier, Grenoble, France*
CARLA W. GRAY • *Department of Molecular and Cell Biology, University of Texas at Dallas, Richardson, TX*
SYLVAIN GUÉRIN • *Laboratory of Molecular Endocrinologie, Centre Hopitalier Universitaire de Québec, Université Laval, Québec, Canada*
JEFFREY J. HAYES • *Department of Biochemistry and Biophysics, University of Rochester Medical Center, Rochester, NY*
HERMANN HEUMANN • *Max Planck Institute of Biochemistry, Martinsried, Germany*
MARK ISALAN • *Laboratory of Molecular Biology, Medical Research Council, Cambridge, UK*
ILIAN JELESAROV • *Biochemisches Institut der Universität Zurich, Zurich, Switzerland*
LINDA JEN-JACOBSON • *Department of Biological Sciences, University of Pittsburgh, Pittsburgh, PA*
BRENDA F. KAHL • *Department of Biochemistry and Molecular Biology, Colorado State University, Fort Collins, CO*
YOUNGGYU KIM • *Waksman Institute and Department of Chemistry, Howard Hughes Medical Institute, Rutgers University, Piscataway, NJ*

G. GEOFF KNEALE • *Biophysics Laboratories, School of Biological Sciences, University of Portsmouth, Portsmouth, UK*
ANNIE KOLB • *Institut Pasteur, Paris, France*
MICHAEL R. KURPIEWSKI • *Department of Biological Sciences, University of Pittsburgh, Pittsburgh, PA*
SIMON LABBÉ • *Department of Biological Chemistry, The University of Michigan Medical School, Ann Arbor, MI*
MARC-ANDRÉ LANIEL • *Laboratory of Molecular Endocrinologie, Centre Hopitalier Universitaire de Québec, Université Laval, Québec, Canada*
OLIVIER LAROCHELLE • *Centre de Recherche en Cancérologie, Université Laval, CHUQ/L'Hotel-Dieu de Québec, Québec, Canada*
LUC LEBEAU • *Faculté de Médecine, Illkirch, France*
BENOIT LEBLANC • *NIDDK, NIH, Bethesda, MD*
HSIAO-HUI LIU • *Department of Biochemistry and Genetics, Medical School, University of Newcastle upon Tyne, Newcastle upon Tyne, UK*
YURI L. LYUBCHENKO • *Departments of Biology and Microbiology, Arizona State University, Tempe, AZ*
IAN W. MANFIELD • *Department of Genetics, University of Leeds, Leeds, UK*
JAMES A. MCCLELLAN • *Biophysics Laboratories, School of Biological Sciences, University of Portsmouth, Portsmouth, UK*
RAINER MEFFERT • *Ministerium für Umwelt und Forsten des Landes Rheinland-Pfalz, Mainz, Germany*
WILLI METZGER • *Ministerium für Umwelt und Forsten des Landes Rheinland-Pfalz, Mainz, Germany*
TOM MOSS • *Centre de Recherche en Cancérologie et départment de Biologie Médicale de l'Université Laval, Centre Hopital Universitaire de Québec, Québec, Canada*
NIKOLAI NARYSHKIN • *Waksman Institute and Department of Chemistry, Howard Hughes Medical Institute, Rutgers University, Piscataway, NJ*
PETER E. NIELSEN • *Department of Medical Biochemistry and Genetics, Laboratory of Biochemistry, The Panum Institute, Copenhagen, Denmark*
ANTHONY W. OLIVER • *Biophysics Laboratories, School of Biological Sciences, University of Portsmouth, Portsmouth, UK*
STÉPHANE OUELLET • *Department de Biologie Médicale, Université Laval, et Unité de Recherche en Génétique Humaine et Moléculaire, Centre de Recherche, Pavilion Saint-Francois d'Assise, Québec, Canada*
ATHANASIOS G. PAPAVASSILIOU • *Department of Biochemistry, School of Medicine, University of Patras, Patras, Greece*
DAMIAN PARRY • *Department of Biochemistry and Genetics, Medical School, University of Newcastle upon Tyne, Newcastle upon Tyne, UK*

MARVIN PAULE • *Department of Biochemistry and Molecular Biology, Colorado State University, Fort Collins, CO*
JIM PERSINGER • *Department of Biochemistry and Molecular Biology, School of Medicine, Southern Illinois University, Carbondale, IL*
SIMON E. PLYTE • *Pharmacia and Upjohn, Milano, Italy*
GUY G. POIRIER • *Unité Santé et Environment, CHUQ, Pavillon CHUL, Québec, Canada*
GABRIELE RATHGEBER • *Merck KGaA, Darmstadt, Germany*
CHRISTOPHER M. READ • *Biophysics Laboratories, School of Biological Sciences, University of Portsmouth, Portsmouth, UK*
VIRGIL RHODIUS • *School of Biochemistry, University of Birmingham, Birmingham, UK*
FRANÇOIS ROBERT • *Whitehead Institute for Biomedical Research, Cambridge, MA*
CHRISTOPH RUSSMANN • *Fachbereich Physik, Universität Kaiserlautern, Germany*
NIGEL SAVERY • *School of Biochemistry, University of Birmingham, Birmingham, UK*
HANS-JOCHEN SCHAFER • *Institute für Biochemie, Johannes Gutenberg-Universität, Mainz, Germany*
PETER SCHICKOR • *Max Planck Institute of Biochemistry, Martinsried, Germany*
DEBORAH B. SCHULMAN • *Department of Molecular Biology and Microbiology, School of Medicine, Case Western Reserve University, Cleveland, OH*
PATRICK SCHULTZ • *Faculté de Médecine, Illkirch, France*
CARL SÉGUIN • *Centre de Recherche en Cancérologie, Université Laval, CHUQ/L´Hotel-Dieu de Québec, Québec, Canada*
DAVID R. SETZER • *Department of Molecular Biology and Microbiology, School of Medicine, Case Western Reserve University, Cleveland, OH*
PETER E. SHAW • *Department of Biochemistry, School of Biomedical Sciences, University of Nottingham, Queen's Medical Center, Nottingham, UK*
LUDA S. SHLYAKHTENKO • *Departments of Plant Biology and Microbiology, Arizona State University, Tempe, AZ*
MARTHA N. SIMON • *Brookhaven National Laboratory, Biology Department, Upton, NY*
A. FRANCIS STEWART • *European Molecular Biology Laboratory, Heidelberg, Germany*
GALE STEWART • *Centre de Recherche en Cancérologie, Université Laval, CHUQ/L´Hotel-Dieu de Québec, Québec, Canada*
PETER G. STOCKLEY • *Department of Genetics, University of Leeds, Leeds, UK*
IAN TAYLOR • *Laboratory of Molecular Biophysics, University of Oxford, Oxford, UK*

Jean-Philippe Therrien • *Division de Pathologie, Department de Biologie Médicale, Université Laval, et Unité de Recherche en Génétique Humaine et Moléculaire, Centre de Recherche, Pavilion Saint-Francois d'Assise, Québec, Canada*

Andrew A. Travers • *Lab Molecular Biology, Medical Research Council, Cambridge, UK*

Joseph S. Wall • *Brookhaven National Laboratory, Biology Department, Upton, NY*

Michelle Webb • *Department of Chemistry, University of Sheffield, Sheffield UK*

Evgeny Zaychikov • *Max Planck Institute of Biochemistry, Martinried, Germany*

Christian Zwieb • *Department of Molecular Biology, The University of Texas Health Center at Tyler, Tyler, TX*

1

Filter-Binding Assays

Peter G. Stockley

1. Introduction

Membrane filtration has a long history in the analysis of protein–nucleic acid complex formation, having first been used to examine RNA–protein interactions *(1)*, before being introduced to DNA–protein interaction studies by Jones and Berg in 1966 *(2)*. The principle of the technique is straightforward. Under a wide range of buffer conditions, nucleic acids pass freely through membrane filters, whereas proteins and their bound ligands are retained. Thus, if a particular protein binds to a specific DNA sequence, passage through the filter will result in retention of a fraction of the protein–DNA complex by virtue of the protein component of the complex. The amount of DNA retained can be determined by using radioactively labeled DNA to form the complex and then determining the amount of radioactivity retained on the filter by scintillation counting. The technique can be used to analyze both binding equilibria and kinetic behavior, and if the DNA samples retained on the filter and in the filtrate are recovered for further processing, the details of the specific binding site can be probed by interference techniques.

The technique has a number of advantages over footprinting and gel retardation assays, although there are also some relative disadvantages, especially where multiple proteins are binding to the same DNA molecule. However, filter binding is extremely rapid, reproducible, and, in principle, can be used to extract accurate equilibrium and rate constants *(3–5)*. We have used the technique to examine the interaction between the *E. coli* methionine repressor, MetJ, and various operator sites cloned into restriction fragments *(6,7, see also* Chapter 15). Results from these studies will be used to illustrate the basic technique.

From: *Methods in Molecular Biology, vol. 148: DNA–Protein Interactions: Principles and Protocols, 2nd ed.*
Edited by: T. Moss © Humana Press Inc., Totowa, NJ

Before discussing the experimental protocols it is important to understand some fundamental properties of the filter-binding assay. The molecular basis of the discrimination between nucleic acids and proteins during filtration is still not fully understood. Care should therefore be taken to characterize the assay with the system under study. Nucleic acid–protein complex retention occurs with differing efficiencies, depending on the lifetime of the complex, the size of the protein component, the buffer conditions, and the extent of washing of the filter. Experiments with the *lac* repressor system have shown that prior filtration of protein followed by passage of DNA containing operator sites does not result in significant retention of the nucleic acid, presumably because filter-bound protein is inactive for further operator binding. The DNA retained on filters is therefore a direct reflection of the amount of complex present when filtration began. Furthermore, incubation of the *lac* repressor with large amounts of DNA that does not contain an operator site followed by filtration also does not lead to significant retention. Because the *lac* repressor (and, indeed, essentially all DNA-binding proteins) binds nonsequence-specifically to DNA, forming short-lived complexes, it is clear that these are not readily retained. The experiments with the *lac* repressor ***(3–5)*** can therefore be used as a guide when designing experimental protocols. The repressor is a large protein (being a tetramer of 38-kDa subunits) but the basic features seem to apply even to short peptides with molecular weights <2 kDa ***(8)***.

In any particular system, the percentage of the DNA–protein complex in solution retained by the filter should ideally be constant throughout the binding curve, and this is known as the retention efficiency. Experimental values range from 30 to >95%. An example of the sort of results obtained with the MetJ repressor is shown in **Fig. 1**.

2. Materials

2.1. Preparation of Radioactively End-Labeled DNA

1. Plasmid DNA carrying the binding site for a DNA-binding protein on a convenient restriction fragment (usually <200 bp).
2. Restriction enzymes and the appropriate buffers as recommended by the suppliers.
3. Phenol: redistilled phenol equilibrated with 100 m*M* Tris–HCl, pH 8.0.
4. Chloroform.
5. Solutions for ethanol precipitation of DNA: 4 *M* NaCl and ethanol (absolute and 70% v/v).
6. Calf intestinal alkaline phosphatase (CIAP).
7. CIAP reaction buffer (10X): 0.5 *M* Tris–HCl, pH 9.0, 0.01 *M* $MgCl_2$, 0.001 *M* $ZnCl_2$.
8. TE buffer: 10 m*M* Tris–HCl, pH 8.0, 1 m*M* ethylenediaminetetraacetic acid (EDTA).
9. 20% w/v Sodium dodecyl sulfate (SDS).
10. 0.25 *M* EDTA, pH 8.0.

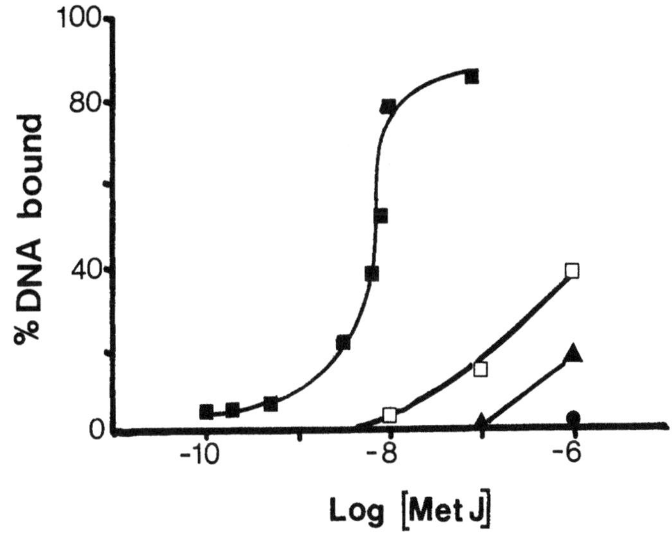

11. T4 polynucleotide kinase (T4 PNK).
12. T4 PNK reaction buffer, 10X: 0.5 *M* Tris–HCl, pH 7.6, 0.1 *M* $MgCl_2$, 0.05 *M* dithiothreitol.
13. Radioisotope: γ-[^{32}P]-ATP.
14. 30% w/v acrylamide stock (29:1 acrylamide: *N,N'*-methylene-bisacrylamide).
15. Polyacrylamide gel elution buffer: 0.3 *M* sodium acetate, 0.2% w/v sodium dodecyl sulfate (SDS), 2 m*M* EDTA.
16. Polymerization catalysts: ammonium persulfate (10% w/v) and *N,N,N',N'*-tetramethylethylene diamine (TEMED).
17. X-ray film, autoradiography cassette and film developer.
18. Plastic wrap and scalpel.

2.2. Filter-Binding Assays

1. Nitrocellulose filters: We use HAWP (00024) filters from Millipore (Bedford, MA), but suitable filters are available from a number of other manufacturers, such as Schleicher and Schuell (Dassel, Germany). Filters tend to be relatively expensive. Some manufacturers produce sheets of membrane that can be cut to size and are thus less expensive.
2. Filter-binding buffer (FB): 100 m*M* KCl, 0.2 m*M* EDTA, 10 m*M* Tris–HCl, pH 7.6.
3. Binding buffer (BB): This is FB containing 50 μg/mL bovine serum albumin (BSA, protease and nuclease free; *see* **Note 1**).

4. Filtration manifold and vacuum pump: We use a Millipore 1225 Sampling Manifold (cat. no. XX27 025 50), which has 12 sample ports.
5. Liquid scintillation counter, vials, and scintillation fluid.
6. Siliconized glass test tubes.
7. TBE buffer: 89 m*M* Tris, 89 m*M* boric acid, 10 m*M* EDTA, pH 8.3.
8. Formamide/dyes loading buffer: 80% v/v formamide, 0.5X TBE, 0.1% w/v xylene cyanol, 0.1% w/v bromophenol blue.
9. Sequencing gel electrophoresis solutions and materials: 19% w/v acrylamide, 1% w/v bis-acrylamide, 50% w/v urea in TBE.
10. Acetic acid (10% v/v).

3. Methods

3.1. Preparation of End-Labeled DNA

1. Digest the plasmid (20 µg in 200 µL) with the restriction enzymes used to release a suitably sized DNA fragment (usually <200 bp). Extract the digest with an equal volume of buffered phenol and add 2.5 volumes of ethanol to the aqueous layer in order to precipitate the digested DNA. If preparing samples for interference assays) only one restriction digest should be carried out at this stage, *see* Chapter 15.
2. Add 50 µL 1X CIAP reaction buffer to the ethanol-precipitated DNA pellet (<50 µg). Add 1 U CIAP and incubate at 37°C for 30 min followed by the addition of a further aliquot of enzyme and incubate for a further 30 min. Terminate the reaction by adding SDS and EDTA to 0.1% (w/v) and 20 m*M*, respectively in a final volume of 200 µL and incubate at 65°C for 15 min. Extract the digest with buffered phenol, then with 1:1 phenol:chloroform, and, finally, ethanol precipitate the DNA from the aqueous phase as above.
3. Redissolve the DNA pellet in 18 µL 1X T_4 PNK buffer. Add 20 µCi γ-[^{32}P]-ATP and 10 U T4 PNK and incubate at 37°C for 30 min. Terminate the reaction by phenol extraction and ethanol precipitation (samples for interference assays should be digested with the second restriction enzyme at this poin)t. Redissolve the pellet in nondenaturing gel loading buffer and electrophorese on a nondenaturing polyacrylamide gel.
4. After electrophoresis, separate the gel plates, taking care to keep the gel on the larger plate. Cover the gel with plastic wrap and in the darkroom, under the safelight, tape a piece of X-ray film to the gel covering the sample lanes. With a syringe needle, puncture both the film and the gel with a series of registration holes. Alternatively, register the film and the gel using fluorescent marker strips. Locate the required DNA fragments by autoradiography of the wet gel at room temperature for several min (approx 10 min). Excise slices of the gel containing the bands of interest using the autoradiograph as a guide. Elute the DNA into elution buffer overnight (at least) at 37°C. Ethanol precipitate the eluted DNA by adding 2.5 vol of ethanol, wash the pellet thoroughly with 70% v/v ethanol, dry briefly under vacuum, and rehydrate in a small volume (approx 50 µL) of TE. Determine the radioactivity of the sample by liquid scintillation counting of a 1-µL aliquot.

3.2. Filter-Binding Assays

3.2.1. Determination of the Equilibrium Constant

1. Presoak the filters in FB at 4°C for several hours before use. Care must be taken to ensure that the filters are completely "wetted." This is best observed by laying the dry filters carefully onto the surface of the FB using blunt-ended tweezers and observing buffer uptake.
2. Prepare a stock solution of radioactively labeled DNA fragment in an appropriate buffer, such as FB. We adjust conditions so that each sample to be filtered contains roughly 20 kcpm. Under these conditions, the DNA concentration is <1 p*M*. Aliquot the stock DNA solution into plastic Eppendorf tubes. It is best at this stage if relatively large volumes are transferred in order to minimize errors caused by pipeting. We use 180 μL/sample. If the DNA-binding protein being studied requires a cofactor, it is best to add it to the stock solution at saturating levels so that its concentration is identical for every sample.
3. Prepare a serially diluted range of protein concentrations diluting into BB. A convenient range of concentrations for the initial assay is between 10^{-11} and 10^{-5} *M* protein.
4. Immediately add 20 μL of each protein concentration carefully to the sides of the appropriately labeled tubes of stock DNA solution. When the additions are complete centrifuge briefly (5 s) to mix the samples and then incubate at a temperature at which complex formation can be observed (37°C for MetJ). For each binding curve it is important to prepare two control samples. The first contains no protein in the 20 μL of BB and is filtered to determine the level of background retention. The second is identical to the first but is added to a presoaked filter in a scintillation vial (*see* **step 6**) and is dried directly without filtering. This gives a value for 100% input DNA.
5. After an appropriate time interval to allow equilibrium to be established, recentrifuge the tubes to return the liquid to the bottom of the tube and begin filtering.
6. The presoaked filters are placed carefully on the filtration manifold ensuring that excess FB is removed and that the filter is not damaged. Cracks and holes are easily produced by rough handling. The sample aliquot (200 μL) is then immediately applied to the filter, where it should be held stably by surface tension. Apply the vacuum. If further washes are used they should be applied as soon as the sample volume has passed through the filter. Remove the filter to a scintillation vial and continue until all the samples have been filtered.
7. The scintillation vials should be transferred to an oven at 60°C to dry the filters thoroughly (approx 20 min) before being allowed to cool to room temperature and 3–5 mL of scintillation fluid added. The radioactivity associated with each filter can now be determined by counting on an open channel (*see* **Note 2**).
8. Correct the value for each sample by subtracting the counts in the background sample (no protein). Calculate the percentage of input DNA retained at each protein concentration using the value for 100% input from the unaltered sample. Plot a graph of percentage retained vs the logarithm of the protein concentration (e.g.,

Fig. 1). The binding curve should increase from left to right until a plateau is reached. This is rarely at 100% of input DNA. The plateau value can be assumed to represent the retention efficiency, and for quantitative measurements, the data points can be adjusted accordingly. There is not enough space here to describe in detail the form of the binding curve or how best to interpret the data. (For an authoritative yet accessible account, *see* **ref. 9**). For our purposes, the protein concentration at 50% saturation can be thought of as the equilibrium dissociation constant.

9. Once an initial binding curve has been obtained, the experiment should be repeated with sample points concentrated in the appropriate region (i.e., the region where the percentage retained is changing most rapidly).

   Control experiments with DNAs that do not contain specific binding sites should also be carried out to prove that binding is sequence-specific. Highly diluted protein solutions appear to lose activity in our hands, possibly because of nonspecific absorption to the sides of tubes, among other things. We therefore produce freshly diluted samples daily. BB can be stored at 4°C for several days without deleterious effect. Ideally, binding curves should be reproducible. However, there is some variability between batches of filters and we therefore recommend not switching lot numbers during the course of one set of experiments.

3.2.2. Kinetic Measurements

Kinetic analysis of the binding reaction depends on prior determination of the equilibrium binding curve, especially the concentration of DNA-binding protein required to saturate the input DNA. This information allows a reaction mixture containing a limiting amount of protein to be set up (e.g., at a protein concentration that produces 75% retention). Both association and dissociation kinetics can be studied. The major technical problem arises because of the relatively rapid sampling rates that are required. However, it is almost always possible to adjust solution conditions such that sampling at 10 s intervals is all that is needed. Dissociation measurements often need to be made over periods of up to 1 h, whereas association reactions are usually complete within several min.

3.2.2.1. Dissociation

Repeat **steps 1** and **2** of **Subheading 3.2.1.** but do not aliquot the stock DNA solution. Add to this sample the appropriate concentration (i.e., which produces approx 75% retention) of stock protein and allow to equilibrate. Add a 20-fold excess of unlabeled DNA fragment containing the binding site and begin sampling (approx 200 µL aliquots) by filtration. Plots of radioactivity retained vs time can then be analyzed to derive kinetic constants. In the simplest case of a bimolecular reaction, a plot of the natural logarithm of the radioactivity retained at time *t* divided by the initial radioactivity vs time yields the first-order dissociation constant from the slope. An important control experiment is to repeat the experiment with DNA that does not contain a specific binding site to show that dissociation is sequence-specific.

A variation of this experiment can be used in which the concentration of protein in the reaction mix is diluted across the range where most complex formation occurs. In this case it is necessary to prepare the initial complex in a small volume (approx 50 µL) and then dilute 100 times with BB, followed by filtering 500 µL aliquots.

3.2.2.2. Association

Set up a stock DNA concentration in a single test tube in (**Subheading 3.2.1.** (**steps 1** and **2**). Incubate both this DNA and the appropriate solution of protein at the temperature at which complexes form. Add the appropriate volume of protein (e.g., 200 µL) to the DNA stock solution (1800 µL) and immediately begin sampling (10 × 200 µL aliquots).

3.2.3. Interference Measurements

Experiments of this type can be used to gain information about the site on the DNA fragment being recognized by the protein. The principle is identical to that used in gel retardation interference assays but has the advantage that the DNA does not have to be eluted from gels after fractionation.

1. Modify the purified DNA fragment radiolabeled (approx 100 kcpm) at a single site with the desired reagent; for example, hydroxyl radicals, which result in the elimination of individual nucleotide groups ***(10)*** (*see* Chapter 16), dimethyl sulfate (DMS) ***(11)*** (*see* Chapter 14), which modifies principally guanines, or ethyl nitrosourea, ENU (*see* Chapter 15), which ethylates the nonesterified phosphate oxygens. The extent of modification should be adjusted so that any one fragment has no more than one such modification. This can be assessed separately in test reactions and monitored on DNA sequencing gels.
2. Ethanol precipitate the modified DNA, wash twice with 70% (v/v) ethanol and then dry briefly under vacuum. Resuspend in 200 µL FB. Remove 20 µL as a control sample. Add 20 µL of the appropriate protein concentration to form a complex and allow equilibrium to be reached. Filter as usual but with a siliconized glass test tube positioned to collect the filtrate. (The Millipore manifold has an insert for just this purpose.) Do not over dry the filter.
3. Place the filter in an Eppendorf tube containing 250 µL FB, 250 µL H_2O, and 0.5% (w/v) SDS. Transfer the filtrate into a similar tube and then add SDS and H_2O to make the final volume and concentration the same as the filter-retained sample. Add an equal volume of buffer-saturated phenol to each tube, vortex, and centrifuge to separate the phases. Remove the aqueous top layers, re-extract with chloroform:phenol (1:1), and then ethanol precipitate. A Geiger counter can be used to monitor efficient elution of radioactivity from the filter, which can be re-extracted if necessary.

4. Recover all three DNA samples (control, filter-retained, and filtrate) after ethanol precipitation and, if necessary, process the modification to completion (e.g., piperidine for DMS modification, NaOH for ENU, and so on). Ethanol precipitate the DNA, dry briefly under vacuum, and then redissolve the pellets in 4 µL formamide/dyes denaturing loading buffer. At this stage, it is often advisable to quantitate the radioactivity in each sample by liquid scintillation counting of 1-µL aliquots. Samples for sequencing gels should be adjusted to contain roughly equal numbers of counts in all three samples.
5. Heat the samples to 90°C for 2 min and load onto a 12% w/v polyacrylamide sequencing gel alongside Maxam–Gilbert sequencing reaction markers *(12)*. Electrophorese at a voltage that will warm the plates to around 50°C. After electrophoresis, fix the gel in 1 L 10% v/v acetic acid for 15 min. Transfer the gel to 3MM paper and dry under vacuum at 80°C for 60 min. Autoradiograph the gel at –70°C with an intensifying screen.
6. Compare lanes corresponding to bound, free, and control DNAs for differences in intensity of bands at each position (*see* **Note 3**). A dark band in the "free fraction" (and a corresponding reduction in the intensity of the band in the "bound fraction") indicates a site where prior modification interferes with complex formation. This is interpreted as meaning that this residue is contacted by the protein or a portion of the protein comes close to the DNA at this point. (*See* Chapters 14–16 for more extensive discussions of interference experiments.)

3.3. Results and Discussion

Figure 1 shows a typical filter-binding curve for the *E. coli* methionine repressor binding to its idealized operator site of $(dAGACGTCT)_2$ cloned into a pUC-polylinker. In the presence of saturating amounts of cofactor (SAM), a sigmoidal binding curve is produced, whereas in the absence of SAM, the binding curve does not saturate in the protein concentration range tested. Similar binding curves have been analyzed to produce Scatchard and Hill plots *(9)* in order to examine the cooperativity with respect to protein concentration *(6)*. However, such multiple binding events should also be studied by gel retardation assays which yield data about the individual complex species (*see* Chapter 2).

Table 1 shows the results obtained for binding to a series of variant operator sites and illustrates the apparent sensitivity of the technique. However, in order to make such comparisons, it is essential to determine the binding curves accurately and with the same batches of protein and filters to minimize minor differences between experiments. **Table 1** lists the affinities of a number of variant *met* operator sites cloned into pUC-polylinkers as determined by filter binding in the presence of saturating levels of corepressor, SAM. The repressor binds cooperatively to tandem arrays of an 8-bp met-box sequence (dAGACGTCT) with a stoichiometry of one repressor dimer per met-box. The variant operators were designed to examine both the tandem binding and the alignment of repressor dimers with the two distinct dyads in tandem met-box sequences *(6)*.

Table 1
The Relative K_ds of a Number of Variant Met Operator Sites

Variant	Operator Sequence	Relative K_d
00045	AGACGTCT	>12.2
00048	AGACGTCTAGACGTCT	1.0
00184	AGACGTCatGACGTCT	2.8
00299	gtctAGACGTCTagac	4.9

Note: The K_d is the concentration of protein that produces 50% binding of input DNA. Values are averages of several experiments and are quoted relative to the two met-box perfect consensus sequences (00048) which, under the conditions used, had an apparent K_d of 82 ± 5 n*M* MetJ monomer. Sequences in capitals represent matches to the consensus *met* box.

Operator variants are as follows:

1. 00045-A single 8-bp met-box or half-site. The binding curve does not saturate because singly bound repressor dimers dissociate very rapidly.
2. 00048-Two perfect met-boxes representing the idealized minimum operator sequence. Repressors bind cooperatively with high affinity.
3. 00184-Two met-boxes with the central T–A step reversed. The crystal structure of the repressor–operator complex shows that the central T–A step is not contacted directly by the repressors, rather the pyrimidine–purine step promotes a sequence-dependent DNA distortion that results in protein–DNA contacts elsewhere in the operator fragment. The A–T step has less tendency to undergo this conformational change and this is reflected in its lowered affinity.
4. 00299-A "shifted" two met-box operator used to define the alignment between the repressor twofold axis and the operator dyads. The low affinity of this construct compared to 00048 confirms that each repressor dimer is centered on the middle of a met-box.

3.3.1. In Vitro Selection Experiments

In recent years, in vitro selection experiments have been used to identify the range of preferred DNA target sequences by DNA-binding proteins ***(13,14)***, (*see also* Chapter 42). The technique depends on the separation of protein-bound DNA sequences from unbound, nonspecific, or low-affinity sites. Filter binding is an attractive option for this selection step because of the speed with which filtration and recovery of the bound fraction can be achieved. However, it is important to be aware that some minor DNA variants can be retained specifically by the filters, thus biasing the selected sequences. One way to avoid this and still retain the advantages of filter binding is to alternate rounds of filter binding with separation by gel retardation (*see* Chapter 2). A detailed discussion of the factors involved in such experiments is beyond the scope of this chapter and the reader is referred to more detailed descriptions (e.g., **ref. *15***).

4. Notes

1. None of the radioactivity is retained by the filter. This again can be caused by a variety of factors. Check that the preparation of DNA-binding protein is still functional (if other assays are available) or that the protein is still intact by SDS-polyacrylamide gel electrophoresis. Check the activity/concentration of the cofactor if required. A common problem we have encountered arises because of the different grades of commercially available BSA. It is always advisable to use a preparation that explicitly claims to be nuclease and protease free.
2. All of the radioactivity is retained by the filter. This is a typical problem when first characterizing a system by filter binding and can have many causes. Check that the filters being used "wet" completely in FB and do not dry significantly before filtration. Make sure that the DNA remains soluble in the buffer being used by simple centrifugation in a bench-top centrifuge. If the background remains high, add dimethyl sulfoxide to the filtering solutions. Classically, 5% (v/v) is used but higher concentrations (approx 20% v/v) have been reported with little, if any, effect on the binding reaction. We have experienced excessive retention when attempting to analyze the effects of divalent metal ions on complex formation, and, in general, it is best to avoid such buffer conditions.
3. Poor recoveries from the filter-retained samples in interference assays, or other problems in processing such samples further, can often be alleviated by addition of 20 µg of tRNA as a carrier during the SDS/phenol extraction step.

Acknowledgment

I am grateful to Yi-Yuan He for providing the data shown in **Table 1** and **Fig. 1**.

References

1. Nirenberg, M. and Leder, P. (1964) RNA codewords and protein synthesis. The effect of trinucleotides upon the binding of sRNA to ribosomes. *Science* **145,** 1399–1407.
2. Jones, O. W. and Berg, P. (1966) Studies on the binding of RNA polymerase to polynucleotides. *J. Mol. Biol.* **22,** 199–209.
3. Riggs, A. D., Bourgeois, S., Newby, R. F., and Cohn, M. (1968) DNA binding of the *lac* repressor. *J. Mol. Biol.* **34,** 365–368.
4. Riggs, A. D., Suzuki, H., and Bourgeois, S. (1970) *lac* repressor-operator interaction. I. Equilibrium studies. *J. Mol. Biol.* **48,** 67–83.
5. Riggs, A. D., Bourgeois, S., and Cohn, M. (1970) The *lac* repressor-operator interaction. III. Kinetic studies. *J. Mol. Biol.* **53,** 401–417.
6. Phillips, S. E. V., Manfield, I., Parsons, I., Davidson, B. E., Rafferty, J. B., Somers, W. S., et al. (1989) Cooperative tandem binding of Met repressor from *Escherichia coli. Nature* **341,** 711–715.

7. Old, I. G., Phillips, S. E. V., Stockley, P. G., and Saint-Girons, I. (1991) Regulation of methionine biosynthesis in the enterobacteriaceae. *Prog. Biophys. Mol. Biol.* **56,**145–185.
8. Ryan, P. C., Lu, M., and Draper, D. E. (1991) Recognition of the highly conserved GTPase center of 23S ribosomal RNA by ribosomal protein L11 and the antibiotic thiostrepton. *J. Mol. Biol.* **221,** 1257–1268.
9. Wyman, J. and Gill, S. J. (1990) In *Binding and Linkage: Functional Chemistry of Biological Macromolecules,* chap. 2, University Science Books, Mill Valley, CA.
10. Siebenlist, U. and Gilbert, W. (1980) Contacts between *Escherichia coli RNA* polymerase and an early promoter of phage T7. *Proc. Natl. Acad. Sci. USA* **77,** 122–126.
11. Hayes, J. J. and Tullius, T. D. (1989) The missing nucleoside experiment: a new technique to study recognition of DNA by protein. *Biochemistry* **28,** 9521–9527.
12. Maxam, A. M. and Gilbert, W. K. (1980) Sequencing end-labelled DNA with base-specific chemical cleavages. *Methods Enzymol.* **65,** 499–560.
13. Tuerk, C. and Gold, L. (1990) Systematic evolution of ligands by exponential enrichment: RNA ligands to bacterophage T4 DNA polymerase. *Science* **249,** 505–510.
14. Ellington, A. D. and Szostak, J. W. (1990) *In vitro* selection of RNA molecules that bind specific ligands. *Nature* **346,** 818–822.
15. Conrad, R. C., Giver, L., Tian, Y. and Ellington, A. D. (1996) *In vitro* selection of nucleic acid aptamers that bind proteins. *Methods Enzymol.* **267,** 336–367.

2

Electrophoretic Mobility Shift Assays for the Analysis of DNA-Protein Interactions

Marc-André Laniel, Alain Béliveau, and Sylvain L. Guérin

1. Introduction

Several nuclear mechanisms involve specific DNA–protein interactions. The electrophoretic mobility shift assay (EMSA, also known as the gel mobility shift or gel retardation assay), first described almost two decades ago *(1,2)*, provides a simple, efficient and widely used method to study such interactions. Its ease of use, its versatility, and especially its high sensitivity (10^{-18} mol of DNA *[2]*) make it a powerful method that has been successfully used in a variety of situations not only in gene regulation analyzes but also in studies of DNA replication, repair, and recombination. Although very useful for qualitative purposes, EMSA has the added advantage of being suitable for quantitative and kinetic analyzes *(3)*. Furthermore, because of its very high sensitivity, EMSA makes it possible to resolve complexes of different protein or DNA stoichiometry *(4)* and even to detect conformational changes.

1.1. Principle of the Method

Electrophoretic mobility shift assay (EMSA) is based on the simple rationale that proteins of differing size, molecular weight, and charge will have different electrophoretic mobilities in a nondenaturing gel matrix. In the case of a DNA–protein complex, the presence of a given DNA-binding protein will cause the DNA to migrate in a characteristic manner, usually more slowly than the free DNA, and will thus cause a change or shift in the DNA mobility visible upon detection.

While the kinetic analysis of EMSA, which has been extensively covered elsewhere (**ref.** *5* and references therein), is not the prime focus of this chapter, it will be useful to understand the basic theory underlying such analyzes. A

From: *Methods in Molecular Biology, vol. 148: DNA–Protein Interactions: Principles and Protocols, 2nd ed.*
Edited by: T. Moss © Humana Press Inc., Totowa, NJ

univalent protein, P, binding to a unique site on a DNA molecule, D, will yield a complex, PD, in equilibrium with the free components:

$$\mathbf{P + D} \underset{k_d}{\overset{k_a}{\rightleftharpoons}} \mathbf{PD}$$

where k_a is the rate of association and k_d is the rate of dissociation. In the case of a strong interaction between protein and DNA, with $k_a > k_d$, two distinct bands are observed, corresponding to the complex PD and to the free DNA. However, because of the dissociation that inevitably occurs during electrophoresis and because the DNA released from a complex during electrophoresis can never catch up with the free DNA, a faint smear may be seen between the two major bands. In contrast, a weak DNA–protein interaction, with $k_a < k_d$, should produce a fainter band corresponding to the complex PD and a more intense smear. However, even weak DNA–protein interactions may lead to distinct bands in EMSA because of their stabilization in the gel matrix as a result of the cage effect ***(6)*** and/or of molecular sequestration ***(7)***. In both cases, the dissociation of the complex is slower within the gel than it is in free solution, but in the cage effect, the gel matrix prevents dissociated components P and D from freely diffusing and thus favors a reformation of the complex PD, whereas in molecular sequestration, the gel matrix isolates complex PD from competing molecules that could promote its dissociation.

As for a single DNA molecule bearing multiple binding sites for a given protein, there will generally be as many mobility shifts formed as there are binding sites. For example, in the case of two independent binding sites on the DNA fragment (D):

$$\begin{array}{ccc} \mathbf{2P + D} & \underset{k_{d1}}{\overset{k_a}{\rightleftharpoons}} & \mathbf{PD1 + P} \\ k_a \upharpoonleft\downharpoonright k_{d2} & & k_a \upharpoonleft\downharpoonright k_{d3} \\ \mathbf{PD2 + P} & \underset{k_{d4}}{\overset{k_a}{\rightleftharpoons}} & \mathbf{P2D} \end{array}$$

this would result in three DNA containing bands: the free DNA (D), the complex with both sites occupied by protein (P2D), and the complexes with only one occupied site (PD1 and PD2, which will generally migrate together).

The kinetics of more complex situations, such as dimerizing protein complexes and multiple DNA–protein interactions, are beyond the scope of this chapter, but some interesting and insightful articles have been recently published ***(4,8)*** in which these questions are expressly addressed.

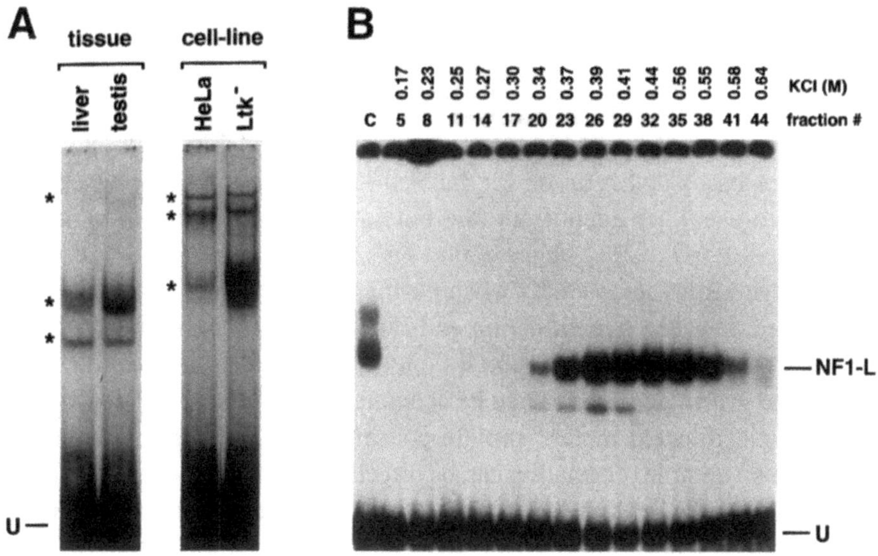

Fig. 1. **Panel A.** Autoradiograph of an EMSA performed using crude nuclear proteins from both whole rat tissues and established tissue-culture cells. A 33-bp synthetic oligonucleotide bearing the DNA sequence from the initiator site of the rat PARP gene promoter was 5' end-labeled and used as a probe in EMSA. It was incubated with crude nuclear proteins *(5* µg) obtained either from fresh rat tissues (liver and testis) or from established tissue-culture cells (HeLa and Ltk⁻). A number of nuclear proteins (indicated by asterisks) were found to bind the rPARP promoter with varying efficiencies and most were common to both the tissues and the cell lines selected. U: Unbound fraction of the labeled probe.

Panel B. Monitoring the enrichment of a nuclear protein by EMSA. Crude nuclear proteins *(*50 mg) of a rat liver extract were prepared and further purified on a heparin–Sepharose column. Nuclear proteins were eluted using a 0.1–1.0 *M* KCl gradient and fractions individually incubated with a 34-bp double-stranded synthetic oligonucleotide bearing the DNA sequence of the rat growth hormone promoter proximal silencer-1 element as the labeled probe. Both the concentration of KCl required to elute the proteins contained in each fraction, as well as the fraction number selected are indicated, along with the position of a major shifted DNA–protein complex corresponding to the rat liver form of the transcription factor NF1 (termed NF1-L). C: control lane in which the silencer-1 labeled probe was incubated with 5 µg crude nuclear proteins from rat liver; U: unbound fraction of the labeled probe.

1.2. Applications of the EMSA

Because EMSA often allows the detection of specific DNA-binding proteins in unpurified protein extracts (*see* **ref. 9** and **Fig. 1A**), the technique has been widely used to analyze crude cell or tissue extracts or partially purified extracts for the presence of protein factors implicated in transcription *(10–13)* and in DNA replication *(9,14)*, recombination *(15)*, and repair *(16)*. The use of unlabeled competitor DNA fragments further aids in identification of DNA-binding proteins (*see* **ref.** ***9***, ***15***, and ***17*** and **Fig. 2A**), and their purification can be easily monitored by EMSA (*see* **ref.** ***9***, and ***13*** and **Fig. 1B**). Moreover, mutation or bases delection on the labeled DNA probe is often an efficient approach to use when identifying the binding site of the protein of interest *(10,12)*.

EMSA yields invaluable data when purified or recombinant proteins are to be analyzed, because quantification and kinetic studies are rapidly achieved *(10,14)*. Parameters of a DNA–protein interaction, such as association, dissociation, and affinity constants, can be accurately measured *(2,3,7,10)*, and the effect of salt, divalent metals, protein concentration and the temperature of incubation on complex formation can be directly observed (*see* **ref.** ***15***, ***20***, and ***21*** and **Fig. 3A,B**). EMSA has also greatly contributed to the elaboration of models of complex assembly in the areas of transcription *(11)*, DNA replication *(14)* and DNA repair *(16)*.

Although EMSA is an informative and versatile method on its own, it becomes more powerful when used in combination with other techniques. Methylation *(23)* and other forms of binding interference studies (*see* Chapters 14 to 16), where a partially modified DNA probe is used, help to define the exact position of the DNA binding site of the protein *(10,24)*. Immunological methods using specific antibodies, as in supershift experiments (*see* **refs.** ***12*** and ***13*** and **Fig. 2B**), are also very helpful in identifying the identity of the protein component of given complexes. However, when analyzing large or multiprotein complexes, supershifts may not be suitable because the supershifted complexes may not be distinguished from the shifted ones or may not identify the different proteins involved. Immunoblotting of EMSA gels *(25)*, "Shift-Western blotting" *(26)* and immunodepletion EMSA *(27)* can be used to resolve such problems. In addition, determination of the molecular weight of the DNA-binding protein(s) identified by EMSA can be achieved by sodium dodecyl sulfate-polyacrylamide gel electrophoresis (SDS-PAGE), either directly *(28)* or following ultraviolet cross-linking of the DNA–protein complex *(29)*.

1.3. Overview of the Procedure

Several components are required for EMSA and may influence the outcome of the procedure.

1.3.1. Nuclear Extract

The choice of protein extract is governed by the objective of the study. Whole-cell or nuclear extracts are very useful in analyzing the regulatory

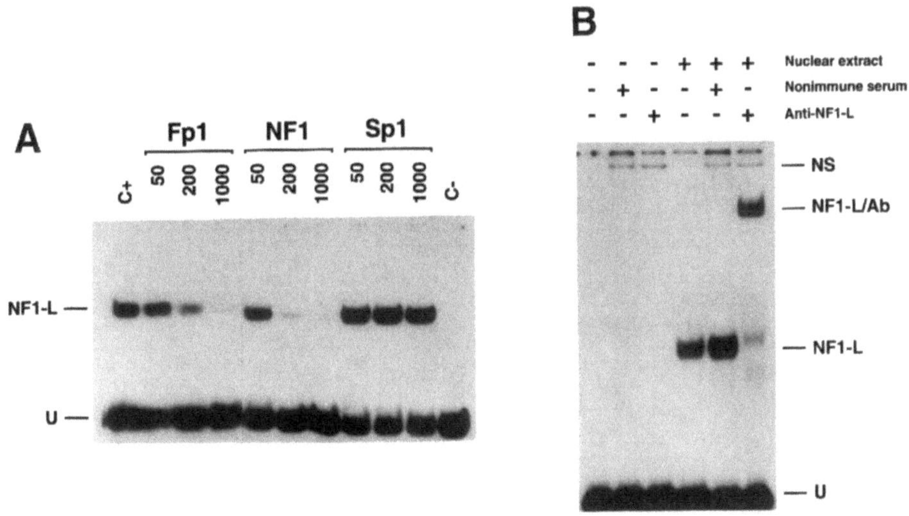

Fig. 2. **Panel A.** Competition in EMSA as a tool to evaluate the specific formation of DNA–protein complexes. A synthetic double-stranded oligonucleotide bearing the NF1 binding site from the Fp1 element of the human CRBP1 gene was 5' end labeled and incubated with 1 μg of a heparin–Sepharose-enriched preparation of rat liver NF1-L. Increasing concentrations (50-, 200-, and 1000-fold molar excess) of unlabeled, double-stranded oligonucleotides containing various DNA binding sites (Fp1, NF1, or Sp1) were added as competitors during the binding assays, and DNA/protein complex formation was analyzed on native 8% polyacrylamide gels. Control lanes containing the labeled probe alone (C–) or incubated with proteins in the absence of any competitor DNA (C+) have also been included. The position of the specifically retarded DNA/protein complex (NF1-L) and that of the free probe (U) is also shown. (Modified from **ref. *18***: reprinted with permission from *Mol. Endocrinol.*, Copyright [1994].)

Panel B. The identity of DNA-binding proteins as revealed by supershift analyses in EMSA. The rGH silencer-1 labeled probe used in **Fig. 1B** was incubated with (+) or without (–) 0.2 μg of a heparin–Sepharose-enriched preparation of NF1-L (*see* panel A) , in the presence of either nonimmune serum (1 μL) or a polyclonal antibody directed against rat liver NF1-L. Formation of DNA/protein complexes was then monitored by EMSA as in **Fig. 1B**. The position of the previously characterized NF1-L DNA/protein complex is shown (NF1-L) along with that of a supershifted complex (NF1-L/Ab) resulting from the specific interaction of the anti-NF1-L antibody with the NF1-L/silencer-1 complex. The position of a nonspecific complex (NS), resulting from the binding of an unknown serum protein to the labeled probe selected, is indicated, as well as the position of the remaining free probe (U). (Modified from **ref. *19***: reprinted with permission from *Eur. J. Biochem.*, Copyright [1994].)

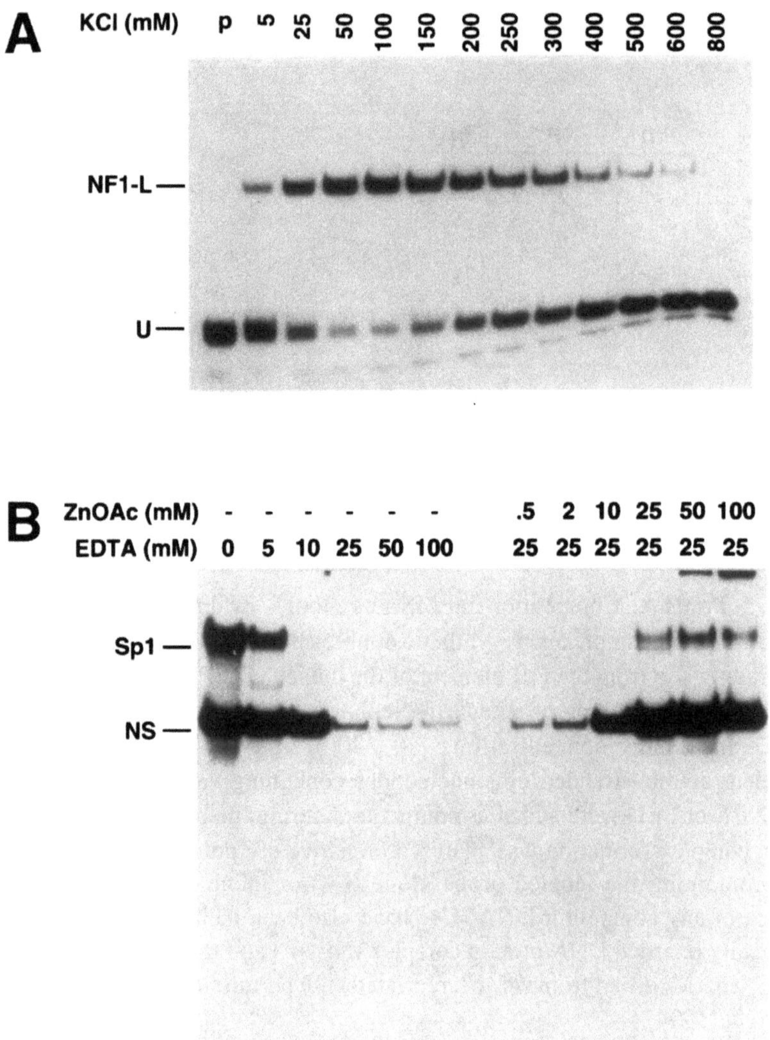

Fig. 3. **Panel A.** Salt-dependent formation of DNA–protein complexes in EMSA. A 5' end labeled 35-bp synthetic double-stranded oligonucleotide bearing the NF1-L binding site of the 5'-flanking sequence of the human CRBP1 gene (and designated Fp5) was incubated in the presence of 1 µg of a heparin–Sepharose-enriched preparation of NF1-L and increasing concentrations of KCl (5 to 800 m*M*) using binding conditions similar to those described in this chapter. Formation of the Fp5/NF1-L DNA–protein complex was then resolved by electrophoresis on a 4% native polyacrylamide gel. Very little free probe (U) is observed in the presence of either 50 or 100 m*M* KCl, providing evidence that optimal binding of NF1-L to its target site in Fp5 is obtained at these salt concentrations. (Modified from **ref. *20***; reprinted with permission from *Biotechniques*, Copyright [1992].)

elements of a DNA fragment such as a gene promoter. Partial protein purification allows further characterization of a DNA–protein interaction and can be achieved by column chromatography on DNA-cellulose or heparin–Sepharose, or by SDS-polyacrylamide gel fractionation and subsequent protein renaturation (*see* **ref. *30*** and **Note 1**). Purified or recombinant proteins give valuable information on protein interactions, competition, dimerization or cooperativity. Whatever protein extract used, its quality is a key factor in EMSA (*see* **Notes 2** and **3**).

1.3.2. DNA Probe

Cloned DNA fragments of 50–400 bp in length or synthetic oligonucleotides of 20–70 nucleotides work very well in EMSA (*see* **ref. *17*** and **Note 4**) and although double-stranded DNA is used most often, single-stranded DNA may also be effective ***(15)***. Although larger DNA fragments usually encompass more extensive regulatory sequences, oligonucleotides will generally contain fewer protein binding sites and thereby yield more specific information, the two approaches often complementing one another. The detection of DNA–protein complexes is usually achieved by labeling of DNA probe (*see* **Note 5**), and this is performed using a [^{32}P]-labeled deoxynucleotide. However, other, less hazardous methods are available (*see* **Note 5**), including labeling with ^{33}P ***(31)***, with digoxygenin ***(32)*** or with biotin ***(33)***.

1.3.3. Gel Matrix

Acrylamide gels (*see* **Note 6**) combine high resolving power with broad size-separation range and provide the most widely used matrix. Alternatively,

---

Panel B. DNA-binding properties of nuclear proteins revealed by EDTA chelation in EMSA. A double-stranded synthetic oligonucleotide bearing the sequence of the rat PARP US-1 binding site for the transcription activation factor Sp1 was 5' end-labeled and incubated with 10 µg crude nuclear proteins from HeLa cells in the presence of increasing concentrations of EDTA (0–100 m*M*) under binding conditions identical to those described in this chapter. Formation of DNA/protein complexes was evaluated by EMSA on a 8% polyacrylamide gel. As little as 10 m*M* EDTA proved to be sufficient to chelate zinc ions and to totally prevent binding of Sp1 to the US-1 element. Similarly, reaction mixtures containing the US-1 labeled probe incubated with 10 µg nuclear proteins from HeLa cells in the presence of 25 m*M* final concentration of EDTA were supplemented with increasing concentrations (0.5–100 m*M*) of zinc acetate (ZnOAc) to evaluate the binding recovery for both Sp1 and the nonspecific DNA–protein complex (NS). A substantial proportion of the DNA-binding capability of both the Sp1 and the NS proteins could be recovered upon further addition of 25 m*M* zinc acetate, providing evidence that both factors probably interact with DNA through the use of a Zn-finger-containing DNA binding domain, a fact that was already known for Sp1. (Modified from **ref. *22***: reprinted with permission from *Eur. J. Biochem.*, Copyright [1993].)

the use of less toxic, commercially available matrices has been reported *(34–36)*. Because of their larger pore size, agarose gels are sometimes used, either alone or in combination with acrylamide, to study larger DNA fragments or multiprotein complexes *(37)*. Gel concentration is also important in EMSA (*see* **Note 7**), however although lower concentration will generally allow the resolution of larger complexes, it may affect their stability *(7)*.

1.3.4. Buffer

Different low-ionic-strength buffers can be used in EMSA (*see* **ref. *36*** and **Note 8**), and can include cofactors such as Mg^{2+} or cAMP, which may be necessary for some DNA–protein interactions *(37)*.

1.3.5. Nonspecific Competitors

To ensure specificity of the DNA–protein interaction, a variety of nonspecific competitors may be used. This is particularly important when using crude protein extracts which contain nonspecific DNA-binding proteins. To avoid nonspecific binding activities interfering with the EMSA, an excess of a nonspecific DNA such as salmon sperm DNA, calf thymus DNA or synthetic DNAs such as poly(dI:dC) is used (*see* **refs. *37*** and ***38*** and **Notes 9** and **10**). The addition of nonionic or zwitterionic detergents *(39)* or nonspecific proteins (e.g., albumin *[40]*) may also increase specific DNA–protein interactions.

2. Materials

2.1. Probe Labeling

1. [γ^{32}P] ATP. Caution: ^{32}P emits high-energy beta radiation. Refer to the rules of your local control radioactivity agency for handling and proper disposal of radioactive materials and waste (*see* **Note 5**).
2. Approximately 25–50 ng of DNA from a 30-bp double-stranded oligonucleotide. For a typical 70-bp probe derived from a subcloned promoter fragment, estimate the amount of the plasmid DNA that is required to end up with about 100–200 ng of the DNA fragment of interest following its isolation from the polyacrylamide gel (*see* **Note 4**).
3. Calf intestinal alkaline phosphatase (CIAP) and 10X CIAP reaction buffer: 0.5 *M* Tris–HCl pH 9.0, 10 m*M* $MgCl_2$, 1 m*M* $ZnCl_2$, 10 m*M* spermidine.
4. T_4 polynucleotide kinase and 10X kinase buffer: 0.5 *M* Tris–HCl pH 7.5, 0.1 *M* $MgCl_2$, 40 m*M* DTT, 1 m*M* spermidine, 1 m*M* EDTA.

2.2. Probe Isolation

1. Standard electrophoresis apparatus for agarose gel.
2. Stock solution of 10X TBE: 0.89 *M* Tris, 0.89 *M* boric acid, and 20 m*M* EDTA.
3. 1% (w/v) agarose in 1X TBE supplemented with 0.5 µg/mL of ethidium bromide from a 10-mg/mL solution. **Caution**: Ethidium bromide is a powerful mutagenic agent (*see* **Note 11**).

4. Restriction enzyme(s) with corresponding buffer(s).
5. For DNA precipitation, a preparation of 1 mg/mL tRNA, a solution of 3 *M* NaOAc (pH 5.2), and a supply of dry ice.
6. Phenol/chloroform: Phenol saturated with 100 m*M* Tris–HCl pH 8.0.
7. 40% (w/v) 29:1 acrylamide–bisacrylamide: 29:1 (w/w) acrylamide and *N′*,*N′*-methylene bis-acrylamide. After complete dissolution of the components, the solution should be filtered using Whatman No. 1 paper and can be stored at room temperature. **Caution**: Acrylamide is a potent neurotoxic agent (*see* **Note 6**).
8. Dialysis tubing: molecular weight cutoff of 3500 and flat width of 18 mm.
9. Plastic wrap.
10. Autoradiography cassettes and film: Kodak XOmat AR.

2.3. Electrophoretic Mobility Shift Assay

1. Standard vertical electrophoresis apparatus for polyacrylamide gels, a gel length of 15 cm is adequate. (*See* **Note 12**.)
2. 40% (w/v) 39:1 acrylamide–bisacrylamide: 39:1 (w/w) acrylamide and *N′*,*N′*-methylene bis-acrylamide. **Caution**: Acrylamide is a potent neurotoxic agent (*see* **Note 6**).
3. 5X Tris–glycine: 250 m*M* Tris, 12,5 m*M* EDTA, and 2 *M* glycine. (*See* **Note 8**.)
4. Extract (crude or enriched) containing cell or tissue nuclear proteins. (*See* **Note 2**.)
5. 2X binding buffer: 20 m*M* HEPES pH 7.9, 20% glycerol, 0.2 m*M* EDTA, 1 m*M* tetrasodium pyrophosphate (*see* **Note 3**) and 0.5 m*M* PMSF.
6. 6X loading buffer: 0.25% bromophenol blue, 0.25% xylene cyanol, and 40% sucrose.
7. Whatman chromatographic paper (3MM) and plastic wrap.
8. Standard gel dryer.
9. Autoradiography cassettes and film: Kodak XOmat AR.

3. Methods

3.1. Probe Labeling

3.1.1. Labeling DNA Fragments Derived from a Subcloned Sequence

1. Select restriction enzymes that produce the shortest DNA fragment containing the sequence of interest. One of these restriction enzymes should produce a protruding 5′ end or blunt end to support labeling with T_4 polynucleotide kinase (*see* **Note 13**). Following the manufacturer's optimal enzymatic conditions, prepare a digestion mix with one of the restriction enzymes in 50 µL to linearize the vector. The initial amount of DNA should be calculated to end up with at least 100–200 ng of DNA after double-restriction enzyme digestion and further isolation of the DNA fragment from the polyacrylamide gel.
2. Before proceeding with dephosphorylation, make sure that digestion is complete by loading a sample (50–100 ng) on a 1% (w/v) agarose minigel. Once complete digestion of the plasmid DNA has been verified, add directly to the digestion reaction mix 1 U of CIP, 10 µL of 10X CIP buffer, and fill to 100 µL with H_2O. Incubate at 37°C for 90 min.

3. To totally eliminate and inactivate CIAP, transfer the reaction mix at 70°C for 10 min and perform a phenol/chloroform followed by a chloroform extraction. Precipitate DNA by adding a 1/10th volume of 3 *M* NaOAc, pH 5.2 and 2 volumes of cold 95% ethanol. Allow DNA to precipitate on dry ice for 30 min, then centrifuge for 15 min.
4. Resuspend DNA in 33 µL of H_2O, add 5 µL of 10X kinase buffer, 10 µL (100 µCi) of [γ^{32}P]ATP and 2 µL of T_4 polynucleotide kinase. Mix and incubate at 37°C for 2 h.
5. Following the labeling procedure, reprecipitate DNA and resuspend in 30 µL of H_2O. Keep a 2-µL sample and digest the remainder with the second restriction enzyme, following manufacturer's conditions.

3.1.2. Labeling Double-Stranded Synthetic Oligonucleotides

1. Mix equal amounts of the complementary strands, heat at 5°C over the specific melting temperature (T_M) of the sequence for 5 min, and let cool to room temperature (RT). When DNA reaches RT, place at 4°C for a few hours prior to use.
2. Use 25–50 ng of the double-stranded oligonucleotide preparation and perform DNA labeling with T_4 polynucleotide kinase as described in **step 4** of **Subheading 3.1.1.** but using 30 µCi of [γ^{32}P] ATP.

3.2. Probe Isolation

3.2.1. For a Typical 70-bp Probe Derived from a Subcloned Promoter Fragment

1. Rigorously clean and dry the polyacrylamide gel apparatus and its accessories prior to use. Gel plates should be cleaned using any good quality commercial soap and then rinsed with 95% ethanol. One plate can be treated with a coat of Sigmacote (chlorinated organopolysiloxane in heptane) to facilitate gel removal from the plates after running.
2. Prepare a 6% polyacrylamide gel ***(41)*** as follows; mix 2.5 mL of 10X TBE, 3.75 mL of 40% acrylamide (29:1) stock solution, and H_2O to 25 mL final volume. Add 180 µL of 10% ammonium persulfate and 30 µL of TEMED. Carefully stir and pour the acrylamide solution between the plates. Insert well-forming comb and allow the gel to set for 30 min., then mount the gel in the electrophoresis tank and fill the chamber with 1X TBE .
3. To the double-digested DNA, add 10 µL of 6X loading buffer and load into two separate wells. For the 2-µL control sample from the single digestion, add 2 µL of loading buffer and load in a free well. Migration should be stopped when bromophenol blue, which is used as a migration marker, reaches two-thirds of the gel length.
4. Carefully disassemble the apparatus and discard the running buffer as radioactive waste. Remove one plate and leave the gel on the remaining plate. Cover the gel with plastic wrap and, in a dark room, place a film over it. It is very important to mark the exact position of the gel on the film as a reference. This can be achieved by using [^{32}P]-labeled black ink. Expose the film for 3 min and develop.

5. If the digestion step with restriction enzymes is complete, two labeled bands resulting from the double digestion should appear on the autoradiogram (provided that each of the restriction enzymes selected initially cut the probe-bearing recombinant plasmid only once). Using a razor blade, cut out from the film the lower band corresponding to the selected probe. Replace the film on the gel (which is still covered with plastic wrap), aligning the reference marks carefully. Using the aperture in the film as guide, remove the probe-containing gel fragment using a scalpel blade.
6. Place the acrylamide fragment in a dialysis tubing closed at one end and add 1 mL of 1X TBE. Remove any remaining air bubbles, close the other end, and place the dialysis tubing in a standard horizontal electrophoresis tank filled with 1X TBE. Run at 100 V for 15 min.
7. Through the action of electrophoretic migration, the labeled probe will pass from the acrylamide fragment to the TBE solution contained in the dialysis tubing. DNA will concentrate as a thin line along the dialysis tubing (on the cathode side) and must be removed by gently rubbing the tubing with a solid object. Using a Pasteur pipet, transfer the labeled probe-containing TBE from the dialysis tubing into three separate microcentrifuge tubes (about 300 µL each). Other procedures may also be selected for extracting the labeled probe from the polyacrylamide gel *(42)*.
8. Repeat **steps 6** and **7** to make sure that all of the probe has been eluted from the acrylamide fragment. At the end of the second elution, recover the TBE again into three other microcentrifuge tubes.
9. Precipitate the probe by adding 1/10th volume of 3 *M* NaOAc pH 5.2 and two volumes of cold 95% ethanol. Allow labeled DNA to precipitate on dry ice for 30 min.
10. Centrifuge and discard the supernatant and resuspend DNA in 50 µL of sterile H_2O. Pool the samples into one microcentrifuge tube and reprecipitate as in **step 9**.
11. Estimate the recovery of labeled DNA by counting the Cerenkov radiation emitted by the pellet using a β counter or by resuspending the DNA in a small volume (*1*00 µL) and counting a 1-µL aliquot in scintillation liquid.
12. Resuspend the labeled DNA in order to obtain 30,000 cpm/µL.

3.2.2. For a Double-Stranded Oligonucleotide Labeled Probe

Proceed as in **Subheading 3.2.1.** except that **steps 1** through **8** should be omitted and replaced by two sequential precipitations in the presence of 5 µg total tRNA as described in **step 9**. (*See* **Note 14.**)

3.3. EMSA

1. Rigorously clean and dry the electrophoresis tank and its accessories prior to use and treat the glass plates as previously described for probe isolation (**step 1**; **Subheading 3.2.1.**).
2. For a typical 70-bp probe, prepare a 6% polyacrylamide gel (*see* **Note 7**) by mixing 2.5 mL of 10X Tris–glycine, 3.75 mL of 40% acrylamide (39:1) stock solution, and H_2O to 25 mL. Add 180 µL of 10% ammonium persulfate and 30 µL of

TEMED. Carefully stir and pour the acrylamide solution between the plates (*see* **Note 15**). Use a comb that has 0.8-cm-width teeth. Allow the gel to set for at least 2 h, then mount gel in the electrophoresis tank and fill the chamber with 1X Tris–glycine (*see* **Note 8**). As soon as the gel is mounted and set, remove the comb and carefully wash the wells with running buffer.

3. Prerun the gel at 4°C and 120 V (8 V/cm) until the current becomes invariant (this takes around 30 min). Prerunning ensures that the gel will remain at a constant temperature from the moment of sample loading.
4. When the gel is ready for loading, prepare samples as follows. For each sample, mix 12 µL of 2X binding buffer, 1 µL of 1 mg/mL poly(dI:dC) (*see* **Notes 9** and **10**), and 0.6 µL of 2M KCl (*see* **Note 10**); then add 30,000 cpm of labeled probe. Where possible, to minimize pipeting errors, prepare a single mix of the common reaction components and distribute equal volumes into the reactions. Finally, add 1–10 µg protein extract and H_2O to a final volume of 24 µL. Mix each tube gently and incubate at RT for 3 min. As a control, prepare a sample without protein extract and add 1 µL of 6X loading buffer containing bromophenol blue and xylene cyanol.
5. Load samples by changing the pipet tip for each sample.
6. Run at 120 V (8 V/cm) and let samples migrate until the free probe reaches the bottom of the gel (*see* **Note 16**). In the case of a 70-bp probe loaded on 6% acrylamide gel, this means 5–6 h of migration.
7. After the gel run, disassemble the apparatus and remove one of the glass plates, place a Whatman paper over the gel, and carefully lift the gel off the remaining plate. Make sure that the gel is well fixed on the Whatman before lifting the gel to avoid gel breakage. Place plastic wrap over the gel and dry at 80°C for 30 min.
8. Place an X-ray film over the gel in an autoradiography cassette and expose at –70°C overnight.

4. Notes

1. Very intense, large or smeary shifted complexes usually result from multiple comigrating DNA–protein complexes that possess nearly identical electrophoretic mobilities in native polyacrylamide gels despite the fact that the proteins they contain usually have distinctive molecular masses on denaturing SDS-PAGE *(43,44)*. An attractive method that helps to distinguish between the proteins yielding these multiple, comigrating complexes is the SDS–polyacrylamide gel fractionation–renaturation procedure *(30)*. This procedure allows recovery and enrichment of specific proteins suitable for further analyzes by EMSA, in addition to providing their approximate molecular masses.
2. When using crude nuclear extracts for detecting DNA–protein complexes in EMSA, the quality of the extract is very critical. Whenever possible, nuclei purification procedures using a sucrose cushion or pad *(45)* is to be preferred in order to eliminate contamination by cytosolic proteins that most often also contain substantial amounts of proteases. Purifying nuclei on sucrose pads has generally yielded high-quality nuclear extract samples. However, such extracts require large quantities of fresh tissue, rendering the approach inappropriate when

limiting amounts of small animal tissues such as spleen, pancreas or prostate are available. In these cases, short microprocedures adapted to prevent protease actions can also be performed ***(46)***. Once the crude extract has been obtained, its quality must be evaluated. An informative way to test extracts is to assess the DNA-binding ability of the ubiquitously expressed transcription factor Sp1. We have found this transcription factor to be particularly sensitive to proteases ***(47)***. Little or no Sp1 binding to its high-affinity binding site (5'-GATCATATCTGCGGGGCGGGGCAGACACAG-3') ***(48)*** is usually indicative of a poor quality nuclear extract. Although such an assay is clearly invaluable when crude extracts are obtained from established tissue-culture cells, caution must be observed when extracts are prepared from whole animal tissues because not all organs express Sp1 at the same level ***(22,47,49)***.

3. The analysis of crude extracts prepared from whole animal tissues by EMSA is somewhat restricted because of the numerous enzymatic activities, such as proteases and deacetylases, these may contain. Degradation of nuclear proteins by endogenous proteases can be prevented by the addition of protease inhibitors. Whole animal tissue extracts are also often contaminated with highly active endogenous phosphatases. Tissues such as liver, kidney, and bone have been reported to be rich in these enzymes ***(50)***, some of which substantially decrease the sensitivity of the EMSA by removing the [^{32}P]-labeled phosphate from the DNA probe. Although addition of phosphatase inhibitors, such as tetrasodium pyrophosphate or sodium fluoride, to the reaction buffer can efficiently prevent dephosphorylation, we have found that the same can also be achieved by simply reducing either the temperature at which the binding reaction is normally performed (30 min of incubation at 4°C) or the time allowed for the DNA–protein interaction to occur (as low as 1 min of incubation at 22°C) ***(21)***. Alternatively, probes labeled by fill-out of unpaired 5' termini using T_4 DNA polymerase or the Klenow fragment of DNA polymerase I and an appropriate [α-^{32}P] dNTP may be used. (*See also* **Note 13**.)
4. When double-stranded oligonucleotides are selected as labeled probes in EMSA, we recommend their size be in the range 20–70 bp. When working with subcloned DNA sequences, optimal signal strength and resolution can be achieved using fragments of 50–250 bp. Although larger fragments may be used, they require longer migration times in order to efficiently resolve the potential DNA–protein complexes. Furthermore, larger labeled probes are likely to bind an increased number of nuclear proteins, which may complicate the interpretation of the results.
5. Handling [γ^{32}P] ATP requires that special care be taken when labeling the DNA probes used in EMSA. The reader is referred to the standard procedures and the guidelines on manipulation of radioactive materials in effect at each research facility. Alternative procedures for nonradioactive probe labeling have been reported for EMSA analyzes ***(32,33)***.
6. Acrylamide is a potent neurotoxic compound that is easily absorbed through skin. Wearing gloves and a mask to avoid direct contact with the skin or inhalation is

therefore required when manipulating dry acrylamide or acrylamide solutions. Similar care should also be taken with polyacrylamide gels, as they may still contain low levels of unpolymerized acrylamide. Acrylamide solution is light sensitive and should be kept away from direct light. It is worth noting that acrylamide and *bis*-acrylamide are slowly converted to acrylic and bisacrylic acid, respectively, upon prolonged storage. To avoid the use of acrylamide, alternative nontoxic gel matrices are available, whose resolution properties are comparable to those of polyacrylamide ***(34,35)***. The use of agarose gels containing a nontoxic synergistic gelling and sieving agent (Synergel™) that helps improve the resolution of DNA–protein complexes has also been reported recently ***(36)***.

7. The concentration of the polyacrylamide gel used in EMSA is primarily dictated by both the size of the labeled probe selected and the resolution of the DNA–protein complexes obtained. It can vary from 4% with large labeled DNA fragments (of over 150 bp in length) to 12% with synthetic oligonucleotides. Two (or more) closely migrating DNA–protein complexes that would normally appear as a single diffuse, smeary complex on a 4% gel can usually be resolved on a 8% gel. However, although increasing the gel concentration usually improved the resolution of DNA–protein complexes, other complexes became unstable in high concentration gels.
8. Although we feel DNA–protein interactions are best revealed using the Tris–glycine buffer system, some complexes may not be detectable under such conditions. The alternative use of other running buffer systems with varying ionic strength, such as Tris–acetate, pH 7.5 or TBE, pH 8.0 ***(23)*** is advisable in order to explore a broader range of DNA–protein complexes.
9. Nonspecific DNA–protein interactions are usually prevented by the addition to the reaction mix of 1–5 µg of a nonspecific competitor DNA. Although this is clearly very effective when crude nuclear extracts are used, such high concentrations of nonspecific competitor DNA were found to compete even for specific DNA–protein complexes when enriched preparations of nuclear proteins are used in EMSA ***(38)***. The more enriched the nuclear protein of interest, the lower the amount of nonspecific competitor required. For example, we routinely use 1–2 µg poly(dI:dC) with crude nuclear proteins, 250 ng when the nuclear extract is enriched on a heparin–Sepharose column, and no more than 25–50 ng with purified or recombinant proteins.
10. The signal strength of a shifted DNA–protein complex can be substantially increased by procedures which favor the interaction between the protein of interest and its target sequence. This can easily be achieved with enriched preparations of nuclear proteins either by increasing the amount of the labeled probe used or by decreasing the concentration of poly(dI:dC), or both. Furthermore, the DNA-binding ability of some nuclear proteins proved to be highly dependent on the salt concentration (usually KCl) of the reaction mix. Transcription factors such as NF1-L and Sp1 interact best with their respective target sequence in the presence of 100 m*M* and 150 m*M* KCl, respectively ***(20)***. It is therefore useful to evaluate the optimum KCl concentration for complex formation on any given DNA probe.

11. Ethidium bromide is a powerful carcinogen that also possesses a moderate toxicity. Wearing gloves is essential when manipulating solutions that contain this DNA dye. Decontamination of ethidium bromide-containing solutions can be achieved using either hypophosphorous acid or potassium permanganate (*see* **ref. *41*** for an overview and detailed protocols).
12. Nearly all vertical electrophoresis apparatus can be used to perform EMSA analyzes. Although gel electrophoresis is performed at room temperature in some EMSA protocols, we recommend 4°C. With some apparatus this can be easily achieved using a specially designed cooling unit. However, for apparatus not equipped with a cooling unit, simply run the gel in a cold room.
13. Although 5' end-labeling of the selected DNA fragment is best done using polynucleotide kinase, very efficient labeling can also be accomplished using alternative procedures, such as filling 5' protruding ends using the Klenow fragment of *E. coli* DNA polymerase I ***(41)***, a particularly attractive alternative when crude nuclear extracts rich in various phosphatases are used (in the event that no phosphatase inhibitors are used in the binding buffer). Larger DNA segments can also be efficiently labeled by PCR.
14. Chemical synthesis of olignonucleotides yields a substantial proportion of intermediate products of progressively decreasing length. This is particularly true for larger oligonucleotides, because the efficiency of each nucleotide addition normally ranges between 98.5% and 99%. For a 40-mer oligonucleotide, this means that 60% of the synthesized products are of the correct length and that the remaining 40% range in size between 1 and 39 nucleotides. Further purification by high-performance liquid chromatography, OPC column, or gel electrophoresis is recommended before annealing oligonucleotides. The loss of even a few bases at the ends of the synthetic duplex may be sufficient to prevent protein binding and therefore reduce the ability to detect the DNA–protein complex in EMSA.
15. We have found that the thickness of the native polyacrylamide gel strongly affects the resolution of shifted DNA–protein complexes; the thinner the gel, the better the resolution. We currently use 0.75-mm-thick gels.
16. Formation of DNA–protein complexes is highly dependent on the voltage selected for their migration into the polyacrylamide gel ***(7)***. We have found that reducing the migration time by running the EMSA at voltage higher than 120 V or 8 V/cm (usually corresponding to 10 mA for a single gel) renders most DNA–protein complexes unstable, preventing their detection.

References

1. Garner, M. M. and Revzin, A. (1981) A gel electrophoresis method for quantifying the binding of proteins to specific DNA regions: application to components of the *Escherichia coli* lactose operon regulatory system. *Nucleic Acids Res.* **9,** 3047–3059.
2. Fried, M. and Crothers, D. M. (1981) Equilibria and kinetics of lac repressor–operator interactions by polyacrylamide gel electrophoresis. *Nucleic Acids Res.* **9,** 6505–6525.

3. Gerstle, J. T. and Fried, M. G. (1993) Measurement of binding kinetics using the gel electrophoresis mobility shift assay. *Electrophoresis* **14,** 725–731.
4. Fried, M. G. and Daugherty, M. A. (1998) Electrophoretic analysis of multiple protein–DNA interactions. *Electrophoresis* **19,** 1247–1253.
5. Cann, J. R. (1998) Theoretical studies on the mobility-shift assay of protein–DNA complexes. *Electrophoresis* **19,** 127–141.
6. Cann, J. R. (1989) Phenomenological theory of gel electrophoresis of protein-nucleic acid complexes. *J. Biol. Chem.* **264,** 17,032–17,040.
7. Vossen, K. M. and Fried, M. G. (1997) Sequestration stabilizes lac repressor-DNA complexes during gel electrophoresis. *Anal. Biochem.* **245,** 85–92.
8. Cann, J. R. (1997) Models of mobility-shift assay of complexes between dimerizing protein and DNA. *Electrophoresis* **18,** 1092–1097.
9. Murakami, Y., Huberman, J. A., and Hurwitz, J. (1996) Identification, purification, and molecular cloning of autonomously replicating sequence-binding protein 1 from fission yeast Schizosaccharomyces pombe. *Proc. Natl. Acad. Sci. USA* **93,** 502–507.
10. Zhang, W., Shields, J. M., Sogawa, K., Fujii-Kuriyama, Y., and Yang, V. M. (1998) The gut-enriched Kruppel-like factor suppresses the activity of the CYP1A1 promoter in an Sp1-dependent fashion. *J. Biol. Chem.* **273,** 17,917–17,925.
11. Tyree, C. M., George, C. P., Lira-DeVito, L. M., Wampler, S. L., Dahmus, M. E., Zawel, L., et al. (1993) Identification of a minimal set of proteins that is required for accurate initiation of transcription by RNA polymerase II. *Genes Dev.* **7,** 1254–1265.
12. Gille, J., Swerlick, R. A., and Caughman, S. W. (1997) Transforming growth factor-β-induced transcriptional activation of the vascular permeability factor (VPF/VEGF) gene requires AP-2-dependent DNA binding and transactivation. *EMBO J.* **16,** 750–759.
13. Roy, A. L., Du, H., Gregor, P. D., Novina, C. D., Martinez, E., and Roeder, R. G. (1997) Cloning of an Inr- and E-box-binding protein, TFII-I, that interacts physically and functionally with USF1. *EMBO J.* **16,** 7091–7104.
14. Ng, J. Y. and Marians, K. J. (1996) The ordered assembly of the ØX174-type primosome. I. Isolation and identification of intermediate protein–DNA complexes. *J. Biol. Chem.* **271,** 15642–15648.
15. Kironmai, K. M., Muniyappa, K., Friedman, D. B., Hollingsworth, N. M., and Byers, B. (1998) DNA-binding activities of Hop**1,** a synaptonemal complex component from Saccharomyces cerevisiae. *Mol. Cell. Biol.* **18,** 1424–1435.
16. Wakasugi, M., and Sancar, A. (1998) Assembly, subunit composition, and footprint of human DNA repair excision nuclease. *Proc. Natl. Acad. Sci. USA* **95,** 6669–6674.
17. Laniel, M. A., Bergeron, M. J., Poirier, G. G., and Guérin, S. L. (1997) A nuclear factor other than Sp1 binds the GC-rich promoter of the gene encoding rat poly(ADP-ribose) polymerase in vitro. *Biochem. Cell Biol.* **75,** 427–434.
18. Eskild, W., Robidoux, S., Simard, J., Hansson, V., and Guérin, S. L. (1994) Binding of a member of the NF1 family of transcription factors to two distinct cis-acting elements in the promoter and 5'-flanking region of the human cellular retinol binding protein 1 gene. *Mol. Endocrinol.* **8,** 732–745.

19. Roy, R. and Guérin, S. L. (1994) The 30-kDa rat liver transcription factor nuclear factor 1 binds the rat growth-hormone proximal silencer. *Eur. J. Biochem.* **219,** 799–806.
20. Robidoux, S., Eskild, W., Kroepelin, C. F., Hansson, V., and Guérin, S. L. (1992) Salt-dependent formation of DNA/protein complexes in vitro, as viewed by the gel mobility shift assay. *Biotechniques* **13,** 354–358.
21. Laniel, M. A. and Guérin, S. L. (1997) Improving sensitivity of the EMSA by restricting tissue phosphatase activities. *Biotechniques* **24,** 964–970.
22. Potvin, F., Roy, R. J., Poirier, G. G., and Guérin, S. L. (1993) The US-1 element from the gene encoding rat poly(ADP-ribose) polymerase binds the transcription factor Sp1. *Eur. J. Biochem.* **215,** 73–80.
23. Ausubel, F. M., Brent, R., Kingston, R. E., Moore, D. D., Seidman, J. G., Smith, J. A., et al. (1992) In *Short Protocols in Molecular Biology,* 2nd ed., Wiley, New York.
24. Bergeron, M. J., Leclerc, S., Laniel, M. A., Poirier, G. G., and Guérin, S. L. (1997) Transcriptional regulation of the rat poly(ADP-ribose) polymerase gene by Sp1. *Eur. J. Biochem.* **250,** 342–353.
25. Granger-Schnarr, M., Lloubes, R., De Murcia, G., and Schnarr, M. (1988) Specific protein–DNA complexes: immunodetection of the protein component after gel electrophoresis and Western blotting. *Anal. Biochem.* **174,** 235–238.
26. Demczuk, S., Harbers, M., and Vennström, B. (1993) Identification and analysis of all components of a gel retardation assay by combination with immunoblotting. *Proc. Natl. Acad. Sci. USA* **90,** 2574–2578.
27. Dyer, R. B. and Herzog, N. K. (1995) Immunodepletion EMSA: a novel method to identify proteins in a protein–DNA complex. *Nucleic Acids Res.* **23,** 3345–3346.
28. Yamamoto, H. (1997) DNA mobility shift assay coupled with SDS-PAGE for detection of DNA-binding proteins. *Biotechniques* **22,** 210–211.
29. Williams, M., Brys, A., Weiner, A. M., and Maizels, N. (1992) A rapid method for determining the molecular weight of a protein bound to nucleic acid in a mobility shift assay. *Nucleic Acids Res.* **20,** 4935–4936.
30. Ossipow, V., Laemmli, U. K., and Schibler, U. (1993) A simple method to renature DNA-binding proteins separated by SDS-polyacrylamide gel electrophoresis. *Nucleic Acids Res.* **21,** 6040–6041.
31. Wolf, S. S., Hopley, J. G., and Schweizer, M. (1994) The application of ^{33}P-labeling in the electrophoretic mobility shift assay. *Biotechniques* **16,** 590–592.
32. Suske, G., Gross, B., and Beato, M. (1989) Non-radioactive method to visualize specific DNA–protein interactions in the band shift assay. *Nucleic Acids Res.* **17,** 4405.
33. Ludwig, L. B., Hughes, B. J., and Schwartz, S. A. (1995) Biotinylated probes in the electrophoretic mobility shift assay to examine specific dsDNA, ssDNA or RNA–protein interactions. *Nucleic Acids Res.* **23,** 3792–3793.
34. Ramanujam, P., Fogerty, S., Heiser, W., and Jolly, J. (1990) Fast gel electrophoresis to analyze DNA–protein interactions. *Biotechniques* **8,** 556–563.
35. Vanek, P. G., Fabian, S. J., Fisher, C. L., Chirikjian, J. G., and Collier, G. B. (1995) Alternative to polyacrylamide gels improves the electrophoretic mobility shift assay. *Biotechniques* **18,** 704–706.

36. Chandrasekhar, S., Souba, W. W., and Abcouwer, S. F. (1998) Use of modified agarose gel electrophoresis to resolve protein–DNA complexes for electrophoretic mobility shift assay. *Biotechniques* **24,** 216–218.
37. Revzin, A. (1989) Gel electrophoresis assays for DNA–protein interactions. *Biotechniques* **7,** 346–355.
38. Larouche, K., Bergeron, M. J., Leclerc, S., and Guérin, S. L. (1996) A careful use of the non-specific synthetic competitor poly(dI–dC):poly(dI–dC) is advised in DNA–protein interaction studies involving enriched nuclear proteins. *Biotechniques* **20,** 439–444.
39. Hassanain, H. H., Dai, W., and Gupta, S. L. (1993) Enhanced gel mobility shift assay for DNA-binding factors. *Anal. Biochem.* **213,** 162–167.
40. Zhang, X. Y., Asiedu, C. K., Supakar, P. C., and Ehrlich, M. (1992) Increasing the activity of affinity-purified DNA-binding proteins by adding high concentrations of nonspecific proteins. *Anal. Biochem.* **201,** 366–374.
41. Sambrook, J., Fritsch, E. F., and Maniatis, T. (1989) In *Molecular Cloning, A Laboratory Manual* , 2nd ed. (Nolan, C., ed.), Cold Spring Harbor Laboratory, Cold Spring Harbor, New York.
42. Harvey, M., Brisson, I., and Guérin, S. L. (1993) A simple apparatus for fast and inexpensive recovery of DNA from polyacrylamide gels. *Biotechniques* **14,** 942–948.
43. Larouche, K., Leclerc, S., Giasson, M., and Guérin, S. L. (1996) Multiple nuclear regulatory proteins bind a single cis-acting promoter element to control basal transcription of the human α4 integrin gene in corneal epithelial cells. *DNA Cell Biol.* **15,** 779–792.
44. Leclerc, S., Eskild, W., and Guérin, S. L. (1997) The rat growth hormone and human cellular retinol binding protein 1 genes share homologous NF1-like binding sites that exert either positive or negative influences on gene expression in vitro, *DNA Cell Biol.* **16,** 951–967.
45. Graves, B. J., Johnson, P. F., and McKnight, S. L. (1986) Homologous recognition of a promoter domain common to the MSV LTR and HSV Tk gene. *Cell* **44,** 565–576.
46. Roy, R., Gosselin, P., and Guérin, S. L. (1991) A short protocol for micropurification of nuclear proteins from whole animal tissues. *Biotechniques* **11,** 770–777.
47. Robidoux, S., Gosselin, P., Harvey, M., Leclerc, S., and Guérin, S. L. (1992) Transcription of the mouse secretory protease inhibitor p12 gene is activated by the developmentally regulated positive transcription factor Sp1. *Mol. Cell. Biol.* **12,** 3796–3806.
48. Dynan,W. S. and Tjian, R. (1983) The promoter-specific transcription factor Sp1 binds to upstream sequences in the SV40 early promoter. *Cell* **35,** 79–87.
49. Saffer, J. D., Jackson, S. P., and Annarella, M. B. (1991) Developmental expression of Sp1 in the mouse. *Mol. Cell. Biol.* **11,** 2189–2199.
50. McComb, R. B., Bowers, G. N., and Posen, S. (1979) *Alkaline Phosphatase.* Plenum, New York.

3

DNase I Footprinting

Benoît Leblanc and Tom Moss

1. Introduction

DNase I footprinting was developed by Galas and Schmitz in 1978 as a method to study the sequence-specific binding of proteins to DNA *(1)*. In the technique, a suitable uniquely end-labeled DNA fragment is allowed to interact with a given DNA-binding protein and then the complex partially digested with DNase I. The bound protein protects the region of the DNA with which it interacts from attack by the DNase. Subsequent molecular-weight analysis of the degraded DNA by electrophoresis and autoradiography identifies the region of protection as a gap in the otherwise continuous background of digestion products; for examples see **Fig. 1**. The technique can be used to determine the site of interaction of most sequence-specific DNA-binding proteins but has been most extensively applied to the study of transcription factors. Because the DNase I molecule is relatively large as compared to other footprinting agents (*see* Chapters 5 and 6 on the use of hydroxy radicals and diethylpyrocarbonate), its attack on the DNA is relatively easily sterically hindered. Thus, DNase I footprinting is the most likely of all the footprinting techniques to detect a specific DNA–protein interaction. This is clearly demonstrated by our studies on the transcription factor xUBF (*see* **Fig. 1B**). The xUBF interaction with the *Xenopus* ribosomal DNA enhancer can be easily detected by DNase I footprinting but has still not been detected by other footprinting techniques.

DNase I footprinting can not only be used to study the DNA interactions of purified proteins but also as an assay to identify proteins of interest within a crude cellular or nuclear extract (e.g., *see* **ref. 2**). Thus, it can serve much the same function as a gel shift analysis (EMSA, Chapter 2) in following a specific DNA-binding activity through a series of purification steps. Because DNase I footprinting can often be used for proteins that do not "gel shift" (UBF,

From: *Methods in Molecular Biology, vol. 148: DNA–Protein Interactions: Principles and Protocols, 2nd ed.*
Edited by: T. Moss © Humana Press Inc., Totowa, NJ

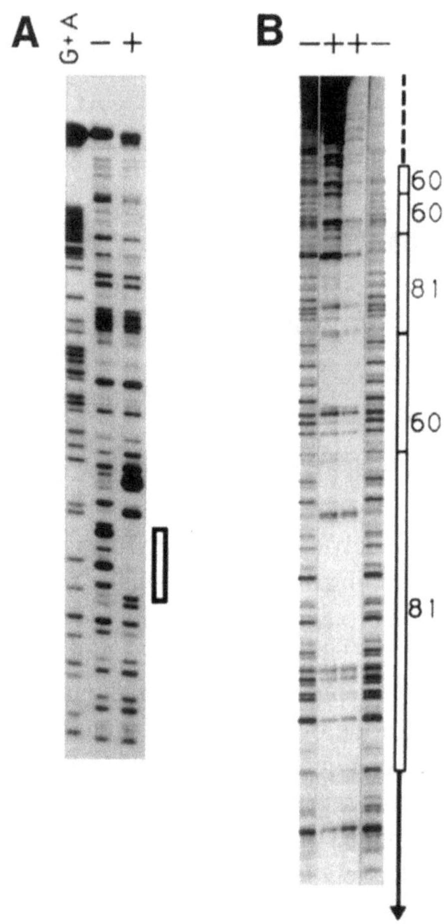

Fig. 1. Examples of DNase I footprints. **(A)** Footprint (open box) of a chicken erythrocyte DNA binding factor on the promoter of the H5 gene *(2)* (figure kindly donated by A. Ruiz-Carrillo). **(B)** Interaction of the RNA polymerase I transcription xUBF with the tandemly repeated 60 and 81b.p. *Xenopus* ribosomal gene enhancers. Both (A) and (B) used 5' end-labeled fragments. Minus and plus refer to naked and complexed DNA fragments, respectively, and G+A to the chemical sequence ladder.

Fig. 1B), it has more general applicability. However, because of the need for a protein excess and the visualization of the footprint by a partial DNA digestion ladder, the technique requires considerably more material than would a gel shift and cannot of itself distinguish individual components of heterogeneous DNA–protein complexes.

DNase I (E.C. 3.1.4.5) is a protein approx 40 Å in diameter. It binds in the minor groove of the DNA and cuts the phosphodiester backbone of both strands

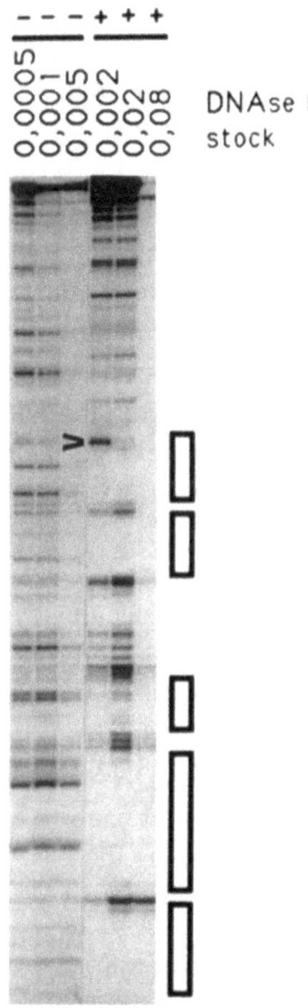

Fig. 2. Course of digestion with increasing amounts of DNase I. Here xUBF was footprinted on the *Xenopus* ribosomal promoter using a 5' end-labeled fragment. The numbers above the tracks refer to the DNase I dilution (in units/μL) employed and minus and plus to the naked and complexed DNAs, respectively. The predominant footprints are indicated by open boxes.

independently *(3)*. Its bulk helps to prevent it from cutting the DNA under and around a bound protein. However, a bound protein will also usually have other effects on the normal cleavage by DNase I, resulting in some sites becoming hypersensitive to DNase I (*see* **Figs. 1** and **2**). It is also not so uncommon to observe a change in the pattern of DNase cleavage without any obvious extended protection (e.g., *see* **Fig. 2**).

Unfortunately, DNase I does not cleave the DNA indiscriminately, some sequences being very rapidly attacked while others remain unscathed even after extensive digestion *(4)*. This results in a rather uneven "ladder" of digestion products after electrophoresis, something which limits the resolution of the technique, see naked DNA tracks in **Figs. 1** and **2**. However, when the protein-protected and naked DNA ladders are run alongside each other, the footprints are normally quite apparent. To localize the position of the footprints, G+A and/or C+T chemical sequencing ladders of the same end-labeled DNA probe *(5)* should accompany the naked and protected tracks (*see* **Note 1**). As a single end-labeled fragment allows one to visualize interactions on one strand only of the DNA, it is usual to repeat the experiment with the same fragment labeled on the other strand. DNA fragments can be conveniently 5' labeled with T_4 polynucleotide kinase and 3' labeled using the Klenow or the T_4 DNA polymerases (fill out) or terminal transferase (e.g., *see* **ref. *6***). A combination of 5' and 3' end labeling allows both DNA strands to be analyzed side by side from the same end of the DNA duplex.

DNase I footprinting requires an excess of DNA-binding activity over DNA fragment used. The higher the percent occupancy of a site on the DNA, the clearer a footprint will be observed. It is therefore important not to titrate the available proteins with too much DNA. This limitation can, in part, be overcome when a protein also generates a gel shift. It is then feasible to fractionate the partially DNase-digested protein–DNA complex by nondenaturing gel electrophoresis and to excise the shifted band (which is then a homogeneous protein–DNA complex) before analyzing the DNA by denaturing gel electrophoresis as in the standard footprint analysis (*see* Chapters 2, 4, 5, and 7 for analogous procedures).

Footprinting crude or impure protein fractions usually requires that an excess of a nonspecific competitor DNA be added. The competitor binds nonspecific DNA-binding proteins as effectively as the specific labeled target DNA fragment and hence, when present in sufficient excess, leaves the main part of the labeled DNA available for the sequence specific protein. Homogeneous and highly enriched protein fractions usually do not require the presence of a nonspecific competitor during footprinting. When planning a footprinting experiment, it is a prerequisite to start by determining the optimal concentration of DNase I to be used. This will be a linear function of the amount of nonspecific DNA competitor but more importantly and less reproducibly, will be a function of the amount and purity of the protein fraction added. As a general rule, more DNase is required if more protein is present in the binding reaction, whether or not this protein binds specifically. Thus, very different DNase concentrations may be required to produce the required degree of digestion on naked and protein-bound DNA. A careful titration of the DNase concentration is therefore essential to optimize the detection of a footprint and can even make the difference between the detection or lack of detection of a given interaction.

The following protocol was developed to study the footprinting of the *Xenopus* ribosomal transcription factor xUBF, which is a rather weak DNA-binding protein with a rather broad sequence specificity. The protocol is not original, being derived from several articles (e.g., **refs.** ***1*** and *7*). It does however represent a very practical approach which can be broadly applied. We recommend that the reader also refer to ***(6)*** for more information on the quantitative analysis of protein–DNA interactions by footprinting.

2. Materials

1. Binding buffer (2X): 20% glycerol, 0.2 m*M* EDTA, 1 m*M* dithiothreitol (DTT) 20 m*M* HEPES pH 7.9 and 4% poly(vinyl alcohol) (*see* **Note 2**).
2. poly dAdT–poly dAdT (Pharmacia LKB): 1 mg/mL solution in TE. Keep at –20°C (*see* **Note 3**).
3. End-labeled DNA fragment of high-specific activity (*see* **Note 1**).
4. Cofactor solution: 10 m*M* $MgCl_2$ and 5 m*M* $CaCl_2$.
5. DNase I stock solution: A standardized vial of DNase I (Sigma, D4263) is dissolved in 50% glycerol, 135 m*M* NaCl, 15 m*M* CH_3COONa pH 6.5 at 10 Kunitz units/µL. This stock solution can be kept at –20°C for many months (*see* **Note 4**).
6. 1 *M* KCl
7. Reaction Stop buffer: 1% sodium dodecyl sulfate (SDS), 200 m*M* NaCl, 20 m*M* EDTA pH 8.0, and 40 µg/mL tRNA (*see* **Note 5**).
8. 10X TBE buffer: 900 m*M* Tris-borate pH 8.3, 20 m*M* EDTA
9. Loading buffer: 7 *M* urea, 0.1X TBE, 0.05% of xylene cyanol and of bromophenol blue.
10. 6% acrylamide, 7 *M* urea, and 1X TBE sequencing gel.
11. Phenol–chloroform (1:1) saturated with 0.3 *M* TNE.

3. Methods

The footprinting reaction is done in three stages: binding of the protein to the DNA, partial digestion of the protein–DNA complex with DNase I, and separation of the digestion fragments on a DNA sequencing gel.

1. The binding reaction is performed in a total volume of 50 µL containing 25 µL of 2X binding buffer, 0.5 µL of 1 mg/mL poly dAdT. poly dAdT, 2–3 ng of end-labeled DNA fragment (approx 15,000 CPM), (*see* **Note 6**), the protein fraction and 1 *M* KCl to bring the final KCl concentration to 60 m*M*. The maximum volume of the protein fraction that can be used will depend on the salt concentration of this solution. The reaction is performed in a 1.5-mL Eppendorf tube.
2. Incubate on ice for 20 min.
3. During the binding reaction, dilute the DNase I stock solution in water at 0°C. We suggest working concentrations of about 0.0005 to 0.1 Kunitz units/µL, depending on the level of protein present (*see* **Note 7** and **step 5**). A good range is 0.0005; 0.001; 0.002; 0.005; 0.02; 0.08 U/µL.

4. After the incubation, transfer the reaction tubes in batches of eight to a rack at room temperature (RT) and add 50 μL of the cofactor solution to each.
5. Add 5 μL of the appropriate DNase I dilution to a tube every 15 s. (0.0005–0.005 for naked DNA; 0.002–0.08 for DNA + proteins).
6. After 2 min digestion, each reaction is stopped by the addition of 100 μL of the stop solution (RT), see **Note 8**.
7. After all the reactions have been processed, phenol–chloroform extract each reaction once.
8. Add two volumes (400 μL) of ethanol (–20°C) and allow nucleic acids to precipitate at –80°C for 20 min.
9. Microcentrifuge for 15 min, at approx 10,000*g* and remove the supernatant with a Pasteur pipet. Check for the presence of a radioactive pellet before discarding the ethanol.
10. Add 200 μL of 80% ethanol (–20°C) and microcentrifuge again. After removing supernatant, dry the pellets in a vacuum desiccator.
11. Resuspend each pellet in 4.5 μL loading buffer, vortex, and centrifuge briefly.
12. A G+A ladder and a molecular-weight marker should be run in parallel with the samples on the sequencing gel (**step 13**) (*see* **Note 9**). The G+A ladder can be prepared as follows ***(5)***: approx 200,000 cpm of end-labeled DNA are diluted into 30 μL H_2O (no EDTA). 2 μL of 1 *M* piperidine formate, pH 2.0, are added and the solution incubated at 37°C for 15 min; 150 μL of 1 *M* piperidine are added directly and the solution incubated at 90°C for 30 min in a well-sealed tube (we use a 500-μL microtube in a thermal cycler). Add 20 μL of 3 *M* CH_3COONa and 500 μL of ethanol and precipitate at –80°C for 10–20 min. Microcentrifuge (10,000*g*, 10 min) and repeat precipitation. Finally, redissolve in 200 μL of H_2O and lyophilize. Resuspend in gel loading buffer and apply about 5000 cpm per track.
13. Prerun a standard 6% acrylamide sequencing gel for 30 min before loading each of the aliquots. Wash the wells thoroughly with a syringe, denature the DNA for 2 min at 90°C, and load with thin-ended micropipet tips. Run the gel hot to keep the DNA denatured (*see* **Note 10**). After the run, wrap the gel in plastic wrap and expose it O/N at –70°C with an intensifying screen (*see* **Note 11**). Several different exposures will probably be required to obtain suitable band densities.

4. Notes

1. Single-stranded breaks in the end-labeled DNA fragment must be avoided as they give false signals indistinguishable from genuine DNase I cleavage and hence can mask an otherwise good footprint. It is therefore advisable to check the fragment on a denaturing gel before use. Always use a freshly labeled fragment (3–4 d at the most), as radiochemical nicking will degrade it.
2. This binding buffer has been shown to work well for the transcription factor NF-1 ***(6)*** and in our lab for both the hUBF and xUBF factors and thus should work for many factors. Glycerol and poly(vinyl alcohol) (an agent used to reduce the available water volume and hence concentrate the binding activity) are not manda-

tory. The original footprinting conditions of Galas and Schmitz *(1)* for the binding of the *lac* repressor on the *lac* operator were 10 m*M* cacodylate buffer pH 8.0, 10 m*M* $MgCl_2$, 5 m*M* $CaCl_2$ and 0.1 m*M* DTT. Particular conditions of pH, cofactors and ionic strength may need to be determined for an optimal binding of different factors to DNA.

3. Because poly dIdC, another nonspecific general competitor, has been shown to compete quite efficiently with G–C-rich DNA sequences, poly dAdT is preferred here. The choice of an appropriate nonspecific competitor (be it synthetic as in this case or natural, for example, pBR322 or calf thymus DNA) may have to be determined empirically for the protein studied. When working with a pure or highly enriched protein, no competitor is usually needed. The DNase I concentration must then be reduced accordingly (to about naked DNA values).
4. These standardized vials allow for very reproducible results. Glycerol will keep the enzyme from freezing, as repeated freeze–thaw cycles will greatly reduce its activity.
5. Do not be tempted to use too much RNA, as it causes a very annoying fuzziness of the gel bands, preventing resolution of the individual bands. The RNA carrier can be completely omitted if care is taken at the precipitation step.
6. The use of 5' end labeling with polynucleotide kinase in the presence of crude protein extracts can sometimes lead to a severe loss of signal because of the presence of phosphatases. In these cases 3' end labeling by "fill out" with Klenow or T_4 DNA polymerases is to be preferred.
7. For naked DNA and very low amounts of protein, dilutions of 0.0005 to 0.005 give a good range of digestion.
8. It is convenient to work with groups of eight sample during the DNase I digestion. The cofactor solution is added to eight samples at a time and then the DNase I digestions are begun at 15-s intervals. Fifteen seconds after adding DNase to the eighth sample, the stop solution is added to sample 1 and then to the other samples at 15-s intervals.
9. In comparing a chemical sequencing ladder with the products of DNase I digestion, one must bear in mind that because the chemical modification and cleavage destroys the target base, each band in the sequencing ladder corresponds to a fragment ending in the base preceding the one read. For example, if a DNase I gel band corresponds in mobility to the sequence ladder band read as G in the sequence ACGT, then the DNase I cleavage occurred between the bases C and G. DNase I cleaves the phosphodiester bond, leaving a 3'-OH, whereas the G+A and C+T sequencing reactions leave a 3'-PO_4, causing a mobility shift between the two types of cleavage ladders. This is a further potential source of error. However, in our experience, the shift is less than a half a base and, hence, cannot lead to an error in the deduced cleavage site.
10. Sequencing gels are not denaturing unless run hot (7 *M* urea produces only a small reduction in the T_m of the DNA). A double-stranded form of the DNA fragment is therefore often seen on the autoradiogram, especially at low levels of DNase I digestion and can sometimes be misinterpreted as a hypersensitive cleav-

age. By running a small quantity of undigested DNA fragment in parallel with the footprint, this error can be avoided.

11. For detection of ^{32}P-labeled DNA in sequencing gels we have used either Cronex Lightning Plus (Dupont) or Kyokko Special (Fuji) intensifying screens, the latter being, in practice, 30% less sensitive but often much less expensive. Fuji-RX or similar films are, in practice, 30% slower than Kodak X-omat AR film, but in North America, at the time of writing, they are five to six times less expensive. Hence, the combinations of screens and films Kyokko/RX:Cronex/RX: Kyokko/AR:Cronex/AR give relative sensitivities of about; 1:1.5:1.5:2. Should the newer slightly higher sensitivity films be used (e.g., Kodak BioMax), it is essential that the appropriate intensifying screens be employed. These newer films are usually most sensitive to green light and do not work well with the Cronex Lightning Plus/Kyokko Special type screens, which emit a blue light.

Acknowledgments

The authors wish to thank Dr. A. Ruiz-Carrillo for providing the autoradiogram in **Fig. 1A**. This work was supported by a project grant from the Medical Research Council (MRC) of Canada. Tom Moss is an MRC of Canada Scientist and a member of the Centre de Recherche en Cancérologie de l'Université Laval which is supported by the FRSQ of Québec.

References

1. Schmitz, A. and Galas, D. J. (1978). DNAase I footprinting: a simple method for the detection of protein-DNA binding specificity. *Nucleic Acids Res.* **5,** 3157–3170.
2. Rousseau, S., Renaud, J., and Ruiz-Carrillo, A. (1989). Basal expression of the histone H5 gene is controlled by positive and negative *cis*-acting sequences. *Nucleic Acids Res.* **17,** 7495–7511.
3. Suck, D., Lalm, A., and Oefner, C. (1988). Structure refined to 2 Å of a nicked DNA octanucleotide complex with DNAse I. *Nature* **332,** 464–468.
4. Drew, H. R. (1984). Structural specificities of five commonly used DNA nucleases. *J. Mol. Biol.* **176,** 535–557.
5. Maxam, A. M. and Gilbert, W. (1980). Sequencing end-labeled DNA with base-specific chemical cleavages, in *Methods in Enzymology, Vol. 65* (Grossman, L. and Moldave, K., eds.), Academic, New York, pp. 499–560.
6. Brown, T. (1987). Analysis of RNA by Northern and slot blot hybridisation, in Current Protocols in Molecular Biology (Ausubel, F. M., Brent, R., Kingston, R. E., Moore, D. D., Seidman, J. G., Smith, J. A., et al., eds.), Greene Publishing Associates/Wiley-Interscience, New York, pp. 4.9.1–4.9.14.
7. Walker, P. and Reeder, R. H. (1988). The *Xenopus laevis* ribosomal gene promoter contains a binding site for nuclear factor-1. *Nucleic Acids Res.* **16,** 10,657–10,668.

4

Footprinting with Exonuclease III

Willi Metzger and Hermann Heumann

1. Introduction

Within the last few years footprinting techniques have become increasingly important in the study of protein–nucleic acid interactions. This is partly the result of a fast-growing number of known nucleic acid-binding proteins but also because of an increase in the available probes that can be chosen in order to tackle a specific problem. There are two major groups of probes—the chemical probes and the enzymatic probes. The enzymatic probes, such as DNase I or exonuclease III, have the advantage of acting specifically on the DNA. Chemical probes are often less specific and may also react with the protein, possibly disturbing the correct interaction of protein with DNA. For the study of very fragile protein–DNA complexes, enzymatic probes are therefore often preferable.

The exploitation of a specific enzymatic function can also be a reason for choosing an enzymatic probe. The exonuclease activity and processivity of exonuclease III makes this enzyme a suitable probe when information about the position of a sequence-specific bound protein is required. A prerequisite for the successful use of exonuclease III as a footprinting probe is, however, that the half-life of the protein–DNA complex should be long compared with the time required for the exonuclease III reaction.

1.1. Enzymatic Activities of Exonuclease III

Exonuclease III (Exo III) is a monomeric enzyme with a molecular weight of 28,000 kDa. It contains several distinct activities: a 3'–5' exonuclease activity, a DNA 3' phosphatase activity, an AP endonuclease activity, and an RNase H activity *(1)*.

From: *Methods in Molecular Biology, vol. 148: DNA–Protein Interactions: Principles and Protocols, 2nd ed.*
Edited by: T. Moss © Humana Press Inc., Totowa, NJ

1.2. Principle of the Procedure

Footprinting with Exo III makes use of the 3'–5' exonuclease activity of this enzyme *(2)*. After a protein has been specifically bound to a DNA fragment containing its recognition site, Exo III is used to remove mononucleotides from both DNA strands in a processive way, beginning from the 3' termini. The specifically bound protein blocks the action of Exo III and leaves double-stranded DNA only in the region bound by the protein (**Fig. 1**). (Any free DNA is fully digested, an advantage of Exo III over other footprinting probes, as there are no background problems caused by the presence of free DNA.) The lengths of the two resultant protected single-strand DNA fragments are determined by electrophoresis on a denaturing sequencing gel using appropriate DNA length standards, the fragments being detected by autoradiography via a 5' radioactive end label. If both termini of the initial DNA duplex are labeled, a decisive association of the protected fragments with the upper and lower strands may be difficult. This problem is overcome by removal of one end-label (e.g., by a restriction enzyme cleavage before Exo III digestion). **Figure 1A** shows the procedure schematically and **Fig. 1B** shows the expected radioactive products (bands) after denaturing gel electrophoresis.

1.3. Interpretation of Results

The region of a DNA duplex protected by a bound protein can be determined from the length of the two single-strand DNA fragments remaining after Exo III treatment. If the initial DNA duplex has a length of k base pairs and the single-strand products have lengths of m and n bases, respectively, the size of the protected region x is given by;

$$x = m + n - k$$

Correct interpretation of the footprinting data, however, requires a critical assessment of the action of the Exo III on the protein–DNA complex. The interpretation is straightforward if the protein is strongly bound to DNA. In this case, because the protein acts as a steric hindrance for Exo III, the protected region gives an upper limit of the size of the DNA segment interacting with the protein. Interpretation of the data is more complicated if the strength of interaction between protein and DNA varies within the interacting domain. Because of the processivity of Exo III, this enzyme may "nibble" into a protein–bound DNA segment for which the binding protein has lower affinity, resulting in an underestimation of the extent of protein–DNA contact. In order to decide if such a process occurs, it is essential to study a time-course of the Exo III digestion.

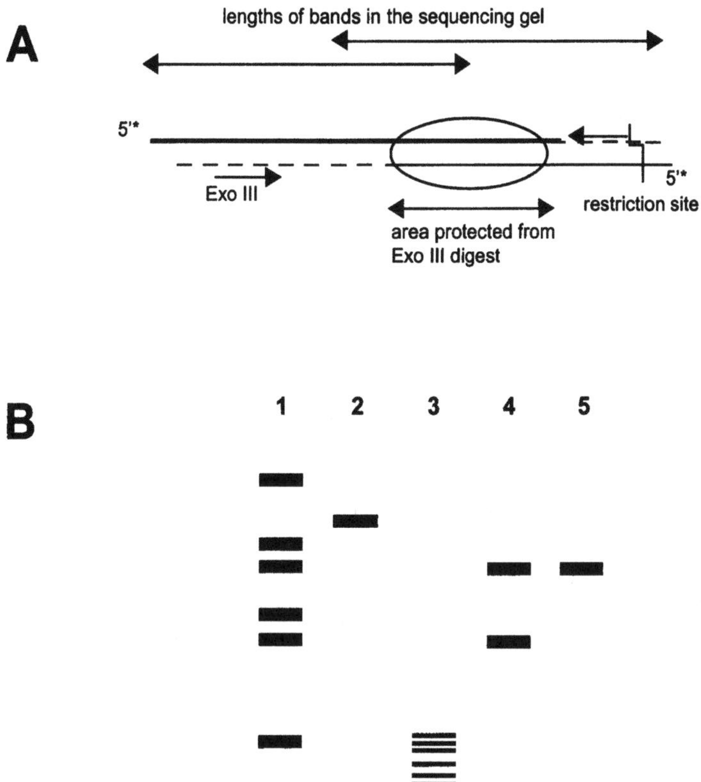

Fig. 1. **(A)** Schematic representation of the Exo III footprinting procedure. **(B)** Exo III footprint of the complex shown in (A). Lane 1: Size markers; lane 2: labeled DNA fragment; lane 3: labeled DNA fragment, after Exo III treatment; lane 4: DNA from DNA–protein complex, after Exo III treatment, both 5' ends of duplex labeled; lane 5: DNA from DNA–protein complex, after Exo III treatment, DNA duplex singly end-labeled label was removed beforehand by cutting with a restriction enzyme as indicated in (A).

1.4. Examples of the Application of Exo III as a Footprinting Probe

Exonuclease III has been used to follow the movement of *E. coli* RNA polymerase during RNA synthesis *(3,4)*. RNA synthesis was arrested at specific positions and the arrested complexes were subjected to Exo III digestion. A set of single–strand products marking the boundaries of RNA polymerase on the DNA at each step of RNA synthesis was observed. Exo III also offers the possibility of detecting specific DNA–protein interactions in crude extracts, because those proteins whose half-lives are greater than the reaction time act as a block for Exo III *(5–7)*.

2. Materials

2.1. Exonuclease III

Exonuclease III is available from BRL (Gaithersburg, MD) or Boehringer Mannheim (Indianapolis, IN). It can be stored for many months at –20°C.

2.2. Sequencing Gel

1. Acrylamide solution: 40% acrylamide, and 0.66% *bis*-acrylamide in H_2O.
2. 10X TBE: 1 *M* Tris-HCl, pH 8.6, 840 m*M* boric acid; and 10 m*M* EDTA.
3. 10% Ammonium persulfate (freshly prepare before use).
4. *N,N,N'N'*-tetramethylethylene diame (TEMED).
5. Preparation of an 8% acrylamide gel: Weigh 21 g of urea, and add 5 mL of 10X TBE solution and 10 mL of acrylamide solution. Dissolve under mild heating. Add double-distilled water to a final volume of 50 mL. Filter and degas the solution (filter pore size 0.2 µm). Add 500 µL of 10% ammonium persulfate solution and 30 µL of TEMED. Pour gel immediately.
6. Formamide loading buffer for the sequencing gel: 100 mL deionized formamide, 30 mg xylenecyanol FF, 30 mg bromophenol blue, and 750 mg EDTA.
7. Electrophoresis buffer: 1X TBE.

2.3. Nondenaturing Gel for the Band-Shift Assay

1. Acrylamide solution: 30% acrylamide and 0.8% *bis*-acrylamide in H_2O.
2. 1 *M* Tris-HCl, pH 7.9.
3. 10% Ammonium persulfate (freshly prepare before use).
4. 5% TEMED (diluted in water).
5. Preparation of nondenaturing gel: Mix 240 µL of 1 *M* Tris-HCl, pH 7.9, 2.75 mL acrylamide solution, and 25.7 mL of H_2O and degas. Add 300 µL of 10% ammonium persulfate and 70 µL of 5% TEMED. Pour gel.
6. Loading buffer for the nondenaturing gel (10X solution): 40% sucrose and 0.1% bromophenol blue.
7. 5% Dichloro-dimethylsilane solution (in chloroform).
8. 0.3% *g*-Methacryl-oxypropyl-trimethoxy-silane and 0.3% acetic acid dissolved in ethanol.
9. Electrophoresis buffer: 8 m*M* Tris-HCl, pH 7.9.

Store the solutions protected from light at 4°C. Dilute buffers with bidistilled water.

2.4. Other Items

1. Sequencing gel apparatus (Pharmacia, Piscataway, NJ).
2. Filters for drop dialysis VS, 0.025 µm (Millipore, Bedford, MA).
3. Peristaltic pump.
4. SpeedVac concentrator.

3. Methods

3.1. Establishing Conditions for Optimum Yield of Specific Protein–DNA Complexes

A very elegant method for establishing the optimum conditions for the formation of a specific protein–DNA complex is acrylamide gel electrophoresis under nondenaturing conditions. This electrophoretic mobility shift assay (EMSA) or "band shift assay" allows one to differentiate between complexed and uncomplexed DNA *(8,9)* and thus to determine the stoichiometry of given complexes and the optimum salt conditions for their formation (*see* Chapter 2). This method can even be applied to high-molecular-weight protein DNA complexes, given the acrylamide concentration is low enough to enable the complex to enter the gel matrix. The gel composition given in **Subheading 2.3.** is optimal for the study of high-molecular-weight complexes (*see* **Note 6**).

To facilitate the handling of low-concentration polyacrylamide gels, the glass gel plates are subjected to a special treatment ensuring the gel will remain bound to only one of the two plates after electrophoresis:

1. Wash the glass plates (20 × 20 cm) with ethanol.
2. Treat one plate with *g*-methacryl-oxypropyl-trimethoxysilane solution, then wash this plate carefully four times with ethanol.
3. Treat the second plate with dichloro-dimethylsilane solution.
4. Form the protein–DNA complex in a volume of about 15 µL under the desired conditions.
5. Dialyze the complex, if necessary, against the electrophoresis buffer by drop dialysis as follows:
   a. Pour the dialysis buffer into a Petri dish.
   b. Place a VS filter (Millipore, *see* **Subheading 2.4.**) with the glossy side upward onto the buffer so that it can float freely.
   c. Place the sample as a drop on the filter. Remove the drop after 1 h when dialysis is complete.
6. Add 1/10 volume of the 10X loading buffer to the dialyzed complex and apply the sample to the nondenaturing polyacrylamide gel.
7. Run the gel at 20 V/cm for approx 2 h. Pump the buffer from the anode to the cathode chambers and back again to avoid pH decrease in the anode chamber (*see* **Note 1**).

3.2. Establishing the Conditions for the Digestion of the DNA

1. Label the DNA at the 5'-ends using T_4 polynucleotide kinase and [γ-^{32}P] ATP. Take an aliquot of end-labeled DNA and remove one 5'-label by asymmetric cleavage with an appropriate restriction enzyme. Use the cleaved and uncleaved labeled DNA in separate analyses as described the following steps.
2. Ensure that the total amount of DNA in an assay of 20 µL is not below 100 ng. The total amount of radioactivity in one assay should be approx 20,000 cpm. Use

the optimal salt conditions optimal for complex. Add 6 m*M* Mg^{2+}, if this is not already present.

3. Add Exo III and incubate at 37°C. In order to establish the optimum conditions, perform a series of experiments using concentrations Exo III between 1 and 200 U per reaction (20 µL) and incubation times varying between 1 and 45 min. Exo III seems to be rather stable over a wide range of ionic strengths, and no large changes in activity were observed in the range between 0 and 100 m*M* NaCl (or KCl) in the incubation mixture (*see* **Note 2**).
4. Add EDTA to a final concentration of 20 m*M* in order to stop the reaction at the appropriate time.
5. Add sodium acetate to a final concentration of 0.3 *M* followed by 2.5 volumes of ice-cold 100% ethanol to precipitate the digested DNA. Keep the solution at –70°C for 20 min.
6. Spin down the solution in a microcentrifuge for 15 min. Wash the pellet with ice-cold 75% ethanol, dry under vacuum, and dissolve in formamide loading buffer.
7. Heat sample at 100°C for 2 min and apply onto a 6–10% sequencing gel. (For the analysis of fragments in the range of 50–150 bases, 8% polyacrylamide is adequate, as described in **Subheading 2.2.**). Use as a length standard a Maxam–Gilbert sequencing reaction of the 5'-end-labeled DNA fragment.
8. Run the gel at 50 W and a temperature of 60°C for 2 h.
9. After electrophoresis, expose the gel overnight at –70°C to X-ray film using an intensifying screen.
10. Ensure that the free DNA is fully digested (usually much shorter single-strand DNA fragments than the predicted half-full-length *[2]* are obtained).

3.3. Exo III Digestion of the Protein–DNA Complex

1. Form the complex using the conditions established under **Subheading 3.1.**
2. Subject the complex to Exo III digestion using the conditions established under **Subheading 3.2.** (*see* **Notes 1** and **3–6**).
3. Add 20 m*M* EDTA, 1% sodium dodecyl fulfate (SDS) (final concentration) to stop the reaction. SDS is necessary in order to destroy the protein–DNA complex.
4. Proceed as described in **Subheading 3.2.** (**steps 5–9**) for recovery and gel electrophoretic analysis of the DNA. For recovery of the DNA, a phenol extraction before the ethanol precipitation is advisable if a crude protein extract has been used for complex formation.

3.4. Modifications of the Procedure

Depending on the kind of protein–DNA complexes being investigated, several modifications of the procedure described in the previous section may be useful or necessary. If the protein–DNA complexes are not homogeneous (e.g., more than one type of protein complex can form or the DNA contains multiple binding sites for a given protein), the desired complex can be purified on a nondenaturing acrylamide gel, as described in **Subheading 3.1.**, provided the complexes show different gel mobilities and have half-lives long enough to survive the electrophoresis. Such a purification step requires the use of 10 times more radioactively labeled DNA.

The procedure is as follows:

1. Form the complex.
2. Subject the complex to digestion with Exo III according to **Subheading 3.3., step 2.**
3. Dialyze the complex by drop dialysis against a low salt buffer (e.g., 10 m*M* Tris-HCl, pH 7.9) in order to avoid salt effects during electrophoresis (**Subheading 3.1., step 5**).
4. Apply the complex to a nondenaturing gel as described in **Subheading 3.1., steps 6** and **7**. The half-life of the complex is in most cases not changed by the Exo III digestion of the DNA.
5. Expose the gel at –70°C to X-ray film using an intensifying screen. A 1-h exposure should be enough to recognize the complexed bands. If not, the recovered DNA will be insufficient for the subsequent sequencing gel analysis.
6. Before removing the film for development mark exactly its position on the gel.
7. Excise the complex bands of interest with a spatula. The band representing the free DNA will be visible as a smear after Exo III digestion.
8. To elute the complexed DNA, put the excised gel slice in 600 µL of bidistilled water. Heat the complex to 90°C for 3 min and shake overnight at room temperature. The effectiveness of the elution can be easily monitored by comparing the radioactivity of the eluate with the radioactivity of the gel slice.
9. Vacuum-dry the eluate in a SpeedVac concentrator.
10. Dissolve the pellet in 10 µL formamide buffer and spin down the gel residue.
11. Transfer the supernatant to a new Eppendorf tube and apply the sample to a sequencing gel as described in **Subheading 3.2., steps 7–9.**

4. Notes

1. The gel concentration has to be adjusted according to the molecular weight of the protein–DNA complex. Here we describe the conditions established for the study of the *E. coli* RNA polymerase (MW 455,000) and a DNA fragment of 130 bp carrying a promoter *(9)*. For some applications, another widely used nondenaturing gel system may be appropriate: 1X TBE buffer, 4% acrylamide, and 0.1% *bis*-acrylamide. Recirculation of the buffer is not necessary here.
2. It has been observed that many batches of commercially available Exo III contain an activity that removes the 5' label. A 5'-phosphatase or a 5'–3' exonuclease activity could account for this phenomenon. Filling in the 5' protruding ends using α-thio-dNTPs as described by some authors *(10,11)* may eliminate the problem. Addition of *E. coli* tRNA can reduce the effect, but will not completely avoid it.

3. Investigation of the complexes of specific binding of proteins and DNA in crude extracts using Exo III requires additional precautions in order to avoid problems caused by endogenous nuclease activities during Exo III exposure. To avoid this problem, sodium-phosphate, tRNA, deoxyoligonucleotides and fragmented phage DNA (e.g., 2 m*M* sodium phosphate, 1 μg of FX 174 DNA cut with *Hae*III, 10 μg of yeast tRNA, and 1 μg mixed p[dN]$_5$) should be added to the assay *(5)*. We find this suppresses nuclease and possibly phosphatase activities contained in the crude extracts (*see also* **Note 2**).
4. Testing different concentrations of Exo III and different incubation periods can provide additional information about the nature of the protein–DNA complex under study. If Exo III is able to "nibble" into a protected area with increasing exposure time, this indicates differences in the strength of protein–DNA interaction *(10)*.
5. Different binding sites for one or more proteins may be detected as distinct stop points for Exo III, as shown in **refs. *12–14***. This applies as much when working with crude extracts as when using purified factors. It is necessary, however, that the ratio of DNA to binding proteins be >1.
6. Heparin, which is often used as a DNA competitor for *E. coli* RNA–polymerase and other DNA-binding proteins, also interacts with Exo III and reduces its activity markedly.

References

1. Rogers, S. G. and Weiss, B. (1980) Exonuclease III of *Escherichia coli* K-12, an AP endonuclease. *Methods Enzymol.* **65,** 201–211.
2. Shalloway, D., Kleinberger, T., and Livingston, D. M. (1980) Mapping of SV 40 DNA replication origin region binding sites for the SV 40 DNA replication antigen by protection against Exonuclease III digestion, *Cell* **20,** 411–422.
3. Metzger, W., Schickor, P., and Heumann, H. (1989) A cinematographic view of *Escherichia coli* RNA polymerase translocation. *EMBO J.* **8,** 2745–2754.
4. Pavco, P. A. and Steege, D. A. (1990) Elongation by *Escherichia coli* RNA polymerase is blocked in vitro by a site specific DNA binding protein. *J. Biol. Chem.* **265,** 9960–9969.
5. Wu, C. (1985) An exonuclease protection assay reveals heat-shock element and TATA box binding proteins in crude nuclear extracts. *Nature* **317,** 84–87.
6. Loh, T. P., Sievert, L. L., and Scott, R. W. (1990) Evidence for a stem cell-specific repressor of Moloney murine leukemia virus expression in embryonic carcinoma cells. *Mol. Cell. Biol.* **10,** 4045–4057.
7. Carnevali, F., La Porta, C., Ilardi, V., and Beccari, E. (1989) Nuclear factors specifically bind to upstream sequences of a *Xenopus laevis* ribosomal protein gene promoter. *Nucleic Acids Res.* **17,** 8171–8184.
8. Fried, M. and Crothers, D. M. (1981) Equilibria and kinetics of lac repressor-operator interactions by polyacrylamide gel electrophoresis. *Nucleic Acids Res.* **9,** 6505–6525.
9. Heumann, H., Metzger, W., and Niehörster, M. (1986) Visualization of intermediary transcription states in the complex between *Escherichia coli* DNA-depen-

dent RNA polymerases and a promoter-carrying DNA fragment using the gel retardation method. *Eur. J. Biochem.* **158,** 575–579.
10. Straney, D. C. and Crothers, D. M. (1987) A stressed intermediate in the formation of stably initiated RNA chains at the *Escherichia coli* lac UV 5 promoter. *J. Mol. Biol.* **193,** 267–278.
11. Straney, D. C. and Crothers, D. M. (1987) Comparison of the open complexes formed by RNA polymerase at the *Escherichia coli* lac UV 5 promoter. *J. Mol. Biol.* **193,** 279–292.
12. Gaur, N. K., Oppenheim, J., and Smith, I. (1991) The Bacillus subtilis sin gene, a regulator of alternate developmental processes, codes for a DNA-binding protein. *J. Bact.* **173,** 678–686.
13. Owen, R. D., Bortner, D. M., and Ostrowski, M. C. (1990) ras Oncogene activation of a VL30 transcriptional element is linked to transformation. *Mol. Cell. Biol.* **10,** 1–9.
14. Wilkison, W. O., Min, H. Y., Claffey, K. P., Satterberg, B. L., and Spiegelman, B. M. (1990) Control of the adipisin gene in adipocyte differentiation. *J. Biol. Chem.* **265,** 477–482.

Further Reading

Kow, Y. W. (1989) Mechanism of action of *Escherichia coli* Exonuclease III. *Biochemistry* **28,** 3280–3287.

5

Hydroxyl Radical Footprinting

Evgeny Zaychikov, Peter Schickor, Ludmilla Denissova, and Hermann Heumann

1. Introduction

The basic principle of the DNA footprinting technique is the measurement of accessibility of the DNA using a probe. The probe can be any enzyme or a chemical reagent that is able to cut the DNA backbone. When the target DNA is a fragment containing a signal sequence for a sequence-specific binding protein, sites on the DNA that interact with the protein are inaccessible to the probe. After electrophoretic separation based on molecular weight, these inaccessible sites appear as blanks in an otherwise regular DNA cleavage pattern, thus revealing the characteristic interaction footprint for the binding protein.

Although the footprinting pattern is a characteristic of the protein–DNA interaction, it is also greatly affected by the type of probe used. Hydroxyl radicals provide DNA footprinting probes, which are very convenient to handle and are distinguished by a number of distinct advantages:

1. Hydroxyl radicals cut the DNA with almost no sequence dependence.
2. Because the probe is very small, the resolution of the footprint is very high (1 bp).
3. The cleavage reaction is effective over a wide range of buffer compositions, salt concentrations, pHs, and temperatures. Only glycerol, a radical scavenger, interferes with the cutting when present at concentrations higher than 0.5%.
4. All chemicals needed are easily available and uncomplicated in their handling.

1.1. Generation and Action of Hydroxyl Radicals

Hydroxyl radicals are generated according to the Fenton reaction by reduction of iron(II) with hydrogen peroxide as follows:

$$Fe^{2+}(EDTA^{4-}) + H_2O_2 \xrightarrow[\text{diothiothreited}]{\text{ascorbate or}} Fe^{3+}(EDTA^{4-}) + OH\cdot$$

From: *Methods in Molecular Biology, vol. 148: DNA–Protein Interactions: Principles and Protocols, 2nd ed.*
Edited by: T. Moss © Humana Press Inc., Totowa, NJ

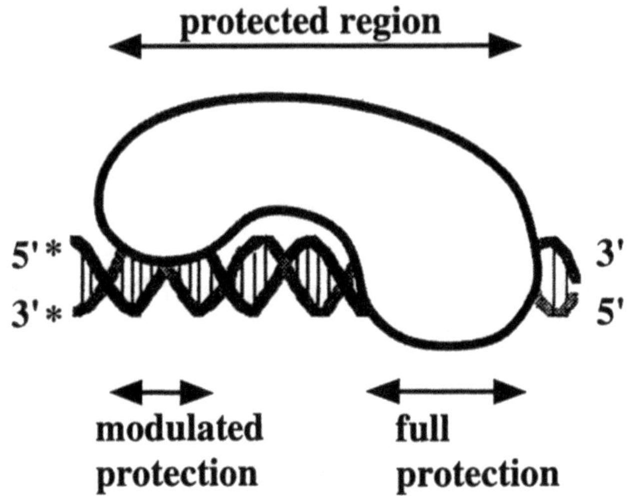

Fig. 1. A putative DNA-binding protein interacts with the DNA over three helical turns. The major portion of the protein interacts with only one side of the DNA over two helical turns. A minor portion of the protein wraps fully around the DNA. The asterisk indicates the position of the radioactive label.

The resulting iron(III) is reduced by ascorbate or dithiothreitol back to iron(II), which can start a new cycle. The use of a negatively charged $[Fe(EDTA)]^{2-}$ complex prevents the iron from interacting electrostatically with DNA, so the only reactant interacting with DNA is the hydroxyl radical generated in solution *(1)*.

An alternative source of hydroxyl radicals is potassium peroxonitrite *(2)*. The radicals are generated via its conjugate acid (ONOOH) when adding a stable alkaline solution of ONOOK in samples buffered at neutral pH:

$$ONOOH \rightarrow NO_2\cdot + OH\cdot$$
$$2NO_2\cdot \rightarrow N_2O_4$$
$$N_2O_4 + H_2O \rightarrow NO_3^- + NO_2^- + 2H^+$$

The exact manner in which hydroxyl radicals act on DNA is still not known. The radicals are thought to abstract an H-atom from the sugar moiety of the DNA backbone, and secondary reactions of the resulting sugar radical cause the backbone to break, leaving a gap in one strand of the double helix with the phosphate groups on either side *(3)*.

1.2. Principle of the Procedure

After formation of the complex of a sequence-specific binding protein and a DNA fragment carrying the binding sequence (*see* **Fig. 1**), the complex is sub-

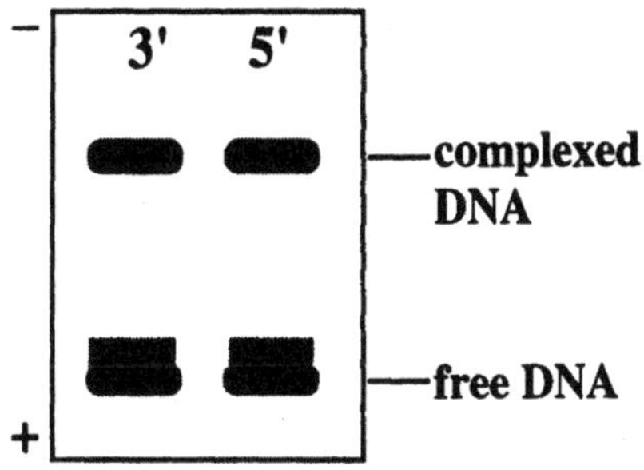

Fig. 2. Protein-DNA complexes are separated from free DNA by nondenaturing gel electrophoresis. The two lanes represent the same complex with a single label at one end of the DNA, at the 3' end, or at the 5' end. Hydroxyl radicals create gaps in the DNA, which cause a significant retardation of the modified fragments within the gel. This effect is visible only in the free DNA and is indicated by the shaded shoulder on the lower band.

jected to hydroxyl radical treatment. Hydroxyl radicals introduce single-base deletions randomly distributed in the DNA. The concentration of the hydroxyl radicals is adjusted so that the yield of deletions is less than one per DNA such that approx 10% of the DNA fragments are affected. Cutting of the DNA is prevented at those sites on the DNA where the protein is bound. This partially cut DNA is applied to a sequencing gel. If the DNA is detected by a unique terminal radioactive label, a DNA ladder is produced that is similar to that obtained by sequence analysis. Blanks within this regular ladder indicate the sites where the protein is bound. This footprint becomes more evident if a reference DNA is included that has been subjected to the same procedure but without previous protein binding. If complex formation is incomplete (i.e., the assay contains free DNA), the footprint will be masked by apparent cleavage within the protein-binding site. This can be avoided by separating the hydroxyl-radical-treated protein DNA complex from free DNA by nondenaturing gel electrophoresis (**Fig. 2**) or by nitrocellulose filter binding before application to a sequencing gel. **Figure 3** shows schematically the footprinting pattern of the protein DNA complex depicted in **Fig. 1**.

1.3. Interpretation of the Footprinting Pattern

Figure 3 shows that a footprinting analysis of both DNA strands includes six DNA ladders (i.e., two DNA sequencing reaction length standards, two free DNAs as reference, and two complexed DNAs). Blanks in the DNA ladder indicate exclusion of radical attack of the DNA because of the presence of the bound protein. These blanks can be assigned to specific sequence positions via the length standards.

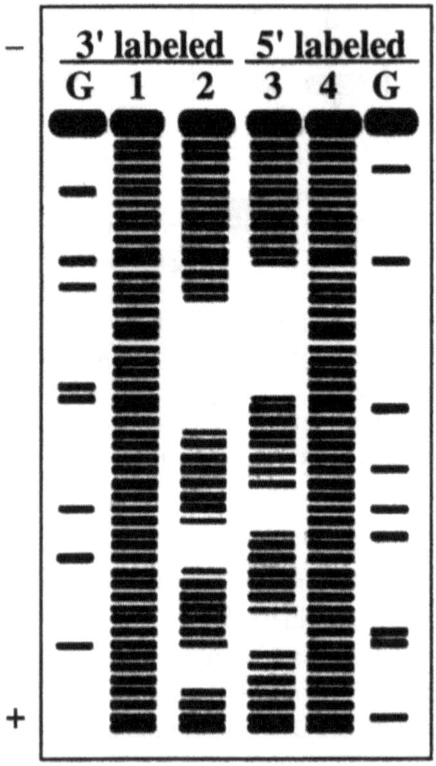

Fig. 3. The bands of the nondenaturing gel in **Fig. 2** containing the complex and the free DNA are eluted and applied (under denaturing conditions) to a sequencing gel. Lanes 1 and 4 show the free DNA labeled respectively at the 3' and the 5' ends. Lanes 2 and 3 show the DNA recovered from the complex. The complex depicted schematically in **Fig. 1** would result in the footprint displayed in lanes 2 and 3. Lanes G contain the length standards obtained by a G-specific Maxam–Gilbert sequencing reaction.

The following information can be extracted from the hydroxyl radical footprinting pattern:

1. The total size of the DNA sequence interacting with the protein can be determined from the extent of the uncleaved or blank positions.
2. A variation of the intensity of the bands within the interacting sequence reflects differences in the modes of interaction. By a comparison of the footprints on both strands, the modes of interaction can often be interpreted:
   a. If both strands show a blank at the same region, it indicates that the protein wraps around the DNA.
   b. A blank on only one strand indicates single-strand formation, with one strand protected by interaction with the protein.

c. Modulation of the intensity of the bands with a regular phasing according to the helix repeat (e.g., 10.3 bp for B-DNA) indicates binding of the protein to one side of the DNA. This interpretation is supported if the complementary strand shows the same pattern but with an offset of two or three bases. This offset is a consequence of the double-helical nature of the DNA, as shown schematically in **Fig. 1**.

1.4. Examples of the Application of Hydroxyl Radicals as Footprinting Probes

1.4.1. Protein DNA Complexes

Numerous Fe^{2+}-dependent hydroxyl radical footprinting studies were performed on protein–DNA complexes. Most useful were those studies of protein–DNA contacts within the transcription machinery.

1. Hydroxyl radicals were used to follow the formation of the transcriptionally active complex between the DNA-dependent RNA polymerase of *E. coli* and T7A1 promoter *(4)*. Temperature-dependent footprinting studies showed that the transcription initiation complex undergoes three different conformations characterized by a specific "footprint" until a transcription competent complex is formed. These conformations could be attributed to the so-called closed, intermediate, and open complex. A bent conformation of DNA in complex with DNA was concluded from the OH radical probing of the lambda P_R promoter complex *(5)*.
2. Site-specific cleavage of the DNA of a RNA polymerase binary complex by both free and EDTA-chelated Fe^{2+} was detected in absence of Mg^{2+} ions *(6,7)*. A phenomenon in FeEDTA-dependent OH radical footprints is hypercleavage of DNA which can be observed in footprints of Mg^{2+}-dependent proteins such as *E. coli* RNA polymerase *(6)* or HIV reverse transcriptase *(8)*. The *E. coli* RNA polymerase footprint contains a hyperreactive cleavage spot within the protected region. This spot could be attributed to a site-specific OH radical cleavage mechanism. This view was supported by footprinting studies using peroxonitrite as an alternative reagent for generating OH radicals. Using this latter reagent, no hyperreactive spot was visible in the OH radical footprint of *E. coli* RNA polymerase, indicating that specifically bound FeEDTA probe was responsible for the hyperreactive cleavage. Free Fe^{2+} ion can replace the catalytically active Mg^{2+} at the polymerization site of RNA polymerase, being chelated with aspartates of the active site *(7)*. The chelated Fe^{2+} generates OH radicals probably according to a mechanism that is analogous to the Fenton reaction and causes a strong local cleavage of both DNA and protein. This hyperreactivity was also observed in Fe^{2+}-dependent OH radical footprints of reverse transcriptase (RT) of the human immunodeficiency virus (HIV-RT). The catalytically active Mg^{2+} of the RNaseH active site of HIV-RT was replaced by Fe^{2+}, leading to site-specific OH radical cleavage of the DNA. It is interesting to note that no hyperreactive cleavage was observed at the other Mg^{2+}-carrying active site of HIV-RT (i.e., at the polymer-

ization site). These different effects of Mg^{2+}/Fe^{2+} substitution in the polymerization sites of HIV-RT and *E. coli* RNA polymerase probably reflect a variation of the redox potentials of the two sites.

3. The movement of *E. coli* RNA polymerase during mRNA synthesis was followed by probing a series of specifically arrested transcribing complexes with hydroxyl radicals ***(9–11)***.

1.4.2. Antibiotic DNA Complexes

Mithramycin, a small antitumor antibiotic drug, was shown to bind to the minor groove of GC-rich DNA sequences, thereby protecting only three bases from hydroxyl radical attack ***(12)***.

1.4.3. DNA Structures

The accessibility of bent DNA was studied using hydroxyl radicals. The bend was induced by A tracts repeated in phase with the helical repeat ***(13)***. Hydroxyl radicals can also be used to measure the number of base pairs per helical turn along any DNA molecule. The DNA is adsorbed onto crystalline calcium phosphate before being subjected to radical treatment. From the variation of the intensity of the bands, the helical periodicity of the DNA can be directly obtained ***(14)***.

1.4.4. RNA Protein Complexes

Splicing-specific ribonucleoprotein complexes were analyzed by hydroxyl radical treatment. These studies revealed that several regions of the 3'-splice site of mRNA precursors are not accessible for hydroxyl radicals, for example, the 3'-intron/exon junction, the polypyrimidine tract, and the site of branch formation were found to be all inaccessible ***(15)***.

1.4.5. RNA Structures

By using hydroxyl radicals, Celander and Cech ***(16)*** demonstrated that at least three magnesium ions are necessary for formation of a catalytically active ribozyme RNA molecule. By using both FeEDTA and ONOOK-generated hydroxyl radicals, information on higher-order structures of tRNA was obtained ***(2)***.

2. Materials

2.1. The Cutting Reaction

Prepare the following solutions separately (*see* **Note 1**):

1. 0.1 *M* dithiotreitol (DTT).
2. 1% Hydrogen peroxide.
3. Iron(II)–EDTA mix: Mix equal volumes of 2 m*M* ammonium iron(II) sulfate hexahydrate ($(NH_4)_2Fe(SO_4)_2 \cdot 6H_2O$) and 4 m*M* EDTA.
4. Stop mix: 4% glycerol, 0.6 *M* sodium acetate, 0.1 mg/mL carrier DNA.

2.2. The Sequencing Gel

1. Urea (ultra pure).
2. 20X TBE: 1 *M* Tris-base, 1 *M* boric acid, and 20 m*M* EDTA.
3. Acrylamide solution: 40% acrylamide and 0.66% bis-acrylamide (*see* **Note 2**).
4. 10% Sodium persulfate (*see* **Note 3**).
5. 10% *N,N,N',N'*-tetramethylethylene diame (TEMED).
6. Sequencing gel (8%): 21 g urea, 2.5 mL of 20X TBE, and 10 mL of the 40% acrylamide solution are made up to 50 mL with bidistilled H_2O and stirred under mild heating until urea is dissolved. The solution is filtered (filter pore size: 0.2 µm) and degassed for 5 min. Immediately before pouring the solution between the glass plates, add 0.3 mL of 10% sodium persulfate and 0.3 mL of 10% TEMED.
7. Loading buffer for the sequencing gel (stock solution): 100 mL formamide (deionized), 30 mg xylenecyanol FF, 30 mg bromophenol blue, and 750 mg EDTA.
8. Electrophoresis buffer: 1X TBE.

2.3. The Nondenaturing Gel for DNA Isolation and Electrophoretic Mobility Shift Assay

1. 20X TBE.
2. Acrylamide solution: 30% acrylamide, and 0.8% *bis*-acrylamide (*see* **Note 2**).
3. 10% Sodium persulfate (*see* **Note 3**)
4. 10% TEMED (aqueous solution).
5. 3% Nondenaturing gel: 1.5 mL of 20X TBE, 3 mL of the acrylamide solution, and 25.5 mL of bidistilled water are mixed and degassed for 5 min. Before pouring the solution between the glass plates add 300 µL of 10% ammonium persulfate and 300 mL of 10% TEMED.
6. Loading buffer for the nondenaturing gel (stock solution): 50% glycerol and 0.1% bromophenol blue.
7. Electrophoresis buffer: 1X TBE.

2.4. Other Items

1. Sequencing gel apparatus.
2. Apparatus for nondenaturing gel electrophoresis.
3. Filters for drop dialysis, VS 0.025 µm (Millipore, Bedford, MA).
4. Filtration device (optional).
5. Nitrocellulose filters (BA85, Schleicher & Schüll) (optional).
6. Peristaltic pump (optional).

3. Method

3.1. Establishing the Conditions for Obtaining Optimum Yield of Specific Protein–DNA Complexes

The method of establishing the conditions for complex formation using the band-shift assay is described in Chapter 4, *see also* Chapter 2.

3.2. Establishing the Conditions for Cutting the DNA by Hydroxyl Radicals

1. End label an aliquot of the DNA fragment of interest under standard conditions *(17)* at the 5'-position, using T_4 polynucleotide kinase and [γ-^{32}P] ATP and end label a second aliquot at the 3'-position, using the Klenow fragment of DNA polymerase I and the appropriate [α-^{32}P] dNTP. In each case, remove one label end by asymmetric cleavage of the DNA fragment with an appropriate restriction endonuclease. Purify your uniquely end-labeled DNA fragment by a non-denaturing gel electrophoresis (*see* **Subheading 2.3.**). The total amount of radioactivity in one assay should be approx 10,000–15,000 cpm. The total amount of DNA in an assay of 20 µL should not be below 100 ng. The optimum length of the DNA fragment is between 100 and 200 base pairs (bp) (*see* **Note 4**).
2. For the cutting reaction, use the buffer conditions and temperature that are optimum for formation of the protein–DNA complex, (*see* **Note 5**). Add to the 20 µL incubation assay 2 µL of each of the previously prepared solutions of DTT, hydrogen peroxide, and the iron(II)–EDTA and mix rapidly. This can be done by placing the individual 2-µL drops separated on the inner wall of the tube and then rapidly mixing them before combining them with the sample using a micropipet.
3. Incubate for variable times (1–5 min is recommended) at the appropriate temperature.
4. Add 25 µL stop mix and 150 µL of ice-cold 100% ethanol to precipitate the DNA. Keep the solution at –70°C for 30 min.
5. Spin down the precipitate in a microcentrifuge for 30 min. Wash the pellet with ice-cold 80% ethanol, dry under vacuum and dissolve the pellet in formamide buffer. Adjust the amount of radioactivity and volume of each sample to approx 5000–6000 cpm in 4–5 µL.
6. Heat the sample for not longer than 2 min at 90°C and place on ice (*see* **Note 6**). Apply the sample onto a 6–10% sequencing gel (for the analysis of fragments in the range of 50–150 bases, a gel consisting of 8% acrylamide is adequate). Use as length standards a Maxam–Gilbert sequencing reaction of the 5'- or 3'-labeled DNA fragment.
7. Run the gel at 50 W at a temperature of 60°C for 1.5–2 h. The gel is ready when the xylenecyanol dye marker is about 3–5 cm above the bottom of the gel.
8. After electrophoresis, expose the gel to an X-ray film using an intensifying screen at –70°C overnight. For subsequent experiments, choose the time of hydroxy radical cleavage that provides an even distribution of bands and leaves around 90% of the DNA uncleaved.

3.3. Footprinting of Protein-DNA Complexes

3.3.1 Preparation of Complexes and Performing Footprinting Reaction

1. Prepare two 15-µL samples of the complexes, one with the DNA labeled at the 3' end and another with the label at the 5' end. Use the conditions established for

optimal complex formation from **Subheading 3.1.** The total amount of radioactivity in one assay should be approx 60,000–80,000 cpm.
2. Pour 30–40 mL of the dialysis buffer containing 8 m*M* Tris–HCl, pH 7.9, into a Petri dish (*see* **Note 5**). Place a Millipore filter (VS 0.025 µm) on the surface of the buffer, shiny side (hydrophobic side) up. Put the samples containing the complexes onto the filter for 1 h in order to remove glycerol and salt (*see* **Note 7**). Remove the samples from the filter using a micropipet and transfer them to a fresh 1.5-mL Eppendorf tube.
3. Subject the two samples to hydroxyl radical treatment as described in **Subheading 3.2., steps 2** and **3** only.

3.3.2. Separation of the Complex and the Free DNA by Nondenaturing Acrylamide Gel Electrophoresis

The conditions for studying a high-molecular-weight complex are described below (*see* **Note 8**). Use the conditions for preparing the nondenaturing gel as described in **Subheading 2.3.** These low-concentration gels are difficult to handle, therefore, the glass plates must be subjected to a special treatment by which the gel is bound to one of the plates. Wash the glass plates (20 × 20 cm) with ethanol. Treat one plate with γ-methacryl-oxypropyl-trimethoxy-silane. Wash this plate carefully four times with ethanol to avoid sticking of the other plate to the gel. Treat the second plate with dichlorodimethylsilane. Use spacers 1–1.5 mm thick. Use combs that allow you to apply amounts of the 50-µL sample.

1. Add loading buffer to the sample (1/10 of the sample volume). Apply the sample onto the nondenaturing acrylamide gel (20 × 20 cm). Run the gel at 20 V/cm for about 2 h. The gel is ready when the dye marker is approx 3–5 cm above the end of the gel.
2. Remove one of the plates by lifting it carefully with a spatula and cover the gel (which sticks to the other glass plate) with plastic wrap. Place an X-ray film on the gel. Put the gel and the film into a cassette and expose for 1–2 h at 4°C. Mark the bands containing the complex and, if visible, the free DNA. This is possible by replacing the film after developing on the gel in exactly the same position. The use of fluorescent marker tapes (e.g., from Stratagene) during exposure of the gel may be very helpful for exact repositioning of the film.
3. Remove the plastic wrap from the gel, cut out the marked bands with a scalpel or spatula, and cut them into small pieces. Put the slices into a eppendorf tube along with 300 µL of elution buffer (0.5% SDS, 1 m*M* EDTA, 0.1 mg/mL carrier DNA) and shake for several hours at room temperature.
4. Spin the tube for a few seconds in a microcentrifuge and transfer the liquid to a new tube. Avoid transferring gel pieces to the new tube. Add 200 µL of water-saturated neutralized phenol, shake 1 min, and centrifuge for 5 min at room temperature. Remove aqueous layer (upper), transfer it to a fresh Eppendorf tube, and add 30 µL of 3 *M* sodium acetate and 1 mL of ice-cold 100% ethanol. Shake the tube for a few seconds and put into –70°C for at least 30 min.

5. Spin the sample in a microcentrifuge for 30 min and remove the supernatant. Wash once with 1 mL of ice-cold 80% ethanol. Dry the sample under vacuum. Dissolve the pellet in a small volume of formamide buffer (usually 5–10 µL, depending on the amount of radioactivity).

*3.3.3. Separation of the Complex and the Free DNA Filtration Through Nitrocellulose Filter (Alternative to **Subheading 3.3.2.**, see **Note 9**)*

The reader may also find it useful to refer to Chapter 1, which discusses the uses of the filter binding assay, and to **Subheading 3.2.3.** of that chapter, which deals with the recovery of DNA from filter-retained complexes.

1. Stop the hydroxyl radical reaction by dilution (at least twofold) with binding buffer containing 2% glycerol. (Alternatively, the sample may be poured directly into the filtration device after completion of the reaction and immediately washed).
2. Mount a nitrocellulose filter (BA85, Schleicher & Schüll) into the filtration device and connect it to a peristaltic pump. (*See* Note 10.)
3. Pore the solution containing the complex treated with radicals onto the filter. Switch on the pump and filter off the solution at the speed of 1–2 mL/min. Wash the filter twice with binding buffer at the same speed. Switch off the pump.
4. Disassemble the filtration device, take off the filter and cut it into small pieces. Place the pieces into an Eppendorf tube.
5. Add 300 µL of solution containing 1% SDS, 0.3 *M* sodium acetate, and 10 µg/mL carrier DNA to the tube and shake for 15 min at room temperature.
6. Carefully transfer the liquid into fresh tube (avoid transferring the nitrocellulose) and add 1 mL of ethanol. Shake the tube for a few seconds and put into –70°C for at least 30 min.
7. Spin the sample in a microcentrifuge for 20 min and remove the supernatant. Wash once with 1 mL of ice-cold 80% ethanol. Dry the sample under vacuum. Dissolve the pellet in a small volume of formamide buffer (usually 5–10 µL, depending on the amount of radioactivity).

3.3.4. Analysis of DNA by Denaturing Gel Electrophoresis

Heat the samples at 90°C for 1–2 min. Proceed as described for free DNA in **Subheading 3.2., steps 7–10**. For both 3' and 5' end-labeled DNA, load the following samples on the gel: the DNA recovered from the complex, the free DNA (either recovered from the gel and/or prepared separately as in **Subheading 3.2.**) and the Maxam–Gilbert reaction ladders as length standard.

4. Notes

1. The iron(II), iron–EDTA mix, and the H_2O_2 solutions should be freshly made before use. The solutions of DTT (0.1 *M*), EDTA (4 m*M*), H_2O_2 (as a 30% stock solution), and the stop mix are stable for months stored at –20°C.

2. The acrylamide solutions are stable for months if protected from light and kept at 4°C.
3. Sodium persulfate has an advantage over routinely used ammonium persulfate of being much more stable in aqueous solution. The 10% sodium persulfate solution may be kept at least 1 mo at +4°C without loss of activity.
4. It is strongly recommended to check the quality of the labeled DNA before use on a sequencing gel. Nicks in the double strand, which could be derived from DNase activities during preparation, will appear as additional bands in the sequencing gel. This admixture of bands will spoil the whole footprint, even when present in only small amounts. Furthermore, it is recommended not to store the pure, labeled DNA longer than 2 wk at –70°C, because the radiation of the label also creates nicks in the DNA.
6. Longer heating or boiling creates additional cuts in the DNA.
5. The buffer conditions can be varied (e.g., the pH), but the ionic strength should not be too high (maximum 50 m*M* NaCl) in order to obtain sharp bands during the nondenaturing electrophoresis to separate complexed and uncomplexed DNA. Many protein–DNA complexes are very stable at low ionic strength (e.g., complexes between RNA polymerase and promoters *[18]*). Therefore, in most cases the stability of the pH in the following electrophoresis is the only limitation to lowering the ionic strength.
7. The purpose of the dialysis is twofold: Removal of glycerol which interferes with the cutting reaction and removal of salt which lowers the quality of the electrophoresis pattern. As a rough approximation, one can remove up to 80–90% of the glycerol and salt present in the sample within 1 h using drop dialysis as described.
8. The gel concentration should be adjusted according to the molecular weight of the protein–DNA complex. Here we have described the conditions established for the study of the *E. coli* RNA polymerase (molecular weight 490,000) and a DNA fragment of 130 bp carrying a promoter *(3)*.
9. Filtration via nitrocellulose is a simpler and faster way to remove unbound DNA, but it does not resolve complexes having different stoichiometries of protein and DNA.
10. A variety of filtration devices may be used allowing handling of relatively small volumes (below 1 mL). Mild vacuum may be used instead of a peristaltic pump.

References

1. Tullius, Th. D., Dombroski, B. A., Churchill, M. E. A., and Kam, L. (1987) Hydroxyl radical footprinting: a high resolution method for mapping protein–DNA contacts. *Methods Enzymol.* **155,** 537–558.
2. Götte, M., Marquet, R., Isel, C., Anderson, V. E., Keith, G., Gross, H., Ehresmann, C., et al. (1996) Probing the higher order structure of RNA with peroxonitrous acid. *FEBS Lett.* **390,** 226–228.
3. Shafer, G. E., Price, M. A., and Tullius, Th. D. (1989) Use of the hydroxyl radical and gel electrophoresis to study DNA structure. *Electrophoresis* **10,** 397–404.

4. Schickor, P., Metzger, W., Werel, W., Lederer, H., and Heumann, H. (1990) Topography of intermediates in transcription initiation of *E. coli*. *EMBO J.* **9,** 2215–2220.
5. Craig, M. L., Suh, W. C., and Record T. M. (1995) HO· and DNase I probing of Eσ70 RNA polymerase–λP_R promoter open complexes: Mg^{2+} binding and its structural consequences at the transcription start site. *Biochemistry* **34,** 15,624–15,632.
6. Zaychikov, E., Denissova L., Meier, T., Götte, M., and Heumann H. (1997) Influence of Mg^{2+} and temperature on formation of the transcription bubble *J. Biol. Chem.* **272,** 2259–2267.
7. Zaychikov, E., Martin, E., Denissova, L., Kozlov, M., Markovtsov, V., Kashlev, M., et al. (1996) Mapping of catalytic residues in the RNA polymerase active center. *Science* **273,** 107–109.
8. Götte, M., Maier, G., Gross, H., and Heumann, H. (1998) Localization of the active site of HIV-1 reverse transcriptase–associated RNase H domain on a DNA template using site-specific generated hydroxyl radicals. *J. Biol. Chem.* **273,** 10,139–10,146.
9. Metzger, W., Schickor, P., and Heumann, H. (1989) A cinematographic view of *Escherichia coli* RNA polymerase translocation. *EMBO J.* **8,** 2745–2754.
10. Zaychikov, E., Denissova, L., and Heumann, H. (1995) Translocation of the *Escherichia coli* transcription complex observed in the registers 11 to 20: "Jumping" of RNA polymerase and asymmetric expansion and contraction of the "transcription bubble". *Proc. Natl. Acad. Sci. USA* **92,** 1739–1743.
11. Heumann, H., Zaychikov, E., Denissova, L., and Hermann, T. (1997) Translocation of DNA-dependent *E. coli* RNA polymerase during RNA synthesis, in *Nucleic Acids and Molecular Biology*, vol. 11, (Eckstein, F. and Lilley, D., eds.), Springer-Verlag, Heidelberg, Germany, pp. 151–177.
12. Cons, B. M. G. and Fox, K. R. (1989) High resolution hydroxyl radical footprinting of the binding of mithramycin and related antibiotics to DNA. *Nucleic Acids Res.* **17,** 5447–5459.
13. Burkhoff, A. M. and Tullius, Th. D. (1988) Structural details of an adenine tract that does not cause DNA to bend. *Nature* **331,** 455–457.
14. Tullius, Th. D. and Dombroski, B. A. (1985) Iron(II) EDTA used to measure the helical twist along any DNA molecule. *Science* **230,** 679–681.
15. Wang, X. and Padgett, R. A. (1989) Hydroxyl radical "footprinting" of RNA: application to pre-mRNA splicing complexes. *Proc. Natl. Acad. Sci. USA* **86,** 7795–7799.
16. Celander, D. W. and Cech, Th. R. (1991) Visualizing the higher order folding of a catalytic RNA molecule. *Science* **251,** 401–407.
17. Maniatis, T., Fritsch, E. F., and Sambrock, J. (1982) *Molecular Cloning, a Laboratory Manual*, Cold Spring Harbor Laboratory, Cold Spring Harbor, NY.
18. Heumann, H., Metzger, W., and Niehörster, M. (1986) Visualization of intermediary transcription states in the complex between *Escherichia coli* DNA-dependent RNA polymerase and a promoter-carrying DNA fragment using the gel retardation method. *Eur. J. Biochem.* **158,** 575–579.

Further Reading

Tullius, Th. D. (1989) Physical studies of protein–DNA complexes by footprinting. *Annu. Rev. Biophys. Biophys. Chem.* **18,** 213–237.

Tullius, Th. D. (1989) Structural studies of DNA through cleavage by the hydroxyl radical, in *Nucleic Acids and Molecular Biology,* vol. 3, (Eckstein, F. and Lilley, D., eds.), Springer-Verlag, Heildelberg, Germany, pp. 1–12.

6

The Use of Diethyl Pyrocarbonate and Potassium Permanganate as Probes for Strand Separation and Structural Distortions in DNA

Brenda F. Kahl and Marvin R. Paule

1. Introduction

In the search for methods to explore the interaction between proteins and DNA, a plethora of footprinting techniques have been developed, many of which are discussed elsewhere in the present work. Most footprinting techniques are based on the simple premise of specific DNA regions being protected from the reagent by the bound protein or molecule of interest. However, a number of studies over the past decade have revealed remarkable distortion of the DNA molecule, including bending and strand separation, in response to the bound protein. These distortions are often within the classical footprint but are rarely detected by these classical techniques. Thus, their detection requires alternative approaches. Unlike enzymatic methods, the chemical probes diethyl pyrocarbonate (DEPC) and potassium permanganate can access and react with the entire sequence of the DNA, distinguishing DNA distortions "under the foot" of a typical footprinting experiment. Combining the information obtained from these chemical probing techniques with the spatial information afforded by traditional footprinting gives an in-depth account of the various ways proteins and other molecules interact with DNA.

Diethyl pyrocarbonate and potassium permanganate are useful probes because of their preferential reactivity with single-stranded vs double-stranded DNA. In addition, unlike typical footprinting techniques where near saturation of the DNA with protein is necessary to observe a clean footprint, potassium permanganate and DEPC probing does not require most of the DNA to be in complex with the protein or molecule of interest. Because of the sensitivity of

From: *Methods in Molecular Biology, vol. 148: DNA–Protein Interactions: Principles and Protocols, 2nd ed.*
Edited by: T. Moss © Humana Press Inc., Totowa, NJ

these chemical probes, it is possible to observe reactive bases even when the population of single-stranded DNA is very small.

The mechanism by which DEPC modifies bases has been investigated, but it is still not clearly understood ***(1–3)***. DEPC predominantly reacts with purine residues, but it may react weakly with cytosine residues as well. DEPC modifies DNA by an out-of-plane attack on several of the nucleophilic centers in purines, leading to the scission of the glycosidic bond. Double-stranded B-form DNA does not undergo modification by DEPC because the close stacking of neighboring bases occludes access to the out-of-plane surfaces. However, under conditions in which the conformation deviates from B form (such as strand separation or bending), purines in the sequence become more accessible to modification by DEPC. Carbethoxylation of the imidazole ring N-7 produces strand scission under alkaline conditions. Thus, DEPC is commonly used to detect purines that are present in melted or distorted DNA sequences. Although DEPC can react with both purines, it shows a marked preference for adenines vs guanines in most instances.

Potassium permanganate reacts with double bonds, oxidizing them to vicinal diols. In nucleic acids, the base thymine is oxidized most vigorously, whereas reaction with C, G, and A is minimal. The mechanism behind this preferential reactivity is believed to arise from an out-of-plane attack on the 5,6 double bond of the thymine ring ***(4–9)***. Although the ring is still intact, the loss of aromaticity resulting from insertion of hydroxyl groups on the 5 and 6 carbons leads to a reduction in hypochromicity. Treatment of the vicinal diol with strong base leads to ring opening and cleavage of the phosphodiester backbone. As with DEPC, stereochemical hindrance from base stacking prohibits reactivity of double-stranded B-form DNA. In DNA, which is denatured or is altered from the B form, the thymine ring becomes susceptible to modification by potassium permanganate. Because of their base preferences, the combined use of DEPC and potassium permanganate allows complete analysis of both GC– and AT–containing sequences in DNA.

Diethyl pyrocarbonate and potassium permanganate modification have been used to detect a number of distorted DNA structures in both prokaryotic and eukaryotic cells, including open complex formation during transcription ***(10–14a)***, steps in promoter clearance ***(14a–17)***, elongation ***(17–19)***, and termination ***(20,21)***, RNA–DNA hybrid structures in transcription elongation complexes ***(22,23)***, drug binding to DNA ***(24–26)***, chromatin positioning ***(27,28)***, recombination events ***(29–31)***, and single-stranded binding protein binding domains ***(32–34)***. On DNA alone, these reagents can reveal sequence-dependent distortions ***(35–37)***, negatively supercoiled DNA ***(38–40)***, cruciform DNA structures ***(41,42)***, DNA hairpins such as those found in triplet expansion diseases like fragile X-syndrome ***(43,44)***, and Z-DNA, H-DNA, or triplex DNA ***(45–48)***.

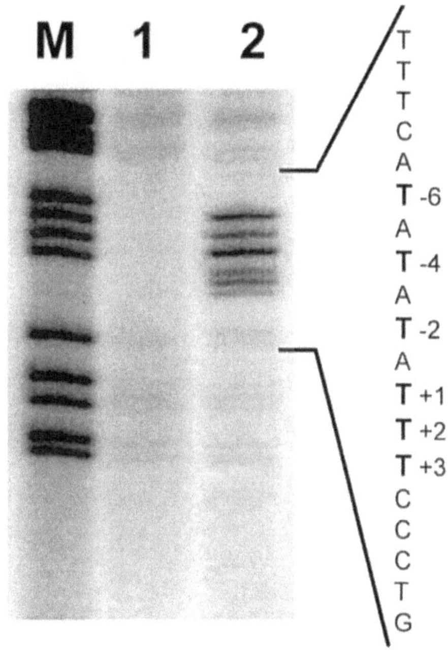

Fig. 1. Potassium permanganate sensitive sites when an open promoter complex is formed on the coding strand by RNA polymerase I from *Acanthamoeba castellanii*. Lane *M* contains a G + A Maxam–Gilbert sequencing ladder, lane 1 contains DNA exposed to $KMnO_4$ treatment in the absence of any proteins; lane 3 contains DNA with the addition of proteins necessary for melting of DNA. Numerical designations refer to the transcription start site. Hypersensitive sites in bold denote regions of strand separation.

Figures 1 and **2** give examples of the data which can be obtained. **Figure 1** shows one of the more common uses of potassium permanganate, the analysis of open promoter complex formation. **Figure 2** shows results for both DEPC and potassium permanganate acting on a stalled transcription elongation complex, showing both the melted transcription bubble and the unreactive RNA–DNA hybrid.

Virtually any sequence that deviates from B-form DNA is a candidate for probing with DEPC and potassium permanganate. Because DEPC and potassium permanganate only modify the susceptible bases without cleaving the phosphodiester backbone, further steps need to be taken to visualize the positions of the modified bases. There are three different methods commonly used: The first, which can only be used for experiments performed in vitro, utilizes 5' or 3' end-labeled DNA fragments. After treatment with DEPC or potassium permanganate, the DNA is treated with piperidine to cleave the

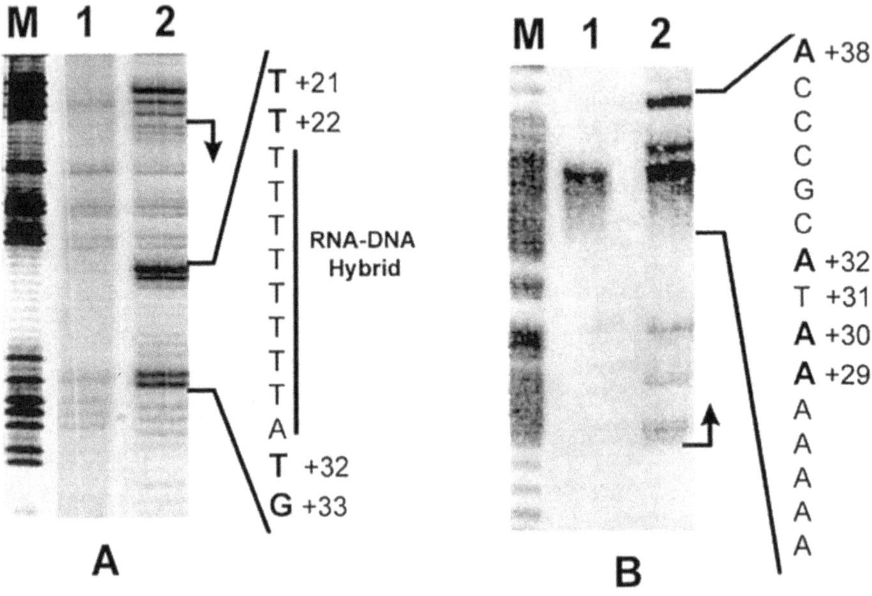

Fig. 2. Potassium permanganate and DEPC probing of stalled transcription complex. **(A)** Potassium permanganate probing of the coding strand when RNA polymerase I is paused at +31 from the transcription start site. Lane *M* contains G + A Maxam-Gilbert sequencing ladder, lane 1 contains DNA alone with $KMnO_4$ treatment; lane 2 displays the permanganate sensitive sites when RNA polymerase from *Acanthamoeba castellanii* is paused at +31 from the transcription start site. The hypersensitive sites in bold at positions +21, +22, and +32, +33 define the transcription bubble on the coding strand. Hyposensitive thymidines in the transcription bubble depict protection due to the formation of an RNA-DNA hybrid. **(B)** DEPC probing on the noncoding strand when RNA polymerase I is paused at +31 from the transcription start site. Lanes are the same as in (A). DEPC-sensitive sites in bold define the leading edge of the transcription bubble on the noncoding strand. Unreactive As in the region may be the result of protein interference or the slightly less reactive nature of DEPC compared to $KMnO_4$.

phosphate backbone on the 3' side of the modified nucleotide. This procedure is useful when comparing the results from multiple footprinting techniques, because the same fragment of labeled DNA may be utilized with each type of experiment. This method also works well when examining short tracts of DNA or DNA that does not amplify well in a thermocycler. The other two methods, primer extension and thermocycle amplification, are used when working in vivo (potassium permanganate only) or with circular pieces of DNA in vitro. Thermocycle amplification is particularly beneficial when working with limited amounts of DNA (in the low nanogram range) or when the ratio of protein–DNA complex to DNA is low. The following materials and methods sections are general guidelines for probing with DEPC and potassium permanganate. Care and attention to detail is necessary for all of these methods, as footprinting with these reagents is generally somewhat more difficult than with many other footprinting reagents. There is usually a need to optimize conditions for each particular application; guidelines for this can be found in the Notes section.

2. Materials

1. Potassium permanganate, DEPC, piperidine, and 2-mercaptoethanol can be purchased from Sigma (St. Louis, MO). DEPC, piperidine, and 2-mercaptoethanol should be stored at 4°C and used with caution in a fume hood. Other reagents listed below should be of the highest quality. Low-adhesion microcentrifuge tubes (siliconized) can be obtained from USA Scientific (Ocala, FL).
2. The following stock solutions can be made, filter sterilized, divided into aliquots and stored at –20°C until ready for use:
   - 1 *M* HEPES pH 7.9
   - 1 *M* $MgCl_2$
   - 0.5 *M* dithiothreitol (DTT)
   - 2 *M* KCl
   - 10 mg/mL bovine serum albumin (BSA), DNase free
   - 0.5 *M* EDTA pH 8.0
   - 3 *M* Sodium acetate pH 5.2
   - 5 mg/mL linear polyacrylamide (acrylamide polymerized without *N,N'*-methylene bisacrylamide)
   - 10 mg/mL Proteinase K
   - 10% sodium dodecyl sulfate (SDS)
   - DEPC stop buffer: 0.2% SDS and 0.6 *M* sodium acetate pH 5.2
   - TE: 10 m*M* Tris–HCl pH 8.0 and 1 m*M* EDTA
   - Electrophoresis loading buffer: 0.1% bromophenol blue, 0.1% xylene cyanol, 10 m*M* EDTA, and 80% deionized formamide
3. A 300 m*M* stock solution of $KMnO_4$ can be made by heating 2.73 g of $KMnO_4$ in 50 mL of deionized water. The solution can be stored at room temperature in a brown bottle for 1 mo.
4. 10X reaction buffer, for example: 200 m*M* HEPES pH 7.9, 100 m*M* $MgCl_2$, 1 m*M* DTT, 1 mg/mL BSA, and 100 m*M* KCl. The choice of this buffer will depend on the conditions needed to form the complex under study.
5. Protein dilution buffer, for example: 20 m*M* HEPES pH 7.9, 1 m*M* EDTA, and 10% glycerol. Choice will depend on protein used.
6. 1 *M* piperidine: 10% v/v in water, freshly made.
7. 0.6 *M* sodium acetate.
8. 0.6 *M* sodium acetate, 20 m*M* EDTA
9. 10X neutralization buffer: 0.5 *M* HEPES pH 7.9, 0.1 *M* $MgSO_4$, and 2 m*M* DTT.
10. dNTP mix: 5 m*M* of each dNTP.

3. Methods

Experiments using DEPC or potassium permanganate follow similar protocols. The only exceptions are the addition of 2-mercaptoethanol to quench reactions using potassium permanganate, the duration of the DNA modification reactions, and the necessity to purify the modified DNA away from DEPC before the cleavage step.

3.1. In Vitro Experiments on Linear DNA Fragments

1. End-labeled DNA (5' or 3') should be separated from excess radiolabeled precursor either by agarose gel electrophoresis or size-exclusion column chromatography, and stored in TE.
2. In a 1.5-mL siliconized microcentrifuge tube, add 4 μL 10X reaction buffer, 40,000 cpm of DNA, and sterile deionized water to a volume of 20 μL. Proteins, diluted in an appropriate buffer, are added to the reaction to give a final volume of 40 μL (*see* **Note 1**). Incubate for the desired amount of time to form DNA–protein complexes.
3. Modifying the susceptible bases (*see* **Notes 2–4**):

   $KMnO_4$ treatment: Add freshly diluted potassium permanganate to give the appropriate concentration and incubate for desired time period. For example, adding 2 μL of 100 m*M* $KMnO_4$ (approximately 9 m*M* final concentration) for 2 min seems to work well for detecting melted DNA in transcription initiation complexes.

   Diethyl procarbonate treatment: Add 1 μL of DEPC to each tube. Mix by briefly vortexing and repeat vortexing every 5 min for 15 min. Vortexing is necessary because DEPC is sparingly soluble in aqueous solutions, so all reactions are run at essentially saturating DEPC and the concentration cannot be altered significantly.
4. Stopping the reactions:

   $KMnO_4$ treatment: Quench the reaction by adding 3 μL of 2-mercaptoethanol, vortex and place on ice. Add 45 μL of 0.2% SDS and 2 mg/mL proteinase K and incubate at 50°C for 1 h. Add 90 μL of 0.6 *M* sodium acetate, 300 μg/mL linear polyacrylamide, and 2.5 volumes 95% ethanol. Mix and centrifuge 30 min at 14,000 rpm in a microfuge. Remove supernatant and wash with 150 μL of 70% ethanol. Centrifuge 5 min, as above. Remove supernatant and dry pellet on medium heat for 5 min in a Speed Vac.

   Diethyl procarbonate treatment: Stop the reaction by adding an equal volume of DEPC stop buffer and phenol–$CHCl_3$ extract. Add 5 μL of 5 mg/mL linear acrylamide, and precipitate with 2.5 volumes of ethanol as above. After centrifugation, rinse with 70% ethanol and centrifuge again. Remove supernatant and dry pellet on medium heat for 5 min in a Speed Vac.
5. Alkaline cleavage: Suspend the pellet in 50 μL of 1 *M* piperidine (10% v/v) and incubate at 90°C for 30 min. Place a lead weight on top of the tubes or use tube locks to prevent the lids from opening.

6. Place tubes on ice to cool and centrifuge briefly. Add 50 µL of 0.6 *M* sodium acetate, 300 µg/mL linearized polyacrylamide, and 250 µL 95% ethanol. Mix and centrifuge 30 min as above. Wash pellet with 150 µL of 70% ethanol. Spin samples for 5 min. Remove supernatant.
7. To remove residual piperidine, add 30 µL sterile deionized water to each sample and dry on medium heat in a Speed Vac (*see* **Note 5**).
8. Add 5 µL electrophoresis loading buffer, vortex samples for 30 s, and heat samples at 95°C for 3 min. Place samples on ice.
9. Load samples on sequencing gel and analyze by standard methods (*see* **Notes 6** and **7**).

3.2. Treatment of DNA In Vivo with KMnO4 and Purification

Potassium permanganate has been used to modify DNA in vivo, followed by analysis of modifications by primer extension or polymerase chain reaction (PCR). DEPC cannot be used for in vivo experiments because of its low solubility. The procedure for in vivo modification is fairly straightforward, however, certain nutrient-rich media can quench permanganate. To avoid this problem, use minimal medium or increase the permanganate concentration so that the reaction mixture does not turn brown in less than 1 min. For some experiments, one can dilute the culture in minimal medium just prior to treatment with potassium permanganate. In vivo modification of mammalian cell cultures usually requires the removal of the growth medium just prior to treatment.

1. To 10 mL of diluted bacterial or yeast culture, add the appropriate amount of $KMnO_4$ (typically in the low millimolar range, depending on the medium, approx 10–20 m*M* for most media) for the desired amount of time (10 s to 5 min) in a shaking water bath. To quench treatment, remove samples from the water bath and pour immediately into prechilled Corex tubes and add 2-mercaptoethanol until the purple color disappears. Centrifuge to pellet cells in a cold Sorvall SS34 rotor for 5 min at 3015*g*. Discard the supernatant.
2. For mammalian cells grown to subconfluence on plastic growth dishes, remove growth medium and wash twice with phosphate-buffered saline or minimal growth medium. Add the desired concentration of potassium permanganate (usually 2–20 m*M*) for the necessary period of time (10 s to 5 min). Stop the permanganate reaction by washing cell monolayers twice with phosphate-buffered saline containing 2% 2-mercaptoethanol and once with phosphate-buffered saline. Cells are then harvested with a rubber policeman or cell scraper.
3. Plasmid and genomic DNA can be isolated by a variety of standard methods ***(49–51)***. Purified modified DNA should be adjusted to a final concentration of approx 15 ng/µL and be free of contaminants that interfere with extension reactions. Extractions involving phenol should be repeated three to four times or until there is no contaminating interphase. Modifications to the DNA can then be visualized by PCR amplification, as detailed in **Subheading 3.4.** An alternative method for identifying modified genomic DNA is the use of ligation mediated PCR (LMPCR) ***(52,53)***.

3.3. Primer Extension Analysis of DNA Treated In Vitro or In Vivo

Because DNA modified in vivo or when in circular form is not end labeled, primer extension is the method of choice to analyze the sites of modification. Modified bases result in extension stop sites because they block the elongating DNA polymerase. For in vitro studies, follow **steps 2–4** of **Subheading 3.1.**, substituting 20–500 ng of purified DNA in place of radiolabeled DNA. For in vivo studies, follow the steps of **Subheading 3.2.**

1. To the isolated, modified DNA (20–500 ng for in vitro and 500 ng for in vivo studies), add (0.3–0.5) × 10^6 cpm of 5' end-labeled primer and dilute to 36 μL with distilled water.
2. Add 4 μL of 0.01 *M* NaOH to each reaction and mix well.
3. Denature DNA by heating to 95°C for 2 min.
4. Add 5 μL of 10X neutralization buffer and mix.
5. Hybridize primer to DNA by heating sample for 3 min at or just under the calculated T_m of the primer.
6. Add 5 μL of a solution containing all four dNTPs at a concentration of 5 m*M* each.
7. Add 0.5–1.0 unit of the Klenow fragment of DNA polymerase I and mix gently. Incubate tube for exactly 10 min at 50°C.
8. Quench by adding an equal volume (~50 μL) of 0.6 *M* sodium acetate, and 20 m*M* EDTA and place on ice.
9. Precipitate DNA by adding 300 μL of 95% ethanol, mix and centrifuge 30 min. Wash the pellet with 150 μL of 70% ethanol. Centrifuge for 5 min, remove supernatant, and dry pellet.
10. Suspend pellet in 5 μL electrophoresis loading buffer and run on a normal sequencing gel. Analyze by standard techniques (*see* **Notes 8** and **9**).

3.4. PCR Amplification of DNA Treated In Vivo or In Vitro

1. In a 0.65-mL microcentrifuge tube, add the following:
   5 μL 10X reaction buffer supplied with the thermostable polymerase
   2 μL dNTP mix
   0.5 × 10^6 cpm end-labeled primer
   Distilled water to a final volume of 49.5 μL
2. Program thermocycler. For example:
   1 Round:
   2 min at 95°C , pause and add 0.5 μL of thermostable polymerase (2.5 units/μL)
   30 s at T_m of primer
   30 s at 72°C
   15–20 Rounds:
   1 m at 95°C
   30 s at T_m of primer
   30 s at 72°C

1 Round:
5 min at 72°C

3. Precipitate DNA by adding 50 µL of 0.6M sodium acetate, 300 mg/mL linear polyacrylamide, and 250 µL 95% ethanol. Mix and centrifuge 30 min at 14,000 rpm in a microfuge. Wash pellet with 150 µL of 70% ethanol and centrifuge for 5 min. Remove supernatant and dry pellet on medium heat for 5 min in a Speed Vac.
4. Suspend pellet in 5 µL loading buffer and run on sequencing gel. Analyze by standard techniques (*see* **Notes 8** and **9**).

4. Notes

1. False positive results can occur from the presence of nucleases in any of the proteins being tested. A necessary control is to incubate each protein with the DNA in the absence of further treatment with the modifying reagent. The DNA isolated from these reactions is run through the remainder of the analysis procedure to reveal any digestion of the DNA by contaminating nucleases.
2. Certain sequences of DNA are sensitive to DEPC and potassium permanganate treatment even in the absence of proteins. It is important to run a control lane of DNA to obtain a background level of sensitive sites.
3. To optimize reaction conditions, a titration of potassium permanganate for varying amounts of time may be necessary. Too little potassium permanganate results in no signal, and too much potassium permanganate can result in a high background. Two to five millimolars of potassium permanganate is typical for in vitro experiments, but for in vivo experiments, where the medium may quench the reagent, concentrations up to 200 m*M* can be used. Times of reaction have been varied from 10 s up to 5 min, but in our hands, there is much less difference in the results obtained with different reaction times than with different potassium permanganate concentrations. Thus, one can set up the experiment for a convenient period of time.
4. Potassium permanganate and DEPC react with proteins as well as DNA, which can impair their function. Thus, negative experiments may result from protein denaturation rather than a lack of DNA modification by the protein.
5. It is important to remove all the piperidine from the DNA following cleavage. If smeared bands are found on the gel, try doing more than one round of drying in the Speed Vac by redissolving the pellet in 30 µL of deionized water and drying as described in **step 7** of **Subheading 3.1.**
6. To obtain the maximum amount of information from the DNA of interest, perform separate experiments with either the template or the RNA-like strand radiolabeled.
7. A Maxam and Gilbert sequencing ladder of the DNA being analyzed run adjacent to the probing reactions is useful to identify specific modified sites.
8. For primer extension and PCR amplification reactions, several factors can affect the observed signal. A loss of signal can be the result of the following:
   a. Improper primer sequence
   b. Annealing temperature higher than T_m

c. Contaminants present in reaction
d. High concentration of magnesium ion

9. Extra bands or smearing can occur if the annealing temperature is suboptimal and allows mispriming. Nonspecific hybridization can also occur if the radiolabled primer has undergone extensive decay. Freshly labeled primer reduces the risk of mispriming events. Supercoiled DNA can cause sequence-induced stopping of the DNA polymerase. Linearizing the plasmid before fill-in or amplification can reduce improper extension.

References

1. Leonard, N. J., McDonald, J. J., and Reichmann, M. E. (1970) Reaction of diethyl pyrocarbonate with nucleic acid components I: Adenine. *Proc. Natl. Acad. Sci. USA* **67,** 93–98.
2. Leonard, N. J., McDonald, J. J., Henderson, R. E. L., and Reichmann, M. E. (1971) Reaction of diethyl pyrocarbonate with nucleic acid components: Adenosine. *Biochemistry* **10,** 3335–3342.
3. Mendel, D. and Dervan, P. B. (1987) Hoogsteen base pairs proximal and distal to echinomycin binding sites on DNA. *Proc. Natl. Acad. Sci. USA* **84,** 910–914.
4. Hayatsu, H. and Ukita, T. (1967) The selective degredation of pyrimidines in nucleic acids by permanganate oxidation. *Biochem. Biophys. Res. Commun.* **29,** 556–561.
5. Howgate, P., Jones, A. S., and Tittensor, J. J. (1968) The permanganate oxidation of thymidine. *J. Chem. Soc.* C, 275–279.
6. Iida, S. and Hayatsu, H. (1971) The permanganate oxidation of thymidine. *J. Biophys. Acta* **240,** 370–375.
7. Rubin, C. M. and Schmid, C. W. (1980) Pyrimidine-specific chemical reactions useful for DNA sequencing. *Nucleic Acids Res.* **8,** 4613–4619.
8. Akman, S. A., Doroshow, J. H., and Dizdaroglu, M. (1990) Base modifications in plasmid DNA caused by potassium permanganate. *Arch. Biochem. Biophys.* **282,** 202–205.
9. Akman, S. A., Forrest, G. P., Doroshaw, J. H., and Dizdaroglu, M. (1991) Mutation of potassium permanganate- and hydrogen peroxide-treated plasmid pZ189 replicating in CV-1 monkey kidney cells. *Mutat. Res.* **261,** 123–130.
10. Sasse-Dwight, S. and Gralla, D. (1988) Probing the *Escherichia coli* glnALG upstream activation mechanism in vivo. *Proc. Natl. Acad. Sci. USA* **85,** 8934–8938.
11. Sasse-Dwight, S. and Gralla, . D. (1989) $KMnO_4$ as a probe for lac promoter DNA melting and mechanism *in vivo*. *Proc. Natl. Acad. Sci. USA* **264,** 8074–8081.
12. Grimes, E., Busby, S., and Minchin, S. (1991) Differential thermal energy requirement for open complex formation by *Escherichia coli* RNA polymerase at two related promoters. *Nucleic Acids Res.* **19,** 6113–6118.
13. Suh, W. C., Ross, W., and Record, M. T., Jr. (1992) Two open complexes and a requirement for magnesium to open the lambda-P-R transcription start site. *Science* **259,** 358–361.
14. Lofquist, A. K., Li, H., Imboden, M. A., and Paule, M. R. (1993) Promoter opening (melting) and transcription initiation by RNA polymerase I requires neither

nucleotide β,γ hydrolysis nor protein phosphorylation. *Nucleic Acids Res.* **21,** 3233–3238.

14a. Kahl, B. F., Li, H., and Paule, M. R. (2000) DNA melting and promoter clearance by eukaryotic RNA polymerase I. *J. Mol. Biol.* **299,** 75–89.

15. Kassavetis, G. A., Blanco, J. A., Johnson, T. E., and Geiduschek, E. P. (1992) Formation of open and elongating transcription complexes by RNA polymerase III. *J. Mol. Biol.* **226,** 47–58.

16. Wong, C. and Gralla, J. D. (1992) A role for the acidic repeat region of transcription factor sigma 54 in setting the rate and temperature dependence of promoter melting in vivo. *J. Biol. Chem.* **267,** 24,762–24,768.

17. Kainz, M. and Roberts, J. (1992) Structures of transcription elongation complexes *in vivo. Science* **255,** 838–841.

18. Ohlsen, K. L. and Gralla, J. D. (1992) Melting during steady-state transcription of the RRNB P–1 promoter *in vivo* and *in vitro. J. Bact.* **174,** 6071–6075.

19. Li, B., Weber, J. A., Chen, Y., Greenleaf, A. L., and Gilmour D. S. (1996) Analysis of promoter-proximal pausing by RNA polymerase II on the hsp70 heat shock gene promoter in a *Drosophila* nuclear extract. *Mol. Cell. Biol.* **16,** 5433–5443.

20. Hartvig L. and Christiansen J. (1996) Intrinsic termination of T7 RNA polymerase mediated by either RNA or DNA. *EMBO J.* **15,** 4767–4774.

21. Komissarova N. and Kashlev, M. (1997) Transcriptional arrest: *Escherichia coli* RNA polymerase translocates backward, leaving the 3' end of the RNA intact and extruded. *Proc. Natl. Acad. Sci. USA* **94,** 1755–1760.

22. Lee, D. N. and Landick, R. (1992) Structure of RNA and DNA chains in paused transcription complexes containing *Escherichia coli* RNA polymerase. *J. Mol. Biol.* **228,** 759–777.

23. Carles-Kinch, K. and Kreuzer, K. N. (1997) RNA-DNA hybrid formation at bacteriophage T4 replication origin. *J. Mol. Biol.* **266,** 915–926.

24. Jeppesen, C. and Nielsen, P. E. (1988) Detection of intercalation-induced changes in DNA structure by reaction with diethyl pyrocarbonate or potassium permanganate. *FEBS Lett.* **231,** 172–176.

25. Fox, K. R. and Grigg, G. W. (1988) Diethyl pyrocarbonate and permanganate provide evidence for an unusual DNA conformation induced by binding of the antitumour antibiotics bleomycin and phleomycin. *Nucleic Acids Res.* **16,** 2063–2075.

26. Bailly C., Gentle, D., Hamy, F., Purcell, M., and Waring, M. J. (1994) Localized chemical reactivity in DNA associated with the sequence-specific bisintercalation of echinomycin. *Biochemical J.* **300,** 165–173.

27. Michelottic, G. A., Michelotti, E. F., Pullner, A., Duncan, R. C., Eick, D., and Levens, D. (1996) Multiple single-stranded cis elements are associated with activated chromatin of the human c-myc gene *in vivo. Mol. Cel. Biol.* **16,** 2656–2669.

28. Hershkovitz, M. and Riggs, A. D. (1997) Ligation-mediated PCR for chromatin-structure analysis of interphase and metaphase chromatin. *Methods* **11,** 253–263.

29. Chiu, S. K., Rao, B. J., Story, R. M., and Radding, C. M. (1993) Interactions of three strands in joints made by RecA protein. *Biochemistry* **32,** 13,146–13,155.

30. Voloshin, O. N. and Camerini-Otero, R. D. (1997) The duplex DNA is very underwound in the three-stranded RecA protein-mediated synaptic complex. *Genes Cells* **2,** 303–314.
31. Plug, A. W., Peters, A. H. F. M., Keegan, K. S., Hoekstra, M. F., De Boer, P., and Ashley, T. (1998) Changes in protein composition of meiotic nodules during mammalian meiosis. *J. Cell Sci.* **111,** 413–423.
32. Duncan, R., Bazar, L., Michelotti, G., Tomonaga, T., Krutzch, H., Avigan, M., et al. (1994) A sequence-specific, single-strain binding protein activates the far upstream element of c-myc and defines a new DNA-binding motif. *Genes Dev.* **4,** 465–480.
33. Sun, W. and Godson, G. N. (1998) Structure of the *Escherichia coli* primase/single-strand DNA-binding protein/phage G4ori-c complex required for primer RNA synthesis. *J. Mol. Biol.* **276,** 689–703.
34. Godson, G. N., Mustaev, A. A., and Sun, W. (1998) ATP cross-linked to *Escherichia coli* single-strand DNA-binding protein can be utilized by the catalytic center of primase as initiating nucleotide for primer RNA synthesis on phage G4ori-c template. *Biochemistry* **37,** 3810–3817.
35. McCarthy, J. G. and Rich, A. (1991) Detection of an unusual distortion in A-tract DNA using potassium permanganate effect of temperature and distamycin on the altered conformation. *Nucleic Acids Res.* **19,** 3421–3430.
36. Matyasek R., Fulnecek, J., Fajkus, J., and Bazdek, M. (1996) Evidence for a sequence-directed conformation periodicity in the genomic highly repetitive DNA detectable with single-strand-specific chemical probe potassium permanganate. *Chromosome Res.* **4,** 340–349.
37. Epplen J. T., Kyas, A., and Maeueler, W. (1996) Genomic simple repetitive DNAs are targets for differential binding of nuclear proteins. *FEBS Lett.* **389,** 92–95.
38. Herr, W. (1985) Diethyl pyrocarbonate: a chemical probe for secondary structure in negatively supercoiled DNA *Proc. Nat. Acad. Sci. USA* **82,** 8009–8013.
39. Voloshin, O. N., Mirkin, S. M., Lyamichev, V. I., Belotserkovskii, B. P., and Frank-Kamenetskii, M. D. (1988) Chemical probing of homopurine-homopyrimidine mirror repeats in supercoiled DNA. *Nature* **333,** 475–476.
40. Bentin, T. and Nielsen, P. E. (1996) Enhanced peptide nucleic acid binding to supercoiled DNA: possible implications for DNA "breathing" dynamics. *Biochemistry* **35,** 8863–8869.
41. Furlong, J. C. and Lilley, D. M. J. (1986) Highly selective chemical modification of cruciform loops by diethyl pryrocarbonate. *Nucleic Acids Res.* **14,** 3995–4007.
42. Scholten, P. M. and Nordheim, A. (1986) Diethylpyrocarbonate: a chemical probe for DNA cruciforms. *Nucleic Acids Res.* **14,** 3981–3993.
43. Balagurumoorthy, P. and Brahmachari, S. K. (1994) Sturcture and stability of human telomeric sequences. *J. Biol. Chem.* **269,** 21,858–21,869.
44. Nadel, Y., Weisman-Shomer, P., and Fry, M. (1995) The fragile X syndrome single strand d(CGG)n nucleotide repeats readily fold back to form unimolecular hairpin structures. *J. Biol. Chem.* **270,** 28,970–28,977.

45. Jiang, H., Zacharia, W., and Amirhaeri, S. (1991) Potassium permanganate as an *in situ* probe for B-Z and Z-Z junctions. *Nucleic Acids Res.* **19,** 6943–6948.
46. Woelfl, S., Wittig, B., and Rich, A. (1995) Identification of transcriptionally induced Z-DNA segments in the human c-myc gene. *Biochim. Biophys. Acta* **1264,** 294–302.
47. Glover, J. N. M., Farah, C. S., and Pulleyblank, D. E. (1990) Structural characterization of separated H-DNA conformers. *Biochemistry* **29,** 11,110–11,115.
48. Haner, R. and Dervan, P. B. (1990) Single-strand DNA triple-helix formation, *Biochemistry* **29,** 9761–9765.
49. Huibregtse, J. M. and Engelke, D. R. (1991) Direct sequence and footprint analysis of yeast DNA by primer extension. *Methods Enzymol.* **194,** 550–562.
50. Holmes, D. S. and Quigley, M. (1981) A rapid boiling method for the preparation of bacterial plasmids. *Anal. Biochem.* **114,** 193–197.
51. Owen, R. J. and Borman, P. (1987) A rapid biochemical method for purifying high molecular weight chromosomal DNA for restriction enzyme analysis. *Nucleic Acids Res.* **15,** 3631.
52. Mueller, P. R. and Wold, B. J. (1989) *In vivo* footprinting of a muscle specific enhancer by ligation mediated PCR. *Science* **246,** 780–786.
53. Garrity, PA. and Wold, B. J. (1992) Effects of different DNA polymerases in ligation-mediated PCR: enhanced genomic sequencing and *in vivo* footprinting. *Proc. Natl. Acad. Sci. USA* **89,** 1021–1025.

7

Footprinting DNA–Protein Interactions in Native Polyacrylamide Gels by Chemical Nucleolytic Activity of 1,10-Phenanthroline-Copper

Athanasios G. Papavassiliou

1. Introduction

The existence of cell-type specific promoter and enhancer elements has been known for several years. However, the mechanisms responsible for the remarkable specificity of such elements, in comparison to the ubiquitously active promoters and enhancers of "housekeeping" genes and DNA tumor viruses, have remained elusive until recently. Although transfection and mutagenesis experiments have taught us a great deal about the structure of cell-type-specific *cis*-acting elements, the breakthrough in understanding the molecular basis for the differential activity of these elements has come from the analysis of their recognition by specific DNA-binding proteins.

Several techniques have been developed for the detection of cell-type-specific DNA-binding activities and the identification of sequence-specific contacts ("footprints") of a protein on DNA. Many of these techniques involve forming DNA–protein complexes (by incubating an asymmetrically labeled double-stranded DNA fragment containing the region of interest with a crude or partly purified protein extract), exposing the complex to enzymatic or to chemical reagents that can cleave or modify the DNA, and determining which bases are protected from attack when the protein(s) is bound. The most widely used reagents are deoxyribonuclease I (DNase I, an endonuclease) and dimethyl sulfate (DMS). Footprinting experiments with DNase I are performed using parallel reactions with free DNA and with DNA–protein complexes, and the nuclease is allowed to digest DNA only to a limited extent *(1)*. The DNA is then purified and denatured, and the single-stranded end-labeled DNA fragments are resolved on a sequencing gel and autoradiographed. Comparing the

From: *Methods in Molecular Biology, vol. 148: DNA–Protein Interactions: Principles and Protocols, 2nd ed.*
Edited by: T. Moss © Humana Press Inc., Totowa, NJ

digestion pattern of the DNA–protein complexes revealed on the autoradiograph with that of free DNA shows a band-free region (footprint) where the bound protein(s) has prevented access of the enzyme to DNA (*see* Chapter 3). In a similar analysis, the DNA is allowed to react mildly with DMS, which methylates primarily deoxyguanosine residues and renders their phosphodiester linkages labile under conditions of Maxam–Gilbert chemistry (*see* Chapter 14). The binding of a protein(s) to a specific DNA region will result in a protection of the corresponding bases from chemical modification ***(2)***.

The suitability of the above assays in determining the binding sequences of proteins on DNA is hindered by several disadvantages. First, the clarity of the footprint is highly dependent on the extent of occupancy of the binding site(s) (i.e., a "clear" footprint is observed only if all DNA molecules are involved in complexes). Unfortunately, this is not always easy to achieve, especially when the concentration and/or purity of the specific binding protein(s) is not satisfactory. Second, DNA–protein complexes formed in crude extracts may often be heterogeneous in terms of both binding specificity and kinetic stability. Therefore, direct footprinting in solution will not correspond to a single species, but, instead, reflect an "integral" of the multiple equilibria operating over the entire region of interest (i.e., the protection pattern will actually represent a composite of more than one complex, with complexes having a very low dissociation rate dominating the footprint). Finally, two different proteins that recognize the same sequence within the probe are most likely to yield indistinguishable footprints. These drawbacks may be overcome by coupling treatment with a footprinting reagent in solution with the electrophoretic mobility shift assay (EMSA; also known as gel retardation assay, *see* Chapter 2) ***(3–5)***. In this approach, the protein and DNA molecules are incubated together, and the equilibrated reaction mixture is exposed to DNase I or DMS, as before. The DNA–protein complexes are subsequently isolated from the free probe by electrophoresis in a nondenaturing polyacrylamide gel. Although the negatively charged free DNA migrates rapidly toward the anode, once it is bound by a specific protein its mobility decreases ***(3,4)***. Following the separation of the free and bound DNA species, the corresponding bands are cut out of the gel, and the DNA eluted and analyzed on a sequencing gel. The region(s) of protection evident in the DNA derived from the complexed fraction, indicates the binding site ***(5)***. Because the complexes are separated from contaminating unbound DNA fragments, their footprints will be free of background cutting, and thus considerably more evident. Similar considerations apply when more than one complex can be formed on the fragment. As long as the DNA-binding proteins differ in their molecular masses and charges, they will cause altered electrophoretic mobilities of the corresponding complexes and, hence, different migration in the native polyacrylamide gel. These complexes can be isolated and run in individual lanes on the sequencing gel. Thus, the exposure of the binding reaction to footprinting reagents, in combination with the fractionation offered by mobility shift gels, permits identification of the regions of DNA bound by protein in different complexes, even if a low percentage of the initial DNA molecules has been complexed.

Although one can substantially increase the sensitivity of DNase I or DMS footprinting experiments in solution by employing the EMSA, several additional problems have still to be faced:

1. DNase I is a relatively bulky molecule (molecular weight [MW] 30,400) that cannot cleave the DNA in the immediate vicinity of a bound protein because of steric hindrance. As a result, the region(s) protected from cutting extends beyond the actual protein-binding site.
2. The nonrandom nature of DNA cleavage by DNase I makes it impossible to assess the involvement in protein binding of nucleotides that lie in an area of the fragment resistant to the endonucleolytic activity of this enzyme (e.g., tracts of A and T residues, or TpA [as opposed to ApT] dinucleotide islands scattered within or adjacent to the binding site), so that binding sites or parts of binding sites may not be detected.
3. The primary site of reaction of DMS with B-DNA is the N-7 atoms of guanine bases, which are located in the major groove. Thus, those guanines in close proximity with the protein will be protected from methylation. However, if a protein primarily makes contacts with a DNA sequence in the minor groove, or if there are no guanine residues in a major groove-binding site, DMS will not reveal these interactions.
4. In many instances, particularly when a complex has a relatively high "off" rate, the bound protein can dissociate from the protected DNA fragment and reassociate to other DNA fragments that have already been nicked by DNase I or modified by DMS. In this case, the DNA-cleavage pattern derived from the complexed fraction will closely resemble that of the uncomplexed DNA, rendering it difficult to observe a footprint. The limitations imposed by the size and the sequence or base specificity of the aforementioned footprinting reagents, as well as the problem of protein exchange from the binding site(s) during treatment, are circumvented by merging the advantages inherent in the EMSA, with the *subsequent* exposure of the gel (hence of the resolved complexes while embedded in the polyacrylamide matrix) to a chemical DNA-scission reagent namely the 1,10-phenanthroline–copper ion (OP–Cu) *(6)*.

1.1. OP–Cu as a Footprinting Agent

1.1.1. Chemistry of DNA Cleavage

1,10 Phenanthroline–copper (**Fig. 1**) is an efficient chemical nuclease that cleaves the phosphodiester backbone of nucleic acids at physiological pH and temperature by oxidation of the deoxyribose (DNA) or ribose (RNA) moiety

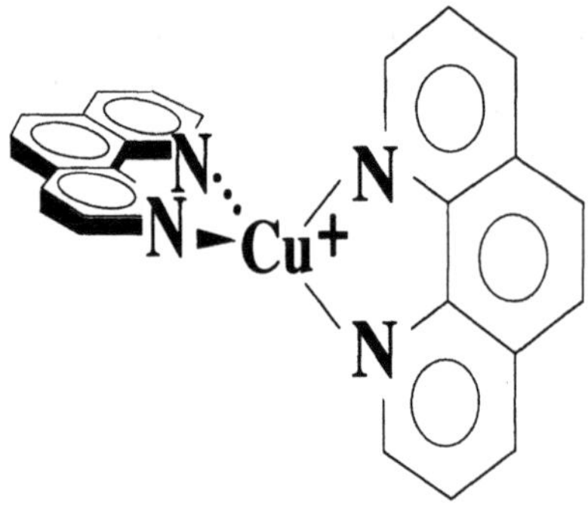

1,10-phenanthroline•Cu(I)

Fig. 1. Structure of 1,10-phenanthroline complexed with copper(I) ion (OP–Cu).

(7). The kinetic scheme of the reaction is summarized in **Fig. 2**. The first step is the formation of the 1,10-phenanthroline-cupric ion coordination complex, under conditions that favor the 2:1 stoichiometry ($[OP]_2Cu^{2+}$). The DNA-scission process is initiated by adding a reducing agent, usually 3-mercapto-propionic acid (a thiol), to the aerobic reaction mixture containing the target DNA. Under these conditions, the 2:1 cupric complex is reduced to the 2:1 cuprous complex ($[OP]_2Cu^+$) that is, in turn, oxidized by molecular oxygen to generate hydrogen peroxide. Hydrogen peroxide is an essential coreactant for the chemical nuclease activity and can be generated as described above or added exogenously *(8)*. The tetrahedral cuprous complex, present at the steady-state concentration defined by the experimental conditions (note the feedback mechanism in **Fig. 2**), then binds reversibly to the minor groove of DNA to form a central intermediate through which the reaction is funneled *(9)*. The DNA-bound cuprous complex undergoes *in situ* a one-electron oxidation by hydrogen peroxide to form a short-lived, highly reactive DNA-bound copper-oxo species that can be written either as a hydroxyl radical coordinated to the cupric ion or as a copper-oxene structure (**Fig. 2**). This species then attacks the H1'-deoxyribose protons of nucleotides, which are accessible in the minor groove; this reaction initiates a series of reactions culminating in cleavage of the phosphodiester backbone *(9)*. Reaction rates at any given sequence position depend on the stability of the intermediate formed between DNA and $(OP)_2Cu^+$ and on the orientation and proximity of the copper-oxo species relative to the C1'-deoxyribose hydrogen in the minor groove. Because both criteria are met satisfactorily in B-DNA sequences, the tetrahedral cuprous complex prefers B-DNA as its substrate. Such stereoelectronic interactions are less efficient in the broad minor groove of A-DNA and not possible in Z-DNA, which

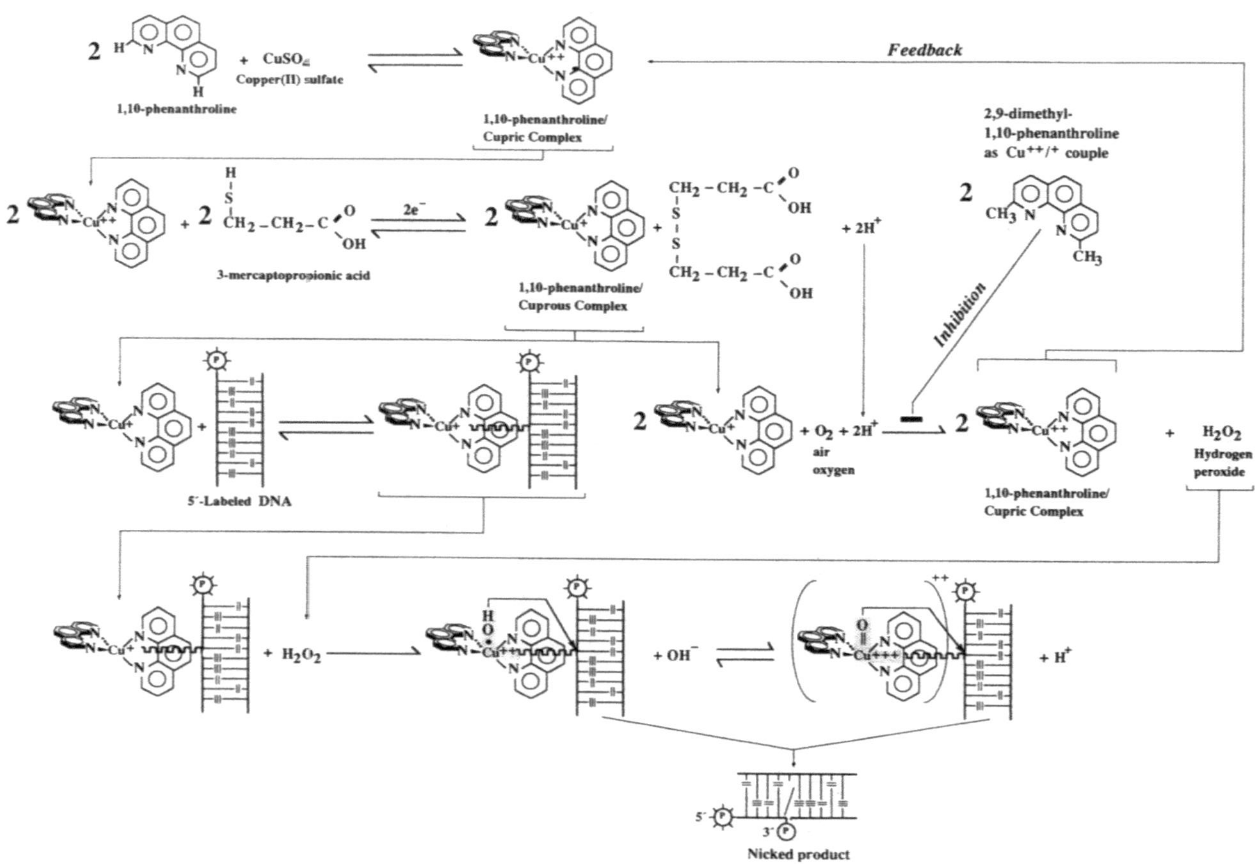

Fig. 2. Schematic representation of the kinetic mechanism for the nuclease activity of 1,10-phenanthroline–copper ion.

has practically no minor groove; as a result, A-DNA is cleaved at 25–33% of the rate with which B-DNA is cleaved and Z-DNA is not cleaved at all ***(9)***. The products of the strand-scission event include the free base, DNA fragments bearing 5'- and 3'-phosphorylated termini, and the deoxyribose oxidation product 5-methylene-2-furanone ***(10)***. The DNA-chain cleavage reaction can be efficiently quenched by adding to the mixture 2,9-dimethyl-1,10-phenanthroline (2,9-dimethyl–OP). This phenanthroline derivative can also chelate copper ions to form a minor groove-associated cuprous complex (thus competing with $[OP]_2Cu^+$), but the reduction potential of the Cu^{2+}/Cu^+ couple is too positive to allow significant nuclease activity under normal assay conditions ***(11)***.

1.1.2. OP–Cu Footprinting Following EMSAs

In as much as the structural and functional properties of DNA are not altered by entrapment in a polyacrylamide gel matrix ***(6)***, the small size and the ready diffusibility of all reaction components in solid supports permit the coupling of OP–Cu footprinting with the EMSA to study DNA–protein interactions ***(12,13)***. In this method, the DNA-binding reaction is performed as usual, electrophoresed under established, nondenaturing conditions, and the entire mobility shift gel is immersed in a footprinting reaction mixture containing 1,10-phenanthroline, cupric ion, and 3-mercaptopropionic acid. Following the reaction quench with 2,9-dimethyl–OP, footprints are obtained after elution of the radioactive free and protein-bound DNA cleavage products from the mobility shift gel and analysis on a sequencing gel (**Fig. 3**). Because the nuclease activity of $(OP)_2Cu^+$ produces 3'-phosphorylated and 5'-phosphorylated ends as cleavage products, sequencing gels can be accurately calibrated with the Maxam–Gilbert sequencing reactions.

1.2. Advantages of OP–Cu over Other Footprinting Agents

1.2.1. General Considerations

The nuclease activity of $(OP)_2Cu^+$ bears several advantages as a footprinting reagent relative to protection analyses using DNase I or DMS as a probe. First, the $(OP)_2Cu^+$ chelate is a small molecule (compared to DNase I) that permits cleavage closer to the edge of the DNA sequence protected by protein binding and, therefore, a more precise definition of it. Second, because the scission chemistry involves attack on the deoxyribose moiety, $(OP)_2Cu^+$ is able to cut at all sequence positions regardless of base. However, the intensity of cutting (rate of cleavage) does depend on local sequence, with attack at adenines of TAT triplets being most preferred (***14***; *see also* legend to **Fig. 3**). Interestingly, a preference for C-3',5'-G steps, rather than T-3',5'-A steps, is observed at a phenanthroline to copper ratio of 1:1, which strongly favors formation of the $OPCu^+$ complex *(15)*. Nevertheless, the cutting patterns obtained with $(OP)_2Cu^+$ are usually sufficiently well-defined to identify protected regions, even though this endonucleolytic agent exhibits some degree of sequence specificity in its rate of cleavage of naked DNA. Third, because $(OP)_2Cu^+$ binds to the minor groove of DNA, it will reveal minor-groove interactions. Because the binding of the coordination complex should be restricted to three base pairs, the complex is more sensitive to local, protein-induced conformational changes than DNase I, which by possessing an extended minor groove-binding site, may be unable to sense. In this context, the complex will also detect binding in the major groove when its approach to its minor groove-binding site is sterically blocked or if the interaction of the protein in the major groove alters the minor groove geometry so that the tetrahedral coordination complex binds poorly (both being frequent features of DNA–protein interactions). Furthermore, because of the difference in their respective mechanisms of cleavage, DNase I and $(OP)_2Cu^+$ probe different aspects of the structure of a DNA–protein complex. DNase I cleavage relies on the accessibility of a particular phosphodiester bond, and thus protection is indicative of an interaction on the outer face of the DNA helix. In contrast, protection from $(OP)_2Cu^+$-mediated cleavage is most likely caused by the inhibition of its binding to the minor groove and implies that a portion of the protein occupies at least the minor groove. Finally, in contrast to other chemical nucleases such as ferrous EDTA (introduces single-stranded nicks in DNA through the generation of diffusible hydroxyl radicals; *see* Chapter 5), the nucleolytic activity of OP–Cu is not inhibited by glycerol, a free radical scavenger, which is present in most protein storage buffers.

1.2.2. Benefits of OP–Cu Footprinting Within Mobility Shift Gels

The major advantage of the combined OP–Cu footprinting procedure arises from the topography of treatment: Preformed DNA–protein complexes are exposed to the chemical nuclease within the gel (i.e., not prior but subsequent to an electrophoretic mobility shift experiment). This characteristic of the technique makes it ideal for protection analysis of kinetically labile complexes *(16)*. At least three factors account for the latter. The first is that the background cleavage is greatly reduced by the separation of unbound DNA from the DNA–protein complex(es) pool. The second factor is the so-called "caging effect" *(3,4)*. The gel matrix forms "cagelike" compartments that prevent a dissociated protein from diffusing away from the DNA, so that by enhancing reassociation, the apparent affinity constant will be higher than the true value. The protein could also interact with the gel matrix, thereby orienting its diffusion toward reassociation. Whatever the mechanism(s), the increase in stabil-

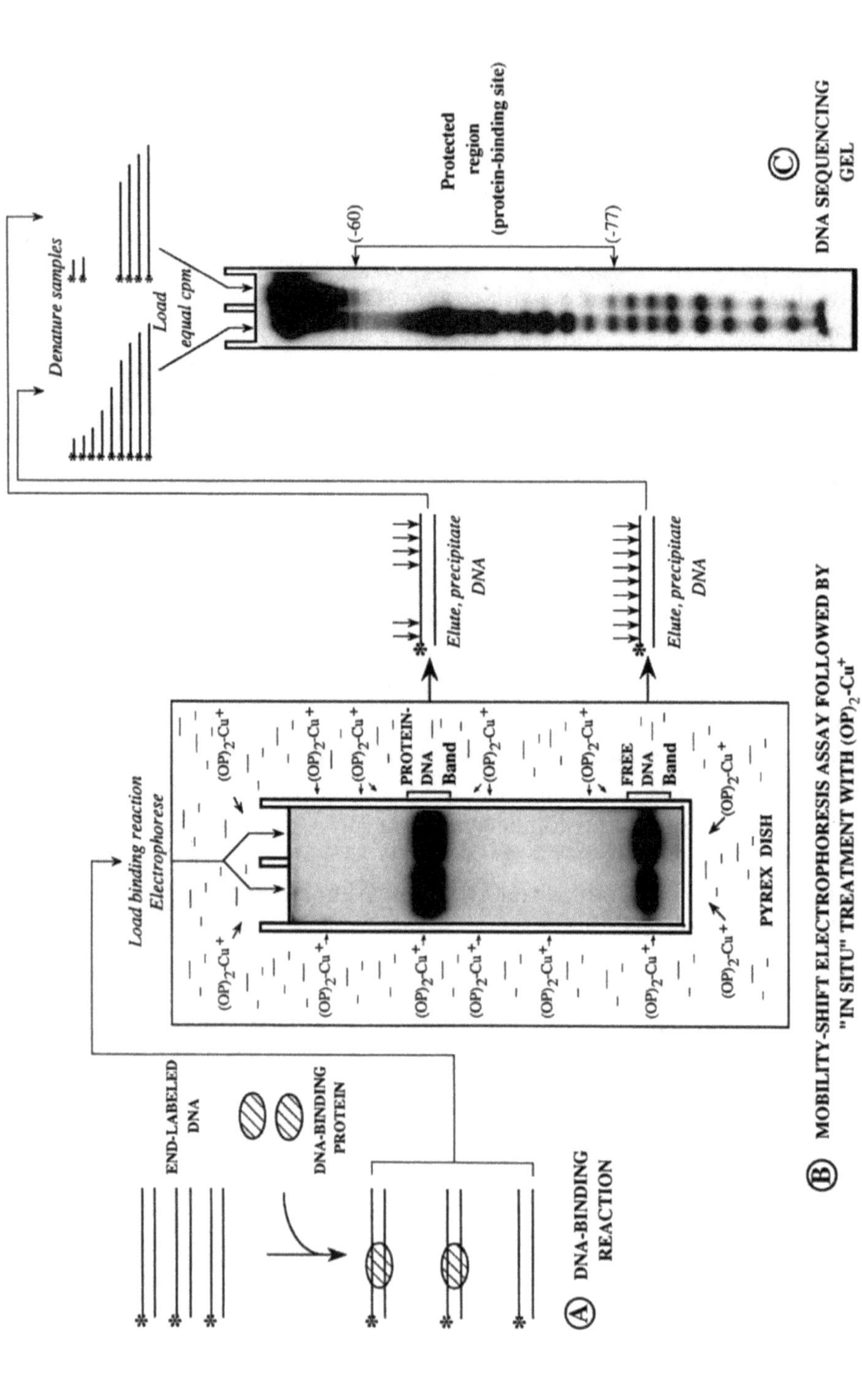

END-LABELED DNA
DNA-BINDING PROTEIN
Ⓐ DNA-BINDING REACTION
Load binding reaction
Electrophorese
(OP)$_2$-Cu$^+$
PROTEIN-DNA Band
FREE DNA Band
PYREX DISH
Ⓑ MOBILITY-SHIFT ELECTROPHORESIS ASSAY FOLLOWED BY "IN SITU" TREATMENT WITH (OP)$_2$-Cu$^+$
Elute, precipitate DNA
Elute, precipitate DNA
Denature samples
Load equal cpm
(-60)
(-77)
Protected region (protein-binding site)
Ⓒ DNA SEQUENCING GEL

ity of the complex contributed by the gel leads to a more efficient blockage of the access of the $(OP)_2Cu^+$ chelate to the protein-binding DNA segment. The third factor comes from the nature and site of action of the cupryl intermediate through which the reaction is funneled, and it acts synergistically with the previous one. Because this highly reactive oxidative species is generated near the surface of the DNA (*in situ*), diffusible radicals, if formed at all, will have a short or restricted diffusive path and, therefore, will be unable to achieve a fast equilibrium distribution along the DNA polymer. Consequently, protein-binding sites exposed during multiple dissociation events will escape the nucleolytic attack most of the time and hence remain intact.

In addition to the fact that discrete complexes with defined stoichiometries and a wide range of kinetic stabilities can be mapped simultaneously, the *in situ* OP–Cu footprinting procedure is superior to oligonucleotide-binding competition assays in the analysis of multiple complexes frequently obtained in electrophoretic mobility shift experiments employing unfractionated extract preparations. For example, multiple retarded bands can arise from protein–protein interactions between a non-DNA-binding transcription factor(s) and a specific DNA-binding protein, or from two proteins binding to distinct DNA sequences in a cooperative manner *(12)*. Although both complexes would be

---

Fig. 3. *(previous page)* Outline of the combined electrophoretic mobility-shift/*in gel* OP–Cu footprinting assay. **(A)** DNA restriction fragments containing a protein-binding site(s) are labeled with ^{32}P at a unique end and incubated with a crude or partially purified extract containing the DNA-binding protein(s) of interest, under optimized binding conditions. **(B)** After equilibration of the DNA-binding reaction, the free and bound DNA fragment populations are separated by electrophoresis through a nondenaturing polyacrylamide gel; the gel is then transferred into a buffer-containing Pyrex dish, and the retarded and unretarded DNA species are exposed *in situ* to the nuclease activity of $(OP)_2Cu^+$. The two DNA fractions are subsequently located by autoradiography of the wet gel, excised and eluted from the gel matrix, precipitated, and recovered in formamide buffer. **(C)** Samples are heat-denatured and equal amounts of radioactivity from the two fractions are electrophoresed on a denaturing polyacrylamide gel (DNA sequencing gel) and autoradiographed. In the sample prepared from the free-DNA band, bands will appear in the gel corresponding to positions of protein binding. For the sample(s) prepared from the protein–DNA band(s), bands will appear at all positions *except* those bound by the protein(s) (protected region). The particular example depicts the OP–Cu mapping of a DNA–protein complex formed between bacterially expressed LFB1 (a liver-specific transcription factor) and an oligonucleotide bearing its binding site within the –95 to –54 region of the α1-antitrypsin promoter. Arrowheads connected by line demarcate the footprinted site. The enhanced cleavage observed within the protein-binding site in the free DNA sample is the result of the presence of repeated TA elements in this sequence (*see* **Subheading 1.2.1.**).

abolished by competition with oligonucleotides, these possibilities can be readily distinguished by direct footprinting within the gel.

1.3. Additional Applications and Outlook

The nucleolytic activity of $(OP)_2Cu^+$ in a polyacrylamide matrix has been also demonstrated to be a viable means of gaining insight into the interactions of RNA-binding proteins with their recognition sequences ***(17,18)***. Application of OP–Cu in this context may be invaluable toward defining structural perturbations in RNA on protein binding and mapping the binding domains of various proteins. Because of the preferential nucleolytic activity of $(OP)_2Cu^+$ toward single-stranded bulge and loop RNA regions (double-stranded stem regions can be cut at elevated concentrations of the chemical nuclease), hypersensitive sites may be obtained on footprinting an RNA–protein complex following a gel retardation assay. Such sites would imply an unwinding of a helical structure on protein binding or perturbations in the minor groove accessibility of the bound RNA molecule.

The *in gel* OP–Cu footprinting methodology has already expanded the "tool box" available to investigators wishing to explore the structure and function relationships of nucleic acid–protein complexes, and emerging improvements in the chemical mechanism (e.g., DNA-strand scission by the coordination complex of OP with a non-redox-active metal) as well as future modifications will likely make this technology even more efficient and broadly useful.

2. Materials

2.1. Analytical and Preparative EMSA

2.1.1. Solutions

1. A variety of binding and gel buffers are commonly employed in EMSA (*see* Chapter 2). A suitable binding buffer is 10 m*M* HEPES pH 7.9, 10% glycerol, 0.1 m*M* EDTA, 0.5 m*M* tetrasodium pyrophosphate, and 0.5 m*M* PMSF. The most common gel buffers are Tris-glycine: 50 m*M* Tris, 2.5 m*M* EDTA, and 0.4 *M* glycine; 0.5X TBE: 45 m*M* Tris, 45 m*M* boric acid, and 1 m*M* EDTA; Tris–acetate: 6.7 m*M* Tris-HCl, pH 7.5, 3.3 m*M* sodium acetate, and 1 m*M* EDTA.
2. Ammonium persulfate (10%; w/v): Weigh out 1 g of ammonium persulfate and put it in a sterile plastic tube containing 10 mL of distilled, deionized water. Vortex vigorously until the salt is completely dissolved. Filter through a 0.22-µm membrane filter. This solution may be stable for a period of a few days at 4°C, but it is recommended that you prepare it freshly for each new gel. Ammonium persulfate is extremely destructive to tissue of the mucous membranes and upper respiratory tract, eyes, and skin. Inhalation may be fatal. Exposure can cause gastrointestinal disturbances and dermatitis. Wear gloves, safety glasses, respira-

tor, and other protective clothing and work in a chemical fume hood. Wash thoroughly after handling.

3. Dye-containing binding buffer: 0.05% (w/v) bromophenol blue in 1X optimized binding buffer (store at 4°C after filtering).

2.1.2. Reagents/Special Equipment

1. Highly purified duplex DNA fragment labeled exclusively at one of its four ends (5' or 3'); use standard procedures for unique labeling ***(19)***. All necessary precautions should be observed to minimize exposure to ionizing radiation during labeling and isolation of the probe. Consult the institutional environmental health and safety office for further guidance in the appropriate use of radioactive materials.
2. Reagents employed in the optimized binding reaction.
3. (16–18) × (16–18)-cm front and back glass gel electrophoresis plates: The plates must be absolutely clean before use. Wash them with warm soapy water; then, holding them by the edges, rinse several times first in tap water and then in deionized water. Finally, rinse with ethanol and let them air-dry. Using a pad of Kimwipes, siliconize the inner side of the back plate with a 2% dimethyldichlorosilane solution in 1,1,1-trichloroethane in a chemical fume hood (this product is particularly toxic; gloves, safety glasses, respirator, and other protective clothing should be worn when handling it.
4. 0.3-cm spacers.
5. Electroresistant plastic tape (e.g., 3M yellow electrical tape).
6. 0.22 and 0.45-μm filters (Millipore, Bedford, MA).
7. *N,N,N',N'*-Tetramethylethylenediamine (TEMED; Bio-Rad, Richmond, CA). TEMED is extremely destructive to tissue of the mucous membranes and upper respiratory tract, eyes, and skin. Inhalation may be fatal. Prolonged contact can cause severe irritation or burns. Wear gloves, safety glasses, respirator, and other protective clothing and work in a chemical fume hood (TEMED is also flammable!). Wash thoroughly after handling.
8. 3-mm gel comb with 10-mm-wide teeth.
9. 10-mL syringe and 18-gage needle.
10. 100- to 200-μL Hamilton syringe.
11. Additional reagents and equipment: Powdered acrylamide and *N,N'*-methylene–*bis*-acrylamide (Bio-Rad); plenty of binder clamps (fold-back spring clips); razor blades; polyacrylamide gel electrophoresis apparatus; constant current power supply; peristaltic pump for recirculating electrophoresis buffer (if required); siliconized 1.5-mL Eppendorf microcentrifuge tubes; spatula. Acrylamide and *N,N'*-methylene–*bis*-acrylamide are potent neurotoxins and are absorbed through the skin. Their effects are cumulative. Wear gloves and a face mask when weighing these substances and when handling solutions containing them. Although polyacrylamide is considered to be nontoxic, it should be handled with care because of the possibility that it might contain small quantities of unpolymerized acrylamide.

2.2. DNA Chemical Cleavage (Footprinting) Reactions within the Gel

2.2.1. Solutions

1. 10 m*M* Tris-HCl, pH 8.0 (store at room temperature after autoclaving).
2. MPA solution (58 m*M* 3-mercaptopropionic acid): Add 100 µL of neat 3-mercaptopropionic acid (Aldrich, Milwaukee, WI) to a sterile 50-mL conical tube containing 19.9 mL of distilled, deionized water. Mix by vortexing. 3-Mercaptopropionic acid is toxic and causes burns in contact with skin and eyes; wear gloves and handle accordingly. Store the liquid reagent in a place protected from light. Dilute immediately prior to use.
3. OP solution (40 m*M* 1,10-phenanthroline): Weigh out 80 mg of 1,10-phenanthroline monohydrate (Aldrich or G. F. Smith) and dissolve (by vortexing and shaking vigorously for 2 min) in 10 mL of absolute ethanol in a sterile 50-mL conical tube. Wear gloves and dust mask when weighing this reagent. Store the powdered reagent in a place protected from light. Prepare just prior to use.
4. Cu^{2+} solution (9 m*M* $CuSO_4$): Weigh out 72 mg of anhydrous copper(II) sulfate (Aldrich or Mallinckrodt, Chesterfield, MO) and dissolve (by vortexing for 1 min) in 50 mL of distilled, deionized water in a sterile 50-mL conical tube. Powdered copper(II) sulfate is irritating to eyes, respiratory system, and skin; wear gloves and eye/face protection when weighing this chemical. Store the powdered chemical sealed in a dry place. Prepare just prior to use.
5. $(OP)_2Cu^+$-STOP solution (28 m*M* 2,9-dimethyl–OP): Weigh out 127 mg of 2,9-dimethyl–1,10-phenanthroline (Neocuproine) monohydrate (Aldrich or G. F. Smith) and dissolve (by vortexing vigorously for 2 min) in 20 mL of absolute ethanol in a sterile 50-mL conical tube. Wear gloves and dust mask when weighing this reagent. Store the powdered reagent in a place protected from light. Prepare just prior to use.

2.2.2. Reagents/Special Equipment

1. 20 × 20-cm Pyrex dish (available in most supermarkets); wash the dish with detergent, water, and then ethanol. Rinse with deionized water and dry with tissues.
2. Sterile 50-mL conical tubes.
3. Additional equipment: protective gloves, glass or plastic beaker, 20-mL glass pipet, and vacuum aspirator.

2.3. Isolation of Free and Complexed DNA Fractions

2.3.1. Direct Elution from the Polyacrylamide Gel Matrix

2.3.1.1. Solutions

1. Gel elution buffer: 0.5 *M* ammonium acetate, pH 7.5 (promotes diffusion of the DNA out of the gel matrix and is readily soluble in ethanol in the subsequent precipitation step), 1 m*M* EDTA, and 0.1% sodium dodecyl sulfate (SDS; w/v; effectively denatures any contaminating DNAase activity). For improved recovery of DNA fragments smaller than 60 bp, the buffer should also include 10 m*M*

magnesium chloride. This stock solution can be stored at room temperature protected from light for several months.

2. Phenol/chloroform/isoamyl alcohol (25:24:1; v/v): Mix 25 vol of phenol (redistilled under nitrogen and equilibrated with 100 m*M* Tris-HCl, pH 8.0, and 1 m*M* EDTA in the presence of 0.1% [w/v] 8-hydroxyquinoline) with 24 vol of chloroform and 1 vol of isoamyl alcohol. Phenol can be stored at 4°C in dark (brown) bottles for up to 2 mo. It is highly corrosive and can cause severe burns. Any areas of skin that come in contact with phenol should be rinsed with a large volume of water or PEG 400 and washed with soap and water (do not use ethanol!). Chloroform is irritating to the skin, eyes, mucous membranes, and respiratory tract. It is also a carcinogen and may damage the liver and kidneys. Wear gloves, protective clothing, safety glasses, and respirator when handling these substances and carry out all manipulations in a chemical fume hood.
3. Chloroform/isoamyl alcohol (24:1; v/v): Mix (in a chemical fume hood and wearing gloves, safety glasses, and respirator) 24 vol of chloroform with 1 vol of isoamyl alcohol. This organic mixture can be stored at room temperature in dark (brown) bottles indefinitely.
4. 70% and 90% (v/v) ethanol.
5. Sequencing-gel loading buffer: 90% (v/v) deionized formamide, 1X TBE (*see* **Subheading 2.3.2.**), 0.025% (w/v) xylene cyanol, and 0.025% (w/v) bromophenol blue. Store at –20°C after filtering. (Preparation of deionized formamide: Combine 200 mL of formamide with 5 g of AG501-X8 [D] [Bio-Rad] in a 250-mL Erlenmeyer flask. Cover the mouth of the flask with Parafilm and gently stir the mixture at room temperature for 30 min. Avoid aeration of the formamide when stirring the mixture. Filter the solution through a coarse-sintered-glass funnel, and store the deionized formamide at –20°C. Formamide is a teratogen; take all safety precautions to avoid contact during the above manipulations.)

2.3.1.2. Reagents/Special Equipment

1. Plastic wrap, such as Saran Wrap® or cling film.
2. Old X-ray film covered with plastic wrap.
3. Glass stirring rod.
4. Small adhesive labels (or Scotch Tape®).
5. Radioactive ink: Mix a small amount of ^{32}P with waterproof black drawing ink, to a concentration of approx 200 cps (on a Geiger counter) per microliter.
6. Fiber-tip pen.
7. Kodak X-Omat AR film.
8. Lightproof cardboard film holder.
9. Aluminum foil.
10. Lab marking pen.
11. Sharp scalpel.
12. Fine-tip waterproof marking pen.
13. Single-edged disposable razor blades.

14. 18-gage syringe needles or sterile forceps.
15. Sterile 3-mL syringes attached to a shortened 18-gage needle (broken with pliers).
16. Siliconized 1.5-mL Eppendorf microcentrifuge tubes.
17. 22-gage syringe needle.
18. Siliconized capless 0.5-mL Eppendorf microcentrifuge tubes.
19. Conformable self-sealing tape (e.g., Parafilm).
20. 1-mL sterile syringes (Becton Dickinson, Heidelberg, Germany).
21. 0.22-µm syringe filters (Millipore).
22. Ice-cold absolute ethanol.
23. Glycogen (from Boehringer, Indianapolis, IN, or Sigma, St. Louis, MO).
24. Drawn-out Pasteur pipets.
25. Elutip™-d mini-columns (Schleicher & Schuell, Dassel, Germany) and low- and high-salt solutions as recommended by manufacturer.
26. Additional equipment: Geiger counter, all equipment for autoradiography, low-speed centrifuge (Beckman J6B or Sorvall RC3), microcentrifuge (Eppendorf or equivalent), 37–42°C shaking incubator, vacuum centrifuge (e.g., SpeedVac, Savant, Hicksville, NY); scintillation counter or Bioscan Quick Count for ^{32}P, and water bath at 68°C.

2.3.2. Electrotransfer of the Entire Gel and Elution from NA-45 Membrane

2.3.2.1. Solutions

1. 10 m*M* EDTA, pH 7.6 (store at room temperature after autoclaving).
2. 0.5 *M* NaOH.
3. 1X TBE buffer: 89 m*M* Tris base, 89 m*M* borate, and 2.5 m*M* EDTA. To prepare 5 L of 5X TBE buffer, dissolve (stirring for at least 1 h) 272.5 g of ultrapure Tris base, 139.1 g of boric acid, and 23.3 g of disodium EDTA dihydrate in 4.5 L of distilled, deionized water, and make up to a final volume of 5 L. It is not necessary to adjust the pH of the resulting solution, which should be about 8.3. Store at room temperature; this stock solution is stable for many months, but it is susceptible to the formation of a precipitate and should occasionally be inspected visually.
4. 20 m*M* Tris-HCl, pH 8.0, and 0.1 m*M* EDTA (store at room temperature after autoclaving).
5. NA-45 membrane elution buffer: 1.0 *M* NaCl, 20 m*M* Tris-HCl, pH 8.0, 0.1 m*M* EDTA (store at room temperature after autoclaving).
6. **Solutions 2–5** of **Subheading 2.3.1.1.**

2.3.2.2. Reagents/Special Equipment

1. NA-45 membrane sheets (0.45-µm pore size; Schleicher & Schuell).
2. Filter paper (Whatman [Clifton, NJ] 3 MM or equivalent).
3. Clean glass plate.
4. **Items 1, 2**, and **4–7** in **Subheading 2.3.1.2.**
5. Metal cassette.
6. **Items 10–14**, **16**, and **22–24** in **Subheading 2.3.1.2.**

7. Additional equipment: protective gloves, electrotransfer unit (high-current [2–3 A] power supply, e.g., Bio-Rad or Hoefer, San Francisco, CA), Kimwipes, Geiger counter, all equipment for autoradiography, microcentrifuge (Eppendorf or equivalent), 55°C water bath with agitation; vacuum centrifuge (e.g., SpeedVac, Savant), scintillation counter or Bioscan Quick Count for ^{32}P, and water bath at 68°C.

2.4. Preparation of the G + A Sequencing Ladder

2.4.1. Solutions

1. Carrier DNA stock solution: salmon sperm DNA extracted sequentially with phenol/chloroform/isoamyl alcohol (25:24:1; v/v) and chloroform/isoamylalcohol (24:1; v/v), precipitated with ethanol, resuspended in distilled, deionized water at 1 mg/mL, and sonicated to an average chain length of 200 bp.
2. 1.0 *M* aqueous piperidine: Add 100 µL of concentrated piperidine (reagent grade; BDH, London, England) into 0.9 mL of distilled, deionized water. Piperidine is somewhat hard to pipet; when diluting rinse the micropipet tip thoroughly by repeated pipeting and then mix the solution well by vortexing. Make dilution in a chemical fume hood just prior to use.
3. 1% sodium dodecyl sulfate (SDS): Prepare a 10% (w/v) solution of SDS in distilled, deionized water (wear dust mask when weighing powdered SDS); heat at 68°C to assist dissolution (do not autoclave). Dilute 1:10 with distilled, deionized water. Store at room temperature.
4. **Solution 5** in **Subheading 2.3.1.1.**

2.4.2. Reagents/Special Equipment

1. Siliconized 1.5-mL Eppendorf microcentrifuge tubes.
2. 20,000 cpm of the end-labeled DNA fragment used in the preparative EMSA.
3. 88% aqueous formic acid.
4. Conformable self-sealing tape (e.g., Parafilm).
5. 1-butanol (*n*-butylalcohol).
6. Drawn-out Pasteur pipet.
7. Additional equipment: wet ice, 37°C water bath with agitation, thermostatted heating block at 90°C, lead weight, microcentrifuge (Eppendorf or equivalent), vacuum centrifuge (e.g., SpeedVac, Savant); water bath at 68°C.

2.5. Analysis of the Chemical Cleavage Products on a DNA Sequencing Gel

2.5.1. Solutions

1. 40% (w/v; 19:1) acrylamide:*N,N'*-methylene–*bis*-acrylamide solution.
2. **Solution 3** in **Subheading 2.3.2.1.**
3. Solution 2 in **Subheading 2.1.1.**
4. Fixing solution: 10% (v/v) acetic acid, and 10% (v/v) methanol.

2.5.2. Reagents/Special Equipment

1. Urea (enzyme grade).
2. 34 × 40-cm front and back glass plates; treat the plates as described in **Subheading 2.1.2.3.**
3. 0.04-cm spacers.
4. 0.4-mm custom-ordered sample comb with 5-mm lanes spaced on 10-mm centers.
5. 10-mL syringe with a 22-gage needle.
6. Calibrated glass capillaries with finely drawn tips or disposable flat-capillary pipet tips (National Scientific Supply Company, Inc., San Rafael, CA).
7. 30-mL syringe with a bent 20-gage needle.
8. Backing paper (Whatman No. 1 or equivalent).
9. Plastic wrap, such as Saran Wrap® or cling film.
10. Filter paper (Whatman 3MM or equivalent).
11. Kodak X-Omat AR film.
12. Large metal autoradiography cassette.
13. Intensifying screen (DuPont [Wilmington, DE] Cronex Lightning Plus).
14. Additional reagents and equipment: TEMED, plenty of binder clamps (fold-back spring clips), sequencing gel electrophoresis apparatus; power supply delivering high voltage (2500–3000 V) (e.g., Bio-Rad, LKB-Pharmacia, Piscataway, NJ); dry-block heater at 90°C, wet ice, aluminum plate of an appropriate size, razor blade, plastic tank at gel dimensions, 10-mL glass pipet, Kimwipes, gel dryer, all equipment for autoradiography.

3. Methods

3.1. Analytical EMSA

1. Perform a preliminary EMSA (*see* Chapter 2 and **Note 1**) to identify conditions for the formation of the complex(es) to be footprinted. Because no universal binding and/or gel system is likely to be found for the study of all DNA–protein interactions, it may be necessary to optimize conditions for formation and adequate resolution of the DNA–protein complex(es) of interest (*see* **Note 1**).
2. If crude or partly fractionated extracts are employed, ascertain the DNA-binding specificity of the resolved complex(es) by performing an analytical competition binding assay (*see* Chapter 2 and **Note 2**).
3. It is advisable, prior to proceeding to the more laborious preparative gel retardation/*in situ* footprinting assay, to perform an additional analytical experiment and obtain a qualitative estimation of the dissociation rates of preequilibrated protein–DNA complexes of interest (*see* **Note 3**). Although the $(OP)_2Cu^+$-mediated cleavage reactions in the gel are relatively insensitive to the kinetic stability of the DNA–protein complex(es) under investigation (*see* **Subheading 1.2.2.**), this information can be used in adjusting the exposure time to the chemical nuclease (*see* **step 6** in **Subheading 3.3.** and **Note 4**), thereby enhancing the clarity of the expected footprint.

3.2. Preparative EMSA

1. Assemble (16–18) × (16–18)-cm front and back glass plates and 0.3-cm spacers for casting a preparative mobility shift polyacrylamide gel (3-mm-thick polyacrylamide gels are preferable, because they are easier to load and give sharper bands than 1.5-mm-thick gels). The plates must be scrupulously clean and free of grease spots to avoid trapping air bubbles while pouring the gel; it is highly recommended to use one glass plate (preferably the back) that has been siliconized on the inner side for ease of removal after electrophoresis is completed. Taking particular care (to prevent leakage), seal the entire length of the two sides and the bottom of the plates with electrician's plastic tape.
2. Prepare, filter (through a 0.45-µm filter), and degas (by applying vacuum) 100 mL of the acrylamide gel solution found during optimization of the analytical assay (**Subheading 3.1, step 1**). Because of the ready permeability of the gel matrix to all reagent and quenching solutions used for the subsequent chemical treatment and the lack of diffusible radicals mediating the DNA-scission reaction, the $(OP)_2Cu^+$ *in situ* footprinting technique is compatible with a broad spectrum of gel porosities (ranging from 3.5% to 6% [w/v], with an acrylamide to *N,N'*-methylene–*bis*-acrylamide molar ratio of 19:1 to 80:1) and gel/running-buffer compositions (from glycerol-containing/low-ionic-strength [pH 7.5–7.9] to high-ionic-strength TBE [pH 8.3] or Tris–glycine [pH 8.5] buffer systems).
3. Add to the solution 0.8 mL of 10% ammonium persulfate and 75 µL of TEMED and swirl the mixture gently.
4. Slowly pour the acrylamide gel mix between the glass plates and quickly insert a 3-mm comb bearing 10-mm-wide teeth. Allow the gel to polymerize (lying flat or nearly flat, to avoid undesirable hydrostatic pressure on the bottom) at room temperature for about 45 min.
5. After polymerization is complete, remove the electrical tape from the bottom of the gel (by cutting with a razor blade) and clamp the gel into place on the electrophoresis apparatus. Fill both chambers of the electrophoresis tank with the buffer used for preparation of the acrylamide gel mix (**step 2**), carefully remove the comb, and immediately rinse the sample wells with reservoir buffer using a 10-mL syringe with an 18-gage needle.
6. Prior to assembling the preparative binding reaction, pre-electrophorese the gel for 60 min at 20 mA, with or without buffer recirculation between the two compartments, depending on the nature of the gel/running-buffer system used (low or high ionic strength). This removes any excess persulfate and unpolymerized acrylamide. Prerunning of the gel should be done at the temperature at which the binding reaction will be performed (known from **step 1** in **Subheading 3.1.**).
7. In a siliconized 1.5-mL Eppendorf tube, scale up the optimized analytical reaction 5- to 10-fold, depending on the relative proportion of the DNA–protein complex(es) obtained. If the detected specific DNA-binding activity(ies) (**Subheading 3.1., step 1**) represents <1% of the total label input, the amount of radioactive probe in the scaled reaction should be at least 250,000 cpm (*see also* **Note 5**).
8. Turn off the electric power. Using a 100- to 200-µL Hamilton syringe, load the preparative binding reaction onto one or two wells (depending on the total vol-

ume of the sample) in the middle of the gel. Raise the tip of the needle as the sample is loaded into the well. Do not attempt to expel all of the sample from the syringe because this almost always produces air bubbles that blow the sample out of the well. If the glycerol concentration in the binding buffer is low (<5%), it is important to load the well gently to prevent dilution (*see also* **Note 6**). Avoid adding bromophenol blue to the binding reaction prior to loading because this dye can rapidly disrupt some DNA–protein complexes. Instead, you may load just dye-containing binding buffer in one of the adjacent lanes to monitor the progress of electrophoresis.

9. Run the gel at 25–35 mA (it may be necessary to adjust the voltage occasionally if a constant power supply is not available) for a time sufficient to allow migration of the free DNA probe to approx 2 cm from the bottom of the gel. Provided the same plate size has been used in establishing the optimal electrophoresis conditions, this can be monitored by the migration of the tracking dye, in correlation with the position of the free probe on the autoradiogram obtained from the optimized analytical assay (**step 1** in **Subheading 3.1.**). If electrophoresis is performed at room temperature, the glass plates should be allowed to become only slightly warm, because excess heating may perturb the equilibrated complexes, or even cause protein denaturation; decrease the current if the plates become any hotter.
10. Following electrophoresis, detach the glass plates from the gel apparatus, and using a spatula, carefully remove the spacers and pry the glass plates apart, taking extreme care not to distort or tear the gel, which should remain attached to only one of the plates (the nonsiliconized front plate).

3.3. DNA Chemical Cleavage (Footprinting) Reactions within the Gel

1. Wear protective gloves and wash your fingers thoroughly in a beaker containing deionized water to remove the talc powder. Immerse the whole gel, still attached to the lower plate (with the gel facing up), in a 20 × 20 cm scrupulously clean Pyrex dish (never use a plastic tray!), containing 200 mL of 10 m*M* Tris-HCl, pH 8.0. Loosen it on its supporting glass plate (omit this step if using Tris–glycine-containing or low-percentage/low-ionic-strength polyacrylamide gels [e.g., a 3.5–4% gel], which are very sticky and extremely difficult to manipulate without fracturing).
2. Prepare the MPA, OP, and Cu^{2+} solutions (*see* **Note 7**).
3. In a sterile 50-mL conical tube, transfer 1 mL of the freshly made OP solution. To this, add 1 mL of the freshly prepared Cu^{2+} solution, and wait 1 min while pipeting up and down (the mixture should turn light blue, indicating efficient formation of the $[OP]_2Cu^{2+}$ chelate). Add 18 mL of distilled, deionized water and vortex the tube. This is the OP/Cu^{2+} solution (1,10-phenanthroline to copper ratio of approx 4.5:1).
4. Add the OP/Cu^{2+} solution (20 mL) to the gel equilibrating in the 200-mL buffer, and shake the Pyrex dish while laying it on an even horizontal surface to distribute evenly.

5. Initiate the chemical nuclease reaction by adding the MPA solution (20 mL); distribute evenly by quickly shaking the Pyrex dish, as earlier. The gel will turn brownish. The appearance of a dark brown precipitate indicates the presence of impurities in the original $CuSO_4$ solution, which will interfere with the cascade leading to DNA-strand scission. It is, therefore, crucial for the assay to use copper(II) sulfate of the best available analytical grade.
6. Incubate for a period of 8–30 min without shaking (do not disturb the equilibrated complexes; the small size and the ready diffusibility of all reaction components within the gel matrix are sufficient for a productive attack on the target DNAs). To obtain an intelligible and homogeneous cleavage pattern of all DNA species in the gel, the exact time of chemical treatment has to be adjusted for each particular case, based on the considerations discussed in **Note 4**.
7. During the last 5 min of the incubation period, prepare the $(OP)_2Cu^+$-STOP solution.
8. Quench the reaction by adding the $(OP)_2Cu^+$ stop solution (20 mL), and wait 2 min while shaking the Pyrex dish (*see* **Note 8**). The gel will turn yellowish, which is diagnostic for the quality of 2,9-dimethyl–OP, and hence for efficient termination of the chemical nuclease action.
9. Using a 20-mL glass pipet, aspirate (staying away from the corners of the gel) all the liquid from the Pyrex dish and carefully rinse the gel (still on the glass plate) with four changes of deionized water. Remove the plate with the gel on it from the Pyrex dish. It is not necessary to take any specific precautions in dispensing the original mixture and the washing material, because all reaction components are oxidatively destroyed. Immerse the Pyrex dish in household bleach for 1 h at room temperature, followed by extensive washing down the drain with tap water.

3.4. Isolation of Free and Complexed DNA Fractions

3.4.1. Direct Elution from the Polyacrylamide Gel Matrix

1. Smoothly wrap the gel and plate with plastic wrap or, preferably, peel off the gel onto a suitable backing material (old X-ray film covered with plastic wrap is best) and wrap it with plastic wrap. Using a glass stirring rod as a rolling pin, remove any air bubbles trapped under the plastic wrap, being careful not to disturb the shape of the gel.
2. To aid accurate subsequent alignment of gel and film, trace three corners of the plastic wrap covering the gel with small adhesive labels (or pieces of Scotch Tape), marked with radioactive ink spots; use an almost equal amount of radioactivity in each ink dot to that in the protein-bound DNA fraction(s) (this can be monitored by a Geiger counter). Use a fiber-tip pen to apply ink of the desired activity to the sticky labels; let the ink dots dry completely before exposing to X-ray film.
3. In the darkroom, tape the sealed gel to a piece of Kodak X-Omat AR film. Enclose the assembly in a lightproof cardboard film holder, exerting an even gentle pressure on the "sandwich", and wrap the entire packet with aluminum foil to ensure a lighttight environment.

4. Expose the film at 4°C for 1–3 h (the length of exposure time depends on the relative abundance of the specific complex[es]) to assess the position of the retarded (bound) and unretarded (free) DNA fragments. The energy of the β particles produced by the ^{32}P decay is sufficient to penetrate several millimeters thickness of hydrated gels, without significant absorption (quenching by the gel is <40%), thus allowing the direct autoradiographic detection of [^{32}P]-labeled DNA embedded in the gel matrix.
5. Develop the film and, using a lab marking pen, encircle the position of the complex(es) to be mapped as well as that of the unretarded probe. Any DNA released by dissociation during the run will trail just above the free DNA band as a smear; do not include this region in your marking.
6. Using a sharp scalpel, cut out the marked rectangles containing the autoradiographic images of the free and bound probe from the X-ray film.
7. Line up the radioactive ink spots on the film with the corresponding markings at the three corners of the plastic wrap. With a fine-tip waterproof marking pen, mark the position of the free and bound probe on the plastic wrap, using the periphery of the rectangular holes on the film as a template.
8. Remove the film and cut through the marks on the plastic wrap with a disposable razor blade for each species. Separate the polyacrylamide slices from the rest of the gel (and from the plastic wrap), using either 18-gage single-use syringe needles or sterile forceps, and transfer them onto a piece of plastic wrap. It is desirable to keep the size of the polyacrylamide strips to a minimum (*see* **Note 5**).
9. Crush the gel slices by extruding them from a sterile 3-mL syringe barrel through a shortened 18-gage needle (broken with pliers) into a siliconized 1.5-mL Eppendorf tube by low-speed centrifugation (5 min at 2500*g*) in a swinging bucket rotor. Alternatively, punch a small hole by forcing a 22-gage sterile needle through the bottom of a siliconized capless 0.5-mL Eppendorf tube, place this tube into another siliconized capped 1.5-mL Eppendorf tube, put the gel slice in the upper tube, and spin at 12,000*g* in a microcentrifuge (minus rotor cover) for 1 min. The gel will be crushed through the hole into the lower tube.
10. To each tube, add enough gel elution buffer to cover the gel paste and mix well by vortexing. The volume of the buffer added depends on the size of gel slice, but, as a guide, 0.5-0.6 mL is used for a slice 10 × 3.5 × 3 mm.
11. Seal each tube with conformable self-sealing tape and allow the DNA fragments to diffuse out by incubating at 37–42°C for 10–16 h in a shaking incubator.
12. Vortex the tubes vigorously and pellet the gel paste by centrifuging at room temperature for 1 min in a microcentrifuge (12,000*g*).
13. Using a micropipet, pipet off the supernatant solution, taking care to avoid polyacrylamide pieces, and transfer it to a 1-mL sterile syringe.
14. Remove any remaining tiny pieces of polyacrylamide by slowly passing the supernatant through a 0.22-μm syringe filter into a fresh siliconized Eppendorf tube (do not use polystyrene tubes to collect the filtrate, as they cannot withstand the subsequent organic extractions). The eluted yield of DNA fragments should be >90%.

15. Extract the filtered supernatant sequentially with an equal volume of phenol/chloroform/isoamyl alcohol (25:24:1; v/v) and chloroform/isoamyl alcohol (24:1; v/v), to eliminate contaminating proteins that might distort DNA fragment migration during subsequent electrophoresis. In both steps, mix the contents of the tube thoroughly by vortexing for 30 s and centrifuge at 12,000*g* (microcentrifuge) for 5 min at room temperature to separate the organic and aqueous phases (*see* **Note 9**).
16. With a micropipet, transfer the aqueous phase (no more than 0.55 mL) to a fresh siliconized Eppendorf tube. Add approx 2 volumes of ice-cold absolute ethanol (no additional salt is required!), vortex well, and precipitate the radioactive DNA fragments by chilling the tube at –20°C for a minimum of 60 min. Although it is generally not necessary to add carrier to aid precipitation (the small acrylamide polymers released from the crushed gel slice will suffice), it is recommended to precipitate the DNA in the presence of glycogen (10 μg/sample, added prior to ethanol) to improve the recovery of DNA even further.
17. Recover DNA by centrifugation at 12,000*g* for 30 min in a microcentrifuge (4°C). Carefully aspirate off the ethanol supernatant with a drawn out Pasteur pipet, taking care not to disturb the faintly visible radioactive pellet (its presence can be monitored by a Geiger counter, and its location identified from the position of the tube in the rotor). (*See also* **Note 10**.)
18. Remove traces of salt trapped in the precipitate (which interfere with subsequent electrophoresis) by rinsing the pellet twice with 1 mL of 70% and 90% (v/v) ethanol, respectively, centrifuging each time at 12,000*g* (microcentrifuge) for 2 min at 4°C. For both washes, invert the tube gently several times; do not vortex.
19. Dry the pellet for 5 min in a vacuum centrifuge.
20. Measure each pellet by Cerenkov counting to determine radioactivity (1500–2000 cpm is sufficient for an overnight exposure with intensifying screen).
21. Resuspend the pellets (by heating at 68°C for 2 min, vortexing vigorously, and repeatedly pipeting) in sequencing-gel loading buffer, so that 5 μL will contain equal Cerenkov cpm from the free and bound DNA fractions. It is important to equalize the number of cpm/μL in the two fractions in order to compare their cleavage patterns accurately. If the sequencing gel is not to be run immediately, the DNA samples can be stored at –70°C.

3.4.2. Electrotransfer of the Entire Gel and Elution from NA-45 Membrane

If the EMSA was performed using Tris–glycine-containing or low-percentage/low-ionic-strength polyacrylamide gels, which behave like poorly set "Jello" and are, therefore, extremely difficult both to manipulate for autoradiography and to handle as polyacrylamide strips in the subsequent DNA elution steps, it is highly recommended to transfer the entire gel electrophoretically onto a sheet of NA-45 membrane (DEAE cellulose in membrane form). Following electroblotting, the NA-45 membrane is exposed to X-ray film, the bands corresponding to free and bound species are cut out, and DNA is eluted. The remaining steps in the procedure, beginning with organic extractions of the eluates, are identical to those described in **Subheading 3.4.1., steps 15–21**.

1. Cut a piece of NA-45 membrane and four pieces of filter paper to the exact size of the gel; cut the membrane between liner sheets wearing gloves.
2. To increase binding capacity, wash the membrane for 10 min in 10 m*M* EDTA, pH 7.6, and for 5 min in 0.5 *M* NaOH, followed by several rapid washes in distilled, deionized water; let the membrane soak in 1X TBE buffer.
3. Remove the plate with gel from the Pyrex dish and place it on a flat surface. Carefully lay two pieces of prewet (in 1X TBE buffer) filter paper onto the surface of the gel, making sure no air bubbles are trapped between the filter paper and gel.
4. Slowly and with extreme care, lift the gel (adhered to the filter paper) and place it (with the gel facing up) on a clean glass plate. Wet the gel with a thin layer of 1X TBE buffer.
5. Wearing gloves, lay the wet membrane sheet over the gel, again being careful not to trap air bubbles beneath the membrane.
6. Complete the "sandwich" by placing the two remaining pieces of prewet (in 1X TBE buffer) filter paper on top of the membrane.
7. Insert the "sandwich" of filter paper/gel/membrane/filter paper into a gel-holder cassette, and load the assembly into one of the center slots in a (wet) transfer apparatus (any of the commercially available electroblot units are suitable), with the NA-45 membrane positioned between the gel and the anode (positive electrode).
8. Fill the transfer apparatus with 1X TBE buffer (precooled at 4°C) and transfer the chemically cleaved double-stranded DNA fragments electrophoretically from the gel to the NA-45 membrane. Electroblotting is performed at 4°C for 3 h, at either 20 V (approx 1 V/cm) if small DNA fragments (40–90 bp) are being transferred or at 35 V (approx 2 V/cm) if fragments >100 bp have been employed in the EMSA.
9. When transfer is completed, turn off the power, remove the gel "sandwich," lift the membrane sheet away from gel while wearing gloves, and rinse it in 20 m*M* Tris–HCl, pH 8.0, and 0.1 m*M* EDTA to remove residual polyacrylamide. Do not let the membrane dry!
10. Place the wet membrane (with transferred DNA face up) on the surface of a used piece of X-ray film wrapped in plastic wrap and cover it with a tightly drawn layer of plastic wrap. With a pad of Kimwipes, push out any trapped air bubbles under the plastic wrap. Efficient transfer can be monitored by checking with a Geiger counter.
11. Follow **steps 2** and **3** in **Subheading 3.4.1.**; use a metal cassette instead of a cardboard film holder to expose the membrane to X-ray film.
12. Autoradiograph the membrane at 4°C for 15–45 min (about one-fourth the time required for the wet gel).
13. Follow steps 5–8 in **Subheading** 3.4.1. Using sterile forceps, transfer the wet NA-45 membrane strips into siliconized 1.5-mL Eppendorf tubes.

14. Add to each NA-45 membrane strip 0.6 mL of NA-45 membrane elution buffer and spin for a few seconds in a microcentrifuge to submerge the whole strip.
15. Incubate at 55°C for 2–3 h in a water bath with agitation.
16. Vortex the tubes vigorously and pellet the NA-45 membrane strips by centrifuging at room temperature for 10 s in a microcentrifuge.
17. Remove the buffer, and place it in a fresh siliconized 1.5-mL Eppendorf tube. Monitor paper for loss of radioactivity; typically, approx 90% of the membrane-bound DNA is released with this technique.
18. Follow **steps 15–21** in **Subheading 3.4.1.**

3.5. Preparation of the G + A Sequencing Ladder

Provided the sequence of the DNA probe is known, you may perform at this stage a Maxam–Gilbert guanine- and adenine-specific modification/cleavage reaction (G + A sequencing ladder) of the end-labeled DNA fragment used in the gel retardation assay. This reaction will be coelectrophoresed with the DNA samples eluted from the free and bound fractions, to identify nucleotides protected from chemical cleavage (protein-contact sites) in the final stage of the footprinting analysis (**Subheading 3.6.**). Below is a fast version (requiring only 1 h) of this otherwise time-consuming reaction.

1. In a siliconized 1.5-mL Eppendorf tube mix successively:
   - 20,000 cpm of the end-labeled DNA fragment used in the EMSA
   - 1.5 µL of carrier DNA stock solution
   - Distilled, deionized water to 10 µL
2. Chill the tube on ice and add 1.5 µL of 88% aqueous formic acid.
3. Incubate at 37°C for 14 min in a water bath.
4. Chill again on ice and add 150 µL of freshly prepared 1.0 *M* aqueous piperidine. Close the tube and wrap the cap tightly with a conformable self-sealing tape.
5. Heat at 90°C for 30 min in a thermostatted heating block, with the wells filled with water. It is necessary to put a lead weight on top of the tube to prevent it from popping open as pressure builds up inside.
6. Cool the tube on ice. Remove the conformable tape and spin for a few seconds in a microcentrifuge; transfer to a fresh siliconized Eppendorf tube.
7. Add 1 mL of 1-butanol. Vortex vigorously until only one phase is obtained.
8. Mark the position of the tube in the rotor and spin at 12,000*g* for 2 min in a microcentrifuge (room temperature).
9. Carefully remove and discard the supernatant using a drawn-out Pasteur pipet, taking care not to disturb the tiny pellet or the area of the tube where the pellet should be located.
10. Add 150 µL of 1% SDS and vortex the tube. Add 1 mL of 1-butanol. Mix well by repeatedly inverting the tube. This step removes remaining traces of piperidine trapped in the precipitate that interfere with the subsequent electrophoretic separation.
11. Resediment the precipitate by centrifuging at 12,000*g* for 2 min in a microcentrifuge and remove the supernatant as in **step 9.**

12. Spin for a few seconds in a microcentrifuge to collect any traces of liquid at the bottom of the tube, and carefully remove it using a micropipet.
13. Dry the pelleted DNA for 5 min in a vacuum centrifuge.
14. Resuspend the samples (as in **step 21** of **Subheading 3.4.1.**) in 5 µL of sequencing-gel loading buffer, and store at –70°C until ready to load onto the sequencing gel.

3.6. Analysis of the Chemical Cleavage Products on a DNA Sequencing Gel

Visualization of the length(s) on the DNA affected by protein(s) binding specifically to it (i.e., the protein-binding site[s], or footprint[s] left by the protein[s] on the DNA) requires electrophoretic fractionation of the single-stranded fragments resulting from the chemical nuclease attack in a denaturing polyacrylamide gel of the type employed in DNA sequencing, followed by autoradiography. The location of the footprint(s) in the known DNA sequence is identified by including the sequencing marker G + A track prepared in **Subheading 3.5.**

1. Assemble and pour a 34 × 40 × 0.04-cm 6–15% (w/v) sequencing polyacrylamide gel, containing 1X TBE buffer and 8.3 *M* urea *(20)*. The percentage of acrylamide depends on the size of the DNA fragments to be separated as well as on the size and location (relative to the labeled end) of the suspected protein-binding site(s). As for the preparative EMSA, you should siliconize the inside surface of the back glass plate to aid pouring into gel mold and removal at the end of electrophoresis. To avoid dispersing of radioactivity across lanes (which might produce significant errors in a subsequent densitometric analysis of the free and bound DNA chemical cleavage patterns on the autoradiogram; *see* **Note 12**), it is recommended to use a 0.4-mm custom-made sample comb with 5-mm lanes and 5-mm spacing. You may wrap the polymerized gel in plastic wrap and keep it at room temperature until use (it can be stored as such for up to 36 h).
2. Attach the gel apparatus to the gel electrophoresis tank. Fill both the top and bottom electrode chambers with 1X TBE buffer and remove the well-forming comb. Check that wells are free from "tails" of polyacrylamide adhering to sides, which may lead to uneven loading of samples and consequently band-shape distortion.
3. Pre-electrophorese the gel for 45–60 min before loading the samples. This removes persulfate ions and heats the gel. Prerunning of the gel is performed at constant temperature (approx 55°C), which is most easily achieved by application of constant power (approx 50–70 W). If the surface temperature becomes too high (>65°C), the glass plates will crack.
4. Thaw the DNA samples (if frozen), heat-denature them (including the G + A sequencing ladder) at 90°C in a dry-block heater for 5 min, and quick-chill in wet ice.
5. Disconnect the power supply, and immediately prior to applying the samples, thoroughly rinse out (using a 10-mL syringe with a 22-gage needle) the wells of

the gel with the upper reservoir TBE buffer; this prevents streaking of the DNA samples caused by urea that has diffused into the wells.

6. Using calibrated glass capillaries with finely drawn tips or, preferably, disposable flat-capillary pipet tips, load (as quickly as possible) 5 μL of each sample (plus the G + A sequencing ladder) onto the wells of the sequencing gel, sweeping the sample evenly from side to side. An untreated naked DNA sample (i.e., not subjected to the gel retardation assay) should always be diluted in sequencing-gel loading buffer, heat-denatured, and coelectrophoresed with the treated samples to verify the integrity of the DNA, as single-strand nicks can mask protein binding sites or even create artificial ones.
7. Remove all bubbles from the bottom of the gel (they may prevent even migration of the samples) using a 30-mL syringe with a bent 20-gage needle.
8. Run the gel under pre-electrophoresing conditions (constant power, approx 50–70 W), taking care not to overheat the glass plates. Uneven migration of fragments ("smiling") caused by an uneven gel temperature can be avoided by clamping an aluminum plate to the front glass plate. It is customary to electrophorese the samples until the bromophenol blue marker dye is about 3–5 cm from the bottom of the gel, but longer electrophoresis times may be required to obtain single-band resolution in the area of the footprint(s). The location of this region(s) depends on the distance between the radioactive label and the suspected protein-binding site(s) as well as on the length of the DNA fragment. Make use of available tables in the literature (referring to the migration of oligodeoxynucleotides in sequencing gels in relation to marker dyes) to determine how long to run your gel in order to achieve the desired electrophoretic resolution in the region of the expected footprint(s) ***(19)***; this will allow you to discern differences between the chemical cleavage patterns of the free DNA and that derived from the complexed fraction(s).
9. After completion of electrophoresis, remove the gel from the apparatus, and with the aid of a razor blade, slowly lift the siliconized plate. The thin polyacrylamide sheet will stick to the unsiliconized plate. Fix the gel for 15–20 min by gently immersing it (still attached to the lower plate) in a tank containing enough fixing solution; this removes excess urea that would otherwise crystallize out.
10. Carefully remove the plate, bringing the gel on it from the tank, and lay it on a flat surface. Place a prewet (in fixing solution) sheet of backing paper (cut slightly bigger than gel dimensions) over the gel, press it gently down on the gel, roll out any air pockets (using a 10-mL glass pipet), and peel it off patiently and with extreme care together with the gel attached.
11. Cover the gel surface (but not the back of the filter paper) with a tightly drawn layer of plastic wrap. With a Kimwipe, push out any trapped air bubbles under the plastic wrap that might interfere with good uniform contact among the film, gel, and screen. Add two sheets of filter paper next to the backing paper as a support pad and put the "sandwich" into a gel dryer (paper pad closest to vacuum source).
12. Dry the gel under vacuum at 80°C for 45–60 min. Do not release the vacuum before the gel is completely dried (sequencing gels with acrylamide concentrations >10% are susceptible to fracturing).

13. In the dark room, place a sheet of Kodak X-Omat AR film against the plastic-covered face of the gel. It is advisable to preflash the X-ray film, that is, exposing the film to a 1-ms flash of light prior to placing it in contact with the sample to an optical density of 0.15 (A_{540}) above the absorbance of the unexposed film. This increases sensitivity (all of the time of exposure to the radioactivity-generated light produces blackening) and linearity of the film response (the degree of blackening above the background is proportional to the amount of radioactivity), which are both essential if a densitometric analysis of the chemical cleavage products is to be performed (*see* **Note 11**). Preflashing requires a photographic flash unit appropriately fitted with filters and adjusted as described ***(21)***.
14. Autoradiograph in a metal cassette containing a single calcium tungstate intensifying screen (place the flashed face of film toward intensifying screen) at –70° to –80°C (to reduce scattering) for 12–16 h. Shorter or longer periods of time may be also required, as the clarity of the footprint depends on the intensity of the bands in the autoradiograph.
15. Immediately remove the film from the cassette and develop it, preferably in an automatic processing machine for X-Omat films. If re-exposing the gel, it is necessary to let the cassette warm up before inserting a second film (cold cassettes will quickly collect moisture from the air).
16. Compare the chemical cleavage pattern of the naked DNA to that of the DNA–protein complex(es). The position of a band in the gel corresponds to the distance between the label and the point at which the DNA has been cleaved by the chemical reagent. Accordingly, bands at the bottom of the gel represent the smallest end-labeled DNA fragments, increasing in size as one reads up the gel until the pattern terminates abruptly in a strongly labeled band, corresponding to the uncleaved full-length probe (*see* **Fig. 3**). The protected region(s) (indicating sequence-specific protein binding) appears as an area resistant to cleavage (footprint), resulting in an almost complete absence of fragments (gap) arising from within the protein-binding site(s) in the chemical cleavage pattern of the complexed DNA. The nucleotides exhibiting protection are identified by aligning the bands in the cutting pattern of the free DNA with positions (bonds) in the sequence of the coelectrophoresed Maxam–Gilbert marker G + A track. In this comparison, it is necessary to note that, regardless of the end of the DNA fragment labeled in the experiment (5' or 3'), the obtained set of products from the chemical cleavage reaction matches the mobilities of the G + A sequencing fragments exactly. This is a consequence of the identical 3' and 5' ends generated by both chemistries at the cleavage points ***(22)***.

4. Notes

1. Optimization of the analytical EMSA. This is generally achieved by assessing binding-reaction parameters and gel electrophoresis conditions. Furthermore, success in interpretation of results from the coupled gel retardation/*in situ* OP–Cu footprinting assay depends critically on some properties of the DNA fragment used in the initial binding reaction.

a. Properties of the DNA fragment used in the binding reaction. An EMSA employing crude extracts works best with short DNA fragments, as these reduce nonspecific interactions of proteins in the extract with sequences flanking the specific binding site(s) and are able to detect binding of large proteins more readily. Optimal sizes range between 100 and 150 bp, with the putative protein-binding site(s) located at approximately the center or at least 20–25 bp away from the radioactive labeled end. If a 20- to 25-bp synthetic oligonucleotide is to be used as a probe, it is advisable to design it in a way that it can be readily subcloned into the polylinker region of a suitable vector, and then labeled and released as a 40- to 45-bp restriction fragment, in order to obtain the desired single-base resolution within and around the expected footprint(s). Although the DNA fragment may be labeled at all ends for EMSAs, the subsequent footprinting analysis requires the DNA fragment to be radioactively labeled (to a high specific activity) at the 3' or 5' end of *one* of the two strands. Klenow enzyme-labeled probes are preferable to kinased probes because some protein extracts contain substantial phosphatase activities. Finally, the labeled probe should be unnicked, because the resulting fragments may obscure the cleavage pattern obtained after the chemical attack in the footprinting analysis. Therefore, sufficient care should be taken to minimize nuclease activities during all steps of preparation, labeling, and isolation. To this end, we have found that purification of singly end-labeled probes from native polyacrylamide gels by "crush-and-soak" methods (similar to that described in **Subheading 3.4.1.**) results in less damage to DNA than electroelution.
b. Binding-reaction parameters. These include binding-buffer composition (pH, ionic strength, metal ion content, and presence or absence of nonionic detergents and/or stabilizer polycations), amount of crude or partly-fractionated extract or purified protein, concentration of labeled DNA probe, type and amount of bulk carrier DNA, and temperature and duration of incubation. Specifically, the following considerations should be evaluated: The optimal ratio of protein to DNA for the assay is best determined by titrating a fixed concentration of the radioactively labeled DNA fragment with increasing amounts of crude or partly-fractionated extracts, or purified protein. Frequently, as protein concentration increases, binding passes through a maximum. Note, however, that increasing amounts of protein to a fixed concentration of DNA will not necessarily increase the yield of specific complex(es) seen. This is because of the fact that whereas a given preparation of any DNA-binding protein(s) tends to be fully active in nonspecific binding, it is typically only fractionally active in site-specific binding activity (the apparent fractional activity varying from 5% to 75%, depending on the particular protein[s] and the individual sample). Therefore, too much protein, particularly with crude extract preparations, leads to occlusion of the binding site(s) by proteins interacting with DNA in a sequence-independent manner. This problem can be minimized by raising simultaneously the concentration of bulk carrier (competitor) DNA (typically of the order of 250- to 5000-fold

[w/w] excess over binding-site DNA); this increases the occupancy of the binding site(s) by sequestering nonspecifically bound proteins, including nonspecific DNA-binding nucleases that may degrade the end-labeled DNA during the binding incubation. Bear in mind, however, that although some proteins are able to locate their target binding site(s) in the presence of vast excesses of nonspecific natural DNAs (sonicated salmon sperm or calf thymus DNA), other proteins cannot tolerate natural DNA carriers, but bind readily in the presence of an excess of synthetic polynucleotides, such as poly d(I-C) · d(I-C) or poly d(A-T) · d(A-T). It is noteworthy in the latter case that the efficacy of competition for nonspecific binding can vary among different batches from the same vendor. On the other hand, if too much carrier DNA is added, it will compete for the specific factor(s) of interest, and the level of complex(es) will decrease. Finally, provided an adequate resolution of DNA-bound species is obtained for a fixed concentration of competitor, increasing the amount of probe increases the fraction of DNA driven into complex(es), until the limit set by the binding constant(s) is reached.

c. Gel electrophoresis conditions. Gel parameters, such as percentage of acrylamide, degree of crosslinking, and pH and type of gel/running-buffer system (high- or low-ionic-strength) dramatically affect the size, aggregation state, and stability of DNA–protein complexes, hence their abundance and quality of separation. Accordingly, it may be necessary to try more than one gel fractionation/buffer system to obtain sufficient formation of the DNA–protein complex(es) of interest. The electrophoresis time has to be optimized for the complex(es) studied and the separation required (if more than one complex has to be mapped). The most promising conditions can then be applied in the subsequent preparative EMSA.

2. Binding competition analysis. If competitor DNA is identical to and in relatively large excess over the labeled DNA, >90% of the radioactive signal should be eliminated from complexes corresponding to protein(s) that interact with the binding-site DNA in a sequence-specific manner. Complexes unaffected or only slightly affected by the addition of competitor are thought to arise from nonspecific, low-affinity binding by abundant proteins that are present in excess to binding-site DNA; DNA derived from these complexes after the *in situ* chemical cleavage reactions can serve as a negative control in the footprinting analysis, because its cutting pattern will closely resemble that of the free (unbound) probe.
3. Assaying relative dissociation rates of DNA–protein complexes. To follow dissociation kinetics, complexes are allowed to form under optimal reaction conditions and, at time zero, exposed to a large excess of an agent that does not perturb their stability (commonly 100- to 250-fold mass excess of nonspecific competitor DNA or, preferably, of the same DNA fragment unlabeled). The "scavenger" molecules sequester the protein(s) as it dissociates from its specific binding site(s), hence preventing it from rebinding. Analysis by the EMSA of aliquots at various times (from a few seconds to 2 h) after quenching the reaction with the sequestering agent shows the amount of free DNA increasing while the

level of protein-bound label diminishes. The experiment can be designed in a way that individual reactions can be started and quenched at different times, such that all reach the point at which they will be applied to the gel more or less simultaneously and electrophoresed for the same period of time. Because dissociation of typical DNA–protein complexes is a first-order process (i.e., independent of the concentration of complexes), the results of the analytical study are also applied to the subsequent preparative assay.

4. Optimizing the time of *in situ* chemical treatment. The length of incubation period is determined by several factors, among which the following are of particular importance:
   a. Kinetic stability of the complex(es) (**step 3** in **Subheading 3.1.**). In principle, the higher the dissociation rate of the complex(es), the lower the time of exposure to the chemical nuclease. However, because of the gel "caging effect" and the *in situ* (on the DNA surface) funneling of the reaction (*see* **Subheading 1.1.1.**), this rule of thumb is applicable only for DNA–protein complexes with half-lives either <<1 min or >>60 min; provided the complex(es) is relatively abundant (*see below*), incubation times <8 min and >30 min, respectively, should be used in these cases.
   b. Relative abundance of the complex(es). The aim of the reaction is to generate, on average, about one chemical cleavage event per DNA strand. Assuming that the cleavage process is governed by Poisson statistics, the product oligonucleotides will statistically be the result of a single cleavage ("single-hit kinetics") when the concentration of the full-length labeled strand is approx 50–70% of its original value (i.e., approx 50–70% of the DNA fragments should be left uncleaved). Accordingly, when the abundance of the target complex(es) is very low (i.e., a bound DNA to free DNA ratio of <<1), early termination of the reactions will be critical for high-"off"-rate complexes, beneficial for intermediate-stability complexes, and safe for extremely stable complexes.
   c. Temperature of incubation. If optimization of the DNA-binding reaction and, consequently, of the EMSA requires that both be performed at low temperature (4°C), you should carry out the chemical nuclease treatment at low temperature as well. However, the amount of dissolved air oxygen in the reaction mixture under these conditions is considerably higher (increases with decreasing temperature); since molecular oxygen catalyzes a rate-limiting step that generates *in situ* hydrogen peroxide (an essential coreactant for the chemical nuclease activity; *see* **Fig. 2**), its presence in more than stoichiometric amounts will shift the subsequent equilibria toward the right side, leading to increased rates of DNA-strand scission. To compensate for this accelerated cleavage kinetics, you should decrease the time of exposure to the chemical nuclease. Inversely, incubation times should be longer than originally established (or you may add hydrogen peroxide exogenously) for chemical treatments performed at bench temperature during hot summer days in non-air-conditioned rooms.

d. Concentration of reagents. Increasing the concentration of OP while holding the concentrations of copper(II) sulfate and 3-mercaptopropionic acid constant increases the overall rate of DNA-strand scission, without significantly affecting the sequence preferences of cleavage and the resulting fragment size distribution *(15)*. On the other hand, substituting 3-mercaptopropionic acid by the same concentration of ascorbic acid reduces the overall rate of DNA cleavage both at low (1:1) and high (>4.5:1) 1,10-phenanthroline/copper molar ratios *(15)*. These observations are of practical significance to estimating the time of chemical treatment, particularly when extremely labile or extraordinary stable complexes are being mapped. A higher concentration of OP and a short incubation time might be used in the former case, whereas ascorbate (as the reducing agent) and prolonged treatments are recommended in the latter.

e. Presence of dithiothreitol (DTT). DTT (often employed in EMSAs to maintain the activity and/or stability of DNA-binding proteins) slows the DNA cleavage rate because this chemical sequesters the copper necessary for the oxidative cleavage to occur. If excessive amounts of DTT (i.e., >1 m*M*) have been used either in the preparative binding reaction or for casting the gel in which the $(OP)_2Cu^+$ cleavage reaction is to be performed, longer incubation times or increased $(OP)_2Cu^+$ concentrations should be used to restore the cleavage efficiency.

5. Trailing of the bands during electrophoresis. Glycerol-containing binding buffers tend to cause significant trailing at the edges of the bands during electrophoresis. If you noticed this trailing under the optimized conditions in the analytical assay, you may substitute glycerol for Ficoll (2.5% [w/v]; Type 400, Pharmacia) in the preparative binding buffer. The presence of Ficoll (a copolymer of sucrose and epichlorohydrin), although not interfering with the thermodynamic and kinetic parameters of the binding reaction, gives rise to straight bands in the gel (by tending not to spread so much because of surface-tension effects when loading the sample), thus minimizing the size of the polyacrylamide strips in the subsequent DNA-elution steps (**Subheading 3.4.1.**).
6. Loading the preparative polyacrylamide gel. If you have problems with the sample not sinking to the bottom of the well (which may be caused by substantial differences between the binding buffer and the electrophoresis buffer, and/or the large volume of the preparative reaction), you can preload the well(s) of the gel with binding buffer or, alternatively, load your sample with the power supply running at 10–15 mA. (Wear dry plastic gloves and use only plastic tips if you do this!)
7. Preparation of the MPA, OP, and Cu^{2+} solutions:

a. Use the recommended suppliers to obtain the liquid and powdered reagents. The care given to the preparation of reagents is crucial. In particular, the water used must be of the highest quality. Our laboratory uses only water purified with a Milli-Q system (Millipore), which removes virtually all organics, ions, and bacteria. Such precautions help prevent spurious reactions of impurities with the reagents.

b. You may try to footprint the complexes using a 1:1 ratio of 1,10-phenanthroline to copper, if your suspected protein-binding site(s) or the adjacent regions are particularly rich in 5'-CG-3' elements (*see* **Subheading 1.2.1.**). In this case, prepare the following solution: 100 m*M* of 3-mercaptopropionic acid, 5 m*M* OP, and 5 m*M* $CuSO_4$.

8. 2,9-Dimethyl–OP is a redox inert analog of OP which acts as a Cu^+-specific chelator; it will bind all the metal ion, preventing further oxidative chemistry and DNA cleavage.
9. Phase inversion during organic extractions of the gel- or membrane-eluted DNA samples. Because of the high-salt content of the elution buffers, the aqueous phase (which normally forms the upper layer) may sometimes be dense enough to form the lower layer. If this is the case, the aqueous phase can be easily identified by monitoring the eluted radioactivity with a Geiger counter or by following the strong yellow color of the organic phase (contributed by 8-hydroxyquinoline that is added to phenol during equilibration as an antioxidant; *see* **Subheading 2.4.1.1.**).
10. Excess acrylamide in the gel-eluted DNA samples. If the DNA pellet is highly contaminated with acrylamide monomers or other impurities from the gel matrix (a problem sometimes encountered when employing low-percentage mobility-shift gels and is usually apparent from the formation of an excessive turbidity during the ethanol-precipitation step), you can further purify it by passing through a pre-equilibrated Elutip™-d mini-column (an ion-exchange column) according to the manufacturer's instructions. Adsorption to and desorption from the column can be followed with a Geiger counter. Alternatively, you may follow the experimental strategy developed by Ragnhildstveit et al. *(22)*.
11. Wondering about the footprint: scanning the autoradiogram. If the effects of protein binding on the cleavage rate of the chemical nuclease are not clearly discernible by eye, or when partial protection is obtained, you may analyze the ladders quantitatively by subtracting the cleavage pattern of the DNA derived from the bound fraction from that derived from the free fraction. This involves calculating the probability of cleavage at each bond (which is related to the amount of radioactivity, or intensity, in the corresponding band in the cutting pattern) and, finally, for each lane, the average number of cuts in the DNA strand, with the aid of automated laser densitometers linked to a computer (available from, for instance, Bio-Rad or LKB-Pharmacia). Because quantification of band intensity is limited within the linear response of Kodak X-Omat AR films to radioactivity (bands with an optical density >0.15 and <1.8 absorbance units), it is essential not to use overexposed films for this type of analysis. Moreover, the film should be free of scratches, fingerprints, or other blemishes that will appear as optical signals indistinguishable from ^{32}P, thus interfering with the calculations. A difference probability plot along the entire length of the binding site(s) and surrounding DNA can be obtained, revealing the protein-binding site(s) much more clearly and reliably than can be done by comparison by eye of the chemical cleavage patterns. If still in doubt however, you may analyze the footprint on the other DNA strand.

12. Appearance of the bands on the autoradiogram. If the band sharpness or shape in the chemical cleavage patterns on the autoradiogram suffers (e.g., narrowing of the bands toward the smallest end-labeled DNA fragments), the DNA samples contain residual proteins, salts, or acrylamide contaminants. These faults should be expected if uneven running and retardation of marker dyes is observed during electrophoresis, and they can be avoided if careful attention is paid during **steps 15–18** in **Subheading 3.4.1.** If necessary, the number of organic extractions (**step 15** in **Subheading 3.4.1.**) and/or ethanol washes (**step 18** in **Subheading 3.4.1.**) can be increased; acrylamide contaminants are efficiently removed by the use of Elutip™-d mini-columns (*see* **Note 10**). Note, however, that smeared or fuzzy bands without apparent electrophoretic problems may also arise from scattering during autoradiography; use a lighttight metal film cassette with a particularly effective closure mechanism, capable of firmly fixing filter/gel and film, and with intensifying screen in place (those produced by Wolf or Picker International Health Care Products [Highland Heights, OH] are best in that).

Acknowledgments

This work has profited greatly from discussions of the author with the students and teachers participating in the 1991–1996 EMBO Courses on "DNA-Protein Interactions."

References

1. Schmitz, A. and Galas, D. J. (1978) The interaction of RNA polymerase and lac repressor with the *lac* control region. *Nucleic Acids Res.* **6,** 111–137.
2. Humayun, Z., Kleid, D., and Ptashne, M. (1977) Sites of contact between λ operators and λ repressor. *Nucleic Acids Res.* **4,** 1595–1607.
3. Garner, M. M. and Revzin, A. (1981) A gel electrophoresis method for quantifying the binding of proteins to specific DNA regions: application to components of the *Escherichia coli* lactose operon regulatory system. *Nucleic Acids Res.* **9,** 3047–3060.
4. Revzin, A. (1989) Gel electrophoresis assays for DNA–protein interactions. *Biotechniques* **7,** 346–355.
5. Topol, J., Ruden, D. M., and Parker, C. S. (1985) Sequences required for in vitro transcriptional activation of a *Drosophila* hsp 70 gene. *Cell* **42,** 527–537.
6. Kuwabara, M. D. and Sigman, D. S. (1987) Footprinting DNA–protein complexes *in situ* following gel retardation assays using 1,10-phenanthroline–copper ion: *Escherichia coli* RNA polymerase–*lac* promoter complexes. *Biochemistry* **26,** 7234–7238.
7. Sigman, D. S., Graham, D. R., D'Aurora, V., and Stern, A. M. (1979) Oxygen-dependent cleavage of DNA by the 1,10-phenanthroline–cuprous complex. Inhibition of *Escherichia coli* DNA polymerase I. *J. Biol. Chem.* **254,** 12,269–12,272.
8. Marshall, L. E., Graham, D. R., Reich, K. A., and Sigman, D. S. (1981) Cleavage of DNA by the 1,10-phenanthroline–cuprous complex. Hydrogen perox-

ide requirement: primary and secondary structure specificity. *Biochemistry* **20,** 244–250.
9. Goyne, T. E. and Sigman, D. S. (1987) Nuclease activity of 1,10-phenan-throline–copper ion. Chemistry of deoxyribose oxidation. *J. Am. Chem.* Soc. **109,** 2846–2848.
10. Pope, L. M., Reich, K. A., Graham, D. R., and Sigman, D. S. (1982) Products of DNA cleavage by the 1,10-phenanthroline copper complex. Identification of *E. coli* DNA polymerase I inhibitors. *J. Biol. Chem.* **257,** 12,121–12,128.
11. Tamilarasan, R., McMillin, D. R., and Liu, F. (1989) Excited-state modalities for studying the binding of copper phenanthrolines to DNA, in *Metal–DNA Chemistry* (Tullius, T. D., ed.), ACS Symposium Series 402, American Chemical Society, Washington, DC, pp. 48–58.
12. Kakkis, E. and Calame, K. (1987) A plasmacytoma-specific factor binds the c-*myc* promoter region. *Proc. Natl. Acad. Sci. USA* **84,** 7031–7035.
13. Flanagan, W. M., Papavassiliou, A. G., Rice, M., Hecht, L. B., Silverstein, S., and Wagner, E. K. (1991) Analysis of the Herpes Simplex Virus type 1 promoter controlling the expression of U_L38, a true late gene involved in capsid assembly. *J. Virol.* **65,** 769–786.
14. Veal, J. M. and Rill, R. L. (1988) Sequence specificity of DNA cleavage by bis(1,10-phenanthroline)copper(I). *Biochemistry* **27,** 1822–1827.
15. Veal, J. M., Merchant, K., and Rill, R. L. (1991) The influence of reducing agent and 1,10-phenanthroline concentration on DNA cleavage by phenanthroline + copper. *Nucleic Acids Res.* **19,** 3383–3388.
16. Papavassiliou, A. G. and Silverstein, S. J. (1990) Interaction of cell and virus proteins with DNA sequences encompassing the promoter/regulatory and leader regions of the Herpes Simplex Virus thymidine kinase gene. *J. Biol. Chem.* **265,** 9402–9412.
17. Darsillo, P. and Huber, P. W. (1991) The use of chemical nucleases to analyze RNA-protein interactions: the TFIIIA-5S rRNA complex. *J. Biol. Chem.* **266,** 21,075–21,082.
18. Papavassiliou, A. G. (1993) *In situ* $(OP)_2$-Cu^+ mapping of electrophoretically resolved RNA-protein complexes. *Anal. Biochem.* **214,** 331–334.
19. Sambrook, J., Fritsch, E. F., and Maniatis, T. (1982) *Molecular Cloning: A Laboratory Manual.* Cold Spring Harbor Laboratory, Cold Spring Harbor, NY.
20. Sambrook, J., Fritsch, E., and Maniatis, T. (1989) *Molecular Cloning. A Laboratory Manual, Second Edition.* Cold Spring Harbor Laboratory, Cold Spring Harbor, NY.
21. Laskey, R. A. (1980) The use of intensifying screens or organic scintillators for visualizing radioactive molecules resolved by gel electrophoresis. *Methods Enzymol.* **65,** 363–371.
22. Maxam, A. and Gilbert, W. (1980) Sequencing end-labeled DNA with base-specific chemical cleavages. *Methods Enzymol.* **65,** 499–560.
23. Ragnhildstveit, E., Fjose, A., Becker, P. B., and Quivy, J. P. (1997) Solid phase technology improves coupled gel shift/footprinting analysis. *Nucleic Acids Res.* **25,** 453–454.

Further Reading

Papavassiliou, A. G. (1995) Chemical nucleases as probes for studying DNA–protein interactions. *Biochem. J.* **305,** 345–357.

Sigman, D. S., Kuwabara, M. D., Chen, C. H., and Bruice, T. W. (1991) Nuclease activity of 1,10-phenanthroline–copper in study of protein–DNA interactions. *Methods Enzymol.* **208,** 414–433.

8

Uranyl Photofootprinting

Peter E. Nielsen

1. Introduction

It has long been known that the uranyl(VI) ion (UO_2^{2+}) forms strong complexes with various inorganic and organic anions, including phosphates, and that the photochemically excited state of this ion is a very strong oxidant *(1)*. For instance, uranyl-mediated photooxidation of alcohols has been studied in detail *(2,3)*. It is also widely recognized that uranyl chemistry and photophysics/photochemistry are very complex. Thus monomeric UO_2^{2+} is only present at low pH (pH approx 2), whereas polynuclear species and various "hydroxides," which often precipitate, form at a higher pH *(4)*.

In spite of this complexity we have found that uranyl-mediated photocleavage of DNA can be used to probe for accessibility of the phosphates in the DNA backbone *(5–8)*. Thus, uranyl is a sensitive probe for protein–DNA–phosphate contacts *(5,6)* as well as for DNA conformation in terms of DNA minor–groove width *(7–10)*. Furthermore, binding sites for divalent metal ions in folded DNA *(11)* or RNA *(12,13,13a,13b)* can be studied by uranyl photocleavage.

The systems that have so far been analyzed by uranyl-mediated DNA photocleavage include the λ-repressor/O_R1 operator complex *(5)*, *E. coli* RNA polymerase/deoP1 promoter transcription initiation open complex *(6)*, transcription factor IIIA (TFIIIA)/*Xenopus* 5S internal control region (ICR) complex *(14)*, catabolite regulatory protein (CRP)/operator DNA complex and the CRP/RNA polymerase/deoP2 promoter initiation complex *(15)*, bent kinetoplast DNA *(7)* and triplex DNA *(16)*. Furthermore, we have found that some drug–DNA complexes (exemplified by mitramycin *[17]* and distamycin *[18]*) may also be studied by uranyl photofootprinting. Finally, divalent metal ion-binding sites in an RNA polymerase–promoter open complex *(6)*, a four way DNA "Holliday junction" *(11)*, a hammer head ribozyme *(12)*, and yeast

From: *Methods in Molecular Biology, vol. 148: DNA–Protein Interactions: Principles and Protocols, 2nd ed.*
Edited by: T. Moss © Humana Press Inc., Totowa, NJ

tRNA^{phe} *(13)* and the tetrahymeria group I intron *(13b)* have been analyzed by uranyl photocleavage. This technique takes advantage of competition of low-affinity cleavage (binding) sites by a chelating agent such as citrate.

The molecular mechanism for uranyl-mediated photocleavage of DNA is not fully understood, but we have shown that uranyl binds to the phosphates of DNA and oxidizes the proximal deoxyriboses, most likely via a direct electron transfer mechanism *(19)*. The main products are 3'-phosphate and 5'-phosphate termini in the DNA and the free nucleobases are liberated in the process *(19)*. Because uranyl binding is to the phosphate groups of the DNA, very little sequence dependence of the photocleavage is seen.

2. Materials

1. Uranyl nitrate ($UO_2(NO_3)_2$), analytical grade: 100 m*M* stock solution in H_2O. This solution was found to be stable for photofootprinting purposes for at least 12 mo, and was diluted to working concentrations immediately prior to use (*see* **Note 1**).
2. ^{32}P end-labeled DNA restriction fragments (*see* **Note 2**).
3. Buffer for formation of protein–DNA complex (*see* **Note 3**).
4. 0.5 *M* Na-acetate, pH 4.5.
5. 96% Ethanol.
6. 70% Ethanol.
7. 2 mg/mL Calf thymus DNA.
8. Gel loading buffer: 80% formamide in TBE buffer, 0.05% bromphenol blue, and 0.05% xylene cyanol.
9. TBE buffer: 90 m*M* Tris–borate, 1 m*M* EDTA, pH 8.3.
10. Polyacrylamide gel: 8% acrylamide, 0.3% *bis*-acrylamide, 7 *M* urea, and TBE buffer. Size: 0.2 mm × 60 cm × 20 cm.
11. λ-Repressor: 1 µg/µL (*see* **Note 4**).
12. $MgCl_2$, 1 m*M* $CaCl_2$, 0.1 m*M* EDTA, and 200 m*M* KCl.
13. DNase I: 1 mg/mL in 10 m*M* Tris-HCl, pH 7.4, and 1 m*M* $MgCl_2$.
14. X-ray film: Agfa Curix RP1.
15. Philips TL 40W/03 fluorescent light tube that fits into standard (20 W) fluorescent light tube sockets if the transformer is changed to 40 W (*see* **Note 5**).

3. Methods

3.1. Uranyl-Photoprobing Protocol

A typical uranyl photoprobing experiment is performed as follows:

1. Form the complex to be analyzed by mixing the ^{32}P end-labeled DNA fragment (>20,000 cpm/sample) (*see* **Note 2**) with the DNA binding ligand in 90 µL of footprinting buffer (*see* **Note 3**) (containing 0.5 µg calf thymus DNA carrier) in a 1.5-mL polypropylene microfuge tube at the desired temperature.
2. Dilute the 100 m*M* uranyl stock solution to 10 m*M* in H_2O.

3. Add 10 µL of this to the sample and mix well (*see* **Note 6**).
4. Place the sample in a thermostated heating/cooling block if the temperature is critical.
5. Irradiate the sample for 30 min at 420 nm by placing the open microfuge tube directly under the fluorescent light tube (*see* **Note 5**).
6. Add 20 µL of 0.5 *M* Na–acetate, pH 4.5 to prevent coprecipitation of uranyl (which will interfere with subsequent gel analysis) and precipitate the DNA by the addition of 250 µL of 96% ethanol.
7. Place the sample on dry ice for 15 min (or overnight at –20°C) and centrifuge 30 min at 20,000*g*.
8. Wash the pellet with 100 µL 70% ethanol, dry in vacuo and redissolve in 4–10 µL 80% formamide gel loading buffer.
9. Heat the sample at 90°C for 5 min.
10. Load 10,000 cpm on a polyacrylamide sequencing gel (0.2–0.4 mm thick, 60 cm long) and run the gel at 2000 V. A sequence ladder (e.g., A+G) is run in parallel (*see* **Subheading 3.2.**).
11. Visualize radioactive bands by autoradiography overnight (or longer) at –70°C using an intensifying screen, or by phosphorimager.
12. Quantitate the results by densitometric scanning of the autoradiograms, if desired (*see* **Note 7**).

3.2. Example Photofootprint

Figure 1 shows a footprinting experiment of the complex between λ-repressor and the O_R1 operator DNA using uranyl and DNase I. Quantitative analysis of these results by densitometric scanning, summarized in **Fig. 2A**, reveals that four predominant regions of the O_R1 operator are protected from photocleavage by uranyl. These regions coincide with those protected from hydroxy radical attack and include the phosphates indicated by ethylation interference and X-ray crystallography to be involved in protein binding *(20)*. However, as discussed in **Subheading 3.3.**, the "footprinting" patterns revealed by these different techniques are not identical. The display of uranyl footprinting data on the double helical model of DNA is often very informative. In the case of the λ-repressor/O_R1 complex such display (**Fig. 2B**) shows that the repressor binds to one face of the helix and that the phosphate contacts predominantly lie either side of the major groove of the DNA, in full accord with binding of the α_3-recognition protein helix of each repressor subunit within the major groove.

3.3. Comparison of Uranyl Photoprobing to Other Techniques

The results obtained by uranyl photofootprinting are comparable to those obtained by hydroxyl radical (EDTA[FeII]) footprinting *(21,22)* and ethylation interference experiments *(23)*. However, because the uranyl ion binds to the phosphates of the DNA, a uranyl-photofootprinting experiment reports on phosphates of the DNA backbone that are accessible to the uranyl and that are

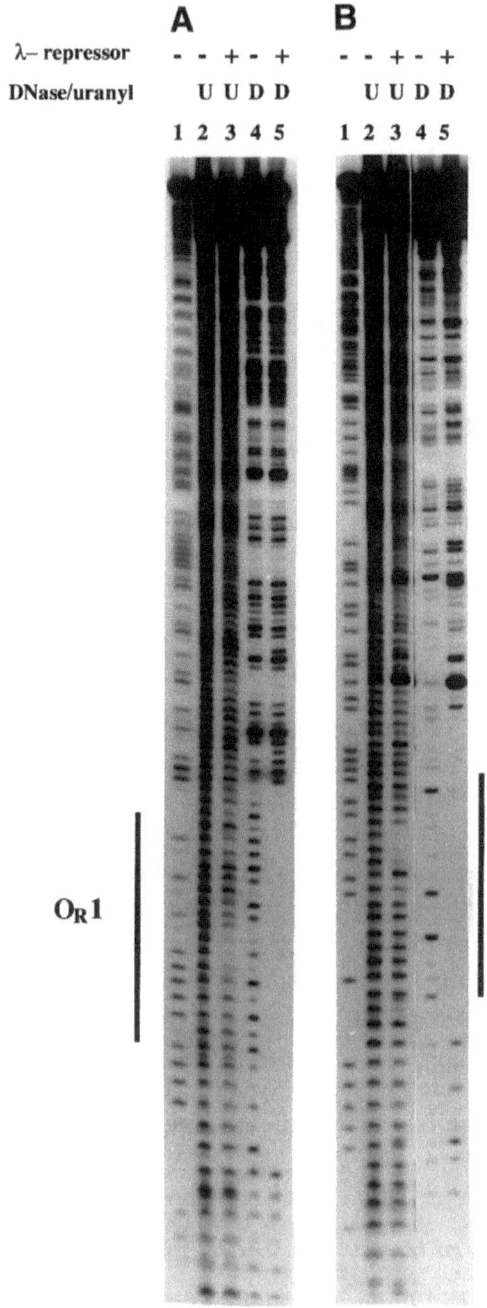

Fig. 1. Uranyl photofootprint and DNase I footprint of the λ-repressor/O_R1 operator complex (*see* **ref. 5** for details). The O_R1 operator sequence was cloned into the *Bam*HI/*Hind*III site of pUC19 and the 225 base pair *Eco*RI/*Pvu*II fragment labeled

therefore not involved in contacts with the bound ligand. On the other hand, hydroxyl radical footprinting reports on the accessibility of the deoxyriboses of the DNA backbone. In the cases studied with both uranyl and hydroxyl radical probing (λ-repressor *[5,21]*, RNA polymerase *[6,24]* and TFIIIA *[14,25]*), the footprint obtained by uranyl involves fewer nucleotides than that obtained by EDTA(FeII).

Interference probing by phosphate ethylation using ethyl-nitroso-urea also reports on the involvement of individual phosphate groups in protein–DNA interaction. However, this is an interference technique; therefore, only phosphate groups that are indispensable for complex formation are detected. Thus, for small complexes (e.g., the λ-repressor-O_R1 complex), ethylation interference and uranyl results are virtually equivalent, whereas for larger complexes (e.g., RNA polymerase–promoter complexes), only some of the contacts detected by uranyl photofootprinting are picked up by ethylation interference. Thus, the data from ethylation interference and hydroxyl radical and uranyl footprinting experiments complement each other, hydroxyl radical attack reflecting the accessibility of individual deoxyriboses, uranyl photofootprinting reporting on the accessibility of individual phosphates, and ethylation interference reporting on phosphates, which are indispensable for complex formation. Furthermore, both hydroxyl radicals and uranyl are able to sense variations in DNA conformation, and for both of these probes, groove width has been implicated as the determinant. Generally speaking, hydroxyl radical cleavage is more intense as the major groove widens, whereas uranyl photocleavage is more intense as it narrows.

UO_2^{2+}, being a divalent cation, may to some extent mimic Mg^{2+} in terms of high-affinity binding sites in protein–nucleic acid complexes and in folded nucleic acids. Consequently, such high-affinity binding sites for the UO_2^{2+} ion can result in hypersensitive cleavage sites. In this respect the uranyl photoprobing can be compared to oxidative cleavage of nucleic acids by the Fe^{2+}/Fe^{3+} ion redox pair *(30,31)*.

Finally, uranyl, being a photochemical technique, has the added, but so far unexplored, potential to be used for temperature and kinetic studies.

---

with ^{32}P at the 3' or 5' end of the *Eco*RI site was used in the experiments. Lanes 1 and 3 are controls without added λ-repressor (0.7 μg/sample). Lanes 1 and 2 are uranyl photofootprints, and lanes 3 and 4 are DNase I (0.5 μg/mL, 5 min at room temperature) footprints. Lanes S are A+G sequence reactions obtained by treating the DNA with 60% formic acid for 5 min at room temperature, and subsequent piperidine treatment. The samples were analyzed on an 8% polyacrylamide gel and run at 2500 V. The gel was subjected to autoradiography for 16 h at –70°C using intensifying screens.

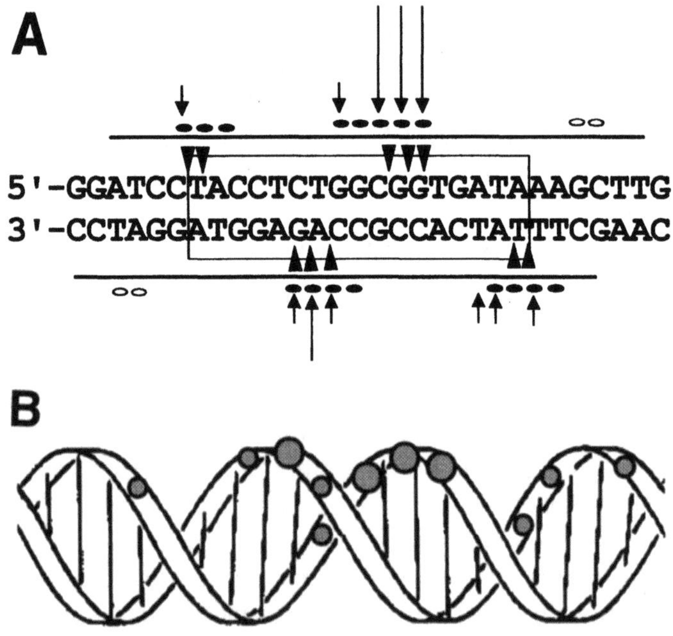

Fig. 2. **(A)** O_R1–operator sequence (box) showing uranyl photofootprint (arrows), EDTA/FeII footprint (dots *[20]*) ethylation interference (arrow heads *[21]*) and DNase I footprint (brackets). **(B)** Display of the uranyl photofootprint on O_R1–DNA double helix. The size of the dots signify the degree of protection.

4. Notes

1. Uranyl acetate ($UO_2[CH_3COO]_2$) gives identical results, but the 100 m*M* stock solution in this case has to be made 50 m*M* in HCl in order for it to be stable.
2. The ^{32}P end-labeled DNA fragments of 50 to 300 base pairs in length are prepared by standard techniques *(28)*. Typically, the plasmid containing the protein-binding site is opened using a restriction enzyme that cleaves at a distance of 20–50 base pairs from the binding site. (This distance is important because the best resolution is obtained in the 20- to 70-bases interval and the bands of uranyl-cleaved DNA fragments become "fuzzy" above 100 bases). The plasmid is labeled either at the 3' end with [α–^{32}P] dNTP and the Klenow fragment of DNA polymerase, or at the 5' end (after dephosphorylation with alkaline phosphatase) with [γ-^{32}P] ATP and polynucleotide kinase. The plasmid is then treated with a second restriction enzyme cutting 50–300 bp from the labeling site and the DNA fragment containing the protein binding site is purified by gel electrophoresis in 5% polyacrylamide, TBE buffer. The DNA fragment is extracted from the excised gel slice with 0.5 *M* NH_4–acetate, 1 m*M* EDTA (16 h, room temperature) and precipitated by addition of 2 vol of 96% ethanol. The pellet is washed with 70% ethanol and dried.
3. Choice of footprinting buffer. The choice of an optimal buffer for a uranyl-photofootprinting experiment is crucial for a successful result. In particular, the

pH of the medium is important. The uranyl-mediated photocleavage of DNA is extremely dependent on pH, cleavage being most efficient at pH 6, less efficient at pH 5 and 7, and virtually absent at pH 8 ***(19)***. Furthermore, as the pH is lowered, a strong modulation of the sequence dependence of the cleavage is observed. In fact, this modulation reflects the conformation of the DNA ***(7–10)***. Thus when conformation-independent DNA cleavage is required, buffers of pH 6.5–7.0 are advantageous, whereas buffers of pH 6.0–6.5 should be chosen for studies of DNA structure. If information about metal-ion-binding sites is desired, citrate (0.1–1 m*M*) should be included in the buffer.

The composition of the buffer and the buffer capacity is also of importance. Because the uranyl solution is acidic, it should be checked that its addition does not significantly affect the pH of the chosen buffer. Furthermore, uranyl-mediated photocleavage of DNA is most efficient in acetate or formate buffers, less efficient in Tris-HCl, very inefficient in HEPES or PIPES buffer, and virtually absent in phosphate buffers (uranyl phosphate precipitates). The ionic strength (as Na^+) is of minor importance and the cleavage is not affected by the presence of Mg^{2+} or dithiothreitol (DTT). Finally, the uranyl photoreaction is not influenced by temperature (0–70°C) ***(19)***. Within these constraints, a buffer that allows protein–DNA binding must be chosen.

4. λ-Repressor was prepared according to **ref. *29*** using an overproducer plasmid: pAE305 in *E. coli.*
5. Light source. Any light source emitting at 300–420 nm can be used. This could be the Philips TL 40 W/03 tube emitting at 420 ± 30 nm. Alternative fluorescent light tubes are Philips TL 20W/12 (300 nm) or TL 20W/09N (365 nm). Lamps emitting below 300 nm are not recommended because of absorption by the DNA bases at these wavelengths. The fluorescent light tubes suggested in this chapter are not very powerful but are quite sufficient for footprinting, they are also inexpensive and do not require sophisticated power supplies. However, if shorter irradiation times are required, uranyl-photofootprinting experiments are quite adequately performed with pyrex-filtered light from high-pressure Hg lamps, xenon lamps, or lasers of the appropriate wavelength (300–420 nm).
6. Order of mixing. It is important that the uranyl be added last because the uranyl–DNA complex is very stable (K_a is estimated to 10^{10} M^{-1} ***[19]***). Uranyl–DNA aggregates that precipitate often form, but this does not adversely affect the outcome of the footprinting reaction. However, if uranyl is added prior to addition of the DNA-binding ligand, the ligand will have limited access to the DNA.

   It is also extremely important that dilution of the uranyl stock solution be performed immediately prior to use since uranyl solutions are not stable at pH 2.0.
7. Examples of densitometric scanning and quantification of footprinting experiments can be found in **refs. *8*, *30***, and ***31***.

References

1. Burrows, H. D. and Kemp, T. J. (1974) The photochemistry of the uranyl ion. *Chem. Soc. Rev.* **3,** 138–165.

2. Azenha, M. E. D. G., Burrows, H. D., Furmosinho, S. J., and Miguel, M. G. M. (1989) Photophysics of the excited uranylionin aqueous solutions. Part 6. Quenching effects of aliphatic alcohols. *J. Chem. Soc. Faraday Trans.* **85,** 2625–2634.
3. Cunningham, J. and Srijaranai, S. (1990) Sensitized photo-oxidations of dissolved alcohols in homogeneous and heterogenous systems. 1. Homogeneous photosentization by uranyl ions. *J. Photochem. Photobiol. A. Chem.* **55,** 219–232.
4. Greenwood, N. N. and Earnshaw, A. (1986) In *Chemistry of the Elements.* Oxford, Pergamon, p. 1478.
5. Nielsen, P. E., Jeppesen, C., and Buchardt, O. (1988) Uranyl salts as photochemical agents for cleavage of DNA and probing of protein-DNA contacts. *FEBS Lett.* **235,** 122–124.
6. Jeppesen, C. and Nielsen, P. (1989) Uranyl mediated photofootprinting reveals strong *E. coli* RNA polymerase-DNA backbone contacts in the +10 region of the deoP1 promoter open complex. *Nucleic Acids Res.* **17,** 4947–4956.
7. Nielsen, P. E., Møllegaard, N. E., and Jeppesen, C. (1990) Uranyl photoprobing of conformational changes in DNA induced by drug binding. *Anti-Cancer Drug Design* **5,** 105–110.
8. Nielsen, P. E., Møllegaard, N. E., and Jeppesen, C. (1990) DNA conformational analysis in solution by uranyl mediated photocleavage. *Nucleic Acids Res.* **18,** 3847–3851.
9. Bailly, C., Møllegaard, N. E., Nielsen, P. E., and Waring, M. J. (1995) The influence of the 2-amino group of guanine on DNA conformation. Uranyl and DNase I probing of inosine/diaminopurine substituted DNA. *EMBO J.* **9,** 2121–2131.
10. Sönnichsen, S. H. and Nielsen, P. E (1996) Enhanced uranyl photocleavage across the minor groove of all $(A/T)_4$ sequences indicates a similar narrow minor groove conformation. *J. Mol. Recogn.* **9,** 219–227.
11. Møllegaard, N. E., Murchie, A. I. H. Lilley, D. M. J., and Nielsen, P. E. (1994) Uranyl photoprobing of a four-way DNA junction. Evidence for specific metal ion binding. *EMBO J.* **13,** 1508–1513.
12. Bassi, G. S. Møllegaard, N. E., Murchie, A. I., von Kitzing, E., and Lilley, D. M. (1995) Ionic interactions and the global conformations of the hammerhead ribozyme. *Nature Struct. Biol.* **2,** 45–55.
13. Nielsen, P. E. and Møllegaard, N. E. (1996) Sequence/structure selective thermal and photochemical cleavage of yeast-$tRNA^{Phe}$ by UO_2^{2+} *J. Mol. Recogn.* **9,** 228–232.
13a. Møllegaard, N. E. and Nielsen, P. E. (2000) Applications of uranyl cleavage mapping of RNA structure. *Meth. Enzymol.* **318,** 43–47.
13b. Wittberger, D., Berens, C., Hamman, C., Westhof, E., and Schroeder, R. (2000) Evaluation of uranyl photocleavage as a probe to monitor ion binding and flexibility in RNAs. J. Mol. Biol. 300, 339–352.
14. Nielsen, P. E. and Jeppesen, C. (1990) Photochemical probing of DNA complexes. *Trends Photochem. Photobiol.* **1,** 39–47.
15. Møllegaard, N. E., Rasmussen, P. B., Valentin-Hansen, P., and Nielsen, P. E. (1993) Characterization of promoter recognition complexes are formed by CRP and CytR for repression and by CRP and RNA polymerase for avtivation of transcription on the *E. coli deo*Cp_2 promoter. *J. Biol. Chem.* **268,** 17,471–17,477.

16. Nielsen, P. E. (1992) Uranyl photofootprinting of triple helical DNA. *Nucleic Acids Res.* **20,** 2735–2739.
17. Nielsen, P. E., Cons, B. M. G., Fox, K. R., and Sommer, V. B. (1990) Uranyl photofootprinting. DNA structural changes upon binding of mithramycin, in *Molecular Basis of Specificity in Nucleic Acid Drug Interactions*, vol. 23 (Pullman, B. and Jortner, J., eds.), Kluwer Academic, Dordrecht, pp. 423–432.
18. Møllegaard, N. E. and Nielsen, P. E. (1997) Uranyl photoprobing of DNA structures and drug-DNA complexes, in *Drug-DNA Interaction Protocols* (Fox, K. R., ed.), Humana, Totowa, NJ, pp. 43–50.
19. Nielsen, P. E., Hiort, C., Buchardt, O., Dahl, O., Sönnichsen, S. H., and Nordèn, B. (1992) DNA Binding and Photocleavage by Uranyl(VI) (UO_2^{2+}) Salts. *J. Amer. Chem. Soc.* **114,** 4967–4975.
20. Schultz, S. C., Shields, G. C., and Steitz, T. A. (1991) Crystal structure of a CAP-DNA complex: the DNA is bent by 90°. *Science* **253,** 1001–1007.
21. Tullius, T. D. and Dombroski, B. A. (1986) Hydroxyl radical "footprinting": high-resolution information about DNA protein contacts and application to lambda repressor and Cro protein. *Proc. Natl. Acad. Sci. USA* **83,** 5469–5473.
22. Burkhoff, A. M. and Tullius, T. D. (1987) The unusual conformation adopted by the adenine tracts in kinetoplast DNA. *Cell* **48,** 935–943.
23. Siebenlist, U., Simpson, R. B., and Gilbert, W. (1980) *E. coli* RNA polymerase interacts homologously with two different promoters. *Cell* **20,** 269–281.
24. O'Halloran, T. V., Frantz, B., Shin, M. K., Ralston, D. M., and Wright, J. G. (1989) The MerR heavy metal receptor mediates positive activation in a topologically novel transcription complex. *Cell* **56,** 119–129.
25. Vrana, K. E., Churchill, M. E. A., Tullius, T. D., and Brown, D. D. (1988) Mapping functional regions of transcription factor TFIIIA. *Mol. Cell. Biol.* **8,** 1684–1696.
26. Berens, C. Streicher, B., Schroeder, R., and Hillen, W. (1998) Visualizing metal-ion-binding sites in group I introns by iron(II)-mediated Fenton reactions. *Chem. Biol.* **5,** 163–175.
27. Zaychikov, E., Martin, E., Denissova, L., Kozlov, M., Markovtsov, V., Kashlev, M., et al. (1996) Mapping of catalytic residues in the RNA polymerase active center. *Science* **273,** 107–109.
28. Maniatis, T. Fritsch, E. F., and Sambrook, J. (1982) *Molecular Cloning. A Laboratory Manual,* Cold Spring Harbor Laboratory, Cold Spring Harbor, NY.
29. Amann, E., Brosins, J., and Ptasne, M. (1983) Vectors bearing a hybrid trp-lac promoter useful for regulated expression of cloned genes in Escherichia coli. *Gene* **25,** 167–178.
30. Jeppesen, C. and Nielsen, P. E. (1989) Photofootprinting of drug-binding sites on DNA using diazo- and azido–9-aminoacridine derivatives. *Eur. J. Biochem.* **182,** 437–444.
31. Dabrowiak, J. C., Kissinger, K., and Goodisman, J. (1989) Quantitative foot–printing analysis of drug–DNA interactions: Fe(III)methidium–propyl–EDTA as a probe. *Electrophoresis* **10,** 404–412.

9

Osmium Tetroxide Modification and the Study of DNA–Protein Interactions

James A. McClellan

1. Introduction

Osmium tetroxide is an foul-smelling chemical used as a fixative in electron microscopy that can also be used to modify thymidine residues within DNA *(1,2)*. The ability of osmium tetroxide to modify DNA is very sensitive to DNA conformation. In particular, osmium tetroxide will attack thymidines that are unstacked, either because they are in a single-stranded region or for some other reason (e.g., because the DNA is bent or because it is overwound *[3]*).

As a footprinting agent, osmium tetroxide has not been widely used. This is principally for historical reasons, and one of the reasons for writing the first version of this chapter was to encourage experimentation with osmium tetroxide as a footprinting agent. Since then, the use of osmium tetroxide in the study of DNA–protein interactions has mushroomed; a search of PubMed using osmium, DNA, and protein as keywords pulled down over 300 references in late 1998. However, it is true that the exquisite sequence specificity of osmium tetroxide is a disadvantage when one wishes to know *all* the contacts that a protein makes in a particular sequence. Additionally, osmium tetroxide has been shown to damage proteins *(4)*, although this is probably a problem common to virtually all chemical footprinting agents rather than a specific disadvantage of osmium tetroxide. On the other hand, osmium tetroxide has definite advantages in the study of certain DNA–protein interactions, specifically those interactions that result in a changed conformation of the DNA double helix. Examples of such interactions include the initiation of transcription *(5)* and the formation of nonstandard conformations in DNA that are known to be recombinogenic and to have the potential to modulate gene expression *(6–8)*.

From: *Methods in Molecular Biology, vol. 148: DNA–Protein Interactions: Principles and Protocols, 2nd ed.*
Edited by: T. Moss © Humana Press Inc., Totowa, NJ

The main advantages of osmium tetroxide are that it is very easy to work with and that the osmium modification reaction takes place under a very wide variety of environmental conditions ***(9)***, including in vivo ***(10–12)***. This makes it possible to study nucleic acid conformation as a function of environmental conditions, such as salt concentration, temperature, and presence or absence of proteins and polyamines. All that is required is that there should be susceptible thymidines (Ts) involved in the conformational change; because most sites where conformational changes occur in DNA are A+T rich, this is not a hard criterion to satisfy.

As an indication of the kinds of information that can be obtained using osmium tetroxide modification, we may cite the results of in vivo modification experiments on AT sequences ***(12)***. Experiments like these provided the first direct evidence that cruciform structures may be seen in vivo. Our experiments also show that previously observed environmentally induced changes in plasmid linking numbers actually do result in changes in torsional stress in vivo, rather than being compensated for by the binding of extra histonelike proteins. However, the data also show that cruciforms are not a normal component of the bacterial genome; we do not see cruciforms if there are significant kinetic barriers to extrusion, and we do not see cruciforms unless we artificially raise the in vivo level of supercoiling by osmotically shocking the cells. Nevertheless, because environmentally induced changes in DNA supercoiling are thought to regulate the bacterial response to osmotic and other stresses ***(13)***, we cannot rule out the possibility that transient formation of unusual DNA structures could be instrumental in such responses. It is particularly interesting that the sequence 5'-ATTATATATATATATATATATATATATAAT-3' is found around the promoter of a key pathogenic gene in *Haemophilus influenzae* ***(14)***. This sequence could attain cruciform geometry at the levels of DNA torsional strain, which we find to be operative inside environmentally stressed bacteria. Moreover, cruciform geometry at artificial promoters is thought to affect transcription ***(15)***.

2. Materials

1. 20 m*M* osmium tetroxide, thawed (*see* **Notes 1** and **2**).
2. 6% v/v aqueous pyridine or 20 m*M* aqueous 2,2' bipyridyl (*see* **Notes 2** and **3**).
3. 10X buffer; 50 m*M* Tris-HCl, 5 m*M* EDTA, pH 8.
4. 1–5 µg of plasmid DNA. Miniprep DNA is adequate; gradient-purified DNA is better.
5. Absolute ethanol.
6. 3 *M* sodium acetate, pH 4.5.
7. Restriction enzymes.
8. Enzymes and radionucleotides for end labeling DNA. In this protocol, the enzyme used is the Klenow fragment of DNA polymerase I. One also needs an [α-P^{32}]

dNTP at specific activity >6000 Ci/m*m*ol, and the other three dNTPs as nonradioactive 2 m*M* stocks.

9. Agarose gel with slots large enough to take >60 mL (they can be made by taping up smaller slots), powerpack and gel tank.
10. 5X loading dye: 10% Ficoll 400, 0.1 *M* sodium EDTA, 1% sodium dodecyl sulfate (SDS), 0.25% bromophenol blue, and 0.25% xylene cyanol.
11. Ethidium bromide (10 mg/mL in water).
12. Long-wavelength (360-nm) ultraviolet (UV) transilluminator.
13. Electroeluter. We use the IBI model UEA.
14. 1 *M* aqueous piperidine, made freshly.
15. High-capacity vacuum pump with trap and desiccator.
16. Alkaline formamide dye; 200 µL deionized formamide, 1 µL of 1 *M* NaOH and xylene cyanol and bromophenol blue to taste. (A needle dipped in the powdered dye and then tapped to remove excess will give quite enough.)
17. Whatmann 3MM paper.

3. Methods

3.1. Osmium Tetroxide Modification

This protocol describes the *in vitro* modification of plasmid DNA and the sequence-level detection of adducts by piperidine cleavage after specific 3-prime end labeling.

1. Solution 1: Combine the plasmid DNA with 5 µL of 10X modification buffer and add distilled water to a total volume of 45 µL.
2. Solution 2: Mix 2.5 µL of 20 m*M* osmium tetroxide with 2.5 µL of 6% pyridine or 20 m*M* bipyridine.
3. Equilibrate solutions 1 and 2 at the desired reaction temperature.
4. Add solution 1 to solution 2, mix well, and incubate for the desired time.
5. Meanwhile, make the stop solution; 180 mL of absolute ethanol and 5 mL of 3 *M* sodium acetate pH 4.5. Chill at –70°C.
6. Add stop solution to reaction, mix well and chill at –70°C for 10 min.
7. Spin at maximum speed in a Eppendorf microfuge for 10 min. Note, non-Eppendorf microcentrifuges are NOT, in general, adequate substitutes. We do not know exactly why, but we suspect it has to do with heating of the rotor during spinning.
8. Pipet off the supernatant and discard as potentially carcinogenic waste. Be careful not to disturb the pellet.
9. Add 1 mL of absolute ethanol to the tube and spin in the microfuge at maximum speed for 5 min.
10. Pipet off the supernatant and discard as in **step 8**.
11. Dry the pellet for 5 min in a vacuum desiccator.
12. Add 44 µL of distilled water.
13. Transfer to a fresh tube. This is important; sometimes residual osmium on the reaction tube can interfere with the subsequent steps (*see* **Note 4**).

14. Add 5 µL of 10X restriction enzyme buffer and 1 µL (about 10 U) of restriction enzyme. The enzyme chosen should give unpaired 5' ends and ideally has a unique site approx 40–200 bases from the site at which modification is expected.
15. Incubate at 37°C for about 1 h.
16. Meanwhile, dry down 5 mCi of a >6000-Ci/m*mol* [α–^{32}P] dNTP. This should be the first nucleotide to be incorporated by DNA polymerase (e.g., if the enzyme used is EcoRI, it should be dATP).
17. To the dried-down radionucleotide, add 2 µL of each 2 m*M* unlabeled stocks of the other dNTPs (i.e., if the labeling is with [α–^{32}P] dATP, add dGTP, dCTP, and dTTP). Mix well.
18. Add this nucleotide mix to the restriction digest. Then, add 1 µL (about 6 U) of the Klenow fragment of *E. coli* DNA polymerase I. We find that Klenow polymerization works well in a variety of restriction enzyme buffers, especially those used for EcoRI and BamHI.
19. Incubate for 1 h (not more) at 37°C.
20. Add 10 µL of 5X loading dye and electrophorese on, for example, a 1% agarose 1X TBE gel until the xylene cyanol (light blue) dye is about halfway down the gel, or, in general until the fragment of interest can be easily excised from the gel. Exactly which gel is chosen will depend on the particular system under study.
21. Stain the gel with 1 µg/mL ethidium bromide. Visualize by long-wave UV and excise the bands of interest from the gel.
22. Electrolelute the labeled DNA into high salt (or use another method).
23. Precipitate and wash the DNA with ethanol. If precipitating from high salt, do not chill!
24. If it is desired to see only the signals on one strand, cut off one of the labeled ends (**Note 5**). After restriction, the DNA should be ethanol precipitated, washed and dried. Do this by repeating **steps 5–11**; it is not necessary to discard the waste as potentially mutagenic.
25. To the dried pellet, add 100 µL of 1 *M* piperidine. Heat at 90°C for 30 min.
26. Transfer to a new tube. Close the tube. Use a needle to punch holes in the cap.
27. Place the tube in liquid nitrogen for about 5 s.
28. Place the tube in a rack in a vacuum desiccator, attach a high-capacity vacuum pump with a trap, and turn it on. Do this quickly, so that the frozen sample does not have time to thaw.
29. Lyophilize for about 2 h.
30. Add 50 µL of water to the tube and repeat lyophilization **step 28** for about 1 h.
31. Repeat **steps 29** and **30**.
32. Resuspend the samples in at least 3 µL of alkaline formamide dye.
33. Heat to 90°C for 5 min and then chill on ice.
34. Load aliquots of about 50 counts per second on a standard TBE–8 *M* urea sequencing gel and electrophorese. Exact details of the sequencing gel will depend on the system under study.
35. Fix the gel in 10% acetic acid for 15 min, transfer to 3MM paper, dry, and autoradiograph.

3.2. Results of Osmium Tetroxide Modification

Figure 1 shows a time-course of osmium tetroxide modification on a 68-bp-long tract of alternating adenines and thymidines within a bacterial plasmid. This sequence is found in the first intron of the frog globin gene. The modification was done at 37°C in the absence of added salt and for the indicated times. A *Bam*HI–*Eco*RI fragment containing the tract was labeled at both ends by Klenow polymerase and [α-^{32}P] dATP. The adducts were cleaved using hot piperidine and electrophoresed on a sequencing gel. The gel was fixed, dried onto paper, and autoradiographed.

The gel shows information from both strands. This is possible because (1) the AT tract is asymmetrically placed on the *Bam*HI–*Eco*RI fragment, and (2) osmium modification is almost completely specific for the AT tract. As can be seen, the modifications are biased toward the label-proximal end of the tract on both strands, and this is intensified at later time-points. In fact, this is also the pattern when the label is placed at the 5' ends (data not shown). Thus, the label-proximal bias of the signals simply indicates multiple modifications rather than an asymmetric structure.

Figure 1 also shows the result of modifying the AT tract in the presence of ions. Here, the patterns are quite different. In the presence of sodium ions, we see modifications at the center of the AT tract and at its ends. In the presence of magnesium ions, we see modifications at the center of the AT tract but not at its ends. These patterns are interpreted as indicating the presence of cruciform structures at the AT tract, either with tight scissor-shaped osmium resistant junctions (in the presence of magnesium) or with floppy square planar osmium sensitive junctions (in the presence of sodium) *(16)*. The central modifications are at what would be the loop of the cruciform, and the modifications at the end of the tract would be at the junction of the cruciform with flanking DNA (**Fig. 2**) (*see* **Note 6**).

The nature of the symmetric structure observed in **Fig. 1** remains obscure; we have termed it the U-structure *(9)*, but this is only a name, not a description. What we know about this structure is that it is easy to interconvert between it and the cruciform, and the conditions under which the U-structure is favored (higher supercoiling and temperature, lower salt concentration) suggest that the U-structure is more unwound than the cruciform. Furthermore, other A+T-rich sequences that form cruciforms can exhibit U-like patterns of modification under the appropriate conditions. One interesting possibility is that the U-structure may be a locally parallel-stranded conformation, in which the A at the 5' end of the top strand makes a reverse Watson–Crick pair to the A at the 5' end of the bottom strand; the Ts following those As pair to each other in the same mode, the As following those Ts pair to each other in the same mode: and

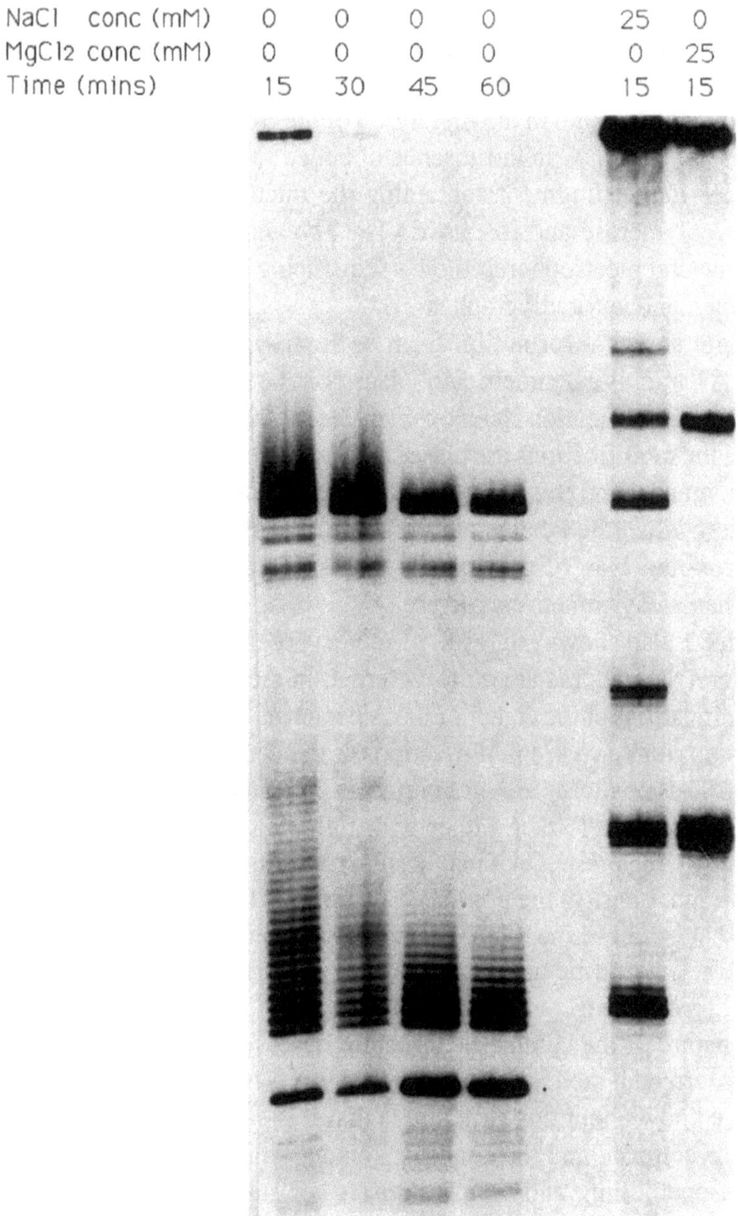

Fig. 1. Time-course of in vitro osmium tetroxide modification of $(AT)_{34}$ tract in plasmid pXG540. The figure shows an autoradiograph of end-labeled piperidine-cleaved *Eco*RI–*Bam*HI fragments from plasmid pXG540 that had been treated with osmium tetroxide under the ambient conditions indicated above each lane (all reactions were done at 20°C). The fragments were separated on a 6% sequencing gel run hot to the touch at constant 70 W.

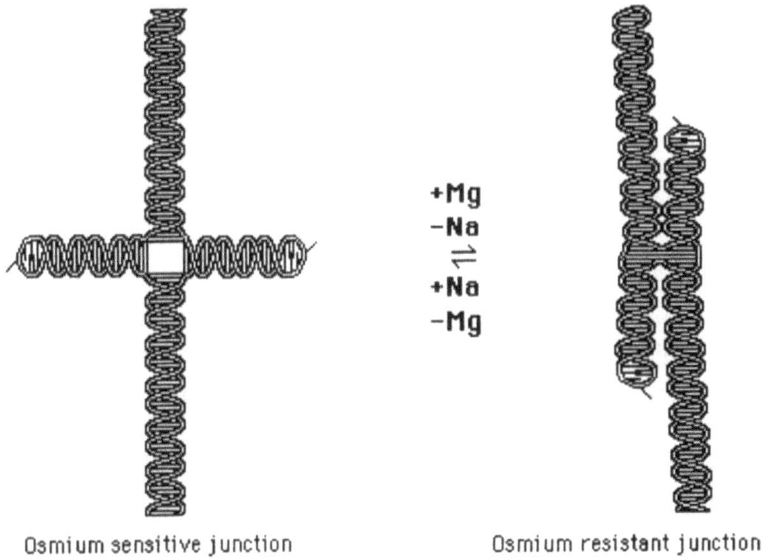

Fig. 2. Two conformations of cruciforms. The figure shows how two different cruciform conformations may be postulated in order to explain the different patterns of osmium modification seen at different ionic strengths.

so on. This is chemically plausible because any base can make a reverse Watson-Crick pair with two hydrogen bonds to another base of the same type (i.e., A pairs to A, T to T, and so on). The proposed structure is shown diagrammatically in **Fig. 3**, and the model is currently being tested.

3.3. In Vivo Osmium Modification

It is possible to modify AT tracts in plasmids inside living bacteria **(10–12)**. **Figure 4** shows the results of one such experiment, in which a number of plasmids with different lengths of AT were modified inside bacterial cells (*E. coli* HB101). The plasmid DNA was recovered by a modification of the Holmes–Quigley boiling method **(12)**; alkaline lysis methods are not used, in order to avoid premature alkaline cleavage of adducts. The DNA was then restricted and end labeled, and the adducts were cleaved by hot piperidine before analysis on sequencing gels, as described earlier.

In vivo modification requires a number of additional tricks if it is to be successful. One has to use bipyridine as the ligand. The number of cells is a critical parameter; the modification reaction should have cells at an optical density (OD) 550-nm of 0.4. If the OD is even twice as high, the experiment is likely to fail. During the modification, the cells should turn a milk chocolate brown; if they do not, discard the experiment and obtain fresh osmium tetroxide. We

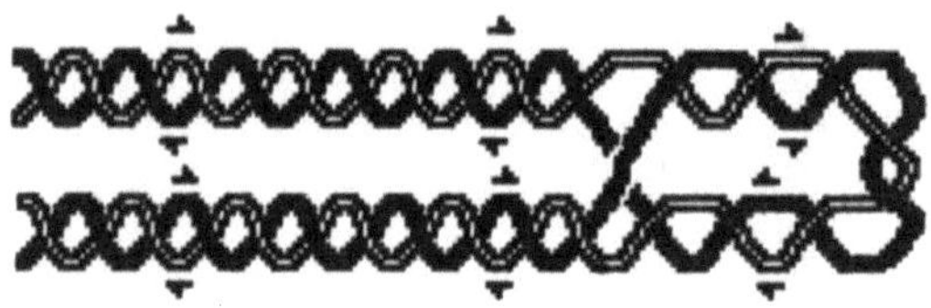

Fig. 3. Proposed locally parallel structure of U-conformation.

usually work on a 50-mL culture scale, and we usually wash the cells and do the modifications in 100 m*M* potassium phosphate buffer, pH 7.4. It is, however, possible to do the modifications in L-broth. After the cells have been boiled the supernatant has to be collected, and it is very viscous indeed. We find that the best way to prepare the supernatant is to spin the lysates at 30,000 rpm for 30 min in a Beckmann table-top ultracentrifuge (120 TS rotor). The supernatant can then be ethanol precipitated or made up as a CsCl/ethidium bromide gradient.

4. Notes

1. Osmium tetroxide is a very powerful oxidizing and crosslinking reagent, which formerly was used for tanning leather. It can react explosively with water. Fumes of osmium tetroxide can damage the cornea of the eye. It is, thus, a chemical to be treated with great respect. To our knowledge, there is no evidence that it is a carcinogen, but this may be because the complex with heterocyclic activators has not been tested as such; this complex is certainly a very powerful and specific covalent modifier of exposed thymidines, and it would be surprising if it was not a carcinogen. It therefore seems prudent to dispose of osmium tetroxide waste as if it were carcinogenic.
2. Osmium tetroxide can be bought from various suppliers, including Sigma and Johnson Matthey. The quality of the reagent varies widely, but, in general, it may be stated that the material bought from Johnson Matthey is superior. One usually buys a 250-mg aliquot, which should cost around £30–40 sterling at 1998 prices. The reagent is shipped inside a metal can. This contains a plastic tube, and inside the tube, wrapped in black paper, is a little sealed glass phial containing yellow crystals. These are osmium tetroxide. One makes up a stock 20 m*M* solution as follows:
   a. The glass tube is plunged into liquid nitrogen and kept there for about 15 s. This treatment prevents the chemical from reacting explosively with water, and also it makes the crystals less sticky and easier to remove from the glass vial.
   b. Inside the fume hood and over a washable tray, the glass vial is broken, ideally with the aid of a diamond knife, and the crystals are tipped into a glass beaker containing 49.2 mL of distilled water. The crystals take some time to dissolve, about 2–3 h.

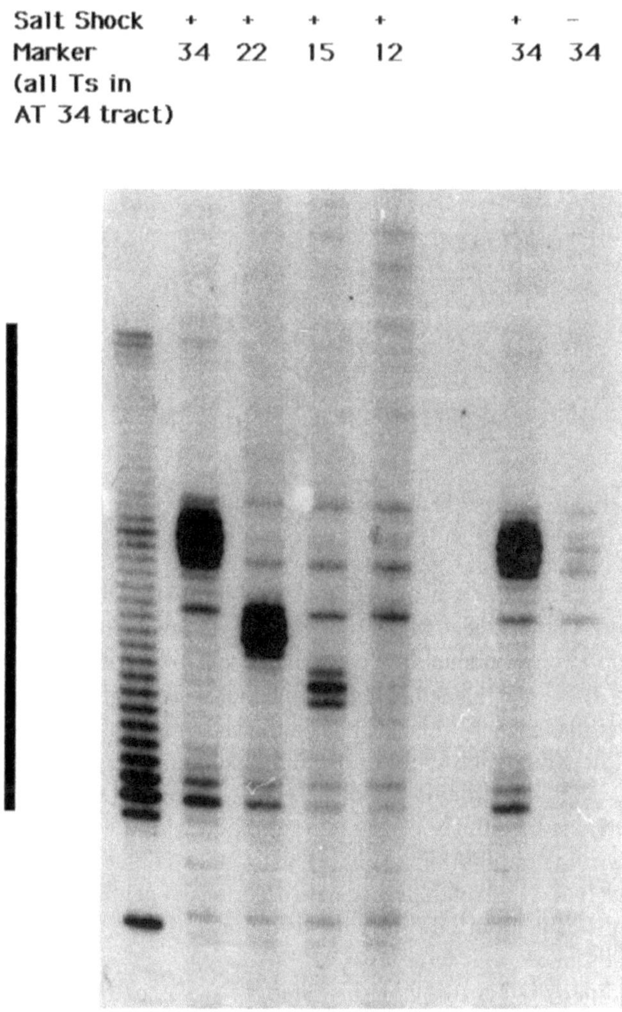

Fig. 4. In vivo modification of AT tracts: cruciform geometry in bacterial plasmids as a consequence of physiological salt shock. Bacterial cells containing isogenic plasmids with various lengths of AT were salt-shocked and treated with osmium tetroxide *in vivo.* Adducts were detected by preparing the DNA, end labeling, and piperidine cleavage, followed by electrophoresis on thin 6% sequencing gels and autoradiography. Strong central modification of the AT tract shows that $(AT)_{34}$ adopts cruciform geometry in salt-shocked but not in control cells, and that $(AT)_{22}$ and $(AT)_{15}$ but not $(AT)_{12}$ also adopt cruciform geometry in salt-shocked cells.

c. Once the crystals have dissolved, the reagent is aliquoted ready for use. Glass containers such as Universals with screw tops should be used, and storage should be at –70°C. It is best to keep the reagent as several separate aliquots

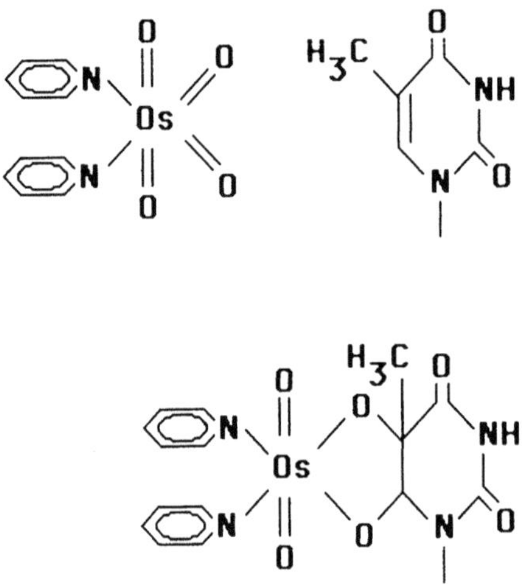

Fig. 5. Stereochemistry of OsO_4 attack on thymidines.

and to use them one at a time. One problem with storage is that the glass containers often crack, which is obviously very dangerous. Volumetric flasks and non-Pyrex containers are particularly sensitive in this regard. The reagent should always be thawed with the container inside a glass beaker in the fume hood. If the reagent is blackish when thawed, discard it; the black color indicates lower oxidation states of osmium, including the metal.

3. By itself, osmium tetroxide is not very reactive with DNA. The species that attacks DNA is a complex of osmium tetroxide with a heterocyclic compound such as pyridine or bipyridine (**Fig. 5**). The attack is on the 5–6 double bond of thymidines (**ref.** ***3*** and references therein). Curiously, neither cytosines nor uracils show significant modifiability with osmium tetroxide, but guanine residues occasionally may.

   Bipyridine is rather insoluble in water. To make a stock solution, it may be necessary to heat the bipyridine and water to 80°C for about 30 min. Once osmium tetroxide is added to pyridine, the mixed reagents should become a straw-yellow color; if they do not, discard them and obtain fresh osmium tetroxide.

   One can change the reactivity of an osmium tetroxide preparation by changing the concentration of osmium and/or ligand, and the overall degree of reaction can be varied from single to multiple hits per molecule by changing the time and/or temperature of the modification reaction. Typical times and temperatures used in our laboratory to obtain single hits are as follows: 45 min at ice temperature; 15 min at 20°C; 5 min at 37°C; 1 min at 40°C.

4. Various methods of detecting osmium tetroxide adducts exist, namely retardation of bands in agarose or acrylamide gels ***(5)***, immunoprecipitation ***(3)***, cleavage by S1 nuclease ***(5)***, cleavage by hot piperidine ***(6)***, and failure of primer extension ***(11)***. In some cases, inhibition of restriction enzyme cleavage can be used ***(10,17)***.

5. A good way to do this is to cut with a frequent cutter that
   a. has a site near to the end from which it is not required to see signals
   b. does not have a site between the end from which it is desired to see signals and the tract where modifications are expected

   For example, the *Eco*RI site of pXG540 has a *Hae*III site 18 bp anticlockwise (in the direction of the Amp promoter). There is no *Hae*III site in the clockwise 170 bp up to the start of the AT tract. So, we often cut and label at *Eco*RI, and then do a limit digest with *Hae*III. This reults in a long labeled fragment that has the information we want on it, and a short fragment (18 bp) that we run off the gel.
6. Osmium tetroxide modification of DNA now has a substantial pedigree, having been used to modify a wide range of sequences in different unusual conformations in vitro, and a narrower range of sequences in different conformations in vivo. **Figure 6** shows some of the actual or proposed structures that have been treated with osmium tetroxide. However, it is very important to be cautious in using osmium modification results, or any other kind of chemical or enzymatic probing, to deduce that a particular structure is forming. This is because osmium does not report on global conformation; it only tells you whether or not a particular T residue has an exposed 5–6 double bond. If a T is exposed, this could be for several reasons:
   a. The T might be in a single-stranded region of the DNA (e.g., a bubble, mismatch *[18]*, B–Z junction or cruciform or H-structure loop).
   b. The T might be in a helix with a shallow or bulging major groove (e.g., at a bend or within a GT/AC tract that was forming Z–DNA).
   c. The T might be in an overwound helix, with exposure of the 5–6 double bond resulting from a loss of base overlap, because of the sharp rotation of each base relative to its neighbors.

   These considerations make it very difficult to conclude on the basis of osmium modification that, for example, Z–DNA is forming; osmium results do not distinguish in any simple way between what we have called the "U"-structure ***(9)***, Z-DNA and a conformation (possibly the eightfold D-helix), which we have observed in locally positively stressed AT tracts ***(19)***. Detecting the loops and junctions of cruciform or H-form DNA can sometimes be done in such a way as to virtually exclude alternative interpretations of the data, but even this is not always possible; for example, a GC tract with ATAT in the middle will react in the same way with osmium tetroxide whether it is in a Z-conformation or cruciform ***(11)***; and there is a published interpretation of chemical modification at H-forming sequences that is radically at odds with the H-structure model ***(20)***.

   In addition to these difficulties, there is a need to be cautious about the effect of the probing chemical on the structural features deduced. Because osmium modification works under a wide range of conditions, it presents fewer such problems of interpretation than some other more fastidious chemicals. Nevertheless, we observe that the ligand used can have an effect on the result obtained; other things being equal, we find that bipyridine is more likely to give a cruciformlike

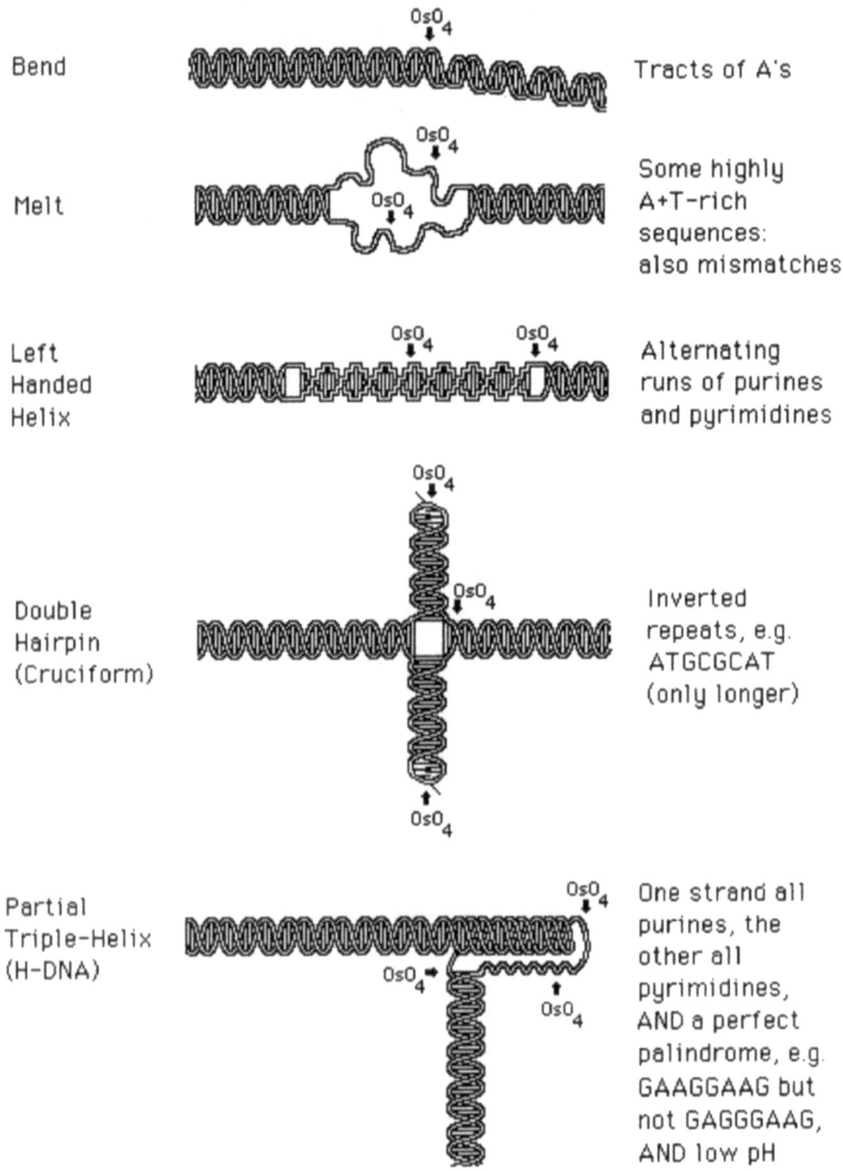

Fig. 6. Unusual DNA structures that react with osmium tetroxide.

pattern of modification and pyridine is more likely to give a U-structure-like pattern when superhelically stressed AT tracts are probed in vitro. This probably has to do with different helix-unstacking potential of the two heterocycles; alternatively, it could be an artifact of contaminating cations in the heterocycles or ion–heterocycle interactions.

References

1. Beer, M., Stern, S., Carmalt, D., and Mohlenrich, K. H. (1966) Determination of base sequence in nucleic acids with the electron microscope. V. The thymine-specific reactions of osmium tetroxide with deoxyribonucleic acid and its components. *Biochemistry* **5,** 2283–2288.
2. Burton, K. and Riley, W. T. (1966) Selective degradation of thymidine and thymine deoxynucleotides. *Biochem. J.* **98,** 70–77.
3. Palecek, E. (1989) Local open DNA structures in vitro and in the cell as detected by chemical probes, in *Highlights of Modern Biochemistry*, (Kotyk, A., Skoda, J., Paces, V., and Kostka, V., eds.), VSP International Science, Zeist, The Netherlands, pp. 53–71.
4. Behrman, E. J. (1988) The chemistry of the interactions of osmium tetroxide with DNA and proteins, in *Symposium on Local Changes in DNA Structure and Their Biological Implications*, Book of Abstracts, p. 6.
5. Buckle, M., Spassky, A., Herbert, M., Lilley, D. M. J., and Buc, H. (1988) Chemical probing of single stranded regions of DNA formed in complexes between RNA polymerase and promoters, in *Symposium on Local Changes in DNA Structure and their Biological Implications*, Book of Abstracts, p. 11.
6. Lilley, D. M. J. and Palecek, E. (1984) The supercoil-stabilised cruciform of ColE1 is hyper-reactive to osmium tetroxide. *EMBO J.* **3,** 1187–1192.
7. Johnston, H. and Rich, A. (1985) Chemical probes of DNA conformation: detection at nucleotide resolution. *Cell* **42,** 713–724.
8. Vojtiskova, M. and Palecek, E. (1987) Unusual protonated structure in the homopurine. homopyrimidine tract of supercoiled and linearised plasmids recognised by chemical probes. *J. Biomol. Struct. Dyn.* **5,** 283–296.
9. McClellan, J. A. and Lilley, D. M. J. (1987) A two-state conformational equilibrium for alternating A-T)*n* sequences in negatively supercoiled DNA. *J. Mol. Biol.* **197,** 707–721.
10. Palecek, E., Boublikova, P., and Karlovsky, P. (1987) Osmium tetroxide recognizes structural distortions at junctions between right- and left-handed DNA in a bacterial cell. *Gen. Physiol. Biophys.* **6,** 593–608.
11. Rahmouni, A. R. and Wells, R. D. (1989) Stabilisation of Z DNA in vivo by localized supercoiling. *Science* **246,** 358–363.
12. McClellan, J. A., Boublikova, P., Palecek, E., and Lilley, D. M. J. (1990) Superhelical torsion in cellular DNA responds directly to environmental and genetic factors. *Proc. Natl. Acad. Sci. USA* **87,** 8373–8377.
17. Palecek, E., Boublikova, P., Galazka, G., and Klysik, J. (1987) Inhibition of restriction endonuclease cleavage due to site-specific chemical modification of the B–Z junction in supercoiled DNA. *Gen. Physiol. Biophys.* **6,** 327–341.
16. Duckett, D. R., Murchie, A. I. H., Diekmann, S., von Kitzing, E., Kemper B., and Lilley, D. M. J. (1988) The structure of the Holliday junction, and its resolution. *Cell* **55,** 79–89.
13. Dorman, C. J. (1991) DNA supercoiling and environmental regulation of gene expression in pathogenic bacteria. *Inf. Immun.* **59,** 745–749.

14. Langermann, S. and Wright, A. (1990) Molecular analysis of the *Haemophilus influenzae* type b pilin gene. *Mol. Microbiol.* **4,** 221–230.
15. Horwitz, M. S. Z. and Loeb, L. A. (1988) An *E. coli* promoter that regulates transcription by DNA superhelix-induced cruciform extrusion. *Science* **241,** 703–705.
18. Cotton, R. G. H., Rodrigues, N. R., and Campbell, R. D. (1988) Reactivity of cytosine and thymine in single-base-pair mismatches with hydroxylamine and osmium tetroxide and its app. lication to the study of mutations. *Proc. Natl. Acad. Sci. USA* **85,** 4397–4401.
19. McClellan, J. A. and Lilley, D. M. J. (1991) Structural alteration in alternating adenine–thymine sequences in positively supercoiled DNA. *J. Mol. Biol.* **219,** 145–149.
20. Pulleyblank, D. E., Haniford, D. B., and Morgan, A. R. (1985) A structural basis for S1 sensitivity of double stranded DNA. *Cell* **42,** 271–280.

10

Determination of a Transcription-Factor-Binding Site by Nuclease Protection Footprinting onto Southwestern Blots

Athanasios G. Papavassiliou

1. Introduction

The interaction of cell-type-specific or inducible transcription factors with regulatory DNA sequences in gene promoters or enhancers is a pivotal step in genetic reprograming during cell proliferation and differentiation and in response to extracellular stimuli. The study of these interactions and the characterization of the factors involved are, therefore, a critical aspect of gene control. Transcription factor–DNA interactions in eukaryotes have been demonstrated by a wide variety of biochemical approaches, including deoxyribonuclease I (DNase I) and chemical nuclease footprinting *(1–3)* (Chapter 3), methylation protection *(4)* (Chapter 14), electrophoretic mobility-shift *(5,6)* (Chapter 2), and Southwestern (SW) assays *(7)* (Chapter 17). Despite their broad applicability, these techniques provide only partial information about the DNA–protein system under investigation. The first three techniques identify either the site(s) of transcription factor binding within the DNA (size and location of nucleotide stretches or atoms on individual bases) or the complexity of the binding pattern (stoichiometry), but do not yield information about the protein(s) involved. On the other hand, the SW assay reveals the relative molecular mass of renaturable (on a membrane support) active species in heterogeneous protein mixtures facilitating their identification, but fails to localize the exact target element within the probing DNA sequence.

Combining SW and DNase I (but also chemical nuclease and methylation protection, *see* below) footprinting methodologies has the dual potential for accurately determining the size of individual DNA-binding transcription factors and precisely mapping their cognate binding sites *(8)* (**Fig. 1**). In addition

From: *Methods in Molecular Biology, vol. 148: DNA–Protein Interactions: Principles and Protocols, 2nd ed.*
Edited by: T. Moss © Humana Press Inc., Totowa, NJ

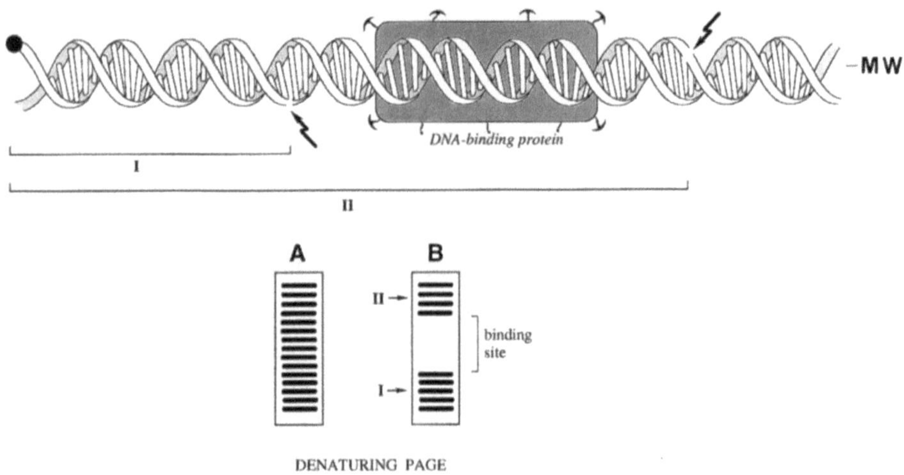

Fig. 1. Rationale in a combined SW–DNase I footprinting procedure. The "hammer"-shaped extensions from the DNA-binding protein indicate immobilization on the blotting membrane. The black-filled sphere indicates the end label in one strand of the DNA footprint probe. MW denotes the molecular size of the membrane-blotted DNA-binding protein (estimated by comparing the position of the active membrane area to the mobilities of coelectroblotted protein MW standards). Arrows mark representative sites of DNase I attack. **(A)** DNase I digestion pattern of free (in solution) probe; **(B)** DNase I digestion pattern of protein-complexed (membrane-bound) probe.

to allowing the detection of only fractional binding (footprints in solution can be obtained only when the binding site[s] is almost completely occupied), coupling *in situ* DNase I footprinting with the SW technique offers the advantage of both resolving the protein component and mapping the binding site of complexes formed by either different factors recognizing an identical sequence within the DNA probe or by two factors interacting in a noncooperative manner with adjacent but distinct sequences. However, this is dependent on the factor being able to bind to DNA as a monomer or a homodimer. If the active form is not a single species (i.e., a heterodimeric or heteromeric complex is required to reconstitute the binding activity), specific DNA binding will not be detected and the procedure will not be applicable. The most critical stage of the coupled assay lies within its first part, namely the SW procedure, and concerns the ability of a transcription factor to efficiently renature into its active form (at least in terms of DNA-binding capacity) on the membrane. Many transcription factors are composed of domains with distinct structural conformations, which aids the process of renaturation. However, because the protein surface immobilized on the membrane poses an impediment on the refolding process, the likelihood of successful renaturation increases with increased size of the DNA-binding transcription factor(s) under study. As a result of the increased kinetic stability of a membrane-immobilized DNA–protein complex (reversible binding to even low-affinity proteins is enhanced because excess unbound DNA has been washed out and hence is not present to compete), reaction parameters such as the size of the DNA probe, the concentration of DNase I, and digestion time are no longer determined by the dissociation rate of the complex, a normal limitation of DNase I protection assays performed in solution.

The fidelity of the combined analytical assay is demonstrated in **Fig. 2**. Evidently, the structural and functional properties of DNA are not altered by entrapment on the blotting membrane surface (i.e., the probe exhibits identical protection pattern and sequence-dependent reactivity with DNase I as it does in solution). Therefore, this coupled assay provides a fast and reliable method that will allow the user who has identified specific regulatory regions in the gene of interest to begin characterizing in detail the transcription factor(s) that bind to them in a certain cellular milieu.

Modifications of the presented DNase I protection analysis on Southwestern blotting membranes (*in situ*) substitute the enzymatic probe for either the chemical nuclease 1,10-phenanthroline–copper ion (OP–Cu) ***(9)*** or dimethyl sulfate (DMS) ***(10)***. Both protocols have the additional advantage of providing information on the nature (i.e., specific–protected versus nonspecific–nonprotected) of several membrane-immobilized DNA–protein species often observed in a SW assay. The molecular and functional properties of DNase I (i.e., its relatively large size, mode of target searching and binding to cleave, and requirement for Mg^{2+} [which often stabilizes both specific and nonspecific complexes]) highly reduce its potential to detect these differences. In addition, these methodologies are useful in rapidly confirming the binding specificity of a protein isolated by screening a cDNA expression library with recognition-site DNA.

2. Materials

2.1. SW Blotting

2.1.1. Solutions

1. Sodium dodecyl sulfate (SDS; 20%; w/v): Dissolve 100 g of SDS (Sigma, St. Louis, MO) in distilled, deionized water to a 0.5 L final volume; heat at 68°C to assist dissolution (do not autoclave). Store at room temperature in a clear bottle. A respirator or dust mask should be worn when handling powdered SDS.
2. Lower (separating) gel buffer (4X stock): 1.5 *M* Tris-HCl, pH 8.8, and 0.4% (w/v) SDS. Filter through a 0.22-µm membrane filter. Store at 4°C.
3. Upper (stacking) gel buffer (4X stock): 0.5 *M* Tris-HCl, pH 6.8, and 0.4% (w/v) SDS. Filter through a 0.22-µm membrane filter. Store at 4°C.

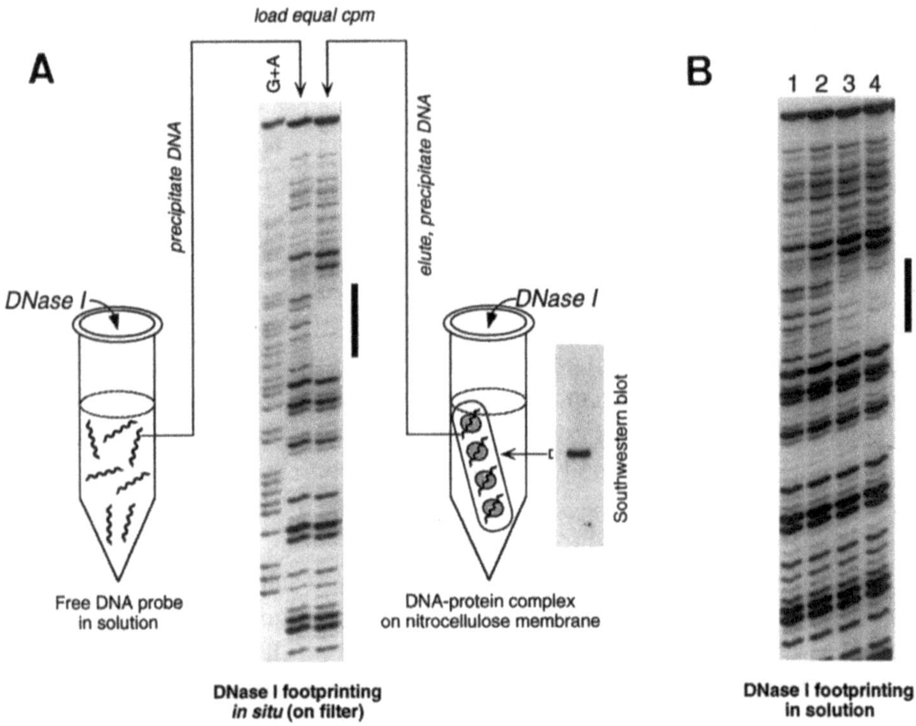

Fig. 2. **(A)** Schematic outline of the manipulations involved in the combined SW–DNase I footprinting procedure. In the example presented, a crude lysate of bacterial cells overexpressing the proto-oncoprotein c-Jun (a component of the transcription factor AP-1) was subjected to a quantitative SW assay utilizing as probe a 134-bp DNA restriction fragment (5'-end labeled in the coding strand) encompassing the AP-1-binding sequence of the human collagenase promoter [5'(-72)TGAGTCA3'(–66)]. **(B)** DNase I footprinting reactions of the same fragment in solution performed with increasing amounts (lanes 2–4) of the c-Jun preparation (lane 1, free-DNA probe). The footprinted region (solid bar) includes in both cases approx 14 bases centered around the AP-1-binding motif *(20)*.

4. Acrylamide/*bis*-acrylamide gel mixture: 30% (w/v) acrylamide, 0.8% (w/v) *N,N'*-methylene–*bis*-acrylamide (Bio-Rad, Richmond, CA). Filter through a 0.22-µm membrane filter. Store at 4°C in the dark. Powdered acrylamide and *N,N'*-methylene-*bis*-acrylamide are potent neurotoxins and are absorbed through the skin. Their effects are cumulative. Wear gloves and a face mask when weighing these substances and when handling solutions containing them (work in a chemical fume hood). Although polyacrylamide is considered to be nontoxic, it should be handled with care because of the possibility that it might contain small quantities of unpolymerized acrylamide.

5. Ammonium persulfate (10%; w/v): Dissolve (by vigorous vortexing) 1 g of ammonium persulfate (Bio-Rad) in 10 mL of distilled, deionized water. Filter through a 0.22-μm membrane filter. Store at 4°C; make fresh weekly. Ammonium persulfate is extremely destructive to tissue of the mucous membranes and upper respiratory tract, eyes, and skin. Inhalation may be fatal. Exposure can cause gastrointestinal disturbances and dermatitis. Wear gloves, safety glasses, respirator, and other protective clothing and work in a chemical fume hood. Wash thoroughly after handling.
6. SDS–polyacrylamide gel electrophoresis (PAGE) running buffer (10X stock): 0.25 *M* Tris base (Sigma) and 1.92 *M* glycine (Sigma). It is not necessary to adjust the pH. The proportions of Tris base and glycine give pH 8.3. Store at room temperature in a large vessel (carboy). Dilute to 1X with distilled, deionized water, then add SDS to a final concentration of 0.1% (w/v).
7. SDS-PAGE sample buffer (4X stock): 0.5 *M* Tris–HCl pH 6.8, 8% (w/v) SDS, 40% (v/v) glycerol (Sigma), and 0.4% (w/v) bromophenol blue (Sigma). Store in aliquots at –20°C. Before use, make a working solution of SDS-PAGE sample buffer by diluting the 4X stock buffer and adding 2-mercaptoethanol (Sigma) to a final concentration of 5% (v/v) (*see* **Note 1**). 2-Mercaptoethanol is harmful if inhaled or absorbed through the skin. High concentrations are extremely destructive to the mucous membranes, upper respiratory tract, skin, and eyes. Use only in a chemical fume hood. Gloves, safety glasses, and respirator should be worn.
8. Western transfer buffer: 50 m*M* Tris base, 380 m*M* glycine, 0.1% (w/v) SDS, and 20% (v/v) methanol (*see* **Note 2**). There is no need to adjust the pH of this buffer by the addition of acid or alkali. Store at room temperature in a large vessel.
9. Dithiothreitol (stock): Prepare a stock solution of 0.5 *M* dithiothreitol (Sigma) in distilled, deionized water and store in aliquots at –20°C. (*See* safety note in **item 7**.)
10. Phenylmethylsulfonyl fluoride (PMSF; stock): Prepare a 100 m*M* stock solution of PMSF (Boehringer, Indianapolis, IN) in absolute ethanol and store in aliquots at –20°C. PMSF is extremely destructive to the mucous membranes of the respiratory tract, the eyes, and the skin. It may be fatal if inhaled, swallowed, or absorbed through the skin. It is a highly toxic cholinesterase inhibitor. Therefore, it should be used in a chemical fume hood and gloves and safety glasses should be worn during handling.
11. 1 *M* KCl.
12. 0.5 *M* $MgCl_2$.
13. SW blocking/renaturation buffer: 3% (w/v) nonfat dried milk (or 5% [w/v] lipid-free bovine serum albumin [BSA]; *see* **Note 3a**), 25 m*M* HEPES·KOH, pH 7.5, 50 m*M* KCl, 6.25 m*M* $MgCl_2$, 1 m*M* dithiothreitol (freshly added from the 0.5 *M* stock solution immediately before use), 10% (v/v) glycerol, 0.1% (v/v) Nonidet P-40 (BDH, London, UK), and 0.2 m*M* PMSF (freshly added from the 100 m*M* stock solution just prior to use). Store at 4°C. Prepare also some SW blocking/renaturation buffer minus dried milk (or BSA).

14. Poly(vinyl alcohol) (stock): Prepare a 10% (w/v) stock solution of poly(vinyl alcohol) (Sigma P 8136; average mol. wt. = 10,000) in distilled, deionized water and store at –20°C.
15. SW binding/washing buffer: 12.5 m*M* HEPES·KOH, pH 7.5, 50 m*M* KCl, 6.25 m*M* $MgCl_2$, 0.5 m*M* dithiothreitol (freshly added from the 0.5 *M* stock solution immediately before use), 2% (w/v) polyvinyl alcohol, 10% (v/v) glycerol, 0.05% (v/v) Nonidet P-40, and 0.2 m*M* PMSF (freshly added from the 100 m*M* stock solution just prior to use). Store at 4°C. (*See* **Note 4b**.)
16. Phenol/chloroform/isoamyl alcohol (25:24:1; v/v): Mix just prior to use 25 vol of phenol with 24 vol of chloroform and 1 vol of isoamyl alcohol. Phenol is highly corrosive and can cause severe burns. Any areas of skin that come in contact with phenol should be rinsed with a large volume of water or PEG 400 (Sigma) and washed with soap and water (*do not use ethanol!*). Chloroform is irritating to the skin, eyes, mucous membranes, and respiratory tract. It is also a carcinogen and may damage the liver and kidneys. Wear gloves, protective clothing, safety glasses, and respirator when handling these substances and carry out all manipulations in a chemical fume hood.
17. Chloroform/isoamyl alcohol (24:1; v/v): Mix 24 vol of chloroform with 1 vol of isoamyl alcohol. This organic mixture can be stored at room temperature in dark (brown) bottles indefinitely.
18. Nonspecific competitor DNA (stock solution): Dissolve salmon/herring sperm or calf thymus DNA in distilled water, deproteinize it by sequential phenol/chloroform/isoamyl alcohol (25:24:1; v/v) and chloroform/isoamylalcohol (24:1; v/v) extractions, sonicate to reduce the mean DNA length to 100–200 bp, precipitate the DNA with ethanol, then resuspend it at 1 mg/mL in distilled, deionized water.

2.1.2. Reagents/Special Equipment

1. *N,N,N',N'*-tetramethylethylenediamine (TEMED; Bio-Rad). TEMED is extremely destructive to tissue of the mucous membranes and upper respiratory tract, eyes, and skin. Inhalation may be fatal. Prolonged contact can cause severe irritation or burns. Wear gloves, safety glasses, respirator, and other protective clothing and work in a chemical fume hood (TEMED is flammable!). Wash thoroughly after handling.
2. Protein extract of interest (e.g., whole-cell-free extract, crude nuclear extract, or partially purified extract), as concentrated as possible.
3. Prestained nonradioactive MW standards (Bio-Rad) or [^{14}C]-methylated protein MW markers (Amersham).
4. Nonfat dried milk powder (e.g., Cadbury's Marvel or Carnation), or fatty acid-free BSA (Boehringer; fraction V).
5. Phenol: Redistilled (under nitrogen) phenol equilibrated with 100 m*M* Tris-HCl, pH 8.0, and 1 m*M* EDTA in the presence of 0.1% (w/v) 8-hydroxyquinoline. Phenol can be stored at 4°C in dark (brown) bottles for up to 2 mo. See the relevant safety note in **Subheading 2.1.1.**
6. Absolute ethanol.

7. Salmon/herring sperm or calf thymus DNA (Sigma or Boehringer).
8. Singly ^{32}P end-labeled DNA probe bearing the binding site(s) of interest (*see* **Note 5**). All necessary precautions should be observed to minimize exposure to ionizing radiation during labeling and isolation of the probe; work behind protective screens whenever possible.
9. Radioactive ink: Mix a small amount of ^{32}P with waterproof black drawing ink, to a concentration of approx 200 cps (on a Geiger counter) per microliter.
10. 0.22-μm membrane filters (Millipore, Bedford, MA).
11. Mini-slab gel electrophoresis apparatus, giving 0.5- to 1.0-mm-thick mini-gels (e.g., Bio-Rad Mini Protean II system), and accompanying equipment.
12. High-current (2–3 A) power supply (e.g., Bio-Rad or Hoefer, San Francisco, CA) and electroblotting apparatus for Western transfer (e.g., Bio-Rad Trans-Blot); additional equipment for Western transfer *(11)*.
13. Nitrocellulose membrane: suitable membranes comprised of unsupported or supported nitrocellulose are available from a number of manufacturers, such as Schleicher & Schuell (Keene, NH; BA85, 0.45 μm), Millipore (Immobilon-NC), and Amersham (Hybond-C/C extra).
14. Plastic trays.
15. Forceps.
16. Plastic wrap such as cling film or Saran Wrap®.
17. X-ray film (e.g., Kodak X-Omat AR, Rochester, NY).
18. Additional equipment: protective gloves/glasses/respirator, 4°C shaking air incubator, sonicator, 25°C shaking air incubator, Geiger counter, Kimwipes, all equipment for autoradiography.

2.2. Exposure of SW Blots to DNase I Treatment

2.2.1. Solutions

1. Eppendorf tube siliconization solution: 2% (v/v) dimethyldichlorosilane in *1,1,1*-trichloroethane (BDH). Eppendorf tubes should be silanized by briefly immersing the opened tubes in a beaker containing this solution, pouring off excess solution, and allowing the tubes to dry in air at room temperature. Dimethyldichlorosilane is particularly toxic. Gloves, safety glasses, respirator, and other protective clothing should be worn when handling it and should only be used in a chemical fume hood.
2. Solutions 1, 12, and 15–17 of **Subheading 2.1.1.**
3. DNase I (stock solution): Dissolve DNase I in 50% glycerol (in distilled, deionized water) to a concentration of 2.5 mg/mL. Store frozen in 10-μL aliquots at –20°C or –70°C. This stock is stable indefinitely.
4. 1 *M* $CaCl_2$.
5. DNase I reaction buffer: 10 m*M* $MgCl_2$ and 5 m*M* $CaCl_2$. Store at room temperature.
6. DNase STOP solution A: 20 m*M* HEPES·KOH, pH 7.5, 20 m*M* EDTA, and 0.5% (w/v) SDS. Store at 4°C.
7. 5 *M* NaCl.
8. DNase-STOP solution B: 60 m*M* HEPES·KOH, pH 7.5, 0.6 *M* NaCl, 60 m*M* EDTA, 1.5% (w/v) SDS. Store at 4°C.

Proteinase K (stock solution): Dissolve Proteinase K in TE (10 m*M* Tris-HCl, pH 7.4, and 1 m*M* EDTA) to a concentration of 2.5 mg/mL. Store in aliquots at –20°C.

10. Probe elution buffer: 20 m*M* HEPES·KOH, pH 7.5, 0.3 *M* NaCl, 3 m*M* EDTA, 0.2% (w/v) SDS, and 50 µg/mL Proteinase K.
11. Glycogen (stock solution): Prepare a stock solution of 10 mg/mL glycogen (Sigma G 0885) in distilled, deionized water and store in aliquots at –20°C; glycogen is used as a carrier to promote the precipitation of nucleic acids.
12. Ice-cold 80% (v/v) ethanol.
13. Formamide loading buffer: 90% (v/v) deionized formamide, 1X TBE (*see* below), 0.025% (w/v) xylene cyanol FF (Sigma), and 0.025% (w/v) bromophenol blue. Store at –20°C after filtering. Formamide is a teratogen; take all safety precautions to avoid contact during manipulations involving this reagent.
14. TBE buffer (5X stock): 445 m*M* Tris base, 445 m*M* borate, 12.5 m*M* EDTA. Dissolve (under stirring for at least 1 h) 272.5 g ultrapure Tris base, 139.1 g boric acid (Sigma), and 23.3 g EDTA·(Na_2) dihydrate (Sigma) in 4.5 L of distilled, deionized water. Make up to a final volume of 5 L. It is not necessary to adjust the pH of the resulting solution, which should be around 8.3. Store at room temperature in a large vessel. This stock solution is stable for many months, but it is susceptible to the formation of a precipitate and should be inspected visually from time to time.
15. Fixing solution: 10% (v/v) acetic acid and 10% (v/v) methanol.

2.2.2. Reagents/Special Equipment

1. DNase I (DPFF grade; Worthington, Freehold, NJ).
2. Proteinase K (Boehringer).
3. Absolute ethanol (at –20°C).
4. Single-edged disposable razor blades.
5. Forceps.
6. Siliconized 0.5-mL Eppendorf microcentrifuge tubes.
7. Siliconized 1.5-mL Eppendorf microcentrifuge tubes.
8. Drawn-out Pasteur pipets.
9. Whatman (Clifton, NJ) 3MM paper.
10. **Items 16** and **17** in **Subheading 2.1.2.**
11. Intensifying screen (e.g., Cronex Lightning Plus; DuPont, Wilmington, DE).
12. Additional equipment: beaker, sharp-tip pencil, wet ice, scintillation vials/counter, vortexer, electronic timer, microcentrifuge (Eppendorf or equivalent), vortexing shaker set at 37°C, Geiger counter, SpeedVac concentrator (Savant, Hicksville, NY), thermostatted heating block at 95°C, plastic tank at gel dimensions (for gel fixing), vacuum gel dryer, and all equipment for autoradiography.

3. Methods

3.1. SW Blotting

The SW protocol involves four steps: electrophoretic separation of proteins by SDS-PAGE, electroblotting of the gel-fractionated proteins onto nitrocellulose (NC) membrane, probing of the blocked blot with the desired DNA probe, and detection of bound DNA by autoradiography of the washed, wet membrane.

1. Prepare and load the sample(s) and protein MW markers onto a standard SDS–polyacrylamide gel ***(11,12)*** and electrophorese at an appropriate voltage until the bromophenol blue dye reaches the bottom of the gel. An 8–10% polyacrylamide separating gel is capable to resolve throughout most of the known DNA-binding transcription factor size range. Because the strength of the final signal obtained by the SW technique is proportional to the quantity of protein electrophoresed on the gel, best results are obtained by running the maximum amount of extract that does not overload the gel. For typical 0.5- to 1-mm-thick protein mini-gels, this is usually 30–150 µg of whole-cell-free, crude nuclear, or partially purified extract protein per lane. (*See* **Note 1**.)
2. Remove the electrophoresed gel from the glass plates. Assemble a Western blot sandwich, place it in the electrophoresis tank containing an appropriate volume of Western transfer buffer, and electroblot the proteins in the electrophoresed gel onto a NC membrane according to standard Western blotting protocols ***(11,13)***. Bear in mind that the best protein transfer is usually achieved by longer transfer times (i.e., 30 V [40 mA] overnight at 4°C). (*See* **Note 2**).
3. When transfer is complete, gently peel the NC membrane off the gel, place it in a plastic tray, and gently wash for 10 min with 20–30 mL of SW blocking/renaturation buffer, omitting dried milk (or BSA).
4. Decant the solution and replace it with a sufficiently large volume of SW blocking/renaturation buffer to completely immerse the membrane (usually 30–50 mL).
5. Incubate overnight at 4°C with gentle rocking or shaking to block nonspecific binding sites on the membrane and to allow renaturation of the filter-immobilized proteins. (*See* **Note 3**.)
6. Using forceps, transfer the membrane to a fresh plastic tray and gently wash for 5 min with 20–30 mL of SW binding/washing buffer. The membrane can be stored in this buffer at 4°C for up to 1 d before incubation with the DNA probe.
7. Immerse the membrane (using forceps) in a fresh plastic tray containing a radioactive probe mixture consisting of the following:
   - SW binding/washing buffer (*see* **Note 4**).
   - 5–10 µg (specific activity approx 10^7 cpm/µg) of an asymmetrically [^{32}P]-labeled DNA fragment bearing the recognition site(s) for the sequence-specific DNA-binding factor(s) of interest (*see* **Note 5**).
   - 20 µg/mL nonspecific competitor DNA (*see* **Note 6**).

   To make the probe concentration as high as possible, the volume of SW binding/washing buffer should be the minimum needed to cover the membrane fully (smallest volumes are achieved if the membrane is sealed in a plastic bag). *Work behind Perspex or glass shields!*
8. Incubate for 2–4 h at room temperature with gentle rocking or shaking. Longer incubation times at a lower temperature may result in a better signal from a poorly binding factor.

9. Carefully remove the radioactive probe mixture and dispose of it safely (*work behind Perspex or glass shields!*).
10. Using forceps, transfer the membrane to a plastic tray and wash for 10 min with 100 mL of SW binding/washing buffer on a platform shaker (room temperature).
11. Repeat **step 10** two to three times or until the radioactive level of the membrane no longer falls appreciably between washes (this can be monitored by a Geiger counter). (*See* **Note 7**.)
12. Place the wet NC sheet between two layers of tightly drawn cling film. With a pad of Kimwipes, push out any trapped air bubbles under the cling film.
13. Expose to X-ray film at 4°C to locate regions of radioactive signal (protein[s]-bound probe). Exposure times of 1–3 h are usually sufficient to detect the radioactive species on the membrane. (*See* **Note 8**.)

3.2. Exposure of SW Blots to DNase I Treatment

The *in situ* footprinting reaction is done in four stages: localization and excision of the areas on the NC sheet corresponding to protein-bound probe, partial digestion of the individual strip-associated and control (free in solution) DNAs with DNase I, extraction of the protein-bound DNA from the strip-immobilized protein–DNA complex, and analysis of the free and complexed DNA digestion products on a DNA sequencing gel.

1. Following autoradiography, align the NC sheet with the autoradiogram (*see* **Note 8**), mark the precise position of radioactive signal(s) with a sharp-tip pencil (this will permit determination of the relative molecular weight of the detected DNA-binding factor[s]), and cut the strip(s) corresponding to radioactive signal(s) with a clean, sharp razor blade.
2. Using forceps, uncover the strip from the cling film and immediately transfer it into a siliconized 0.5-mL Eppendorf tube containing 200 µL of SW binding/washing buffer.
3. Allow the strip to equilibrate for 15 min on ice. Meanwhile, thaw an aliquot of the 2.5-mg/mL DNase I stock solution.
4. Bring the tube to room temperature, place it in a scintillation vial, and determine cpm of the probe retained on the strip by Cerenkov counting.
5. Transfer an equal amount of radioactivity of free probe to a separate, siliconized 0.5-mL Eppendorf tube containing 200 µL of SW binding/washing buffer, and subject it to the same manipulation as its membrane-associated counterpart (**step 3**).
6. Prepare an appropriate dilution of DNase I in ice-cold distilled, deionized water. Mix thoroughly by inversion and gentle vortexing.
7. Add to both samples 200 µL of DNase I reaction buffer and mix by flicking. Let the tubes stand for 1 min at room temperature. It is not necessary to close the caps on the tubes until addition of the DNase stop solutions (**step 10**).
8. Add dilute DNase I to a final concentration of 25 ng/mL and quickly distribute by flicking. It is helpful—especially when footprinting several strips—to have all necessary items (pipetmen, buffers, timer, DNase I) in close proximity. The

smoother this procedure goes, the better (and more reproducible!) the footprint(s) will be.

9. Incubate for 1 min at room temperature (*see* **Note 9**). As with solution footprinting protocols, the exposure time for the free-DNA control reaction can be titrated to achieve the cutting intensity profile of the membrane-bound form. Nevertheless, we have found (employing probes of various lengths) that the above combination of digestion time and DNase I concentration generates sufficient cleavage for a good signal-to-noise ratio.
10. Following treatment, remove the strip (with forceps) from the tube and rapidly immerse it in 500 µL of ice-cold DNase-STOP solution A. Leave on ice for 2 min (*do not vortex!*). Terminate the control reaction (free probe) by adding 200 µL of ice-cold DNase-STOP solution B; vortex thoroughly, spin briefly, transfer into a siliconized 1.5-mL Eppendorf tube, and proceed directly to **step 15**.
11. Place the strip (using forceps) in a siliconized 0.5-mL Eppendorf tube containing 200 µL of probe elution buffer. Spin briefly in a microcentrifuge to submerge the entire strip.
12. Incubate the tube on a vortexing shaker at 37°C for 2 h.
13. Add 100 µL of distilled, deionized water and vortex the tube vigorously (2 min).
14. Microcentrifuge for 2 min to pellet the NC strip, and transfer the supernatant into a siliconized 1.5-mL Eppendorf tube. A Geiger counter can be used to monitor efficient recovery of radioactivity; typically, >90% of the membrane-bound probe is released by this process.
15. Extract once with phenol/chloroform/isoamyl alcohol (25:24:1; v/v) and once with chloroform/isoamyl alcohol (24:1; v/v). In both cases, mix by vortexing (for 10 s) and spin for 5 min.
16. Transfer the aqueous (top) layer to a fresh siliconized 1.5-mL Eppendorf tube. Precipitate the DNA with cold (–20°C) ethanol in the presence of 10 m*M* $MgCl_2$ (added from the 0.5 *M* stock solution; $MgCl_2$ aids in the recovery of small DNA fragments) and 10 µg carrier glycogen; mix by inversion, spin at 12,000*g* for 15 min.
17. Remove and discard the supernatant with a drawn-out Pasteur pipet, being careful not to aspirate the DNA (the bottom of the tube can be held to a Geiger counter to check that the DNA pellet remains). Add 800 µL of ice-cold 80% ethanol and rinse the pellet by gently rolling the microcentrifuge tube. Respin for 3 min.
18. Remove and discard the supernatant using a drawn-out Pasteur pipet, taking extreme care not to disturb the tiny, whitish pellet or the area of the tube where the pellet should be located (the DNA pellet frequently adheres only loosely to the walls of the tube). Dry the pellet in a SpeedVac rotary concentrator. Do not allow the drying procedure to continue past the point of dryness because the sample may be difficult to resuspend.
19. Dissolve the pellet in 6–8 µL of formamide loading buffer; pipet the loading buffer onto the upper, inside surface of tube and tap the tube to drop the droplet onto the DNA pellet. Vortex briefly at high speed for approx 15 s and microcentrifuge for 30 s to collect all of the solution to the bottom of the tube.

20. Transfer to fresh siliconized 1.5-mL Eppendorf tubes and determine the total radioactivity recovered by Cerenkov counting each sample for 1 min in a scintillation counter.
21. Heat samples to 95°C for 3 min to denature DNA and immediately chill in wet ice.
22. Spin briefly to bring the liquid to the bottom of the tubes. Samples can be electrophoresed immediately or stored at –70°C for no more than 24 h after footprinting.
23. Load 1500–2000 cpm of each DNA digestion product (adjust volumes accordingly if necessary; it is essential that a consistent volume of sample be loaded on each lane) onto a pre-electrophoresed, 5–10% denaturing urea (sequencing) gel (*see* **Notes 9b** and **10a**). Electrophorese in 1X TBE buffer at 60–70 W constant power (for a 34 × 40 cm, 0.4-mm-thick gel) until the marker dye fronts have migrated the appropriate distance in order to visualize the DNA region of interest ***(11)***. To determine the location of the transcription factor-binding site(s), the DNase I digests are electrophoresed alongside a Maxam–Gilbert G + A sequencing ladder prepared from the end-labeled footprint probe ***(14)***. *See* Chapter 7 for a fast protocol for preparing such a ladder.
24. Disassemble gel apparatus, carefully lift off one glass plate and soak the gel (still on the second glass plate) in fixing solution for 15 min (*see* **Note 10b**).
25. Drain briefly, overlay the gel with two sheets of 3MM paper, and carefully peel it off the glass plate. Cover the gel surface with plastic wrap and dry under vacuum at 80°C for approx 1 h.
26. Expose the dry gel to X-ray film overnight at –70°C with an intensifying screen (a piece of paper placed between the gel and the film will prevent spurious exposure of the film resulting from static electricity). Several different exposures may be required to obtain suitable band densities.
27. Compare lanes corresponding to free and protein-bound DNAs to identify the region(s) of protection; the region(s) of the DNA fragment that is bound by the factor appears as a blank stretch (footprint) in the otherwise continuous background of digestion products.

4. Notes

1. The diversity of properties characteristic of DNA-binding transcription factors imposes an empirical determination of the conditions under which the sample(s) is prepared for SDS-PAGE. In some cases, the reducing agent (2-mercaptoethanol) should be omitted from the sample buffer and, in others, the SDS concentration should be lowered to 0.5%. Furthermore, some DNA-binding proteins may not withstand the sample boiling before loading on the gel.
2. The presence of methanol in the Western transfer buffer may cause a problem during the electrophoretic transfer of some bulky DNA-binding proteins (gel shrinkage reduces pore size).
3. a. The commercially available nonfat dried milk preparations from various suppliers contain large amounts of protein kinase/phosphatase activities; these activities can potentially interfere with binding of the DNA probe to transcription factors whose DNA-binding capacity is known, or suspected, to be subject to

regulation by inducible phosphorylation/dephosphorylation events. If this is the case, substitute nonfat dried milk in the SW blocking/renaturation buffer for the recommended grade of lipid-free BSA; this particular grade contains only trace amounts of the aforementioned activities and should be preferred as blocking agent ***(15,16)***. Moreover, the use of lipid-free BSA results in an even background throughout and enhances the specific signal-to-noise ratio in the DNA-probing step ***(17)***.

b. By manipulating the conditions for renaturation of the membrane-immobilized proteins (e.g., by incorporating a cycle of protein denaturation and renaturation ***[16]***), the method may be extended to the analysis of proteins resolved in two-dimensional gels ***(18)***. This offers a powerful and convenient means for studying cell cycle/type/stimulus-dependent DNA–transcription factor interactions and their regulatory roles in gene activity.

4. a. Inclusion of poly(vinyl alcohol) (a molecular crowding agent [volume excluder]) in the buffer decreases the amount of small ions/water available for hydration of any probe dissociated from the immobilized protein matrix and renders the aqueous environment unfavorable for the unbound DNA. Consequently, the effective concentration of the probe is increased and interactions with low binding constants are stabilized.

   b. It may be important in some cases to supplement this buffer with $ZnSO_4$ (Aldrich, Milwaukee, WI; final concentration 10 μ*M*) if the transcription factor(s) under study is known, or suspected, to contain a zinc-finger domain(s).

5. a. The DNA chosen for probing the protein blot can be a *cis*-acting regulatory (promoter/enhancer) DNA restriction fragment in the size range of 125–250 bp, with the putative transcription-factor-binding sites located no less than 20–25 bp from the labeled end. This is to ensure that the region of DNA to be investigated for the presence of footprints is capable of being accurately resolved on a sequencing gel.

   b. A prerequisite for the subsequent DNase I footprinting analysis is the use of a DNA probe that has been labeled on only one strand of the DNA duplex. The labeling of only one strand of a promoter/enhancer restriction fragment can be achieved in a number of ways, such as using T_4 polynucleotide kinase and [γ-^{32}P] ATP (5'-end labeling), the large (Klenow) fragment of *E. coli* DNA polymerase I and [α-^{32}P] dNTPs (3'-end labeling ["filling in"]), or the polymerase chain reaction (PCR) amplification ***(19)***. Preparation of radiolabeled DNA employing any of these methodologies requires about 8 h. A combination of 5' and 3' end-labeled DNA probes allows both strands to be analyzed side by side from the same end of the DNA duplex.

   c. For optimal sensitivity in the SW procedure, the probe should be of as high a specific activity as possible and highly purified. The latter can be assured by using a nucleic acid-specific, ion-exchange column, such as the Elutip™-d (Schleicher & Schuell) or the NACS prepac cartridge (BRL, Gaithersburg, MD) ***(11)***. It is recommended not to store the pure, labeled DNA probe longer than 2–4 d, because the radiation creates nicks in the DNA that will appear as additional bands in the sequencing gel.

6. Synthetic alternate copolymers, such as poly[dI–dC] · poly[dI–dC] or poly[dA–dT] · poly[dA–dT] (Boehringer or Pharmacia), at similar final concentrations may be more suitable competitors for some DNA-binding transcription factors.
7. Longer washing times at room temperature are detrimental resulting in dissociation of the bound probe, but longer washing times with cold (4°C) SW binding/washing buffer can reduce background without significant signal loss. To reduce possible low-specificity DNA–protein complexes, the final wash can be performed in cold SW binding/washing buffer with higher salt concentration (i.e., 100–200 m*M* KCl).
8. If prestained nonradioactive protein MW standards have been used, the edges of the plastic wrap should be marked with pieces of tape labeled with radioactive (or fluorescent) ink (let the ink dots dry completely before exposing to X-ray film!). These marks allow the autoradiogram to be aligned with the protein size markers on the NC sheet (**Subheading 3.2., step 1**), facilitating calculations on the relative molecular weight of the specific DNA-bound protein species (obtained radioactive signal[s]). If [^{14}C]-methylated protein MW markers have been used, their position will be apparent on the X-ray film without the need to use the radioactive (or fluorescent) ink procedure.
9. a. Although longer digestion times do not enhance background cutting (uncomplexed DNA is minimized), they may lead to substantial deviations from the required "single-hit kinetics" (i.e., on average, each DNA molecule is nicked at most once; this corresponds to nicking approx 30–50% of the DNA molecules). b. Intense bands due to uncleaved, full-length probe should be visible at the top of each lane in the DNA sequencing gel. This aids in determining whether single-hit kinetics are operative and whether equal amounts of total radioactivity are loaded in each lane.
10. a. If the DNA probe is relatively long (i.e., >175 bp) and multiple transcription-factor-binding sites are to be resolved, a gradient or wedge sequencing gel can be used. b. Wedge-shaped gels must be soaked in fixing solution, followed by 5% glycerol for 10 min prior to drying.

References

1. Galas, D. J. and Schmitz, A. (1978) DNase footprinting: A simple method for the detection of protein-DNA binding specificities. *Nucleic Acids Res.* **5,** 3157–3170.
2. Tullius, T. D. and Dombroski, B. A. (1986) Hydroxyl radical "footprinting": high-resolution information about DNA–protein contacts and application to λ repressor and Cro protein. *Proc. Natl. Acad. Sci. USA* **83,** 5469–5473.
3. Kuwabara, M. D. and Sigman, D. S. (1987) Footprinting DNA–protein complexes *in situ* following gel retardation assays using 1,10-phenanthroline–copper ion: *Escherichia coli* RNA polymerase–*lac* promoter complexes. *Biochemistry* **26,** 7234–7238.
4. Johnsrud, L. (1978) Contacts between *Escherichia coli* RNA polymerase and a *lac* operon promoter. *Proc. Natl. Acad. Sci. USA* **75,** 5314–5318.
5. Garner, M. M. and Revzin, A. (1981) A gel electrophoresis method for quantifying the binding of proteins to specific DNA regions: application to compo-

nents of the *Escherichia coli* lactose operon regulatory system. *Nucleic Acids Res.* **9,** 3047–3060.
6. Fried, M. and Crothers, D. M. (1981) Equilibria and kinetics of lac repressor-operator interactions by polyacrylamide gel electrophoresis. *Nucleic Acids Res.* **9,** 6505–6525.
7. Miskimins, W. K., Roberts, M. P., McClelland, A., and Ruddle, F. H. (1985) Use of a protein-blotting procedure and a specific DNA probe to identify nuclear proteins that recognize the promoter region of the transferrin receptor gene. *Proc. Natl. Acad. Sci. USA* **82,** 6741–6744.
8. Smith, S. E. and Papavassiliou, A. G. (1992) A coupled Southwestern–DNase I footprinting assay. *Nucleic Acids Res.* **20,** 5239–5240.
9. Polycarpou-Schwarz, M. and Papavassiliou, A. G. (1993) Probing of DNA–protein complexes immobilized on protein-blotting membranes by the chemical nuclease 1,10-phenanthroline (OP)–cuprous ion. *Methods Mol. Cell. Biol.* **4,** 22–26.
10. Polycarpou-Schwarz, M. and Papavassiliou, A. G. (1993) Distinguishing specific from nonspecific complexes on Southwestern blots by a rapid DMS protection assay. *Nucleic Acids Res.* **21,** 2531–2532.
11. Sambrook, J., Fritsch, E. F., and Maniatis, T. (1989) *Molecular Cloning: A Laboratory Manual,* Cold Spring Harbor Laboratory, Cold Spring Harbor, NY.
12. Laemmli, U. K. (1970) Cleavage of structural proteins during the assembly of the head of bacteriophage T4. *Nature* **227,** 680–685.
13. Towbin, H., Staehelin, T., and Gordon, J. (1979) Electrophoretic transfer of proteins from polyacrylamide gels to nitrocellulose sheets: procedure and some applications. *Proc. Natl. Acad. Sci. USA* **76,** 4350–4354.
14. Maxam, A. and Gilbert, W. (1980) Sequencing end-labeled DNA with base-specific chemical cleavages. *Methods Enzymol.* **65,** 499–560.
15. Papavassiliou, A. G., Bohmann, K., and Bohmann, D. (1992) Determining the effect of inducible protein phosphorylation on the DNA-binding activity of transcription factors. *Anal. Biochem.* **203,** 302–309.
16. Polycarpou-Schwarz, M. and Papavassiliou, A. G. (1995) Protein–DNA interactions revealed by the Southwestern blotting procedure. *Methods Mol. Cell. Biol.* **5,** 152–161.
17. Papavassiliou, A. G. and Bohmann, D. (1992) Optimization of the signal-to-noise ratio in south-western assays by using lipid-free BSA as blocking reagent. *Nucleic Acids Res.* **20,** 4365–4366.
18. Moreland, R. B., Montross, L., and Garcea, R. L. (1991) Characterization of the DNA-binding properties of the polyomavirus capsid protein VP1. *J. Virol.* **65,** 1168–1176.
19. Lakin, N. D. (1993) Determination of DNA sequences that bind transcription factors by DNA footprinting, in *Transcription Factors: A Practical Approach* (Latchman, D. S., ed.), IRL, Oxford, pp. 27–47.
20. Angel, P., Imagawa, M., Chiu, R., Stein, B., Imbra, R. J., Rahmsdorf, H. J., et al. (1987) Phorbol ester-inducible genes contain a common *cis* element recognized by a TPA-modulated *trans*-acting factor. *Cell* **49,** 729–739.

11

Diffusible Singlet Oxygen as a Probe of DNA Deformation

Malcolm Buckle and Andrew A. Travers

1. Introduction

The DNA double helix is highly malleable and, when constrained, either as a small circle or by the action of a protein, can be readily distorted from its energetically favored conformation. Such distortions may be relatively moderate, as exemplified by smooth bending (which maintains base stacking), or more extreme when this stacking can be disrupted. Deformations of this latter type include kinks, where the direction of the double helical axis is changed abruptly at a single-base step, and localized strand separation, which may be a direct consequence of protein-induced unwinding or of high negative superhelicity in free DNA. Both kinks and localized unwinding can arise transiently during the enzymatic manipulation of DNA by recombinases and by protein complexes involved in the establishment of unwound regions during the initiation of transcription or of DNA replication.

The detection of localized lesions in the DNA double helix requires that the bases at the site of the lesion be accessible to a chemical reagent only when the DNA is distorted. Further, because protein-induced distortions are not necessarily sequence dependent, it is desirable that any reagent used for detection possesses minimal selectivity with respect to the base. Additionally the reagent itself should ideally be noninvasive; that is, it should not form a stable noncovalent complex with DNA and thereby possess the potential to perturb the local conformation of the double helix. Other highly desirable attributes for reagents used for this purpose is that they should possess short half-lives and can be generated on demand *in situ*. These latter characteristics permit the study and detection of transient intermediates in the processes leading to the

From: *Methods in Molecular Biology, vol. 148: DNA–Protein Interactions: Principles and Protocols, 2nd ed.*
Edited by: T. Moss © Humana Press Inc., Totowa, NJ

establishment of complexes competent to initiate replication or transcription or to catalyze recombination.

Chemical reagents so far described that specifically target bases in DNA that is locally deformed include dimethyl sulfate, diethyl pyrocarbonate, osmium tetroxide *(1)*, and potassium permanganate *(2)* (*see* Chapters 6, 9, and 14). However all of these reagents react selectively with specific bases and are also relatively long-lived. Another reagent used extensively for the detection of locally unwound regions of DNA is copper-*o*-phenanthroline *(3)* (*see* Chapter 7). This compound is a minor-groove ligand, which cleaves the sugar–phosphate backbone as a consequence of free-radical attack on a deoxyribose residue close to the site of binding.

One reagent that lacks these shortcomings is oxygen in the singlet state, of which there are two forms with energies of 155 and 92 kJ, respectively. The latter state has a much longer lifetime and can oxidize a variety of unsaturated organic substrates. Typically such a reaction may involve a Diels–Alder-like addition to a 1,3-diene. This highly reactive form of oxygen can be generated by the photochemical excitation of appropriate heterocyclic ring systems that can then promote the conversion of dissolved oxygen in the triplet state to a singlet form. In solution, the singlet state generated in this way has a half-life of approximately 4 ms and can react with accessible DNA bases to form an adduct across a double bond. Once formed, such an adduct sensitizes the sugar–phosphate backbone to alkaline hydrolysis by piperidine, thus permitting the identification of the site of modification *(4)*.

1.1. The Reaction of Singlet Oxygen with DNA

The use of singlet oxygen as a reagent for analyzing DNA structure has been pioneered by the groups of Hélène and Austin, who have introduced two general methods of targeting the DNA. In the first case, a DNA ligand is used as a sensitizer for singlet-oxygen production in a manner analogous to the use of copper-*o*-phenanthroline for the generation of free radicals. Such ligands include methylene blue, which intercalates at sites where the DNA is relatively unwound *(4)* and also a porphyrin ring covalently linked to a defined DNA sequence designed to target a selected region of double helix *(5)*. In these examples photochemical excitation produces singlet oxygen at the site of the bound ligand and reaction is confined to the immediate proximity of the ligand. A second approach is to use singlet oxygen as a freely diffusible reagent. This use is similar in principle to that of a hydroxyl radical produced by the Fenton reaction *(6)*. In such experiments the singlet oxygen is generated by the irradiation of a complex of eosin with Tris (**Fig. 1**) and is then free to diffuse. The eosin is irreversibly oxidized in the course of this reaction. Because, however, the half-life of the singlet oxygen is very short, the concentration of the eosin–

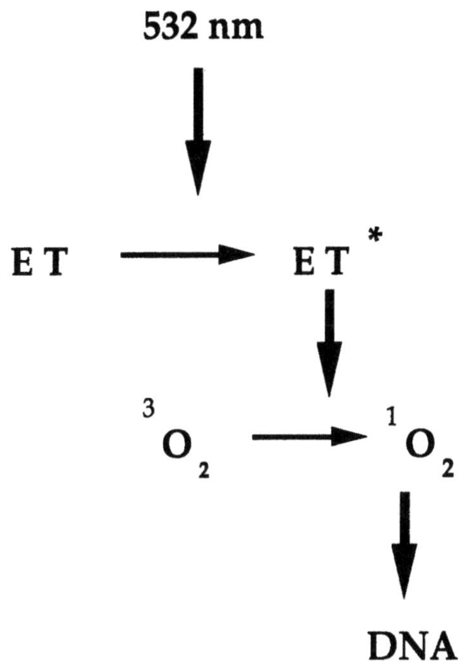

Fig. 1. Reaction cascade for the modification of DNA by singlet oxygen.

Tris complex must be sufficiently high to ensure that singlet oxygen can access potentially reactive sites in the DNA before its reversion to the triplet state.

As with any chemical reagent reacting with a set of chemically distinct targets, the rate of reaction of singlet oxygen with the different bases varies. Notably, guanine as the free base reacts up to 100-fold more rapidly than the other nucleic acid bases. However, the rate of reaction of diffusible singlet oxygen with duplex DNA appears not to be primarily determined by the nature of the bases at the target site but rather by their accessibility. In normal B-form DNA, the bases are generally tightly stacked so as to preclude the entry, and hence the reaction, of the reagent between the base pairs. In the structures of DNA oligomers, it is unusual for the average planes of adjacent base pairs to be separated by a roll angle of greater than 10°. This tight structure is reflected in the relative lack of reactivity of DNA in solution toward singlet oxygen. By contrast, when bound by protein, the DNA can be locally unwound *(7)* or can be kinked so that adjacent base-pairs planes can be inclined by up to 43° relative to each other *(8)*. This deformation of the DNA structure by bound protein would, in principle, be expected to increase the accessibility to singlet oxygen as has been observed both for core nucleosome particles associated with DNA of mixed sequence *(4)* and for the ternary complex of RNA polymerase and catabolite regulatory protein (CRP) with the lac regulatory region *(9)*. In the latter case, which is the only example for which information is so far available, reaction is observed with all four DNA bases, although there are insufficient sites documented to preclude some base selectivity of the reagent. If the local structure of the DNA is the principal determinant of reactivity, the reagent should be able to access the bases through both the major and minor grooves, as is indeed observed *(9)*. However, the precise range of DNA structures available for reaction with singlet oxygen remains to be established, as does the possible influence of bound protein on sensitizing or quenching the reactivity of the bases.

A major advantage of a photoactivated reaction is the ability to produce the reactive species under highly controlled conditions. Both porphyrins and methylene blue can be activated by a continuous laser beam. However, eosin absorbs maximally at 523.2 nm, a wavelength that is close to the 532-nm output from a neodymium–YAG laser. This fortuitous proximity is of particular utility, as it permits the production of singlet oxygen either from an effectively continuous output or from a discrete number of pulses, each of approximately 7 ns duration. This method of activation utilizes the high-energy output of the Nd–YAG laser and also allows the kinetics of the protein-induced structural alterations in DNA to be followed with high precision. It should be noted that because the half-life of singlet oxygen in solution is only 4 ms *(10)*, the time available for reaction with the DNA is essentially limited by the time of irradiation. This short half-life also means that it is unnecessary to terminate any reaction by the addition of a quenching reagent.

2. Materials

1. A neodymium–yttrium–aluminium garnet (Nd-YAG) laser (Spectra-Physics DCR-11) set up as illustrated in **Fig. 2** is used to generate a beam of polarized coherent light at a wavelength of 1064 nm. A doubling crystal correctly aligned in the beam path produces a mixture of light at 1064 and 532 nm. A dichroic mirror arranged so that the 532-nm beam is deflected down onto a thermostatted Eppendorf tube containing 20 μL of the sample to be irradiated subsequently separates this mixture. Alternatively, if the volume of the irradiated solution is small, the different wavelengths can be separated by an appropriate arrangement of prisms and the 532-nM beam directed into an Eppendorf tube held horizontally in a metal block maintained at the required temperature. It is essential to obtain an adequate separation of these two wavelengths because even a relatively low proportion of the primary emission at 1064 nm could result in a rapid heating of the sample. The DCR-11 functions at a frequency of 10 Hz each pulse of about 7 ns duration delivering 160 mJ of energy. Other NdYAG lasers are obtainable that can deliver up to twice this energy per pulse.
2. Eosin isothiocyanate is obtained from Molecular Probes.
3. 10 m*M* Tris-HCl, pH 7.9.

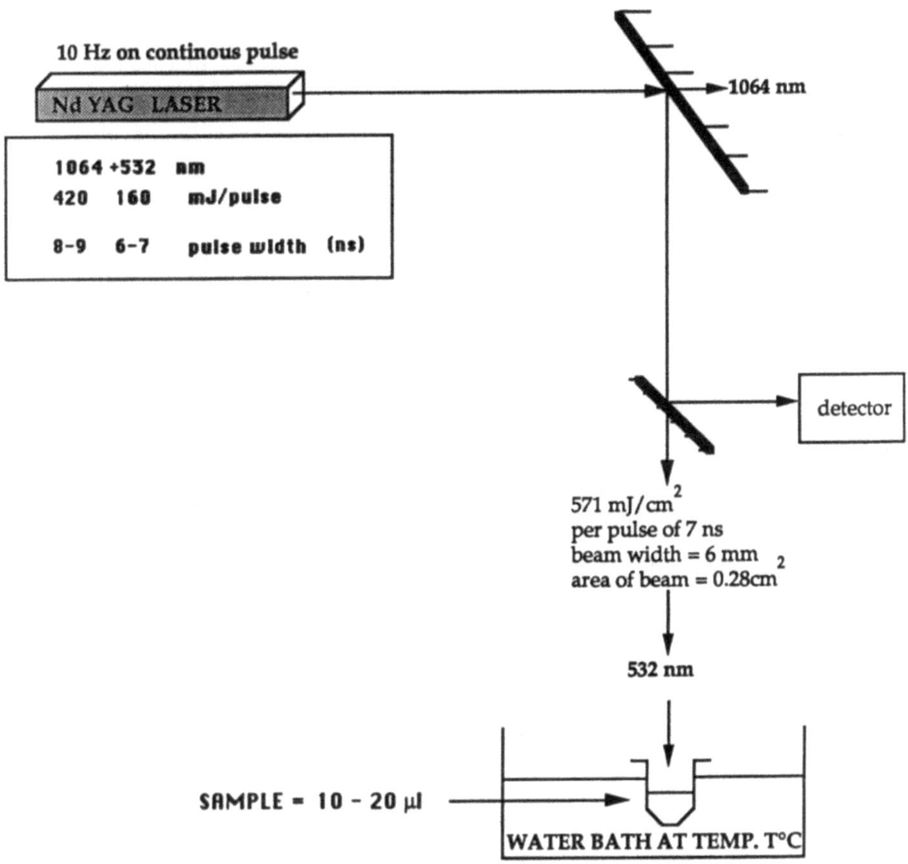

Fig. 2. Use of a Nd-YAG laser for irradiation of a 20-µL reaction mixture.

4. DNA fragment containing the protein-binding site. The fragment should be labeled at one end with ^{32}P using, for example, end filling with either the Klenow fragment of DNA polymerase or reverse transcriptase *(11)*. The concentration of the stock solution of fragment is typically in the region of 100 µg/mL.
5. An appropriate buffer that is suitable for the protein–DNA complexes is under investigation. Avoid the use of reducing agents.
6. Bovine serum albumin: stock solution 10 mg/mL.
7. Phenol: equilibrated with an equal volume of 0.1 *M* Tris-HCl, pH 8.0.
8. Absolute ethanol.
9. Piperidine: 0.1 *M* piperidine is prepared by dilution of redistilled piperidine (10.1 *M*).
10. 10X TBE: 108 g Tris base, 55 g boric acid, and 40 mL of 0.5 *M* EDTA (pH 8.0) in 1:1 solution.
11. Polyacrylamide gel: For a 200 bp DNA fragment, a 40 × 20 × 0.04-cm 8% denaturing polyacrylamide gel is used. The gel is prepared by mixing 10 mL of 40%

acrylamide solution (380 g DNA-sequencing grade acrylamide, 20 g *N,N'*-methylenebisacrylamide in 1:1 solution), 5 mL 10X TBE, 23 g urea, deionized water to 50 mL. When the urea is fully dissolved, add 100 µL of 10% ammonium persulfate and mix with rapid stirring, Then, add 60 µL TEMED (*N,N,N'N'*-tetramethylethylene diamine) and mix rapidly. Pour the gel solution between the sealed gel plates and insert a comb with 0.25-cm teeth into the gel solution. Allow to set, remove comb, and clean slots with gel buffer using a Hamilton syringe.

12. Gel running buffer, 1X TBE: 90 m*M* Tris, 90 m*M* borate, pH 8.3, and 10 m*M* ethylenediaminetetracetate (EDTA) pH 8.0 made by dilution of 10X TBE.
13. Gel loading buffer: 95% formamide, 10 m*M* ethylenediaminetetracetate (pH 8.0), 1 mg/mL xylene cyanol FF, and 1 mg/mL bromophenol blue.
14. X-ray film.
15. 10 m*M* dithiothreitol (DTT).

3. Methods

3.1. Preparation of Eosin–Tris Complex

1. A 10 m*M* stock solution of this complex is formed by incubating 10 m*M* eosin isothiocyanate in 10 m*M* Tris-HCl, pH 7.9 for 2 h at 37°C, taking care to avoid exposure to light. The eosin isocyanate is diluted from a freshly prepared 100 m*M* stock solution.
2. The concentration of eosin is estimated from the absorption of the solution at 525 nm ($\varepsilon_{523.2}$ = 25.6 for a 1 *M* solution) and adjusted by the addition of the appropriate volume of double-distilled water.

3.2. Formation of Nucleoprotein Complexes

1. The first step in the detection of protein-induced deformation of DNA is the formation of a nucleoprotein complex. For example, RNA polymerase (100 n*M*) and end-labeled fragments of DNA containing the *lac* UV5 promoter (4 n*M*) are mixed in a buffer containing 100 m*M* KCl, 10 m*M* $MgCl_2$, 20 m*M* HEPES (pH 8.4), 3% glycerol, and 100 µg/mL bovine serum albumin in a total volume of 20 µL *(9)*.
2. This mixture is then incubated for 30 min at 37°C.

3.3. Irradiation of the Nucleoprotein Complex

1. Once the nucleoprotein complex has been formed a fresh stock of the eosin–Tris complex is added at an appropriate concentration to 20 µL of the target solution in a small Eppendorf tube. Typically, a final concentration of 50 m*M* is used (*see* **Note 1**).
2. Immediately after the addition of the eosin–Tris complex the whole mixture is irradiated for 20 s at 10 Hz. This corresponds to a total energy dose of 115 J/cm^2 (*see* **Note 2**). Successful activation of the eosin–Tris complex is apparent by a detectable change in the color of the solution consequent upon a shift in the absorption maximum from 523 to 514 nm on oxidation.

3. Although the half-life of singlet oxygen is sufficiently short to obviate the need to remove excess reagent, it is advisable to add a quenching agent such as dithiothreitol immediately on cessation of irradiation to minimize any secondary radical reactions. Typically, 1 μL of 10 m*M* dithiothreitol is added to the irradiated solution.

3.4. Detection of Sites of Reaction with Singlet Oxygen

1. After irradiation and quenching, 30 μL double-distilled water is added to each sample.
2. Fifty microliters of a phenol freshly prepared by equilibration of melted phenol with an equal volume of 0.1 *M* Tris-HCl is then added.
3. After mixing with a vortex mixer, the samples are centrifuged for 1 min in a bench-top microcentrifuge at 5000*g* to separate the aqueous and organic layers.
4. The upper aqueous layer is removed with an automatic pipet to a clean Eppendorf tube.
5. Three to four volumes of ethanol at 0°C are then added and the samples placed in a dry ice/ethanol bath for 1 h.
6. Centrifuge the samples for 15 min at 5000*g* in a bench-top centrifuge.
7. Remove the ethanol using an automatic pipet.
8. Dry the samples in a centrifugal evaporator.
9. Resuspend in 100 μL of freshly prepared piperidine solution.
10. Heat at 90°C for 30 min (*see* **Note 6**).
11. Sites of cleavage are determined by separation on polyacrylamide gels (typically 40 cm, run at 60 W constant power until the xylene cyanol FF marker has migrated 23 cm into gel) followed by autoradiography (typically 2–24 h exposure depending on the specific activity of the labeled DNA fragment). A typical result is shown in **Fig. 3**. Cleavage at a particular site results in the generation of a band of defined length. Standard Maxam and Gilbert sequencing reactions *(12)* can be performed and loaded on the same gel to identify the cleavage sites.

4. Notes

1. Because the half-life of singlet oxygen is short, the average path length for diffusion is also short. Consequently, to ensure an adequate rate of reaction, the eosin–Tris concentration must be sufficiently high to ensure that all potential targets in the DNA are accessible to the reactive entity.
2. Although the energy dose used during irradiation may appear to be substantial, it should be borne in mind that even with a high-intensity laser, each pulse delivers only 160 mJ and consequently, full saturation of the system requires a considerable pulse repetition rate.
3. For optimum reactivity, it is essential that radical scavengers such as mercapto–groups should, as far as possible, be rigorously excluded from the reaction mixture, as they would effectively prevent any singlet oxygen from arriving at its target site. For the same reasons, the concentrations of alcohols such as glycerol should be kept as low as possible.
4. Ideally, protein–DNA complexes should be insensitive to the presence of the nonirradiated eosin–Tris complex. However, it has been observed that the half-

Fig. 3. Reaction of singlet oxygen with a binary complex of *E. coli* RNA polymerase with the *lac* UV_5 promoter. The figure shows an autoradiograph of the pattern of reactivity on the transcribed strand in the presence and absence of RNA polymerase. Note that the bands visible in the DNA only lane result from piperidine cleavage at $(Py)_3$ sequences and their occurrence is independent of both irradiation and the presence of eosin–Tris.

life of certain complexes, in particular the binary CAP–DNA and RNA polymerase–DNA complexes, is reduced by approximately an order of magnitude with the sensitizer present *(9)*. For stable complexes in which the protein has a long residence time this effect does not significantly interfere with the detection of DNA deformations because the time of irradiation is short relative to the stability of the complex. At this time, it is unclear whether this effect is general or is restricted to particular complexes. Nevertheless, it is essential to determine the stability of complexes under study under the precise conditions corresponding to those prevailing during irradiation. The method of choice is gel retardation (electrophoretic mobility shift assay).

5. The short time of irradiation allows the use of the singlet-oxygen reaction in kinetic studies. Here again, to prevent any perturbation of an enzymatic manipulation of DNA, it would be necessary to add the eosin–Tris complex immediately prior to irradiation after the reaction under study had proceeded for the required time. For this purpose, a rapid-mixing device would be necessary.
6. To obtain sharp bands on polyacrylamide gels, it is advisable for the samples to be transferred to clean Eppendorf tubes immediately prior to the evaporation of piperidine. Removal of piperidine in a centrifugal evaporator should also be carried out as rapidly as possible, and any form of heating should be avoided, as this increases the nonspecific background cleavage of DNA by piperidine.

References

1. Lilley, D. M. J. and Palecek, C. (1984) The supercoil-stabilised cruciform of colE1 is hyper-reactive to osmium tetroxide. *EMBO J.* **3,** 1187–1195.
2. Sasse-Dwight, S. and Gralla, J. D. (1989) KMn04 as a probe for *lac* promoter DNA melting and mechanism *in vivo*. *J. Biol. Chem.* **264,** 8074–8081.
3. Sigman, D. S., Spassky, A., Rimsky, S., and Buc, H. (1985) Conformational analysis of lac promoters using the nuclease activity of 1,10-phenanthroline–copper ion. *Biopolymers* **24,** 183–197.
4. Hogan, M. E., Rooney, T. F., and Austin, R. H. (1987) Evidence for kinks in DNA folding in the nucleosome. *Nature* **328,** 554–557.
5. Le Doan, T., Perrouault, L., Hélène, C., Chassignol, M., and Thuong, N. T. (1986) Targeted cleavage of polynucleotides by complementary oligonucleotides covalently linked to iron-porphyrins. *Biochemistry* **25,** 6736–6739.
6. Tullius, T. D., Dombroski, B. A., Churchill, M. E. A., and Kam, L. (1987) Hydroxylradical footprinting: a high resolution method for mapping protein–DNA contacts. *Methods Enzymol.* **155,** 537–558.
7. Ansari, A. Z., Chael, M. L., and O'Halloran, T. V. (1992) Allosteric underwinding of DNA is a critical step in positive control of transcription by Hg-MerR. *Nature* **355,** 87–89.
8. Schultz, S. C., Shields, S. C., and Steitz, T. A. (1991) Crystal structure of a CAP–DNA complex: the DNA is bent by 90°. *Science* **253,** 1001–1007.
9. Buckle, M., Buc, H., and Travers, A. A. (1992) DNA deformation in nucleoprotein complexes between RNA polymerase, cAMP receptor protein and the *lac* UV_5 promoter probed by singlet oxygen. *EMBO J.* **11,** 2619–2625.

10. Rougée, M. and Bensasson, R. V. (1986) Détermination des constantes de vitesse de désactivation de l'oxygène singulet ($^1D_7$) en presénce de biomolécules. *Comp. Rend. Acad. Sci. Paris* **302,** 1223–1226.
11. Travers, A. A., Lamond, A. I., Mace, H. A. F., and Berman, M. L. (1983) RNA polymerase interactions with the upstream region of the *E. coli tyrT* promoters. *Cell* **35,** 265–273.
12. Maxam, A. M. and Gilbert, W. (1980) Sequencing end-labeled DNA with base-specific chemical cleavages. *Methods Enzymol.* **155,** 560–568.

12

Ultraviolet-Laser Footprinting

Johannes Geiselmann and Frederic Boccard

1. Introduction

1.1. Measurement of DNA–Protein Interactions In Vitro

A large number of processes within the cell, in particular the regulation of gene expression, rely on the binding of proteins to specific sites on the DNA. A primary ingredient to understanding these processes is the characterization of the protein–DNA interaction *(1)*. Such a characterization consists in determining the position of the binding site on the DNA and measuring the affinity of the protein for this recognition site. A wide variety of footprinting techniques can accomplish this task *(2,3)*.

1.2. Footprinting Techniques

These techniques involve the reaction of a footprinting reagent (in the largest sense) with DNA and the subsequent localization and quantification of the resultant DNA modification. Commonly used footprinting reagents include DNase I, $KMnO_4$, or dimethylsulfate (DMS) *(2,3)*; *see* Chapters 3–14. Most of these methods require extended incubation times of the reagent with the DNA–protein complex, which may lead to artifacts if the footprinting reagent modifies the complex. For example, DMS may react with the protein; DNase I relaxes a supercoiled plasmid, which may destabilize the DNA–protein complex under investigation. UV-laser footprinting eliminates several of these disadvantages, but also creates others (*see* **Subheading 1.4.**).

1.3. Principle of UV-Laser Footprinting

The principle of ultraviolet (UV)-laser footprinting is schematized in **Fig. 1**. The sample containing the nucleoprotein complex is irradiated with a short (less than 10 ns) pulse of UV-laser light. An identical sample, but lacking the

From: *Methods in Molecular Biology, vol. 148: DNA–Protein Interactions: Principles and Protocols, 2nd ed.*
Edited by: T. Moss © Humana Press Inc., Totowa, NJ

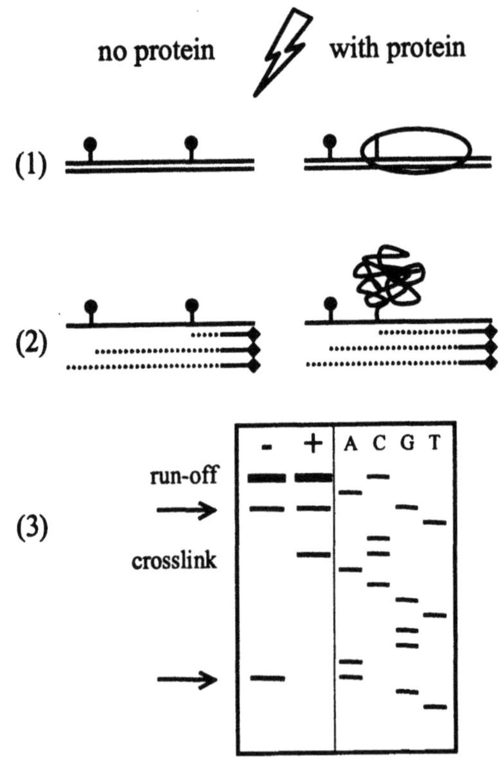

Fig. 1. The principle of UV-laser footprinting. The double-stranded DNA is represented by the double line, and the protein by the ellipse. **(1)** A sample of DNA alone, or the DNA–protein complex is irradiated with one pulse of UV-laser light. The bases can undergo intramolecular photoreactions, react with the solvent (symbolized by the circle), or form a crosslink with the bound protein (symbolized by the line connecting DNA and protein). Only the reactions of the top strand are shown in the figure. **(2)** The samples are denatured, a radioactively labeled primer (line with diamond) is annealed to one stand of the DNA (the top strand) and extended using a DNA polymerase (dotted line). The primer extension stops at damaged bases or at the end of the fragment. **(3)** The primer extension products are analyzed on a sequencing gel, along with a dideoxy-sequencing reaction using the same primer. The location of the photoreactions are marked with an arrow, the crosslink and the runoff are indicated.

protein, is treated in parallel in the same way. The conditions are adjusted such that the number of photons delivered onto the sample exceeds the number of absorbing molecules. The nucleic acid bases are excited and undergo photoreactions, the nature of which depend exquisitely on the local environment of the bases. The possible reactions include intrastrand reactions of consecutive bases (the formation of thymine dimers is the most prominent such reaction), interstrand reactions (although their quantum efficiency is too low to contribute to footprinting signals), reactions with solvent molecules (e.g., with H_2O), and crosslinks with the protein *(4–8)*.

In general, it is not possible, but neither is it necessary, to determine the nature of the photoreaction at a particular base. Because the photoreactions are highly sensitive to the local environment of the DNA, binding of a protein changes this environment and thus produces a footprint (i.e., a difference between the DNA photoreactivity in the absence and presence of the protein). In the example of **Fig. 1**, the photoreaction at the left extremity of the DNA fragment remains constant because the protein does not change the local environment of this base. However, the presence of the protein prevents a photoreaction toward the right end of the fragment (perhaps by excluding water from the vicinity of the base) and favors a new photoreaction with an amino acid side chain, resulting in a covalent bond, a crosslink, between the protein and the DNA. All such reactions modify the nucleotide base and, hence, impede the progression of DNA polymerase during subsequent replication of the damaged DNA strand. Arrest of the polymerase is detected in a primer extension reaction by the appearance of shortened replication products, which then serve to localize the modified base.

1.4. Advantages and Disadvantages of UV-Laser Footprinting

Ultraviolet-laser footprinting circumvents many potential artifacts of more classical techniques by trapping the complex under investigation at the time of irradiation. The signal is acquired very rapidly (on the order of microseconds) (i.e., faster than typical rearrangements of a DNA–protein complex [on the order of milliseconds]), and the footprint thereby freezes the initial state of the complex *(9)*. Laser light, as opposed to ordinary UV light, is needed in order to limit the "incubation time" to several ns while providing a sufficient number of photons to excite all nucleotide bases of the sample. The rapidity of signal acquisition allows one to obtain kinetic structural signals by irradiating a complex at different times after mixing the components. The technique has the further advantage that it can be transposed to in vivo experiments in a straightforward manner because UV light readily penetrates bacterial cells. However, because of the limited sample size of in vivo experiments, the present version of the technique is only suitable for studying DNA–protein interactions in bacteria when using a binding site carried on a multicopy plasmid.

Using the UV-laser technology, it is possible to follow the kinetics of the assembly of DNA–protein complexes. This technique can be used to determine the order in which protein–DNA contacts are established in a multiprotein–DNA complex. Proteins and DNA are mixed at time zero and the sample is irradiated at a specific time interval after mixing *(10)*. Modern mixing techniques allow a time resolution on the order of a several milliseconds *(11)*.

The main disadvantage of UV-laser footprinting is the unpredictability of the footprinting signal. It should be noted that the footprint is *not* the result of the protein shielding the DNA from the UV irradiation. The presence of the protein merely changes the probability of certain photoreactions by excluding solvent, changing the conformation of the DNA, or juxtaposing a reactive amino acid and a particular base. A structural interpretation of the footprinting signal is therefore, in general, not possible. A more practical limitation of the technique is the need for a relatively expensive laser.

We describe the experiment for the particular case of the binding of an *Escherichia coli* protein, the integration host factor (IHF), to one of its specific binding sites, the *yjbE* site ***(12)***, in vitro and in vivo. IHF is a small, heterodimeric protein (molecular weight [MW] of the dimer is approx 19 kDa) that binds to specific sites on the DNA (for a review, *see* **ref. *13***). Upon binding to DNA IHF bends the DNA by about 180° ***(14)***. This DNA bending gives rise to a very strong UV-laser footprinting signal, probably because two consecutive pyrimidines are brought into optimal alignment for the formation of a pyrimidine dimer ***(15)***. Variations of this basic protocol, applicable to the study of *any* DNA–protein interaction, are described in **Subheading 4.**

2. Materials

1. Ultraviolet-laser: The most commonly used lasers are YAG lasers, e.g., the Spectra-Physics Quanta-Ray GCR series lasers, which emit infrared light of 1064 nm. Two consecutive passes through a frequency-doubling crystal yields high-intensity light of 266 nm, which is sufficiently close to the absorption maximum of nucleic acid bases. Frequency doublers are a standard add-on for virtually all commercially available YAG lasers. The laser power should be between 30 and 50 mJ per pulse at 266 nm and a pulse duration of 5 ns is standard. The energy of one 30-mJ pulse represents about 4×10^{16} photons (i.e., 67 nmol photons).
2. Power meter to measure the energy of the laser beam.
3. Thermostated water bath.
4. Spectrophotometer.
5. Water pump and 0.45-µ*M* filter device for washing *E. coli* cells.
6. Phosphoimager: A phospho-storage device is ideal for quantifying sequencing gels. The most commonly used instruments are sold by Molecular Dynamics, Fuji, or by Bio-Rad.
7. IHF binding buffer: 50 m*M* Tris-HCl, pH 7.5, 70 m*M* KCl, 7 m*M* $MgCl_2$, 3 m*M* $CaCl_2$, 1 m*M* EDTA, 10% glycerol, 200 mg/mL bovine serum albumin (BSA), and 1 m*M* β-mercaptoethanol.
8. IHF: working stock solution at 3–5 µ*M* in 50 m*M* Tris-HCl, pH 7.4, 800 m*M* KCl, 40 m*M* K-phosphate, 2 mg/mL BSA, and 10% glycerol.
9. Plasmid pBluescriptII (Strategene) containing the IHF binding site cloned into the multicloning site (MCS). Stock solution in water at 100 n*M*.

10. Primer extension reaction. Annealing buffer: 1 *M* Tris-HCl, pH 7.6, 100 m*M* $MgCl_2$, and 160 m*M* dithiothreitol (DTT). Elongation mix for 2 µL: 2 U of T7 DNA polymerase in 2.4 m*M* of each deoxyribonucleotide, 3 m*M* Tris-HCl, pH 7.5, 0.75 m*M* DTT, 15 mg/mL BSA, and 0.75% glycerol (*see* **Subheading 3.** for the annealing and elongation steps). Stop solution: 95% formamide, 20 m*M* EDTA, 0.05% bromophenol blue, and 0.05% xylene cyanol.
11. Sequencing reaction. Enzyme dilution buffer: 20 m*M* Tris-HCl, pH 7.5, 5 m*M* DTT, 0.1 mg/mL BSA, and 5% glycerol. Stop solution: 95% formamide, 1 m*M* EDTA, 0.05% bromophenol blue, 0.05% xylene cyanol.
12. Sequencing gel: Sequencing reactions are analyzed on 40-cm-long denaturing (7 *M* urea) 8% polyacrylamide (ratio acrylamide:*bis* acrylamide 19:1) gels in 90 m*M* Tris–borate, and 2 m*M* EDTA (TBE), and TBE used as running buffer. Gels were transferred onto Whatman (3MM Chr) paper and dried at 80°C for 40 min in a gel dryer (Bio-Rad model 583) linked to a vacuum pump.
13. Extraction of plasmid DNA. Solution I: 100 m*M* Tris-HCl, pH 7.5, 10 m*M* EDTA, 400 µg/mL RNase I. Solution II: 0.2 *N* NaOH, and 1% sodium dodecyl sulfate (SDS) made freshly. Solution III: 3 *M* potassium, and 5 *M* acetate solution, made by adding 11.5 mL of glacial acetic acid and 28.5 mL of H_2O to 60 mL of 5 *M* potassium acetate.
14. LB and minimal M9 media are used to grow and wash *E. coli* cells, respectively ***(16)***.

3. Methods

3.1. In Vitro UV-Laser Footprinting

1. Arrange the laser beam, using appropriate mirrors, such that it is directed vertically into a water bath.
2. Align, and fix firmly, an Eppendorf holder in the water bath such that the laser beam enters precisely in the center of an open Eppendorf tube. A piece of black paper stuck into the bottom of the Eppendorf tube can help align the tube with the laser beam; the impact of the laser light is very audible and "burns" the site of impact, whitening the otherwise black paper.
3. Operate the laser in repetition mode (*see* **Note 1**) for at least 10 min and adjust the doubling crystals to obtain a laser power at 266 nm of at least 30 mJ per pulse. This laser power is measured during the warm-up period with an appropriate power meter, before the actual footprinting reaction.
4. Incubate IHF at the desired concentration with 5 n*M* plasmid DNA in 40 µL binding buffer for 20 min at 25°C (*see* **Note 2**). It is best to use a flat-bottomed Eppendorf tube, but regular 1.5-mL Eppendorf tubes are adequate. The laser beam generally has a diameter of 5 mm, and for maximal use of the light energy, the sample should have roughly the same dimensions. Care should be taken to ensure that all of the sample is irradiated by the laser beam.
5. Place the sample under the laser beam and irradiate with one pulse of UV-laser light. It is best to operate the laser in repetition mode (i.e., continuously emitting around 10 pulses per second). Most lasers also have the possibility to emit a single pulse of light. However, the power of such a pulse is not very well con-

trolled, because the yield of the doubling crystals is extremely sensitive to temperature. Continuous emission of laser pulses at a frequency of about 10 Hz ensures a constant temperature of the crystals and therefore a stable pulse energy. An electronically controlled shutter is used to obstruct the beam. The shutter opening is coordinated with the emission of the laser pulses to ensure that only one pulse of laser light passes for each opening of the shutter.

6. After irradiation, remove the protein from the irradiated DNA by incubating with 50 µg/mL proteinase K for 15 min at 50°C. Extract the samples with half a volume of a phenol/chloroform solution (made by adding equal volumes of phenol pH 8 and chloroform), precipitate with 2 vol of ethanol, and resuspend the DNA in 18.5 µL of H_2O.
7. Primer extension (*see* **Note 3**): Add 2 µL of a solution of 0.2 µ*M* radiolabeled primer (5'-[^{32}P] labeled using T4-kinase) to 18.5 mL of DNA. (Increasing the primer concentration beyond this twofold excess over template will increase the strength of the primer extension signals only marginally.) Denature the samples by heating to 100°C for 3 min and chill on ice for 5 min. After the addition of 2.5 µL of annealing buffer, incubate the samples for 3 min at 50°C (to anneal primer) and chill again for 5 min on ice. Add 2 µL of the elongation mix (*see* **Note 4**) and incubate the reaction for 10 min at 37°C. The mix is prepared freshly but can be kept on ice for several hours.
8. Precipitate the DNA by adding 150 µL of ethanol and incubating for 10 min at –20°C. Centrifuge for 10 min at full speed (about 12,000*g*) in a microcentrifuge. Resuspend the DNA in 10 µL of loading dye.
9. Analyze 3 µL by gel electrophoresis on a denaturing 8% polyacrylamide sequencing gel. After electrophoresis, transfer the gel onto Whatman paper and vacuum dry at 80°C for 40 min.
10. To determine precisely the location of the footprint, a reference ladder is generated by sequencing the same plasmid DNA using the same radiolabeled primer (*see* **Note 5**). Denature 15 n*M* of plasmid DNA in a volume of 8 µL by heating to 100°C for 2 min. Chill on ice for 5 min. Add 1 µL of radiolabeled primer (0.25 µ*M*) and 1 µL of annealing buffer. Incubate for 3 min at 50°C. Chill annealing reaction on ice for 5 min. Add 2.8 µL of each Deaza G/A T7 Sequencing™ Mixes (Pharmacia) to four termination reaction tubes and prewarm at 37°C. Dilute 1 µL of T7 DNA polymerase with 4 µL of enzyme dilution buffer. Add 2 µL of diluted T7 DNA polymerase to the annealing reaction. Dispense 2.8 µL of annealing reaction in each of the termination tubes. Incubate for 5 min at 37°C. Add 4 µL of the stop solution.
11. Expose the dried gel to a phospoimager screen overnight and scan the screen using the phospho-imager and its associated software (e.g., ImageQuant from Molecular Dynamics).
12. Deduce a line profile of the different lanes using the Phospho-Imager software.
13. Transfer the data to Microsoft Excel and superimpose the scans on the same graph in order to visualize and quantify the footprint (*see* **Notes 6–11**).

3.2. In Vivo UV-Laser Footprinting

1. Grow cultures of IHF$^+$ (W3110) and IHF$^-$ (W3110 *hip*) strains *(12)* transformed with the plasmid carrying the *ihf* site in LB medium to the desired OD_{600} (0.6 or to saturation).
2. Wash the cells in minimal M9 medium (optically transparent buffer), resuspend in minimal M9 medium to a final OD_{600} of 1, and incubate the cells at 37°C (*see* **Note 12**).
3. Irradiate as described above a large number (40–60), of 50-µL cell aliquots (*see* **Note 13**). Freeze the cells immediately after irradiation in a dry-ice bath.
4. Pool the cells and extract plasmid DNA from 2–3 mL of cells using the following alkaline lysis procedure. Centrifuge the bacteria for 5 min in a tabletop Eppendorf centrifuge. Resuspend the pellet in 100 µL of solution I and add 100 µL of solution II. Mix by inversion several times. Add 100 µL of solution III and mix by inverting the tube several times. Centrifuge the tubes at full speed in a tabletop Eppendorf centrifuge (>10,000*g*) for 5 min. Transfer the supernatant to a clean tube and add 210 µL of isopropanol. Incubate for 5 min at room temperature and centrifuge the tubes at full speed for 10 min. Wash the pellet with 70% ethanol and dissolved the DNA in 37 µL of H_2O.
5. For primer extension, add 2 µL of a solution of 0.2 µ*M* primer (5'-[^{32}P] labeled) to 18.5 mL of DNA and process and analyze the samples in the same way as for the in vitro reactions, **steps 7–13** of **Subheading 3.1.** (*see* **Notes 10** and **11**).

4. Notes

1. Laser setup. As mentioned in the **Subheading 3.** in order to obtain a stable laser power it is best to operate the laser in repetition mode. If a single-pulse mode is used the energy of a particular pulse is ill-defined and the absence of a footprinting signal may simply be the consequence of diminishing laser power.

   All lasers provide an electrical signal that allows external equipment to be coordinated with the laser pulse. To our knowledge, shutters are not commercially available. However, it is an easy task for a good mechanics shop to construct such a shutter. We used a shutter made of a small sheet of blackened Teflon obstructing a hole of approx 8-mm in diameter through which the laser beam had to pass in order to reach the sample. Any material can be used, but it should be kept in mind that the laser will eventually burn a hole into the material and that it is best to use a black material in order to minimize hazardous reflections of the laser light. A simple electronic circuit controlled the opening of the shutter by activating an electomagnet that pulled the Teflon sheet (via an attached piece of metal) away from the hole. After the light had passed, the current to the magnet was cut and a spring pulled the sheet back over the hole.
2. The binding buffer described is the standard buffer used for measuring DNA binding of IHF. Other proteins may require different buffer conditions. The buffer may be adjusted with certain limitations. The salt concentration should not be too low in order to avoid nonspecific binding of the protein to DNA. The buffer should not include a high concentration of reagents that absorb at 266 nm. For

example, the interaction of the cyclic AMP receptor protein (CRP) with DNA requires the presence of cAMP in the buffer *(17)*. Keep the concentration of such nucleotides below 100 μ*M*. As mentioned in **Subheading 3.**, one laser pulse contains the equivalent of about 100 nmol in photons. A typical reaction volume is 50 μL; therefore 100 μ*M* ATP absorbs about 5 nmol of photons. A more physiological concentration of ATP in the millimolar range would dramatically decrease the yield of the photoreaction.

3. The procedure describes the footprinting reaction for only one strand of DNA. Evidently, the other strand can be analyzed in the same way using the appropriate primer. The primers should be chosen such that the region of interest is within 200 nucleotides from the primer. Most photoreactions have a quantum efficiency below 1%, the formation of thymine dimers reaching several percent. Therefore, on average, there will be roughly 1 photoreaction per 100 base pairs (bp). Considering only the 100-bp region downstream of a primer assures single-hit conditions. If the DNA carries too many photoreactions, the primer extension will stop at the first defect and the signal of a photoreaction further downstream will pass undetected.
4. The detection of photoreactions. All DNA polymerases are very sensitive to damaged bases. It is not important to use T7 DNA polymerase in the primer extension reaction, any other DNA polymerase (e.g., Klenow, Taq) gives equivalent signals. However, signals obtained with different polymerases may not necessarily be identical because some may be more sensitive to particular photodamaged bases than others. Even RNA polymerases can be used if the template harbors an appropriate promoter. We have successfully used T7 RNA polymerase to transcribe the region of interest from a T7 promoter located on the vector DNA. The major signal of the IHF footprint remained unchanged, but, instead of a single band, T7 RNA polymerase generates a doublet of bands *(15)*.
5. To determine the location of termination sites precisely, we generated a reference DNA ladder consisting of a sequencing reaction of the same DNA region (*see* **Subheading 3.**). In general, it is assumed that the primer extension reaction of the irradiated DNA stops just before the modified base. For example, if the sequence of the DNA read on the gel is 5'-GGAC-3' and the primer extension reaction of the irradiated DNA shows a band at the position of the A in the sequencing reaction (run in parallel on the gel), then the photoreaction most likely took place on the following base pair (the C in the above sequence). Because the primer extension reaction reads the opposite strand, the photoreaction actually damaged the G marked with an asterisk:

   5'-GGAC-3'
   3'-CCTG*-5'

6. There are several possible reasons for not detecting a UV-laser footprinting signal. The most obvious problem is that the protein does not bind to the DNA. Generally, we verify binding by a gel retardation assay. The second reason may be that protein binding does not change the photoreactivity of the bases in the recognition site. This is the case, for example, for CRP, which yields very weak

signals in UV-laser footprinting despite a strong interaction measured by other techniques (J. Geiselmann, unpublished results). Because photoreactivity depends on the sequence, a particular site might not be photosensitive; for example, the main IHF signal had not been observed for the *ssb* site, probably because the sequence of this site does not contain the highly reactive TC pyrimidine doublet *(15)*.

7. A trivial reason for not detecting a photoreaction is that the laser did not hit the sample. The best control, and a control to include in all laser footprinting reactions, is to analyze the primer extension reaction of DNA alone. Two DNA-alone samples should always be included: one irradiated sample and an *identical* sample that has not been exposed to UV-laser light. The irradiated sample should yield readily visible bands all along the lane (**Fig. 2**, lane f), whereas the nonirradiated sample should show no elongation arrests (**Fig. 2**, lane e).
8. The irradiated DNA does not give any elongation arrests with T7 DNA polymerase. Verify the primer extension mix. Perform a control primer extension reaction using the nonirradiated plasmid template, but cut about 100 bp downstream of the primer with a convenient restriction enzyme. Primer extension using this template should give a very strong band corresponding to the elongation reaction reaching the end of the fragment.
9. Artifactual footprinting signals. A contaminated template preparation or a damaged DNA template could be misinterpreted as giving a footprinting signal. It is very important to verify that the nonirradiated template does not produce any bands during the primer extension reaction. This is particularly important for in vivo reactions because the plasmid preparation could partially damage the plasmid and lead to artifactual bands (**Fig. 2**, lanes c and d).
10. Comparing signals obtained under different conditions in vitro, or comparing in vitro to in vivo signals. The efficiency of sample preparation or of the primer extension reaction can vary from sample to sample. In order to compare different lanes we run a small portion (10%) of the primer extension reactions on a sequencing gel and quantify the lanes using a phosphoimager. We then load the same samples on a second sequencing gel, but equilibrating the amounts loaded according to the signal intensities on the first gel. For example, if one of the reactions was only half as efficient as the other ones, we load two times more of this sample on the second gel. This readjustment should not be allowed to exceed a factor of 2.
11. Quantifying lanes. Once an intensity equilibrated gel exposure is obtained, we obtain line profiles of all lanes using the Phospo-Imager software and compare lanes by superimposing the line profiles in Microsoft Excel. Remaining small (several percent) differences in the intensities of the lanes should be normalized by multiplying the scans with a scaling factor between 0.9 and 1.1 that is determined subjectively by the user in such a way that the global patterns superimpose. Lanes containing a high background of nonspecific radioactivity cannot be quantified with confidence.
12. It is important to keep in vivo samples at the physiological temperature (37°C for *E. coli*) to ensure optimal binding.

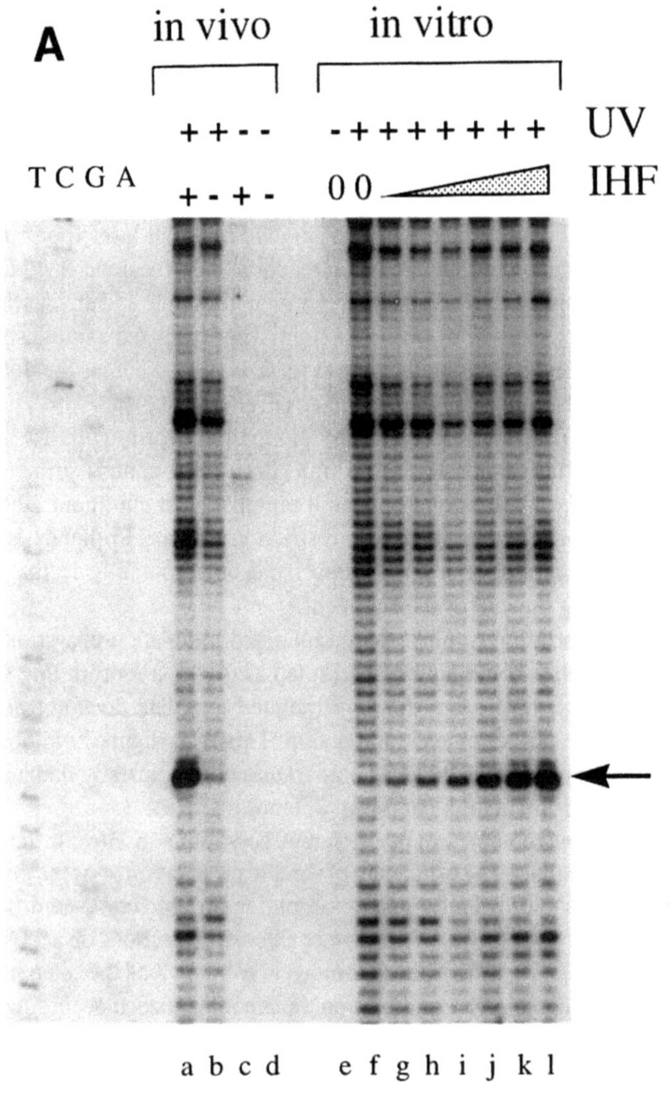

Fig. 2. UV-laser footprinting. **(A)** Primer extension profile of a UV-laser footprinting experiment showing the binding of IHF to a specific binding site. The four lanes on the left, labeled TCGA, are a sequencing reaction using the same primer as the one used for the primer extension of the UV-laser footprinting reaction (only the T lane is clearly visible on the picture). Increasing amounts of IHF (0–200 n*M*, lanes f to l) are incubated with 5 n*M* plasmid and footprinted in vitro, as described in **Subheading 3.** The arrow points to the major footprinting signal. Lane e is identical to lane f, but the DNA has not been irradiated. The in vivo reactions are carried out as described in the protocol. Lane a is derived from wt cells expressing IHF, lane b is a footprinting reaction from a strain lacking IHF.

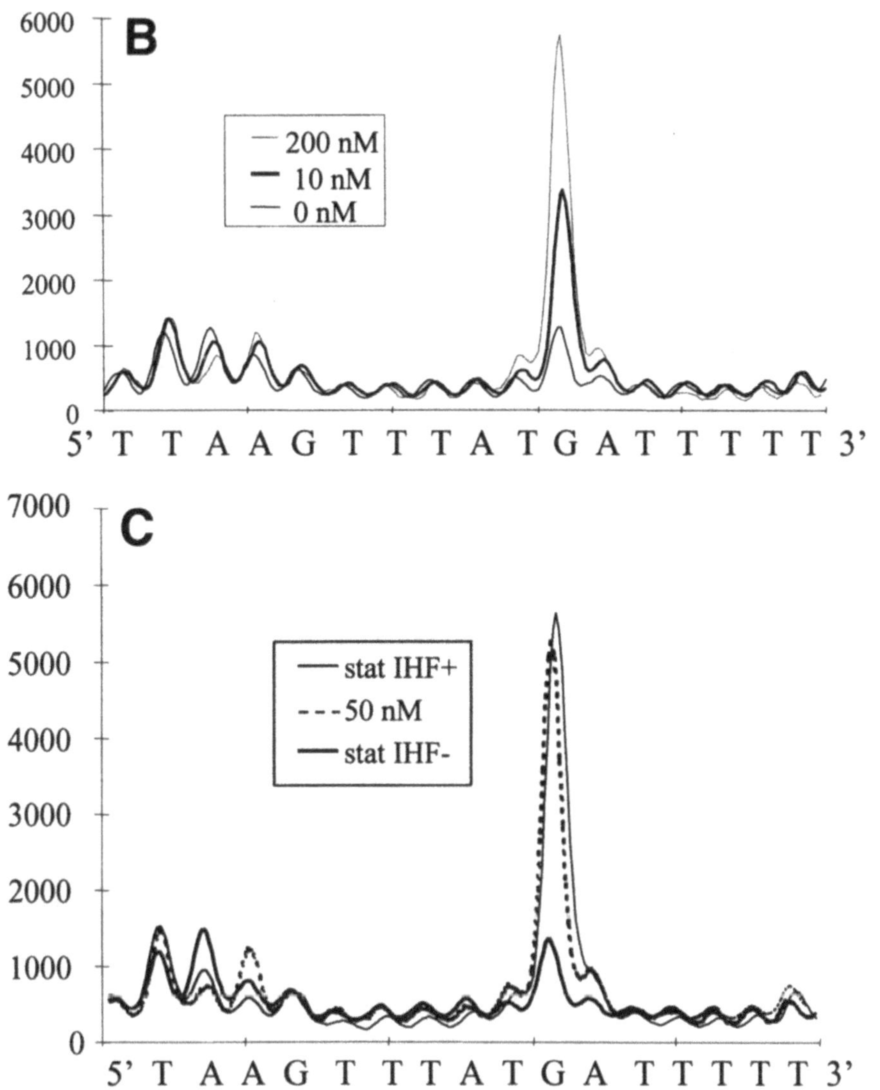

The control lanes c and d show that the preparation of the plasmid DNA is sufficiently clean for an efficient primer extension and that the bands seen in lanes a and b are due to the UV irradiation. **(B)** Superposition of line profiles from lanes f, h, and l, corresponding to the indicated concentrations of IHF. The intensities of the bands are in arbitrary units. **(C)** An equivalent superposition of the in vivo profiles and the in vitro profile corresponding to the 50-n*M* IHF lane shows that *E. coli* contains roughly the same amount of free IHF in stationary-phase cells as was present in the in vitro sample using 50 n*M* (total) IHF. As expected, the footprint in a strain lacking functional IHF shows the same profile as DNA alone in the in vitro reaction.

13. In vivo footprinting. Primer extension of the irradiated template can only be performed in vitro. It is therefore necessary to extract the irradiated DNA from the bacteria. Our current technology allows the measurement of protein binding to specific binding sites carried on a multicopy plasmid. In principle, a primer extension reaction on chromosomal DNA should work as well. In practice the signals obtained from chromosomal DNA are too weak. Increasing the number of samples irradiated does not remedy the problem. Because of the large excess of chromosomal DNA with respect to the primer extension product, we observe abnormal migration of the band in the sequencing gel.

    The sample for in vivo UV-laser footprinting must be prepared such that a single pulse of the laser (typically about 30 mJ per pulse, corresponding to 4×10^{16} photons [i.e., 67 nmol of photons]) delivers more photons than there are absorbing molecules in the sample. For an in vivo experiment, the absorbing molecules are mostly made up of cellular DNA and RNA, as well as free nucleoside phosphates. An upper estimate of the concentration of absorbing molecules within an *Escherichia coli* cell is about 100 m*M*, corresponding to 6×10^8 absorbers per cell. Because a single pulse delivers 4×10^{16} photons and because we want an excess of photons over absorbers, we want to irradiate less than about 10^8 *E. coli* cells per pulse. This numbers corresponds to about 100 µL of a suspension at 1 OD_{600}. A large number of 50-µL samples are therefore irradiated and the cells are frozen immediately after irradiation.

References

1. von Hippel, P. H. and Berg, O. G. (1986) On the specificity of DNA–protein interactions. *Proc. Natl. Acad. Sci. USA* **83(6),** 1608–1612.
2. Sauer, R. T. (1991) Protein–DNA interactions, in *Methods in Enzymology*, vol. 208, Academic Press, San Diego, CA.
3. Jost, J.-P. and Saluz, H. P. (1991) A laboratory guide to in vitro studies of protein-DNA interactions, in *Biomethods*, vol. 5, Birkhäuser Verlag, Basel.
4. Hockensmith, J. W., Kubasek, W. L., Vorachek, W. R., and von Hippel, P. H. (1993) Laser cross-linking of proteins to nucleic acids. I. Examining physical parameters of protein–nucleic acid complexes. *J. Biol. Chem.* **268,** 15,712–15,720.
5. Pashev, I. G., Dimitrov, S. I., and Angelov, D. (1991) Crosslinking proteins to nucleic acids by ultraviolet laser irradiation. *Trends Biochem. Sci.* **16,** 323–326.
6. Panyutin, I. G., Kovalsky, O. I., and Budowsky, E. I. (1989) Irradiation of the template with high-intensity (pulse-laser) ultraviolet light results in DNA–polymerase termination events at deoxyguanosine residues. *FEBS Lett.* **258,** 274–276.
7. Menshonkova, T. N., Simukova, N. A., Budowsky, E. I., and Rubin, L. B. (1980) The effect of high intensity ultraviolet irradiation on nucleic acids and their components. Cleavage of *N*-glycosidic bond in thymidine, adenosine and 2'-deoxyadenosine. *FEBS Lett.* **112,** 299–301.
8. Matsunaga, T., Hieda, K., and Nikaido, O. (1991) Wavelength dependent formation of thymine dimers and (6–4) photoproducts in DNA by monochromatic ultraviolet light ranging from 150 to 365 nm. *Photochem. Photobiol.* **54,** 403–410.

9. Hockensmith, J. W., Kubasek, W. L., Vorachek, W. R., Evertsz, E. M., and von Hippel, P. H. (1991) Laser cross-linking of protein-nucleic acid complexes. *Methods Enzymol.* **208,** 211–236.
10. Eichenberger, P., Dethiollaz, S., Buc, H., and Geiselmann, J. (1997) Structural kinetics of transcription activation at the *malT* promoter of *Escherichia coli* by UV laser footprinting. *Proc. Natl. Acad. Sci. USA* **94,** 9022–9027.
11. Buckle, M., Pemberton, I. K., Jacquet, M. A., and Buc, H. (1999) The kinetics of sigma subunit directed promoter recognition by *E. coli* RNA polymerase. *J. Mol. Biol.* **285,** 955–964.
12. Murtin, C., Engelhorn, M., Geiselmann, J., and Boccard, F. (1998) A quantitative UV laser footprinting analysis of the interaction of IHF with specific binding sites: re-evaluation of the effective concentration of IHF in the cell. *J. Mol. Biol.* **284,** 949–961.
13. Nash, H. A. (1996) The HU and IHF proteins: accessory factors for complex protein-DNA assemblies, in *Regulation of Gene Expression in* Escherichia coli (Lin, E. E. C. and Lynch, A. S., eds.), R.G. Landes Company, Austin, TX, pp. 149–179.
14. Rice, P. A., Yang, S., Mizuuchi, K., and Nash, H. A. (1996) Crystal structure of an IHF-DNA complex: a protein-induced DNA U-turn. *Cell* **87,** 1295–1306.
15. Engelhorn, M., Boccard, F., Murtin, C., Prentki, P., and Geiselmann, J. (1995) In vivo interaction of the *Escherichia coli* integration host factor with its specific binding sites. *Nucleic Acids Res.* **23,** 2959–2965.
16. Miller, J. H. (1992) *A Short Course in Bacterial Genetics.* Cold Spring Harbor Laboratory, Cold Spring Harbor, NY.
17. Buckle, M., Buc, H., and Travers, A. (1992) DNA deformation in nucleoprotein complexes between RNA polymerase, cAMP receptor protein and the lac UV5 promoter probed by singlet oxygen. *EMBO J.* **11,** 2619–2625.

13

In Vivo DNA Analysis

Régen Drouin, Jean-Philippe Therrien, Martin Angers, and Stéphane Ouellet

1. Introduction

The in vivo analysis of DNA–protein interactions and chromatin structure can provide several kinds of critical information regarding regulation of gene expression and gene function. For example, DNA sequences spanned by nuclease-hypersensitive sites or bound by transcription factors often correspond to genetic regulatory elements. Using the ligation-mediated polymerase chain reaction (LMPCR) technology it is possible to map such DNA sequences and to demonstrate the existence of unusual DNA structures directly in living cells. LMPCR analyses can thus be used as a primary investigative tool to identify the regulatory sequences involved in gene expression. Once specific promoter sequence sites are shown to be bound by transcription factors in living cells, it is often possible to establish the identity of these factors simply by comparison with the consensus binding sites of known factors such as Sp1, AP-1, NF-1, and so forth. The identity of each factor can then be confirmed using in vitro gel shift (electrophoretic mobility shift assay [EMSA]) or footprinting assays.

Clearly, gene promoters are best studied in their natural state in the living cell and, thus, it is not surprising that in vivo DNA footprinting is one of the most accurate predictors of the state of transcriptional activity of genes *(1–3)*. The native state of a gene and most of the special DNA structures are unavoidably lost when DNA is cloned or purified *(1–4)*. Hence, the commonly used in vitro methods, such as in vitro footprinting and EMSAs, cannot demonstrate that a given DNA–protein interaction actually occurs within the cells of interest. With the advent of in vivo DNA footprinting, in vitro studies have been extended to the situation in living cells, revealing the cellular processes impli-

From: *Methods in Molecular Biology, vol. 148: DNA–Protein Interactions: Principles and Protocols, 2nd ed.*
Edited by: T. Moss © Humana Press Inc., Totowa, NJ

cated in the regulation of gene expression. LMPCR is the method of choice for in vivo footprinting and DNA structure studies because it can be used to investigate complex animal genomes, including that of human. The quality and usefulness of the information obtained from any in vivo DNA analysis, however, depends on three parameters: (1) the integrity of the native chromatin substrate used in the experiment, (2) the structural specificity of the chromatin probe, and (3) the sensitivity of the assay. The ideal chromatin substrate is, of course, that found inside intact cells. However, a near-ideal chromatin substrate is still to be found in permeabilized cells, allowing the application of a wider range of DNA cleavage agents, including DNase I.

In vivo footprinting assesses the local reactivity of modifying agents on the DNA of living cells as compared to that on purified DNA (*see* **Figs. 1–4**). Two steps characterize an in vivo footprinting analysis: (1) the treatment of purified DNA and of cells with a given DNA modifying agent and (2) the visualization of nucleotide modifications on a DNA sequencing gel. The latter step requires that the modifying agent either directly induces DNA strand breaks or modifies DNA nucleotides such that strand breaks can subsequently be induced in vitro. A comparison is then made between the modification frequency on purified DNA and that on the DNA in living cells. For example, each guanine residue of purified DNA has a near-equivalent probability of being methylated by dimethylsulfate (DMS) and, thus, the cleavage pattern of in vitro modified DNA appears on a sequencing gel as a ladder of bands of roughly equal intensity. However, as a result of the presence of DNA-binding proteins, all guanine residues do not show the same accessibility to DMS in living cells (**Fig. 1**). Thus, differences between banding patterns obtained from in vitro and in vivo modified DNA can be used to infer the sites of protein binding in living cells. As will be seen, it is always advisable to validate such interpretations using more than one footprinting agent.

The step of visualizing in vivo footprints has historically been problematic because of the dilute nature of the sequences of interest and the complexity of the genomes of higher eukaryotes. The development of an extremely sensitive and specific technique, such as LMPCR, was thus necessary. The LMPCR tech-

---

Fig. 1. *(opposite page)* Overall scheme for in vivo DNA analysis using DMS. The methylation of guanine residues following DMS treatment of purified DNA (in vitro) and cells (in vivo) is shown by vertical arrows and methylated residues (Me). When purified DNA is treated with DMS, every guanine residue has a similar probability of being methylated. However, the guanine residue in intimate contact with a sequence-specific DNA-binding protein illustrated by the dotted oval is protected from DMS methylation, whereas the guanine residues localized close to the boundary of a DNA–protein contact that modifies DNA structure, allowing a better accessibility to DMS, is

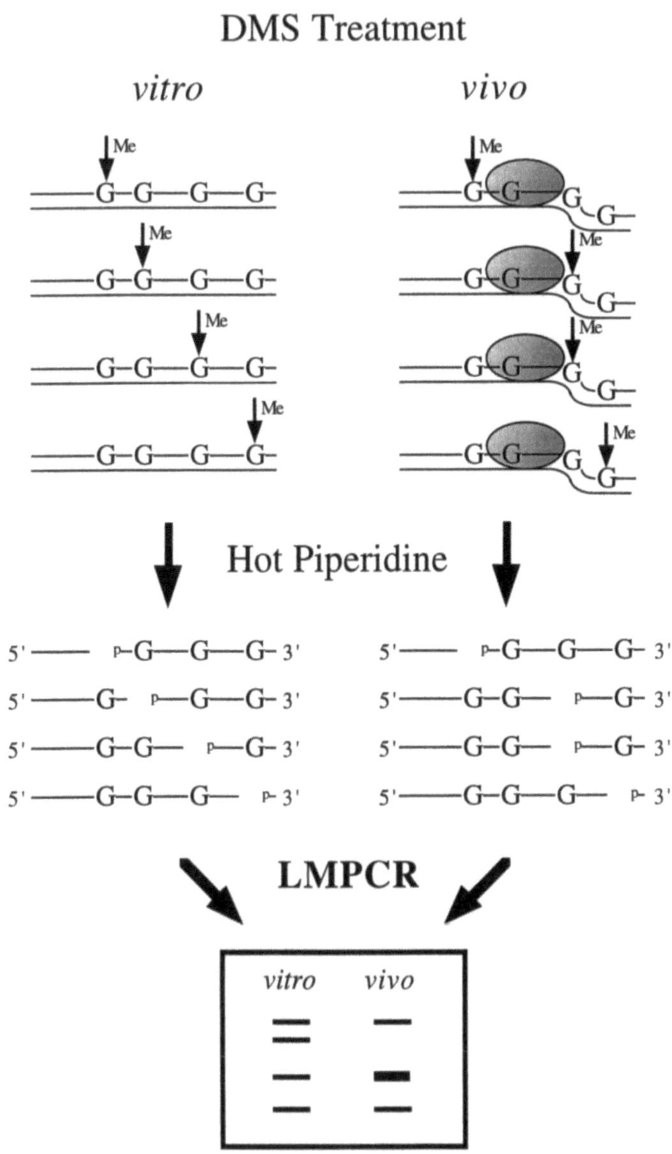

nethylated more frequently. The methylated guanine residues are cleaved by hot pip-
eridine leaving phosphorylated 5' ends. On the sequencing ladder following LMPCR, guanine residues that are protected from methylation appear as missing or less intense bands when compared with the sequencing ladder from the same DNA sequence obtained after DMS treatment of purified DNA. On the other hand, guanine residues that undergo enhanced DMS methylation appear as darker bands in the sequencing ladder relative to the purified DNA control.

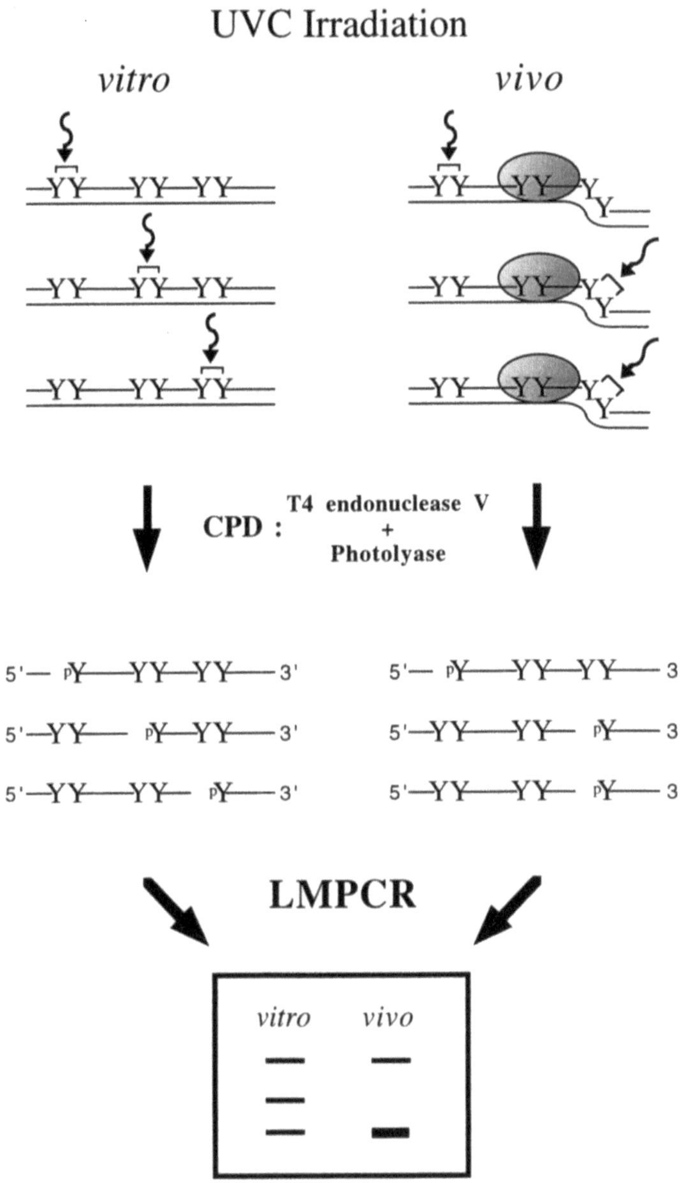

Fig. 2. Overall scheme for in vivo DNA analysis using UVC and CPD formation. The CPD formation following UVC exposure of purified DNA (in vitro) and cells (in vivo) is shown with curved arrows and brackets linking two adjacent pyrimidines (Y). When purified DNA is irradiated with UVC, the frequency of CPD formation at dipyrimidine sites is determined by the DNA sequence. However, the presence of a sequence-specific DNA-binding protein illustrated by the dotted oval as well as DNA structure can prevent (negative photofootprint) or enhance (positive photofootprint)

nique quantitatively maps single-strand DNA breaks having phosphorylated 5' ends within single-copy DNA sequences. It was first developed by Mueller and Wold *(5)* for DMS footprinting, and, subsequently, Pfeifer and colleagues adapted it to DNA sequencing *(6)*, methylation analyses *(1,6,7)*, DNase I footprinting *(2)*, nucleosome positioning *(2)*, and UV photofootprinting *(4,8)*. LMPCR can be combined with a variety of DNA-modifying agents used to probe the chromatin structure in vivo. It is our opinion that no single technique can provide as much information on the DNA–protein interactions and DNA structures existing within living cells as can LMPCR.

1.1. General Overview of LMPCR

Genomic sequencing techniques such as that developed by Church and Gilbert *(9)* can be used to map strand breaks in mammalian genes at nucleotide resolution. However, by incorporating an exponential amplification step, LMPCR (outlined in **Fig. 5**) constitutes a genomic sequencing method orders of magnitude more sensitive than the direct technique of Church and Gilbert. It uses 20 times less DNA than this latter technique to obtain a nucleotide-resolution banding pattern and allows short autoradiographic exposure times. The unique aspect of LMPCR is the blunt-end ligation of an asymmetric double-stranded linker (5' overhanging to avoid self-ligation or ligation in the wrong direction) onto the 5' end of each cleaved blunt-ended DNA molecule *(5,6)*. The blunt end is created by the extension of a gene-specific primer (primer 1 in **Fig. 5**) until a footprinting strand break is reached. Because the generated breaks will be randomly distributed along the genomic DNA and thus have 5' ends of unknown sequence, the asymmetric linker adds a common and known sequence to all 5' ends. This then allows exponential PCR amplification from an adjacent genomic sequence to that of the generated breaks using the longer oligonucleotide of the linker (linker-primer) and a second nested gene-specific primer (primer 2, *see* **Fig. 5**). After 20–22 cycles of PCR, the DNA fragments are size-fractionated on a sequencing gel. LMPCR preserves the quantitative representation of each fragment in the original population of cleaved molecules *(10–13)*, allowing quantification on a phosphorimager *(14–17)*. Thus, the band intensity pattern obtained by LMPCR directly reflects the frequency distribu-

---

CPD formation. The CPDs are cleaved by T_4 endonuclease V digestion and photolyase photoreactivation leaving phosphorylated 5' ends. On the sequencing ladder following LMPCR, the negative photofootprints appear as missing or less intense bands when compared with the sequencing ladder from the same DNA sequence obtained after UVC irradiation of purified DNA. On the other hand, positive photofootprints appear as darker bands in the sequencing ladder relative to the purified DNA control.

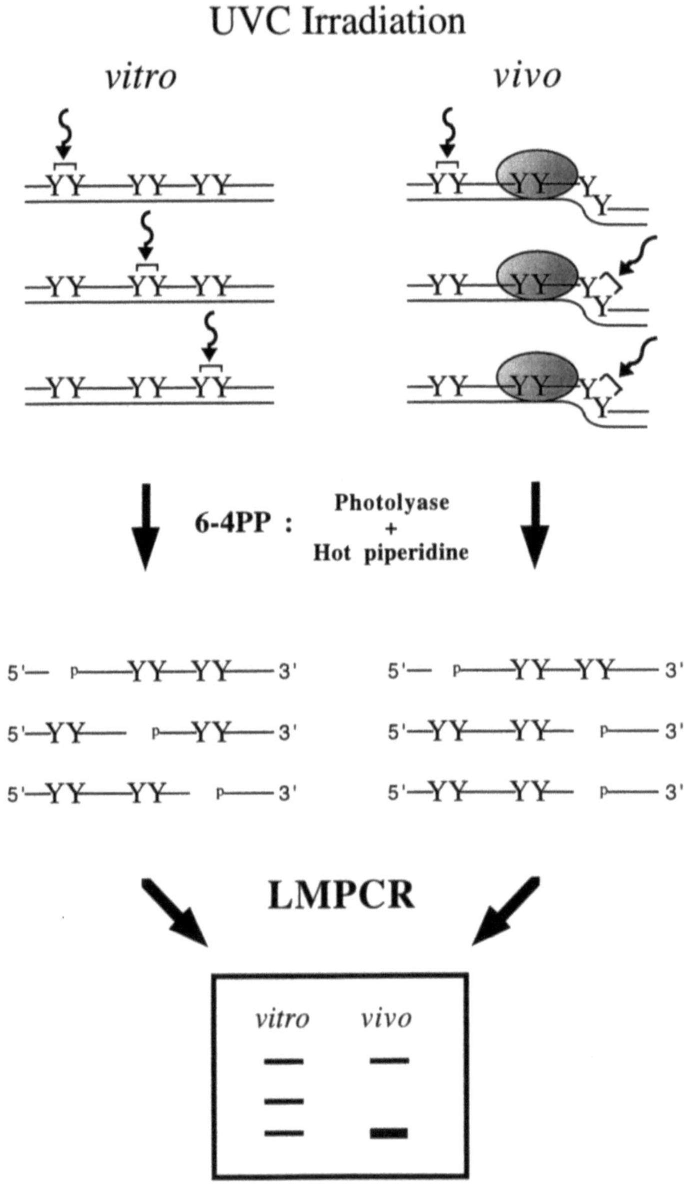

Fig. 3. Overall scheme for in vivo DNA analysis using UVC and 6–4PP formation. The 6–4PP formation following UVC exposure of purified DNA (in vitro) and cells (in vivo) is shown with curved arrows and brackets linking two adjacent pyrimidines (Y). When purified DNA is irradiated with UVC, the frequency of 6–4PP formation at dipyrimidine sites is determined by the DNA sequence. However, the presence of a sequence-specific DNA-binding protein illustrated by the dotted oval as well as DNA structure can prevent (negative photofootprint) or enhance (positive photofootprint)

tion of 5'-phosphoryl DNA breaks along a 200-bp sequence adjacent to the nested primer.

Two methods exist to reveal the sequence and footprinting ladders created by LMPCR. Pfeifer and colleagues *(6)* took advantage of electroblotting DNA onto a nylon membrane followed by hybridization with a gene-specific probe to reveal sequence ladders, otherwise known as "indirect end labeling". On the other hand, Mueller and Wold *(5)* used a nested third radiolabeled primer for the last one or two cycles of the PCR amplification step. We find Pfeifer's method much more sensitive than Mueller and Wold's (unpublished data). In this chapter, we will describe our LMPCR protocol as modified from the protocol of Pfeifer and colleagues.

In summary, LMPCR is the method of choice to study the in vivo structure of promoters with respect to the positions of DNA–protein interactions, of special DNA structures and of chromatin structures such as nucleosomes. To perform in vivo DNA analysis, three probing agents are regularly combined with LMPCR: DMS, ultraviolet (UV) and DNase I (**Figs. 1–4, Table 1**). These probing agents provide complementary information and each has its associated advantages and drawbacks (**Table 2**). To best characterize DNA–protein interactions, it is often necessary to use two or even all three of these methods. Treatments with any probing agents must produce either strand breaks or modified nucleotides that can be converted to DNA strand breaks with a 5'-phosphate in vitro (**Figs. 1–4, Table 3**). In this chapter, we describe protocols routinely used in our laboratory for DMS, UV, and DNase I in vivo treatments as well as the associated LMPCR technology. These protocols may also be adapted to footprinting with other probing agents, such as $KMnO_4$ and OsO_4 (*see* Chapters 6 and 9), although a detailed description is beyond the scope of the present chapter.

1.2. In Vivo Dimethylsulfate (DMS) Footprint Analysis (Fig. 1)

Dimethylsulfate is a small, highly reactive molecule that easily diffuses through the outer cell membrane and into the nucleus. It preferentially methylates not only the N7 position of guanine residues via the major groove but, to a lesser extent, also the N3 position of adenine residues via the minor groove.

6–4PP formation. First, CPDs are photoreactivated by photolyase and then 6–4PPs are cleaved by hot piperidine treatment leaving phosphorylated 5' ends. On the sequencing ladder following LMPCR, the negative photofootprints appear as missing or less intense bands when compared with the sequencing ladder from the same DNA sequence obtained after UVC irradiation of purified DNA. On the other hand, positive photofootprints appear as darker bands in the sequencing ladder relative to the purified DNA control.

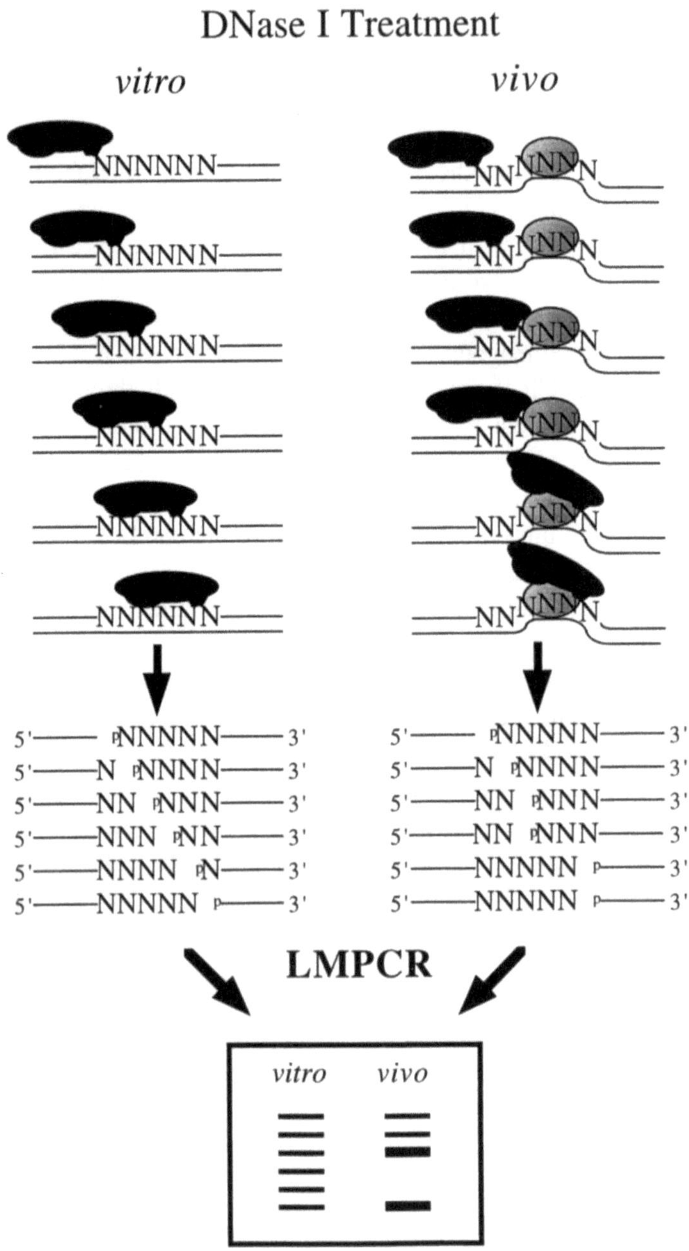

Fig. 4. Overall scheme for in vivo DNA analysis using DNase I. The DNase I enzyme (the solid black) digestion of purified DNA (in vitro) and cells (in vivo) is shown. When purified DNA is digested with DNase I, the cleavage pattern shows that sites of the nucleotide sequence have similar probabilities of being cleaved. However, the presence of a sequence-specific DNA-binding protein illustrated by the dotted oval

The most significant technical advantage of in vivo DMS footprinting is that DMS can be simply added to the cell culture medium, requiring no cell manipulation (*see* **Table 2** for advantages and drawbacks). Each guanine residue of purified DNA displays about the same probability of being methylated by DMS. Because DNA inside living cells forms chromatin and is often found associated with a number of proteins, it is expected that its reactivity toward DMS will differ from purified DNA. **Figures 6** and **7** show in vivo DMS treatment patterns compared to the treatment of purified genomic DNA. Proteins in contact with DNA either decrease accessibility of specific guanines to DMS (protection) or, as frequently observed at the edges of a footprint, increase reactivity (hyperreactivity) *(1)*. Hyperreactivity can also indicate a greater DMS accessibility of special in vivo DNA structures *(19)*. Hot piperidine cleaves the glycosylic bond of methylated guanines and adenines, leaving a ligatable 5'-phosphate *(20)*.

Genomic footprinting using DMS reveals DNA–protein contacts located in the major groove of the DNA double helix (**Table 1**). However, it should be noted that in vivo DNA studies using DMS alone may not detect some DNA–protein interactions *(21)*. First, no DNA–protein interaction will be detected in the absence of guanine residues. Second, some proteins do not affect DNA accessibility to DMS. Third, certain weak DNA–protein contacts could actually be disrupted because of the high reactivity of the DMS. Thus when using DMS, it is often important to also apply alternative footprinting approaches *(21,22)*.

1.3. Photofootprint Analysis (Figs. 2 and 3)

Ultraviolet light (UVC: 200–280 nm; UVB: 280–320 nm) can also be used as a modifying agent for in vivo footprinting *(4,8,23–25)*. When cells are subjected to UV light (UVC or UVB), two major classes of lesions may be introduced into the DNA at dipyrimidine sequences (CT, TT, TC, and CC): the cyclobutane pyrimidine dimer (CPD) and the pyrimidine (6–4) pyrimidone photoproduct (6–4PP) *(26)*. CPDs are formed between the 5,6 bonds of any two adjacent pyrimidines, whereas a stable bond between positions 6 and 4 of two adjacent pyrimidines characterizes 6–4PPs. 6–4PP are formed at a rate 15–30% of that of CPDs *(27)* and are largely converted to their Dewar valence

---

as well as DNA structure can prevent (protection) or enhance (hypersensitive) DNase I cleavage. The DNase I cleavage leaves phosphorylated 5' ends. On the sequencing ladder following LMPCR, DNA sequences that are protected from DNase I cleavage appear as missing or less intense bands when compared with the sequencing ladder from the same DNA sequence obtained after DNase I digestion of purified DNA. On the other hand, hypersensitive sites that undergo enhanced DNase I cleavage appear as darker bands in the sequencing ladder relative to the purified DNA control.

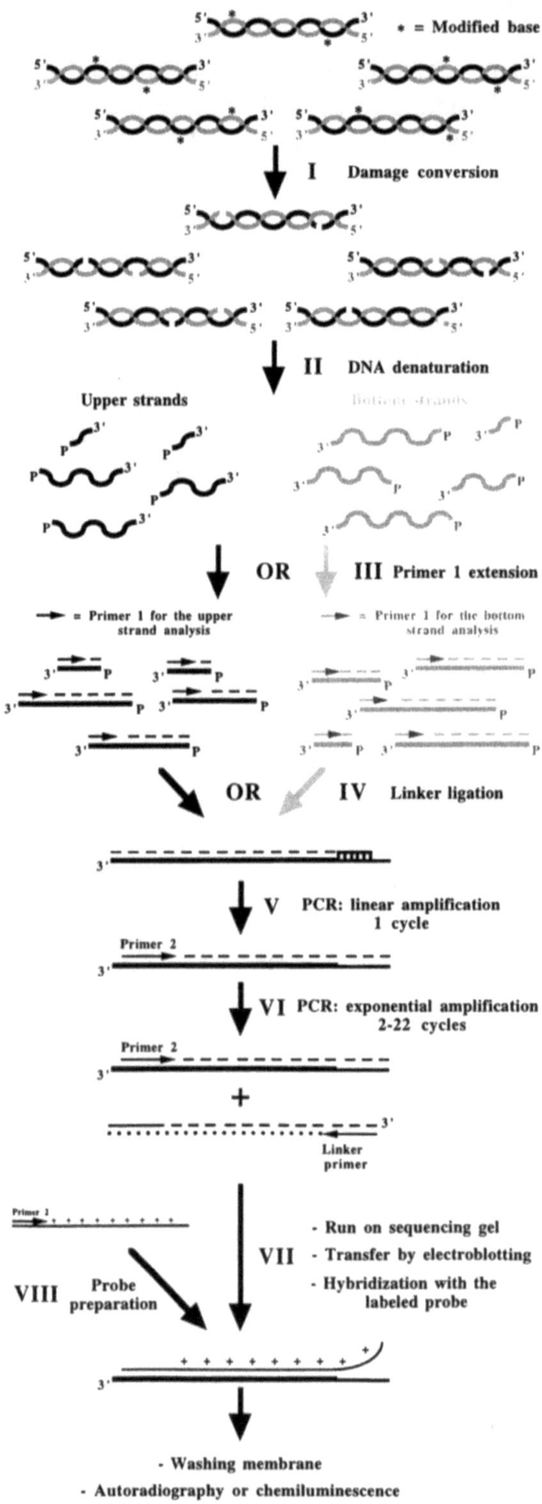
* = Modified base
I Damage conversion
II DNA denaturation
Upper strands
Bottom strands
OR
III Primer 1 extension
= Primer 1 for the upper strand analysis
= Primer 1 for the bottom strand analysis
OR
IV Linker ligation
V PCR: linear amplification 1 cycle
Primer 2
VI PCR: exponential amplification 2-22 cycles
Primer 2
+
Linker primer
VII
- Run on sequencing gel
- Transfer by electroblotting
- Hybridization with the labeled probe
VIII Probe preparation
- Washing membrane
- Autoradiography or chemiluminescence

isomers by direct secondary photolysis (photoisomerization) *(27)*. In living cells, the photoproduct distribution is determined both by sequence context and chromatin structure *(28)*. In general, CPDs and 6–4PPs appear to form preferentially in longer pyrimidine runs. Because UVB and UVC radiation are primarily absorbed in the cell by the DNA, there are relatively few perturbations of other cellular processes, and secondary events that could modify the chromatin structure or release DNA–protein interactions. Furthermore, intact cells are exposed for a short period of time only to a high-intensity UV irradiation. Thus, UV irradiation is probably one of the least disruptive footprinting method and, hence, truly reflects the in vivo situation (**Table 2**). As for DMS, DNA-binding proteins influence the distribution of UV photoproducts in a significant way *(23)*. When the photoproduct spectrum of irradiated purified DNA is compared with that obtained after irradiation of living cells, some striking differences become apparent. These are referred to as "photofootprints" *(23)*. The photoproduct frequency within sequences bound by sequence-specific DNA-binding proteins (transcription factors) is suppressed or enhanced in comparison to purified DNA *(4,8,29)*. Effects of chromatin structure may be significant in regulatory gene regions that bind transcription factors (***Fig. 6***). Mapping of CPDs at the single-copy gene level can reveal positioned nucleosomes because CPDs are modulated in a 10-bp periodicity within nucleosome core DNA *(30,31)*. 6–4PPs form more frequently in linker DNA than in core DNA *(32)*.

Photofootprints reveal variations in DNA structure associated with the presence of transcription factors or other proteins bound to the DNA. UV light has the potential to reveal all DNA–protein interactions provided there is a dipyrimidine sequence on either DNA strand within a putative protein-binding sequence. Because photofootprints can be seen outside protein-binding sites,

---

Fig. 5. *(previous page)* Outline of the LMPCR procedure. Step I: specific conversion of modified bases to phosphorylated 5' single-strand breaks; Step II: denaturation of genomic DNA; Step III: annealing and extension of primer 1 (although both strands can be studied, each LMPCR protocol only involves the analysis of either the nontranscribed strand or the transcribed strand); Step IV: ligation of the linker; Step V: first cycle of PCR amplification, this cycle is a linear amplification because only the gene-specific primer 2 can anneal; Step VI: cycles 2 to 22 of exponential PCR amplification of gene-specific fragments with primer 2 and the linker primer (the longer oligonucleotide of the linker); Step VII: separation of the DNA fragments on a sequencing gel, transfer of the sequence ladder to a nylon membrane by electroblotting, and visualization of the sequence ladder by hybridization with a labeled single-stranded probe; Step VIII: preparation and isotopic or nonisotopic labeling of single-stranded probe.

Table 1
Purposes of the Three Main In Vivo Footprinting Approaches

Approaches	Activities
1. Dimethylsulfate (DMS)	i. Localizes in vivo DNA–protein contacts located in the major groove of the DNA double helix ii. Can detect special DNA structures
2. UV irradiation (UVB or UVC)	i. Localizes in vivo DNA–protein interactions and shows how DNA structure is affected in the presence of transcription factors ii. Can detect special DNA structures iii. Can show evidence of positioned nucleosomes
3. DNase I	i. Localizes in vivo DNA–protein contacts ii. Precisely maps in vivo DNase I hypersensitive sites iii. Shows evidence of nucleosomes and their positions; can differentiate core DNA from linker DNA

UV light should not be used as the only in vivo footprinting agent. The precise delimitations of the DNA–protein contact are difficult to determine with the simple in vivo UV probing method.

The distribution of UV-induced CPDs and 6–4PPs along genomic DNA can be mapped at the sequence level by LMPCR following conversion of these photoproducts into ligatable 5'-phosphorylated single-strand breaks. CPD are enzymatically converted by cleavage with T_4 endonuclease V followed by UVA (320–400 nm) photoreactivation of the overhanging pyrimidine using photolyase (**Fig. 2**) *(8)*. Because the 6–4PPs and their Dewar isomers are hot alkali-labile sites, they can be cleaved by hot piperidine (**Fig. 3**) *(29)*.

1.4. In Vivo DNase I Footprint Analysis (Fig. 4)

DNase I treatment of permeabilized cells gives clear footprints when the DNase I-induced breaks are mapped by LMPCR *(2)*. Both living cells (in vivo) and purified DNA (in vitro) are treated with DNase I. As with DMS and UV, footprint analyses are obtained by comparing in vivo DNase I digestion patterns to patterns obtained from the digestion of purified genomic DNA (**Fig. 7**). When compared to purified DNA, permeabilized cells show protected bands at DNA–protein interaction sequences and DNase I hypersensitive bands in regions of higher-order nucleoprotein structure *(2)*. Compared to DMS, DNase I is less base selective, is more efficient at detecting minor groove DNA–protein contacts, provides more information on chromatin structure, displays larger and clearer footprints, and better delimits the boundaries of DNA–protein interactions (**Fig. 7**). The nucleotides covered by a protein are almost completely protected on both strands from DNase I nicking, allowing a better

Table 2
Advantages and Drawbacks of the Three Main In Vivo Footprinting Approaches

Approaches	Advantages	Drawbacks
DMS	Treatment is technically easy to carry out; the DMS is a small molecule that penetrates very easily into living cells with little disruption.	1. Requires guanines, therefore is sequence dependent. 2. Does not detect all DNA–protein interactions.
UV irradiation (UVB or UVC)	1. Treatment is technically easy to carry out; UV light penetrates through the outer membrane of living cells without disruption. 2. Detects many DNA–protein interactions. 3. Very sensitive to particular and DNA structures.	1. Requires two adjacent pyrimidines, therefore is sequence dependent. 2. The interpretation of the results is sometime difficult; to differentiate between DNA–protein interactions. and special DNA structures can be very difficult.
DNase I	1. Little sequence dependency. 2. No conversion of modified bases required. 3. Detects all DNA-protein contacts. 4. Very sensitive to particular DNA structures.	1. Technically difficult to carry out; reproducibility is often a problem. 2. DNase I is a protein that can penetrate in living cells only following membrane permeabilization, thus causing some cell disruption.

Table 3
Mapping Schemes Used with the Three Main In Vivo Footprinting Approaches

Approaches	Strand breaks	Modified bases	Conversion of modified bases to DNA single-strand breaks
DMS	Few	Guanine: methylated guanines at N7 position Adenine: to a much lesser extent, methylated adenines at N3 position	Hot piperidine
UV irradiation (UVB or UVC)	Very few	(i) Cyclobutane pyrimidine dimers (ii) 6–4 Photoproducts	(i) T_4 endonuclease V followed by photolyase (ii) Photolyase followed by hot piperidine
DNase I	Yes	None	No conversion required

delimitation of the boundaries of DNA–protein contacts. However, it should be underlined that the relatively bulky DNase I molecule cannot cleave the DNA in the immediate vicinity of a bound protein because of steric hindrance. Consequently, the regions protected from cutting can extend beyond the actual DNA–protein contact site. On the other hand, when DNA is wrapped around a nucleosome-size particle, DNase I cutting activity is increased at 10-bp intervals and no footprint is observed (**Tables 1** and **2**).

DNase I, a relatively large 31-kDa protein, cannot penetrate cells without previous cell-membrane permeabilization. Cells can be efficiently permeabilized by lysolecithin *(2)* or Nonidet P40 *(33)*. It has been shown that cells permeabilized by lysolecithin remain intact, replicate their DNA very efficiently, and show normal transcriptional activities *(34,35)*. There are numerous studies showing that lysolecithin-permeabilized cells maintain a normal nuclear structure to a greater extent than isolated nuclei, because the chromatin structure can be significantly altered during the nuclear isolation procedures *(2)*. Indeed, DNase I footprinting studies using isolated nuclei can be flawed because transcription factors are lost during the isolation of nuclei in polyamine containing buffers *(2)*. Even though other buffers may be less disruptive, factors can still be lost during the isolation procedure, leading to the loss of footprints or partial loss of footprints.

DNase I digestion of DNA leaves ligatable 5'-phosphorylated breaks, but the 3'-ends are free hydroxyl groups. Pfeifer and colleagues *(2,36)* observed that these genomic 3'-OH ends can be used as primers and be extended by the DNA polymerases during the initial extension and/or PCR steps of LMPCR, thereby reducing significantly the overall efficiency of LMPCR and giving a

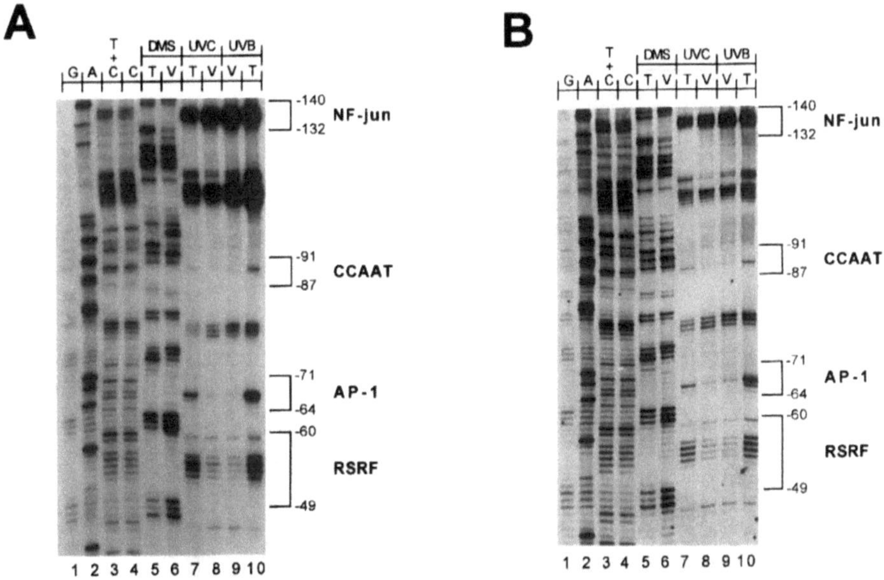

Fig. 6. LMPCR analysis of methylated guanines and CPD along the nontranscribed strand of the c-jun promoter following DMS treatment, and UVB and UVC irradiation respectively. **(A)** The membrane was hybridized with an isotopic [^{32}P]-dCTP-labeled probe. The membrane was exposed on film between two intensifying screens for 25 min at –70°C. **(B)** The membrane was hybridized with a digoxigenin-labeled probe and exposed on film for 40 min at room temperature. For this experiment, one LMPCR protocol was carried out and only one gel was run on which all the samples (20 in total) were loaded symmetrically in duplicate. Each symmetrical well of each set of samples was loaded with exactly the same amount of DNA. Lanes 1–4: LMPCR of DNA-treated with chemical cleavage reactions. These lanes represent the sequence of the c-*jun* promoter analyzed with JD primer set ***(18)***. Lanes 5–6: LMPCR of DMS-treated naked DNA (T: in vitro) and fibroblasts (V: in vivo) followed by hot piperidine treatment. Lanes 7–10: LMPCR of UVC- and UVB-irradiated naked DNA (T) and fibroblasts (V) followed by T_4 endonuclease V/photolyase digestion. On the right, the consensus sequences of transcription factor binding sites are delimited by brackets. The numbers indicate their positions relative to the major transcription initiation site.

background smear on sequencing gels. To avoid the nonspecific priming of these 3'-OH ends, three alternative solutions have been applied: (1) blocking these ends by the addition of a dideoxynucleotide ***(2,36)***; (2) enrichment of fragments of interest by extension product capture using biotinylated gene-specific primers and magnetic streptavidin-coated beads ***(18,37–39)***; and (3) performing primer 1 hybridization and primer 1 extension at a higher temperature (52–60°C vs 48°C, and 75°C vs 48°C, respectively) using a thermostable

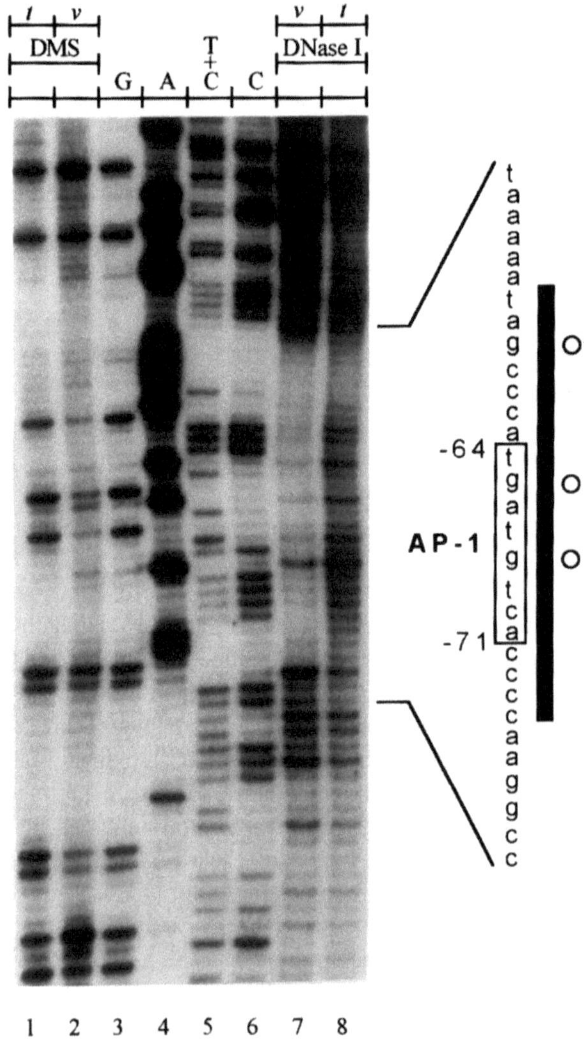

Fig. 7. LMPCR analysis of methylated guanines and DNA strand breaks along the transcribed strand of the c-jun promoter following DMS treatment and DNase I digestion, respectively. The membrane was hybridized with an isotopic [^{32}P]-dCTP-labeled probe. Lanes 1–2: LMPCR of DMS-treated purified DNA (t: in vitro) and fibroblasts (v: in vivo) followed by hot piperidine treatment. Lanes 3–6: LMPCR of DNA-treated with chemical cleavage reactions. These lanes represent the sequence of the c-*jun* promoter analyzed with JC primer set *(18)*. Lanes 7–8: LMPCR of DNase I-digested permeabilized fibroblasts (v) and purified DNA (t). As a reference, a small portion of the chemically derived sequence is shown on the right of the autoradiogram, the AP-1-like binding sequence is enclosed by a box, and the numbers indicate its position relative to the major transcription initiation site. Open circles represent guanines

enzyme such as *Vent* exo⁻ DNA polymerase and cloned *Pfu* DNA polymerase (*3,40–42*). Although effective, the first two alternatives involve additional manipulations that are time-consuming. Because of its simplicity, we select primer 1 with a higher T_m (52–60°C) and use the cloned *Pfu* DNA polymerase for the primer 1 extension.

1.5. Choice of DNA Polymerases for LMPCR

Ligation-mediated PCR involves the PCR amplification of a mixture of genomic DNA fragments of different size. During the LMPCR procedure, DNA polymerases are required for two steps: primer extension (PE) and PCR amplification. For the PE step, the best DNA polymerase would be one that (1) is thermostable and very efficient, (2) has no terminal transferase activity, (3) is able to efficiently polymerize about 0.75 kb of DNA even when the DNA is very GC rich, and (4) is able to polymerize through any DNA secondary structures. For the PCR step, the best DNA polymerase would be (1) thermostable, (2) very efficient, (3) able to amplify indiscriminately a mixture of DNA fragments of different lengths (between 50 and 750 bp) and of varying GC-richness (from 5% to 95%), and (4) able to efficiently resolve DNA secondary structures. We find cloned *Pfu* DNA polymerase that corresponds to *Pfu* exo⁻ is the best enzyme for the PE and PCR steps of LMPCR (*42*). In this chapter, LMPCR protocols using cloned *Pfu* DNA polymerase for PE and PCR steps will be described in detail. However, because the more frequently used combination of DNA polymerases is Sequenase™ 2.0 for the PE step and *Taq* DNA polymerase for the PCR step, a description of an alternative LMPCR protocol using Sequenase 2.0 and *Taq* DNA polymerase will also be included.

2. Materials

2.1. DNA Purification (for 10^7 to 10^8 cells)

1. Any types of cells (i.e., fibroblasts, lymphocytes, etc.).
2. Trypsin–EDTA (Gibco-BRL).
3. Hank's Balanced Salt Solution (HBSS) (Gibco-BRL).
4. Buffer A: 300 m*M* sucrose, 60 m*M* KCl, 15 m*M* NaCl, 60 m*M* Tris-HCl, pH 8.0, 0.5 m*M* spermidine, 0.15 m*M* spermine, and 2 m*M* EDTA. Store at –20°C.
5. Buffer A + 1% Nonidet P40. Store at –20°C.
6. Conical tissue culture tubes, 50 mL.
7. Buffer B: 150 m*M* NaCl and 5 m*M* EDTA pH 7.8.

---

that are protected against DMS-induced methylation (negative DMS footprints) in vivo. The black bar shows the protected sequence against DNase I-induced cleavage in vivo. Thus, in vivo DNase I footprinting analysis delimits much better the DNA–protein interactions.

8. Buffer C: 20 m*M* Tris-HCl, pH 8.0, 20 m*M* NaCl, 20 m*M* EDTA, and 1% sodium dodecyl sulfate (SDS).
9. Proteinase K from *Tritirachium album* (Roche Molecular Biochemicals).
10. RNase A from bovine pancreas (Roche Molecular Biochemicals).
11. Phenol, equilibrated with 0.1 *M* Tris-HCl, pH 8.0 (Roche Molecular Biochemicals, cat. no. 108-95-2).
12. Chloroform.
13. 5 *M* NaCl.
14. Precooled absolute ethanol (–20°C).
15. Precooled 70% ethanol (–20°C).
16. *N*-2-Hydroxyethylpiperazine-*N*'-2-ethanesulfonic acid (HEPES).
17. 4'–6-Diamidino-2-phenylindole (DAPI).
18. Nanopure H_2O should be used in making any buffers, solutions, and dilutions, unless otherwise specified.

2.2. Chemical Cleavage for DNA-Sequencing Products

1. Potassium tetrachloropalladate(II) (K_2PdCl_4, Aldrich).
2. K_2PdCl_4 solution: 10 m*M* K_2PdCl_4 and 100 m*M* HCl, pH 2.0 (adjusted with NaOH). Store at –20°C.
3. K_2PdCl_4 stop: 1.5 *M* sodium acetate, pH 7.0, and 1 *M* β-mercaptoethanol.
4. Dimethylsulfate (DMS, 99+%, Fluka). Considering its toxic and carcinogenic nature, DMS should be manipulated in a well-ventilated hood. DMS is stored under nitrogen at 4°C and should be replaced every 12 mo. DMS waste is detoxified in 5 *M* NaOH.
5. DMS buffer: 50 m*M* sodium cacodylate and 1 m*M* EDTA pH 8.0. Store at 4°C.
6. DMS stop: 1.5 *M* sodium acetate pH 7.0 and 1 *M* β-mercaptoethanol. Store at –20°C.
7. Hydrazine (Hz, anhydrous, Aldrich). Considering its toxic and carcinogenic potentials, Hz should be manipulated in a well-ventilated hood. Hz is stored under nitrogen at 4°C in an explosion-proof refrigerator and the bottle should be replaced at least every 6 mo. Hz waste is detoxified in 3 *M* ferric chloride.
8. Hz stop: 300 m*M* sodium acetate pH 7.0 and 0.1 m*M* EDTA. Store at 4°C.
9. 5 *M* NaCl.
10. 3 *M* Sodium acetate pH 7.0.
11. Precooled absolute ethanol (–20°C).
12. Precooled 80% ethanol (–20°C).
13. Dry ice.
14. Piperidine (99+%, Fluka or Sigma): 10 *M* stock diluted to 2 *M* with H_2O just before use by adding 250 µL stock under 1 mL H_2O in a 1.5-mL microtube on ice. Cap immediately to minimize evaporation. Considering its toxic and carcinogenic potentials, piperidine should be manipulated in a well-ventilated hood. Piperidine 10 *M* is stored at 4°C under nitrogen atmosphere.
15. Teflon tape.
16. Lock caps.
17. 3 *M* Sodium acetate pH 5.2.
18. 20 µg/µL glycogen.
19. Vacuum concentrator (SpeedVac concentrator, Savant).

2.3. Treatment of Purified DNA and Living Cells with Modifying Agents

2.3.1. DMS Treatment

1. DMS (99+%, Fluka).
2. Trypsin–EDTA (Gibco-BRL).
3. Hank's Balanced Salt Solution (HBSS).

2.3.2. 254-nm UV and UVB Irradiation

1. Germicidal lamp (254 nm) for UVC irradiation (Philips G15 T8, TUV 15W).
2. UVB light for UVB irradiation (Philips, FS20T12/UVB/BP).
3. UVX digital radiometer (Ultraviolet Products, Upland, CA).
4. 0.9% NaCl.
5. UV irradiation buffer: 150 m*M* KCl, 10 m*M* NaCl, 10 m*M* Tris-HCl, pH 8.0, and 1 m*M* EDTA.
6. Buffer A + 0.5% Nonidet P40. Store at –20°C.
7. Scraper.

2.3.3. DNase I Treatment

1. Deoxyribonuclease I (DNase I, Worthington biochemical corporation; 45A134).
2. Trypsin–EDTA (Gibco-BRL).
3. Hank's Balanced Salt Solution (HBSS).
4. L-α-Lysophosphatidylcholine (L-α-Lysolecithin).
5. Nonidet P40.
6. Solution I: 150 m*M* sucrose, 80 m*M* KCl, 35 m*M* HEPES, pH 7.4, 5 m*M* $MgCl_2$, and 0.5 m*M* $CaCl_2$.
7. Solution II: 150 m*M* sucrose, 80 m*M* KCl, 35 m*M* HEPES, pH 7.4, 5 m*M* $MgCl_2$, and 2 m*M* $CaCl_2$.
8. Conical tubes, 15 and 50 mL.
9. Buffer B: 150 m*M* NaCl and 5 m*M* EDTA, pH 7.8.
10. Buffer C: 20 m*M* Tris-HCl, pH 8.0, 20 m*M* NaCl, 20 m*M* EDTA, and 1% SDS.
11. Proteinase K from *Tritirachium album* (Roche Molecular Biochemicals).
12. RNase A from bovine pancreas (Roche Molecular Biochemicals).
13. Phenol (*see* **Subheading 2.1., item 11**).
14. Chloroform.
15. 5 *M* NaCl.
16. Precooled absolute ethanol (–20°C).
17. Precooled 80% ethanol (–20°C).

2.4. Conversion of Modified Bases to DNA Single-Strand Breaks

2.4.1. DMS-Induced Base Modifications

Piperidine (99+%, *see* **Subheading 2.2., item 14**).

2.4.2. UV-Induced Base Modifications

1. 10X dual buffer: 500 m*M* Tris-HCl, pH 7.6, 500 m*M* NaCl, and 10 m*M* EDTA.
2. 1 *M* 1,4-Dithiothreitol (DTT, Roche Molecular Biochemicals).
3. 5 mg/mL nuclease-free bovine serum albumine (BSA, Roche Molecular Biochemicals).
4. T_4 endonuclease V enzyme (Epicentre Technologies). The saturating amount of T_4 endonuclease V enzyme can be estimated by digesting UV-irradiated genomic DNA with various enzyme quantities and separating the cleavage products on alkaline agarose gel ***(43)***. The saturating amount of the enzyme is the next to the minimum quantity that produces the maximum cleavage frequency as evaluated on the alkaline agarose gel.
5. *E. coli* photolyase enzyme (Pharmingen). The saturating amount of photolyase can be estimated by photoreactivating UV-irradiated genomic DNA with various enzyme quantities, digestion with T_4 endonuclease V, and separating the cleavage products on alkaline agarose gel ***(43)***. The saturating amount of photolyase is the next to the minimum enzyme quantity which produces no cleavage following T_4 endonuclease V digestion as evaluated on the gel. Because photolyase is light sensitive, all steps involving photolyase should be carried out under yellow light.
6. UVA black light (UV F15T8BLB 360 nm, Philips, 15W).
7. Plastic film (plastic wrap).
8. 0.52% SDS solution.
9. Phenol (*see* **Subheading 2.1., item 11**).
10. Chloroform.
11. 5 *M* NaCl.
12. Precooled absolute ethanol (–20°C).
13. Precooled 80% ethanol (–20°C).
14. Piperidine (99+%, *see* **Subheading 2.2., item 14**).

2.5. Ligation-Mediated Polymerase Chain Reaction Technology

2.5.1. Primer Extension (Steps II and III, Fig. 5)

1. A gene-specific primer (primer 1) is used to initiate primer extension. The primer 1 used in the first-strand synthesis are 15- to 22-mer oligonucleotides and have a calculated melting temperature (T_m) of 50–60°C. They are selected using a computer program (Oligo 4.0 software, National Biosciences) ***(44)*** and, optimally, their T_m, as calculated by a computer program (GeneJockey software), should be about 10°C lower than that of subsequent primers (*see* **Note 1**) ***(45)***. The first-strand synthesis reaction is designed to require very little primer 1 with a lower T_m so that this primer does not interfere with subsequent steps ***(11–13,46)***. The primer 1 concentration is set at 50 μ*M* in H_2O and then diluted 1:100 in H_2O to give 0.5 pmol/μL.
2. Siliconized microtubes (0.625 μL) (National Scientific Supply Co, Inc.).
3. Thermocycler (PTC™, MJ research, Inc.).

4. 10X cloned *Pfu* buffer: 200 m*M* Tris-HCl, pH 8.8, 20 m*M* $MgSO_4$, 100 m*M* NaCl, 100 m*M* $(NH_4)_2SO_4$, 1% (v/v) Triton X-100, and 1 mg/mL nuclease-free BSA (*see* **Note 2**).
5. Cloned *Pfu* mix: 1.5 m*M* of each dNTP and 1.5 U cloned *Pfu* DNA polymerase, also named *Pfu* exo^- (2.5 U/µL, Stratagene).
6. 5X Sequenase buffer: 200 m*M* Tris-HCl, pH 7.7, and 250 m*M* NaCl.
7. Mg–dNTPs mix: 20 m*M* $MgCl_2$, 20 m*M* DTT, and 0.375 m*M* of each dNTP.
8. T7 Sequenase V.2 (Amersham).
9. 310 m*M* Tris-HCl, pH 7.7.

2.5.2. Ligation (Step IV, **Fig. 5**)

1. The DNA molecules that have a 5' phosphate group and a double stranded blunt end are suitable for ligation. A DNA linker with a single blunt end is ligated directionally onto the double-stranded blunt end of the extension product using T_4 DNA ligase. This linker has no 5' phosphate and is staggered to avoid self-ligation and provide directionality. Also, the duplex between the 25-mer (5' GCGGTGACCCGGGAGATCTGAATTC) and 11-mer (5' GAATTCAGATC) is stable at the ligation temperature, but denatures easily during subsequent PCR reactions ***(5,46)***. The linker was prepared in aliquots of 500 µL by annealing in 250 m*M* Tris-HCl, pH 7.7, 20 pmol/µL each of the 25-mer and 11-mer, heating at 95°C for 3 min, transferring quickly at 70°C, and cooling gradually to 4°C over a period of 3 h. Linkers are stored at –20°C and thawed on ice before use. Linker: L25 (60 pmol/µL, 5'-GCGGTGACCCGGGAGATCTGAATTC), L11 (60 pmol/µL, 5'-GAATTCAGATC), 2 *M* Tris-HCl, pH 7.7, and 1 *M* $MgCl_2$.
2. T_4 DNA ligase (1 U/µL, Roche Molecular Biochemicals).
3. Ligation mix: 30 m*M* DTT, 1 m*M* ATP, 83.3 µg/mL of BSA, 100 pmol of linker, and 3.25 U/microtube of T_4 DNA ligase. If cloned *Pfu* DNA polymerase was used for primer extension (step III, **Fig. 5**), the ligation mix is prepared by adding per microtube: 1.35 µL of 1 *M* DTT, 0.5 µL of 100 m*M* ATP, 0.15 µL of 5 µg/µL BSA, 1.1 µL of Tris-HCl, pH 7.4, 5.0 µL of 20 pmol/µL linker, 3.25 µL of 1 U/µL T_4 ligase, and 33.65 µL of H_2O. If Sequenase was used for primer extension (step III, **Fig. 5**), the ligation mix is prepared by adding per microtube: 1.35 µL of 1 *M* DTT, 0.5 µL of 100 m*M* ATP, 0.75 µL of 5 µg/µL BSA, 5.0 µL of 20 pmol/µL linker, 3.25 µL of 1 U/µL T_4 ligase, and 34.15 µL of H_2O.
4. 7.5 *M* ammonium acetate.
5. 0.5 *M* EDTA, pH 8.0.
6. 20 µg/µL glycogen.
7. Precooled absolute ethanol (–20°C).
8. Precooled 80% ethanol (–20°C).

2.5.3. Polymerase Chain Reaction (Steps V and VI, **Fig. 5**)

1. At this step, gene-specific fragments can be exponentially amplified because primer sites are available at each end of the target fragments (i.e., primer 2 on one end and the longer oligonucleotide of the linker on the other end). Primer 2 may

or may not overlap with primer 1. The overlap, if present, should not be more than seven to eight bases ***(11–13,46)***. Primer 2 is diluted in H_2O to give 50 pmol/µL.

2. 10X cloned *Pfu* buffer: 200 m*M* Tris-HCl, pH 8.8, 20 m*M* $MgSO_4$, 100 m*M* NaCl, 100 m*M* $(NH_4)_2SO_4$, 1% (v/v) Triton X–100, and 1 mg/mL nuclease-free BSA (*see* **Note 2**).
3. Cloned *Pfu* DNA polymerase, also named *Pfu* exo^- (2.5 U/µL, Stratagene).
4. Cloned *Pfu* DNA polymerase mix per microtube: 2X cloned *Pfu* buffer, 0.5 m*M* of each dNTP, 10 pmol of LP25 (Linker Primer), 10 pmol of primer 2, and 3.5 U of cloned *Pfu* DNA polymerase.
5. Mineral oil.
6. Cloned *Pfu* DNA polymerase stop: 1.56 *M* sodium acetate pH 5.2 and 20 m*M* EDTA.
7. Formamide loading dye: 94% formamide, 2 m*M* EDTA, pH 7.7, 0.05% xylene cyanole FF, and 0.05% bromophenol blue ***(11–13)***. The formamide loading dye is freshly premixed by adding 1 part H_2O to 2 parts formamide loading dye.
8. 5X *Taq* buffer: 50 m*M* Tris-HCl, pH 8.9, 200 m*M* NaCl, and 0.05% [w/v] gelatin (*see* **Note 2**).
9. *Taq* DNA polymerase (5 U/µL, Roche Molecular Biochemicals).
10. *Taq* DNA polymerase mix per microtube: 2X *Taq* buffer, 4 m*M* $MgCl_2$, 0.5 m*M* of each dNTP, 10 pmol LP25 (Linker Primer), 10 pmol primer 2, and 3 U *Taq* DNA polymerase.
11. *Taq* DNA polymerase stop: 1.56 *M* sodium acetate pH 5.2 and 60 m*M* EDTA.
12. Phenol (*see* **Subheading 2.1., item 11**) premixed with chloroform in a ratio of 92 µL of phenol for 158 µL of chloroform.
13. Precooled absolute ethanol (–20°C).
14. Precooled 80% ethanol (–20°C).

*2.5.4. Gel Electrophoresis and Electroblotting (Step VII, **Fig. 5**)*

1. 60-cm-long × 34.5-cm-wide sequencing gel apparatus (Owl Scientific).
2. Spacers (0.4-mm thick).
3. Plastic well-forming comb (0.4-mm thick, Bio-Rad).
4. 5X (0.5 *M*) Tris–borate–EDTA (TBE) buffer: 500 m*M* Tris, 830 m*M* boric acid, and 10 m*M* EDTA, pH 8.3. Use this stock to prepare 1X (100 m*M*) TBE buffer.
5. 8% Polyacrylamide, to prepare 1 L, add 77.3 g of acrylamide, 2.7 g of *bis*-acrylamide, 420.42 g of urea, and 200 mL of 0.5 *M* TBE dissolved in H_2O. Polyacrylamide solution should be kept at 4°C.
6. Gel preparation: Mix 100 mL of 8% polyacrylamide with 1 mL of 10% ammonium persulfate (APS) and 30 µL of *N,N,N′,N′*-tetra-methylethylenediamide (TEMED). This mix is prepared immediately before pouring the solution between the glass plates. Without delay, take the gel mix into a 50-mL syringe and inject the mix between the plates, maintaining a steady flow. During pouring, the plates should be kept at a 30° angle and tilted to the side into which the mix is injected. Any air bubbles should be avoided and removed if they form. The gel should be left to polymerize for a minimum of 3 h before use. If the gel is to be left over-

night, 45 min after pouring, place a moistened paper tissue over the comb, and cover the upper end of the assembly with a plastic film to prevent the gel from drying out.
7. Flat gel loading tips (National Scientific Supply Co).
8. Power supply (Bio-Rad PowerPac 3000).
9. Electroblotting apparatus (HEP3, Owl Scientific Inc.) used according to the manufacturer's instructions.
10. Whatman 3MM Chr paper (Fisher Scientific).
11. Plastic film (plastic wrap).
12. Whatman 17 Chr papers (Fisher Scientific).
13. Nylon membrane, positively charged (Roche Molecular Biochemicals, cat. no. 1 417 240).
14. Power supply (Bio-Rad, model 200/2.0).
15. UVC (254 nm) germicidal lamp.

2.5.5. Hybridization (Step VII, **Fig. 5**)

The hybridization is performed in a rolling 8-cm-diameter × 22 cm long borosilicate glass hybridization tubes in a hybridization oven (Hoefer). The nylon membrane is soaked in 100 m*M* TBE and, using a 25-mL pipet, placed in the tube so that the membrane sticks completely to the wall of the hybridization tube. Following hybridization and washing, the membranes are placed in an autoradiography cassette FBAC 1417 (Fisher Scientific) and exposed to Kodak X-ray film (XAR-5, 35 × 43 cm, Kodak Scientific Imaging Film) with intensifying screens (35 × 43 cm, Fisher Scientific, cat. no. FB-IS-1417) at –70°C when a radiolabeled probe has been hybridized and without intensifying screens at room temperature when a digoxigenin-labeled probe has been hybridized.

2.5.5.1. Radiolabeled Probe

1. Hybridization buffer: 250 m*M* sodium phosphate pH 7.2, 1 m*M* EDTA, 7% SDS, and 1% BSA.
2. Radiolabeled probe diluted in 6–7 mL of hybridization buffer.
3. Washing buffer I: 20 m*M* sodium phosphate pH 7.2, 1 m*M* EDTA, 0.25% BSA, and 2.5% SDS.
4. Washing buffer II: 20 m*M* sodium phosphate pH 7.2, 1 m*M* EDTA, and 1% SDS.
5. Plastic film (plastic wrap).

2.5.5.2. Digoxigenin-Labeled Probe

1. Prehybridization buffer: 5X SSC (750 m*M* NaCl and 75 m*M* sodium citrate pH 7.0), 1% casein, 0.1% *N*-lauroylsarcosin, and 0.02% SDS.
2. Digoxigenin-labeled probe diluted in 15 mL of prehybridization buffer (use only 7.5 mL for hybridization).
3. 2X washing solution: 2X SSC and 0.1% SDS.

4. 0.5X washing solution: 0.5X SSC and 0.1% SDS.
5. Buffer 1: 150 m*M* NaCl and 100 m*M* maleic acid, pH 7.5.
6. Buffer 2: buffer 1 + 1% (w/v) casein.
7. Antidigoxigenin antibodies (Roche Molecular Biochemicals).
8. Buffer 1 + 0.3% Tween-20.
9. Buffer 3: 100 m*M* Tris-HCl, pH 9.5, 100 m*M* NaCl, and 50 m*M* $MgCl_2$.
10. CSPD® [disodium 3-(4-methoxyspiro {1,2-dioxetane–3,2'-(5' chloro)tricyclo [3.3.1.$1^{3,7}$]decan}–4-yl)phenyl phosphate] substrate (Roche Molecular Biochemicals, cat. no. 1 655 884).
11. Acetate sheets.
12. Doubleseal (Model 855, Decosonic).

2.6. Preparation of Single-Stranded Hybridization Probes (Step VIII, Fig. 5)

2.6.1. Template Preparation: PCR Products

2.6.1.1. PCR Amplification

1. 5X *Taq* buffer: 50 m*M* Tris-HCl, pH 8.9, 200 m*M* NaCl, and 0.05% (w/v) gelatin (*see* **Note 2**).
2. *Taq* DNA polymerase (5 U/µL, Roche Molecular Biochemicals).
3. One primer 2 (50 pmol/µL) for each strand of the DNA fragment to be amplified distant from 150 to 450 bp.
4. *Taq* DNA polymerase mix per microtube: 2X *Taq* buffer, 4 m*M* $MgCl_2$, 0.4 m*M* of each dNTP, 10 pmol of each primer 2, and 3 U *Taq* DNA polymerase.
5. Mineral oil.
6. *Taq* DNA polymerase stop: 1.56 *M* sodium acetate, pH 5.2, and 60 m*M* EDTA.
7. Phenol (*see* **Subheading 2.1., item 11**) premixed with chloroform in a ratio of 92 µL of phenol to 158 µL of chloroform.
8. Precooled absolute ethanol (–20°C).
9. Precooled 80% ethanol (–20°C).
10. 5X TAE loading buffer: 5X TAE (200 m*M* Tris base, 100 m*M* glacial acetic acid, and 5 m*M* EDTA, pH 8.0), 0.025% bromophenol blue, 30% Ficoll 400, and 2% SDS.

2.6.1.2. Purification and Quantification of PCR Products

1. Agarose.
2. 1X TAE buffer: 40 m*M* Tris base, 20 m*M* glacial acetic acid, and 1 m*M* EDTA, pH 8.0.
3. DNA size standards (ϕX 174 RF, Canadian life technologies, cat. no. 15611-015).
4. Ethidium bromide.
5. Siliconized microtubes (0.625 mL) and 1.5-mL microtubes.
6. Glass wool.
7. 3 *M* sodium acetate, pH 7.0.
8. Precooled absolute ethanol (–20°C).

9. Precooled 80% ethanol (–20°C).
10. Low DNA mass ladder (Gibco-BRL, cat. no. 10068-013).
11. 5X universal neutral loading buffer: 0.25% bromophenol blue, 0.25% xylene cyanol FF, and 30% glycerol in H_2O. Store at 4°C.

2.6.2. Labeling of Single-Strand Hybridization Probes

2.6.2.1. Isotopic Labeling

1. Siliconized microtubes (0.625 mL) and 1.5-mL microtubes.
2. 5X *Taq* buffer: 50 m*M* Tris-HCl, pH 8.9, 200 m*M* NaCl, and 0.05% (w/v) gelatin (*see* **Note 2**).
3. 100 m*M* $MgCl_2$.
4. DNA templates: PCR products (10 ng/µL) or DNA plasmids (20 ng/µL).
5. Primer 2 (50 pmol/µL).
6. dNTP (dATP, dGTP, dTTP) mix (200 µ*M* of each).
7. dNTP (dATP, dGTP, dTTP) mix diluted 1:10 in H_2O. This mix is changed every 2 wk.
8. *Taq* DNA polymerase (5 U/µL, Roche Molecular Biochemicals).
9. α-[^{32}P]dCTP (3000 Ci/mmol, New England Nuclear).
10. 7.5 *M* ammonium acetate.
11. 20 µg/µL glycogen
12. Precooled absolute ethanol (–20°C)
13. Geiger counter.
14. TE buffer pH 8.0: 10 m*M* Tris-HCl, pH 8.0, and 1 m*M* EDTA, pH 7.8.
15. Hybridization buffer: 250 m*M* sodium phosphate, pH 7.2, 1 m*M* EDTA, 7% SDS, and 1% BSA.

2.6.2.2. Digoxigenin (Nonisotopic) Labeling

1. Siliconized microtubes (0.625 mL) and 1.5-mL microtubes.
2. 5X *Taq* buffer: 50 m*M* Tris-HCl, pH 8.9, 200 m*M* NaCl, and 0.05% (w/v) gelatin (*see* **Note 2**).
3. 100 m*M* $MgCl_2$.
4. DNA templates: PCR products (10 ng/µL) or DNA plasmids (20 ng/µL).
5. Primer 2 (50 pmol/µL).
6. dNTP mix (A:G:C:T = 25 m*M* : 25 m*M* : 25 m*M* : 20 m*M*).
7. dNTP mix diluted 1:8.3 in H_2O. This mix is changed every 2 wk.
8. 1 m*M* digoxigenin-11-dUTP (Roche Molecular Biochemicals) diluted 1:2 in H_2O.
9. *Taq* DNA polymerase (5 U/µL, Roche Molecular Biochemicals).
10. 7.5 *M* ammonium acetate.
11. 20 µg/µL glycogen.
12. Precooled absolute ethanol (–20°C).
13. TE buffer pH 8.0: 10 m*M* Tris-HCl, pH 8.0, and 1 m*M* EDTA, pH 7.8.
14. Prehybridization buffer: 5X SSC (750 m*M* NaCl and 75 m*M* sodium citrate pH 7.0), 1% casein, 0.1% *N*-lauroylsarcosine, and 0.02% SDS.

3. Methods

3.1. DNA Purification (for 10^7 to 10^8 Cells)

1. Detach cells using trypsin (if needed) and sediment the cell suspension by centrifugation in 50-mL conical tubes.
2. Resuspend the cells in 5–15 mL of buffer A.
3. Add 1 volume (5–15 mL) of buffer A containing 1% Nonidet P40.
4. Incubate on ice for 5 min.
5. Sediment nuclei by centrifugation at 4500*g* for 10 min at 4°C.
6. Remove the supernatant. Resuspend nuclei in 1–10 mL of buffer A by gentle vortexing. Resediment nuclei at 4500*g* for 10 min at 4°C.
7. Remove supernatant. It is recommended to leave a small volume (100–500 μL) of buffer A to facilitate resuspension of nuclei.
8. Dilute the nuclei in 1–2 mL of buffer B.
9. Add an equivalent volume of buffer C and proteinase K to a final concentration of 450 μg/mL.
10. Incubate at 37°C for 3 h, shake occasionally (*see* **Note 3**).
11. Add RNase A to a final concentration of 150 μg/mL.
12. Incubate at 37°C for 1 h.
13. Purify DNA by extraction with 1 vol phenol (one to two times as needed), 1 vol phenol:chloroform (one to two times as needed), and 1 vol chloroform. Phenol extraction and phenol-chloroform extraction should be repeated if the aqueous phase is not clear (*see* **Note 3**).
14. Precipitate DNA in 200 m*M* NaCl and 2 vol of precooled absolute ethanol. Ethanol should be added slowly and to facilitate DNA recovery, rock the tube very gently.
15. Recover DNA by spooling the floating DNA filament with a micropipet tip. If DNA is in small pieces or not clearly visible, recover DNA by centrifugation (5000*g* for 30 min at 4°C), but expect RNA contamination (*see* **Note 4**). RNA contamination does not cause any problems for LMPCR. RNase digestion can be repeated if needed.
16. Remove supernatant and wash DNA once with 10 mL of 70% ethanol.
17. Centrifuge the DNA (5000*g* for 10 min at 4°C).
18. Remove supernatant and air-dry DNA pellet.
19. Dissolve DNA in 10 m*M* HEPES, pH 7.4, and 1 m*M* EDTA (HE buffer) at an estimated concentration of 60–100 μg/mL. The quantity of DNA can be estimated based upon the number of cells that were initially used for DNA purification. About 6 μg of DNA should be purified from 1×10^6 cells.
20. Carefully measure DNA concentration by spectrophotometry at 260 nm. Alternatively, DNA can be measure by fluorometry after staining with DAPI. Only double-strand DNA concentration has to be measured, be careful if there is RNA contamination (*see* **Note 5**).

3.2. Chemical Cleavage for DNA Sequencing Products

In vivo DNA analysis using LMPCR requires complete DNA sequencing ladders from genomic DNA. Base-specific chemical modifications are performed according to Iverson and Dervan *(47)* for the A reaction and Maxam and Gilbert for the G, T+C, and C reactions. DNA from each of these base-modification reactions is processed by LMPCR concomitantly with the analyzed samples and loaded in adjacent lanes on the sequencing gel to allow the identification of the precise location and sequence context of footprinted regions. The chemical modifications induced by DMS, Hz, and K_2PdCl_4 and cleaved by piperidine destroy the target base. Therefore, one must bear in mind that when analyzing a chemical-sequencing ladder, each band corresponds to a DNA fragment ending at the base preceding the one read. In this section, we will describe the chemical sequencing of genomic DNA. The cleavage protocol below works optimally with 10–50 μg of genomic DNA per microtube. Before chemical sequencing, the required amount of DNA per microtube is ethanol precipitated and the pellet is air-dried. For each base-specific reaction, we usually carried out the treatment in three microtubes containing 50 μg of genomic DNA for three different incubation times with the modifying agent in order to obtain low, medium and high base-modification frequencies.

3.2.1. A Reaction

1. Add 160 μL of H_2O and 40 μL of K_2PdCl_4 solution to the DNA pellet and carefully mix on ice using a micropipet.
2. Incubate at 20°C for 5, 10, or 15 min.
3. Add 50 μL of K_2PdCl_4 stop.
4. Add 750 μL of precooled absolute ethanol.

3.2.2. G Reaction

1. Add 5 μL of H_2O, 200 μL of DMS buffer, and 1 μL of DMS to the DNA pellet and carefully mix on ice using a micropipet.
2. Incubate at 20°C for 30, 45, or 60 s.
3. Add 50 μL of DMS stop.
4. Add 750 μL of precooled absolute ethanol.

3.2.3. T+C Reaction

1. Add 20 μL of H_2O and 30 μL of Hz to the DNA pellet and carefully mix on ice using a micropipet.
2. Incubate at 20°C for 120, 210, or 300 s.
3. Add 200 μL of Hz stop.
4. Add 750 μL of precooled absolute ethanol.

3.2.4. C Reaction

1. Add 5 μL of H_2O, 15 μL of 5 *M* NaCl, and 30 μL of Hz to the DNA pellet and carefully mix on ice using a micropipet.
2. Incubate at 20°C for 120, 210, or 300 s.

3. Add 200 µL of Hz stop.
4. Add 750 µL of precooled absolute ethanol.

All samples are processed as follows:

1. Mix samples well and place on dry ice for 15 min.
2. Centrifuge for 15 min at 15,000*g* at 4°C.
3. Remove supernatant, then recentrifuge for 1 min and remove all the liquid using a micropipet.
4. Carefully dissolve pellet in 405 µL of H_2O.
5. Add 45 µL of 3 *M* sodium acetate pH 7.0.
6. Add 1 mL of precooled absolute ethanol.
7. Leave on dry ice for 15 min.
8. Centrifuge for 15 min at 15,000*g* at 4°C.
9. Take out supernatant and then respin.
10. Wash with 1 mL of precooled 80% ethanol; spin 5 min at 15,000*g* in a centrifuge at 4°C.
11. Remove the supernatant, spin quickly, remove the liquid with a micropipet and air-dry pellet.
12. Dissolve pellet in 50 µL of H_2O, add 50 µL of freshly prepared 2 *M* piperidine, and mix well using a micropipet.
13. Secure caps with Teflon™ tapes and lock the caps with "lock caps".
14. Incubate at 82°C for 30 min.
15. Pool all three microtubes of the same chemical reaction in a new 1.5-mL microtube.
16. Add H_2O until a volume of 405 µL is reached, then add 10 µL of 3 *M* sodium acetate pH 5.2, 1 µL of glycogen, and 1 mL of precooled absolute ethanol.
17. Leave on dry ice for 15 min.
18. Spin 10 min at 15,000*g* at 4°C.
19. Take out the supernatant and wash twice with 1 mL of precooled 80% ethanol, then respin for 1 min and remove all the liquid using a micropipet.
20. Add 200 µL of H_2O and remove traces of remaining piperidine by drying the sample overnight in a Speedvac concentrator.
21. Dissolve DNA in H_2O to a concentration of 0.5 µg/µL.
22. Determine the DNA strand break frequency by running the samples on a 1.5% alkaline agarose gel ***(43)***. The size range of the fragments should span 100–500 bp.

3.3. Treatment of Purified DNA and Cells with Modifying Agents

3.3.1. DMS Treatment

1. If cells are grown to confluence as monolayer, replace the cell culture medium with a freshly prepared serum-free medium containing 0.2% DMS and incubate at room temperature for 6 min. If cells are grown in suspension, sediment the cells by centrifugation and remove the cell culture medium. The cells are diluted

in a freshly prepared serum-free medium containing 0.2% DMS and are then incubated at room temperature for 6 min.

2. Remove the DMS-containing medium and quickly wash the cell monolayer with 10–20 mL of cold HBSS. Sediment cells by centrifugation if they are treated in suspension and remove the DMS-containing medium and wash the cells with 10 mL of cold HBSS.
3. Detach cells using trypsin for cells grown as monolayer.
4. Nuclei are isolated and DNA purified as described in **Subheading 3.1.**
5. Purified DNA obtained from the same cell type is treated as described in **Subheading 3.2.2.** Usually, a DMS treatment of 45 s should give a break frequency corresponding to that of the in vivo treatment described in this section. This DNA is the in vitro treated DNA used to compare with DNA DMS-modified in vivo (*see* **Notes 5** and **6**).

3.3.2. 254-nm UV and UVB Irradiation

1. If cells are grown as monolayer in Petri dishes, replace cell culture medium with cold 0.9% NaCl. If cells are grown in suspension, sediment the cells by centrifugation and remove the cell culture medium. The cells are diluted in cold 0.9% NaCl at a concentration of 1×10^6 cells/mL (*see* **Note 7**) and, to avoid cellular shielding, a thin layer of the cell suspension is placed in 150-mm Petri dishes.
2. Expose the cells to 0.5–2 kJ/m^2 of UVC (254-nm UV) or 25–100 kJ/m^2 of UVB. The cells should be exposed on ice in uncovered Petri dishes. The UV intensity is measured using a UVX digital radiometer.
3. Remove the 0.9% NaCl by aspiration for cells grown as monolayer in Petri dishes or by sedimentation for cell suspensions.
4. If cells were irradiated in suspension; follow the procedure described in **Subheading 3.1.** to isolate nuclei and purify DNA. After DNA purification, DNA is dissolved in H_2O at a concentration of 0.2 µg/µL. For cells cultured in Petri dishes, add in each dish 8 mL of buffer A containing 0.5% Nonidet P40.
5. Incubate on ice for 5 min.
6. Scrape the cells and transfer them in a conical 50-mL tube. In the same conical 50-mL tube, pool cells from Petri dishes that undergo the same procedure.
7. Wash the dishes twice with 8 mL of buffer A + 0.5% Nonidet P40 per each of three identical Petri dishes.
8. Continue from **step 5** of **Subheading 3.1.** After DNA purification, DNA is dissolved in H_2O at a concentration of 0.2 µg/µL.
9. Expose purified DNA to the same UVC or UVB dose as the cells. Purified DNA should be irradiated on ice and diluted in the UV irradiation buffer at a concentration of 60–75 µg/mL (*see* **Note 6**). Purified DNA should be obtained from the same type of cells as the type irradiated in vivo (*see* **Note 8**). This DNA is used as control DNA to compare with DNA UV modified in vivo (*see* **Notes 6** and **7**).
10. Following UV irradiation, DNA is ethanol precipitated and DNA is resuspended in H_2O at a concentration of 0.2 µg/µL.

3.3.3. DNase I Treatment

Genomic footprinting with DNase I requires cell permeabilization (*see* **Note 9**). Cells grown as a monolayer can be permeabilized while they are still attached to the Petri dish or in suspension following trypsinization. Here, we will describe cell permeabilization using lysolecithin applied to monolayer cell cultures (steps labeled **a**). For monolayer cultures, cells are grown to about 80% of confluency. Alternatively, we describe cell permeabilization using lysolecithin or Nonidet P40 applied to cells in suspension (steps labeled **b**). For cells in suspension, cells are diluted at a concentration of approx 1×10^6 cells/mL. To permeabilize the vast majority of cells in suspension, they must not be clumped and not form aggregates during the permeabilization step and subsequent DNase I treatment. To achieve this, we gently flick the microtubes during permeabilization and DNase I treatment and keep the cell concentration below 2×10^6/µL.

1a. For cells in monolayers, permeabilize the cells by treating them with 4 mL of 0.05% lysolecithin in solution I (prewarmed) at 37°C for 1–2 min ***(48)***.
2a. Remove the lysolecithin and wash with 10 mL of solution I. Add 3 mL of DNase I (2–4 U/mL) to solution II and incubate at room temperature for 3–5 min. DNase I concentration and incubation times may have to be adjusted for different cell types. During this incubation, no more than 10% of the cells should be released from the dish.
3a. Stop the reaction and lyse the cells by removal of the DNase I solution and addition of 1.5 mL of buffer C containing 600 µg/mL of proteinase K. Add 1.5 mL of buffer B and mix gently by rocking the flask or the Petri dish. Transfer lysis solution to a 15-mL tube (then continue to **step 4**).

Alternatively:

1b. Sediment the cell suspension by centrifugation. Wash the cells with HBSS. Resuspend the cells in solution II at a concentration of 20×10^6/mL and aliquote by transferring 100 µL of the cell suspension per 1.5-mL microtube. Add to each microtube 100 µL of solution II prewarmed at 37°C containing 0.1% lysolecithin or 0.25% Nonidet P40. Mix gently by flicking. Incubate at room temperature for 3 min.
2b. Quickly spin to pellet the cells. Add 50 µL of 2000 U/mL DNase I and mix gently by flicking. Incubate at room temperature for 5 min.
3b. Quickly spin and remove supernatant, resuspend the cells in 1.5 mL of buffer B, and, using a pipet, rapidly transfer to a 15-mL tube in which there are and 1.5 mL of buffer C containing 600 µg/mL of proteinase K (then continue to **step 4**).

4. Incubate at 37°C for 3 h, shake occasionally.
5. Add RNase A to a final concentration of 200 µg/mL and incubate at 37°C for 1 h.
6. Purify DNA by phenol-chloroform extraction (*see* **Subheading 3.1., step 13**).
7. Precipitate DNA in 200 m*M* NaCl and 2 volumes of precooled absolute ethanol.

8. Leave on dry ice for 20 min.
9. Recover DNA by centrifugation (5000*g* for 30 min at 4°C), but expect RNA contamination. RNA contamination does not cause any problems for LMPCR. RNase A digestion can be repeated if needed.
10. Remove supernatant and wash DNA once with 10 mL of precooled 80% ethanol.
11. Centrifuge the DNA (5000*g* for 10 min at 4°C). Remove supernatant and air-dry DNA pellet.
12. Dissolve DNA in H_2O and carefully measure DNA concentration (*see* **Subheading 3.1., step 20**).
13. To obtain purified DNA controls (*see* **Notes 6** and **8**), digest 50 µg of purified DNA in solution II with 4–8 U/µL of DNase I at room temperature for 10 min. Stop the reaction by adding 400 µL of phenol. Extract once with phenol–chloroform and once with chloroform. Dissolve DNA in H_2O at a concentration of 0.5 µg/µL.

3.4. Conversion of Modified Bases to DNA Single-Strand Breaks

When purified DNA or cells are treated with DMS and UV, DNA base modifications are induced (**Table 3**). These modifications must be converted to single-strand breaks before running LMPCR. Following UV exposure, CPDs and 6–4PPs are converted individually because they use different conversion procedures (**Table 3**). On the other hand, DNase I digestion generates DNA strand breaks suitable for LMPCR without any conversion procedures. Before running LMPCR, the DNA strand break frequency must be determined by running the samples on a 1.5% alkaline agarose gel *(43)*. The size range of the fragments should span 200–2000 bp (*see* **Note 6**).

3.4.1. DMS-Induced Base Modifications (see *Fig. 1*)

1. Dissolve DNA (10–50 µg) in 50 µL H_2O, add 50 µL of 2 *M* piperidine and mix well using a micropipet.
2. Samples are processed as described in **Subheading 3.2., steps 13–20**.
3. Dissolve DNA in H_2O to a concentration of 0.2 µg/µL.

3.4.2. UV-Induced Base Modifications

3.4.2.1. CPD (*see* Fig. 2)

1. To specifically cleave CPDs, dissolve 10 µg of UV-irradiated DNA in 50 µL H_2O, add 50 µL of a solution containing 10 µL of 10X dual buffer, 0.1 µL of 1 *M* DTT, 0.2 µL of 5 mg/mL BSA, a saturating amount of T_4 endonuclease V and complete with H_2O to a final volume of 50 µL. Mix well by flicking the microtube and quick spin.
2. Incubate at 37°C for 1 h.
3. To perform the photolyase digestion to remove the overhanging dimerized base that would otherwise prevent ligation *(8)*, add 10 µL of the following mix: 1 µL of 10X dual buffer, 1 µL of 1 *M* DTT, 0.2 µL of 5 mg/mL BSA, a saturating amount of photolyase, and complete with H_2O to a final volume of 10 µL. Mix well by flicking the microtube and quick spin.

4. Preincubate the microtubes at room temperature for 3–5 min under yellow light with their caps opened.
5. Leaving their caps open, cover the microtubes with a plastic film to prevent UVB-induced damage and place open ends 2–3 cm from a UVA black light for 1 h.
6. Add 290 μL of 0.52% SDS, mix well, and extract DNA using 1 vol (400 μL) phenol, 1 vol phenol:chloroform, and 1 volume chloroform.
7. To precipitate DNA, add 18 μL of 5 *M* NaCl and 1 mL of precooled absolute ethanol.
8. Leave 15 min on dry ice, spin 20 min at 15,000*g* in a centrifuge at 4°C.
9. Wash once with 1 mL of precooled 80% ethanol.
10. Spin 8 min at 15,000*g* in a centrifuge at 4°C.
11. Air-dry the pellet and dissolve DNA in H_2O to a concentration of 0.2 μg/μL.

3.4.2.2. 6–4PP (*See* **Fig. 3**)

1. Dissolve DNA (10–50 μg) in 50 μL of H_2O, add 50 μL of 2 *M* piperidine and mix well using a micropipet.
2. Samples are processed as described in **Subheading 3.2., steps 13–20**.
3. Dissolve DNA in H_2O to a concentration of 0.2 μg/μL.

3.5. Ligation-Mediated Polymerase Chain Reaction Technology

The LMPCR protocol using cloned *Pfu* DNA polymerase for primer extension and PCR steps is labeled with **a** in **Subheadings 3.5.1.** and **3.5.3.** An alternative LMPCR protocol using Sequenase for primer extension steps and *Taq* DNA polymerase for PCR steps is labeled **b** in **Subheadings 3.5.1.** and **3.5.3.** Aside from the ligation mix (*see* **Subheading 2.5.2.**), the ligation step (**Subheading 3.5.2.**) is identical with both enzyme combinations. The primer extension, ligation, and PCR steps are carried out in siliconized 0.625-mL microtubes and a thermocycler is used for all incubations.

3.5.1. Primer Extension (Steps II and III, *Fig. 5*)

1a. Mix 0.5–2 μg of genomic DNA, 3 μL of cloned *Pfu* buffer, and 1 pmol of primer 1 in a final volume of 25 μL.
2a. Denature DNA at 98°C for 3 min.
3a. Incubate for 20 min at 45°C to 55°C, depending of the T_m of the primer 1.
4a. Cool to 4°C.
5a. Add 5 μL of the cloned *Pfu* mix. Flick and quick spin.
6a. Incubate the samples at the annealing temperature for 30 s, then increase the temperature to 75°C at a rate of 0.3°C/s and incubate at 75°C for 10 min. Finally, the samples are cooled to 4°C.

Alternatively:

1b. Mix 0.5–1.6 μg of DNA in Sequenase buffer with 1 pmol of primer 1 in a final volume of 15–18 μL.
2b. Denature DNA at 98°C for 3 min.

3b. Incubate for 20 min at 45°C to 50°C, depending of the T_m of the primer 1.
4b. Cool to 4°C.
5b. Add 9 µL of the following mix: 7.5 µL of Mg–dNTP mix, 1.1 µL of H_2O, and 0.4 µL of T7 Sequenase V.2. Flick and quick spin.
6b. Incubate at 48°C for 5 min, 50°C for 1 min, 51°C for 1 min, 52°C for 1 min, 54°C for 1 min, 56°C for 1 min, 58°C for 1 min, and 60°C for 1 min. Then, the samples are cooled to 4°C.
7b. Add 6 µL of cold 310 m*M* Tris-HCl, pH 7.7.
8b. Incubate at 67°C for 15 min to inactivate the Sequenase, then cool to 4°C.

3.5.2. Ligation (Step IV, **Fig. 5**)

1. To the primer extension reaction, add 45 µL of the ligation mix and mix well with the pipet. Note that the composition of the ligation mix (*see* **Subheading 2.5.2.**) is different whether Sequenase or cloned *Pfu* DNA polymerase was used for the primer extension (Section 3.5.1 and Step III in **Fig. 5**).
2. Incubate at 18°C overnight.
3. On ice, precipitate DNA by adding 28.75 µL of 7.5 *M* ammonium acetate, 0.25 µL of 0.5 *M* EDTA, pH 8.0, 1 µL of 20 µg/µL glycogen, and 275 µL of precooled absolute ethanol.
4. Leave 15 min on dry ice and spin 20 min at 15,000*g* in a centrifuge at 4°C.
5. Wash once with 500 µL of precooled 80% ethanol.
6. Spin 8 min at 15,000*g* in a centrifuge at 4°C.
7. Air-dry DNA pellets and dissolve DNA pellets in 50 µL of H_2O.

3.5.3. Polymerase Chain Reaction (Steps V and VI, **Fig. 5**)

1a. Add 50 µL of the cloned *Pfu* DNA polymerase mix and mix with a pipet. The reaction mix is overlaid with 50 µL of mineral oil.
2a. Cycle 22 times as described in **Table 4** for cloned *Pfu* DNA polymerase. The last extension should be done for 10 min to fully extend all DNA fragments.
3a. Add 25 µL of cloned *Pfu* DNA polymerase stop under the mineral oil layer. Then, continue to **step 4**.

Alternatively:

1b. Add 50 µL of the *Taq* DNA polymerase mix and mix with the pipet. The reaction is overlaid with 50 µL of mineral oil.
2b. Cycle 22 times as described in **Table 4** for *Taq* DNA polymerase. The last extension should be done for 10 min to fully extend all DNA fragments.
3b. Add 25 µL of *Taq* DNA polymerase stop mix under the mineral oil layer. Then, continue to **step 4**.

4. Extract with 250 µL of premixed phenol–chloroform (92 µL :158 µL) and transfer to 1.5-mL microtubes.
5. Add 400 µL of precooled absolute ethanol.
6. Leave 15 min on dry ice; spin 20 min at 15,000*g* in a centrifuge at 4°C.

Table 4
Exponential Amplification Steps Using Cloned *Pfu* DNA Polymerase or *Taq* DNA Polymerase

Cycle	Denaturation T in °C for D in s) *Pfu*	*Taq*	Annealing (*T* is the T_m of the oligonucleotide for D in s) *Pfu* or *Taq*	Polymerization (D in s) *T* is the same for all cycles: 75°C for *Pfu* and 74°C for *Taq* —
0	—	93 for 120	—	—
1	98 for 300	98 for 150	T_m for 180	180
2	98 for 120	95 for 60	T_m –1°C for 150	180
3	98 for 60	95 for 60	T_m –2°C for 120	180
4	98 for 30	95 for 60	T_m –3°C for 120	180
5	98 for 20	95 for 60	T_m –4°C for 90	150
Repeat cycle 5, 13 more times (add 5 s per cycle for annealing and polymerization)				
19	98 for 20	95 for 60	T_m –3°C for 240	240
20	98 for 20	95 for 60	T_m –2°C for 240	240
21	98 for 20	95 for 60	T_m –1°C for 240	240
22	98 for 20	95 for 60	T_m for 240	600

Note: Temperature (T) and duration (D) of the denaturation, annealing and polymerization steps.

7. Wash once with 500 µL of precooled 80% ethanol.
8. Spin 8 min at 15,000*g* in a centrifuge at 4°C.
9. Air-dry DNA pellets.
10. Dissolve DNA pellets in 7.5 µL of premixed formamide loading dye in preparation for sequencing gel electrophoresis. For the sequence samples G, A, T+C, and C, it is often advisable to dissolve DNA pellets in 15 µL of premixed formamide loading dye.

*3.5.4. Gel Electrophoresis and Electroblotting (Step VII, **Fig. 5**)*

The PCR-amplified fragments are separated by electrophoresis through a 8% polyacrylamide/7 *M* urea gel, 0.4 mm thick and 60–65 cm long, then transferred to a nylon membrane by electroblotting ***(11–13)***.

1. Prerun the 8% polyacrylamide gel until the temperature of the gel reaches 50°C. Running buffer is 100 m*M* TBE. Before loading the samples, wash the wells thoroughly using a syringe.
2. To denature DNA, heat the samples at 95°C for 2–3 min, then keep them on ice prior to loading.
3. Load an aliquot of 3–3.5 µL using flat tips.
4. Run the gel at the voltage or power necessary to maintain the temperature of the gel at 50°C. This will ensure that the DNA remains denatured.

5. Stop the gel when the green dye (xylene cyanole FF) reaches 1–2 cm from the bottom of the gel.
6. Separate the glass plates using a spatula, then remove one of the plates by lifting it carefully. The gel should stick to the less treated plate (*see* **Note 10**).
7. Cover the lower part of the gel (approx 40–42 cm) with a clean Whatman 3MM Chr paper, carefully remove the gel from the glass plate and cover it with a plastic film (*see* **Note 10**).
8. On the bottom plate of the electroblotter, individually layer three sheets of Whatman 17 CHR paper, 43 cm × 19 cm, presoaked in 100 m*M* TBE, and squeeze out the air bubbles between the paper layers by rolling with a bottle or pipet.
9. Add 150 mL of 100 m*M* TBE on the top layer and place the gel quickly on the Whatman 17 CHR papers before TBE is absorbed. Before removing the plastic film, remove all air bubbles under the gel by gentle rolling with a 10-mL pipet.
10. Remove the plastic film and cover the gel with a positively charged nylon membrane presoaked in 100 m*M* TBE, remove all air bubbles by gently rolling a 10-mL pipet, then cover with three layers of presoaked Whatman 17 CHR paper and squeeze out air bubbles with rolling bottle. Paper sheets can be reused several times except for those immediately under and above the gel.
11. Place the upper electrode onto the paper.
12. Electrotransfer for 45 min at 2 A. The voltage should settle at approximately 10–15 V.
13. UV-crosslink (1000 J/m^2 of UVC) the blotted DNA to the membrane, taking care to expose the DNA side of the membrane. If probe stripping and rehybridization are planned, keep the membrane damp.

*3.5.5. Hybridization (Step VII, **Fig. 5**)*

3.5.5.1. Radiolabeled Probe

1. Prehybridize with 15 mL of hybridization buffer at 60–68°C for 20 min. The prehybridization temperature is based on the T_m of the primer used to prepare the probe.
2. Decant the prehybridization buffer and add the labeled probe in 6–8 mL of hybridization buffer.
3. Hybridize at 60–68°C (2°C below the calculated T_m of the probe) overnight.
4. Wash the membrane with prewarmed washing buffers. The buffers should be kept in an incubator or water bath set at a temperature of 4°C higher than the hybridization temperature. The membrane is placed into a tray on an orbital shaker. Wash with buffer I for 10 min and with buffer II three times for about 10 min each time.
5. Wrap the membrane in plastic film (*see* **Note 10**). Do not let the membrane become dry if stripping and rehybridization are planned after exposure to the film.
6. Expose membrane to X-ray films with intensifying screens at –70°C. Although longer exposure might be necessary, an exposure of 0.5–8 h is usually enough to produce a sharp autoradiogram. Nylon membranes can be rehybridized if more than one set of primers have been included in the primer extension and amplification reactions *(11–13)*. Probes can be stripped by soaking the membranes in boiling 0.1% SDS solution twice for 5–10 min each time.

3.5.5.2. Digoxigenin-Labeled Probe

1. Prehybridize with 20 mL of prehybridization buffer at 60–68°C for at least 3 h.
2. Decant the prehybridization buffer and add 7.5 mL of digoxigenin-labeled probe in prehybridization buffer.
3. Hybridize at 60–68°C (2°C below the calculated T_m of the probe) overnight.
4. Wash the membrane twice with 20 mL of 2X washing solution for 5 min each at room temperature, followed by two washes with 20 mL of 0.1X washing solution for 15 min each at 65°C. The membrane is placed into a rolling 8-cm-diameter × 22-cm-long borosilicate glass hybridization tube in a hybridization oven. Manipulate the membrane exclusively with tweezers (*see* **Note 11**) and do not let it dry following the hybridization step.
5. Wash the membrane with 50 mL of buffer 1 for 1 min at room temperature.
6. Transfer the membrane to a new hybridization tube and incubate with 20 mL of buffer 2 for 1 h at room temperature.
7. Replace buffer 2 with 20 mL of buffer 2 containing the antidigoxigenin antibody diluted 1:10,000 (prepared 5 min before use) and incubate for 30 min at room temperature.
8. Remove the antibody solution and wash the membrane with 20 mL of buffer 1.
9. Transfer the membrane to a new hybridization tube and incubate with 20 mL of buffer 1 containing 0.3% Tween-20 for 15 min at room temperature.
10. Replace the solution with 20 mL of buffer 3 and incubate for 5 min at room temperature.
11. Place the membrane between two cellulose acetate sheets and pour 0.5 mL:100 cm^2 of CSPD® diluted 1:100 in buffer 3 onto the membrane between the acetate sheet sandwich. Carefully remove the air bubbles and seal the acetate sheets using heat (Doubleseal). Incubate the membrane for 15 min at 37°C.
12. Expose membrane to X-ray films for 40 min at room temperature (*see* **Note 11**).

3.6. Preparation of Single-Stranded Hybridization Probes (Step VIII, Fig. 5)

The [^{32}P]-dCTP or digoxigenin-labeled single-stranded probe is prepared by 30 cycles of repeated linear primer extension using *Taq* DNA polymerase. Primer 2 (or primer 3, *see* **Note 12**) is extended on a double-stranded template which can be a plasmid or a PCR product. The latter is produced using two opposing primers 2, separated by a distance of 150–450 bp. Alternatively, any pair of gene specific primers suitable for amplifying a DNA fragment containing a suitable probe sequence (*see* **Note 12**) can be employed.

3.6.1. Template Preparation: PCR Products

3.6.1.1. PCR Amplification

1. To 50 µL of purified genomic DNA (100 ng) in H_2O, add 50 µL of the *Taq* DNA polymerase mix and mix with the pipet. The reaction is overlaid with 50 µL of mineral oil.

2. Cycle 35 times at 95°C for 1 min (97°C for 3 min for the first cycle), 61–73°C (1–2°C below the calculated T_m of primer 2 with the lowest T_m) for 2 min, and 74°C for 3 min. The last extension should be done for 10 min.
3. Add 25 µL of *Taq* DNA polymerase stop under the mineral oil layer.
4. Extract with 250 µL of premixed phenol–chloroform (92 µL:158 µL) and transfer to 1.5-mL microtubes.
5. Add 400 µL of precooled absolute ethanol.
6. Leave 15 min on dry ice, spin 20 min at 15,000*g* in a centrifuge at 4°C.
7. Wash once with 1 mL of precooled 80% ethanol.
8. Spin 8 min at 15,000*g* in a centrifuge at 4°C.
9. Air-dry DNA pellets.
10. Resuspend DNA pellets in 25 µL of 1X TAE loading buffer.

3.6.1.2. Purification and Quantification of PCR Products

1. Load 25 µL of PCR products per well along with an appropriate DNA mass ladder.
2. Migrate the PCR products on a neutral 1.2–1.5% agarose gel.
3. Stain the gel with ethidium bromide and recover the band containing the DNA fragment of expected molecular weight on a UV transilluminator using a clean scalpel blade. Minimize the size of the slice by removing as much extraneous agarose as possible.
4. Crush the slice and put it in a 0.625-mL microtube pierced at the bottom and containing a column of packed dry glass wool (*see* **Note 13**).
5. Insert the 0.625-mL microtube containing the column in a 1.5-mL microtube and spin 15 min at 7000*g*. Transfer the flowthrough to a new 1.5-mL microtube. If there is still some visible agarose remaining, repeat **step 5**.
6. Add 50 µL of H_2O to wash the column of any remaining DNA by spinning 8 min at 7000*g*. Pool all of the flowthrough contents in one 1.5-mL microtube.
7. Complete the volume to 405 µL with H_2O, add 45 µL of 3 *M* sodium acetate pH 7.0, and 1 mL of precooled absolute ethanol to precipitate DNA. Leave 15 min on dry ice, spin 20 min at 15,000*g* in a centrifuge at 4°C.
8. Wash once with 1 mL of precooled 80% ethanol and spin 8 min at 15,000*g* in a centrifuge at 4°C.
9. Air-dry DNA pellet.
10. Dissolve DNA pellets in 103 µL of H_2O.
11. Load aliquots of 1 and 2 µL of the DNA template dissolved in 1X universal neutral loading buffer along with a quantitative DNA molecular-weight ladder and electrophorese on a neutral 1.5% agarose gel.
12. Stain the gel with ethidium bromide and photograph on a UV transilluminator. The DNA concentration of the aliquots is estimated by comparison with the DNA ladder band intensities and H_2O is added to obtain a final concentration of template DNA of 3 ng/µL. The DNA template is aliquoted and stored at –20°C.

3.6.2. Labeling of Single-Strand Hybridization Probes

3.6.2.1. Isotopic Labeling

1. Prepare 150 µL of the following mix: 30 µL of 5X *Taq* buffer, 3 µL of 100 m*M* $MgCl_2$, 1 µL of dNTPs mix diluted 1:10 in H_2O, 20–40 ng of plasmid or 10–20 ng

of PCR products, 1.5 µL of 50 pmol/µL primer 2, 2.5 U of *Taq* DNA polymerase, and 10 µL of α-[^{32}P]-dCTP (3000 ci/mmol).
2. Cycle 30 times at 95°C for 1 min (97°C for 3 min for the first cycle), 60–68°C for 2 min, and 74°C for 3 min.
3. Transfer the mixture to a conical 1.5-mL microtube with screw cap.
4. Precipitate the probe by adding 50 µL of 10 *M* ammonium acetate, 1 µL of glycogen, and 420 µL of precooled absolute ethanol.
5. Leave 5 min at room temperature and spin 5 min at 15,000*g* in a centrifuge at room temperature.
6. Transfer the supernatant to into a new 1.5-mL microtube. Using a Geiger counter, compare the counts per minute between the pellet (probe) and the supernatant, count from the probe should be more than or equal to the supernatant for optimal results.
7. Dissolve the probe in 100 µL of TE buffer.
8. Add the probe to 6–8 mL of hybridization buffer and keep the probe at 65°C.

3.6.2.2. Digoxigenin (Nonisotopic) Labeling

1. Prepare 150 µL of the following mix: 30 µL of 5X *Taq* buffer, 3 µL of 100 m*M* $MgCl_2$, 1 µL of dNTP mix diluted 1:8.3, 20–40 ng of plasmid or 10–20 ng of PCR products, 1.5 µL of 50 pmol/µL primer 2, 2.5 U of *Taq* DNA polymerase, and 1.2 µL of 0.5 m*M* digoxigenin–11-dUTP.
2. Cycle 30 times at 95°C for 1 min (97°C for 3 min for the first cycle), 60–68°C for 2 min, and 74°C for 3 min.
3. Precipitate the probe by adding 50 µL of 10 *M* ammonium acetate, 1 µL of glycogen, and 420 µL of precooled absolute ethanol. Spin for 10 min at 15,000*g* in a centrifuge at room temperature.
4. Check the incorporation of the digoxigenin-labeled nucleotide by a dot blot (*see* **Note 14**).
5. Resuspend the probe in 100 µL of TE buffer.
6. Add the probe to 15 mL of prehybridization buffer.

4. Notes

1. Primers should be selected to have a higher T_m at the 5' end than in the 3' end. This higher annealing capacity of the 5' end lowers false priming, thus allowing a more specific extension and less background ***(49)***. A guanine or a cytosine residue should also occur at the 3' end. This stabilizes the annealing and facilitates the initiation of the primer extension. It is important that the selected primer does not have long runs of purines or pyrimidines, does not form loops or secondary structure, and does not anneal with itself. If primer dimerization occurs, less primer will be available for annealing and polymerization will not be optimal. The purity of the primers is verified on a 20% polyacrylamide gel (to prepare a 500-mL mix, dissolve 96.625 g of acrylamide, 3.375 g of *bis*-acrylamide, 210.21 g of urea corresponding to 7 *M*, in 100 m*M* TBE); if more than one band is found, the primer is reordered. The primers are also tested in a conventional PCR to prepare the template for the probe synthesis (*see* **Note 12**).

2. Originally, *Pfu* and *Taq* buffers were prepared using KCl, which was, however, shown to stabilize secondary DNA structures, thus preventing an optimal polymerization ***(50)***. The use of NaCl prevents, to some extent, the ability of DNA to form secondary structures. This is particularly helpful when GC-rich regions of the genome are being investigated.
3. The genomic DNA used for LMPCR needs to be very clean and undegraded. Any shearing of the DNA during preparation and handling before the first primer extension must be avoided. After an incubation of 3 h, if clumps of nuclei are still visible, proteinase K at a final concentration of 300 μg/mL should be added and the sample reincubated at 37°C for another 3 h.
4. If no DNA can be seen, add glycogen (1–2 μg) to the DNA solution and put the DNA on dry ice for 20 min and centrifuge the DNA (5000*g* for 20 min at 4°C). This should help DNA recovery but increases the probability of RNA contamination.
5. Because in vivo DNA analysis is based on the comparison of DNA samples modified in vivo with DNA control modified in vitro, given the quantitative characteristic and high sensitivity of LMPCR technology, the DNA concentrations should be as accurate as possible. Indeed, it is critical to start LMPCR with similar amounts of DNA in every sample to be analyzed. The method used to evaluate DNA concentration should measure only double-stranded nucleic acids. RNA contamination does not affect LMPCR, although it can interfere with the precise measurement of the DNA concentration.
6. The DNA frequency of DNA breakage is even more critical than the DNA concentration. For DMS and UV, the base-modification frequency determines the break frequency following conversion of the modified bases to single-strand breaks, whereas for DNase I, the frequency of cleavage is exactly the break frequency. The break frequency must be similar among the samples to be analyzed. It should not average more than one break per 150 bp for in vivo DNA analysis, the optimal break frequency varying from one break per 200 bp to one break per 2000 bp. When the break frequency is too high, we typically observe dark bands over the bottom half of the autoradiogram and very pale bands over the upper half, reflecting the low number of long DNA fragments. In summary, to make the comparison of the in vivo modified DNA sample with a purified DNA control easily interpretable and valid, the amount of DNA and the break frequency must be similar between the samples to be compared. On the other hand, it is not so critical that the break frequency of the sequence ladders (G, A, T+C, and C) be similar to that of the samples to be studied. However, to facilitate sequence reading, the break frequency should be similar between the sequence reactions. It is often necessary to load less DNA for the sequence ladders.
7. If the cell density is too high, multiple cell layers will be formed and the upper cell layer will obstruct the lower ones. This will result in an inhomogeneous DNA photoproduct frequency.
8. It is imperative that the purified DNA samples used as DNA control and the in vivo DNA samples come from the same cell type. For instance, differing cytosine

methylation patterns of genomic DNA from different cell types affect photoproduct formation (**17,29**) and give altered DNase I cleavage patterns *(2)*.

9. A nearly ideal chromatin substrate can be maintained in permeabilized cells. Nonionic detergents such as lysolecithin *(48)* and Nonidet P40 *(33)* permeabilize the cell membrane sufficiently to allow the entry of DNase I. Conveniently, this assay can be performed with cells either in a suspension or in a monolayer. One concern is that permeabilized cells will lyse after a certain time in a detergent, thus care must be taken to monitor cell integrity by microscopy during the course of the experiment. A further difficulty with the permeabilization technique concerns the relatively narrow detergent concentration range over which the treatment can be performed. Each cell type appears to require specific conditions for the detergent cell permeabilization. Furthermore, the DNase I concentration must be calibrated for each cell type to produce an appropriate cleavage frequency. Optimally, the in vivo DNase I protocol works better if the enzyme has cleaved the DNA backbone every 1.5–2 kb. Cutting frequencies greater than 1 kb are associated with higher LMPCR backgrounds because the number of 3'-OH ends is much higher, making suppression of the extension of these ends more difficult.
10. To facilitate sequencing gel removal following migration, it is crucial to siliconize the inner face of both glass plates prior to pouring the gel. For security, cost-effectiveness, efficiency, and time-saving, we recommend treating the glass plates with RAIN-AWAY® solution (Wynn's Canada, product no. 63020). We apply 0.75 mL on one plate and 1.5 mL on the other before each five utilizations as specified by the manufacturer. In this way, the gel is easier to pour and tends to stick on the less siliconized plate.

    Whenever plastic film is needed, we recommend plastic wrap brand. This brand was found to be less permeable to liquids and more resistant to tears than other brands. This is particularly important when membranes are exposed on the phosphorimager plate in order to avoid the moistening of the plate and irreversibly damaging it.
11. We adapted the nonisotopic digoxigenin-based probe labeling method and chemiluminescent detection system (Roche Molecular Biochemicals) to reveal DNA sequence ladders after LMPCR amplification, sequencing gel electrophoresis, and electroblotting. Compared to the isotopic method, the nonisotopic method has a higher specificity, higher sensitivity, lower background, and lower cost, and is therefore a highly recommendable alternative. As shown in **Fig. 6B**, the sequence ladder revealed by nonisotopic labeling was clearer, sharper and presented lesser background compared to the isotopic labeling method (**Fig. 6A**). Unlike isotopic probes, digoxigenin-labeled probes are innocuous, can be easily disposed of, can be stored for long periods, and can even be reused. It is worth noting, however, that this nonisotopic detection method requires some minor precautions. First, the nylon membrane used for this type of detection must bear a specific density and homogeneous distribution of positive charge. Among membranes we tested, the one sold by Roche Molecular Biochemicals, unquestionably gave the best results. Secondly, care should be taken with the manipulation

of the membrane. The use of tweezers is strongly recommended in order to reduce nonspecific spots and background. As seen in **Fig. 6B**, in spite of taking every precaution, some small spots are still observed on the "chemiluminogram." These might be explained by the powder from gloves. An alternative explanation for these spots could be the presence of nondissolved crystals in the antibody solution (to minimize this problem, this solution can be spun for 15–30 s before use) or in the detection buffer. However, the use of an appropriate membrane and meticulous manipulations can produce very good results with the nonisotopic detection method.

12. To avoid long probes, (i.e., greater than 200 bp), plasmid DNA is cut with an appropriate restriction enzyme (e.g., *see* **ref.** ***16***). If a third primer (primer 3) is used to make the probe, it should be selected from the same strand as the amplification primer (primer 2) just 5' to primer 2 sequence and with no more than seven to eight bases of overlap on this primer, and should have a T_m of 60–68°C. As first reported by Hornstra and Yang ***(41,51,52)***, we simply use the primer 2 employed in the amplification step and we produce the probe from PCR products. Such probes cost less (no primer 3), are more convenient (the preparation of the PCR products permits the testing of primers) and simplify the assay because no cloning requirement is needed as long as the sequence is known.
13. The bottom of a capless 0.625-mL siliconized microtube can be easily pierced with a heated needle. It is important to emphasize that the hole should be made as small as possible for the column to efficiently retain agarose. The pierced microtube is packed with wetted glass wool. Three successive centrifugation steps of 1 min each at 16,000*g* are necessary to compact and dry the glass wool. The water is recuperated in a capless 1.5-mL microtube. If glass wool is found with the effluent, the column should be discarded. A final 5-min centrifugation at 16,000*g* should be carried out to ensure the glass wool is fully compacted and dry. The glass wool column is stored at room temperature in a new capless 1.5-mL microtube and covered with a plastic film to protect the column from dust. In this way, the column can be stored indefinitely until it is used.
14. To verify whether digoxigenin was incorporated in the probe, use an aliquot of 1 µL from the 100-µL probe preparation and pipet onto a small piece of positively charged membrane (*see* **Subheading 2.5.4., item 13**). Expose the membrane to 1000 J/m^2 of 254-nm UV to crosslink the probe onto the membrane. Place the membrane in a Petri dish and add 15 mL of buffer 1 (*see* **Subheading 2.5.5.2., item 5**). Discard the buffer 1, add 20 mL of buffer 2 (*see* **Subheading 2.5.5.2., item 6**), and place the dish on a shaker for 10 min at room temperature. Discard the buffer 2, add 20 mL of digoxigenin-antibody coupled with a peroxidase (anti-Digoxigenin-AP, Roche Molecular Biochemicals, cat. no. 1 093 274) diluted 1:5 000 in buffer 2. Incubate 10 min at room temperature. Add 20 mL of buffer 1 in a new Petri dish, transfer the membrane to this new dish and wash the membrane for 10 min at room temperature. Always manipulate the membrane with tweezers (*see* **Note 11**). Remove the buffer 1 and add 20 mL of buffer 3 (*see* **Subheading 2.5.5.2., item 9**). Wait 5 min to allow the membrane to reach the appropriate pH

(pH 9.5) for the detection. During this time, prepare the detection solution by adding 90 µL of NBT (4-Nitroblue tetrazolium chloride, Roche Molecular Biochemicals, cat. no. 1 383 213) and 70 µL of BCIP (X-phosphate/5-bromo-4-chloro-3-indolyl-phosphate, Roche Molecular Biochemicals, cat. no. 1 383 221) to 20 mL of buffer 3. Discard buffer 3 and add the detection solution to the dish containing the membrane and place it in a dark room. Check occasionally and monitor the appearance of staining. If no staining appears after 1 h, this means that the incorporation of DIG was not efficient. The detection solution is very toxic, manipulate it carefully and eliminate this solution as toxic waste.

Acknowledgments

The authors wish to thank Dr. Elliot A. Drobetsky for his precious help in editing this text and for exciting LMPCR discussions. We are grateful to Mrs. Nancy Dallaire, Isabelle Paradis, and Nathalie Bastien for their technical assistance and valuable contribution to the development of the LMPCR technology. This work was supported by the Medical Research Council of Canada (MRC) and the Canadian Genetic Diseases Network (MRC/NSERC NCE program). R. Drouin is presently a research scholar ("Chercheur-boursier") of the "Fonds de la Recherche en Santé du Québec" (FRSQ).

References

1. Pfeifer, G. P., Tanguay, R. L., Steigerwald, S. D., and Riggs, A. D. (1990) *In vivo* footprint and methylation analysis by PCR-aided genomic sequencing: comparison of active and inactive X chromosomal DNA at the CpG island and promoter of human PGK–1. *Genes Dev.* **4,** 1277–1287.
2. Pfeifer, G. P. and Riggs, A. D. (1991) Chromatin differences between active and inactive X chromosomes revealed by genomic footprinting of permealized cells using DNase I and ligation-mediated PCR. *Genes Dev.* **5,** 1102–1113.
3. Chen, C.-J., Li, L. J., Maruya, A., and Shively, J. E. (1995) *In vitro* and in vivo footprint analysis of the promoter of carcinoembryonic antigen in colon carcinoma cells: effects of interferon γ treatment. *Cancer Res.* **55,** 3873–3882.
4. Tornaletti, S. and Pfeifer, G. P. (1995) UV light as a footprinting agent: modulation of UV-induced DNA damage by transcription factors bound at the promoters of three human genes. *J. Mol. Biol.* **249,** 714–728.
5. Mueller P. R. and Wold, B. (1989) *In vivo* footprinting of a muscle specific enhancer by ligation mediated PCR. *Science* **246,** 780–786.
6. Pfeifer, G. P., Steigerwald, S. D., Mueller, P. R., Wold, B., and Riggs, A. D. (1989) Genomic sequencing and methylation analysis by ligation mediated PCR. *Science* **246,** 810–813.
7. Pfeifer, G. P., Steigerwald, S. D., Hansen, R. S., Gartler, S. M., and Riggs, A. D. (1990) Polymerase chain reaction-aided genomic sequencing of an X chromosome-linked CpG island: methylation patterns suggest clonal inheritance, CpG site autonomy, and an explanation of activity state stability. *Proc. Natl. Acad. Sci. USA* **87,** 8252–8256.

8. Pfeifer, G. P., Drouin, R., Riggs, A. D., and Holmquist, G. P. (1992) Binding of transcription factors creates hot spots for UV photoproducts in vivo. *Mol. Cell. Biol.* **12,** 1798–1804.
9. Church, G. M. and Gilbert, W. (1984) Genomic sequencing. *Proc. Natl. Acad. Sci. USA* **81,** 1991–1995.
10. Pfeifer, G. P. (1992) Analysis of chromatin structure by ligation-mediated PCR. *PCR Methods Appl.* **2,** 107–111.
11. Pfeifer, G. P. and Riggs, A. D. (1993) Genomic footprinting by ligation mediated polymerase chain reaction, in *PCR Protocols: Current Methods and Applications* (White, B., ed.), Humana, Totowa, NJ, pp. 153–168.
12. Pfeifer, G. P. and Riggs, A. D. (1993) Genomic sequencing, in *DNA Sequencing Protocols* (Griffin, H. G. and Griffin, A. M., eds.), Humana, Totowa, NJ, pp. 169–181.
13. Pfeifer, G. P., Singer-Sam, J., and Riggs, A. D. (1993) Analysis of methylation and chromatin structure. *Methods Enzymol.* **225,** 567–583.
14. Gao, S., Drouin, R., and Holmquist, G. P. (1994) DNA repair rates mapped along the human PGK1 gene at nucleotide resolution. *Science* **263,** 1438–1440.
15. Tornaletti, S. and Pfeifer, G. P. (1994) Slow repair of pyrimidine dimers at p53 mutation hotspots in skin cancer. *Science* **263,** 1436–1438.
16. Rodriguez, H., Drouin, R., Holmquist, G. P., O'Connor, T. R., Boiteux, S., Laval, J., Doroshow, J. H., and Akman, S. A. (1995) Mapping of copper/hydrogen peroxide-induced DNA damage at nucleotide resolution in human genomic DNA by ligation-mediated polymerase chain reaction. *J. Biol. Chem.* **270,** 17,633–17,640.
17. Drouin, R. and Therrien, J.-P. (1997) UVB-induced cyclobutane pyrimidine dimer frequency correlates with skin cancer mutational hotspots in p53. *Photochem. Photobiol.* **66,** 719–726.
18. Rozek, D. and Pfeifer, G. P. (1993) In vivo protein–DNA interactions at the c-jun promoter: preformed complexes mediate the UV response. *Mol. Cell. Biol.* **13,** 5490–5499.
19. Cartwright, I. L. and Kelly, S. E. (1991) Probing the nature of chromosomal DNA–protein contacts by in vivo footprinting. *BioTechniques* **11,** 188–203.
20. Maxam, A. M. and Gilbert, W. (1980) Sequencing end-labeled DNA with base-specific chemical cleavages. *Methods Enzymol.* **65,** 499–560.
21. Chin P. L., Momand, J., and Pfeifer, G. P. (1997) *In vivo* evidence for binding of p53 to consensus binding sites in the *p21* and *GADD45* genes in response to ionizing radiation. *Oncogene* **15,** 87–99.
22. Angers, M., Drouin, R., Bachvarova, M., Paradis, I., Marceau, F., and Bachvarov, D. R. (2000) *In vivo* protein–DNA interactions at the kinin B1 receptor promoter: no modification upon interleukin-1 beta or lipopolysaccharide induction. *J. Cell. Biochem.* **78,** 278–296.
23. Becker, M. M. and Wang, J. C. (1984) Use of light for footprinting DNA in vivo. *Nature* **309,** 682–687.
24. Pfeifer, G. P. and Tornaletti, S. (1997) Footprinting with UV irradiation and LMPCR. *Methods* **11,** 189–196.

25. Pfeifer, G. P., Chen, H. H., Komura, J., and Riggs, A.D. (1999) Chromatin structure analysis by ligation-mediated and terminal transferase-mediated polymerase chain reaction. *Methods Enzymol.* **304**, 548–571.
26. Cadet, J., Anselmino, C., Douki, T., and Voituriez, L. (1992) Photochemistry of nucleic acids in cells. *J. Photochem. Photobiol. B: Biol.* **15,** 277–298.
27. Mitchell, D. L. and Nairn, R. S. (1989) The biology of the (6–4) photoproducts. *Photochem. Photobiol.* **49,** 805–819.
28. Holmquist, G. P. and Gao, S. (1997) Somatic mutation theory, DNA repair rates, and the molecular epidemiology of p53 mutations. *Mutat. Res.* **386,** 69–101.
29. Pfeifer, G. P., Drouin, R., Riggs, A. D., and Holmquist, G. P. (1991) *In vivo* mapping of a DNA adduct at nucleotide resolution: detection of pyrimidine (6–4) pyrimidone photoproducts by ligation-mediated polymerase chain reaction. *Proc. Natl. Acad. Sci. USA* **88,**1374–1378.
30. Gale, J. M. and Smerdon, M. J. (1990) UV induced (6–4) photoproducts are distributed differently than cyclobutane dimers in nucleosomes. *Photochem. Photobiol.* **51,** 411–417.
31. Gale, J. M., Nissen, K. A., and Smerdon, M. J. (1987) UV-induced formation of pyrimidine dimers in nucleosome core DNA is strongly modulated with a period of 10.3 bases. *Proc. Natl. Acad. Sci. USA* **84,** 6644–6648.
32. Mitchell, D. L., Nguyen, T. D., and Cleaver, J. E. (1990) Nonrandom induction of pyrimidine–pyrimidone (6–4) photoproducts in ultraviolet-irradiated human chromatin. *J. Biol. Chem.* **265,** 5353–5356.
33. Rigaud, G., Roux, J., Pictet, R., and Grange, T. (1991) *In vivo* footprinting of rat TAT gene: dynamic interplay between the glucocorticoid receptor and a liver-specific factor. *Cell* **67,** 977–986.
34. Miller, M. R., Castellot, J. J., and Pardee, A. B. (1978) A permeable animal cell preparation for studying macromolecular synthesis. DNA synthesis and the role of deoxyribonucleotides in S phase initiation. *Biochemistry* **17,** 1073–1080.
35. Contreras, R. and Fiers, W. (1981) Initiation of transcription by RNA polymerase II in permeable SV40-infected CV-1 cells; evidence of multiple promoters for SV40 late transcription. *Nucleic Acids Res.* **9,** 215–236.
36. Tanguay, R. L., Pfeifer, G. P., and Riggs, A. D. (1990) PCR-aided DNase I footprinting of single-copy gene sequences in permeabilized cells. *Nucleic Acids Res.* **18,** 5902.
37. Törmänen, V., Pfeifer, G. P., Swiderski, P. M., et al. (1992) Extension product capture improves genomic sequencing and DNase I footprinting by legation-mediated PCR. *Nucleic Acids Res.* **20,** 5487–5488.
38. Tornaletti, S., Bates, S., and Pfeifer, G. P. (1996) A high-resolution analysis of chromatin structure along p53 sequences. *Mol. Carcinogen.* **17,** 192–201.
39. Szabo, P. E., Pfeifer, G. P., and Mann, J. R. (1998) Characterization of novel parent-specific epigenetic modifications upstream of the imprinted mouse *H19* gene. *Mol. Cell. Biol.* **18,** 6767–6776.
40. Garrity, P. A. and Wold, B. J. (1992) Effects of different DNA polymerases in ligation-mediated PCR: enhanced genomic sequencing and in vivo footprinting. *Proc. Natl. Acad. Sci. USA* **89,** 1021–1025.

41. Hornstra, I. K. and Yang, T. P. (1994) High resolution methylation analysis of the human hypoxanthine phosphoribosyltransferase gene 5' region on the active and inactive X chromosomes: correlation with binding sites for transcription factors. *Mol. Cell. Biol.* **14,** 1419–1430.
42. Angers, M., Cloutier, J.-F., and Drouin, R. (2000) The effectiveness of *Pfu* exo$^-$ DNA polymerase in ligation-mediated PCR is mainly modulated by the ratio of DNA molecules per unit of enzyme. Submitted.
43. Drouin, R., Gao, S., and Holmquist, G. P. (1996) Agarose gel electrophoresis for DNA damage analysis, in *Technologies for Detection of DNA Damage and Mutations* (Pfeifer, G. P., ed.), Plenum, New York, pp. 37–43.
44. Rychlik, W. and Rhoads, R. E. (1989) A computer program for choosing optimal oligonucleotides for filter hybridization, sequencing and in vitro amplification of DNA. *Nucleic Acids Res.* **17,** 8543–8551.
45. Drouin, R., Rodriguez, H., Holmquist, G. P., and Akman, S. A. (1996) Ligation-mediated PCR for analysis of oxidative DNA damage, in *Technologies for Detection of DNA Damage and Mutations* (Pfeifer, G. P., ed.), Plenum, New York, pp. 211–225.
46. Mueller PR, Wold, B. (1991) Ligation-mediated PCR: applications to genomic footprinting. *Methods* **2,** 20–31.
47. Iverson, B. L. and Dervan, P. B. (1987) Adenine specific DNA chemical sequencing reaction. *Nucleic Acids Res.* **19,** 7823–7830.
48. Zhang, L. and Gralla, J. D. (1989) *In situ* nucleoprotein structure at the SV40 major late promoter: melted and wrapped DNA flank the start site. *Genes Dev.* **3,** 1814–1822.
49. Rychlik, W. (1993) Selection of primers for polymerase chain reaction, in *PCR Protocols: Current Methods and Applications* (White, B., ed.), Humana, Totowa, NJ, pp. 31–40.
50. Fry, M. and Loeb, L. A. (1994) The fragile X syndrome d(CGG)n nucleotide repeats form a stable tetrahelical structure. *Proc. Natl. Acad. Sci. USA* **91,** 4950–4954.
51. Hornstra, I. K. and Yang, T. P. (1992) Multiple in vivo footprints are specific to the active allele of the X-linked human hypoxanthine phosphoribosyltransferase gene 5' region: implications for X chromosome inactivation. *Mol. Cell. Biol.* **12,** 5345–5354.
52. Hornstra, I. K. and Yang, T. (1993) *In vivo* footprinting and genomic sequencing by ligation-mediated PCR. *Anal. Biochem.* **213,** 179–193.

14

Identification of Protein–DNA Contacts with Dimethyl Sulfate

Methylation Protection and Methylation Interference

Peter E. Shaw and A. Francis Stewart

1. Introduction

Dimethyl sulfate (DMS) is an effective and widely used probe for sequence-specific protein–DNA interactions. It is the only probe routinely used both for in vitro (methylation protection, methylation interference) and in vivo (DMS genomic footprinting) applications because it rapidly reacts with DNA at room temperature and readily penetrates intact cells *(1)*. DMS methylates predominantly the 7-nitrogen of guanine and the 3-nitrogen of adenine. Thus reactivity with G residues occurs in the major groove and with A residues in the minor groove. In standard Maxam and Gilbert protocols *(2)*, the methylated bases are subsequently converted to strand breaks and displayed on sequencing gels.

Methylation protection and interference are essentially combinations of the gel retardation assay or electrophoretic mobility shift assay (EMSA) *(3,4)* (Chapter 2) with the DMS reaction of the Maxam and Gilbert sequencing procedure. Protein–DNA interactions are reflected either as changes in DMS reactivities caused by bound protein (methylation protection) or as selective protein binding dictated by methylation (methylation interference).

In methylation protection, protein is first bound to DNA that is uniquely end labeled and the complex is reacted with DMS. DMS reactivities of specific residues are altered by bound protein either by exclusion, resulting in reduced methylation, or by increased local hydrophobicity, resulting in enhanced methylation, or by local DNA conformational changes, such as unwinding, resulting in altered reactivity profiles *(5–7)*. After the DMS reaction, free DNA is separated from protein-bound DNA by gel retardation and both DNA

From: *Methods in Molecular Biology, vol. 148: DNA–Protein Interactions: Principles and Protocols, 2nd ed.*
Edited by: T. Moss © Humana Press Inc., Totowa, NJ

fractions are recovered from the gel. Methylated residues are converted into strand scissions and the free and bound DNA fractions are compared on a sequencing gel. A complete analysis requires the examination of both strands. This is accomplished by preparing two DNA probes, each uniquely labeled at one end, and carrying both probes through the protocols. A binding site characterized by methylation protection will therefore appear as a cluster of altered DMS reactivities.

In methylation interference *(8,9)*, DNA is first reacted with DMS, purified and then presented to protein. Under the reaction conditions used methylation is partial, yielding approximately one methylated base per DNA molecule. Thus, the protein is presented with a mixture of DNA molecules that differ with respect to the positions of methyl groups. Some methyl groups will interfere with protein binding because they lie in or near the binding site. Gel retardation separates the mixture into two fractions: free DNA, which, as long as DNA is in excess over binding activity, represents the total profile of methylation reactivity, and bound DNA, which will not contain any molecules with methyl groups incompatible with binding. Both free and bound DNA fractions are recovered, methylated residues are converted to strand scissions, and the fractions are compared on a sequencing gel. The binding site is observed as the absence of bands in the bound sample corresponding to the positions where methylation interferes with binding.

It is obvious that these two uses of DMS may not deliver identical results. For example, **Fig. 1** presents a comparison obtained from experiments with the serum response element binding factors $p67^{SRF}$ $p62^{TCF}$ and their binding site in the human c-fos promoter (SRE). Because the use of DMS in vivo for genomic footprinting is limited to the equivalent of methylation protection, a direct comparison between in vivo and in vitro data excludes the more widely used methylation interference assay.

The two techniques are, however, very similar in practical terms and thus are presented together. Both techniques rely on preestablished conditions that permit a protein–DNA complex to be resolved in a gel retardation assay (Chapter 2) and on chemical DNA sequencing methodologies, for which the reader is advised to consult **ref.** *2* for a detailed treatment.

2. Materials

1. Dimethyl sulfate (DMS) (Merck), analytical grade.
2. Piperidine (Sigma), analytical grade; use freshly made 1:10 dilution in double-distilled water.
3. Phenol/chloroform 50% v/v, buffered with 50 m*M* Tris-HCl, pH 8.0.
4. NA45 paper (Schleicher & Schuell).
5. 3MM paper (Whatman) or GB 002 paper (Schleicher & Schuell).

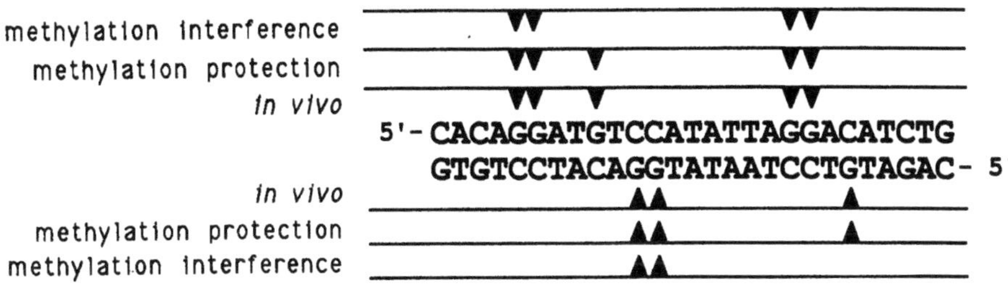

Fig. 1. Comparison of methylation interference and protection patterns formed by factors binding at the c-fos serum response element (SRE) in vitro and in vivo. G residues identified by methylation interference *(9)*, methylation protection and in vivo genomic footprinting *(7)* are indicated. An additional G on both the upper and lower strands is implicated in the protein–DNA interaction by methylation protection.

6. Electrophoresis equipment suitable for gel retardation or EMSA.
7. Electroblotting apparatus for Western transfer (e.g., Bio-Rad Trans-Blot).
8. Standard DNA sequencing gel electrophoresis equipment.
9. Vacuum gel drier (optional).
10. TBE buffer: 89 m*M* Tris base, 89 m*M* boric acid, and 2 m*M* EDTA. Make 10X stock as 108 g Tris base, 55 g boric acid, and 40 mL of 0.5 *M* EDTA pH 8.3 per liter.
11. NA45 elution buffer: 10 m*M* Tris-HCl, pH 8.0, 1 m*M* EDTA, and 1 *M* NaCl.
12. Carrier DNA: Salmon testis DNA or calf thymus DNA (Sigma), dissolved at 3 mg/mL in 10 m*M* Tris-HCl, pH 8.0, and 1 m*M* EDTA, and sheared.
13. Sequencing loading buffer: 90% formamide, 10 m*M* EDTA, and 0.1% (w/v) bromophenol blue, 0.1% (w/v) xylene cyanol blue.
14. Gel retardation loading buffer: 20% Ficoll, 20 m*M* EDTA, 0.1% w/v bromophenol blue.
15. 2X DMS buffer: 120 m*M* NaCl, 20 m*M* Tris-HCl, pH 8.0, 20 m*M* $MgCl_2$, and 2 m*M* EDTA.
16. DMS stop buffer: 1.5 *M* NaAc, pH 7.0, and 1 *M* 2-mercapto-ethanol, store frozen.
17. X-ray film (e.g., Kodak X-Omat) or imaging plate for phosphorimager.

3. METHOD

3.1. Methylation Protection

1. Incubate 300,000 cpm of uniquely end-labeled DNA probe (*see* **Note 1**) and a corresponding amount of protein together in a total volume of 100 µL, as previously optimized for gel retardation analysis.
2. Add 1 µL of DMS and incubate at room temperature (the incubation time depends on the length of the DNA probe and is empirical; as a guide for a 200-bp fragment, 1.5 min, for a 50-bp oligonucleotide duplex, 3 min).
3. Add 1/10 vol of 250 m*M* dithiothreitol (DTT), mix gently, add 1/10 vol of gel retardation loading buffer, mix gently, load onto a 2-mm-thick retardation gel in

1X TBE (or an alternative buffer as determined to be best for the given complex) and run as optimized for analytical gels. The load may need to be spread over up to 10 times more well area as compared with an analytical retardation assay (*see* **Note 2**).

4. After electrophoresis, separate the glass plates carefully so that the gel adheres to one plate and cover the gel with cling film. Expose to X-ray film long enough to reveal complexes clearly (i.e., 6 h to overnight). The alignment of the film to the gel must be reliably marked.
5. Put the developed film on a light box. Remove the gel from the cling film and realign it on the X-ray film. Cut pieces of NA45 paper sufficiently large to cover individual complexes in the gel yet small enough to fit into 1.5-mL tubes when rolled up. Wet the paper pieces in retardation gel running buffer and, with the help of tweezers, position one over each complex of interest in the gel, as visualized from the underlying film. Also position a similar sized piece of paper over (some of) the uncomplexed DNA. NA45 paper can be labeled with pencil before wetting.
6. Carefully cover the gel and paper pieces with two sheets of 3MM paper wetted in 1X TBE (or alternative gel running buffer from **step 3**). Lay a ScotchBrite pad from the electroblotting apparatus on top of the paper and turn the gel over. Carefully remove the second glass plate, cover the other side of the gel with 3MM paper and ScotchBrite as before and insert the package into an electrotransfer apparatus as described in the manufacturer's instructions with the NA45 paper toward the anode. Transfer in 1X TBE (or the alternative retardation gel buffer) at 80 V for 1.5 h (*see* **Note 3**).
7. Stop the transfer, unpack the gel carefully with the NA45 paper on top and transfer each piece to a labeled 1.5-mL tube containing 600 µL of elution buffer (check that the radioactivity has transferred to the paper). Incubate at 70°C for 1 h.
8. Remove NA45 paper from each tube, check that at least half the radioactivity has eluted into the buffer (do not expect quantitative elution, but at least 50% should come off), add 20 µg carrier DNA, extract with phenol/chloroform and precipitate with 1 volume of isopropanol. Wash precipitate once in 70% ethanol and dry under vacuum. (*See* **Notes 4** and **5**.)

9a. To reveal modified Gs: Dilute piperidine 1:10 in water and add 50 µL to each pellet. Vortex briefly and incubate at 90°C for 30 min (tubes must be clamped or weighted down to prevent the lids opening); then dry under a good vacuum. Take up the samples in 100 µL of water and repeat the drying process. This strand scission protocol should not convert methylated A residues into strand breaks. It is often observed, however, that breakages at As do occur with reasonable efficiency.

9b. To reveal modified As and Gs: The following modification will produce efficient cleavage at both methyl-G and methyl-A residues. After the preparative retardation gel, resuspend the dried, purified DNA in 30 µL of 10 m*M* sodium phosphate pH 6.8, and 1 m*M* EDTA. Incubate for 15 min at 92°C. Then add 3 µL of 1 *M* NaOH and incubate for 30 min at 92°C, followed by 320 µL of 500 m*M* NaCl,

50 µg/mL carrier DNA, and 900 µL ethanol. Chill and centrifuge to pellet the radioactivity. Wash the pellet in 70% ethanol and dry.

10. Measure the Cerenkov counts in each tube, then redissolve the samples in water (e.g., 10 cpm/µL) and transfer equivalent counts (1000 cpm in each case would be optimal) from each into fresh tubes. Dry down and redissolve in 5 µL sequencing loading buffer.
11. Prepare and pre-electrophorese a standard sequencing gel (5–12% acrylamide, depending on probe length). Denature probes at 95°C for 5 min, snap cool in ice and load onto the gel. Run the gel until optimal separation of sequence is achieved. (*See* **Notes 6–8**.)
12. Stop electrophoresis, remove the gel from the tank and lift off one glass plate. Fix the gel in 20% ethanol and 10% acetic acid for 10 min. Drain briefly and then overlay the gel with two sheets of 3MM paper and carefully peel it off the glass plate. Cover the gel surface with cling film and dry on a vacuum gel drier (*see* **Note 9**). Expose the dry gel to X-ray film with intensifying screens as necessary, or to an imaging plate.

3.2. Methylation Interference

1. Mix 300,000 cpm of end-labeled probe (*see* **Note 1**); 100 µL of 2X DMS buffer and water to 200 µL. Add 2 µL of DMS and incubate at room temperature (the same guidelines as given in **Subheading 3.1.2.** apply for the reaction time). Stop the reaction by the addition of 50 µL cold DMS stop mix and precipitate with 850 µL cold ethanol. Redissolve in 200 µL cold 0.3 *M* NaAc pH 7.0, add 700 µL cold ethanol, and reprecipitate. Wash twice in 80% ethanol and redissolve the probe in water or binding buffer (about 20,000 cpm/µL).
2. Incubate the probe with protein for gel retardation as previously optimized for gel retardation analyses of the complexes in question in a total volume of 100 µL.
3. Add 1/10 vol of gel retardation loading buffer, mix gently and load onto a 2-mm-thick retardation gel in 1X TBE (or alternative buffer); then, run as optimized for analytical gels. However, the load should be spread over up to 10 times more well area (*see* **Note 2**).
4. Continue with **step 4** and all subsequent steps as described for methylation protection (**Subheading 3.1.**).

4. Notes

1. To have sufficient counts to complete the procedure, proceed with at least 10 times the amount of material required for a simple gel retardation analysis (i.e., at least 300,000 cpm).
2. A common difficulty with these methods is the persistence of contaminants that accompany DNA after the preparative retardation gel. These contaminants interfere with the migration of DNA on the sequencing gel, producing blurred and distorted patterns. In order to minimize this problem it is worth ascertaining the load limit of the retardation gel so that the protein–DNA complex will not smear but be well resolved and therefore concentrated in the gel before elution.

3. It is also possible to use a semidry electrotransfer apparatus (e.g., Bio-Rad Trans-blot SD) to transfer the DNA from the gel retardation gel onto NA45 paper. In this case, both the transfer time and potential are reduced.
4. In some instances, it may prove difficult to elute the DNA from the NA45 paper, in which case raising the salt concentration or the temperature may improve elution. (Extending the incubation time does not seem to help.) If not, the batch of NA45 may be to blame or it is even conceivable that the DNA–protein complex in question is adsorbed too tightly onto the paper. It is not possible to phenol extract the NA45 paper in order to remove bound protein–DNA.
5. Retain the isopropanol supernatants until you are sure the samples have precipitated quantitatively. Add more carrier DNA if required.
6. It is similarly advisable to load as little material onto the sequencing gel as practicable. With the advent of the phosphorimager, the lower limit for the sequencing gel is well under 1000 cpm/lane.
7. If the end-labeled DNA fragment is relatively long and multiple binding sites are to be resolved, a gradient or wedge sequencing gel can be used in **step 11** of **Subheading 3.1.**
8. An appropriate complement for the final result is to perform the Maxam and Gilbert G+A reactions on the end-labeled probe. On the sequencing gel, these reactions should provide unambiguous sequence information and, in case difficulties are encountered, clues as to the steps that are problematic.
9. It is not essential to dry down the sequencing gel because after one glass plate has been removed, it can be covered with cling film and exposed to X-ray film at –70°C with one screen. This alternative should only be considered if the signal is sufficiently strong or if a gel drier is not available.

References

1. Church, G. M. and Gilbert, W. (1984). Genomic sequencing. *Proc. Natl. Acad. Sci. USA* **81,** 1991–1995.
2. Maxam, A. and Gilbert, W. (1980) Sequencing end-labelled DNA with base-specific chemical cleavages. *Methods Enzymol.* **65,** 499–560.
3. Fried, A. and Crothers, D. M. (1981) Equilibria and kinetics of lac repressor–operator interactions by polyacrylamide gel electrophoresis. *Nucleic Acids Res.* **9,** 6505–6525.
4. Garner, M. M. and Revzin, A. (1981) A gel electrophoresis method for quantifying the binding of protein to specific DNA regions: application to components of the *E. coli* lactose operon regulatory system. *Nucleic Acids Res.* **9,** 3047–3059.
5. Gilbert, W., Maxam, A., and Mirzabekov, A. (1976) Contacts between the LAC repressor and DNA revealed by methylation. in *Control of Ribosome Biosynthesis, Alfred Benzon Symposium IX* (Kjelgaard, N. O. and Maaloe, O., eds.), Academic, New York, pp. 139–148.
6. Johnsrud, L. (1978) Contacts between *Escherichia coli* RNA polymerase and a lac operon promoter. *Proc. Natl. Acad. Sci. USA* **75,** 5314–5318.

7. Herrera, R. E., Shaw, P. E., and Nordheim, A. (1989). Occupation of the c-fos serum response element in vivo by a multi-protein complex is unaltered by growth factor induction. *Nature* **340,** 68–70.
8. Siebenlist, U. and Gilbert, W. (1980) Contacts between *E. coli* RNA polymerase and an early promoter of phage T7. *Proc. Natl. Acad. Sci. USA* **77,** 122–126.
9. Shaw, P. E., Schröter, H., and Nordheim, A. (1989). The ability of a ternary complex to form over the serum response element correlates with serum inducibility of the human c-fos promoter. *Cell* **56,** 563–572.

15

Ethylation Interference

Iain W. Manfield and Peter G. Stockley

1. Introduction

Structural studies of DNA–protein complexes have now made it clear that specific sequence recognition in these systems is accomplished in two ways, either directly by the formation of hydrogen bonds to base-pair edges from amino acid side chains located on a DNA-binding motif, such as a helix–turn–helix, or indirectly as a result of sequence-dependent distortions of the DNA conformation *(1)*. These contacts occur in the context of oriented complexes between macromolecules that juxtapose the specific recognition elements. As part of these processes, proteins make a large number of contacts to the phosphodiester backbone of DNA, as was predicted from biochemical assays of the ionic strength dependence of DNA binding.

Contacts to phosphate groups can be inferred by the ethylation interference technique *(2)*. Ethylnitrosourea (EtNU) reacts with DNA to form, principally, phosphotriester groups at the nonesterified oxygens of the otherwise phosphodiester backbone. Minor products are the result of the reactions of EtNU with oxygen atoms in the nucleotide bases themselves (*see* **Note 1**). Under alkaline conditions and at high temperature, the backbone can be cleaved at the site of the modification to form a population of molecules carrying either 3'-OH or 3'-ethylphosphate groups.

The length of the ethyl group (approx 4.5 Å) means that at a number of positions along a DNA molecule encompassing the binding site for a protein, complex formation will be inhibited by the presence of such a modification. At other sites, outside the binding site, no interference with protein binding at the specific site will be observed. Addition of the DNA-binding protein to a randomly ethylated DNA sample, followed by some procedure to separate the complexes formed from unbound DNA, will fractionate the DNA sample into

From: *Methods in Molecular Biology, vol. 148: DNA–Protein Interactions: Principles and Protocols, 2nd ed.*
Edited by: T. Moss © Humana Press Inc., Totowa, NJ

those molecules able to bind protein with high affinity and those for which the ethylation has lowered the affinity (**Fig. 1**). In practice, modification at different sites produces molecules with a spectrum of affinities for the protein. It is, therefore, not possible to prove conclusively that a particular phosphate is contacted by the protein, but only that ethylation at that site interferes with complex formation.

Only occasionally are large amounts of pure protein readily available for in vitro biochemical assay of DNA-binding activity and, often, only small amounts of crude nuclear extracts are available. In many commonly used assays, complex formation could not easily be detected in such situations. For example, using DNase I or hydroxyl radicals, a high level of binding-site occupancy is required for a footprint to be detected. Fractional occupancy is readily detected by gel retardation of complexes (electrophoretic mobility shift [EMSA]; *see* Chapter 2) but offers only limited characterization of the details of the protein–DNA interaction. Interference techniques, such as the ethylation and hydroxyl radical interference (*see* Chapter 16) techniques ***(3)***, do allow the molecular details of complex formation to be studied even when only small amounts of crude protein are available ***(4)***. Whatever the level of saturation, DNA fragments modified at sites reducing the affinity of protein for DNA are less likely to form complexes. Therefore, the bound fraction on gel retardation assays will always give an indication of the sites that do not inhibit complex formation when modified. The groups on the DNA recognized by the protein can then be inferred. Ethylation can also be used to analyze RNA–protein complexes ***(5,6)***.

We have used the ethylation interference technique to probe the interaction of the *Escherichia coli* methionine repressor, MetJ, with its binding site in vitro. Binding sites for MetJ consist of two, or more, immediately adjacent copies of an 8-bp site with the consensus sequence 5'-AGACGTCT-3', which has been termed a "met box." X-ray crystallography has been used to determine the structure of the MetJ dimer, the complex with corepressor, *S*-adenosyl methionine (SAM), and a complex of the holorepressor with a 19-mer oligonucleotide containing two met boxes ***(7,8)***. The structure of the protein–DNA complex in the crystal reveals two MetJ dimers (one per met box) binding to the DNA by insertion of a β-ribbon into the major groove. The general features of this model are corroborated by the results of the ethylation interference experiments and by data from a range of other footprinting techniques.

2. Materials

2.1. Preparation of Radioactively End-Labeled DNA

1. Plasmid DNA carrying the binding site for a DNA-binding protein on a convenient restriction fragment (usually <200 bp).

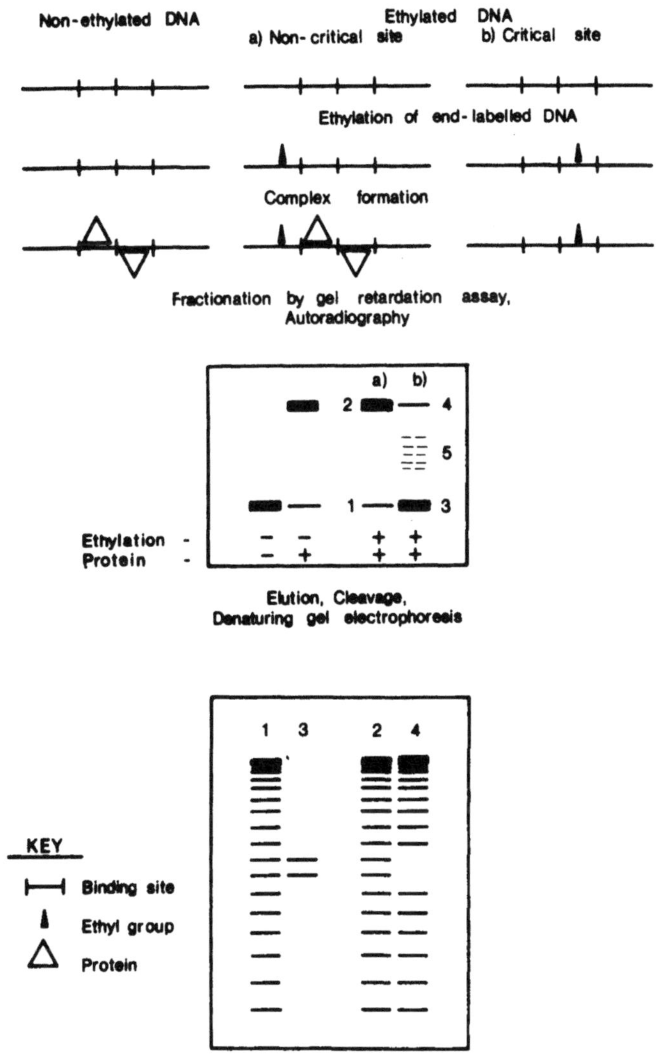

Fig. 1. Diagrammatic outline of the ethylation interference experiment. The upper section shows ethylation of end-labeled DNA (ethyl groups represented by small solid triangles) and complex formation with protein (represented by large open triangles). The expected mobility of each species on nondenaturing gels is shown in the middle section. In idealized form, the pattern of cleavage products that might be expected from recovery of each species after gel retardation assay is shown in the lower section. In practice, samples represented by 1 + 3 and 2 + 4 migrate to the same position on the retardation gel and are, therefore, not separated. The resultant pattern is shown in **Fig. 2**.

2. Restriction enzymes and the appropriate buffers as recommended by the suppliers.
3. Phenol: redistilled phenol equilibrated with 100 m*M* Tris-HCl, pH 8.0.
4. Chloroform.

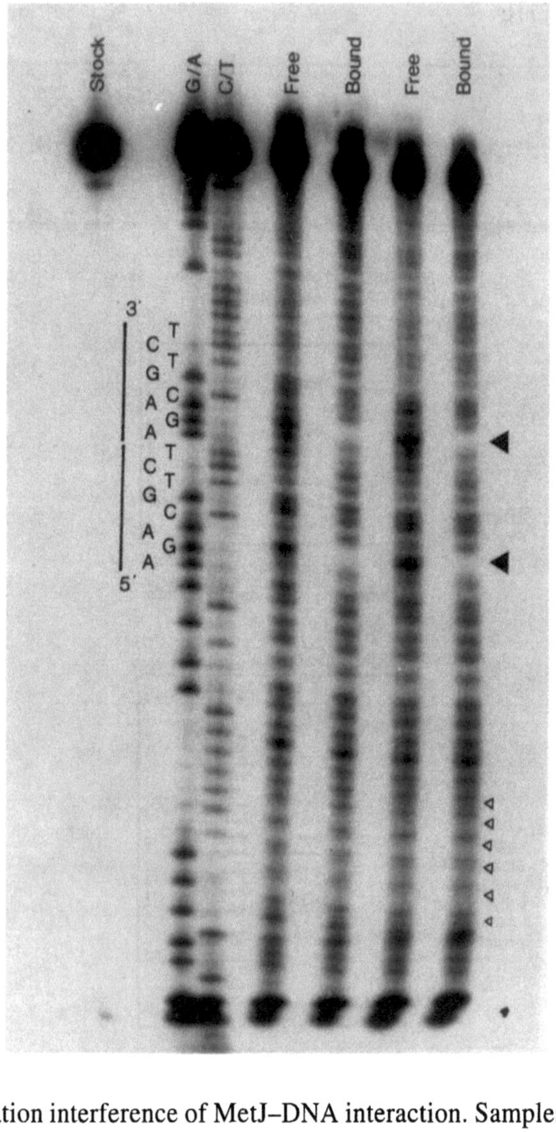

Fig. 2. Ethylation interference of MetJ–DNA interaction. Samples for denaturing gel electrophoresis were prepared following the methods given here. "Stock" indicates nonethylated DNA that has been through the cleavage reaction. "G/A" and "C/T" are the products of the purine- and pyrimidine-specific Maxam–Gilbert chemical cleavage reactions, respectively. "Free" and "Bound" are the DNA fractions that can and cannot form complexes, respectively. The two sets of data represent results obtained with DNA ethylated for 30 min (left-hand lanes) or 60 min (right-hand lanes). The sequence of the MetJ binding site is indicated along the side of the autoradiograph. The phosphate ethylation that interferes most strongly with complex formation is indicated by a large solid triangle. Small, open triangles indicate the minor reaction products of the cleavage reaction, which, for small fragments, are resolved on these gels.

5. Solutions for ethanol precipitation of DNA: 4 *M* NaCl and ethanol (absolute and 70% v/v).
6. Alkaline phosphatase (AP), from calf intestine (CIAP) or shrimp.
7. 10X AP reaction buffer: 0.5 *M* Tris-HCl, pH 9.0, 0.1 *M* $MgCl_2$, and 0.001 *M* $ZnCl_2$.
8. TE buffer: 10 m*M* Tris-HCl, pH 8.0, and 1 m*M* ethylenediaminetetraacetic acid (EDTA).
9. Sodium dodecyl sulfate (SDS) 20% w/v.
10. EDTA (0.25 *M*, pH 8.0).
11. T4 polynucleotide kinase (T4-PNK).
12. 10X T4-PNK reaction buffer: 0.5 *M* Tris-HCl, pH 7.6, 0.1 *M* $MgCl_2$, and 0.05 *M* dithiothreitol (DTT).
13. Radioisotope: [γ-^{32}P]-ATP.
14. 30% w/v Acrylamide stock: 29:1 :: acrylamide: *N,N'*-methylene–*bis*-acrylamide.
15. Polyacrylamide gel elution buffer: 0.3 *M* sodium acetate, 0.2% (w/v) SDS, and 2 m*M* EDTA.
16. Polymerization catalysts: ammonium persulfate (APS) (10% w/v) and *N,N,N',N'*-tetramethylethylenediamine (TEMED).

2.2. Ethylation Modification and Fractionation of DNA

1. End-labeled DNA in TE: 250,000 cpm are required per ethylation reaction to be performed (roughly 20 ng DNA fragment/reaction). Ethylation of more DNA allows a range of protein concentrations to be used when protein–DNA binding reactions are prepared. Standard safety procedures should be used when handling radiolabeled DNA (e.g., work behind Lucite shields).
2. Cacodylate buffer: 50 m*M* sodium cacodylate (used without adjustment of the pH, which is usually close to 8.0). Cacodylate buffer contains arsenic and therefore should be handled with caution. It is only necessary to prepare small volumes (approx 10 mL) of this solution. Passing the solution through a 0.45-µm filter is the preferred method of sterilization.
3. Ethylnitrosourea (EtNU): This reagent is readily synthesized ***(9)*** provided, the appropriate safety measures are used. Commercial suppliers do not appear to provide this material at present. The solid material should be stored at –20°C and allowed to warm to room temperature before use. EtNU should be handled in a fume hood and contaminated waste stored there until disposal. Wear two pairs of latex gloves when handling samples containing EtNU.
4. tRNA (1 mg/mL).
5. Solutions for ethanol precipitation of DNA: 4 *M* NaCl and ethanol (absolute and 70% v/v).
6. Complex buffer: for MetJ, for example, 10 m*M* Tris-HCl, pH 7.4, 150 m*M* NaCl, and 1 m*M* *S*-adenosyl methionine *p*-toluene sulfonate salt (SAM). Add glycerol to this buffer to 10% (v/v).
7. Purified DNA-binding protein or protein extract.
8. Nondenaturing gel acrylamide stock solution: 30% (w/v) (37:1 acrylamide: *N,N'*-methylene–*bis*-acrylamide).

9. Electrophoresis buffer stock solution: 1.0 *M* Tris-HCl, pH 8.0, and solid SAM.
10. Ammonium persulfate (APS) (10% w/v).
11. TEMED.
12. Glass plates: 150 × 150 × 1.5 mm.
13. Peristaltic pump capable of recirculating buffer at 5–10 mL/min.
14. X-ray film.
15. Autoradiography cassette.
16. X-ray film developer and fixer.
17. Plastic wrap.
18. Scalpel blade.
19. Syringe needle.
20. Polyacrylamide gel elution buffer: 0.3 *M* sodium acetate, 0.1% (w/v) SDS, and 1 m*M* EDTA.
21. 10 m*M* sodium phosphate, pH 7.0, and 1 m*M* EDTA.
22. 1.0 *M* NaOH (freshly prepared).
23. Acetic acid solutions: 1.0 *M* and 10% v/v.
24. Sequencing gel loading buffer: 80% (v/v) formamide, 0.5X TBE, 0.1% (w/v) xylene cyanol, 0.1% (w/v) bromophenol blue.
25. Acrylamide stock solutions for sequencing gel: 19% (w/v) acrylamide, 1% (w/v) *N,N'*-methylene–*bis*-acrylamide, and 50% (w/v) urea in TBE.
26. TBE (1X): 89 m*M* Tris, 89 m*M* boric acid, and 10 m*M* EDTA, pH 8.3.

2.3. Maxam–Gilbert Chemical DNA Sequencing Reactions

There is insufficient space to cover these methods in detail here, but extensive information, materials, methods, and troubleshooting guides are readily available in published literature ***(10,11)***.

3. Methods

3.1. Preparation of End-Labeled DNA

1. Digest the plasmid (e.g., 10 μg of the plasmid in a reaction volume of 200 μL with one of the pair of restriction enzymes used to release a suitably sized DNA fragment [usually <200 bp]). Extract the digest with an equal volume of buffered phenol and then add 2.5 vol of ethanol to the aqueous layer to precipitate the DNA.
2. Add 50 μL 1X AP reaction buffer to the ethanol-precipitated DNA pellet (<50 μg). Add 1 U AP, incubate at 37°C for 30 min, followed by addition of a further aliquot of enzyme, and incubate for a further 30 min. Terminate the reaction by adding SDS and EDTA to 0.1% (w/v) and 20 m*M*, respectively, in a final volume of 200 μL and incubate at 65°C for 15 min. Extract the digest with an equal volume of buffered phenol, then with 1:1 phenol:chloroform, and finally ethanol precipitate the DNA from the aqueous phase by addition of 2.5 vol of ethanol.
3. Redissolve the DNA pellet (~2.5 μg) in 18 μL 1X T4-PNK buffer. Add 20 μCi [γ–^{32}P]-ATP and 10 U T4-PNK, and incubate at 37°C for 30 min. Terminate the

reaction by phenol extraction (followed by ethanol precipitation and a second restriction enzyme digest) or by addition of nondenaturing gel loading buffer, and electrophoresis on a nondenaturing polyacrylamide gel. We use 12% (w/v) polyacrylamide gels (19:1, acrylamide:*bis*-acrylamide) with 1X TBE as the electrophoresis buffer.

4. After electrophoresis, locate the required DNA fragments by autoradiography of the wet gel. Excise slices of the gel containing the bands of interest using the autoradiograph as a guide. Elute the DNA into 500 µL elution buffer overnight (at least) at 37°C. Ethanol precipitate the DNA, wash the pellet thoroughly with 70% (v/v) ethanol, dry under vacuum, and rehydrate in a small volume of TE buffer (e.g., 50 µL). Determine the radioactivity of the sample by liquid scintillation counting of a 1 µL aliquot.

3.2. Ethylation Modification and Protein-Binding Assay

3.2.1. Ethylation Reaction

1. Dispense the required volume containing approx 250,000 cpm of radiolabeled DNA solution into an Eppendorf tube, add cacodylate buffer to a final volume of 100 µL and heat the solution to 50°C in a heating block. Prepare the minimum volume of EtNU-saturated ethanol (at 50°C) for the required reactions, add 100 µL of this to the DNA, mix, and incubate at 50°C for 60 min *(see* **Notes 1** and **2**).
2. Add to the sample 5 µL of 4 *M* NaCl, 2 µg of tRNA, and 150 µL of ethanol. Mix and place at –20°C for 60 min or in a dry ice/ethanol bath for 15 min. Pellet the DNA by centrifugation in a microfuge for 15 min and remove the supernatant (store separately to be destroyed by incineration). Add 500 µL of 70% ethanol, mix thoroughly, recentrifuge, and remove the supernatant. Dry the pellet briefly under vacuum.

3.2.2. Fractionation of DNA by Gel Retardation Assay

The following procedure allows the separation of free and bound DNA by means of gel retardation. The precise conditions will depend on the protein under investigation *(see* **Notes 3–6**).

1. Mix 10 mL of nondenaturing acrylamide stock solution, 1.5 mL of 1.0 *M* Tris-HCl, pH 8.0, 29.5 mL of distilled water, 0.2 mL of APS, 15 µL of TEMED, and 1.5 mg of SAM, and pour into gel frame. Insert the well former and leave to polymerize for 1–2 h.
2. When polymerized, insert the gel into the tank and connect peristaltic pump tubing such that buffer is pumped in both directions (i.e., top to bottom and bottom to top reservoirs). Pre-electrophorese gel for 30 min at 100 V.
3. Redissolve the pellet of ethylated DNA in complex buffer plus glycerol *(see* **Note 6**). Set aside 10% of the ethylated DNA sample, which will be used as an unfractionated control to indicate the variation in level of modification at each residue. Dry this sample under vacuum and store at –20°C until **step 1** of **Subheading 3.2.4.** Add DNA-binding protein to a concentration that would saturate

unmodified DNA and incubate at 37°C for 15 min to allow complex formation. **Note:** Exact conditions will vary depending on the protein being studied.

4. Load the DNA–protein complex solution onto the gel and electrophorese into the gel at 250 V for 2 min. Reduce the voltage to 100 V and continue electrophoresis until a small amount of bromophenol blue dye loaded into an unused lane has reached the bottom of the gel.

3.2.3. DNA Recovery

1. After electrophoresis, remove one plate and wrap the gel and remaining plate in clear plastic wrap film (we use Saran Wrap™). Cut a piece of X-ray film large enough to cover the lanes used on the gel and fix it to the gel firmly with masking tape. Using a syringe needle, make a number of holes through the film and gel that will serve to orient the developed film with respect to the gel. Alternatively, align the film and the gel using fluorescent marker strips.
2. Place the assembly in an autoradiography cassette and leave at 4°C overnight.
3. Take the film off the gel and develop as usual. When dry, align the film and gel using the holes created previously. Using a syringe needle, make a series of holes into the gel around the fragments of interest using the bands on the film as a guide. Remove the film and excise the marked regions of polyacrylamide. Place gel fragments (10 × 5 mm) in Eppendorf tubes, add 600 µL of gel elution buffer, and incubate at 37°C overnight (at least).
4. Transfer 400 µL of the eluate to a fresh tube, add 2 µg of tRNA and 1 mL of ethanol, mix, and place at –20°C for 60 min. Pellet DNA by centrifugation in a microfuge for 15 min. Discard the supernatant (check for absence of radioactivity), wash the pellet with 500 µL of 70% ethanol, recentrifuge, and discard the supernatant. Dry the pellet briefly under vacuum. If the DNA does not pellet readily, incubate the sample at –20°C for longer or recentrifuge at 4°C.

3.2.4. Phosphotriester Cleavage and DNA Sequencing

1. Redissolve each pellet in 15 µL sodium phosphate buffer and add 2.5 µL of 1 *M* NaOH. Seal tube with plastic film (e.g., Parafilm) and incubate at 90°C for 30 min. Centrifuge samples briefly to collect any condensation. Add 2.5 µL of 1.0 *M* acetic acid, 2 µL of 4 *M* NaCl, 1 µg of tRNA, and 70 µL of ethanol. Leave samples at –20°C for 60 min. Pellet DNA as described in **step 4** of **Subheading 3.2.3.**
2. Redissolve the pellet in 4 µL of sequencing gel loading buffer. Heat to 90°C for 2 min and load samples onto a 12% w/v polyacrylamide sequencing gel alongside Maxam–Gilbert sequencing reaction markers. Electrophorese at a voltage that will warm the plates to around 50°C. After electrophoresis, fix the gel in 1 L of 10% (v/v) acetic acid for 15 min. Transfer the gel to 3MM paper and dry under vacuum at 80°C for 60 min. Autoradiograph the gel at –70°C with an intensifying screen.
3. Compare lanes corresponding to bound, free, and control DNA for differences in intensity of bands at each position. A dark band in the "free fraction" (and a

corresponding reduction in the intensity of the band in the "bound fraction") indicates a site where ethylation interferes with complex formation. This is interpreted as meaning that this residue is contacted by the protein or a portion of the protein comes close to the DNA at this point. For 5' end-labeled DNA, the ethylation reaction products migrate slightly more slowly than the Maxam–Gilbert chemical sequencing products (*see* **Notes 7** and **8**, and also **Fig. 1**).

3.3. Results and Discussion

The result of an ethylation interference experiment with MetJ is shown in **Fig. 2** along with densitometer traces showing quantitative comparisons of the distribution of products in bound and free fractions (**Fig. 3**) *(13)*. Visual examination of the autoradiograph shows that ethylation at 5'-pG2 results in total exclusion of such fragments from protein–DNA complexes. Densitometry indicates that ethylation at other sites inhibits complex formation to varying degrees and that there are more sites in the 3' half-site than in the 5' half-site for which ethylation inhibits complex formation. These data can be interpreted in terms of the MetJ-DNA crystal structure *(8)*. The site at which ethylation completely inhibits complex formation corresponds to the phosphate 5' to the guanine at position 2 of each met box. The crystal structure shows a contact to this phosphate from the N-terminus of the repressor B-helix. Indeed, at the center of the operator site, sequence-dependent distortions of the oligonucleotide fragment away from B-DNA result in displacement of this 5' G2 phosphate by up to 2 Å in the direction of the protein. Thus, this site of complete inhibition of complex formation corresponds to an important contact between the DNA and a secondary structural element in the protein, which presumably is unable to adjust to the presence of a bulky ethyl group. The other sites of ethylation interference effects are contacted by amino acid side chains and peptide backbone groups in extended loops of the repressor. It might be expected that side chains and loops could be flexible enough to reorient in order to reduce steric hindrance between the protein and the ethyl group, thus explaining the partial interference effects. Similar good correlations between the contacts identified by ethylation interference experiments and those seen in crystals have been demonstrated in other systems, such as the phage 434 repressor *(14)* and the phage Lambda repressor *(15)*.

MetJ shows a high level of sequence specificity; therefore, clear interference effects are observed. However, we have observed that proteins with low-sequence specificity show weak interference effects. Densitometry and quantitation may be necessary to resolve these effects from the background of cleavage from fragments shifted by nonspecifically bound proteins. Additionally, proteins binding to DNA asymmetrically (e.g., to sites which do not show the dyad symmetry common to prokaryotic repressor proteins) will show dif-

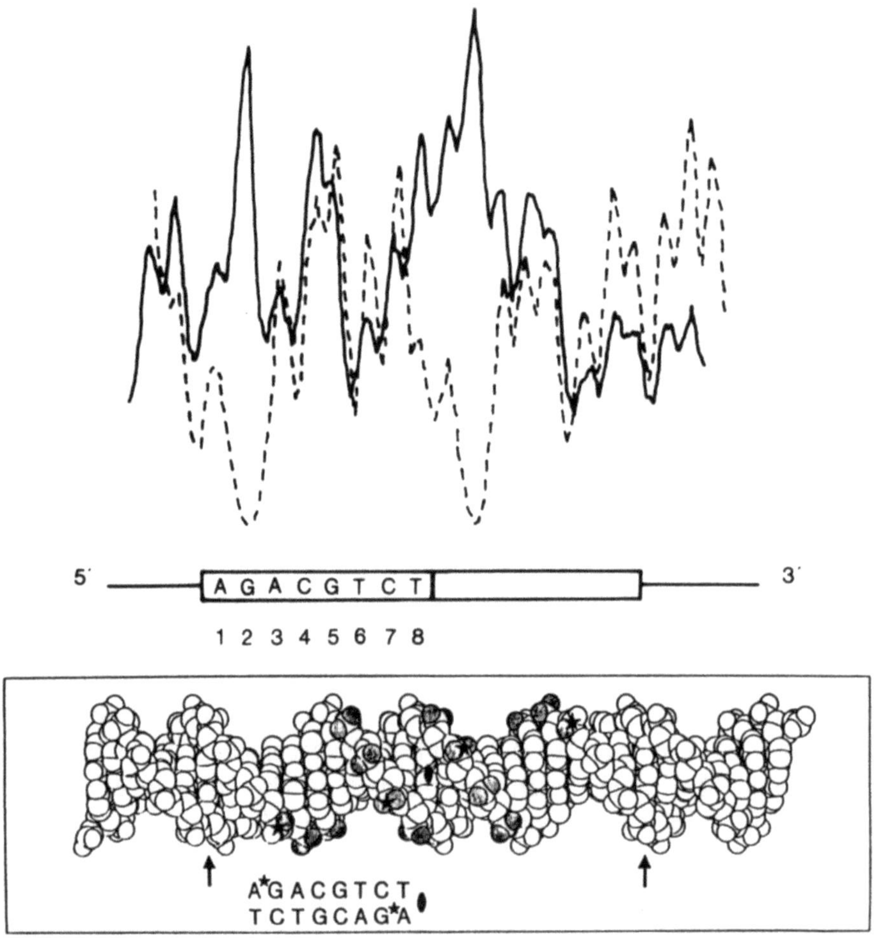

Fig. 3. Summary of the results. **Top:** Densitometer traces of ethylated DNA from the bound and free fractions. DNA from the free fraction is indicated by the trace with solid lines and DNA from the bound fraction by the trace with dashed lines. The position of each binding site is indicated at the bottom of the trace. The sequence within each box is 5'-AGACGTCT–3'. Note that the greatest inhibitory effect is at the second residue (G2) of each box and that the number of positions, ethylation of which inhibits complex formation, is greater in the 3' box than in the 5' box. **Bottom:** Space filling representation of the operator site showing the positions where ethylation results in complete inhibition of complex formation (starred phosphates).

ferent interference effects on each strand, again providing valuable information on the structure of the protein–DNA complex.

The two nonesterified oxygens at each phosphate are nonequivalent diastereoisomers, one of which projects toward the major groove while the other projects toward the minor groove. Although, upon modification, both isomers will show the same charge-neutralization effects, the steric effects on protein–nucleic acid complex formation will be markedly different. The effect has been investigated by incorporation of unique diastereoisomers of ethyl and methyl phosphonates into synthetic oligonucleotides. These modified target sites have then been used in binding assays for restriction enzymes and repressors using standard gel retardation or filter binding assays. Ethylated phosphotriesters have been shown to reduce the T_m of a d[GGAA(Et)TTCC] duplex with the Rp–Rp duplex melting 11°C lower than the unmodified duplex, as a result of steric clashes ***(16)***. Therefore, incorporation of methyl phosphonates may be a better approach and has been used to probe lac repressor–operator interactions. Modification of phosphates outside the binding site had no effect on complex formation, whereas sites shown to be important by biochemical assays showed different effects for each isomer. In one case, the alkyl group blocked direct interaction with the phosphate or close approach of the protein, whereas in the other case, ionic or hydrogen bonded interactions with phosphate remained possible ***(17)***.

4. Notes

1. Modification at secondary (nonphosphate) sites: Early work on the reaction of alkylating agents with DNA showed that a number of products are obtained. With EtNU, phosphotriester groups comprise 60–65% of the reaction products ***(9)***. The remaining products are the result of reactions at base oxygen groups with relative abundance in the order thymine O2 = guanine O6 > thymine O4 >> cytosine O2 for double-stranded DNA ***(12)***. To our knowledge, the effects of such modifications on DNA binding by proteins have not been addressed in the literature on ethylation interference experiments. However, some prediction of the effects might be made based on a knowledge of the structure of DNA. Thymine O2 and cytosine O2 are in the minor groove, and guanine O6 and thymine O4 are in the major groove. For sequence-specific DNA-binding proteins interacting with DNA via the major groove, it would be expected that the presence of an ethyl group would inhibit complex formation, but that a modification in the minor groove would be less inhibitory.
2. Level of modification of DNA by EtNU: A similar intensity of each band (subject to the position-dependent variation of reactivity observed with EtNU) is the required level of modification. We have used a single batch of "home-produced" EtNU for all our ethylation interference experiments. The conditions of time and temperature used with this batch give an appropriate level of probe modification. There may be variations in the reactivity of EtNU from different sources leading to undermodification or overmodification. The correct modification conditions can readily be determined by performing a test ethylation on a small amount of DNA, such as 20,000 cpm, followed by alkaline cleavage (omitting the gel fractionation step), sequencing gel electrophoresis, and autoradiography. Overmodification will produce a bias toward short fragments.

3. Choice of fractionation method: The original report of the use of the phosphate ethylation reaction used the filter binding assay to fractionate protein-bound and protein-free DNA *(2)*; *see also* Chapter 1. Filter-bound DNA is then eluted by washing the filter in a high-salt buffer containing SDS. For other experiments, we find that TE + 0.1 % (w/v) SDS efficiently elutes DNA from nitrocellulose filters. The rapid recovery of DNA from filters is an advantage of using the filter binding assay compared with the gel retardation assay. Despite this, fractionation by gel retardation assay has proven to be by far the most popular method in the literature. The advantage of the gel retardation assay is that both the presence and amounts of multiple complexes can be determined, and these complexes recovered separately, something not possible by filter binding. Parallel binding reactions with ethylated and unethylated DNA and separation on nondenaturing polyacrylamide gels readily demonstrate any differences in the mobility of complexes formed with each DNA sample.
4. Cofactor requirements: High-affinity DNA binding in our system (MetJ) is dependent on the presence of *S*-adenosyl methionine at millimolar concentrations. This cofactor is present in binding reactions and is included in the gel mix, but it would be prohibitively expensive also to include it in the electrophoresis buffer. This does not seem to affect the results obtained by this technique. Electrophoresis for extended times does deplete the lower region of the gel of corepressor, leading to some complex dissociation. Another feature specific to this system is the hydrolysis of the corepressor to presumably inactive products. For this reason, gels were not left to polymerize for more than 2 h.

   In other systems where a cofactor is not required or in which the cofactor is cheap and/or stable over extended periods, the conditions used for the gel-retardation-assay fractionation step should be optimized by a consideration of the specific features of the system under study. For further details, consult Chapter 2.
5. Effects of salt precipitates: During ethanol precipitation of DNA, salt is often also precipitated. This white crystalline precipitate is readily distinguished from the almost clear nucleic acid pellets. The interaction of DNA-binding proteins with their sites is strongly ionic strength dependent, and therefore the presence of a high concentration of salt following ethanol precipitations will inhibit complex formation in addition to any ethylation interference effects. After the cleavage reaction at modified sites, another ethanol precipitation is performed. The presence of a large amount of salt at this stage will prevent complete dissolution of the pellet and will interfere with subsequent electrophoresis. A dark background between each band was often observed on autoradiographs. It is believed that this is caused by the presence of salt in the sample. Reprecipitation as described below helps to reduce this problem.

   To remove any salt precipitate, the pellet can be dissolved in a small volume of TE (e.g., 100 µL) and precipitated by addition of 2 volumes of ethanol without addition of further amounts of salt. The DNA can be pelleted as described in **step 4** in **Subheading 3.2.3.**

6. Recommended controls for the gel-retardation-assay (EMSA) fractionation step: With a precomplex formation modification reaction such as ethylation, it is important to perform appropriate control reactions especially for the gel retardation step. For such binding reactions, we use 20,000-cpm unmodified DNA in the presence and absence of MetJ at the concentration used in the binding reaction with ethylated DNA samples. The specific activity of protein samples may vary from batch to batch. The control reaction outlined above will confirm that the protein sample used is active for DNA binding. In gel retardation assays, the mobility of complexes is a function of a number of properties of the system, such as charge and molecular weight of the protein, stoichiometry of the complex, and bending of the DNA induced by binding of the protein. Demonstration that the mobility of complexes formed with ethylated and unethylated DNA is the same is probably good evidence that there are no significant differences between the complexes.

   The protein concentration used in the fractionation step dictates how many interfering sites are reported. Because the ethyl groups at different sites affect protein binding to differing degrees, increasing the protein concentration can mask any weak interference effects such that only the most strongly interfering sites will be detected. Comparison of the yield of complex on modified DNA with the control complex on unmodified DNA will show the level of binding-site saturation and, therefore, indicate the level of discrimination between strongly and weakly interfering sites that can be expected. Using a range of protein concentrations in the binding reaction with aliquots of the ethylated DNA should therefore allow the strength of the inhibitory effect at each site to be placed in rank order. This is valuable structural information, as it might be expected that the strongest effects will be observed at sites that are in closest contact to the protein in the complex.
7. The presence of multiple cleavage products at each phosphate: The products of the cleavage reaction at phosphotriester groups carry either 3'-OH or 3'-ethylphosphate groups. For large fragments on low-percentage polyacrylamide gels, these two species are not resolved. However, for short fragments on high-percentage gels, two bands are observed at each residue. In practice, this does not produce problems with data analysis.
8. Troubleshooting: We have experienced few problems with this technique. Most problems have been associated with the specific properties of the proteins we have studied. However, it is possible to envisage a number of potential problems and explanations for these, and remedies are presented here.

   Of the available structures of DNA-binding proteins complexed to DNA fragments, there are none in which the protein does not make some contacts to the phosphodiester backbone. Thus, it is expected that because of the size of the ethyl group, an interference effect will always be observed. In the event that no inhibition of complex formation is observed, it should be confirmed that ethylation has occurred by performing a titration of the ethylation reaction as discussed in **Note 2**.

It is possible that an increase in free DNA is observed on the gel retardation (EMSA) assay but that no cleavage products are observed on the sequencing gel, although the full-length DNA is present. This could be caused by an error with the buffer used to resuspend the eluted DNA pellet, the NaOH solution used to perform strand scission, or the temperature of the reaction, all of which can be readily checked.

References

1. Otwinowski, Z., Schevitz, R. W., Zhang, R.-G., Lawson, C. L., Joachimiak, A., Marmorstein, R. Q., et al. (1988) Crystal structure of the trp repressor/operator complex at atomic resolution. *Nature* **335,** 321–329.
2. Siebenlist, U. and Gilbert, W. (1980) Contacts between *Escherichia coli* RNA polymerase and an early promoter of phage T7. *Proc. Natl. Acad. Sci. USA* **77,** 122–126.
3. Hayes, J. J. and Tullius, T. D. (1989) The missing nucleoside experiment: a new technique to study recognition of DNA by protein. *Biochemistry* **28,** 9521–9527.
4. Damante, G., Fabbro, D., Pellizzari, L., Civitareale, D., Guazzi, S., Polycarpouschwartz, M., et al. (1994) Sequence-specific DNA recognition by the thyroid transcription factor–1 homeodomain. *Nucleic Acids Res.* **22,** 3075–3083.
5. Calnan, B. J., Tidor, B., Biancalana, S., Hudson, D., and Frankel, A. D. (1991) Arginine-mediated RNA recognition—the arginine fork. *Science* **252,** 1167–1171.
6. Li, H. L. and Nicholson, A. W. (1996) Defining the enzyme binding domain of a ribonuclease III processing signal. Ethylation interference and hydroxyl radical footprinting using catalytically inactive RNase III mutants. *EMBO J.* **15,** 1421–1433.
7. Rafferty, J. B., Somers, W. S., Saint-Girons, I., and Phillips, S. E. V. (1989) Three dimensional crystal structures of *Escherichia coli met* repressor with and without corepressor. *Nature* **341,** 705–710.
8. Somers, W. S. and Phillips, S. E. V. (1992) Crystal structure of the met repressol-operator complex at 28Å resolution: DNA recognition by β-strands. *Nature* **359,** 387–393.
9. Jensen, D. E. and Reed, D. J. (1978) Reaction of DNA with alkylating agents. Quantitation of alkylation by ethylnitrosourea of oxygen and nitrogen sites on poly [dA-dT] including phosphotriester formation. *Biochemistry* **17,** 5098–5107.
10. Maxam, A. M. and Gilbert, W. (1977) A new method for sequencing DNA. *Proc. Natl. Acad. Sci. USA* **74,** 560–564.
11. Maxam, A. M. and Gilbert, W. (1980) Sequencing end-labelled DNA with base-specific chemical cleavages. *Methods Enzymol.* **65,** 499–560.
12. Singer, B. (1976) All oxygens in nucleic acids react with carcinogenic ethylating agents. *Nature* **264,** 333–339.
13. Phillips, S. E. V., Manfield, I., Parsons, I., Davidson, B. E., Rafferty, J. B., Somers, W. S., et al. (1989) Cooperative tandem binding of Met repressor from *Escherichia coli. Nature* **341,** 711–715.

14. Bushman, F. D., Anderson, J. E., Harrison, S. C., and Ptashne, M. (1985) Ethylation interference and X-ray crystallography identify similar interactions between 434 repressor and operator. *Nature* **316,** 651–653.
15. Ptashne, M. (1992) A genetic switch: Phage [lambda] and higher organisms. Cell and Blackwell Scientific, Cambridge, MA; 2nd edition.
16. Summers, M. F., Powell, C., Egan, W., Byrd, R. A., Wilson, W. D., and Zon, G. (1986) Alkyl phosphotriester modified oligodeoxyribonucleotides. VI. NMR and UV spectroscopic studies of ethyl phospotriester (Et) modified Rp–Rp and Sp–Sp duplexes, {d[GGAA(Et)TTCC]}2. *Nucleic Acids Res.* **14,** 7421–7436.
17. Noble, S. A., Fisher, E. F., and Caruthers, M. H. (1984) Methylphosphonates as probes of protein–nucleic acid interactions. *Nucleic Acids Res.* **12,** 3387–3304.

16

Hydroxyl Radical Interference

Peter Schickor, Evgeny Zaychikov, and Hermann Heumann

1. Introduction

Interference studies are just the inverse approach of "footprinting" experiments. In one type of experiment, the effect of a chemical modification of a single base on the subsequent binding of a sequence-specific protein is determined, whereas in the other, it is the accessibility of the protein-bound DNA to modification that is determined. Thus, the experiments necessarily differ in the order in which the protein binding and DNA modification steps occur. The "interference" approach is characterized first by chemical modification of the DNA and by subsequent protein binding. Such studies provide information on the change of the binding strength following single-base modification. This change can either be positive or negative and can be quantified by the gel shift assay *(1)* (*see* Chapters 2 and 5).

Here, we describe the use of hydroxyl radicals as the modifying reagent. This probe has a number of advantages compared to the most commonly used probes, such as dimethylsulfate (DMS) or ethylnitrosourea. Hydroxyl radicals cleave the backbone of DNA with almost no sequence dependence, whereas most other reagents react in a highly sequence-dependent manner with the bases of the DNA. Furthermore, hydroxyl radicals modify the DNA by elimination of a nucleoside, producing a "gap" in one DNA strand (for details of the chemical reaction, *see* Chapter 5). This allows the study of two kinds of effect:

1. The effect on protein binding caused by missing contacts. Other reagents prevent binding by introducing bulky groups into a base of the DNA. Whether this reflects the importance of a base for protein–DNA interaction or whether the bulky group is just a steric hindrance is difficult to determine.
2. The effect on the DNA structure because of the eliminated base. The missing nucleoside is a center of enhanced flexibility in the DNA. Therefore, the effect of

From: *Methods in Molecular Biology, vol. 148: DNA–Protein Interactions: Principles and Protocols, 2nd ed.*
Edited by: T. Moss © Humana Press Inc., Totowa, NJ

DNA flexibility on the protein–DNA interaction can be studied. The modification of the DNA by hydroxyl radicals is a very fast and highly reproducible experiment in contrast to other methods. The reagents needed are easily available.

1.1. Generation and Action of Hydroxyl Radicals

Hydroxyl radicals introduce randomly distributed nucleoside eliminations and associated backbone cleavages in the DNA. The generation and action of hydroxyl radicals is described in detail in Chapter 5.

1.2. Principle of the Procedure

A DNA fragment labeled either at the 3' or the 5' end is subjected to hydroxyl radical treatment. The concentration of the hydroxyl radicals is adjusted so that the number of base eliminations is less than one per DNA, this means that only approx 10% of the DNA fragments will be modified. This population of randomly modified DNA molecules is incubated with the protein under study (*see* **Fig. 1**). Those DNA molecules that are still able to bind the protein can be separated from those DNA molecules that are no longer able to bind the protein by nondenaturing gel electrophoresis or electrophoretic mobility shift. The bands containing free DNA and the complexed DNA (**Fig. 2**) are eluted and, after denaturation, are applied on a sequencing gel in order to determine the positions of the modifications (**Fig. 3**). The relative effect on the binding strength caused by a single-base elimination can be quantified by determining the intensity change of the different bands by densitometric scanning.

1.3. Interpretation of the Interference Pattern

The interpretation of an electrophoresis pattern obtained by hydroxyl radical interference studies (**Fig. 3**) is not as straightforward as hydroxyl radical footprinting studies. The reason is that a single-base elimination generated by hydroxyl radicals can cause two effects on protein binding:

1. Lack of specific contacts between the protein and the DNA. This leads to a decrease in the binding strength.
2. Enhanced flexibility of the DNA at the position of the base elimination and backbone cleavage. This can lead to an increase or a decrease of the binding strength.

A quantitative interpretation of the interference pattern is not always possible, as the intensity of the bands of the interference pattern reflects the sum of the different effects contributing to the binding strength. Additional information concerning the protection of the DNA by the protein (e.g., by using hydroxyl radical footprinting) is necessary in order to differentiate between the two effects. The examples in the following subheadings may be used as a guideline for interpretation.

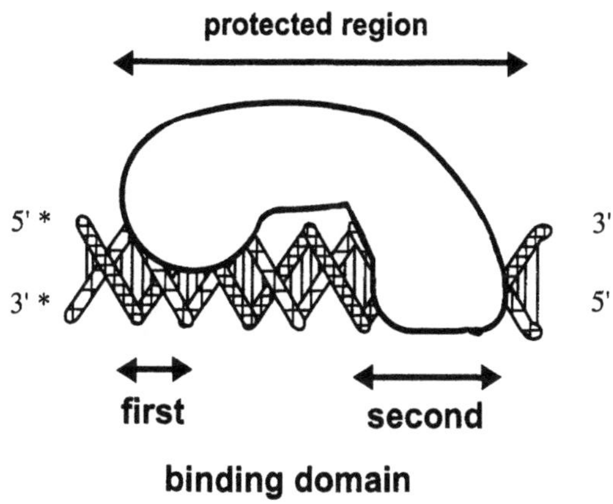

Fig. 1. A schematic representation of a protein–DNA complex with two distinct interaction sites. At the first binding site, the protein interacts with only one side of the DNA. At the second binding site, the protein wraps fully around the DNA. The asterisk indicates the position of the radioactive label.

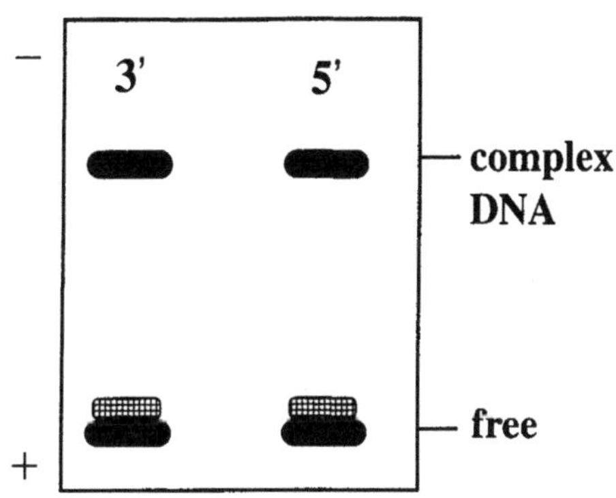

Fig. 2. Nondenaturing gel electrophoresis of the complex with hydroxyl radical pretreated DNA as target for the protein binding. The two lines show the same complex labeled at the 3' and the 5' ends, as indicated in **Fig. 1**. **Note:** The single-base elimination leads to an enhanced flexibility of the DNA indicated by the "smear" of the band representing the free DNA.

1.4. Examples of the Application of Hydroxyl Radicals as Interference Probes

1.4.1. Transcription Factors

The interference of binding of some eubacterial and eukaryotic transcription factors was investigated using hydroxyl radical pretreated DNA-binding sequences. In all cases, the single-base elimination led to a decrease of the

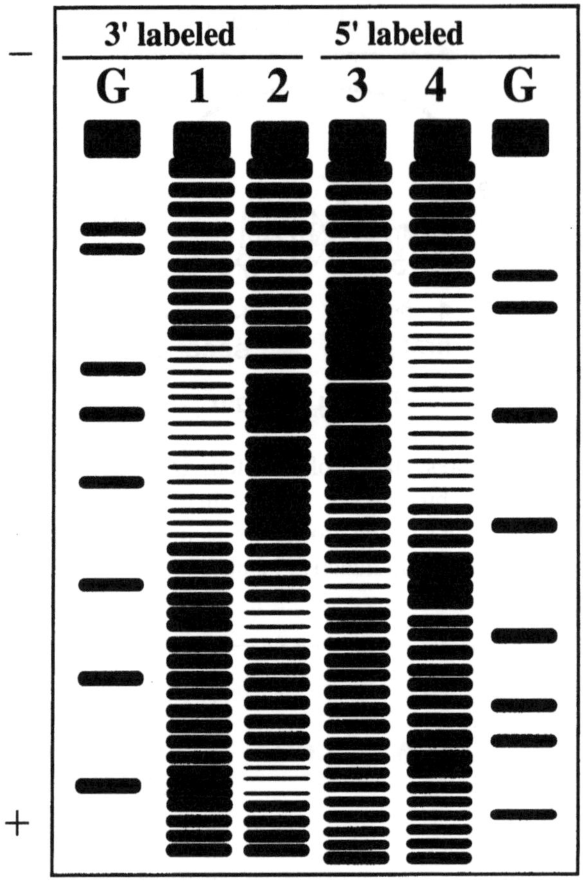

Fig. 3. A putative interference pattern of the protein DNA complex (**Fig. 1**) obtained after separation of "free" DNA and "complexed" DNA by a nondenaturing gel electrophoresis (**Fig. 2**). Lanes I and 4 show the pattern of the "free" DNA labeled at the 3' and the 5' ends, respectively. Lanes 2 and 3 show the pattern of the "complexed" DNA. Lanes G contain the length standards obtained by a G-specific Maxam–Gilbert sequencing reaction.

binding strength indicating a loss of contacts. Examples are the progesterone receptor *(2)*, λ-repressor, cro-protein *(3)*, necrosis factor-κB *(4)*, and GCN4 transcription factor *(5)*.

1.4.2. RNA Polymerase-Promoter Interaction

A strong eubacterial promoter was subjected to hydroxyl radical treatment in order to investigate the influence of single-base eliminations on the binding of *Escherichia coli* RNA polymerase *(6)*. This study revealed three patterns of interaction that could be attributed to different sites of the promoter:

1. Direct base contact with the template strand in the "–35 region": This was concluded from the reduced affinity of the polymerase for a promoter having a base eliminations in this region together in combination with the results of hydroxyl radical footprinting studies *(7)*, which revealed close contacts between the bases of this sequence and the protein. This is an example in which a base elimination leads to missing contacts between protein and DNA.
2. A DNA-structure-dependent interaction in the "–10 region": This was inferred from the increased binding affinity of the polymerase for a promoter having base eliminations in this region in conjunction with footprinting studies that revealed that this region is in close contact with the protein *(7)*. This is an example in which a base elimination leads to enhanced flexibility of the DNA favoring protein binding and suggests that the DNA adopts a particular conformation when bound to the protein.
3. An interaction that is based on a defined spatial relationship between the "–35 region" and the "–10 region" domains. This conclusion was drawn from the following findings:
   a. Base elimination within the promoter region between the two domains reduces the binding affinity of RNA polymerase.
   b. The eliminated bases had no contact with the protein (as shown by their accessibility to hydroxyl radical footprinting studies *[7]*).
   c. The effect was observed for both DNA strands. This is an example in which base elimination leads to a loss of the defined spatial relationship between two functionally important sites because of an enhancement in DNA flexibility.

2. Materials

2.1. The Cutting Reaction

Prepare the following solutions separately (*see* **Note 1**):

1. 0.1 *M* dithiotreitol (DTT).
2. 1% Hydrogen peroxide.
3. Iron(II)–EDTA-mix: Mix equal volumes of 2 m*M* ammonium iron(II) sulfate hexahydrate ($[NH_4]_2Fe[SO_4]_2 \cdot 6H_2O$) and 4 m*M* EDTA.
4. Stop mix: 4% glycerol, and 0.6 *M* sodium acetate.

2.2. The Sequencing Gel

1. Urea (ultra pure).
2. 20X TBE: 1 *M* Tris base, 1 *M* boric acid, and 20 m*M* EDTA.
3. Acrylamide solution: 40% acrylamide, and 0.66% *bis*-acrylamide (*see* **Note 2**).
4. 10% Sodium persulfate (*see* **Note 3**).
5. 10% TEMED.
6. Sequencing gel (8%): 21 g urea, 2.5 mL of 20X TBE, and 10 mL of 40% acrylamide solution are made up to 50 mL with bidistilled H_2O and stirred under mild heating until urea is dissolved. The solution is filtered (filter pore size: 0.2 µm)

and degassed for 5 min. Then, 0.3 mL of 10% sodium persulfate and 0.3 mL of 10% TEMED are added immediately before pouring the solution between the glass plates.

7. Loading buffer for the sequencing gel (stock solution): 100 mL formamide (deionized), 30 mg xylenecyanol FF, 30 mg bromophenol blue, and 750 mg EDTA.
8. Electrophoresis buffer: 1X TBE.

2.3. The Nondenaturing Gel for DNA Isolation and Electrophoretic Mobility Shift Assay

1. 20X TBE: as described in **Subheading 2.2.**
2. Acrylamide solution: 30% acrylamide, and 0.8% *bis*-acrylamide (*see* **Note 2**).
3. 10% Sodium persulfate (*see* **Note 3**)
4. 10% *N,N,N',N'*-tetramethylethylene diamine (TEMED) (aqueous solution).
5. 3% Nondenaturing gel: 1.5 mL of 20X TBE, 3 mL of the acrylamide solution, and 25.5 mL of bidistilled water are mixed and degassed for 5 min. Then 300 µL of 10% ammonium persulfate and 300 µL of 10% TEMED are added before pouring the solution between the glass plates.
6. Loading buffer for the nondenaturing gel (stock solution): 50% glycerol, and 0.1% bromophenol blue.
7. Electrophoresis buffer: 1X TBE.

2.4. Other Items

1. Sequencing gel apparatus.
2. Apparatus for nondenaturing gel electrophoresis.
3. Filters for drop dialysis, VS, 0.025 µm (Millipore, Bedford, MA).
4. SpeedVac concentrator.
5. 1X TE: 10 m*M* Tris-HCl, pH 7.9, and 1 m*M* EDTA.

3. Methods

3.1. Establishing the Conditions for Obtaining Specific Protein–DNA Complexes

The method of establishing the conditions for complex formation using the electrophoretic mobility-shift assay is described in Chapter 4. The enzyme to DNA ratio should be adjusted so that the ratio of complexed to free DNA is about 1:1 (*see* **Note 4**).

3.2. Hydroxyl Radical Base Elimination

1. End-label an aliquot of the DNA fragment of interest under standard conditions *(8)* at the 5'-position, using T_4 polynucleotide kinase and [γ-^{32}P] ATP and end label a second aliquot at the 3'-position, using the Klenow fragment of DNA polymerase I and the appropriate [α-^{32}P] dNTP. In each case remove one label end by asymmetric cleavage of the DNA fragment with an appropriate restriction

endonuclease. Purify the uniquely end-labeled DNA fragments by nondenaturing gel electrophoresis. The total amount of DNA in an assay of 20 µL should not be below 100 ng (*see* **Note 5**).

2. Dissolve each end-labeled DNA preparation in 20 mL of 1X TE buffer. Add to both samples 2 µL each of the previously prepared solutions of DTT, hydrogen peroxide, and the iron(II)–EDTA mix by putting single drops of each solution on the inner wall of the tube and rapidly mixing the three drops before combining them with the sample using a micropipet.
3. Incubate for 3–4 min at room temperature.
4. Add 25 mL stop mix and 150 µL of ice-cold 100% ethanol to precipitate the DNA. Keep the solution at –70°C for 30 min.
5. Recover the DNA by microcentrifugation for 30 min. Wash the pellet with ice-cold 80% ethanol, dry the pellet under vacuum, and redissolve the pellet in 20–50 mL of 1X TE buffer.
6. Heat the sample for not longer than 2 min at 90°C and place on ice (*see* **Note 5**). Apply the sample onto a 6–10% sequencing gel (for the analysis of fragments in the range of 50–150 bases a gel consisting of 8% acrylamide is adequate). Use as length standards a Maxam–Gilbert sequencing reaction of the 5'- or 3'- labeled DNA fragment.
7. Run the gel at 50 W at a temperature of 60°C for 1.5–2 h. The gel is ready when the xylenecyanol dye marker is about 3–5 cm above the bottom of the gel.
8. After electrophoresis, expose the gel to an X-ray film using an intensifying screen at –70°C overnight. For subsequent experiments choose the time of hydroxy radical cleavage that provides an even distribution of bands and leaves around 90% of the DNA uncleaved.

3.3. Interference Studies on Protein-DNA Complexes

1. Prepare two 20-µL samples of the complex to be studied using DNA labeled respectively at the 3' and the 5' ends and hydroxy radical treated as described in **Subheading 3.2.** Use conditions established for optimal complex formation (*see* **Subheading 3.1.**). The total amount of radioactivity in one assay should be around 60,000–80,000 cpm.
2. Pour 30–40 mL of the dialysis buffer containing 8 m*M* Tris-HCl, pH 7.9, into a Petri dish (*see* **Note 6**). Place a Millipore filter (*see* **Subheading 2.4.**) on the surface of the buffer, shiny side (hydrophobic side) up. Put the samples containing the complexes onto the filter for 1 h in order to remove salt (*see* **Note 7**). Remove the samples from the filter by a micropipet and transfer them to a fresh 1.5-mL Eppendorf tube.
3. Separate the complex and the free DNA on a nondenaturing acrylamide gel electrophoresis. The method for separation of complexes from free DNA is described in **Subheading 3.3.2.** in Chapter 5.
4. The whole procedure for the recovery of DNA from the gel is described in **Subheading 3.3.2.** in Chapter 5.

5. Adjust the amount of radioactivity and volume in each sample to about 5000–6000 cpm in 4–5 µL. Heat the samples for not longer than 2 min at 90°C and put them on ice (*see* **Note 8**).
6. Analyze the DNA by denaturing gel electrophoresis. Apply the samples to a 6–10% sequencing gel (for the analysis of fragments in the range of 50–150 bases a gel consisting of 8% acrylamide is adequate). Load the following samples on the gel: end-labeled DNA, the Maxam–Gilbert reaction as length standard, the free DNA recovered from the gel, and the DNA recovered from the complex.
7. Run the gel for 1.5–2 h at about 50 W to obtain a temperature of 60°C.
8. Expose the gel after electrophoresis to an X-ray film using an intensifying screen at –70°C overnight.

4. Notes

1. The iron(II), iron-EDTA mix, and the H_2O_2 solutions should be freshly made before use. The solutions of DTT (0.1 M), EDTA (4 m*M*), H_2O_2 (as a 30% stock solution), and the stop mix are stable for months being stored at –20°C.
2. The acrylamide solutions are stable for months if protected from light and kept at 4°C.
3. Sodium persulfate has an advantage over routinely used ammonium persulfate of being much more stable in aqueous solution. The 10% sodium persulfate solution may be kept at least 1 mo at 4°C without loss of activity.
4. It is advisable to keep the enzyme-to-DNA ratio <1 in order to assure stringent sequence selection conditions.
5. It is strongly recommended to check the quality of the labeled DNA on a sequencing gel before use. Nicks in the double strand, which could result from DNase activities during the preparation procedure, will appear as additional bands in the sequencing gel. This admixture of bands would spoil the whole footprint, even when present in only small amounts. Furthermore, it is recommended not to store the freshly labeled DNA longer than 2 wk because of the danger of radiation damage to the DNA.
6. The buffer conditions can be varied (e.g., the pH), but the ionic strength should not be too high (a maximum of 50 m*M* NaCl) in order to obtain sharp bands during the following electrophoresis. Many protein–DNA complexes are very stable at low ionic strength (e.g., complexes between RNA polymerase and promoters *[6]*). Therefore, in most cases the stability of the pH in the following electrophoresis is the only limitation to lowering the ionic strength.
7. The purpose of the dialysis is the removal of salt, the presence of which would lower the quality of the electrophoresis pattern. As a rough approximation, one can remove up to 80–90% of the salt present in the sample within 1 h of drop dialysis.
8. Longer heating or boiling creates additional cuts in the DNA.

References

1. Heumann, H., Metzger, W., and Niehörster, M. (1986) Visualization of intermediary transcription states in the complex between *Escherichia coli* DNA-depen-

dent RNA polymerase and a promoter-carrying DNA fragment using the gel retardation method. *Eur. J. Biochem.* **158,** 575–579.
2. Chalepakis, G. and Beato, M. (1989) Hydroxyl radical interference: a new method for the study of protein–DNA interaction. *Nucleic Acids Res.* **17,** 1783.
3. Hayes, J. J. and Tullius, Th. D. (1989) The missing nucleoside experiment: a new technique to study recognition of DNA by protein. *Biochemistry* **28,** 9521–9527.
4. Schreck, R., Zorbas, H., Winnacker, E. L., and Baeuerle, P. A. (1990) The nf-kappa-b transcription factor induces DNA bending which is modulated by its 65-kd subunit. *Nucleic Acids Res.* **18,** 6497–6502.
5. Gartenberg, M. R., Ampe, C., Steitz, T. A., and Crothers, D. M. (1990) Molecular characterisation of the GCN4-DNA complex. *Proc. Natl. Acad. Sci. USA* **87,** 6034–6038.
6. Werel, W., Schickor, P., and Heumann, H. (1991) Flexibility of the DNA enhances promoter affinity of *Escherichia coli* RNA polymerase. *EMBO J.* **10,** 2589–2594.
7. Schickor, P., Metzger, W., Werel, W., Lederer, H., and Heumann, H. (1990) Topography of intermediates in transcription initiation of *E. coli. EMBO J.* **9,** 2215–2220.
8. Maniatis, T., Fritsch, E. F., and Sambrock, J. (1982) *Molecular Cloning. A Laboratory Manual.* Cold Spring Harbor Laboratory, Cold Spring Harbor, NY.

17

Identification of Sequence-Specific DNA-Binding Proteins by Southwestern Blotting

Simon Labbé, Gale Stewart, Olivier LaRochelle, Guy G. Poirier, and Carl Séguin

1. Introduction

Southwestern blotting was first described by Bowen et al. *(1)* and was used to identify DNA-binding proteins that specifically interact with a chosen DNA fragment in a sequence-specific manner. In this technique, mixtures of proteins such as crude nuclear extracts or partially purified preparations are first fractionated on a sodium dodecyl sulfate (SDS) denaturing gel; the gel is then equilibrated in a SDS-free buffer to remove detergent and the proteins transferred by electroblotting to an immobilizing membrane. During the transfer the proteins renature and hence DNA-binding proteins may subsequently be detected on the membrane by their ability to bind radiolabeled DNA. Fractionation of crude nuclear extracts on an SDS gel followed by electroblotting and analysis for sequence-specific DNA binding directly on the blot combines the advantages of a high-resolution fractionation step with the ability to rapidly analyze for a large number of different DNA-binding specificities.

The successful identification of specific DNA-binding proteins by the Southwestern technique largely depends on the renaturation of the proteins after their separation by SDS-polyacrylamide gel electrophoresis (PAGE). The ease with which renaturation can be achieved after treatment with SDS varies from protein to protein. Some DNA-binding proteins may be inefficiently renatured and thus be unable to bind DNA once they are adsorbed onto membranes. In addition, any multimeric protein that requires a combination of different molecular-weight subunits to bind DNA will be missed. Proteins requiring a cofactor(s) in order to show their ability to specifically interact with DNA may also be difficult to detect.

From: *Methods in Molecular Biology, vol. 148: DNA–Protein Interactions: Principles and Protocols, 2nd ed.*
Edited by: T. Moss © Humana Press Inc., Totowa, NJ

Site-specific protein–DNA interactions may also be obscured by the large number of nonspecific DNA-binding proteins present in a typical crude nuclear extract. Conditions for DNA binding, such as pH, ionic strength, and divalent cation requirement, as well as the type and amount of nonspecific DNA used as competitor, should be optimized for each protein under investigation. Special attention should also be given DNA-binding proteins frequently present in crude nuclear extracts and known to copurify with sequence-specific DNA-binding proteins (*see* **Note 1**). Despite these shortcomings, Southwestern blotting has, over the last few years, been used to identify and characterize several sequence-specific DNA-binding proteins, histones, and nonhistone proteins; for examples, *see* **Fig. 1** and **refs. *2–14***, as well as RNA-binding proteins ***(15–17)***. In addition, preparative Southwestern blots have been used to define DNA sequences recognized by a given DNA-binding protein ***(18)***, whereas Southwestern screening of cDNA libraries has been used to isolate several sequence-specific DNA-binding proteins ***(19,20)***.

2. Materials

All solutions are made with double-distilled or Milli Q water.

1. Crude nuclear extracts or protein chromatographic fractions.
2. 1.5 *M* Tris-HCl, pH 8.8.
3. 0.5 *M* Tris-HCl, pH 6.8.
4. Sodium dodecyl sulfate (SDS) 10% and 2-bis-mercaptoethanol.
5. Tris–glycine buffer 10X: 250 m*M* Tris base and 1.92 *M* glycine pH 8.3. To make this up, dissolve 30 g Tris base, and 144 g glycine in 1 L of water (*see* **Note 2**).
6. Electrophoresis running buffer: 2X Tris–glycine buffer containing 1% SDS.
7. Transfer buffer: Tris–glycine buffer 1X.
8. Binding buffer 1X: 20 m*M* HEPES, pH 7.9, 5 m*M* $MgCl_2$, 50 m*M* NaCl, and 1 m*M* dithiothreitol (DTT) (*see* **Notes 3** and **4**).
9. Sample loading buffer 2X: 125 m*M* Tris, pH 6.8, 4% (w/v) SDS, 20% (v/v) glycerol, 10% (v/v) 2-*bis*-mercaptoethanol, and 0.025% (w/v) bromophenol blue. Make 1-mL aliquots and store at –20°C.
10. Acrylamide (50:1) and (30:0.8):acrylamide:*N',N'*-methylene–*bis*-acrylamide made up in water (*see* **Note 5**). Deionize the acrylamide solution with a mixed-bed resin ***(21)*** and store at 4°C. **Note:** Acrylamide is a potent neurotoxin. Wear gloves and a facemask when handling the dry powder.
11. 10% Ammonium persulfate (w/v) in water (prepared weekly and stored in aliquots at –20°C), and *N,N,N',N'*-tetramethylethylenediamine (TEMED).
12. Methanol and isobutanol.
13. Blocking buffer–5%: 5% (w/v) nonfat dried milk and 0.01% (v/v) Antifoam A emulsion (Sigma) in 1X binding buffer.
14. Blocking buffer–0.25%: 0.25% (w/v) nonfat dry milk, 0.01% (v/v) Antifoam A emulsion in 1X binding buffer.

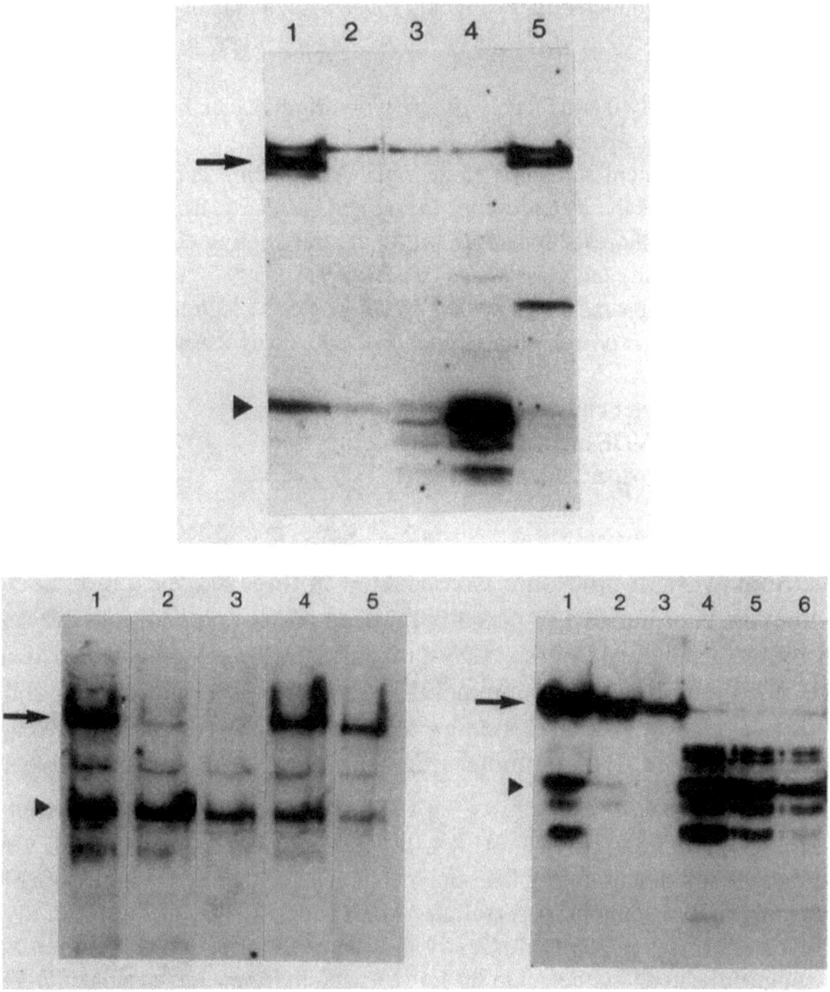

Fig. 1. Southwestern analysis of mouse L-50 cell crude nuclear extracts. Approximately 200 μg of extracted protein was resolved by SDS-PAGE, transferred to a PVDF membrane, and incubated with radiolabeled DNAs. **Upper panel:** Binding of labeled oligos corresponding to control and mutant metal regulatory elements (MREs) of the mouse metallothionein 1 gene promoter. Lanes: 1, control MREd; 2, Mutant-a; 3, Mutant-b; 4, Mutant-c; 5, Mutant-d. **Left lower panel:** Analysis of the binding specificity of the MREd oligo to nuclear proteins using MREd as the probe and cold MREd (lanes 1–3) or Mutant-a (lanes 4 and 5) as competitor. **Right lower panel:** Competition experiments using heterologous probe and nonspecific nucleic acids as competitors. Probes: lanes, 1–3, MREd; lanes 4–6, a 309 bp pBR322 *Msp*I fragment. Nonspecific nucleic acid competitor cocktail was added as follows: lanes 1 and 4, none; lanes 2 and 5, 0.8 μg/mL; lanes 3 and 6, 4 μg/mL. The arrow corresponds to M_r 108,000 and the arrowhead indicates the position of Mr 45,000. (From **ref. *21*.**)

15. End-labeled oligonucleotides (oligos) or DNA fragments of high specific activity. Standard procedures ***(21)*** can be followed in the preparation of DNA samples (*see* **Note 6**).
16. Poly(dA–dT)·poly(dA–dT), poly(dI–dC)·poly(dI–dC), calf thymus DNA, salmon sperm DNA, *Escherichia coli* DNA.
17. ^{14}C-Labeled protein standards (5 µCi/mL, Amersham) and prestained protein standards such as the Kaleidoscope Prestained Standards from Bio-Rad.
18. Vertical electrophoresis apparatus with a central cooling core (e.g., Protean II slab cell, Bio-Rad) and glass plates (*see* **Note 7**).
19. Electroblotting apparatus with cooling coil (e.g., Trans-blot transfer cell, Bio-Rad).
20. Electrophoresis DC power supply capable of 200 V and 2 Amp (e.g., model 200/2.0 of Bio-Rad).
21. Recirculating water chiller apparatus.
22. Immobilon™–PVDF membrane (Millipore).
23. Whatman 3MM paper.

3. Method

The Southwestern procedure is conducted in three stages: Crude nuclear protein extracts or purified or partially purified protein preparations are separated by electrophoresis on an SDS–polyacrylamide gel, transferred to the PVDF membrane, and then the immobilized proteins assessed for their ability to bind a labeled oligo corresponding to a given DNA cis-acting regulatory element or a labeled DNA promoter fragment.

1. Prepare an 8% acrylamide–*bis*-acrylamide (50:1) separating gel (*see* **Note 4**). Mix 25 mL 1.5 *M* Tris-HCl, pH 8.8, 15.7 mL acrylamide, and 57.2 mL water. Filter the solution through a 0.45-µm membrane and then add 0.5 mL 10% SDS, 0.5 mL of 10% ammonium persulfate (APS), and 35 µL TEMED. Mix gently by inversion and pour the gel. Gently overlayer the acrylamide solution with isobutanol and allow the gel to set for at least 1 h. Remove the isobutanol, rinse gel surface thoroughly with water. Prepare the stacking gel by mixing 6.25 mL of 0.5 *M* Tris-HCl, pH 6.8, 3.35 mL acrylamide–*bis*-acrylamide (30:0.8), 15 mL water. Filter the solution and add 0.25 mL of 10% SDS, 0.25 mL of 10% APS and 17 µL TEMED. Mix gently, pour over the separating gel, add a comb, and allow the stacking gel to set for at least 1 h. Just before using, carefully wash the wells with running buffer using a syringe.
2. Dilute equal volumes of protein preparation and 2X loading buffer. We typically use 100–400 µg (protein) of a crude nuclear extract (prepared according to Dignam ***[22]***) in the presence of a cocktail of four protease inhibitors (i.e., leupeptin, polymethylsulfonyl fluoride [PMSF], antipain, and chemostatin A). Lower amounts are required when using purified or partially purified protein preparations. Mix loading buffer and samples just before loading to avoid SDS precipitation. We do not boil or otherwise purposely denature the protein preparations (*see* **Note 8**). Include 15 µL of ^{14}C-labeled protein markers in a separate

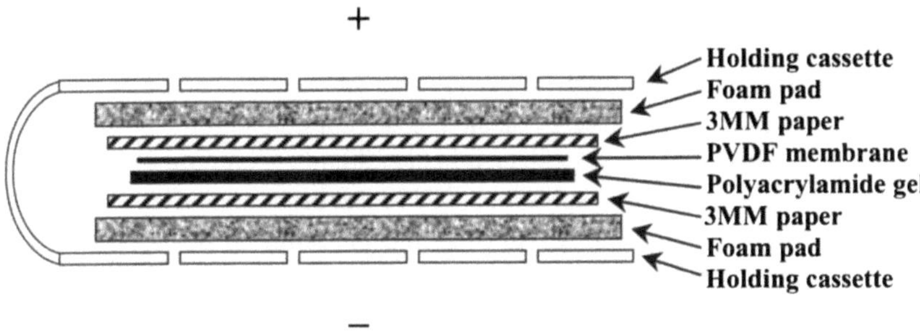

Fig. 2. Schematic representation of the protein blotting sandwich assemblage used in Southwestern experiments. The orientation of the sandwich in relation to the cathode (+) and the anode (–) is indicated.

lane. Run gel in electrophoresis electrode buffer (2X Tris–glycine buffer, 1% SDS) overnight at 75 V (constant voltage). Do not prerun the gel. We perform the electrophoresis at 4°C by connecting the cooling coil provided with the electrophoresis apparatus to a recirculating water chiller (*see* **Note 9**).

3. Electrophoresis gels should be pre-equilibrated in transfer buffer prior to commencement of electrophoretic transfer. Pre-equilibration will aid the removal of SDS, contaminating the electrophoresis salts from the proteins, and will facilitate subsequent renaturation to the native conformation (*see* **Note 10**). In addition, the pretransfer step will allow any changes in the size of the gel resulting from the swelling or shrinking to occur at this stage rather than during the transfer. Thus, after the gel is run, dismantle the apparatus, remove the gel from the plates, and soak it in transfer buffer for 60 min with gentle agitation. We usually perform this operation in the cold room (*see* **Note 9**). Wear clean disposable gloves.
4. Lay the gel onto 3MM paper prewetted with transfer buffer. Always keep the gel wet by regularly pouring buffer over it.
5. Prewet the PVDF membrane in methanol for few seconds and transfer it to water.
6. Begin mounting the protein-blotting sandwich by placing one sponge (provided with the blotting transfer apparatus) into the holder, followed by the 3MM support paper and placing the gel (*see* **Fig. 2** and **Note 11**).
7. Lay the prewetted PVDF membrane over the gel.
8. Carefully remove any air bubbles trapped between the gel and the membrane. We use a 10-mL glass pipet gently rolled over the membrane.
9. Place a second prewetted (with transfer buffer) 3MM paper and repeat **step 8**. Place the second foam pad onto the sandwich and close the holder. Introduce the holder into the transfer apparatus, previously filled with cold transfer buffer. Be sure that the membrane is toward the positive electrode. Place the transfer apparatus on the stirring plate and connect it to a circulating cooler unit (*see* **Note 12**).
10. It is important that a stirring bar be placed inside the transfer apparatus and that the transfer buffer be stirred during the course of the experiment. This will help to maintain uniform conductivity and temperature during electrophoretic

transfer. Failure to do so will result in poor transfer of proteins and may pose a safety hazard.

11. Transfer at 50 V constant for 3 h.
12. After the transfer is completed, dismantle the apparatus, remove the membrane from the gel, and successively place it in (1) 200 mL of binding buffer in a plastic dish for 15 min with gentle agitation, (2) 250 mL blocking buffer–5% for a minimum of 1 h with agitation, (3) and 150 mL blocking buffer–0.25% containing 10^6 cpm/mL of probe DNA and leave overnight with agitation (*see* **Note 13**). Never let the membrane dry between the transfer and the binding steps.
13. After the transfer, the gel may be transferred in 50% methanol for subsequent silver staining to visualize untransferred proteins (*see* **Note 14**).
14. Wash the membrane by successively soaking it under continuous agitation in (1) blocking buffer–0.25% for 30 s, (2) blocking buffer–0.25% for 20 min, (3) blocking buffer–0.25% for 20 min, and (4) Binding buffer 1X for 20 min. Recover wash liquids because they contain ^{32}P.
15. Let the membrane dry on 3MM paper at room temperature for 1 h, then autoradiograph using an X-ray film and intensifying screen if necessary.

4. Notes

1. Crude nuclear extracts prepared from human HeLa and mouse L cells contain three common nonspecific DNA-binding proteins that often contaminate preparations of affinity-purified factors. These are poly (ADP–ribose) polymerase (PARP), which has an M_r of 116,000 (as well as a typical proteolytic fragment of M_r-60,000) ***(27)***, the Ku antigen, which consist of two polypeptides of M_r-70,000 and 80,000 ***(28)***, and replication protein A (RP-A) of M_r-74,000 ***(29)***. The heat-shock protein hsp70 (M_r-70,000), which can stick rather tenaciously to proteins that are improperly folded, may also cause problems. Thus, when Southwestern analyses are performed with affinity-purified nuclear proteins, it is recommended to be suspicious of positive polypeptides of M_r-60,000, 70,000–74,000, 80,000, and 110,000, and to perform control experiments. For instance, in the case of PARP, if a polypeptide with an M_r in the range of this protein is detected in the Southwestern procedure, a Western analysis could be performed on an aliquot of the protein preparation using anti-PARP antibodies available commercially (Roche Diagnostics/Boehringer). Alternatively, a chromatographic step with Red Agarose (Bio-Rad), which binds PARP with high affinity, can be performed to remove of this possible contaminant ***(27)***. One should, however, bear in mind that PARP has recently been shown to act as a coactivator for a number of transcription factors, including AP-2 ***(30)***, E47 ***(31)***, and TEF-1 ***(32)***; thus, its detection using a specific DNA probe corresponding to the binding site of a known transcription factor could actually be functionally relevant.
2. Tris–glycine buffers should not be pH adjusted.
3. Dithiothreitol should be stored at –20°C, thawed on ice, and added when the solution is ice-cold.

4. Several DNA-binding proteins require zinc ions to efficiently bind DNA. As a precaution, we routinely add 5 μ*M* $ZnCl_2$ to all solutions (loading, electrophoresis, transfer, binding, and wash buffers).
5. The percentage of gel and the ratio of acrylamide to bis-acrylamide will depend on the M_r of the protein of interest. We have used a ratio of 50:1 to increase resolution of the high-molecular-weight protein species. A ratio of 30:0.8 or 29:1 may be more appropriate for average protein species. Plates are cleaned with phosphate-free soap and rinsed with deionized water, Milli Q water, and ethanol 95%. We usually do not silanize the plates with dichlorodimethylsilane, as we observed that it could interfere with the subsequent steps.
6. Synthetic oligos or DNA fragments can be conveniently 5' labeled with T_4 kinase ***(21)***. A radiolabeled probe with a high specific activity increases the sensitivity of detection and reduces background ***(23)***. A nonradioactive Southwestern procedure, using chemiluminescent detection, has also been described ***(24)***.
7. The procedure can both be carried out using a standard electrophoresis and transfer apparatus or a small-size apparatus such as the Mini-protean II and mini-Trans-Blot transfer cell of Bio-Rad ***(2)***.
8. We do not boil samples before electrophoresis in order to avoid irreversibly denaturing proteins or to affect the DNA-binding properties of more sensitive protein species ***(4)***.
9. Depending of the biochemical properties of the protein under investigation and of its stability at room temperature, it is generally recommended to perform the whole procedure at 4°C. Particular conditions of temperature may need to be determined for an optimal binding of different factors to DNA.
10. Although the extent of renaturation is not known, this wash, by removing SDS from the proteins, permits functional binding to the DNA recognition sequence. Triton X-100 ***(1,12)***, urea ***(3,5,8,11)*** or guanidium HCl ***(9,10)*** can be included in the transfer buffer to promote the removal of SDS and to facilitate subsequent renaturation of the proteins. Low-molecular-weight proteins (≤10 kDa) may diffuse more readily out of gels during the pre-equilibration step. To avoid this, one can shorten the pre-equilibration period, changing the buffer several times to aid equilibration. Methanol (20%) can also be added to the transfer buffer ***(5,9,25)***. However, this organic solvent may affect the DNA-binding properties of the proteins. The choice of renaturation procedure and buffers should be determined empirically and will depend on the factors studied.
11. Each layer of the sandwich is thoroughly prewetted with buffer and then carefully positioned on top of the previous layer, taking care to avoid trapping any air bubbles which would distort the resulting transfer.
12. Placing the trans-blot apparatus in the cold room is an inadequate means of controlling transfer buffer temperature. Efficient heat removal is obtained by connecting the transfer apparatus to a recirculating cooler. Transfer is a function of molecular weight, with the largest proteins being transferred more slowly. The precise transfer conditions should take into account the molecular-weight range of the proteins under investigation and the composition of the gel.

13. It is important to block any remaining free binding sites on the membrane in order to reduce nonspecific binding of the DNA probe. The binding buffer should, thus, be designed to optimize the specific binding of the radiolabeled probe while keeping nonspecific binding to a minimum. Nonspecific binding can be reduced by the addition of an excess of unlabeled DNA such as poly(dI–dC)·poly(dI–dC) or poly(dA–dT)·poly(dA–dT), or sheared *E. coli*, salmon sperm, or calf thymus DNA. Nonspecific binding can also be reduced by increasing salt concentrations in the binding reaction and/or in the wash buffer. In addition, the use of lipid-free bovine serum albumin (BSA) instead of nonfat dry milk has been reported to optimize the signal-to-noise ratio ***(26)***. Although specific high-affinity DNA–protein interactions will not normally be competed by nonspecific competitor DNA, the addition of a large excess of competitor DNA may eventually do so. Thus, the amount of competitor DNA and the salt concentration of buffers will depend on the factor of interest. The specificity and affinity of binding should be further examined by adding an unlabeled target sequence or a nonfunctional mutated target sequence DNA as a competitor to the blocking and/or binding solutions. Alternatively, to determine the specificity of an interaction, the binding of a radiolabeled mutant target DNA can be compared with the binding of the unmutated sequence (*see* **Fig. 1**) *(2)*.
14. Some proteins may not retain their ability to bind selectively with their target DNA after immobilization of the protein on the membrane. Furthermore, the membrane will not necessarily retain all of the proteins that have migrated out the gel. It is, thus, useful to determine which proteins have been transferred to the membrane. Detection of the transferred proteins may be achieved after autoradiography by staining the membrane with either Coomassie brilliant blue R-250, Amido black, or Ponceau rouge. Proteins remaining in the gel after transfer can be visualized by silver staining.

Acknowledgments

We thank R. Hancock for comments on the manuscript and S. Desnoyers for helpful suggestions. S.L. is the recipient of a scholarship from FRSQ, and O.L. of a Studentship from NSERC. This work was supported by a Grant from NSERC to C.S.

References

1. Bowen, B., Steinberg, J., Laemmli, U. K., and Weintraub, H. (1980) The detection of DNA-binding proteins by protein blotting. *Nucleic Acids Res.* **8,** 1–20.
2. Séguin, C. and Prévost, J. (1988) Detection of a nuclear protein that interacts with a metal regulatory element of the mouse metallothionein 1 gene. *Nucleic Acids Res.* **16,** 10,547–10,560.
3. Jack, R. S., Brown, M. T., and Gehring, W. J. (1983) Protein blotting as a means to detect sequence-specific DNA-binding proteins. *Cold Spring Harb. Symp. Quant. Biol.* **47**(Pt 1), 483–491.

4. Miskimins, W. K., Roberts, M. P., McClelland, A., and Ruddle, F. H. (1985) Use of a protein-blotting procedure and a specific DNA probe to identify nuclear proteins that recognize the promoter region of the transferrin receptor gene. *Proc. Natl. Acad. Sci. USA* **82,** 6741–6744.
5. Silva, C. M., Tully, D. B., Petch, L. A., Jewell, C. M., and Cidlowski, J. A. (1987) Application of a protein-blotting procedure to the study of human glucocorticoid receptor interactions with DNA. *Proc. Natl. Acad. Sci. USA* **84,** 1744–1748.
6. Hughes, E. N., Engelsberg, B. N., and Billings, P. C. (1992) Purification of nuclear proteins that bind to cisplatin-damaged DNA. Identity with high mobility group proteins 1 and 2. *J. Biol. Chem.* **267,** 13,520–13,527.
7. Wegenka, U. M., Buschmann, J., Lutticken, C., Heinrich, P. C., and Horn, F. (1993) Acute-phase response factor, a nuclear factor binding to acute-phase response elements, is rapidly activated by interleukin-6 at the posttranslational level. *Mol. Cell. Biol.* **13,** 276–288.
8. Kwast-Welfeld, J., Debelle, I., Walker, P. R., Whitfield, J. F., and Sikorska, M. (1993) Identification of a new cAMP response element-binding factor by southwestern blotting. *J. Biol. Chem.* **268,** 19,581–19,585.
9. Dhawan, P., Chang, R., and Mehta, K. D. (1997) Identification of essential nucleotides of the FP1 element responsible for enhancement of low density lipoprotein receptor gene transcription. *Nucleic Acids Res.* **25,** 4132–4138.
10. Villafuerte, B. C., Zhao, W., Herington, A. C., Saffery, R., and Phillips, L. S. (1997) Identification of an insulin-responsive element in the rat insulin-like growth factor-binding protein-3 gene. *J. Biol. Chem.* **272,** 5024–5030.
11. Wu, J., Jiang, Q., Chen, X., Wu, X. H., and Chan, J. S. (1998) Identification of a novel mouse hepatic 52 kDa protein that interacts with the cAMP response element of the rat angiotensinogen gene. *Biochem. J.* **329,** 623–629.
12. Melkonyan, H., Hofmann, H. A., Nacken, W., Sorg, C., and Klempt, M. (1998) The gene encoding the myeloid-related protein 14 (MRP14), a calcium-binding protein expressed in granulocytes and monocytes, contains a potent enhancer element in the first intron. *J. Biol. Chem.* **273,** 27,026–27,032.
13. Carrion, A. M., Mellstrom, B., and Naranjo, J. R. (1998) Protein kinase A-dependent derepression of the human prodynorphin gene via differential binding to an intragenic silencer element. *Mol. Cell. Biol.* **18,** 6921–6929.
14. Chen, A. and Davis, B. H. (1999) UV irradiation activates JNK and increases alphaI(I) collagen gene expression in rat hepatic stellate cells. *J. Biol. Chem.* **274,** 158–164.
15. Katsu, Y., Yamashita, M., and Nagahama, Y. (1997) Isolation and characterization of goldfish Y box protein, a germ-cell- specific RNA-binding protein. *Eur. J. Biochem.* **249,** 854–861.
16. Johnston, K. A., Polymenis, M., Wang, S., Branda, J., and Schmidt, E. V. (1998) Novel regulatory factors interacting with the promoter of the gene encoding the mRNA cap binding protein (eIF4E) and their function in growth regulation. *Mol. Cell. Biol.* **18,** 5621–5633.
17. Hamann, S. and Stratling, W. H. (1998) Specific binding of Drosophila nuclear protein PEP (protein on ecdysone puffs) to hsp70 DNA and RNA. *Nucleic Acids Res.* **26,** 4108–4115.

18. Keller, A. D. and Maniatis, T. (1991) Selection of sequences recognized by a DNA binding protein using a preparative Southwestern blot. *Nucleic Acids Res.* **19,** 4675–4680.
19. Singh, H., Clerc, R. G., and LeBowitz, J. H. (1989) Molecular cloning of sequence-specific DNA binding proteins using recognition site probes. *BioTechniques* **7,** 252–261.
20. Stuempfle, K. J. and Floros, J. (1997) Caution is advised when cDNA expression libraries are screened by southwestern methodologies. *BioTechniques* **22,** 260–264.
21. Sambrook, J., Fritsch, E. F., and Maniatis, T. (1989) *Molecular Cloning. A Laboratory Manual* (Ford, N., Nolan, C., and Ferguson, M., eds.), Cold Spring Harbor Laboratory, Cold Spring Harbor, NY.
22. Dignam, J. D. (1990) Preparation of extracts from higher eukaryotes. *Methods Enzymol.* **182,** 194–203.
23. Handen, J. S. and Rosenberg, H. F. (1997) An improved method for southwestern blotting. *Front. Biosci.* **2,** c9–c11.
24. Dooley, S., Welter, C., and Blin, N. (1992) Nonradioactive Southwestern analysis using chemiluminescent detection. *BioTechniques* **13,** 540–543.
25. Schaufele, F., Cassill, J. A., West, B. L., and Reudelhuber, T. (1990) Resolution by diagonal gel mobility shift assays of multisubunit complexes binding to a functionally important element of the rat growth hormone gene promoter. *J. Biol. Chem.* **265,** 14,592–14,598.
26. Papavassiliou, A. G. and Bohmann, D. (1992) Optimization of the signal-to-noise ratio in south-western assays by using lipid-free BSA as blocking reagent. *Nucleic Acids Res.* **20,** 4365–4366.
27. Mazen, A., Ménissier-de Murcia, J., Molinete, M., Simonin, F., Gradwohl, G., Poirier, G., et al. (1989) Poly(ADP-ribose) polymerase: a novel finger protein. *Nucleic Acids Res.* **12,** 4689–4698.
28. Mimori, T., Hardin, J. A., and Steitz, J. A. (1986) Characterization of the DNA-binding protein antigen Ku recognized by autoantibodies from patients with rheumatic disorders. *J. Biol. Chem.* **261,** 2274–2278.
29. Wold, M. S. (1997) Replication protein A: A heterotrimeric, single-stranded DNA- binding protein required for eukaryotic DNA metabolism. *Annu. Rev. Biochem.* **66,** 61–92.
30. Kannan, P., Yu, Y., Wankhade, S., and Tainsky, M. A. (1999) PolyADP-ribose polymerase is a coactivator for AP-2-mediated transcriptional activation . *Nucleic Acids Res.* **27,** 866–874.
31. Dear, T. N., Hainzl, T., Follo, M., Nehls, M., Wilmore, H., Matena, K., and Boehm, T. (1997) Identification of interaction partners for the basic-helix-loop-helix protein E47. *Oncogene* **14,** 891–898.
32. Butler, A. J. and Ordahl, C. P. (1999) Poly(ADP-ribose) polymerase binds with transcription enhancer factor 1 to MCAT1 elements to regulate muscle-specific transcription. *Mol. Cell. Biol.* **19,** 296–306.

18

A Competition Assay for DNA Binding Using the Fluorescent Probe ANS

Ian A. Taylor and G. Geoff Kneale

1. Introduction

Fluorescence spectroscopy is a useful technique for investigating the interaction of DNA-binding proteins with DNA. Generally, use is made of the intrinsic fluorescence of the protein arising from the aromatic amino acids, which is frequently perturbed in a DNA–protein complex (*see* Chapter 33). In some cases, however, changes in the intrinsic fluorescence emission of a protein arising from its interaction with nucleic acid may not be detectable. For example, if tryptophan and/or tyrosine residues are not located in the proximity of the DNA-binding site, the emission spectrum may not be perturbed by the interaction. Furthermore, in the presence of a large number of tryptophan and tyrosine residues, a relatively small perturbation in the overall emission spectrum brought about by DNA binding may be masked.

To overcome these problems, an alternative approach is to add an extrinsic fluorescence probe to the system that competes with DNA for the binding site of the protein. One can then measure the change in the fluorescence emission spectrum of the probe as DNA is added. If the fluorescence characteristics of the free and bound probe differ, displacement of the probe by DNA can then be observed.

The fluorescent probe 1-anilinonaphthalene-8-sulfonic acid (ANS) and its derivatives have long been used to study protein structure *(1)* and, more recently, to study protein–nucleic acid interactions *(2–4)*. ANS has the property that its fluorescence emission spectrum undergoes a 50-nm blue shift along with an approx 100-fold enhancement when transferred from an aqueous environment to a less polar solvent, such as methanol (*see* **Fig. 1**). ANS will bind weakly to hydrophobic patches on protein molecules with an average dissocia-

From: *Methods in Molecular Biology, vol. 148: DNA–Protein Interactions: Principles and Protocols, 2nd ed.*
Edited by: T. Moss © Humana Press Inc., Totowa, NJ

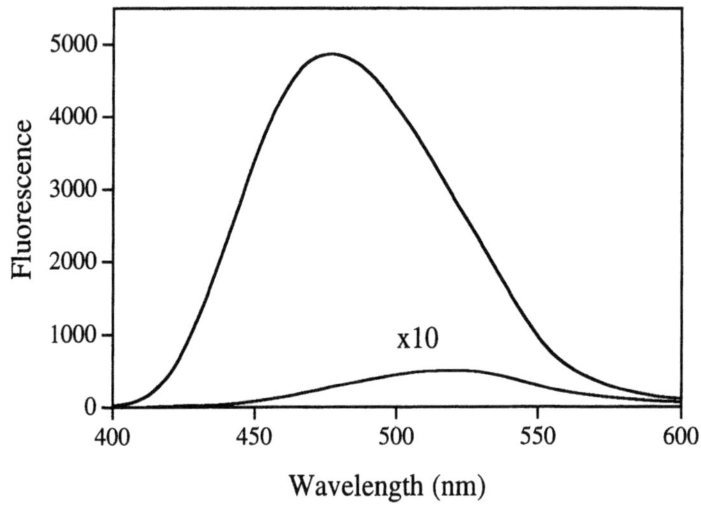

Fig. 1. The effect of solvent polarity on the fluorescence emission spectrum of ANS. Fluorescence emission spectra (λ_{ex} = 370 nm) of 50 μ*M* ANS in 100% methanol (upper curve) and in aqueous buffer (lower curve) are shown.

tion constant of 100 μ*M (3)*. When molecules of ANS are bound to protein, enhancement and shifting of the fluorescence spectrum similar to that observed in apolar media often occurs (*see* **Fig. 2**). Thus, bound molecules of ANS fluoresce much more a strongly and at shorter wavelength than ANS molecules in an aqueous solvent.

The precise reason why ANS molecules bind at DNA-binding sites is not entirely clear, as such sites are not particularly hydrophobic. Nevertheless, ANS is a planar aromatic molecule that will have some properties in common with the DNA bases despite the lack of hydrogen-bonding capacity. Furthermore, the negatively charged sulfonate group of ANS may mimic the phosphate group of the DNA backbone.

The protocol described has been successfully applied to an investigation of the DNA-binding properties of a type I modification enzyme (M.*Eco*R124I *[4]*). It was possible to demonstrate differential binding affinity for an oligonucleotide containing the canonical recognition sequence and one that differs by just one base pair in a nonspecific spacer sequence in the enzymes recognition site. It remains to be seen how general the method is because there have been very few instances of its application to DNA-binding proteins. Since it can be established fairly rapidly whether ANS binds to a given protein, and whether there is some release of ANS by the addition of DNA, it is a technique worth investigating. **Subheadings 3.1.** and **3.2.** deal with preliminary experiments. If these are encouraging, then accurate fluorescence titrations can be

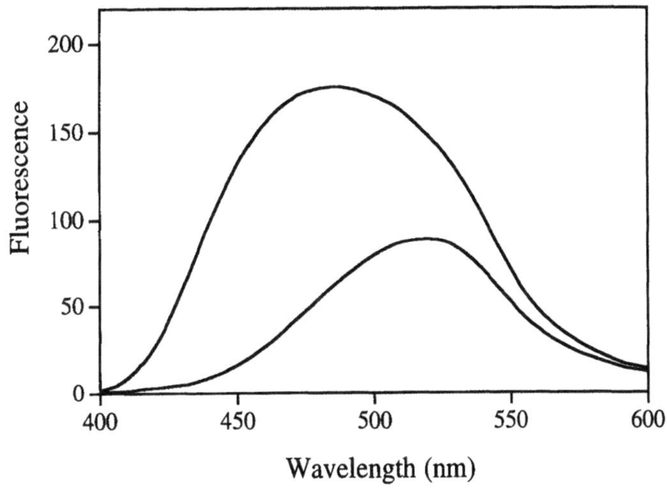

Fig. 2. Enhancement and spectral shift of ANS fluorescence by the addition of protein. Fluorescence emission spectrum (λ_{ex} = 370 nm) of 100 μ*M* ANS alone (lower curve) and with the addition of 2.2 μ*M* M.*Eco*R124I (upper curve).

undertaken to investigate the DNA binding characteristics of the protein in more detail (**Subheading 3.3.**).

2. Materials

1. A fluorimeter is required that is capable of scanning with both emission and excitation monochromators and in which the emission and excitation slit widths can be varied. In our laboratory, we routinely use a Perkin-Elmer LS50B. The fluorimeter is controlled by a PC using software written by the manufacturer.
2. It is desirable that the cell holder compartment be thermostatically controlled to ±0.1°C, and some models of fluorimeter have a built-in temperature control. Alternatively, this can be achieved by circulating water through the cell holder. In this case the temperature is controlled by a programmable circulating water bath (e.g., Neslab RTE-100).
3. Good quality quartz cuvets (preferably stoppered) with all four faces polished are required. A 1 × 0.4-cm (semimicro) cuvet is preferred, because this minimizes the inner filter effect compared to the standard 1 × 1-cm (3-mL) cuvets (*see* **Subheading 3.1.**). The excitation beam should pass through the 0.4-cm path because absorption of the emission beam should be negligible.
4. All buffers should be prepared from the highest-quality reagents and ultrapure water. The buffers should be degassed and filtered to remove any particulate contaminants. In the example provided the assay was carried out in 10 m*M* Tris-HCl, pH 8.2, 100 m*M* NaCl, and 5 m*M* $MgCl_2$ (*see* **Note 1**).
5. High-purity ANS free of contaminating bis-ANS may be obtained from Molecular Probes (Eugene, OR). A 1 m*M* solution of ANS should be made fresh just prior to use.

6. A 50-μ*M* stock solution of the purified DNA-binding protein of interest.
7. A 100-μ*M* stock of an oligonucleotide duplex containing the DNA-binding site to be investigated (*see* **Note 2**).

3. Methods

3.1. Titration of Protein with ANS

To find optimal conditions for the use of ANS in a DNA-binding experiment, it is advisable to first titrate the protein of interest with ANS to check the extent to which the protein binds the fluorescent probe. The precise concentrations of the reagents and composition of the buffer used here work well for M.*Eco*R124I and its subsequent binding to a 30-bp DNA duplex containing a single recognition site. It may be necessary to vary the conditions for other systems.

1. Prepare 250 mL of a degassed and filtered standard buffer that will be used throughout the set of experiments (e.g., 10 m*M* Tris-HCl, pH 8.2, 100 m*M* NaCl, and 5 m*M* $MgCl_2$). Dialyze or desalt (*see* **Note 3**) the DNA-binding protein into this same standard buffer and prepare approx 2 mL of a 1 μ*M* solution.
2. Prepare 50 mL of a 1 m*M* ANS solution again in the same buffer. The concentration of ANS can be determined from its UV absorption spectrum ($\varepsilon_{370\ \text{aqueous}}$ = 5500/*M*/cm; *see* **Note 4**).
3. Adjust the excitation and emission slits on the fluorimeter to 2.5 nm (wider slits can be used if the signal is weak) and set the desired temperature. Allow time for the machine to "warm up" and also for the cell holder to temperature equilibrate.
4. Place 1 mL of buffer in the fluorimeter cuvet and record the fluorescence emission spectrum between 400 and 600 nm, using an excitation wavelength (λ_{ex}) of 370 nm. Add 2 μL aliquots of the 1 m*M* ANS solution to the cuvet and mix gently. After each addition record the fluorescence emission spectrum as before. At the end of the titration measure, the OD_{370} of the sample using the same optical path length as used in the fluorescence titration.
5. In a clean cuvet, place 1 mL of the solution of 1 μ*M* DNA-binding protein and record the fluorescence emission spectrum between 400 and 600 nm (λ_{ex} = 370 nm). Repeat the ANS titration as in **step 4** and again record the OD_{370} of the sample at the end of the titration.
6. To analyze the titration data, choose the wavelength in the emission spectrum that shows the largest difference between the two titrations. This will probably be in the vicinity of 480 nm, although shorter wavelengths can be used to minimize the background signal from free ANS (*see* **Fig. 2**).
7. To obtain the corrected fluorescence intensity (F_{corr}), first correct for any dilution (if significant, *see* **Note 5**), then correct the observed fluorescence intensity (F_{obs}) by application of **Eq. 1**. which accounts for any nonlinearity resulting from inner filter effects (*see* **Note 6**).

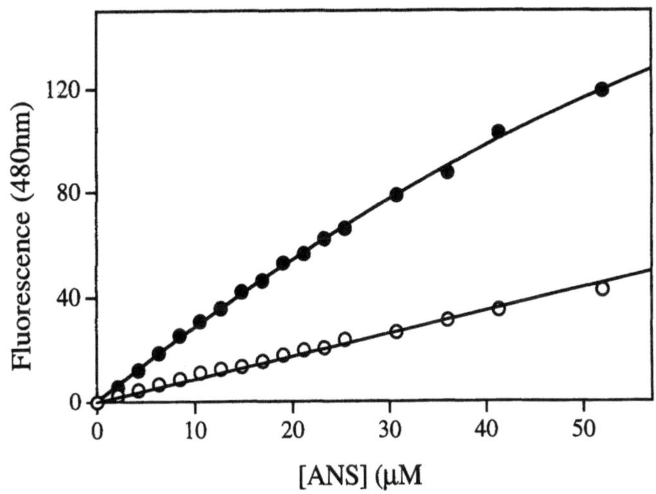

Fig. 3. Fluorescence titration of M.*Eco*R124I (1.5 μ*M*) with ANS (upper curve). The lower curve shows the fluorescence increment in the absence of protein. Both titrations have been corrected for dilution and the inner filter effect.

$$F_{corr} = F_{obs} \times 10^{(Aex/2)} \quad (1)$$

Apply **Eq. 1** to the data from each point in the titration. A_{ex} is the absorbance of the sample at the excitation wavelength (370 nm). This absorbance can be calculated from the ANS extinction coefficient and the known ANS concentration, taking into account the appropriate path length, but it should be checked against the OD_{370} value measured at the end of each titration.

8. Plot the corrected fluorescence at the chosen wavelength against the ANS concentration for each titration. A typical case is shown in **Fig. 3**. The titration curve in the absence of protein should be linear if the corrections for inner filter and dilution have been correctly applied.
9. Subtraction of the curve of ANS added to buffer from the curve of ANS added to the protein solution yields the binding curve of ANS to the protein, as illustrated in **Fig. 4**. The shape of the curve will depend on the ANS binding properties of the particular protein under study (*see* **Note 7**). In the case shown in **Fig. 4**, this curve is fairly representative of the situation in which several ANS molecules are associated weakly.

3.2. Preliminary investigation of the Displacement of ANS by DNA

Once satisfied that the protein under investigation binds ANS, one must ascertain if any of the bound ANS molecules are located in the DNA-binding site of the protein. If so, their fluorescence will change upon being displaced by DNA.

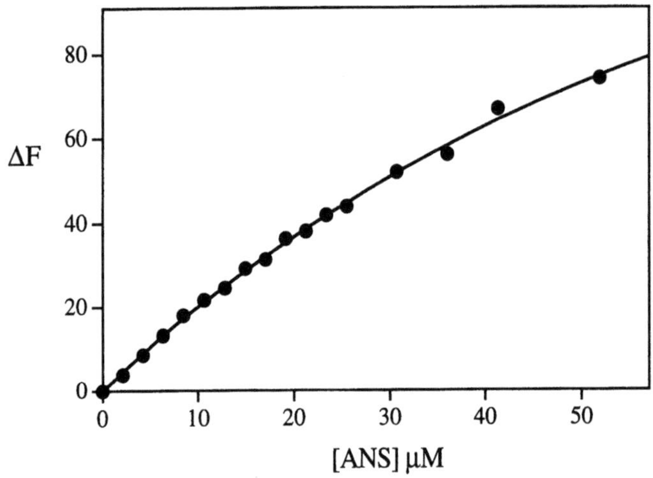

Fig. 4. The binding curve for ANS to M.*Eco*R124I, generated by subtraction of the lower curve from the upper curve in **Fig. 3**.

1. Make up a solution of 100 μ*M* ANS in buffer (*see* **Note 8**). Measure its fluorescence emission spectrum between 400 and 600 nm (using λ_{ex} = 370 nm).
2. Prepare an identical solution of ANS containing 1 μ*M* DNA. Measure the emission spectrum as in **step 1**. These two spectra should be effectively identical. There should be no observable interaction between the nucleic acid and the ANS.
3. Make up a 100 μ*M* ANS solution containing 1 μ*M* protein, and an identical solution containing 1 μ*M* DNA in addition. Measure the fluorescence emission spectra of these two samples as in **step 1**.

The presence of protein in the ANS solution should cause a change in the shape and intensity of the emission spectrum. An increase in quantum yield accompanied by a blue-shifted spectrum should be observed, When the nucleic acid is present and bound to the protein, any ANS (which is weakly bound) in the DNA-binding site of the protein will be displaced and change the form of the spectrum toward that of free ANS (*see* **Fig. 5**).

3.3 Fluorescence Competition Assay

Once it has been established that the DNA fragment of interest shows a measurable displacement of ANS, further investigation of the DNA-binding characteristics of the protein can be conducted. Titrations can be done in a number of different ways, but we have found it more reproducible to titrate the protein into a solution of ANS in the presence and absence of DNA. The difference in fluorescence at each point then represents the amount of ANS displaced (i.e., the amount of DNA bound). The concentrations used should be those found to be optimal from the earlier experiments. Because the concentra-

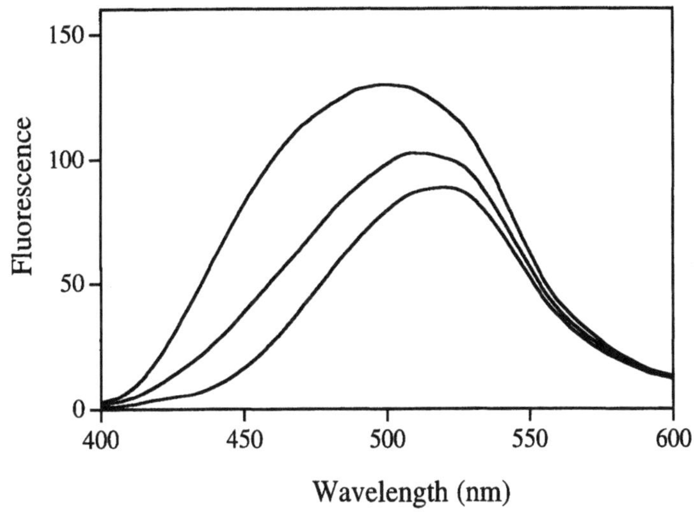

Fig. 5. Fluorescence emission spectra (λ_{ex} = 370 nm) of 100 μ*M* ANS in buffer (lower), in a solution of 1 μ*M* M.*Eco*R124 (upper), and in a solution of 1 μ*M* M.EcoR124I and 1 μ*M* DNA (middle).

tion of ANS is constant throughout, the absorbance at 370 nm should remain unchanged and no inner filter correction need be applied.

1. Place 1 mL of a 100 μ*M* ANS solution in the cuvet and record the emission spectrum between 400 and 600 nm (λ_{ex} = 370 nm).
2. Add small (2–10 μL) aliquots of the 50 μ*M* stock protein solution to the cell up to a final concentration of 3 μ*M*. After each addition record the fluorescence emission spectrum.
3. Make up 1 mL of 100 μ*M* ANS, this time containing 1 μ*M* DNA, and record the fluorescence emission spectrum as in **step 1**.
4. Titrate the same small aliquots of the stock 50 μ*M* protein solution into the ANS/DNA mixture and record the emission spectrum after each addition.
5. Select an appropriate wavelength (e.g., 480 nm; *see* **step 6** of **Subheading 3.1.**) and plot the fluorescence intensity at each point in the titration against the protein concentration for both experiments (*see* **Fig. 6**).
6. To obtain a binding curve for the protein–nucleic acid interaction, subtract the fluorescence intensities at each point in the two experiments (with and without DNA) and replot against protein concentration (*see* **Fig. 7**). This difference represents the amount of bound DNA, because DNA is solely responsible for the decrease in fluorescence through displacement of ANS from the binding site. For a high-affinity DNA protein interaction (with K_d substantially less than the concentration of DNA used in the titration), competition from the ANS will be negligible and a stoichiometric binding curve will be produced; the sharp break in the curve at the stoichiometric point indicates the point at which all the DNA is

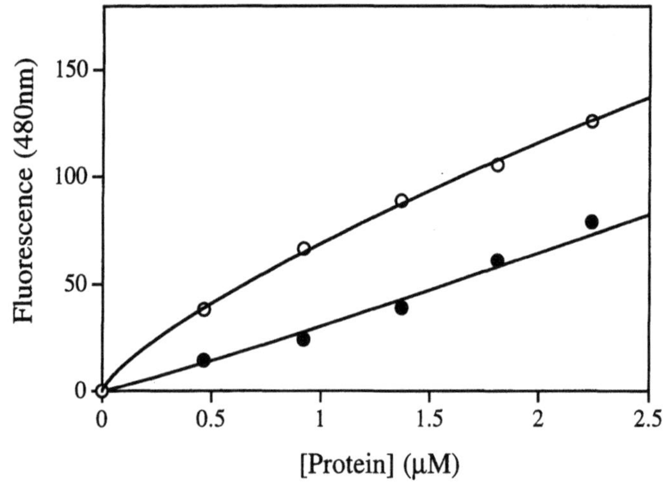

Fig. 6. Data from an ANS competition assay. The upper set of points represent the fluorescence increase resulting from successive additions of M.*Eco*R124I to a 100 μ*M* solution of ANS. The lower set of points represent the fluorescence increase when the same titration is carried out in the presence of 1 μ*M* DNA.

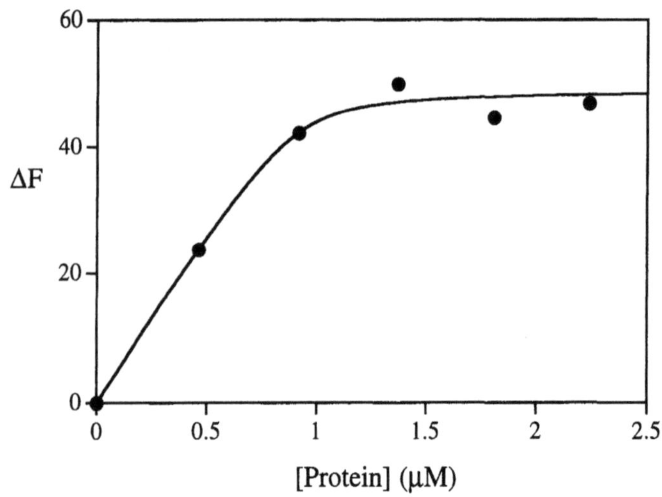

Fig. 7. Binding curve for the interaction of M.*Eco*R124I with an oligonucleotide containing its recognition sequence. This curve is generated by the subtraction of the lower curve from the upper curve in **Fig. 6**.

bound. Curvature of the plot around the stoichiometric point represents a lower-affinity interaction. For a more detailed discussion of DNA binding curves, *see* Chapter 33.

4. Notes

1. The exact composition of the binding buffer will be dependent on the DNA-binding protein being studied. Also, it is advisable to avoid the presence of strongly absorbing compounds and/or quenchers that may interfere with fluorescence measurements. If the interaction has been characterized by another method (e.g., by gel retardation assay), the fluorescence experiment should initially be carried out in the binding buffer used in these studies.
2. In our laboratory, titrations are carried out using short synthetic DNA duplexes (30-mers) that contain the protein's recognition sequence. Short oligonucleotides have the advantage that relatively large amounts of highly pure material are readily obtainable. However, the same protocol should be applicable to the use of longer nucleic acids, such as restriction fragments or polynucleotides.
3. It is important that all the components in the titration are "optically matched." Preparation of a matched protein sample is best achieved by either dialysis or buffer exchange. If the protein sample is limiting, small amounts can be prepared by buffer exchange using NAP5 columns (Pharmacia) or, alternatively, by dialysis using Slide-A-lyzers (Pierce).
4. The value of $\varepsilon_{370\ \text{aqueous}}$ for ANS in buffer was derived by comparison of the OD_{370} of two equimolar solutions of ANS, one in a buffer and the other in 100% methanol for which ε_{370} is known (6800/*M*/cm; Molecular Probes). If the buffer used differs significantly from the one used here, then it is advisable to recalculate the ε_{370}.
5. If the stock solution of ANS is available at high concentration, then the volume of sample in the cuvet during the titration can be assumed to be constant. If a more dilute stock solution is used, there will be significant change in volume (>5%). It is necessary to account for this when calculating the ANS concentration at each point in the titration.
6. See Chapter 33 for a more detailed discussion of the inner filter effect. As a guide, for excitation at 370 nm in an aqueous buffer, 52 μ*M* ANS has an OD_{370} of 0.11 in a 0.4-cm path-length cuvet. This gives rise to an inner filter correction of 1.14 using **Eq. 1**.
7. The binding curve generated for ANS can, in principle, take many forms. The shape will depend on the number and relative affinity of ANS binding sites on the protein. If the protein contains high-affinity sites, the curve may be biphasic and may allow the stoichiometry of the strong interaction to be determined. A more likely situation is that there will be numerous ANS binding sites with differing but weak affinities ($K_d > 100\ \mu M$). The result of this is a curved plot similar to **Fig. 3**.
8. The concentration of the ANS solution used in the titration must be determined empirically from the previous experiments. It should be high enough to ensure that a good fraction of the ANS binding sites on the protein are occupied (as determined in **Subheading 3.1.**),
9. As well as direct excitation of the fluorescent probe (i.e., with an excitation wavelength of 370 nm for ANS), it may be possible to investigate energy transfer

effects between aromatic amino acid residues in the protein and the bound probe. If the excitation is performed at 280 nm to excite both tyrosine and tryptophan, energy transfer to ANS will be apparent from the emission spectrum between 400 and 600 nm. In principle this effect could also be useful to for following displacement of ANS in a titration with DNA.

10. A further extension of the fluorescent probe approach is to employ the covalent probe 5-((((2-iodoacetyl) amino) ethyl) amino) naphthalene-1-sulfonic acid (1,5-IAEDANS) (**5**). This reagent reacts with accessible cysteine residues in the protein and has a higher quantum yield than ANS in aqueous solution. One can look at the emission spectrum of the bound probe, or it may be possible to observe energy transfer from aromatic residues in the protein. Any of these fluorescence characteristics could change when the DNA is bound if the probe is located near the DNA-binding site. We have used this technique to study the interaction of M.*Eco*R124I with DNA and found that energy transfer from the protein to the bound probe decreased by over 30% when DNA was bound. As long as the presence of the probe does not inhibit binding, then titrations with DNA can be used to produce DNA-binding curves. However if the probe does inhibit, this can also be informative; the labelled cysteine(s) can be identified by peptide mapping by analogy with the methods reported in Chapter 20.

References

1. Brand, L. and Gohlke, J. R. (1972) Fluorescence probes for structure. *Annu. Rev. Biochem.* **41,** 843–868.
2. Secnik, J., Wang, Q., Chang, C. M., and Jentoft, J. E. (1990) Interactions at the nucleic acid binding site of the avian retroviral nucleocapsid protein: studies utilizing the fluorescent probe 4,4'-bis(phenylamino)(1,1'-binaphthalene)-5,5'-disulfonic acid. *Biochemistry* **29,** 7991–7997.
3. York, S. S., Lawson, R. C., Jr., and Worah, D. M. (1978) Binding of recrystallized and chromatographically purified 8-anilino-1-naphthalenesulfonate to *Escherichia coli* lac repressor. *Biochemistry* **17,** 4480–4486.
4. Taylor, I., Patel, J., Firman, K., and Kneale, G. (1992) Purification and biochemical characterization of the EcoR124 type I modification methylase. *Nucleic Acids Res.* **20,** 179–186.
5. Kelsey, D. E., Rounds, T. C., and York, S. S. (1979) lac repressor changes conformation upon binding to poly[dA-T)]. *Proc. Natl. Acad. Sci. USA* **76,** 2649–2653.

19

Site-Directed Cleavage of DNA by Linker Histone-Fe(II) EDTA Conjugates

David R. Chafin and Jeffrey J. Hayes

1. Introduction

The ordered and regular packaging of eukaryotic DNA within the chromatin complex allows the efficient utilization of this substrate for nuclear processes such as DNA replication, transcription, recombination, and repair *(1,2)*. Thus, an understanding of the organization of protein–DNA interactions and associations within the chromatin complex is a prerequisite for a complete molecular description of these processes. For example, the linker histone protein plays crucial roles in the stability and organization of the chromatin fiber *(3,4)*. This multidomain protein undoubtedly makes complicated and diverse but poorly understood interactions within the chromatin fiber *(1)*. Currently, there is disagreement regarding the site of association of the globular domain of this protein within the nucleosome proper *(5,6)*. Moreover, the molecular interactions and chemical activities of its N- and C-terminal tails are most likely modulated by the multiple posttranslational phosphorylation events known to occur within these domains *(1,2)*. Thus, the linker histone tails represent critical points for signal transduction within the chromatin complex likely to be manifested as structural alterations within chromatin.

To elucidate the multiple interactions between the linker histone protein and several model chromatin complexes, we have opted for a site-directed chemical cleavage methodology originally introduced by Ebright and colleagues and Fox and colleagues *(7,8)*. This method relies on targeting a DNA cleavage reagent via the unique nucleophilic properties of a cysteine sulfhydryl engineered at rationally selected locations within the protein of interest. The linker histone protein is a particularly suitable candidate for this type of approach because only in one rare instance *(9)* has this protein been found to contain a

From: *Methods in Molecular Biology, vol. 148: DNA–Protein Interactions: Principles and Protocols, 2nd ed.*
Edited by: T. Moss © Humana Press Inc., Totowa, NJ

cysteine residue. The single sulfhydryl group within the protein is modified with a bifunctional reagent that contains a cysteine-specific moiety at one end and an iron(II)-based DNA cleavage reagent at the other (**Fig. 1**) *(7,8,10–12)*. The protein is then assembled into the chromatin complex of interest and DNA cleavage is initiated by standard Fenton chemistry *(7,8)*. The DNA from such complexes is prepared and the location of DNA cleavage is mapped to single-base-pair resolution on DNA sequencing gels.

2. Materials

2.1. Construction of Cysteine Substituted Protein

2.1.1. Point Mutation by PCR

1. Oligonucleotide primers: Two oligonucleotide primers complimentary to the 5' and 3' ends of the sequence to be amplified are needed. In addition, if the codon to be altered is located more than approx 10–15 nucleotides from the end of the coding sequence, one additional primer is needed that must contain the sequence substitutions for the altered codon flanked by 12–15 nucleotides of complementary sequence on each side. Store at –20°C.
2. 10X stock containing all four dNTPs at 10 m*M* concentration each.
3. A source of clean reliable 18 meg Ω water, free of chemical contaminants.
4. 10X polymerase chain reaction (PCR) buffer; can be obtained commercially from the supplier of the PCR enzyme of choice.
5. Vent or *Taq* DNA polymerase; can be obtained from commercial sources.

2.1.2. Ligation and Transformation of PCR Insert into DH5α or BL21 Cells

1. DH5α or BL21 cells can be obtained commercially or prepared in competent form; store at –70°C.
2. Luria–Bertani (LB) medium, sterile.
3. 1000X stock of ampicillin (100 mg/mL).
4. LB–agar plates containing 0.1 mg/mL ampicillin.

2.1.3. Overexpression and Purification of Mutant Histone Proteins

1. 100X (0.2 *M*) stock of isopropyl β-D-thiogalactopyranoside (IPTG).
2. Luria–Bertani (LB) medium, sterile.
3. 10 mg/mL lysozyme solution.
4. 1000X PMSF (phenylmethylsulfonyl fluoride): 100 m*M* in ethanol.
5. Triton-X100 detergent.
6. A 2-*M* solution of NaCl.
7. A 50% (v/v) slurry of Bio-Rex 70, 100-mesh chromatography resin (Bio-Rad).
8. 10 m*M* Tris-HCl, pH 8.0, and 1 m*M* EDTA solutions containing 0.5 *M*, 0.6 *M*, 1.0 *M*, and 2.0 *M* NaCl.

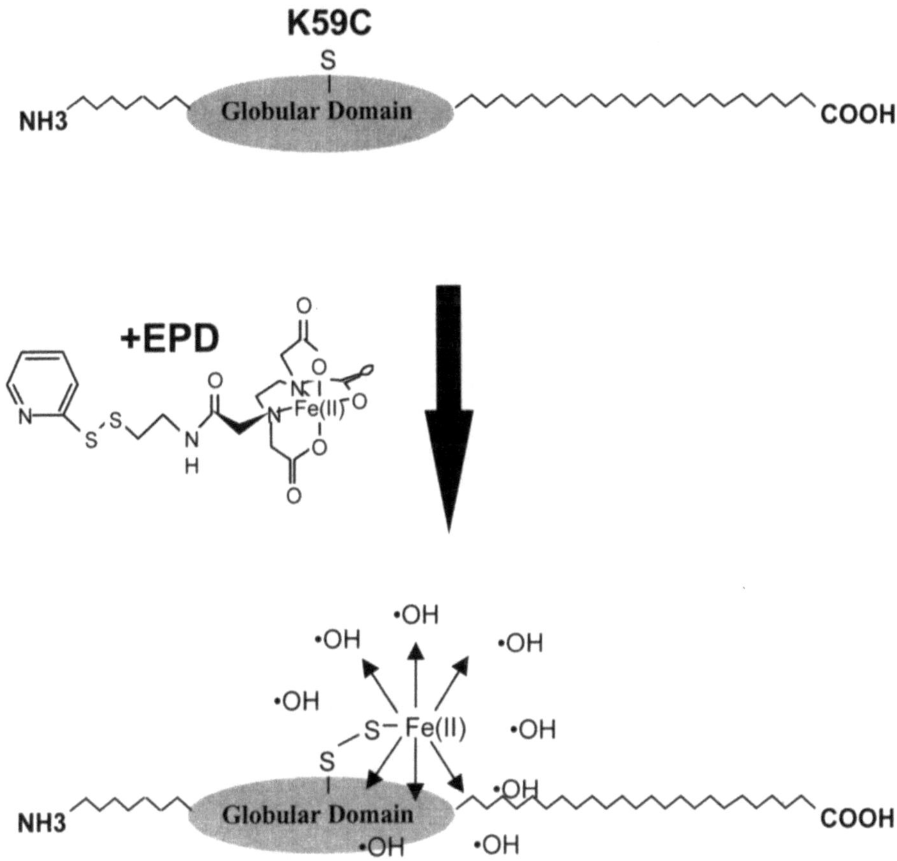

$$Fe(II) + H_2O_2 \rightleftharpoons \bullet OH + {}^-OH + Fe(III)$$

Fig. 1. Lysine at position 59 within the globular domain of histone H1 was changed to a cysteine residue (K59C, **top**). The free sulfhydryl group on the cysteine residue was coupled to the DNA cleavage reagent EPD (**middle**). Hydroxyl radicals were produced from the Fe^{2+} center contained within the EPD moiety by standard Fenton chemistry (**bottom** and equation).

2.2. Reduction and Modification of Cysteine-Substituted Proteins with EDTA-2-aminoethyl 2-pyridyl disulfide (EPD)

1. 1 *M* stock of DTT (dithiothreitol), made fresh; store at –20°C.
2. A 50% slurry of Bio-Rex 70, 100–200 mesh chromatography resin (Bio-Rad).
3. 10 m*M* Tris-HCl, pH 8.0 solutions containing 0.5 *M*, 0.6 *M*, 1.0 *M*, and 2.0 *M* NaCl.

4. 0.3 *M* stock of EPD synthesized according to **refs.** *7* and *8*. Alternatively, iodoacetamido-1,10 phenanthroline–Cu^{2+} can be employed in place of EPD (Molecular Probes, Eugene, OR).
5. Disposable 10 mL plastic chromatography columns (Bio-Rad).
6. Coomassie blue stain: 45% methanol, 10% acetic acid, and 0.25% coomassie brilliant blue R250.

2.3. Radioactive End-Labeling of a Purified DNA Restriction Fragment

1. Linear DNA fragment with convenient restriction sites on either end, previously phosphatased.
2. 10X T_4 polynucleotide kinase buffer (supplied with enzyme).
3. [γ-^{32}P]dATP, 6000 Ci/mmol.
4. T_4 polynucleotide kinase 10,000 units/mL (Promega).
5. 2.5 *M* ammonium acetate.
6. 95% Ethanol, –20°C.
7. 70% Ethanol, –20°C.
8. 10% SDS stock solution.
9. Alkaline phosphatase from calf intestine (Boehringer Mannheim)
10. TE buffer: 10 m*M* Tris-HCl, pH 8.0, and 1 m*M* EDTA.

2.4. In Vitro Reconstitution of Nucleosomes

1. 10 m*M* Tris-HCl, pH 8.0, and 1 m*M* EDTA solutions containing 1.2 *M*, 1.0 *M*, 0.8 *M*, and 0.6 *M* NaCl.
2. TE buffer.
3. Stock of 6000–8000 molecular-weight cutoff dialysis tubing, 14 mm in diameter.
4. Stock of sonicated calf thymus (CT) DNA, approx 1–2 mg/mL.
5. Stock of 5 *M* NaCl.
6. Source of purified core histone proteins H2A/H2B and H3/H4, ours were purified from chicken erythrocytes (*see* **Note 7**).

2.5. Maxim–Gilbert G-Specific Reaction

1. 10X G-specific reaction buffer: 0.5 *M* NaCacodylate and 10 m*M* EDTA.
2. Dimethylsulfate (DMS) (Sigma).
3. G-reaction stop buffer: 1.5 *M* Na acetate, 1 *M* β-mercaptoethanol and 0.004 µg/µL sonicated calf thymus DNA.
4. Piperidine (neat, 10 *M* stock) (Sigma).

2.6. Chemical Mapping of Protein–DNA Interactions with EPD

1. 0.7% agarose made with 0.5X TBE. (**Note:** Treat all solutions with chelex 100 resin [Bio-Rad] to remove adventitious redox-active metals.)
2. Histone dilution buffer: 10 m*M* Tris-HCl, pH 8.0, and 50 m*M* NaCl.
3. Stock solution of 20 m*M* sodium ascorbate, store frozen at –20°C.
4. Stock solution of 1 m*M* Fe(II) 2 m*M* EDTA; store frozen at –20°C.

5. Solution of 0.15% H_2O_2, freshly made.
6. Stop solution: 50% glycerol and 10 m*M* EDTA.
7. Microcentrifuge filtration devices (Series 8000 can be obtained from Lida Manufacturing Corporation).
8. 10 m*M* Tris-HCl, pH 8.0, and 0.1% SDS.
9. Microcentrifuge pestles, can be obtained from Stratagene.
10. 95% and 70% ethanol solutions, cooled to –20°C.
11. 3 *M* sodium acetate.

2.7. Sequencing Gel Analysis

1. Solid urea; molecular biology grade.
2. 5X TBE.
3. 40% Acrylamide (19:1 acrylamide:*bis*-acrylamide).
4. 20% APS (ammonium persulfate).
5. TEMED (*N,N,N',N'*-tetramethylethylenediamine).
6. 1 mL formamide loading buffer: 100% formamide, 0.05% bromophenol blue, 0.05% xylene cyanol, and 1 m*M* EDTA.

3. Methods

3.1. Overexpression and Purification of Single-Cysteine-Substituted Proteins

The following methods work well for incorporating a single amino acid substitution into any protein of interest. Standard three or four primer PCR methods are used.

1. Standard PCR methods are used to amplify a DNA fragment containing a cysteine codon in place of the wild-type codon. If the codon to be changed is near the end of the amplified coding region, then only two primers are necessary with the change incorporated into one of these "parent" primers. If more central to the sequence, then a 3 primer technique is used with the change incorporated into an internal primer, amplify with the internal primer and one of the parent primers and then use the short amplified fragment as a primer with the remaining parent primer and the original DNA as the template. Finally, if this method fails, two complementary internal primers with the intended change are used to amplify overlapping short fragments using the appropriate parent primers; these two fragments are then combined with the parent primers and the entire insert amplified without an additional template added.
2. Gel purify the DNA insert of interest. We typically use the electroelution technique after separating the PCR DNA on a 1% agarose gel. A slice of agarose containing the insert of interest is placed into a dialysis membrane with enough TBE to cover the agarose. Both ends of the dialysis membrane are clipped shut and placed into a standard DNA electrophoresis apparatus containing 1X TBE buffer; the DNA electroeluted is for 15 min at 150 V. Remove the TBE buffer from the dialysis membrane containing the PCR DNA and precipitate.

3. Ligate the insert containing the single cysteine substitution into the appropriate expression vector. We typically use the pET expression system (Novagen). Both DNAs must be digested with the same restriction endonucleases. Incubate equimolar amounts of insert DNA and pET3d DNA in 1X T_4 ligation buffer. Add 400 U of T_4 DNA ligase (Bio-Labs) and incubate at 4°C overnight (*see* **Note 2**).
4. Check the efficiency of the ligation by transforming a small amount of the ligation sample into DH5a cells. Place the transformation on LB–ampicillin plates and incubate at 37°C overnight.
5. Prepare DNA from several colonies by placing a single colony into 3–5 mL of LB–ampicillin medium and grow at 37°C. Isolate the DNA from these cultures by standard DNA mini-prep techniques (*see* **Note 3**).
6. Digest part of the isolated plasmid with the original restriction endonucleases used for ligation to liberate the DNA fragment corresponding to the original insert. The plasmids that contain correct inserts can be used to transform BL21 cells for overexpression.
7. Transform the pET plasmid containing the insert into BL21 cells in the same manner as for the DH5a cells (*see* **step 4**).
8. Place one BL21 colony from the LB–ampicillin plate into 200 mL of LB–ampicillin medium.
9. Grow the culture in the absence of IPTG at 37°C to an optical density of 0.6 at 595-nm wavelength light. Add IPTG to a final concentration of 0.2 mg/mL and return the culture to 37°C for approx 2–4 h (*see* **Note 4**).
10. After 2–4 h, pellet the bacteria by centrifugation at 4000*g* for 15 min.
11. Decant the supernatant and resuspend the pellet in 5–10 mL of TE buffer.
12. Add 10 mg/mL lysozyme to a final concentration of 0.2 mg/mL. Then, add Triton-X 100 to a final concentration of 0.2% and incubate for 30 min at room temperature.
13. Dilute the bacteria twofold with 2 *M* NaCl to a final concentration of 1 *M* NaCl. Transfer the bacteria to Oakridge centrifuge tubes on ice.
14. Sonicate the bacterial slurry for 6 min total in two 3-min sonication steps (*see* **Note 5**).
15. Pellet the cell debris by centrifugation at 10,000*g* for 30 min at 4°C.
16. Add PMSF to a final concentration of 0.1 m*M*. Dilute the supernatants twofold with TE buffer to bring the NaCl concentration to 0.5 M.
17. Incubate the diluted supernatant with 12.5 mL of a 50% suspension of Bio-Rex 70-mesh beads for 4 h at 4°C with rotation. Linker histones and most other proteins will bind directly to the Bio-Rex 70 beads. However, core histone proteins must first be incubated with their dimerization partner proteins before they will bind to the chromatography matrix (i.e., H2A with H2B) (*see* **Note 6**).
18. After 4 h collect the beads in a plastic 10-mL disposable chromatography column. Collect the flowthrough fraction in a 50-mL conical tube and freeze.
19. Wash the column with 2–3 column volumes of 10 m*M* Tris-HCl, pH 8.0 containing 0.6 *M* NaCl. Collect the first 10 mL of the wash fraction in a 15-mL conical tube and freeze.

20. Elute the bound proteins with two separate 1-column volume elution steps of 10 m*M* Tris-HCl, pH 8.0 containing 1.0 *M* NaCl. Collect the 1.0 *M* elution steps in separate 15 mL conical tubes and freeze.
21. After elution, wash the column with one column volume of 10 m*M* Tris-HCl, pH 8.0 containing 2.0 *M* NaCl. Collect the 2.0 *M* elution step in a 15-mL conical tube and freeze.
22. Check 10 µL of each fraction for protein by loading a small amount on a 12% or 18% SDS–polyacrylamide gel.

3.2. Reduction and Modification of Cysteine-Substituted Proteins

3.2.1. Reduction of Cysteine-Substituted Proteins

1. Incubate protein of interest in a 15-mL conical tube with 50 m*M* DTT final concentration for 1 h on ice.
2. Dilute the protein sample twofold with TE; this dilutes the NaCl concentration to 500 m*M* NaCl.
3. Add 0.8 mL of a 50% slurry of Bio-Rex 70 chromatography resin and incubate for 2 h at 4°C with rotation. We have found that Bio-Rex 70 can bind approx 0.5 mg protein/g resin.
4. Pour slurry into a 10-mL plastic, disposable chromatography column and collect the flowthrough fraction. Disposable chromatography columns are commercially available from Bio-Rad or other manufacturers.
5. Wash the column with 3–5 column volumes of buffer containing 10 m*M* Tris-HCl, pH 8.0, and 0.5 *M* NaCl. Immediately remove 20 µL of the freshly eluted wash sample into a separate Eppendorf tube for analysis later on a 12% SDS–polyacrylamide gel and immediately freeze the larger sample in case the protein did not bind the resin. Aliquoting the sample in this manner ensures that the sample does not need to be thawed for analysis.
6. An intermediate wash of the column with buffer containing 0.6 *M* NaCl is performed to remove proteins that are less well-bound because of partial degradation. Aliquots of these samples are obtained in the same manner as the previous wash step.
7. Linker histone protein can be eluted with 1-column volume steps of the same buffer except with 1.0 *M* NaCl. Typically, five separate 1-column volume 1.0 *M* NaCl elution steps are performed and collected separately. Usually, only 5 µL of the elution steps needs be aliquoted for SDS–polyacrylamide gel analysis. As above, the large elution fractions are frozen immediately. A final elution with buffer containing 2.0 *M* NaCl buffer will ensure that all of the protein has been eluted from the column.
8. Check the protein content of each aliquot obtained from the wash and elution steps on a 12% SDS–polyacrylamide gel. After separation, incubate the protein gel in enough Coomassie blue stain to cover the protein gel. Stain for approx 1 h at room temperature and destain with 45% methanol and 10% acetic acid until the background of the gel is clear.

3.2.2. Modification of Cysteine Substituted Proteins with EPD

1. Thaw the fraction containing the reduced protein to be modified with EPD on ice. Working as quickly as possible, add a 1.1 fold molar excess of EPD to 60 μL (approx 30 μg) of reduced protein. Incubate for 1 h at room temperature in the dark.
2. Removal of excess cleavage reagent requires one more round of Bio-Rex 70 chromatography, identical to that in **step 3** of **Subheading 3.2.1.** except that 60 μL of the 50% slurry is added to the protein. In addition, the slurry is poured into a column made from a blue 1-mL pipet tip fitted with glass wool at the opening. Wash and elute as in **step 5** of **Subheading 3.2.1.** except all elution volumes are scaled according to the resin amount. Aliquots for protein analysis are exactly the same size as previously indicated.
3. Postmodification labeling with ^{14}C-NEM (N-[ethyl–1–^{14}C]-maleimide) (New England Nuclear) can be used to quantitatively determine the extent of modification with the DNA cleavage reagent. Add 0.25–0.5 μCi of ^{14}C-NEM to each protein aliquot made from the elution fractions of the Bio-Rex column. After 10 min, add 2 volumes of 2X protein loading buffer and separate the proteins on a 12% SDS-polyacrylamide gel. Stain and destain the gel as in **step 8** of **Subheading 3.2.1.** and dry the gel onto a piece of Whatman filter paper. Visualize the labeled proteins by exposing the dried gel to ultra sensitive Bio-Max autoradiography film.
4. A protein gel at this step performs two functions. (1) It determines which fractions contain the protein of interest and (2) it determine the extent of modification with the DNA cleavage reagent.

3.3. Radioactive End-Labeling of a Purified DNA Restriction Fragment

1. Treat approx 5 μg of plasmid DNA or 1 μg of a purified DNA fragment with the appropriate restriction endonuclease in the manufacturer's buffer.
2. Precipitate the DNA by adjusting the solution to 0.3 *M* sodium acetate and adding of 2.5 vol of cold ethanol.
3. Resuspend the DNA in phosphatase buffer and treat with alkaline phosphatase for 1 h at 37°C.
4. Adjust the solution to 0.1% SDS, phenol extract the solution, and then precipitate the aqueous phase twice with ethanol and sodium acetate.
5. Resuspend the DNA in 10 μL TE and add 2.5 μL of 10X T_4 polynucleotide kinase buffer.
6. Add 50 μCi of [γ-^{32}P]dATP and adjust the volume to 24 μL with water.
7. Start the reaction by adding 10 units of T_4 polynucleotide kinase and incubate for 30 min at 37°C.
8. Stop the kinase with 200 μL of 2.5 *M* ammonium acetate (NH_4Oac) and 700 μL of cold 95% ethanol.
9. Pellet the DNA in a microcentrifuge for 30 min at room temperature.

10. Wash the DNA pellet briefly with cold 70% ethanol and dry the DNA in a Speedvac concentrator.
11. Dissolve the DNA in 34 µL of TE buffer.
12. Digest the DNA fragment with a second restriction endonuclease that liberates the fragment of interest and yields fragments that can be easily separated on a native 6% polyacrylamide gel.
13. After separation, wrap the gel tightly in plastic wrap and apply fluorescent markers onto various portions of the gel for alignment purposes (can be obtained from Stratagene) or accurately mark the position of the gel on the film. Expose the wet gel to the autoradiography film for 1 min, which is sufficient to detect the specific band containing the labeled fragment.
14. Excise the band of interest from the polyacrylamide gel and place into a clean Eppendorf tube. Crush the acrylamide gel slice with a Eppendorf pestle and add 700 µL of TE buffer. The labeled fragment will elute overnight with passive diffusion.
15. Split the sample equally into two Series 8000 Microcentrifuge Filtration Devices and spin for 30 min in a microcentrifuge.
16. Precipitate the eluted DNA and dissolve in TE buffer pH 8.0. Add enough TE buffer so that the labeled DNA is approx 1000 cpm/µL (*see* **Note 7**).

3.4. Reconstitution of Nucleosomes by Salt Step Dialysis

The method described here for the reconstitution of nucleosomes allows for large quantities of nearly homogeneous core particles in 12 h ***(13)***. Moreover, reconstituted nucleosomes are known to bind linker histone in a physiologically relevant manner according to multiple criteria. Virtually any piece of DNA 147 bp or longer can be used. However to obtain nucleosomes with only one translational position, the DNA sequence should contain nucleosome positioning sequences such as that from the *Xenopus borealis* somatic 5S rRNA gene ***(14–16)***. The DNA can be labeled on the 5' or 3' end with commercially available enzymes after phosphatase treatment as described above.

1. Add approx 5–8 µg of unlabeled calf thymus DNA, 200,000–400,000 cpm of singly labeled *Xenopus borealis* 5S ribosomal DNA, purified chicken erythrocyte core histone protein fractions (H2A/H2B and H3/H4) (*see* **Note 1**), 160 µL of 5 *M* NaCl (2.0 *M* final), and TE buffer to a final of volume 400 µL.
2. Place the reconstitution mixture into a 6 to 8 kDa molecular weight cut-off dialysis bag. All subsequent dialysis steps are for 2 h at 4°C against 1 L of dialysis buffers unless specified. The first dialysis buffer is 10 m*M* Tris-HCl, pH 8.0, 1.2 *M* NaCl and 1 m*M* EDTA. Subsequent dialyses steps are with fresh buffer containing 1.0 *M*, 0.8 *M*, and then 0.6 *M* NaCl. The procedure is completed with a final dialysis against TE buffer overnight. Nucleosomes at this stage can be used for gel-shift experiments where EDTA does not interfere.
3. For DNA cleavage experiments with EPD, two additional dialysis steps are required. First dialyze the reconstitutes against 10 m*M* Tris-HCl, pH 8.0 several

hours to remove the EDTA. A second dialysis against fresh 10 m*M* Tris-HCl, pH 8.0 removes trace amounts of EDTA and prepares the samples for chemical mapping with EPD.

3.5. Maxim–Gilbert G-Specific Reaction

The G-specific reaction used in the Maxim–Gilbert sequencing method provides an easy and quick method to identify the exact location of bases within any known sequence on sequencing gels. It is used here to determine the sites of DNA to base-pair resolution. Because this method is not generally used any longer, the steps are outlined as follows:

1. Add approx 20,000 cpm of singly labeled DNA (same DNA used to reconstitute nucleosomes).
2. Add 20 µL of 10X G-specific reaction buffer.
3. Add water to a final volume of 200 µL.
4. Start by adding 1 µL of straight dimethylsulfate (DMS) to the tube. Mix immediately and spin briefly in a microfuge (do this in a hood; be careful not to get any DMS on your skin or on standard laboratory gloves. Store DMS in a tightly capped brown glass bottle at 4°C).
5. Add 50 µL of G-reaction stop solution and mix immediately.
6. Add 2.5 vol of –20°C 95% ethanol to precipitate the DNA.
7. Wash the DNA with –20°C ethanol; dry and dissolve the DNA in 90 µL of H_2O.
8. Add 10 µL of piperidine and incubate at 90°C for 30 min.
9. Dry the DNA solution in a Speedvac to completion.
10. Dissolve the DNA in 20 µL of water and repeat the drying step. Repeat this step one more time.
11. Dissolve DNA in 100 µL TE buffer and store at 4°C.

3.6. Site-Directed Hydroxyl Radical Cleavage of DNA

3.6.1. Binding Single-Cysteine-Substituted Linker Histone Proteins to Reconstituted Nucleosomes

1. The exact amount of each mutant linker histone protein needed to stoichiometrically bind the nucleosome needs to be determined empirically. Increasing amounts of the linker histone are titrated to a fixed volume of reconstituted nucleosomes (typically 5000 cpm) and analyzed via a gel-shift procedure ***(17)***. This is typically scaled up 10-fold for the site-specific cleavage reaction.
2. Add 5% glycerol final to the binding reaction (analytical scale only, site-specific cleavage reactions contain 10-fold less glycerol).
3. Add 50 m*M* NaCl final to the binding reaction (*see* **Note 8**).
4. Incubate the binding reactions for 15 min at room temperature.
5. Separate the complexes on a 0.7% agarose and 0.5X TBE gel. After drying the gel, expose to autoradiograpic film and determine the amount of protein necessary for good complex formation.

6. Several assays for the correct binding of linker histones to DNA have been performed ***(17,18)***. One of the easiest involves a brief digestion with micrococcal nuclease in the chromatosome stop assay ***(13)***.

3.6.2. Site-Directed Hydroxyl Radical Mapping of Linker Histone-DNA Interaction

1. Scale up the binding reaction to include 40,000–50,000 cpm of labeled reconstituted nucleosomes and add enough modified mutant linker histone to form H1–nucleosome complexes.
2. Add glycerol to 0.5% final concentration (*see* **Note 9**).
3. Add sodium ascorbate to a final concentration of 1 m*M*.
4. Add H_2O_2 to a final concentration of 0.0075%.
5. Incubate for 30 min at room temperature in the dark.
6. After 30 min, add 1/10 vol of 50% glycerol and 10 m*M* EDTA solution.
7. Load samples immediately onto a running (90 V) preparative 0.7% agarose and 0.5X TBE gel.
8. Separate the samples so that the H1–nucleosome complexes are well resolved from tetramer and free DNA bands.
9. Next, wrap the gel tightly with plastic wrap so that the gel cannot move within the plastic. Lay fluorescent markers onto various portions of the gel for alignment purposes (can be obtained from Stratagene) or accurately mark the position of the gel on the film.
10. Expose the wet gel for several hours at 4°C.
11. Next, develop the autoradiograph and overlay onto the wet gel, lining up the fluorescent markers.
12. Cut and remove the agarose containing the H1–nucleosome complexes or bands of interest and place them into Series 8000 Microcentrifuge Filtration Devices.
13. Freeze the filtration tubes containing the agarose plugs on dry ice for 15 min.
14. Spin down the agarose in a microcentrifuge at maximum speed for 30 min at room temperature. The fluid from the agarose matrix will be collected in the 2-mL centrifuge tube surrounding the filtration device.
15. Gently remove the agarose plug from the bottom of the filtration device and place into a clean Eppendorf tube. Save the centrifugation devices for use later.
16. Using a microcentrifuge pestle, crush the agarose pellet and add 500 µL of 10 m*M* Tris-HCl, pH 8.0, and 0.1% SDS and continue to crush the agarose.
17. After the agarose is crushed into tiny pieces, place all samples at 4°C overnight.
18. Place the crushed agarose into the same centrifugation device and pellet. Spin down the agarose in a microcentrifuge at maximum speed for 30 min at room temperature.
19. Combine identical samples from both spins and precipitate the DNA.
20. Dissolve the DNA in 15 µL of TE buffer.
21. Heat the samples to 90°C for 2 min to denature.

3.7. Sequencing Gel Analysis of H1°C-EPD Cleavage

3.7.1. Sequencing Gel Electrophoresis

1. Add equal numbers of counts from each sample, including the G specific reaction, to clean eppendorf tubes.
2. Place the samples into a Speedvac concentrator and dry to completeness.
3. Dissolve the samples in 4 µL of formamide loading buffer.
4. Place samples directly onto ice to prevent renaturation.
5. Separate samples on a 6% polyacrylamide and 8 *M* urea sequencing gel running at constant 2000 V.

3.7.2 Example of Site-Directed Cleavage of Nucleosomal DNA by EPD

An example of a linker histone site-directed DNA cleavage reaction is presented in **Fig. 2**. A schematic of the 5S mononucleosome is shown in the left panel. The thick black line represents the 5S ribosomal DNA fragment that contains the transcriptional coding sequence for this gene (gray arrow). The 5S ribosomal gene fragment was used because it contains a nucleosomal positioning sequence that precisely wraps the DNA around the core histones and provides a homogeneous population of nucleosomes. Furthermore, because one major translational position predominates within this population of nucleosomes, the precise orientation of the DNA as it wraps around the core histones is known. This enables us to determine to base-pair resolution, the sites of cleavage by EPD with respect to the nucleosome structure. A singly end-labeled 5S DNA fragment was incorporated into nucleosomes via the salt dialysis procedure detailed earlier. Labeled mononucleosomes were bound by an EPD-modified linker histone containing a single cysteine substitution for the lysine residue at position 59, referred to as K59C-EPD. After allowing hydroxyl radical cleavage for 30 min, the protein–DNA complexes were separated on a 0.7% agarose gel (**Fig. 2A**) and the labeled DNA fragments corresponding to the H1–nucleosome complexes were purified. These purified DNAs were then analyzed on a 6% sequencing gel (**Fig. 2B**, right panel).

Fig. 2 *(opposite page)*. **(A)** Various DNAs from control or hydroxyl radical cleavage reactions were separated on a 0.7% agarose gel. The wet gel was exposed to autoradiographic film for 3–4 h according to **Subheading 3.6.2.** Lanes 1 and 2 contain free DNA (FD) and bulk nucleosomes (Nuc.), respectively, not exposed to hydroxyl radical cleavage. Lanes 3–5 contain nucleosomes or H1–nucleosome complexes (H1–Nuc.) subjected to hydroxyl radical cleavage. Nucleosomes–K59C (lane 3), nucleosomes with unmodified K59C (lane 4) or nucleosomes with K59C–EPD (lane 5). **(B)** A linear schematic of the 5S nucleosome is shown (**left**). The DNA (black line)

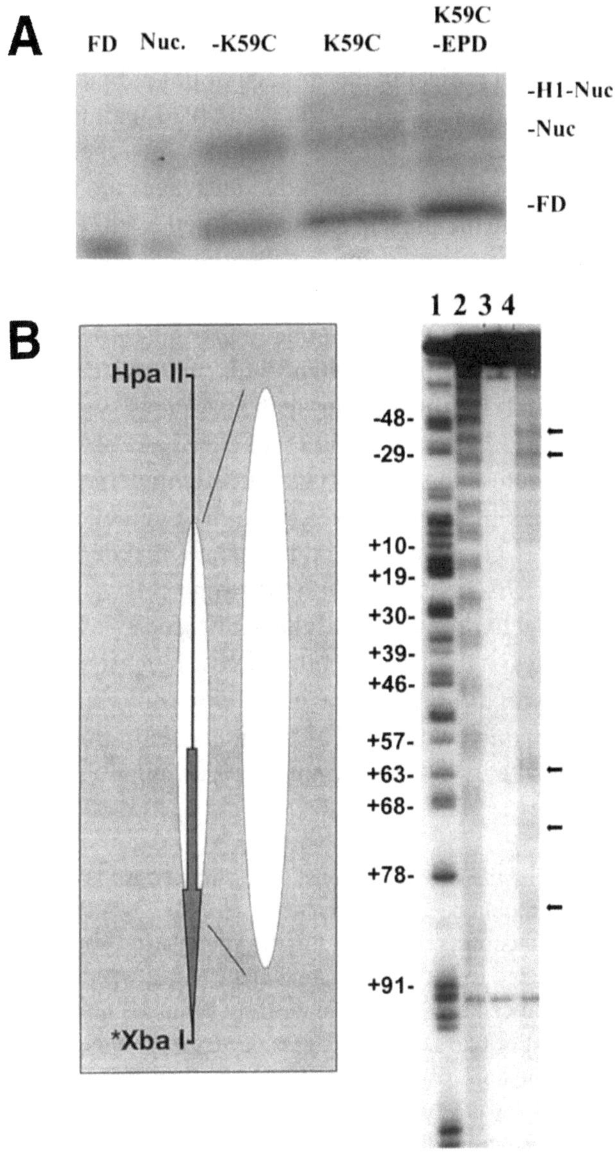

was restricted with the restriction enzymes shown and radiolabeled (*) at the *Xba*I site. The position of the 5S nucleosome (white oval) is shown with respect to the start site of transcription (gray arrow) and with respect to the size of DNAs in the sequencing gel (larger white oval). Labeled DNAs from hydroxyl radical cleavage reactions were analyzed on a 6% sequencing gel (**right**). Lane 1: Maxim–Gilbert G-specific reaction; lane 2: hydroxyl radical cleavage in the absence of histone H1; lanes 3 and 4: hydroxyl radical cleavage with K59C (lane 3) or K59C–EPD (lane 4). Specific cleavages are shown to the right of the gel (black arrows).

The gel reveals that the reconstituted mononucleosomes used for this experiment give a characteristic 10-bp protection when footprinted by general cleavage with hydroxyl radical in the absence of linker histone ***(15)***. This indicates that the 5S DNA has been properly assembled with the histone octamer into a nucleosome (**Fig. 2B**, lane 2). The linker histone-directed cleavage experiments are shown in lanes 3 and 4. No cleavages are observed when the reaction is carried out in the presence of unmodified K59C (**Fig. 2B**, lane 3). In contrast, when the cleavage reagents are added in the presence of K59C-EPD bound nucleosomes, two sets of cleavages are evident (**Fig. 2B**, lane 4). The cleavages at +62, +72, and +82 correspond to the end of the nucleosome where the DNA exits. This result is consistent with previous data suggesting that the linker histone binds the nucleosome at the periphery, tucked inside a superhelical gyre of DNA ***(12,19)***. A second set of cleavages occurs at –29 and –39. These cleavages occur on the DNA strand directly underneath that of the +62/+72/+82 cleavages as the DNA makes one full superhelical turn around the histone octamer. It is possible that amino acid 59 makes close contacts with both strands of the DNA, consistent with the strong cleavages seen at each site. It is also possible that hydroxyl radicals have diffused away from the –29/–39-cleavage site and cleave the DNA in other areas. Inconsistent with this, glycerol, a very good hydroxyl radical scavenger, does not seem to have an effect on the cleavages obtained with K59C-EPD at concentrations known to eliminate hydroxyl radical cleavage (Chafin and Hayes, unpublished results).

4. Notes

1. A complication of the in vitro reconstitution procedure is that purified histone proteins are often obtained in two fractions, H2A/H2B and H3/H4 ***(22)***. Thus, in addition to total histone mass, the ratio between these two substituents must be empirically adjusted to yield maximum octamer–DNA complexes ***(13)***.
2. Many ligation procedures are available from primary literature or commercial sources. Ligation of two DNA fragments occurs more rapidly at room temperature or 37°C if the base-pair overlap is sufficiently stable.
3. Many DNA mini-prep procedures are described in detail in **ref.** ***(20)***. The DNA isolated for the techniques described here were from the boiling DNA mini-prep procedure ***(20)***.
4. Before proceeding, it is recommended that a small amount of the culture be checked for overexpression of the protein of interest. This can be done by removing 1 mL of the culture before and after induction by IPTG.
5. Sonication techniques tend to increase the temperature of the sample quickly, which could induce proteolysis of the proteins. The sample must therefore be cooled before and during sonication. Allow several min between sonication runs to keep the sample as cold as possible.
6. Histones H2A or H2B do not bind to Bio-Rex 70 when purified individually. However, we have found that when allowed to heterodimerize, they bind to the

column and elute off consistently in 1 *M* NaCl *(13)*. This characteristic could be the result of the fact that histone H2A and H2B are completely unfolded when separated from each other *(21)*.

7. Storing labeled DNA in a concentrated form is not advised, as autodegradation of the DNA takes place. DNA can be stored for several weeks at approx 5000 cpm/µL.
8. Several methods can be used for the incorporation of linker histones into reconstituted mononucleosomes. The method described here involves direct addition of linker histones to mononucleosis in 50 m*M* NaCl. Linker histones are folded in low-salt solutions in the presence of DNA *(23)*. Indeed, we find that linker histones can be directly mixed to nucleosomes in either 5- or 50-m*M* NaCl solutions and these proteins then bind in a physiologically relevant manner *(17)*.
9. Glycerol is a good scavenger for hydroxyl radicals and will generally inhibit hydroxyl-radical-based cleavage if added at a final concentration over 0.5% and therefore should be avoided. However, adding small concentrations of glycerol will allow hydroxyl radical cleavage to occur if the EPD moiety is in close proximity to the DNA backbone but not cleavage from sites farther away.

References

1. Wolffe, A. P. (1995) *Chromatin: Structure and Function.* Academic, London.
2. van Holde K. E. (1989) *Chromatin.* Springer-Verlag, New York.
3. Thoma, F., Koller, T., and Klug, A. (1979) Involvement of histone H1 in the organization of the nucleosome and the salt-dependent superstructures of chromatin. *J. Cell Biol.* **83,** 403–427.
4. Carruthers, L. M., Bednar, J., Woodcock, C. F. L., and Hansen, J. C. (1998) Linker histones stabilize the intrinsic salt-dependent folding of nucleosomal arrays: mechanistic ramifications for higher-order folding. *Biochemistry* **37,** 14,776–14,787.
5. Crane-Robinson, C. (1997) Where is the globular domain of linker histone located on the nucleosome? *Trends Biochem. Sci.* **22,** 75–77.
6. Zhou, Y.-B., Gerchman, S. E., Ramakrishnan, V., Travers, Andrew, and Muyldermans, S. (1998) Position and orientation of the globular domain of linker histone H5 on the nucleosome. *Nature* **395,** 402–405.
7. Ebright, Y. W., Chen, Y., Pendergrast, P. S., and Ebright, R. H. (1992) Incorportation of an EDTA–metal complex at a rationally selected site within a protein: application to EDTA–iron DNA affinity cleaving with catabolite gene activator protein (CAP) and Cro. *Biochemistry* **31,** 10,664–10,670.
8. Ermacora, M. R., Delfino, J. M., Cuenoud, B., Schepartz, A., and Fox, R. O. (1992) Conformation-dependent cleavage of staphlyococcal nuclease with a disulfide-linked iron chelate. *Proc. Natl. Acad. Sci. USA* **89,** 6383–6387.
9. Neelin, J. M., Neelin, E. M., Lindsay, D. W., Palyga, J., Nichols, C. R., and Cheng, K. M. (1995) The occurrence of a mutant dimerizable histone H5 in Japanese quail erythrocytes. *Genome* **38,** 982–990.
10. Chen, Y. and Ebright, R. H. (1993) Phenyl-azide-mediated photocrosslinking analysis of Cro–DNA interaction. *J. Mol. Biol.* **230,** 453–460.

11. Lee, K.-M. and Hayes, J. J. (1997) The N-terminal Tail of Histone H2A Binds to Two Distinct Sites Within the Nucleosome Core. *Proc. Natl. Acad. Sci. USA* **94,** 8959–8964.
12. Hayes, J. J. (1996) Site-directed cleavage of DNA by a linker histone–Fe(II)EDTA conjugate: localization of a globular domain binding site within a nucleosome. *Biochemistry* **35,** 11,931–11,937.
13. Hayes, J. J. and Lee, K.-M. (1997) In vitro reconstitution and analysis of mononucleosomes containing defined DNAs and proteins. *Methods* **12,** 2–9.
14. Rhodes, D. (1985) Structural analysis of a triple complex between the histone octamer, a *Xenopus* gene for 5S RNA and transcription factor IIIA. *EMBO J.* **4(13A),** 3473–3482.
15. Hayes, J. J., Tullius, T. D., and Wolffe, A. P. (1990) The structure of DNA in a nucleosome. *Proc. Natl. Acad. Sci. USA* **87,** 7405–7409.
16. Simpson, R. T. (1991) Nucleosome positioning: occurrence, mechanisms and functional consequences. *Prog. Nucleic Acids Res. Mol. Biol.* **40,** 143–184.
17. Hayes, J. J. and Wolffe, A. P. (1993) Preferential and asymmetric interaction of linker histones with 5S DNA in the nucleosome. *Proc. Natl. Acad. Sci. USA* **90,** 6415–6419.
18. Allan, J., Hartman, P. G., Crane-Robinson, C., and Aviles, F. X. (1980) The structure of histone H1 and its location in chromatin. *Nature* **288,** 675–679.
19. Pruss, D., Bartholomew, B., Persinger, J., Hayes, J. J., Arents, G., Moudrianakis, E. N., et al. (1996) A new model for the nucleosome: a binding site for linker histone inside the DNA gyre. *Science* **274,** 614–617.
20. Sambrook, J., Fritsch, E. F., and Maniatis, T. (1989) *Molecular Cloning: A Laboratory Manual,* 2nd ed., Cold Spring Harbor Laboratory, Cold Spring Harbor, NY.
21. Karantza, V., Baxevanis, A. D., Freire, E., and Moudrianakis, E. N. (1995) Thermodynamic studies of the core histones: Ionic strength and pH dependence of H2A–H2B dimer stability. *Biochemistry* **34,** 5988–5996.
22. Simon, R. H. and Felsenfeld, G. (1979) A new procedure for purifing histone pairs H2A + H2B and H3 + H4 from chromatin using hydroxyapatite. *Nucleic Acids Res.* **6,** 689–696.
23. Clark, D. J. and Thomas, J. O. (1986) Salt-dependent co-operative interaction of histone H1 with linear DNA. *J. Mol. Biol.* **187,** 569–580.

20

Nitration of Tyrosine Residues in Protein–Nucleic Acid Complexes

Simon E. Plyte

1. Introduction

Chemical modification is a powerful tool for investigating the accessibility and function of specific amino acids within folded proteins. It has provided significant information regarding the role of different amino acids at the binding sites of numerous enzymes and DNA-binding proteins. The identification of such residues by chemical modification has then often be used to plan subsequent site-directed mutagenesis experiments. These data complement those from crystallographic and nuclear magnetic resonance (NMR) studies in determining the residues located at the active site; thus, one needs to consider all these techniques when elucidating protein structure and function. For example, chemical modification of leukotriene A4 hydrolase, 3-hydroxyisobutyrate dehydrogenase, and lactate dehydrogenase *(1–3)* have contributed significantly to the understanding of active-site mechanisms in these proteins and in elucidating the mechanisms of DNA binding in the Fd and Pf1 gene 5 proteins *(4–5)*.

Reagents exist to modify cysteine, methionine, histidine, lysine, arginine, tyrosine and carboxyl groups selectively. However, in this chapter, we are only concerned with the selective modification of tyrosine residues (for reagents and conditions for the modification of the other amino acids, *see* **ref.** ***6***). The side chain of tyrosine can react with several compounds, the most commonly used being N-acetylimidizole and tetranitromethane (TNM). N-acetylimidizole will O-acetylate tyrosine residues in solution *(7)*, and this reagent has been used to modify numerous proteins including the Fd gene 5 protein *(4)*. However, this reagent can also *N*-acetylate primary amines, and in the study on the Fd gene 5 protein *(4)* in addition to acetylation of three tyrosine residues, all

From: *Methods in Molecular Biology, vol. 148: DNA–Protein Interactions: Principles and Protocols, 2nd ed.*
Edited by: T. Moss © Humana Press Inc., Totowa, NJ

five lysine residues were found to be modified. Tetranitromethane is a reagent highly specific for tyrosine residues and reacts under mild conditions to form the substitution product 3-nitrotyrosine *(8)*. The modified tyrosine has a characteristic adsorption maximum at 428 nm, and this can be used to quantitate the number of tyrosine residues modified *(8)*. However, under harsher conditions, there have been some reports of modification of sulfhydryl groups and limited cases of reaction with histidine and tryptophan *(9)*.

1.1. Strategies

1.1.1. Tyrosine Accessibility

The general strategy employed in chemical modification experiments is to determine the accessibility of the target residues within the native protein and the extent of protection offered by the bound substrate. Peptide mapping of the labeled protein then allows the roles of the individual residues to be assessed. First, the free protein is nitrated and then digested into fragments by proteolysis. These peptides are then separated to enable identification of the modified residue(s). The nucleoprotein complex is also nitrated and the modified residues identified in a similar way. From a comparison of these data, the extent of protection at each site can be established.

For peptide mapping, a protease should be chosen that, on digestion of the target protein, will place each tyrosine in a separate peptide. However, this is not essential if the modified residues are identified by N-terminal sequencing. It is possible that tyrosine modification may affect the efficiency of α-chymotrypsin digestion and this enzyme should be avoided if possible. The peptides can be separated by reverse-phase high-performance liquid chromatography (HPLC), and those containing tyrosine purified for further analysis. The tyrosine-containing peptide can be easily identified directly after HPLC purification by the characteristic fluorescence emission maximum of 3-nitrotyrosine at 305 nm (when excited at 278 nm). A particular tyrosine residue can then be identified by N-terminal sequence analysis.

The identification of nitrated tyrosine residues in the free protein provides information concerning the solvent accessibility of these residues in the protein and indicates which residues are likely to be buried within the protein. DNA protection studies will indicate which of these residues may be involved in protein–DNA interactions. However, the protection from nitration by bound DNA is only an indication of a functional role for a particular residue. The bound DNA may confer protection to a residue several angstroms away or may induce protein oligomerization (cooperative binding) which protects the tyrosine by protein–protein interactions. Consequently, functional studies need to be performed to further determine the role of the protected residue(s). The situation is analogous to the two types of analysis frequently used in the investigation of the DNA bases involved in complexes: "footprinting" and "interference" techniques. The data obtained from chemical modification and protection studies can then be used to design site-directed mutagenesis experiments to look at the function of an individual residue by observing the effects of its replacement with other amino acids.

1.1.2. Functional Studies

A protocol for functional studies will not be described in this chapter, but some general considerations will be mentioned here. One should nitrate the free protein and determine whether the modified protein still binds to DNA. This information should indicate whether the residues protected in the nucleoprotein complex are implicated in DNA binding. However, with proteins that bind cooperatively to DNA, a reduction in DNA-binding affinity may result from a disruption of protein–protein interactions rather than from protein–DNA interactions. A possible way to resolve this ambiguity is to bind the native and modified proteins to short oligonucleotides where the cooperativity factor is negligible. Modification of residues involved in protein–protein interactions should not significantly affect the intrinsic binding of the modified protein to DNA, when compared to the native protein.

Tyrosine residues can interact with DNA either by hydrophobic interactions via stacking with DNA bases or by hydrogen-bonding with the nucleotide through the phenolic OH group ***(10)***. Nitrotyrosine has a pK_a of 8.0, which may disrupt H-bonding as well as base stacking interactions. However, the addition of sodium dithionate reduces 3-nitrotyrosine to 3-amino tyrosine (which has a pK_a similar to that of native tyrosine) and may restore H-bonding interactions ***(11)***. Reduction with this reagent may provide further information concerning the nature of the tyrosine–nucleic acid interaction.

1.1.3. Rates of Modification

Nitration of a protein will initially report on the accessibility of specific tyrosine residues in the presence and absence of DNA. However, if the modified tyrosine residues can be analyzed individually, one can look at the nitration rates of the tyrosines and determine the degree of accessibility of each residue. This is achieved by removing aliquots of protein (at various time intervals) from a nitration experiment and determining the percentage nitration of each tyrosine for a given time-point. This can be done by quantitating the nitrated and unnitrated products after digestion, either by measuring the peak areas (recorded at 214 nm), taken directly from the HPLC profile ***(5)***, or by amino acid analysis of the purified peptides.

2. Reagents

All chemicals should be of AnalaR grade or higher and dissolved in double-distilled water. For HPLC analysis, trifluoroacetic acid (TFA), water and acetonitrile should be of HPLC grade. Buffers for HPLC should be filtered (0.2 µm) and degassed before use.

1. Tetranitromethane (TNM) stock solution: a 300 m*M* stock solution of TNM in ethanol. Store in the dark at 4°C. Note that TNM can cause irritation to the skin and lungs, and the solution should be made up in the fume hood. Additionally, TNM can be explosive in the presence of organic solvents such as toluene.
2. Nitration buffer: 150 m*M* NaCl and 10 m*M* Tris, pH 8.0.
3. Desalting column: Disposable "10DG" Econo columns (Bio-Rad, Richmond, CA) are preferred.
4. µBondapak C_{18} HPLC column (Waters Associates, Milford, MA) or a similar reverse-phase column.
5. Trypsin (TPCK treated).
6. Standard sodium dodecyl sulfate–polyacrylamide gel electrophoresis (SDS-PAGE) equipment with DC power supply capable of 150 V.
7. SDS–polyacrylamide gel stock solutions:
   Solution A: 152 g acrylamide and 4 g *bis*-acrylamide. Make up to 500 mL.
   Solution B: 2 g SDS and 30 g Tris base, pH 8.8. Make up to 500 mL.
   Solution C: 2 g SDS and 30 g Tris base, pH 6.8. Make up to 500 mL.
   When making up these solutions, they should all be degassed and filtered using a Buchner filter funnel. They should be stored in lightproof bottles; they will keep for many months.
8. 10% ammonium persulfate (APS): dissolve 0.1 mg in 1 mL of dH_2O.
9. 15% SDS–polyacrylamide gel: Mix together 8.0 mL of solution A, 4.0 mL of solution B, and 3.9 mL of dH_2O. Add 150 µL of 10% APS and 20 µL of *N,N,N',N'*-tetramethylethylene diamine (TEMED). Mix well and then pour between the plates. Immediately place a layer of dH_2O (or butanol) on top of the gel to create a smooth interface with the stacking gel. When the resolving gel has set, pour off the water and prepare the stacking gel. This is done by adding 750 mL of solution A and 1.25 mL of solution C to 3.0 mL of dH_2O. Finally, add 40 µL of APS and 10 µL of TEMED, pour on the stacking gel and insert the comb. To avoid the gel sticking to the comb, remove the comb as soon as the gel has set.
10. 10X SDS running buffer: 10 g SDS, 33.4 g Tris base, and 144 g glycine made up to 1 L.
11. High methanol protein stain: Technical-grade methanol 500 mL, 100 mL glacial acetic acid and 0.3 g PAGE 83 stain (Coomassie blue), made up to 1 L.
12. Destain solution: 100 mL methanol and 100 mL glacial acetic acid, made up to 1 L.
13. 2X SDS-PAGE loading buffer: 4% (w/v) SDS, 60 m*M* Tris-HCl, pH 6.8, 20% glycerol, 0.04% (w/v) bromophenol blue, and 1% (v/v) β-mercaptoethanol.

3. Methods

The method is a fairly general one for protein-nucleic acid complexes. However, precise details of conditions for nucleoprotein complex dissociation and peptide mapping will vary with the system under investigation. As an example of the technique, nitration of the Pf1 gene 5 protein and nucleoprotein complex will be described *(5)*.

3.1. Nitration

1. Desalt the protein or nucleoprotein complex into nitration buffer to a concentration between 0.5 and 5 mg/mL (*see* **Note 1**). For initial determination of nitrated residues, 0.5 mg of protein should be sufficient. However, if a time-course experiment is performed, larger amounts of protein are required.
2. To 1 mL of sample, add a 10-fold molar excess of 300 m*M* TNM (in ethanol) and incubate at room temperature for 1 h, stirring gently (*see* **Note 2**). The reaction is stopped by the addition of acid (add HCl to pH 2.0) or by rapid desalting into 10 m*M* Tris–HCl pH 8.0 (*see* **Note 3**).
3. Run an aliquot of the modified protein–nucleoprotein complex on an SDS gel, together with native protein, to determine whether there has been any TNM-induced crosslinking (*see* **Note 4**). If analyzing the free protein proceed to **step 5**; if modifying the nucleoprotein complex, proceed to **step 4**.
4. Dissociate the nucleoprotein complex by the addition of salt (*see* **Note 5**). Large DNA fragments can be removed by ultracentrifugation, whereas smaller fragments can be either digested with nucleases or removed by gel filtration. The protein is then dialyzed or desalted into the appropriate protease digestion buffer.
5. Digest the protein to completion with the desired protease(s) and then lyophilize the peptides for separation by HPLC. The peptides may be stored at –20°C.

In the example provided, the Pf1 gene 5 protein and nucleoprotein complex were incubated at room temperature in the presence of a 64 *M* excess of TNM (300 m*M* in ethanol) for 3 h. The reaction was stopped by desalting the protein (and nucleoprotein complex) into 10 m*M* Tris-HCl, pH 8.0. The nucleoprotein complex was dissociated by the addition of $MgCl_2$ to 1 *M* and the phage genomic DNA was then removed by ultracentrifugation at 221,000g (in a Beckman L8 ultracentrifuge; 70.1 Ti rotor) for 2.5 h. The protein was desalted into 10 m*M* Tris-HCl, pH 8.0 for proteolysis and digestion with trypsin (Sigma [St. Louis, MO], TCPK treated) at an enzyme substrate ratio of 1:25 (w/w) for 3 h at 37°C. Phenylmethane sulfonyl fluoride was added to a final concentration of 1 m*M* and the sample was lyophilized overnight. This procedure results in the complete separation of the three tyrosine-containing tryptic peptides.

3.2. Peptide Mapping

1. Peptides can usually be separated by reverse-phase HPLC on a C_{18} column. Generally the peptides are applied to the column in 8 *M* urea and 2% β-mercapto-

ethanol and separated in an acetonitrile gradient in the presence of 0.05–0.1% TFA. The acetonitrile gradient must be determined empirically for each particular protein.
2. Determine separation conditions for peptides from the native protein (*see* **Note 6**) and identify tyrosine-containing peptides (*see* **Note 7**).
3. Apply peptides from the nitrated protein and initially elute under the same conditions that were used for the native protein (*see* **Note 8**). If necessary, change the acetonitrile gradient to achieve separation of the tyrosine-containing peptides and their nitrated counterparts.

In the example provided, the tryptic peptides from a native gene 5 protein were resuspended in 200 µL of 8 *M* urea and 2% β-mercaptoethanol and clarified prior to HPLC analysis (**Fig. 1**, top). Tyrosine-containing peptides were initially detected by their fluorescence properties (*see* **Note 7**) and then identified by automated Edman degradation on an Applied Biosystems 477A pulsed liquid amino acid sequencer. Nitrated peptides were applied to the C_{18} column and separated under identical conditions (**Fig. 1**, bottom). The nitrated peptides were initially detected by their altered retention times and by virtue of their yellow color in 10 m*M* Tris-HCl, pH 8.0. The identity of the nitrated peptides was subsequently confirmed by N-terminal sequencing (*see* **Note 9**).

3.3. Functional Studies

As discussed in Subheading 1., one should check whether nitration of the protein impairs the ability to bind DNA (other chapters in this volume can be consulted for possible approaches such as EMSA [Chapter 2] or DNaseI footprinting [Chapter 3]). The protein isolated from the nitrated nucleoprotein complex should be checked for DNA binding. Because the target amino acid residues in contact with the DNA should have been protected from modification, the protein from the nitrated complex would be expected to retain DNA-binding activity.

4. Notes

1. As an alternative to desalting, the protein can be dialyzed into nitration buffer.
2. The molar excess of TNM can be increased to ensure maximal modification (e.g., the Pf1 gene 5 protein was nitrated in a 64-fold molar excess of TNM in the example provided). Note, however, that at high concentrations, protein insolubility can become a problem.
3. One can desalt the protein into a buffer appropriate for proteolysis or dissociation of the nucleoprotein complex at this stage, as required.
4. Tetranitromethane-induced crosslinking has been widely reported, and an SDS gel should be run to check for the appearance of adducts. Reducing the concentration of portion or molar excess of TNM may help to limit adduct formation.

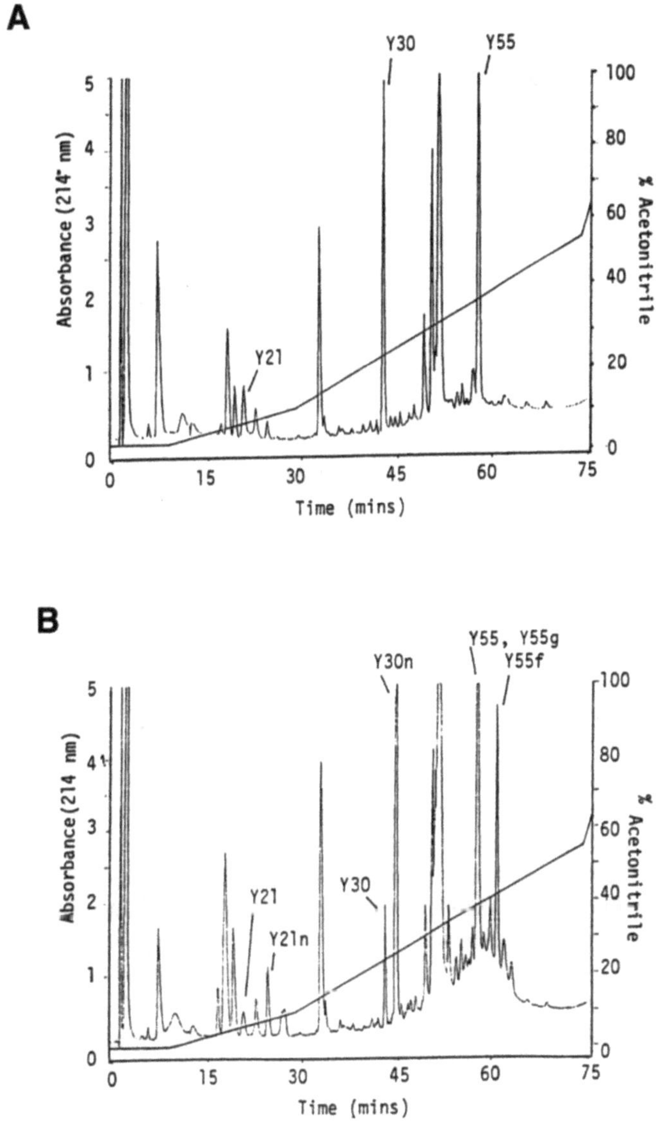

Fig. 1. High-performance liquid chromatographic elution profile of tryptic peptides of the Pf1 gene 5 protein for **(top)** native protein and **(bottom)** nitrated protein. Peaks Y21, Y30, and Y55 correspond to tryptic peptides containing tyrosine 21, 30, and 55, respectively. n denotes a nitrated peptide. Aliquots (100 μL) were applied to a μBondapak C18 HPLC column (Waters Associates) (300 × 4.6 mm inner diameter) fitted with a C18 guard column. The HPLC buffers for this experiment were as follows: buffer A: 0.05% TFA/H_2O; buffer B: 0.05% TFA/acetonitrile. Peptides were separated in the following gradient at a flow rate of 2 mL/min: O% B for 5 min; 0–10% B in 20 min; 10–55% B in 45 min; 55–90% B in 5 min; 90% B for 5 min; 90–100% B in 5 min.

Gel filtration is another way of removing the adducts prior to peptide mapping (*see also* **Note 10**).

5. Usually, protein–nucleic acid interactions can be disrupted by the addition of NaCl or $MgCl_2$ to 1–2 *M*. The conditions required to effect separation will vary with the nature of the complex.
6. It is not essential to have complete separation of all fragments, only separation of the tyrosine-containing peptides and their nitrated counterparts. The HPLC conditions should be adjusted to achieve this.
7. Tyrosine residues have a characteristic fluorescence emission maximum at 303 nm when excited at 278 nm. This phenomenon can be used initially to determine which peptides contain a tyrosine residue (this may not be possible, however, if there is a tryptophan residue present in the same peptide as a result of energy transfer). If on-line fluorescence detection is not available, the fractions can be taken directly from the HPLC and analyzed in a fluorimeter. The peptides should then be unambiguously identified by either N-terminal sequencing or amino acid analysis.
8. The addition of a nitrate group to the tyrosine should alter the hydrophobicity and, hence, the retention time of that particular peptide in an acetonitrile gradient. This should allow immediate identification of the nitrated peptides. However, it is possible that a nitrated peptide comigrates with another unmodified peptide. Therefore, freeze-dry all peptides from HPLC and resuspend in 10 m*M* Tris-HCl, pH 8.0: The nitrated peptides will have a faint yellow color (absorbance maximum at 428 nm).
9. For peptides sequenced on an applied biosystems 477A pulse liquid amino acid sequencer (fitted with a 120A separation system for the analysis of PTH-derivitized amino acids), PTC-3-nitrotyrosine elutes just after DTPU.
10. TNM induced oligomerization has been observed on the nitration of numerous proteins including several DNA-binding proteins ***(4,12,13)***. This is usually considered undesirable, and steps are often taken to reduce the crosslinking and remove adducts before analysis (e.g., by gel filtration). However, advantage can be taken of this crosslinking ability; Martinson and McCarthy ***(14)*** used TNM as a reagent to crosslink histones specifically. On nitration with TNM, we have also shown that the Pf1 gene 5 protein forms an SDS-stable dimer ***(13)***. Initial analysis of the peptide adduct in this case suggested that tyrosine 55 from one monomer was crosslinked to phenylalanine 76 from the other monomer (forming an interdimer crosslink rather than an intradimer crosslink). The adducts are thought to form via a free-radical mechanism, resulting in zero-length crosslinks between residues in close proximity ***(14,15)*** Thus, if adduct formation is limited to one or two species, additional structural information can be obtained from the experiment. The crosslinked proteins should be digested and the peptide adduct purified by HPLC. N-Terminal sequencing, amino acid analysis, and gas chromatograpy–mass spectroscopy (of the hydrolyzed peptide) should enable unambiguous identification of the two residues participating in the crosslink and provide structural information concerning the relative positions of these residues in the protein.

References

1. Mueller, M. J., Samuelson, B., and Haeggstrom, J. Z. (1995) Chemical modification of leukotriene A4 hydrolase. Indications for essential tryosyl and arginyl residues at the active site. *Biochemistry* **34,** 3536–3543.
2. Hawes, J. W., Crabb, D. H., Chan, R. M., Rougraff, P. M., and Harris, A. (1994) Chemical modification and site-directed mutagenesis studies of rat 3-hydroxy-isobuyrate dehydrogenase. *Biochemistry* **34,** 4231–4237.
3. Kochhar, S., Hunziker, P. E., Leong-Morgenthaler, P., and Hottinger, H. (1992) Primary strucure, physiochemical properties and hemical modification of NAD+-dependent *d*-lactate dehydrogenase. *J. Biol. Chem.* **267,** 8499–8513.
4. Anderson, R., Nakashima, Y., and Coleman, J. (1975) Chemical modification of functional residues of the Fd gene 5 DNA-binding protein. *Biochemistry* **14,** 907–917.
5. Plyte, S. E. and Kneale, G. G. (1991) Mapping the DNA binding site of the Pf1 gene 5 protein. *Protein Eng.* **4(5),** 553–560.
6. Lundblad, R. and Noyes, M. (1984) *Chemical Reagents for Protein Modification I and II*, CRC, Boca Raton, FL.
7. Riordan, J., Sokolovsky, M., and Vallee, B. (1967) Environmentally sensitive tyrosine residues. Nitration with tetranitromethane. *Biochemistry* **6,** 358.
8. Sokolovsky, M., Riordan, J., and Vallee, B. (1966) Tetranitromethane. A reagent for the nitration of tyrosyl residues in proteins. *Biochemistry* **5,** 3582–3589.
9. Sokolovsky, M., Harell, G., and Riordan, J. (1969) Reaction of tetranitromethane with sulphydryl groips in proteins. *Biochemistry* **8,** 4740–4745.
10. Dimicoli, J. and Helene, C. (1974) Interaction of aromatic redisues of proteins with nucelic acids I and II. *Biochemistry* **13,** 714–730.
11. Sokolovsky, M. Riordan, J., and Vallee, B. (1967) Conversion of 3-nitrotyrosine to 3-amino-tyrosine in peptides and proteins. *Biochem. Biophys. Res. Commun.* **27,** 20.
12. Anderson, R. and Coleman, J. (1975) Physiochemical properties of DNA-binding proteins: gene 32 protein of T4 and *Escherichia coli* unwinding protein. *Biochemistry* **1,** 5485–5491.
13. Plyte, S. E. (1990) The biochemical and biophysical characterization of the Pf1 gene 5 protein and its complex with nucelic acids, Ph.D. thesis, Portsmouth University, Portsmouth, UK.
14. Martinson, H. and McCarthy, B (1975) Histone–histone associations within chromatin. Crosslinking studies using tetranitromethane. *Biochemistry* **14,** 1073–1078.
15. Williams, J. and Lowe, J. (1971) The crosslinking of tyrosine with tetranitromethane. *Biochem. J.* **121,** 203–209.

21

Chemical Modification of Lysine by Reductive Methylation

A Probe for Residues Involved in DNA Binding

Ian A. Taylor and Michelle Webb

1. Introduction

The basic side chains of lysine residues often play essential roles in DNA–protein recognition. They are able to contribute to the overall affinity of an interaction through nonspecific charge–charge interactions with the phosphate backbone and contribute substantially to the specificity of the interaction by forming direct hydrogen bonds with functional groups on the edges of the bases. This dual role and their almost ubiquitous presence in the interface of DNA–protein complexes make them very attractive targets for chemical modification experiments.

Numerous chemical reagents to chemically modify lysine side chains in proteins are available *(1)*. Unfortunately, the conditions required for such procedures are often harsh and result in total denaturation of the protein. Furthermore, many reagents are not entirely specific to lysine and often react with other residues such as cysteine, histidine, and tyrosine. Two methods, that are able to specifically modify lysine residues under native conditions and have been applied successfully to the investigation of protein nucleic acid interactions, are amidination with imidoesters *(2)* and, the subject of this chapter, reductive alkylation *(3)*.

Reductive alkylation has become a widespread and well-established technique for the specific modification of lysine residues and a variety of reagents have been used to produce this chemical modification *(4)*. A reductive methylation reaction using formaldehyde and the reducing agent sodium cyanoborohydride *(5)* is particularly useful because under mild solution condi-

From: *Methods in Molecular Biology, vol. 148: DNA–Protein Interactions: Principles and Protocols, 2nd ed.*
Edited by: T. Moss © Humana Press Inc., Totowa, NJ

$$\text{Pr-NH}_2 \overset{\text{CH}_2\text{=O}}{\rightleftharpoons} \text{Pr-NH-CH}_2\text{-OH} \xrightarrow{-\text{H}_2\text{0}} \overset{\text{(A)}}{\text{Pr-N=CH}_2} \xrightarrow{\text{NaCNBH}_3} \overset{\text{(B)}}{\text{Pr-NH-CH}_3} \qquad \text{(I)}$$

$$\text{Pr-NH-CH}_3 \overset{\text{CH}_2\text{=O}}{\rightleftharpoons} \text{Pr-N(CH}_3\text{)-CH}_2\text{-OH} \xrightarrow{-\text{H}_2\text{0}} \text{Pr-N(CH}_3\text{)=CH}_2 \xrightarrow{\text{NaCNBH}_3} \overset{\text{(C)}}{\text{Pr-N(CH}_3\text{)}_2} \qquad \text{(II)}$$

Fig. 1. Reductive methylation of lysine. The reaction scheme *(6)* occurs in two distinct phases. Adduction of formaldehyde to the lysine ε-amino group generates the Schiff base (**A**), which is then reduced by sodium cyanoborohydride to ε-*N*-mono-methyl-lysine (**B**). A second round of the reaction generates the final product ε-*N,N*-dimethyl-lysine (**C**).

tions (aqueous buffer, pH 7–8), the accessible lysine residues on proteins are completely converted to the ε-*N,N*-dimethyl derivatives. The reaction (**Fig. 1**) occurs in two distinct phases. Initially, the ε-amino group of the lysine forms an adduct with the formaldehyde to produce a Schiff base. This then undergoes reduction by sodium cyanoborohydride to the monomethylamine derivative in the second part of the reaction. A further round of the reaction, which occurs more rapidly than the first produces the dimethyl derivative.

The attractiveness of this modification is that dimethylation of the lysine side chain is a relatively small chemical change. The pK_a is only slightly affected and may even remain unchanged *(5)*. Because of this, the residue maintains its ionization properties and the potential for the formation of the same ion-pair interactions as in the unmodified protein remains, although there is some loss of hydrogen-bonding capacity. Nevertheless, the modification is a mild one and unlikely to significantly perturb the native enzyme structure.

A major use of any chemical modification procedure is the incorporation of isotopic labels at specific positions in proteins. Reductive methylation experiments that incorporate ^{13}C have been used to probe the environment of lysine side chains in proteins using nuclear magnetic resonance (NMR) spectroscopy *(7)*. The incorporation of radiolabels into proteins by reductive methylation enables the number of lysines that are accessible to be determined. In the case of DNA-binding proteins, the labeling is carried out on the free protein and the DNA–protein complex. This immediately provides information about the number of lysines present in the DNA-binding site. Subsequent peptide mapping strategies allow identification of specific residues, the degree of labeling at particular sites is then used to derive the location of that residue within the protein. In this way, [^3H] formaldehyde has been used as a source of radiolabel to probe the role of the lysine residues of the core histones in the nucleosome

(8), the interaction of the linker histone H5 with nucleosomes in long chromatin *(9)*, and the role of lysine residues in a specific DNA–protein complex between the type I DNA methylase M.*Eco*R124I *(10)* and a short DNA duplex containing its recognition site.

Subheading 3. contains a set of protocols to enable quantification of the number of lysine residues susceptible to reductive methylation in a protein and a DNA–protein complex. The peptide mapping and N-terminal protein-sequencing procedures required to identify and determine the extent of modification at individual residues are also described. Because the modification reaction is sensitive to contaminants in commercial grades of sodium cyanoborohydride (most likely cyanide), a method to recrystallize the reagent *(6)* is described in **Subheading 3.1.** In order to obtain quantitative information about the total extent of protein modification and/or determine the relative accessibility of individual residues, it is vital that the specific activity of the [^{3}H] formaldehyde be determined accurately. **Subheading 3.2.** describes a protocol to do this using a simple peptide substrate α-melanocyte-stimulating hormone (α-MSH). The data from this experiment then allow quantitative conclusions to be drawn from the subsequent protein protection and peptide mapping experiments that are described in **Subheadings 3.3.** and **3.4.**

2. Materials

2.1. Reagents and Materials

1. Reagents were obtained as follows: urea, formic acid, hydrochloric acid and glacial acetic acid (ARISTAR grade), dichloromethane (dried, AnalaR), glycine, Tris (hydroxymethyl) methylamine, sodium chloride and Na_2EDTA (AnalaR), trifluoroacetic acid and water (HiPerSolv grade), acetonitrile (far UV HiPerSolv grade), and sodium cyanoborohydride (Schuchardt), all from Merck; dithiothreitol (DTT) and HEPES (both molecular biology grade), α-MSH and silica gel (Type III, indicating) from Sigma [^{3}H] formaldehyde (approx 100 Ci/mol). NEN radiochemicals; filter paper (No. 1) from Whatman; dialysis membrane (Slide-A-Lyser) from Pierce; scintillation fluid (Ecoscint™ H) from National Diagnostics); trypsin (sequencing grade) and Pefabloc™ from Boehringer Mannheim.
2. DNA-binding protein (approx 10 mg).
3. Stock solution of oligonucleotide duplex, 300 μ*M* (1 μmol synthesis, high-performance liquid chromatography [HPLC] purified).
4. Formaldehyde (Sigma): 37% solution and 15% methanol stabilizer.
5. 8 *M* urea and 50 m*M* acetic acid.
6. 10 m*M* HEPES, pH 7.5.
7. 1 *M* Na-glycine, pH 7.0.
8. 50 m*M* TBS: 10 m*M* Tris-HCl, 50 m*M* NaCl, and 1 m*M* EDTA, pH 7.5.
9. 2 *M* Tris base.

2.2. Equipment

1. HPLC: binary gradient formation using high-pressure-mixing, variable-wavelength ultraviolet (UV) absorbance detector and fraction collection are required.
2. Reverse-phase HPLC columns: semipreparative C_3 (9.4 × 250 mm), analytical C_3 (4.6 × 250 mm), and analytical C_{18} (4.6 × 250 mm). Zorbax 300SB or an equivalent wide-pore packing is recommended.
3. High-resolution anion-exchange column (e.g., TSK-GEL, DEAE-NPR [4.6 × 35 mm]).
4. Freeze-dryer.
5. Scanning (UV/visible) spectrophotometer.
6. Liquid scintillation counter.
7. Protein-sequencing facilities.

3. Methods

3.1. Recrystallization of Sodium Cyanoborohydride

1. Dissolve 44 g of sodium cyanoborohydride in 100 mL of acetonitrile and remove any undissolved material by centrifugation.
2. Add 600 mL of dichloromethane to the mixture place in a sealed container and allow the sodium cyanoborohydride to precipitate overnight at 4°C.
3. Filter the mixture (Whatman filter paper No. 1) to collect the precipitate and wash with an additional 100 mL of cold dichloromethane.
4. Allow the powder to dry and store in a vacuum desiccator containing silica gel. Prepare aqueous stock solutions immediately before use.

3.2. Determination of the Effective Specific Activity of ^{3}H Formaldehyde

There is a substantial variation in the effective specific activity of different batches of [^{3}H] formaldehyde. In order to accommodate this, it is necessary to determine the effective specific activity of each individual batch. A convenient way to do this is to use a simple peptide substrate for which the number of accessible amino groups is known. The effective specific activity can then be calculated from the amount of radioactivity that can be incorporated into the fully modified peptide. For this purpose the peptide hormone α-MSH (Ac-SYSMEHFRWGKPV-NH_2) is recommended (*see* **Note 1**).

1. Redissolve the contents of a 1-mg vial of α-MSH in 1 mL of 8 *M* urea and 50 m*M* acetic acid.
2. Equilibrate a C_{18} reverse-phase column in 2% acetonitrile and 0.05% (v/v) trifluoroacetic acid (TFA) at a flow rate of 1 mL/min and set the UV absorbance detector to 280 nm. Apply the peptide in approx 200-μg aliquots and elute with a 0–60% increasing gradient of acetonitrile, 0.05% TFA (v/v) over 60 min.
3. Collect the peptide-containing fractions from multiple runs, pool, and lyophilize them. Store the purified peptide at –20°C in a box containing silica gel.

4. Dissolve 100 nmol of purified α-MSH in 0.5 mL of 10 m*M* HEPES, pH 7.5. Add 10 m*M* sodium cyanoborohydride (freshly made) followed by 2 m*M* [^{3}H] formaldehyde (*see* **Notes 2** and **3**) and incubate at 24°C for 3 h. After this time add a second aliquot of the reagents and allow the reaction to continue for an additional 3 h.
5. Adjust the sample to 8 *M* urea and 50 m*M* acetic acid and purify the modified peptide by reverse-phase chromatography using the same gradient as in **step 2**. Pool the fractions, lyophilize, and store at –20°C in a box containing silica gel.
6. To determine the specific activity of the modified peptide, redissolve the labeled peptide in 1 mL of water and determine its concentration from the absorbance at 280 nm (ε_{280} = 7000/*M*/cm). Add 10 µL to 1 mL of liquid scintillant (Ecoscint H or equivalent), mix well, and determine the amount of incorporated radioactivity by liquid scintillation counting. The specific activity of the labeled peptide ($\sigma_{\alpha\text{-MSH}}$) (in nCi/nmol) is calculated from **Eq. 1**. The effective specific activity of the [^{3}H] formaldehyde ($\sigma_{[^3\text{H}]\ \text{formaldehyde}}$) is simply half this value (**Eq. 2**) (*see* **Note 4**):

$$\sigma_{\text{a-MSH}} = (\text{dpm})/(\text{No. nmoles counted} \times 2220) \quad (1)$$

$$\sigma_{[^3\text{H}]\ \text{formaldehyde}} = {}^1/_2\ \sigma_{\alpha\text{-MSH}} \quad (2)$$

3.3. Surface Labeling of Protein and DNA–Protein Complex by Reductive Methylation and Quantification of Residues Modified

Prior to mapping the positions of modified lysine residues, it is necessary to determine the number of residues susceptible to reductive methylation and the proportion of these that are protected by the presence of DNA.

1. Prepare approx 3.0 mL of a solution of the DNA-binding protein in 10 m*M* HEPES, 50 m*M* NaCl, and 1 m*M* EDTA (pH 7.5) by either dialysis or buffer exchange. The concentration of DNA-binding protein in this solution should be 1–2 mg/mL.
2. Take 1.5 mL of protein and add an equimolar amount, or slight excess, of the DNA duplex from a concentrated stock solution to form the DNA–protein complex.
3. Withdraw a 300-µg aliquot from the free protein and the DNA–protein complex samples and add glycine to 50 m*M* from a neutral 1 *M* stock solution. Dilute the samples to 1 mL with 50 m*M* TBS and load into dialysis cassettes (*see* **Note 5**). Dialyze against 2 L of 50 m*M* TBS at 4°C. These samples are controls for efficiency of the whole procedure and also serve as the "zero" time points (*see below*).
4. To the remaining solutions, add a 30-fold molar excess of sodium cyanoborohydride over the total lysine content of the protein. Then, add a 10-fold molar excess of [^{3}H] formaldehyde (*see* **Note 6**) and incubate at 24°C.
5. At timed intervals (10 min up to 5 h) withdraw 300 µg samples of protein from each time-course. Quench the reaction by the addition of 50 m*M* glycine and

dilute to 1 mL with 50 m*M*TBS. Load the samples into dialysis cassettes and dialyze overnight against 2 L of 50 m*M* TBS using at least 3 changes.

6. After extensive dialysis, remove all the samples from the cassettes. For free-protein samples, continue from **step 7** onward. Samples from the DNA–protein complex time-course need to be processed to remove the DNA as follows. Equilibrate a small-volume (1 mL or less) high-performance anion exchange (*see* **Note 7**) HPLC column (analytical TSK-GEL, DEAE-NPR [4.6 × 35 mm] or an equivalent) in 50 m*M* TBS at a flow rate of 1 mL/min. Set the UV detector to 280 nm. Apply each sample from the time-course to the column and collect any flowthrough. Elute the protein (if bound) and the DNA by application of an increasing NaCl gradient from 0.05 *M* to 1.0 *M* over 50 column volumes. Collect the protein-containing fractions and proceed.
7. Determine the molar concentration of the samples from each time-course (including the "zero") using the absorbance at 280 nm. Add 10 µL of each sample to 1 mL of liquid scintillant, mix well, then determine the amount of incorporated radioactivity at each time-point by liquid scintillation counting.
8. Calculate the specific activity (in nCi/nmol) of the labeled protein ($\sigma_{protein}$) at each time-point of the reaction using **Eq. 3**. Then, using the value for the effective specific activity of the [^{3}H] formaldehyde determined in **Subheading 3.2.** calculate the number of lysine residues modified at each point in the time-course using **Eq. 4** (*see* **Note 8**):

$$\sigma_{protein} = (\text{dpm})/(\text{No. nmoles counted} \times 2220) \tag{3}$$

$$\text{No. modified lysines} = \sigma_{(protein)} = 2\sigma_{([3H]\ formaldehyde)} \tag{4}$$

9. Plot the number of lysine residues modified against time and fit the data to a single exponential process using **Eq. 5** (**Fig. 2**). In most cases, the data should fit well to this model (*see* **Note 9**) and the total number of modifiable lysine residues is then given by the limit value (*L*). Fitting the data in this manner also allows a rate constant (*k*) for the incorporation of radiolabel to be derived. Both of these parameters can be affected by DNA binding.

$$\text{No. modified lysines} = L(1-e^{-kt}) \tag{5}$$

10. Compare the fitted curves of the time-course for the reaction of free protein and for the DNA–protein complex (**Fig. 3**). Formation of the DNA–protein complex may well reduce *L*, indicating the presence of a population of strongly protected lysines. At the same time, differences in *k* are likely to arise from an overall lowering of the rate of modification because of decreased accessibility of lysine residues in the presence of DNA.

3.4. Pulse Chase Labeling of Proteins

If the fraction of lysine residues protected by DNA is large, as determined in **Subheading 3.3.**, then a pulse-labeling procedure carried out on the free protein will reveal which lysine residues are surface accessible and likely to be

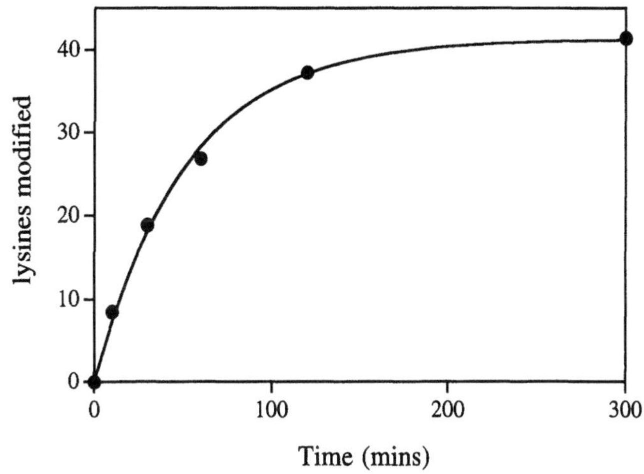

Fig. 2. Time-course of reductive methylation of the type I DNA methyltransferase M.*Eco*R124I. The curve is the best fit of the data to a single exponential ($L = 41$, $k = 0.019$/min).

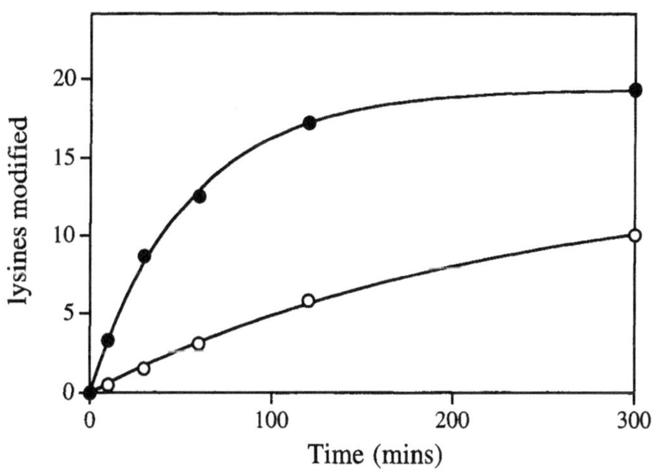

Fig. 3. The effect of DNA binding on reductive methylation of the DNA recognition subunit (HsdS) from M.*Eco*R124I. The upper curve is a best fit to the data from a time-course for the modification reaction of free protein ($L = 19$, $k = 0.019$/min). The lower curve is the best fit to the data from a time-course for modification of the protein in the DNA–protein complex ($L = 14$, $k = 0.004$/min).

involved in DNA binding. If only a small number of lysine residues are protected by DNA, then a modification to the procedure should be undertaken (*see* **Note 10**).

1. Prepare 1 mL of DNA-binding protein at a concentration of 1–2 mg/mL in 50 m*M* TBS by either dialysis or buffer exchange.
2. Add sodium cyanoborohydride to a 30-fold molar excess over the total lysine content of the protein and then initiate the chemical modification reaction by the addition of [^{3}H] formaldehyde at a 10-fold molar excess. Incubate at 24°C, the length of time will depend on the results from the experiments in **Subheading 3.3.** Aim to modify for a length of time when the reaction is about 50% complete (*see* **Note 11**). This will probably be between 10 and 60 min.
3. At the end of the pulse, quench the reaction by the addition of 50 m*M* glycine and dialyze overnight against 1 L of 50 m*M* TBS. Change the dialysis buffer at least three times.
4. Equilibrate a semipreparative C_3 reverse-phase column (9.4 × 250 mm) in 5% acetonitrile and 0.05% (v/v) TFA at a flow rate of 3 mL/min and set the UV detector to 225 nm. Remove the protein from the dialysis cassette, add urea to a final concentration of 8 *M* and DTT to 50 m*M*. Incubate the sample briefly at room temperature, acidify by the addition of 100 m*M* acetic acid, and apply the protein to the column. Elute with a 5–65% gradient of acetonitrile and 0.05% (v/v) TFA over 60 min. Collect the protein containing fractions and lyophilize them.
5. Redissolve the modified protein in 1 mL of 8 *M* urea and 10 m*M* HEPES (pH 7.5) and determine the protein concentration from the absorbance at 280 nm. At this point, the extent of label incorporation should be determined as in **Subheading 3.3., steps 7** and **8.**
6. To "chase" the reaction with unlabeled reagent, add a 30-fold molar excess of sodium cyanoborohydride over the total lysine content followed by a 10-fold excess of unlabeled formaldehyde. Incubate for 3 h at 24°C, then add a second aliquot of these reagents and continue the reaction for a further 3 h.
7. Add DTT to a final concentration of 50 m*M*, incubate briefly at room temperature then acidify with 100 m*M* acetic acid. Purify the fully modified protein by reverse-phase chromatography as in **step 4**. Lyophilize the fractions containing protein and store in aliquots of approx 2 nmol at –20°C in a box containing silica gel.
8. Redissolve a 2 nmol aliquot of modified protein in 100 µL of 0.9% formic acid (*see* **Note 12**). Ensure that the sample is fully dissolved then dilute to 500 µL with dH_2O and titrate to pH 8 by the addition of 35 µL of 2 *M* Tris base.
9. Dissolve the contents of a vial of sequencing grade trypsin (*see* **Note 13**) in 1 m*M* HCl to give a concentration of 1 mg/mL. Add the trypsin to the protein to give an enzyme to a substrate ratio of 1:10 (w/w) and incubate at 37°C for approx 18 h. To increase the efficiency of cleavage add the trypsin in three aliquots at roughly 4-h intervals. Terminate the digest by the addition of 1 m*M* Pefabloc and store at –20°C until required.
10. Equilibrate an analytical C_3 reverse-phase HPLC column (4.6 × 250 mm) in 2% acetonitrile and 0.05% (v/v) TFA at a flow rate of 1 mL/min and set the UV absorbance detector to 214 nm. Adjust the tryptic digests to 8 *M* urea and 50 m*M* DTT and incubate briefly at room temperature. Acidify the mixture by the addition of 100 m*M* acetic acid and apply to the column. Elute the peptides with an increasing gradient of acetonitrile collecting 250 µL fractions. For a complex

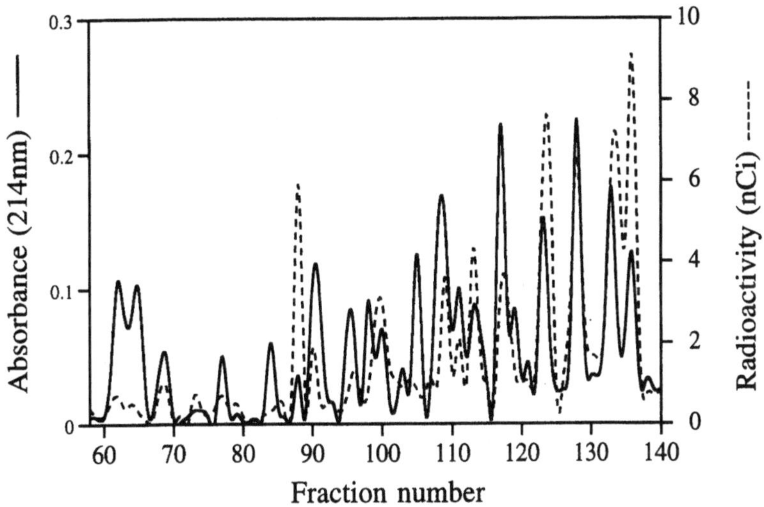

Fig. 4. Separation of reductively methylated peptides from pulse-labeled HsdS by C_4 reverse-phase chromatography. The absorbance at 214 nm and the amount of radioactivity (nCi) are plotted for each fraction.

mixture of peptides, the following gradient works well: 2–35% in 45 min followed by 35–60% in 20 min. It may be necessary to alter this for the particular protein under investigation.

11. Remove 25 µL from each fraction, add 1 mL of liquid scintillant, and determine the level of radioactivity by liquid scintillation counting. Pool the fractions across each peak and lyophilize them. Store at –20°C in a box containing silica gel until required.
12. Construct an overlaid chromatogram as in **Fig. 4** and use this to select peaks with an apparently high specific activity (*see* **Note 14**). These peaks require further fractionation by C_{18} reverse-phase chromatography (*see* **Note 15**) before ultimately submitting them for N-terminal amino acid sequencing.
13. Equilibrate an analytical C_{18} reverse-phase column in 2% acetonitrile and 0.05% TFA at a flow rate of 1 mL/min and set the UV detector to 214 nm. Redissolve each selected peak in 500 µL of 8 *M* urea and 100 m*M* acetic acid and apply to the column. Elute the bound peptides with a linear gradient of acetonitrile (2–50% in 70 min) and collect 250-µL fractions.
14. Remove 25 µL from fractions across each peak and determine the incorporated radioactivity as in **step 11**. Lyophilize the remainder and store at –20°C in a box containing silica gel until required.
15. Analyze each purified peptide using automated N-terminal amino acid sequencing. The objective is to determine the number of pmoles of each amino acid released at each cycle of the sequencing reaction and also to determine the amount of radioactivity associated with the residue. A suggested method for doing this is described in **Note 16**.

3.5. Data Analysis

The data from the N-terminal sequencing are used to determine the specific activity of each modified lysine as follows. Plot the number of pmoles of each residue released at each cycle pmole(*n*) versus the cycle number (*n*); then, fit these data to **Eq. 5** (*see* **Fig. 5A**). *E* is the efficiency of the sequencing process (usually around 90%) and pmole(0) is the amount of starting material.

$$\text{pmole}(n) = \text{pmole}(0) \times E^n \tag{5}$$

As PTH-dimethyl-lysine is not a standard amino acid, the number of pmoles of modified lysine is not determined directly from integration of the HPLC trace. Instead, the value can be calculated by interpolation of the fitted curve for the cycle at which the dimethyl-lysine was released. Plot the amount of radioactivity released at each cycle versus the cycle number (**Fig. 5B**). Significant quantities of radioactivity should only be present at a cycle where a modified lysine is present. Combine the data from the two plots and use **Eq. 1** to calculate the specific activity (σ_{lysx}); then, determine the fractional modification ($\sigma_{lysx}/\sigma_{\alpha\text{-MSH}}$) for each dimethyl-lysine in a peptide. The value of this ratio is proportional to the accessibility of the residue during the pulse part of the chemical modification reaction. Values close to unity indicate a high degree of accessibility, whereas values close to zero indicate a residue that is inaccessible to chemical modification.

These data can be used to build up a picture of the protein surface and identify clusters of lysines that are potential surfaces for DNA binding. Residues identified by these methods are then targets for site-directed mutagenesis experiments.

4. Notes

1. The α-MSH peptide contains only a single lysine and has a blocked N-terminus. The presence of the two aromatic residues allow its concentration to be determined accurately from its UV absorbance at 280 nm (ε_{280} = 7000/*M*/cm). We have extensively characterized the reductive methylation reaction with this peptide using NMR and mass spectroscopy ***(10)***. Under the reaction conditions used in **Subheading 3.2.**, >95% of the product is dimethylated and the remainder monomethylated. In principle, any peptide substrate with a known number of free amino groups and for which the concentration can be measured accurately could be used to determine the effective specific activity. However, if a different peptide substrate is used, it is advisable to extensively characterize the reaction in the same way. On a routine basis, if access can be gained to a mass spectrometer, it may be worthwhile checking the completeness of the reaction in this way.
2. [^{3}H] Formaldehyde from NEN is supplied as a 0.3 *M* aqueous solution in snap-off glass vials. Make sure the whole contents of the vial are at the bottom and then leave on ice for 10 min before breaking the seal. After opening, transfer the

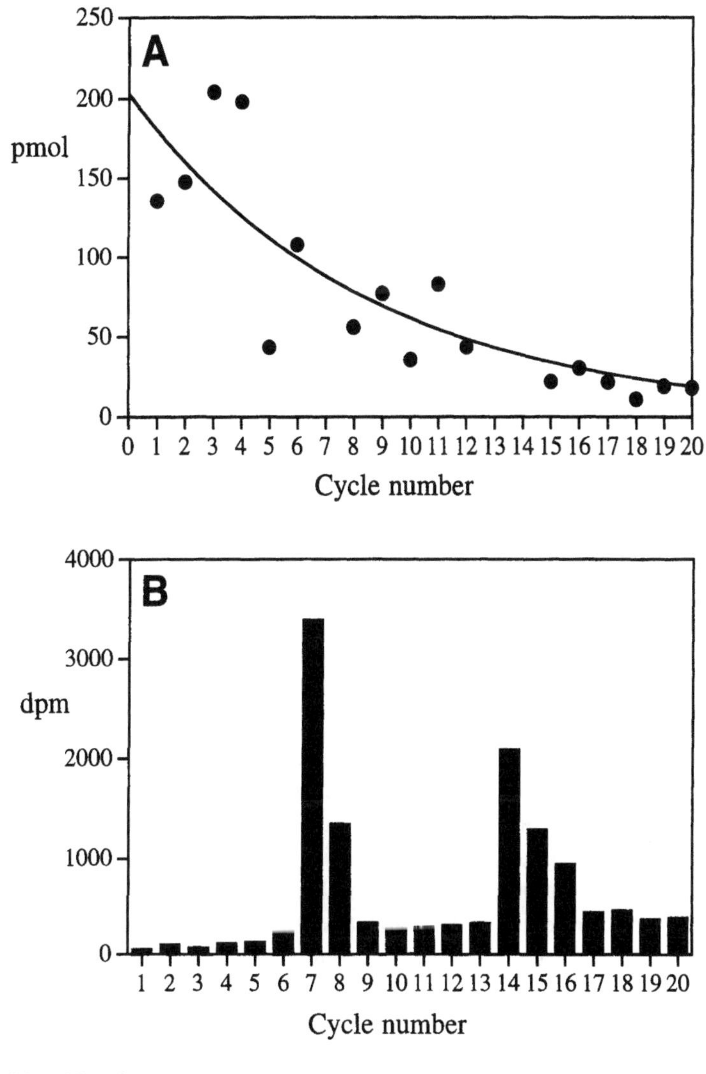

Fig. 5. Identification of radiolabeled lysine residues in the tryptic peptide D190-N220 from the reductively methylated pulse-labeled HsdS subunit from M.*Eco*R124I. **(A)** The pmole yield at each cycle of the Edman degradation sequencing reaction. Fitting of the data to **Eq. 5** allows the yield of dimethyl-lysine to be determined by interpolation. **(B)** A histogram showing the amount of radioactivity released at each cycle of the same sequencing reaction. The combination of radioactivity and picomole released at each cycle allows estimation of the specific activity of individual residues.

contents to a screw-cap microfuge tube and store at 4°C. If possible, use all of the reagent within 1–2 d of opening.

3. The efficiency of the reductive methylation of proteins is greatly reduced by the presence of amines in the solution because of competitive inhibition. Thus, com-

monly used buffers such as Tris–HCl and triethanolamine have to be avoided. HEPES and phosphate are good alternatives. A further problem is the potential for the formaldehyde–lysine adducts to undergo side reactions leading to unwanted protein crosslinking. To prevent this, the sodium cyanoborohydride should be added to the protein solution prior to addition of the formaldehyde.

4. The stoichiometry of the reductive methylation reaction dictates that two molecules of formaldehyde are required to complete the modification of a lysine residue to ε-*N,N*-dimethyl-lysine (*see* **Fig. 1**). Because of this, the effective specific activity of the [^{3}H] formaldehyde is half the value determined for the fully dimethylated peptide.
5. For dialysis of small volumes, (0.5–2 mL) Slide-A-Lyser cassettes (Pierce) are extremely useful. We find that a 1-mL sample volume is easy to inject and recover from the cassette without large losses of sample and without large dilution. If necessary, a smaller volume could be used.
6. The amount and the exact ratio of the reagents used for the reaction are somewhat empirical. The major concern is the prevention of side reactions resulting from reactive formaldehyde–lysine adducts. For a detailed account, *see* **ref. *6***. Briefly, the concentration of formaldehyde needs to be at an excess over the number of lysine residues to drive the reaction to completion, but not so high as to favor protein crosslinking. The other requirement is that the sodium cyanoborohydride be in excess over the formaldehyde to ensure efficient reduction of the Schiff bases.
7. To separate the DNA from protein, DEAE or Q ion-exchange columns are the method of choice. DNA oligonucleotides will bind very strongly to these matrices and the protein either can be recovered from the flowthrough or will elute earlier in a NaCl gradient. An alternative is to use heparin–Sepharose or, for basic proteins, a cation-exchange resin. If the chromatographic separation of the DNA from the protein is problematic, treat each sample with DNase I (FPLCpure, Pharmacia) before application to the column.
8. The N-terminus of the protein can also be reductively methylated. If the protein is relatively small or the total number of modified residues is low, then it is worthwhile to consider this when calculating the extent of modification.
9. If the time-course is extended to 5 h incubation, the reaction should be complete and the data will usually fit well to a single exponential process. Occasionally, this is not the case—for instance, if the protein contains several distinct populations of lysines with different kinetics. In this case, a more complex model will be needed to deconvolve the various classes of reacting species.
10. The pulse-labeling method involves treating the protein with a short pulse of labeled formaldehyde followed by a "cold" chase. This will identify all the surface lysines and provide information about their accessibility. If a large proportion of the total number of modifiable lysine residues are protected by DNA, then a pulse chase procedure of this kind will identify residues likely to be involved in DNA binding. However, if only a small proportion of lysines are protected by DNA, then an initial "cold" labeling should be performed on the DNA–protein

complex, followed by separation of the protein from the DNA before the pulse with labeled reagents is applied.

11. This is a compromise between getting enough label into the protein to allow the sites of modification to be easily determined and providing for differential labeling at individual sites so that information about the accessibility of each lysine can be obtained.
12. There may be some difficulty in redissolving the lyophilized protein in the aqueous buffers required for tryptic digestion (e.g., 20 m*M* Tris-HCl, pH 8.0). The formic acid strategy described works well in some cases but is not guaranteed. An alternative is to dissolve the sample in 100 m*M* Tris-HCl and 8 *M* urea pH 8.0 and then to add an equal volume of trypsin in 1 m*M* HCl such that the final urea concentration is 4 *M* and the trypsin concentration 1:10 (w/w).
13. The objective of the proteolytic digest is to produce peptides of an optimal length (5–30 amino acids) for quantitative analysis by automated Edman degradation. A tryptic digest of a reductively methylated protein will produce an arginine specific digest, as ε-*N,N*-dimethyl-lysine residues are not substrates for tryptic cleavage ***(11)***. Such a digest will yield some peptides that are suitable for N-terminal sequencing but will probably not cover the entire protein. In order to produce further peptides of suitable length, other proteases and chemical cleavage reagents should be investigated. The usefulness of these agents can vary substantially. In general, the best enzymatic alternatives are chymotrypsin and V8 protease. Cyanogen bromide is the best alternative for chemical cleavage.
14. The main criterion for selection of peaks is an apparent high specific activity, indicating that surface accessible lysines are present in peptides eluted within the peak. Additional information about the location of buried lysines within the protein can be gained by sequencing peptides which show very low or apparently no label incorporation. Although this could yield valuable data, one should be aware that some of these "cold" peaks are likely to be peptides derived from trypsin.
15. After an initial separation of a complex mixture of reductively methylated peptides, by C_3 reverse-phase chromatography, a further fractionation of peptides using either C_8 or C_{18} is highly recommended. Often a peak with an apparently high specific activity taken from an initial C_3 separation will resolve into multiple components on C_8 or C_{18}, only some of which are labeled. Avoid loading peptides eluted from a C_3 column at high acetonitrile concentrations (>50%) onto columns with longer alkyl-chain-bonded phases as the interaction between the sample and the bonded phase may be too strong for efficient recovery. For these larger, more hydrophobic peptides, it may be better to reapply to a C_3 column and then elute with a different gradient to the one used initially. Alternatively, redigest with a different enzyme and separate the products on a C_{18} column.
16. A simple and effective way to quantify the degree of incorporated label at individual lysine residues involves splitting the peptide sample during the automated Edman degradation sequencing reaction; most sequenators are equipped with this facility. After extraction of each 2-analino–5-thiazolinone-derivatized amino

acid, split the sample and divert 50% to a fraction collector. Convert the remainder to the phenylthiohydantoin derivative and identify the residue by on-line HPLC analysis in the usual way. Calculate the number of picomoles of the newly released residue by integration of the HPLC peak. Determine the amount of radioactivity associated with the residue by liquid scintillation counting of the material that was diverted to the fraction collector. This method may require modification, depending on the configuration of the available sequenator.

References

1. Lundblad, R. L. and Noyes, C. M. (1984) *Chemical Reagents for Protein Modification.* CRC, Boca Raton, FL.
2. Hunter, M. J. and Ludwig, M. L. (1962) *J. Am. Chem. Soc.* **84,** 3491–3497.
3. Means, G. E. and Feeney, R. E. (1968) Reductive alkylation of amino groups in proteins. *Biochemistry* **7,** 2192–2201.
4. Means, G. E. and Feeney, R. E. (1995) Reductive alkylation of proteins. *Anal. Biochem.* **224,** 1–16.
5. Jentoft, N. and Dearborn, D. G. (1983) Protein labelling by reductive alkylation. *Methods Enzymol.* **91,** 570–579.
6. Jentoft, N. and Dearborn, D. G. (1979) Labelling of proteins by reductive methylation using sodium cyanoborohydride. *J. Biol. Chem.* **254,** 4359–4365.
7. Zhang, M., Thulin, E., and Vogel, H. J. (1994) Reductive methylation and pKa determination of the lysine side chains in calbindin D9k. *J. Protein Chem.* **13,** 527–535.
8. Lambert, S. F. and Thomas, J. O. (1986) Lysine-containing DNA-binding regions on the surface of the histone octamer in the nucleosome core particle. *Eur. J. Biochem.* **160,** 191–201.
9. Thomas, J. O. and Wilson, C. M. (1986) Selective radiolabelling and identification of a strong nucleosome binding site on the globular domain of histone H5. *EMBO J.* **5,** 3531–3537.
10. Taylor, I. A., Webb, M., and Kneale, G. G. (1996) Surface labeling of the type I methyltransferase M.EcoR124I reveals lysine residues critical for DNA binding. *J. Mol. Biol.* **258,** 62–73.
11. Poncz, L. and Dearborn, D. G. (1983) The resistance to tryptic hydrolysis of peptide bonds adjacent to *N*-epsilon, *N*-dimethyllysyl residues. *J. Biol. Chem.* **258,** 1844–1850.

22

Limited Proteolysis of Protein–Nucleic Acid Complexes

Simon E. Plyte and G. Geoff Kneale

1. Introduction

Limited Proteolysis is a useful structural probe for investigating the globular nature of proteins by preferentially digesting the more accessible regions often found between domains. Generally, proteases require a small region of polypeptide chain possessing conformational flexibility for accommodation in the active site *(1)*. The regions of a protein possessing conformational flexibility are often found between tightly folded domains and are, therefore, preferential sites for proteolysis. In practice, limited proteolysis is achieved by dilution of the enzyme sufficiently so that it will only digest the most accessible regions leaving the domains intact. Digestion of protein–nucleic acid is often advantageous in that the DNA may provide steric protection of the DNA-binding domain not afforded by the free protein. The generation of domains by limited proteolysis relies directly on the tertiary structure of the protein under investigation and provides much firmer evidence for their existence than that provided by sequence homology.

An increasing number of nucleic-acid-binding proteins are known in which regions of their polypeptide chain are folded separately into compact globular domains, each possessing a distinctive function. For example, digestion of the A1 heterogeneous nuclear ribonucleoprotein (A1 hnRNP) with *Staphylococcus aureus* V8 protease produces two discrete domains, both capable of binding single-stranded nucleic acids *(2,3)*. Similarly, digestion of the *Pf1* gene 5 nucleoprotein complex results in the production of a 12-kDa domain that retains much of the single-stranded DNA-binding ability of the intact protein *(4)*. Further, using limited proteolysis, a cryptic DNA-binding domain was revealed in the COOH terminus of yeast TFIIIB70 and a core ssDNA-binding domain was

From: *Methods in Molecular Biology, vol. 148: DNA–Protein Interactions: Principles and Protocols, 2nd ed.*
Edited by: T. Moss © Humana Press Inc., Totowa, NJ

generated, possessing increased binding affinity, in human replication protein A *(5,6)*. In addition to its use for the analysis of domain structure, limited proteolytic fragments from *Escherichia coli* DNA gyrase B, for example, permitted the successful crystallization and structure determination of one of its domains *(7)*.

1.1. Strategy

The strategy adopted for the limited proteolysis of nucleoprotein complexes can be considered in four parts: optimization of the proteolysis, characterization of the proteolysed complex, purification of the DNA-binding domains, and sequence characterization of the fragment(s).

1.1.1. Proteolysis of Nucleoprotein Complex

The nucleoprotein complex should be digested with various proteases to establish which conditions are optimal for generating a protease-resistant domain. We routinely vary two parameters (enzyme/substrate ratio and time of digestion) when determining the best conditions for limited proteolysis. However, other parameters, such as temperature, ionic strength, and pH may also be varied. To determine the appropriate enzyme/substrate ratio for a particular protease, the nucleoprotein complex is digested at several enzyme/substrate ratios, removing samples at regular time intervals for sodium dodecyl sulfate–polyacrylamide gel electrophoresis (SDS-PAGE) analysis. The appearance of a discrete domain, resistant to further degradation (even if only transiently), is evidence for the existence of a domain, although not necessarily one that binds DNA. Choice of protease is often critical (*see* **Table 1**). Initially, it is best to try a relatively nonspecific enzyme (e.g., papain) because this decreases the likelihood of activity being dependent on primary sequence rather than tertiary structure.

1.1.2. Preliminary Characterization of DNA-Binding Properties of the Proteolyzed Nucleoprotein Complex

An initial indication of DNA binding can be found during the proteolysis experiment by removing two aliquots for gel analysis that can be run on polyacrylamide or agarose gels appropriate for the size of complex in the presence and absence of the denaturant SDS. A retardation in the mobility of the DNA (seen under ultraviolet [UV] light) in the absence of SDS implies that the fragment is still associated with DNA and constitutes a DNA-binding domain. However, this does not prove that the proteolyzed fragment is a discrete DNA-binding domain; it is possible that the nucleoprotein complex has only been "nicked" by the protease and maintains its native tertiary structure by noncovalent interactions. Therefore, it is necessary to purify the domain and fully characterize its DNA-binding properties.

Table 1
Useful Enzymes for Limited Proteolysis

Enzyme	Substrate specificity	Inhibitors
α-Chymotrypsin	Preferentially cuts C-terminally to aromatic amino acids	Aprotinin, PMSF, DFP, TPCK, cymostatin
Elastase	Cuts C-terminally to aliphatic noncharged amino acids (e.g., Ala, Val, Leu, Ile, Gly, Ser)	PMSF, DFP
Endoproteinase Arg-C	Cuts C-terminally to arginine residues	DFP, TLCK
Endoproteinase Lys-C	Cuts C-terminally to lysine residues	Aprotinin, DFP
Papain	Nonspecific protease but shows some preference for bonds involving Arg, Lys Gln, His, Gly, and Tyr	PMSF, TPCK, TLCK, leupeptin, heavy metal ions
Pepsin	Nonspecific protease	Pepstatin
Subtilisin	Nonspecific protease	DFP, PMSF
Trypsin	Cuts C-terminally to lysine and arginine residues	DFP, PMSF, TLCK
Endoproteinase Glu-C (V8 protease)	Cuts C-terminally to glutamic acid and or aspartic acid residues[a]	DFP

Abbreviations used: DFP, diisopropyl fluorophosphate (extremely toxic!); PMSF, phenylmethyl sufonyl fluoride; TPCK, *N*-tosyl-l-phenylalanine chloromethyl ketone; TLCK, Nα-*p*-tosyl-l-lysine chloromethly ketone.

[a]Will cut C-terminally to glutamic acid residues in ammonium bicarbonate, pH 8.0, or ammonium acetate, pH 4.0; will cut C-terminally to glutamic and aspartic acid residues in phosphate buffer, pH 7.8.

1.1.3. Purification of the DNA-Binding Domain

Purification of the fragment can make use of the fact that it will still be associated with DNA. Ultracentrifugation of the proteolyzed nucleoprotein complex (if large fragments of DNA are used) concentrates the domain and removes residual protease and small proteolytic fragments. The proteolyzed nucleoprotein complex can then be dissociated and the domain further purified if necessary. Alternatively, the DNA-binding fragment can be purified by affinity chromatography on DNA agarose. Several techniques are available to determine whether the purified domain binds DNA (*discussed in several chapters*) and include gel retardation assay, a variety of footprinting techniques, fluorescence spectroscopy, and circular dichroism.

1.1.4. Determination of the Amino Acid Sequence of the Domain

N-Terminal sequencing and amino acid analysis of the purified DNA-binding domain should be sufficient to establish the sequence of the domain, if the native amino acid sequence is known. Alternatively, N-terminal sequencing and mass spectroscopy should enable unambiguous identification of the domain. If certain proteases have been used to generate the domain (e.g., trypsin, α-chymotrypsin, endoproteinase Arg-C, and so forth), the C-terminal amino acid may also be known. If there are still ambiguities, carboxypeptidase digestion of the fragment can also be used to help identify the C-terminal residues, although this is not always reliable. If this still does not yield an unambiguous result, one must resort to amino acid sequencing of the entire fragment.

2. Materials

1. Spectra-Por dialysis membrane washed thoroughly in double-distilled water.
2. All proteases should be of the highest grade available and treated for contaminating protease activity, if necessary. A list of useful enzymes and their inhibitors is given in **Table 1**.
3. Buffers should be AnalR grade or higher and made up in double-distilled water.
4. SDS–polyacrylamide gel stock solutions:

   Solution A: 152 g acrylamide and 4 g *bis*-acrylamide. Make up to 500 mL.

   Solution B: 2 g SDS and 30 g Tris base, pH 8.8. Make up to 500 mL.

   Solution C: 2 g SDS, 30 g Tris base, pH 6.8. Make up to 500 mL.

   When making up these solutions, they should all be degassed and filtered using a Buchner filter funnel. They should be stored in lightproof bottles and will keep for many months.
6. 10% ammonium persulfate (APS): dissolve 0.1 mg in 1 mL of dH_2O.
7. 15% SDS–polyacrylamide gel: Mix together 8.0 mL of solution A, 4.0 mL of solution B, and 3.9 mL of dH_2O. Add 150 µL of 10% APS and 20 µL of *N,N,N',N'*-tetramethylethylene diamine (TEMED). Mix well and then pour between the plates. Immediately place a layer of dH_20 (or butanol) on top of the gel to create a smooth interface with the stacking gel. When the resolving gel has set, pour off the water and prepare the stacking gel. This is done by adding 750 µL of solution A and 1.25 mL of solution C to 3.0 mL of dH_2O. Finally, add 40 µL of APS and 10 µL of TEMED, pour on the stacking gel, and insert the comb. Remove the comb as soon as the gel has set to avoid the gel sticking to the comb.
8. 10X SDS running buffer: 10 g SDS, 33.4 g Tris base, and 144 g glycine made up to 1 L.
9. High-methanol protein stain: technical-grade methanol 500 mL, 100 mL glacial acetic acid and 0.3 g PAGE 83 stain (Coomassie blue), made up to 1 L.
10. Destain solution: 100 mL methanol and 100 mL glacial acetic acid, made up to 1 L.
11. 2X SDS-PAGE loading buffer: 4% (w/v) SDS, 60 m*M* Tris-HCl, pH 6.8, 20% glycerol, 0.04% (w/v) bromophenol blue, and 1% (v/v) β-mercaptoethanol.
12. 6X agarose gel loading buffer: 0.25% (w/v) bromophenol blue, 0.25% (w/v) xylene cyanol, and 30% glycerol.
13. 6X agarose gel loading buffer plus SDS: as in **item 12** plus 12% SDS (w/v).
14. TE buffer: 10 m*M* Tris-HCl, pH 7.5, and 1 m*M* EDTA.
15. 5 *M* NaCl or $MgCl_2$ (or other concentrated salt solutions for dissociation of the nucleoprotein complex [e.g., NaSCN]).

3. Methods

The method given here covers the first three objectives outlined in **Subheading 1.1.** Experimental details for the determination of the amino acid sequence of the fragment can be found in any standard text on protein chemistry. The following protocol was used for the generation of an 11-kDa DNA-binding domain from the *Pf1* gene 5 protein *(8)*. This protein binds cooperatively to ssDNA to produce a nucleoprotein complex of several million Daltons. Different nucleoprotein complexes will require different conditions of digestion and purification, but the basic principles remain the same.

3.1. Limited Proteolysis

1. Dialyze the nucleoprotein complex into the appropriate digestion buffer (*see* the manufacturer's recommendations for the buffer, temperature of reaction, and inhibitor). We routinely digest the nucleoprotein complex at approx 1 mg/mL, but the concentration is not too critical.
2. Prepare 40 tubes containing 5 mL of 2X SDS loading buffer plus 1 µL of the appropriate protease inhibitor. Leave on ice.
3. Pipet 55 µL of the nucleoprotein complex (55 mg) into each of four tubes labeled 1:100, 1:200, 1:500, 1:1000, respectively. Place on ice until needed.
4. Dissolve the protease in digestion buffer to a concentration that will give an enzyme/substrate ratio of 1:100 (w/w) when 1 µL of the protease is added to 50 µL of nucleoprotein complex (i.e., 0.5 mg/mL).
5. Prepare three dilutions of the protease. In this case, the protease is diluted 1:2, 1:5, and 1:10 with digestion buffer that will result in a final enzyme substrate ratio of 1:200, 1:500, and 1:1000 (w/w).
6. Remove 5 µL of the nucleoprotein complex from each of the four tubes and add to 1 of the 40 tubes containing 2X loading buffer (plus inhibitor) and place on ice. This is the time = 0 tube and should be labeled accordingly.
7. Add 1 µL of the protease to the appropriate nucleoprotein solution (e.g., protease diluted 1:5 to the nucleoprotein solution marked 1:500) and incubate at the specified temperature.
8. Remove 5 µL samples every 15 min and add to 2X loading buffer (in the appropriately marked tube) and then place on ice.
9. At the end of the experiment, boil the samples and run an SDS polyacrylamide gel. The presence of a degraded fragment(s), resistant to further proteolysis, is evidence for a discrete domain (*see* **Note 1**).
10. Adjustment of the enzyme/substrate ratios, time course, and choice of enzymes is often necessary. The optimum conditions must be found by trial and error.

3.2. Purification of the DNA-Binding Domain

1. Digest a large quantity (several milligrams) of the nucleoprotein complex under the optimized conditions determined in **Subheading 3.1.** to produce the DNA-binding domain (*see* **Note 2**). Add the appropriate inhibitor and run a sample on SDS-PAGE to check the digestion.

2. For very large nucleoprotein complexes, the proteolyzed complex can be purified away from the protease and small proteolytic fragments by ultracentrifugation. Spin the nucleoprotein complex at 229,000*g* (Beckman 70.1 Ti rotor) for 3 h at 4°C (*see* **Note 3**). Carefully wash the centrifuge tube with 4 mL of TE buffer, discard the washings, and resuspend the nucleoprotein complex in 2 mL of TE buffer on ice. Another ultracentrifugation step can be performed to remove all traces of the protease. For smaller nucleoprotein complexes, the DNA can be immobilized on a large resin (e.g., DNA cellulose) prior to interaction with the DNA-binding protein. Low-speed centrifugation can then be used to purify the DNA-associated domain. Sometimes, limited proteolysis can generate several fragments that bind DNA. These may arise from the same region of the protein; if so, this can be overcome by allowing the proteolysis to proceed further or by increasing the amount of protease.
3. Dissociate the proteolyzed nucleoprotein complex by the addition of salt to the appropriate concentration (*see* **Note 4**). The DNA can then be removed by ultracentrifugation (if sufficiently large) or nuclease digestion. If the DNA was originally bound on a solid support, then it can be removed by low-speed centrifugation (*see* **Note 5**).
4. Remove the high-salt buffer by desalting or dialysis. If the sample contains several different domains or a residual undigested protein, it will be necessary to purify the domains to homogeneity. Various chromatographic techniques are available to further purify the domains, including chromatofocusing, ion exchange, affinity, and gel filtration chromatography. These techniques permit recovery of the domain in a native state for further biochemical analysis. Alternatively, if the fragment is only to be used for sequence analysis, the mixture can be applied to a C_3 reverse-phase high-performance liquid chromatography column and separated in an acetonitrile gradient.
5. If the sequence of the native protein is known, then the sequence of the DNA-binding domain can be established by N-terminal sequencing and amino acid analysis. Additionally, the mass of the fragment (determined by mass spectroscopy) should help locate the sequence of the DNA-binding domain.

4. Notes

1. Often during the experiment, a protease-resistant fragment is only transiently formed during complete digestion of the protein. If this occurs, vary some of the parameters (enzyme dilution, temperature, etc.) to try and prolong the lifetime of the fragment.
2. Scaling up of the digestion is not generally a problem and we routinely digest several milligrams (>10) of nucleoprotein complex if necessary.
3. The speed and duration of centrifugation will vary depending on the size of the nucleoprotein complex. For smaller complexes, ultracentrifugation may not be appropriate.
4. In many cases, a NaCl concentration between 1 *M* and 2 *M* is sufficient to dissociate the nucleoprotein complex. However, some nucleoprotein complexes

remain associated above 2 *M* NaCl and require 1 *M* $MgCl_2$ or 1 *M* NaSCN for dissociation *(8)*. The appropriate salt concentration can be determined by SDS-PAGE analysis of the pellet and supernatant after ultracentrifugation at different ionic strengths.

5. The DNA can also be removed by DNase digestion followed by gel fitration (i.e., desalting column) or by extensive dialysis against TE buffer.

References

1. Vita, C., Dalzoppo, D., and Fontana, A (1987) Limited proteolysis of globular proteins: molecular aspects deduced from studies on thermolysin, in *Macromolecular Biorecognition* (Chaiken, I., Chaiancone, E., Fontana, A., and Veri, P., eds.), Humana Press, Clifton, NJ.
2. Merril, B., Stone, K., Cobianchi, F., Wilson, S., and Williams, K. (1988) Phenylalanines that are conserved among several RNA-binding proteins form part of a nucleic acid binding pocket in the heterogeneous nuclear ribonucleoprotein. *J. Biol. Chem.* **263,** 3307–3313.
3. Bandziulis, R., Swanson, M., and Dreyfuss, G. (1989) RNA binding proteins as developmental regulators. *Genes Dev.* **4,** 431–437.
4. Plyte, S.E. and Kneale, G.G. (1993) Characterization of the DNA-binding domain of the Pf1 gene 5 protein. *Biochemistry* **32,** 3623–3628
5. Huet, J., Conesa, C., Carles, C., and Sentenac, A. (1997) A cryptic DNA-binding domain at th COOH terminus of TFIIIB70 affects formation, stability and function of perinitiation complexes. *J. Biol. Chem.* **272,** 18,341–18,349.
6. Bochareva, E., Frappier, L., Edwards, A., and Bochareva, A. (1998) The RPA32 subunit of human replication protein A contains a single stranded DNA-binding domain. *J. Biol. Chem.*, **273,** 3932–3936.
7. Wigley, D., Davies, G., Dodson, E., Maxwell, A., and Dodson, G. (1991) Crystal structure of an N-terminal fragment of the DNA gyrase B protein. *Nature* **351,** 624–629.
8. Kneale, G.G. (1983) Dissociation of the Pf1 nucleoprotein assembly complex and characterization of the DNA-binding protein. *Biochem. Biophys. Acta* **739,** 216–224.

23

Ultraviolet Crosslinking of DNA–Protein Complexes via 8-Azidoadenine

Rainer Meffert, Klaus Dose, Gabriele Rathgeber, and Hans-Jochen Schäfer

1. Introduction

In biological systems, photoreactive derivatives have been widely applied to study specific interactions of receptor molecules with their ligands by photoaffinity labeling *(1–3)*. While the receptors are generally proteins (e.g., enzymes, immunoglobulins, or hormone receptors), the ligands differ widely in their molecular structure (e.g., sugars, amino acids, nucleotides, or oligomers of these compounds).

The advantage of photoaffinity labeling compared with affinity labeling, or chemical modification with group-specific reagents is that photoactivatable nonreactive precursors can be activated at will by irradiation (**Fig. 1**). These reagents do not bind covalently to the protein unless activated. On irradiation of the precursors, highly reactive intermediates are formed that react indiscriminately with all surrounding groups. Therefore, after activation, a photoaffinity label, interacting at the specific binding site, can label all the different amino acid residues of the binding area. Today, aromatic azido compounds are mostly used as photoactivatable ligand analogs. They form highly reactive nitrenes upon irradiation because of the electron sextet in the outer electron shell of these intermediates (**Fig. 2**).

In addition to the azido derivatives, photoreactive precursors forming radicals or carbenes on irradiation can be used as photoaffinity labels. All of these intermediates (nitrenes, e.g.) vigorously try to complete an electron octet (**Fig. 3**).

To produce covalent crosslinks between proteins and DNA, various methods have been applied *(4–11)*: ultraviolet (UV) irradiation, γ-irradiation, chemi-

From: *Methods in Molecular Biology, vol. 148: DNA–Protein Interactions: Principles and Protocols, 2nd ed.*
Edited by: T. Moss © Humana Press Inc., Totowa, NJ

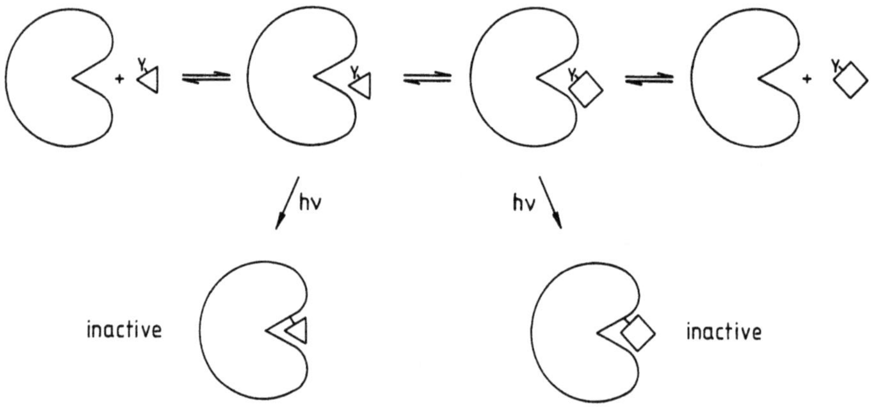

Fig. 1. Photoaffinity labeling of receptor proteins (e.g., enzymes) by photoactivatable ligand analogs (e.g., substrate analog/product analog). In the dark (upper line), the biological interactions of the protein with the ligand analog can be studied. On irradiation (lower line), the protein (enzyme) is labeled and inactivated by the substrate analog/product analog.

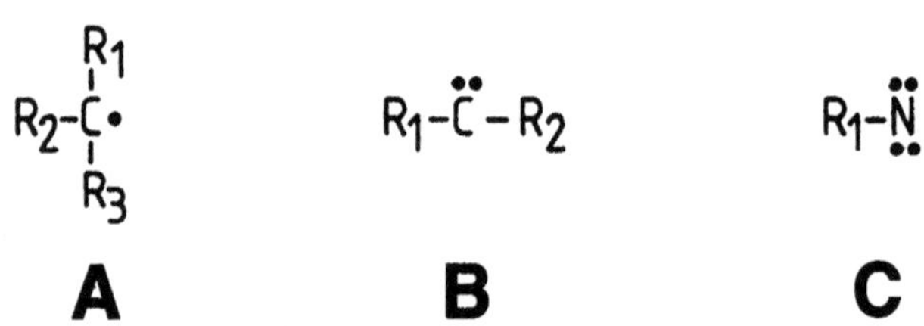

Fig. 2. Highly reactive photogenerated intermediates: radical (**A**), carbene (**B**), and nitrene (**C**).

cal methods, and even vacuum or extreme dryness. Besides these methods, photoaffinity labeling and photoaffinity crosslinking are helpful tools for the study of specific interactions between proteins and deoxyribonucleic acids. To date, many successful attempts have been made to photocrosslink proteins to nucleic acids using different photoactivatable deoxynucleotides. 5-bromo-, 5-iodo-, 5-azido-, and 5-[*N*-(*p*-azidobenzoyl)-3-aminoallyl]-2'-deoxyuridine-5'-monophosphate ***(12–18)***, 4-thio-2'-deoxythymidine-5'-monophosphate ***(19)***, and 8-azido-2'-deoxyadenosine-5'-monophosphate ***(20,21)*** have been incorporated into deoxyribonucleic acids to bind DNA covalently to adjacent proteins (for a review *see* **ref.** ***22***).

Here, we describe the synthesis of 8-azido-dATP (8-N_3dATP), its incorporation into DNA by nick translation, and the procedure to photocrosslink azido-modified DNA to proteins ***(20,21)***.

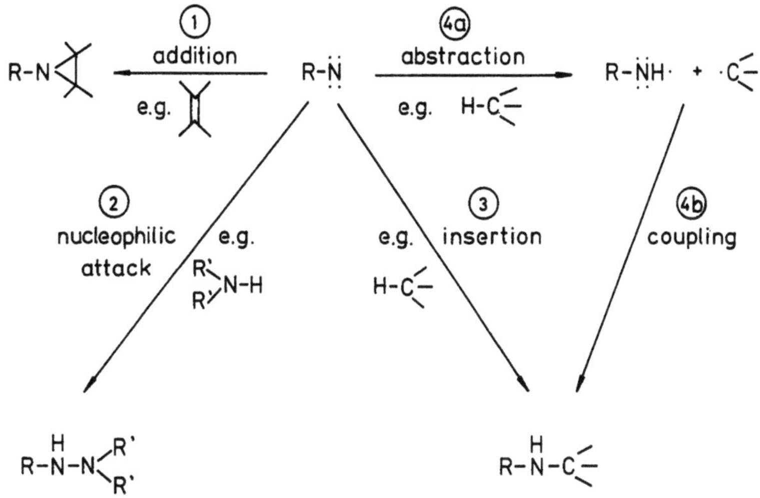

Fig. 3. Reactions of nitrenes. Cycloaddition to multiple bonds forming three-membered cyclic imines (1), addition to nucleophiles (2), direct insertion into C-H bonds yielding secondary amines (3), and hydrogen atom abstraction followed by coupling of the formed radicals to a secondary amine (4a, 4b).

2. Materials

2.1. Synthesis of 8-N_3dATP

1. dATP (disodium salt, Boehringer Mannheim, Mannheim, Germany).
2. Potassium acetate buffer: 1 *M*, pH 3.9.
3. Bromine.
4. Sodium disulfite ($Na_2S_2O_5$).
5. Ethanol.
6. DEAE–Sephadex A-25.
7. Triethylammonium bicarbonate buffer: 0.7 *M*, pH 7.3.
8. Dimethylformamide.
9. Hydrazoic acid: 1 *M* in benzene.
10. Triethylamine.

2.2. Characterization of 8-N_3dATP

1. Silica gel plates F_{254} (Merck, Darmstadt, Germany).
2. Cellulose plates F (Merck).
3. Isobutyric acid/water/ammonia (66:33:1 v/v).
4. *n*-Butanol/water/acetic acid (5:3:2 v/v).

2.3. Preparation of Azido-Modified DNA

1. DNA (e.g., pBR 322 or pWH 106).
2. Deoxyribonucleotides (dATP, dGTP, dCTP, dTTP, [α-^{32}P]-dCTP).

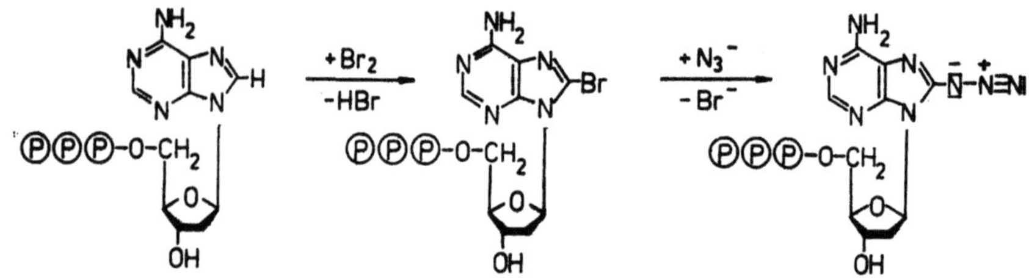

Fig. 4. Synthesis of 8-N_3dATP.

3. DNase I (*Escherichia coli,* 2000 U/mg, Boehringer Mannheim) in 0.15 *M* NaCl and 50% glycerol.
4. 50 m*M* Tris-HCl, pH 7.2.
5. Magnesium sulfate ($MgSO_4$).
6. Bovine serum albumin.
7. DNA polymerase I (*E. coli,* Boehringer Mannheim, No. 104493, purchased containing definite amounts of DNase I).
8. Ethylenediaminetetraacetic acid disodium salt (EDTA).
9. Sephadex A-25.

2.4. Photocrosslinking

An ultraviolet lamp (e.g., Mineralight handlamp UVSL 25 at position "long wave") emitting UV light at wavelengths of 300 nm and longer.

3. Methods

3.1. Synthesis of 8-N_3dATP

The synthesis of 8-N_3dATP (**Fig. 4**) is performed principally by analogy to the synthesis of 8-N_3ATP *(23)* (*see* **Note 1**). In the first step, bromine exchanges the hydrogen at position 8 of the adenine ring. Then, the bromine is substituted by the azido group.

1. Dissolve 0.2 mmol (117.8 mg) of dATP in 1.6 mL of potassium acetate buffer (1 *M*, pH 3.9) and add 0.29 mmol (15 µL) of bromine. Keep the reaction mixture in the dark at room temperature for 6 h (the absorption maximum shifts from 256 nm to 262 nm; *see* **Note 2**).
2. Reduce excessive bromine by addition of traces of (approx 5 mg) $Na_2S_2O_5$ until the reaction mixture looks colorless or pale yellow. Pour the reaction mixture into 20 mL of cold ethanol (–20°C) and allow to stand for at least 30 min at –20°C in the dark.
3. Collect the precipitated deoxynucleotide by centrifugation and redissolve the residue in 0.5 mL of double-distilled water. Further purification is achieved by ion-exchange chromatography over DEAE–Sephadex A-25 column (50 × 2 cm) with a linear gradient of 1000 mL each of water and triethylammonium bicarbonate (0.7 *M*, pH 7.3).

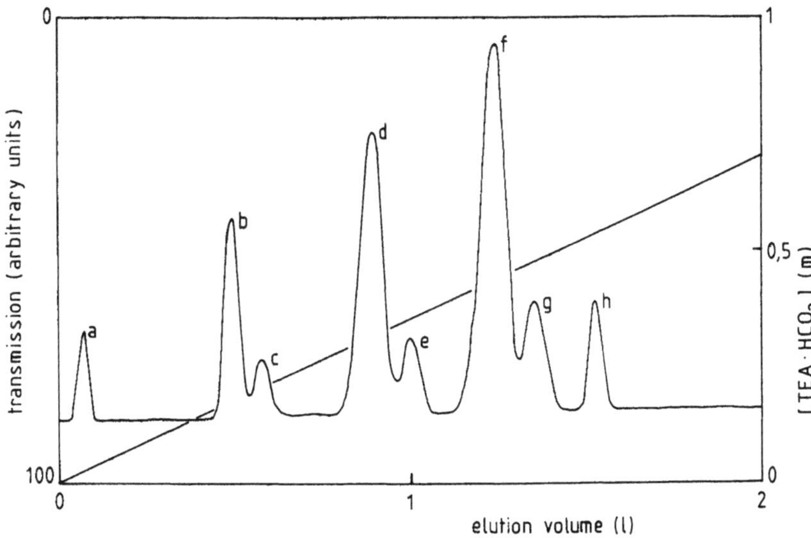

Fig. 5. Elution profile (anion-exchange chromatography on DEAE–Sephadex A 25; elution buffer: linear gradient of 1000 mL each of water and 0.7 *M* triethylammonium bicarbonate (pH 7.3) of the reaction products of 8-N_3dATP synthesis: front **(a)**, 8-N_3dAMP **(b)**, 8-BrdAMP **(c)**, 8-N_3dADP **(d)**, 8-BrdADP **(e)**, 8-N_3dATP **(f)**, 8-BrdATP **(g)**, and probably a higher phosphorylated 8-azidoadenosine derivative **(h)**.

4. Combine the fractions containing 8-bromo-dATP (8-BrdATP) (main peak of the elution profile) and dry by lyophilization. 8-BrdATP is obtained as the triethylammonium salt. The expected yield should be 65% (spectroscopically).
5. Dissolve 0.1 mmol (87.3 mg) of dried 8-BrdATP (triethylammonium salt) in 3 mL of freshly distilled dimethylformamide (*see* **Notes 3** and **4**). Add a dried solution of 0.8 mmol (34.4 mg) of hydrazoic acid (HN_3) in 800 µL of benzene and 0.8 mmol (111.3 µL) of freshly distilled triethylamine. Keep the reaction mixture in the dark at 75°C for 7 h (the absorption maximum shifts from 262 nm to 280 nm).
6. Evaporate the solvents under vacuum and redissolve the residue in 1 mL of water. Further purification is achieved by ion-exchange chromatography over DEAE–Sephadex A-25 as described in step 3. **Figure 5** shows the elution profile of the chromatography (*see* **Notes 4** and **5**).
7. Combine the fractions containing 8-N_3dATP and dry the solution by lyophilization. 8-N_3dATP is obtained as the triethylammonium salt. Yield: 30% (spectroscopically). 8-N_3dATP can be stored at –20°C in the dark freeze-dried (*see* **Notes 6–8**) or frozen in aqueous solution, pH 7.0.

3.2. Characterization of 8-N_3dATP

1. Thin-layer chromatography (TLC). TLC is carried out on silica gel plates F_{254} or cellulose plates F. The development is performed in either isobutyric acid/water/ammonia (66:33:1 v/v) or *n*-butanol/water/acetic acid (5:3:2 v/v).

2. Ultraviolet absorbance. Record the UV absorbance spectrum of 8-N_3dATP. It shows a maximum at 280 nm. The UV absorbance of 8-N_3dATP is pH dependent (*see* **Note 9**).
3. Photoreactivity. The photoreactivity of 8-N_3dATP is tested by two different methods. It can either be demonstrated by the spectroscopic observation of the photolysis (**Fig. 6**; *see* **Note 10**) or by the ability of the photolabel to bind irreversibly to cellulose on thin-layer plates on UV irradiation (Mineralight handlamp UVSL 25) prior to the development of the chromatogram. After development, most of the irradiated label is detected at the origin of the chromatogram in contrast to the nonirradiated control, which has completely migrated.

3.3. Preparation of Azido-Modified DNA

Azido-modified and [^{32}P]-labeled DNA are prepared by nick translation. For this purpose, the detailed and exact composition of the reaction medium depends strongly on the size as well as on the amount of the DNA to be modified. The optimal ratio of DNA, DNase I, and DNA polymerase I (Kornberg enzyme) should be tested in preliminary experiments (*see* **Notes 11** and **12**).

Here, we describe the well-tested reaction conditions for the modification of plasmid pBR322 (4363 bp). The preparation of azido-modified pWH106 (4970 bp) can be performed analogously.

1. Add 17.3 pmol of pBR322 to a mixture of 50 nmol of dGTP, 50 nmol of dTTP, 50 nmol of 8-N_3dATP, and 500 pmol of dCTP; prepare on ice.
2. Add 370 kBq of [α-^{32}P]-dCTP (110 TBq/mmol) and 20 pg of DNase I (freshly prepared out of a stock solution of 1 mg of DNase I in 1 mL of 0.15 *M* NaCl and 50% glycerol).
3. Adjust the reaction medium to an end concentration of 50 *mM* Tris-HCl, pH 7.2, 10 *mM* $MgSO_4$, and 50 mg/mL of bovine serum albumin (standard reaction volume: 100 µL).
4. Start the nick translation reaction by adding 100 U of DNA polymerase I from *E. coli*.
5. Incubate for 1 h at 15°C in the dark.
6. Stop the reaction by adding EDTA (final concentration: 20 m*M*).
7. Separate the unincorporated deoxyribonucleotides from photoreactive [^{32}P]-labeled pBR322 by gel filtration over Sephadex A-25 column using a 1-mL syringe.
8. Precipitate photoreactive pBR322 by adding two volumes of cold ethanol and redissolve the precipitate in double-distilled water. Store the aqueous solution at –20°C in the dark.

Control nonphotoreactive DNA can be prepared analogously replacing the 8-N_3dATP by 50 nmol of dATP.

3.4. Photocrosslinking

1. Prepare 20–30 µL aqueous solutions containing the photoreactive DNA (0.5 pmol) and the protein (1–25 pmol) to be cross-linked (*see* **Notes 13** and **14**).

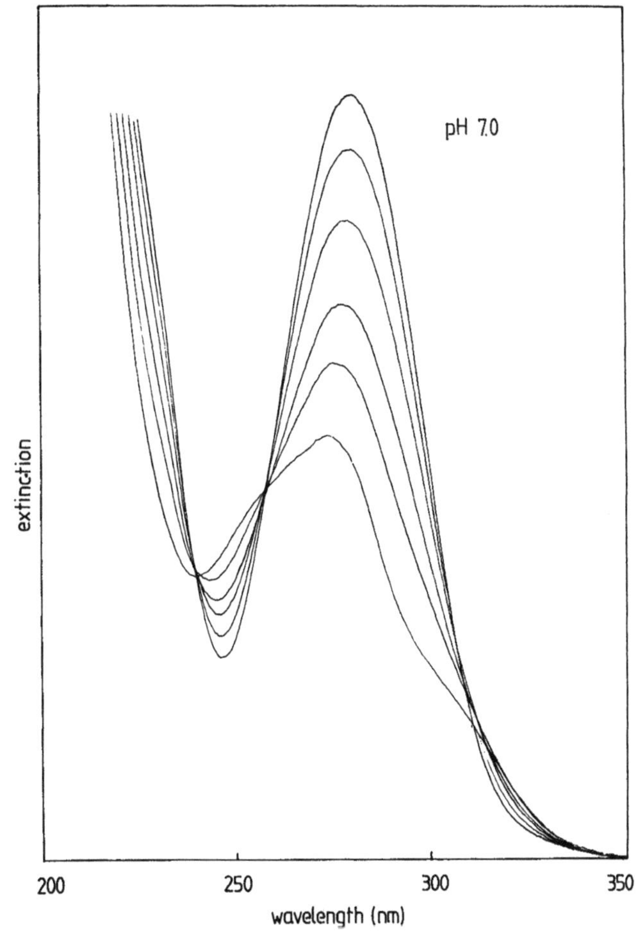

Fig. 6. Change of the optical absorption spectrum of 8-N_3dATP on UV irradiation in Tris-HCl buffer (0.01 *M*, pH 7.0, 20°C). The irradiation time between two subsequent absorption spectra was 2 min initially. It was increased up to 10 min toward the end of photolysis. The final spectrum was taken after 30 min of irradiation. During the photolysis, the absorbance at 280 nm decreased; two new absorbance maxima at 248 and 305 nm are formed.

2. Incubate the reaction mixture for 10 min at 37°C in the dark.
3. Expose the sample to UV irradiation (*see* **Notes 15** and **16**). The irradiation times can be chosen in a range from 1 s to 60 min (*see* **Note 17**).
4. Keep the solutions in the dark before and after photolysis (*see* **Note 6**).

3.5. Analysis of DNA-Protein Adducts

Analysis of DNA–protein adducts can be made, for example, by polyacrylamide gel electrophoresis of the irradiated samples followed by autoradiography. SDS–polyacrylamide gel electrophoresis should be performed immediately after photocrosslinking according to Laemmli ***(24)*** with some variations. After the addition of 20 mg/mL of bromophenol blue, the samples are loaded onto a SDS–polyacrylamide gel of 5% polyacrylamide (separating gel) with an overlay of 3.5% polyacrylamide (stacking gel) containing 1% SDS. After the electrophoretic separation, the gels are silver-stained according to Adams and Sammons ***(25)***, dried, and exposed to X-ray film at –70°C. A quantitative determination of the DNA–protein adducts is possible by densitometric measurement of the autoradiogram ***(26)***. **Figure 7** shows a typical result on photocrosslinking of *Eco*RI-digested plasmid pWH106 with a specific interacting protein (Tet repressor).

Another possibility to detect the DNA–protein adducts is the application of the nitrocellulose filter binding assay according to Braun and Merrick ***(27)***, *see also* Chapter 1.

4. Notes

1. Experiments to synthesize [α-^{32}P] or [U-^{14}C]-labeled 8-N_3dATP by starting the synthesis with [α-^{32}P] or [U-^{14}C]dATP, respectively, failed. This is most probably because of the formation of bromine radicals induced by radiation. These radicals could react unspecifically with the deoxyribonucleotide, suppressing the very specific electrophilic substitution of the hydrogen in position 8 of the adenine ring by the bromine ion.
2. Do not stop the reaction of dATP with bromine before 6 h even if the absorption maximum is near 262 nm after 1 or 2 h, otherwise a significant reduction of the yield of 8-BrdATP may occur.
3. 8-BrdATP obtained as triethylammonium salt is soluble in dimethylformamide in contrast to the alkali salts of this nucleotide. This is advantageous for the following substitution of bromine by the azido group yielding 8-N_3dATP.
4. The exchange reaction of bromine by the azido group requires absolute dryness. However, the formation of 8-N_3dAMP and 8-N_3dADP is usually observed, resulting from a limited hydrolytic cleavage of 8-N_3dATP (*see* **Fig. 5**).
5. Besides the three azidoadenine deoxyribonucleotides, minor amounts of unreacted 8-bromoadenine deoxyribonucleotides are eluted as well (*see* **Fig. 5**).
6. Because of the photoreactivity of azido compounds, samples containing 8-N_3dATP should always be kept in the dark if possible. However, short exposure of azido compounds to normal daylight in our laboratory never falsified the results obtained.
7. 8-N_3dATP can be stored frozen at –20°C in aqueous solution, pH 7.0, in the dark for at least 2 yr without significant loss of photoreactivity, as demonstrated by subsequent photocrosslinking experiments.
8. Exclude dithiothreitol from any buffers or other solutions that contain 8-N_3dATP. It is well-known that dithiothreitol reduces azido groups to the corresponding amines ***(28)***. In addition, the UV absorbance of dithiothreitol resembles that

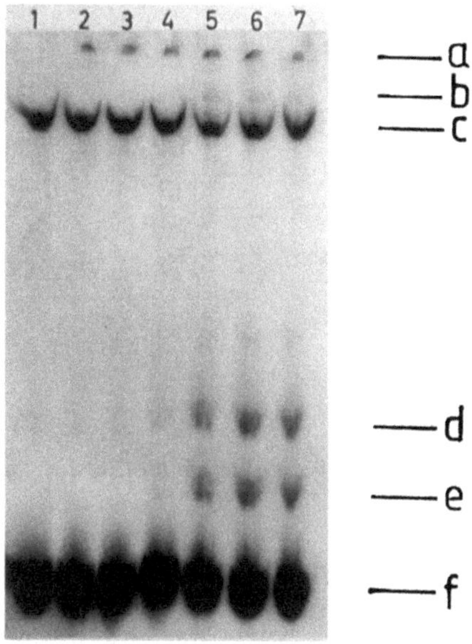

Fig. 7. Photocrosslinking of proteins to DNA (pWH106). Autoradiogram of a denaturing 5% SDS-polyacrylamide gel electrophoresis showing photocrosslinking of Tet repressor to azido-activated 187-bp and 3848-bp fragments of pWH106 (radioactive labeled by ^{32}P). Each 187-bp fragment contains two *tet* operator sequences, the 3848-bp fragment none. UV irradiation of azidomodified 187-bp fragment in the presence of Tet repressor results in a reduced migration of the 187-bp fragment because of covalent crosslinking of the DNA to one or two Tet repressor dimers. In each of lanes 1–7, 0.06 pmol pWH106 (cleaved by *Eco*RI) and 20 pmol Tet repressor were applied. Lane 1: photoactive fragments of pWH 106 without protein (30' UV); lane 2: nonphotoactive fragments of pWH106 with Tet repressor (30' UV); lanes 3–7: photoactive fragments of pWH106 with Tet repressor (0', 1', 4', 10', 30' UV). Fractions: Origin of sample loading (**a**), traces of 3848-bp fragment covalently crosslinked (unspecifically) to Tet repressor (**b**); 3848-bp fragment (no Tet repressor bound) (**c**); 187-bp fragment covalently crosslinked to two Tet repressor dimers (**d**); 187-bp fragment covalently crosslinked to one Tet repressor dimer (**e**); 187-bp fragment (no Tet repressor bound) (**f**).

of 8-N_3dATP because of the formation of disulfide bonds by oxidation on storage in aqueous solution. This results in a reduced rate of photocrosslinking by the UV irradiation.

9. The UV absorption of 8-N_3dATP shows a maximum at 280 nm. The absorbance at 280 nm increases with decreasing pH value (*see* **step 2** of **Subheading 3.2.**). A second absorption maximum at 219 nm shifts to 204 nm in acidic solution. Both effects are a result of the protonation at N^1 of the purine ring *(29)*. The UV absorption spectrum of 8-N_3dATP resembles that of 8-N_3ATP *(23)*.

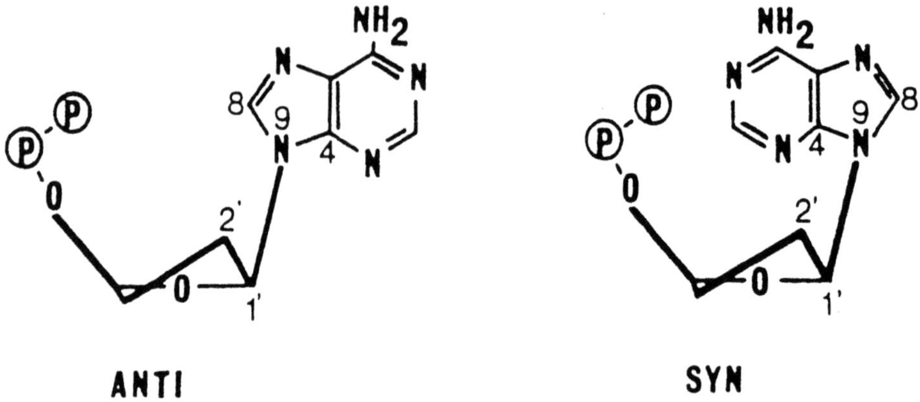

Fig. 8. Conformation of adenine nucleotides. (From **ref.** *23*.)

10. When testing its photoreactivity, take into account that the photolysis of 8-N_3dATP is pH dependent. Exhaustive irradiation in neutral solution yields new absorption maxima at 248 and 305 nm, whereas in acidic or basic solution, the destruction of the purine ring is observed, as indicated by the disappearance of the absorbance between 240 and 310 nm (data not shown).
11. By analogy with 8-azidoadenine nucleotides, 8-N_3dATP should prefer the *syn* conformation (**Fig. 8**) as a result of the bulky substituent in position 8 of the purine ring *(30)*. This, however, seems to be contradicted by our results indicating that DNA polymerase I *(E. coli)* accepts 8-N_3dATP in the nick translation reaction (it has been suggested that this enzyme only interacts with 2-deoxynucleoside triphosphates in the *anti* conformation *[31]*). The discrepancy may be explained in two ways: First, the steric requirements for the binding of 8-N_3dATP by DNA polymerase I are less restrictive than assumed *(32)* or second, 8-N_3dATP interacts in the *anti* conformation with the binding site of the enzyme. This could be demonstrated for the interaction of 8-N_3ATP with the F_1ATPase from mitochondria *(33)*.
12. The preparation of azido-modified and [^{32}P]-labeled DNA by nick translation is critically dependent on the ratio of DNA, DNase I, and DNA polymerase I in the reaction medium. High concentrations of DNase I, on the one hand, result in a very efficient incorporation rate of the azido-modified and radioactive labeled deoxynucleotides, but on the other hand, the degradation of the DNA probes by DNase I has to be evaluated. Application of too small amounts of DNase I results in inefficient incorporation of the photoactivatable deoxynucleotides and in an insufficient photocrosslinking to the interacting proteins during subsequent irradiation of the azido-modified DNA.
13. Tris-HCl buffer, 50 *mM*, pH 7.2, can be used instead of double-distilled water without any significant effect on the photocrosslinking efficiency.
14. The amount of protein planned to be photocrosslinked to photoreactive DNA can be varied over a wide range. Too high an excess of proteins, however, should be

avoided because the absorbance maximum of proteins at 280 nm will lead to inefficient crosslinking rates.

15. One way to expose the samples to UV light is to deposit the probes (typically 30–50 μL) in plastic wells (normally used for radioimmunoassay or enzyme-linked immunosorbent assay tests). The UV lamp is positioned directly above the samples; thus more than one probe can be irradiated simultaneously.
16. The emitted light of the UV lamp used for photocrosslinking should not contain light of shorter wavelengths than 300 nm because of the possibility of photodamaging DNA or protein. For example, the Mineralight handlamp UVSL 25 (long wave) emits UV light of mainly 366 nm. The small portion of UV light of wavelengths between 300 and 320 nm emitted allows the photoactivation of the azido group without any significant photodamage of DNA or protein.
17. Ultraviolet irradiation times for photocrosslinking can be chosen over a wide range (*see* **step 3** of **Subheading 3.4.**). Optimal UV irradiation rates (flux per unit area) must be tested. In our experiments (using the Mineralight handlamp UVSL 25 fixed in a position resulting in a fluence rate of 4 $J/m^2/s$ at the position of the sample) DNA–protein adducts are first detectable after irradiation times of 10–30 s; irradiation periods longer than 15–20 min do not improve the yield of photocrosslink products.

Acknowledgments

The authors thank Dr. Marianne Schüz (Wiesbaden) for editing the manuscript. This work was supported by the Bundesministerium für Forschung und Technologie (07QV8942) and by the Deutsche Forschungsgemeinschaft (Scha 344/1-3).

References

1. Bayley, H. and Knowles, J. R. (1977) Photoaffinity labeling. *Methods Enzymol.* **46,** 69–114.
2. Bayley, H. (1983) Photogenerated reagents in biochemistry and molecular biology, in *Laboratory Techniques in Biochemistry*, vol. 12 (Work, T. S. and Burdon, R. H., eds.), Elsevier, Amsterdam.
3. Schäfer, H.-J. (1987) Photoaffinity labeling and photoaffinity crosslinking of enzymes, in *Chemical Modifications of Enzymes, Active Site Studies* (Eyzaguirre, J., ed.), Ellis Horwood, Chichester, pp. 45–62.
4. Smith, K. C. (1962) Dose dependent decrease in extractability of DNA from bacteria (by UV-light). *Biochem. Biophys. Res. Commun.* **8,** 157–163.
5. Shetlar, M. D. (1980) Crosslinking of proteins to nucleic acids by UV-light. *Photochem. Photobiol. Rev.* **5,** 105–197.
6. Welsh, J. and Cantor, C. R. (1984) Protein–DNA crosslinking. *Trends Biochem. Sci.* **9,** 505–508.
7. Ekert, B., Giocanti, N., and Sabattier, R. (1986) Study of several factors in RNA–protein crosslink formation induced by ionizing radiations within 70S ribosomes of *E. coli* MRE 600. *Int. J. Radiat. Biol.* **50,** 507–525.

8. Lesko, S. A., Drocourt, J. L., and Yang, S. U. (1982) DNA–protein and DNA interstrand crosslinks induced in isolated chromatin by H_2O_2. and Fe-EDTA-chelates. *Biochemistry* **21,** 5010–5015.
9. Wedrychowsky, A., Ward, W. S., Schmidt, W. N., and Hnilica, L. S. (1985) Chromium-induced crosslinking of nuclear proteins and DNA. *J. Biol. Chem.* **260,** 7150–7155.
10. Summerfield, F. W. and Tappel, A. L. (1984) Crosslinking of DNA in liver and testis of rats fed 1,3-propanediol. *Chem. Biol. Interact.* **50,** 87–96.
11. Dose, K., Bieger-Dose, A., Martens, K.-D., Meffert, R., Nawroth, T., Risi, S., et al. (1987) Survival under space vacuum—biochemical aspects. *Proc. 3rd Eur. Symp. Life Sci. Res. in Space (ESA SP–271)*, pp. 193–195.
12. Lin, S. Y. and Riggs, A. D. (1974) Photochemical attachment of lac repressor to bromodeoxyuridine-substituted lac operator by UV radiation. *Proc. Natl. Acad. Sci. USA* **71,** 947–951.
13. Evans, R. K., Johnson, J. D., and Haley, B. E. (1986) 5-Azido-2'-deoxyuridine-5'-triphosphate: a photoaffinity labeling reagent and tool for the enzymatic synthesis of photoactive DNA. *Proc. Natl. Acad. Sci. USA* **83,** 5382–5386.
14. Bartholomew, B., Kassavetis, G. A., Braun, B. R., and Geiduschek, E. P. (1990) The subunit structure of *Saccharomyces cerevisiae* transcription factor IIIC probed with a novel photocrosslinking reagent. *EMBO J.* **9,** 2197–2205.
15. Lee, D. K., Evans, R. K., Blanco, J., Gottesfeld, J., and Johnson, J. D. (1991) Contacts between 5 S DNA and *Xenopus* TFIIIA identified using 5-azido-2'-deoxyuridine-substituted DNA. *J. Biol. Chem.* **266,** 16,478–16,484.
16. Blatter, E. E., Ebright, Y. W., and Ebright, R. H. (1992) Identification of an amino acid-base contact in the GCN4-DNA complex by bromouracil-mediated photocrosslinking. *Nature* **83,** 650–652.
17. Willis, M. C., Hicke, B. J., Uhlenbeck, O. C., Cech, T. R., and Koch, T. H. (1993) Photocrosslinking of 5-iodouracil-substituted RNA and DNA to proteins. *Science* **262,** 1255–1257.
18. Hicke, B. J., Willis, M. C., Koch, T. H., and Cech, T. R. (1994) Telomeric protein–DNA point contacts identified by photo-cross-linking using 5-bromodeoxyuridine. *Biochemistry* **33,** 3364–3373.
19. Bartholomew, B., Braun, B. R., Kassavetis, G. A., and Geiduschek, E. P. (1994) Probing close DNA contacts of RNA–polymerase III transcription complexes with the photoactive nucleoside 4-thiodeoxythymidine. *J. Biol. Chem.* **269,** 18,090–18,095.
20. Meffert, R. and Dose, K. (1988) UV-induced crosslinking of proteins to plasmid pBR322 containing 8-azidoadenine 2'-deoxyribonucleotides. *FEBS Lett.* **239,** 190–194.
21. Meffert, R., Rathgeber, G., Schäfer, H.-J., and Dose, K. (1990) UV-induced crosslinking of Tet repressor to DNA containing tet operator sequences and 8-azidoadenines. *Nucleic Acids Res.* **18,** 6633–6636.
22. Bartholomew, B., Tinker, R. L., Kassavetis, G. A., and Geiduschek, E. P. (1995) Photochemical crosslinking assay for DNA tracking by replication proteins. *Methods Enzymol.* **262,** 476–494.

23. Schäfer, H.-J., Scheurich, P., and Dose, K. (1978) Eine einfache darstellung von 8-N_3ATP: ein Agens zur Photoaffinitätsmarkierung von ATP-bindenden Proteinen. *Liebigs Ann. Chem.* **1978,** 1749–1753.
24. Laemmli, U. K. (1970) Cleavage of structural proteins during the assembly of the head of bacteriophage T4. *Nature* **227,** 680.
25. Adams, L. D. and Sammons, D. W. (1981) A unique silver staining procedure for color characterization of polypeptides. *Electrophoresis* **2,** 155–165.
26. Westermeier, R., Schickle, H., Thesseling, G., and Walter, W. W. (1988) Densitometrie von gelelektrophoresen. *GIT Labor-Medizin* **4/88,** 194–202.
27. Braun, A. and Merrick, B. (1975) Properties of UV-light-mediated binding of BSA to DNA. *Photochem. Photobiol.* **21,** 243–247.
28. Staros, J. V., Bayley, H., Standring, D. N., and Knowles, J. R. (1978) Reduction of aryl azides by thiols: implications for the use of photoaffinity reagents. *Biochem. Biophys. Res. Commun.* **80,** 568–572.
29. Koberstein, R., Cobianchi, L., and Sund, H. (1976) Interaction of the photoaffinity label 8-azido-ADP with glutamate dehydrogenase. *FEBS Lett.* **64,** 176–180.
30. Vignais, P. V. and Lunardi, J. (1985) Chemical probes of the mitochondrial ATP synthesis and translocation. *Annu. Rev. Biochem.* **54,** 977–1014.
31. Czarnecki, J. J. (1978) Ph.D. Thesis, University of Wyoming, Laramie, WY.
32. Englund, P. T., Kelly, R. B., and Kornberg, A. (1969) Enzymatic synthesis of DNA: binding of DNA to DNA polymerase. *J. Biol. Chem.* **244,** 3045–3052.
33. Garin, J., Vignais, P. V., Gronenborn, A. M., Clore, G. M., Gao, Z., and Bäuerlein, E. (1988) 1H-NMR studies on nucleotide binding to the catalytic sites of bovine mitochondrial F_1-ATPase. *FEBS Lett.* **242,** 178–182.

24

Site-Specific Protein–DNA Photocrosslinking

Analysis of Bacterial Transcription Initiation Complexes

Nikolai Naryshkin, Younggyu Kim, Qianping Dong, and Richard H. Ebright

1. Introduction

1.1. Site-Specific Protein-DNA Photocrosslinking

In work carried out in collaboration with the laboratory of D. Reinberg (University of Medicine and Dentistry of New Jersey), we have developed a site-specific protein–DNA photocrosslinking procedure to define positions of proteins relative to DNA in protein–DNA and multiprotein–DNA complexes *(1–3)*. The procedure has four parts (**Fig. 1**):

1. Chemical *(4–6)* and enzymatic *(7)* reactions are used to prepare a DNA fragment containing a photoactivatible crosslinking agent and an adjacent radiolabel incorporated at a single, defined DNA phosphate (with a 9.7 Å linker between the photoreactive atom of the crosslinking agent and the phosphorus atom of the phosphate, and with an approximately 11 Å maximum "reach" between potential crosslinking targets and the phosphorus atom of the phosphate).
2. The multiprotein–DNA complex of interest is formed using the site-specifically derivatized DNA fragment, and the multiprotein–DNA complex is ultraviolet (UV)-irradiated, initiating covalent crosslinking with proteins in direct physical proximity to the photoactivatible crosslinking agent.
3. Extensive nuclease digestion is performed, eliminating uncrosslinked DNA and converting crosslinked DNA to a crosslinked, radiolabeled 3–5 nucleotide "tag."
4. The "tagged" proteins are identified.

The procedure is performed in systematic fashion, with preparation and analysis of at least 10 derivatized DNA fragments, each having the photoactivatible crosslinking agent incorporated at a single, defined DNA phosphate

From: *Methods in Molecular Biology, vol. 148: DNA–Protein Interactions: Principles and Protocols, 2nd ed.*
Edited by: T. Moss © Humana Press Inc., Totowa, NJ

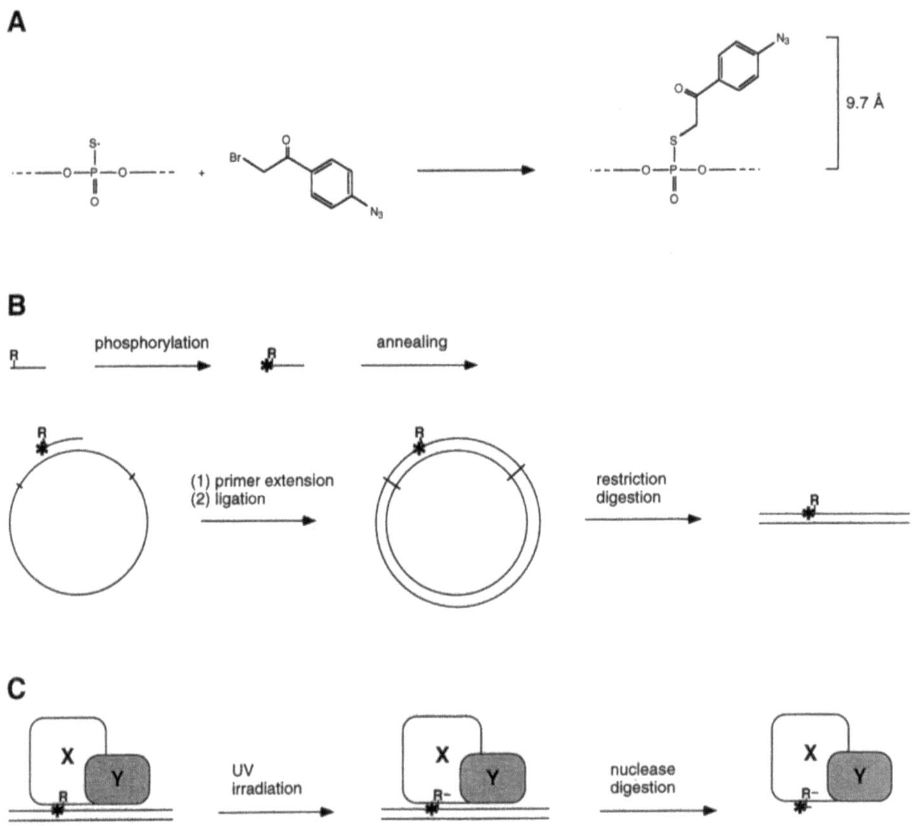

Fig. 1. Site-specific protein–DNA photocrosslinking ***(1–3)***. **(A,B)** Chemical and enzymatic reactions are used to prepare a full-length-promoter DNA fragment with a phenyl-azide photoactivatible crosslinking agent (R) and an adjacent radioactive phosphorus (∗) incorporated at a single, defined site. Based on the chemistry of incorporation, the maximum distance between the site of incorporation and the photoreactive atom is 9.7 Å; the maximum distance between the site of incorporation and a cross-linked atom is approx 11 Å. **(C)** UV irradiation of the derivatized protein–DNA complex initiates crosslinking. Nuclease digestion eliminates uncrosslinked DNA and converts crosslinked, radiolabeled DNA to a crosslinked, radiolabeled 3–5 nucleotide "tag."

(typically each second DNA phosphate—each 12 Å—on each DNA strand spanning the region of interest ***(1–3,8)***.

The results of the procedure define the translational positions of proteins relative to the DNA sequence. Plotted on a three-dimensional representation of a DNA helix, the results also define the rotational orientations of proteins relative to the DNA helix axis, and the groove orientations of proteins relative to the DNA major and minor grooves ***(1–3,8)***.

The procedure has been validated in experiments with three multiprotein–DNA complexes for which crystallographic structures are available (i.e., the TBP–DNA complex, the TBP-TFIIA-DNA complex, and the TBP–TFIIB–DNA complex ***(1,9–13)***. In each case, there was a one-to-one correspondence between sites at which strong crosslinking was observed and sites that in the crystallographic structure were within 11 Å of crosslinked proteins ***(1,9–13)***. The procedure also has been applied to multiprotein-DNA complexes for which crystallographic structures are not available ***(1–3,8)***, including a eukaryotic transcription complex containing 16 distinct polypeptides and having a molecular mass in excess of 800 kDa (the RNAPII–TBP–TFIIB–TFIIF–DNA complex ***[2]***) and a eukaryotic transcription complex containing 27 distinct polypeptides and having a molecular mass in excess of 1700 kDa (the RNAPII–TBP–TFIIB–TFIIE–TFIIF–TFIIH–DNA complex ***[2a]***).

The procedure is related to a procedure developed by Geiduschek and co-workers (***14–17***; *see also* **refs. *18–23***), but offers important advantages. First, because the photoactivatible crosslinking agent is incorporated into DNA chemically, it can be incorporated at a single, defined site. (In the procedure of Geiduschek and co-workers, this is true only at certain DNA sequences.) Second, because the photoactivatible crosslinking agent is incorporated on the DNA phosphate backbone, it can be incorporated at any nucleotide: A, T, G, or C. Third, since the photoactivatible crosslinking agent is incorporated on the DNA phosphate backbone, it probes interactions both in the DNA minor groove and in the DNA minor groove.

1.2. Bacterial Transcription Initiation Complexes

Escherichia coli RNA polymerase holoenzyme (RNAP) consists of two copies of an α-subunit (36.5 kDa), one copy of a β-subunit (151 kDa), one copy of a β'-subunit (155 kDa), and one copy of a σ-subunit (70.3 kDa for the principle σ subunit species, σ^{70}) ***(24)***. RNAP is a molecular machine that carries out a complex series of reactions in transcription initiation ***(24–26)***. Formation of a catalytically competent transcription initiation complex involves three steps ***(24–26)***:

1. RNAP binds to promoter DNA, interacting solely with DNA upstream of the transcription start, to yield an RNAP–promoter closed complex (RP_c; also referred to as RP_{c1}).
2. RNAP then wraps promoter DNA around its circumference, capturing and interacting with DNA downstream of the transcription start, and RNAP undergoes a protein conformational change, clamping tightly onto DNA, to yield an RNAP–promoter intermediate complex (RP_i; also referred to as RP_{c2} and I_2).
3. RNAP then melts approx 14 bp of promoter DNA surrounding the transcription start, rendering accessible the genetic information in the template strand of DNA, to yield a catalytically competent RNAP–promoter open complex (RP_o).

In the case of the *E. coli lacPUV5* promoter, the RNAP–promoter intermediate complex (RP_i) and the RNAP–promoter open complex (RP_o) can be trapped by formation at 14–19°C in the absence of NTPs, and formation at 37°C in the absence of NTPs, respectively ***(27,28)***. Electrophoretic mobility shift, DNA footprinting, fluorescence anisotropy, and 2-aminopurine fluorescence experiments suggest that the trapped complexes are stable and homogeneous (***28–31***; A. Kapanidis, X. Shao, N.N., Y.K., and R.H.E., unpublished data). Kinetic experiments suggest that the trapped complexes correspond to bona fide on-pathway intermediates ***(27,28)***.

In current work, we are using systematic site-specific protein–DNA photocrosslinking to define RNAP–promoter interactions in RNAP–promoter intermediate and open complexes. We are constructing a set of 110 derivatized DNA fragments, each containing a photoactivatible crosslinking agent incorporated at a single, defined position of the *lacPUV5* promoter (positions –79 to +30). For each derivatized DNA fragment, we are forming RNAP–promoter intermediate and open complexes, isolating complexes using nondenaturing polyacrylamide gel electrophoresis, UV-irradiating complexes *in situ*—in the gel matrix—and identifying crosslinked polypeptides. We are performing experiments both with wild-type RNAP and with RNAP derivatives having discontinuous β and β' subunits ("split-β RNAP" and "split-β' RNAP;" reconstituted in vitro from recombinant α, recombinant σ^{70}, and sets of recombinant fragments of β and β'; ***32,33***). Use of split-β and split-β' RNAP permits unambiguous assignment of crosslinks to β and β' (which are not well-resolved in SDS–polyacrylamide gel electrophoresis) and permits rapid, immediate mapping of crosslinks to segments of β and β' (e.g., N-terminal segment, central segment, or C-terminal segment) (**Fig. 2**).

In this chapter, we present protocols for preparation of derivatized *lacPUV5* promoter DNA fragments, formation of RNAP–promoter intermediate and open complexes, UV irradiation of complexes, and identification of crosslinks. In addition, we present support protocols for preparation of wild-type RNAP, split-β RNAP, and split-β' RNAP.

2. Materials

2.1. Preparation of Derivatized DNA Fragment, Chemical Reactions

1. Azidophenacyl bromide (Sigma).
2. Tetraethylthiuram disulfide/acetonitrile (PE Biosystems).
3. dA-CPG, dC-CPG, dG-CPG, T-CPG (1 μmol, 500 Å) (PE Biosystems).
4. dA, dC, dG, T β-cyanoethylphosphoramidites (PE Biosystems).
5. Reagent kit for oligodeoxyribonucleotide synthesis (0.02 *M* iodine) (PE Biosystems).

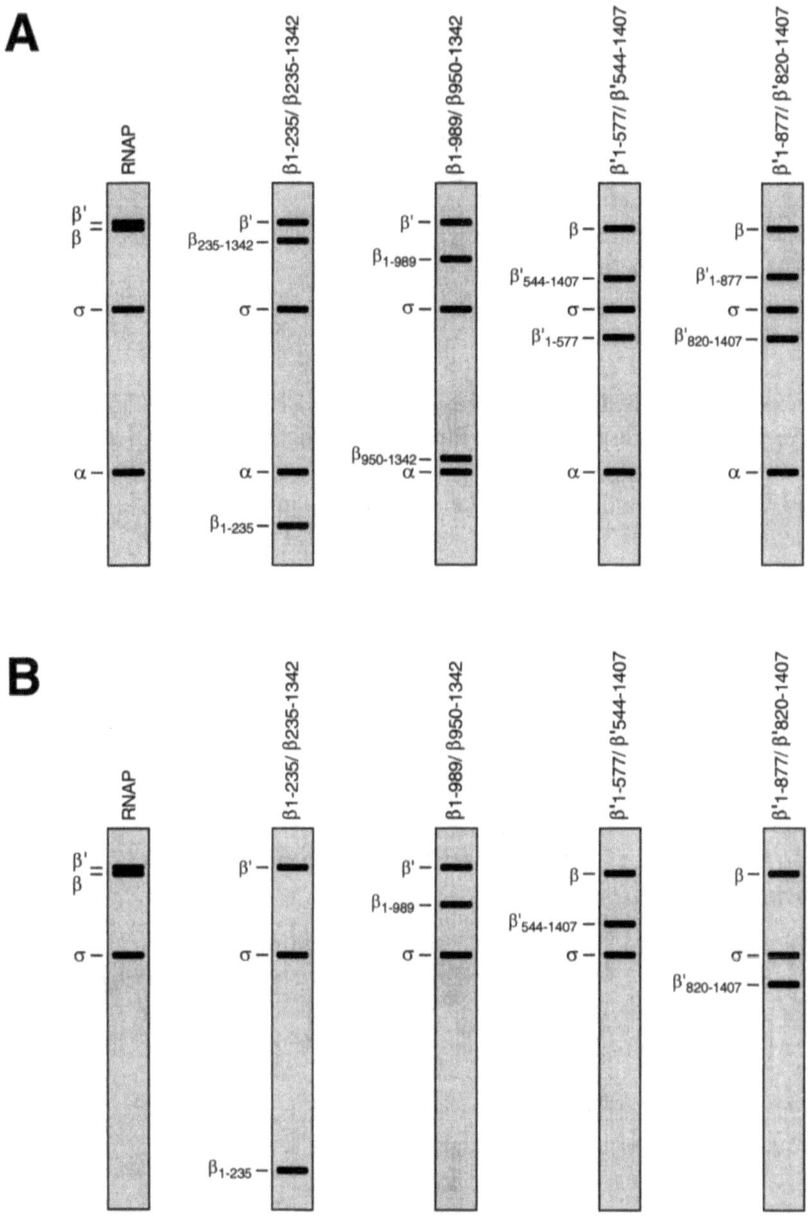

Fig. 2. Use of split-subunit RNAP derivatives *(32,33)* permits unambiguous assignment of crosslinks to RNAP subunits and permits rapid mapping of crosslinks to segments of RNAP subunits. **(A)** Subunit compositions of RNAP, two split-β RNAP derivatives, and two split-β' RNAP derivatives (idealized Coomassie-stained SDS-PAGE gels). **(B)** Results of site-specific protein–DNA photocrosslinking experiments using the RNAP derivatives of panel A and a DNA fragment derivatized at a site close to or in contact with residues 1–235 of β, residues 821–1407 of β', and σ^{70} in the RNAP–promoter complex (idealized autoradiographs of SDS-PAGE gels).

6. Denaturing loading buffer: 0.3% bromophenol blue, 0.3% xylene cyanol, and 12 m*M* EDTA, in formamide.
7. 0.5X TBE: 45 m*M* Tris-borate, pH 8.3, and 1 m*M* EDTA.
8. TE: 10 m*M* Tris-HCl, pH 7.6, 1 m*M* EDTA.
9. 50 m*M* triethylammonium acetate, pH 7.0 (Prime Synthesis).
10. 1 *M* potassium phosphate, pH 7.0.
11. 3 *M* sodium acetate, pH 5.2.
12. 100% ethanol (store at –20°C).
13. 70% ethanol (store at –20°C).
14. Dichloromethane (anhydrous) (PE Biosystems).
15. Acetonitrile (anhydrous) (PE Biosystems).
16. Acetonitrile (high-performance liquid chromatographic [HPLC] grade) (Fisher).
17. Formamide (Sigma).
18. 12% polyacrylamide (29:1 acrylamide:*bis*-acrylamide), 8 *M* urea, 0.5X TBE slab gel (10 × 7 × 0.075 cm).
19. Oligonucleotide purification cartridge (OPC) (PE Biosystems).
20. LiChrospher 100 RP–18 reversed-phase HPLC column (5 μm) (Merck).
21. Autoradiography intensifying screen (Sigma).
22. 254-nm germicidal lamp.
23. ABI392 DNA/RNA synthesizer (PE Biosystems).
24. Varian 5000 HPLC (Varian).
25. L-3000 diode-array HPLC UV detector (Hitachi).
26. Speedvac evaporator (Savant).

2.2. Preparation of Derivatized DNA Fragment, Enzymatic Reactions

1. Derivatized oligodeoxyribonucleotide (**Subheading 3.1.**).
2. M13mp2*(ICAP-UV5)* or M13mp2*(ICAP-UV5)*-rev ssDNA (*see* **Notes 1** and **2**).
3. T_4 polynucleotide kinase (10 U/μL) (New England Biolabs, cat. no. M0201L).
4. T_4 DNA polymerase (3 U/μL) (New England Biolabs, cat. no. M0203L).
5. T_4 DNA ligase (5 U/μL) (Roche Molecular Biochemicals, cat. no. 799009).
6. *Hae*III (40 U/μL)(Roche Molecular Biochemicals, cat. no. 1336029).
7. *Pvu*II (40 U/μL)(Roche Molecular Biochemicals, cat. no. 899216).
8. [γ^{32}P]-ATP (10 mCi/mL, 6000 Ci/mmol) (NEN).
9. 100 m*M* ATP (Amersham Pharmacia Biotech).
10. 100 m*M* dNTPs (Amersham Pharmacia Biotech).
11. Upstream primer (5'-CGGTGCGGGCCTCTTCGCTATTAC–3').
12. 10X phosphorylation buffer: 500 m*M* Tris-HCl, pH 7.6, 100 m*M* $MgCl_2$, 15 m*M* β-mercaptoethanol.
13. 10X annealing buffer: 400 m*M* Tris-HCl, pH 7.9, 500 m*M* NaCl, and 100 m*M* $MgCl_2$.
14. 10X digestion buffer: 100 m*M* Tris-HCl, pH 7.9, 500 m*M* NaCl, and 100 m*M* $MgCl_2$, (*see* **Note 3**).
15. Elution buffer: 0.5 *M* ammonium acetate, 10 m*M* magnesium acetate pH 7.5, and 1 m*M* EDTA.

16. Denaturing loading buffer: 0.3% bromophenol blue, 0.3% xylene cyanol, and 12 m*M* EDTA, in formamide.
17. Nondenaturing loading buffer: 0.3% bromophenol blue, 0.3% xylene cyanol, and 30% glycerol, in water.
18. 0.5X TBE: 45 m*M* Tris-borate pH 8.3, and 1 m*M* EDTA.
19. TE: 10 m*M* Tris-HCl, pH 8.0, and 1 m*M* EDTA.
20. Low-EDTA TE: 10 m*M* Tris-HCl, pH 8.0, and 0.1 m*M* EDTA.
21. 0.5 *M* EDTA, pH 8.0.
22. 10% sodium dodecyl sulfate (SDS).
23. 100% ethanol (store at –20°C).
24. 70% ethanol (store at –20°C).
25. 12% polyacrylamide (29:1 acrylamide:*bis*-acrylamide), 8 *M* urea, and 0.5X TBE slab gel (10 × 7 × 0.075 cm).
26. 7.5% polyacrylamide (29:1 acrylamide:*bis*-acrylamide), and 0.5X TBE slab gel (10 × 7 × 0.15 cm).
27. CHROMA SPIN+TE–10 spin column (Clontech).
28. CHROMA SPIN+TE–100 spin column (Clontech).
29. Spin-X centrifuge filter (0.22 µm, cellulose acetate) (Fisher).
30. PicoGreen dsDNA quantitation kit (Molecular Probes, cat. no. P-7589).
31. Disposable scalpels (Fisher).
32. Autoradiography markers (Stratagene).
33. Light box (VWR).
34. Speedvac evaporator (Savant).

2.3. Preparation of RNAP and RNAP Derivatives

1. *E. coli* strain XL1-blue (Stratagene, cat. no. 200249).
2. *E. coli* strain BL21(DE3) pLysS (Novagen, cat. no. 69388-3).
3. Plasmids encoding RNAP subunits (*see* **Table 1**).
4. Plasmids encoding fragments of RNAP subunits (*see* **Table 2**).
5. LB broth: 10 g/L tryptone, 5 g/L yeast extract, and 10 g/L NaCl; autoclave sterilized.
6. TYE agar plates containing 200 µg/mL ampicillin and 35 µg/mL chloramphenicol: 10 g/L tryptone, 5 g/L yeast extract, 8 g/L NaCl, and 15 g/L agar; autoclave sterilized without antibiotics; supplemented with antibiotics after cooling to 55°C; poured into sterile 100 × 15 mm Petri plates at approx 25 mL/plate.
7. TYE agar plates containing 200 µg/mL ampicillin and 20 µg/mL tetracycline.
8. TYE agar plates containing 40 µg/mL kanamycin and 35 µg/mL chloramphenicol.
9. TYE agar plates containing 40 µg/mL kanamycin and 20 µg/mL tetracycline.
10. 100 mg/mL ampicillin (filter sterilized) (Sigma).
11. 35 mg/mL chloramphenicol in ethanol (filter sterilized) (Sigma).
12. 40 mg/mL kanamycin (filter sterilized) (Sigma).
13. 20 mg/mL tetracycline in methanol (filter sterilized) (Sigma).
14. 1 *M* IPTG (filter sterilized) (Roche Molecular Biochemicals).
15. Buffer A: 20 m*M* Tris-HCl, pH 7.9, 500 m*M* NaCl, and 5 m*M* imidazole.
16. Buffer B: 20 m*M* Tris-HCl, pH 7.9, 6 *M* guanidine chloride, and 500 m*M* NaCl.

Table 1
Plasmids Encoding RNAP Subunits

Plasmid	Relevant Characteristics	Ref.
pHTT7f1-NHα	Ap^R; ori-pBR322; ori-f1; *ϕ10P-rpoA(H6,Nter)*[a]	***34***
pMKSe2	Ap^R; ori-pBR322; *lacP-rpoB*	***35***
pT7β'	Ap^R; ori-pBR322; *ϕ10P-rpoC*	***36***
pHTT7f1-σ	Ap^R; ori-pBR322; ori-f1; *ϕ10P-rpoD*	***34***

[a]*rpoA(H6,Nter)* is a derivative of *rpoA* having a nonnative hexahistidine coding sequence immediately after the *rpoA* start codon.

Table 2
Plasmids Encoding Fragments of RNAP Subunits

Plasmid	Relevant characteristics	Ref.
$p\beta_{1-235}$	Ap^R Km^R; ori-pBR322; *lacP-rpoB(1–235)*	***32***
$p\beta_{235-1342}$	Ap^R; ori-pBR322; *lacP-rpoB(235–1342)*	***32***
$p\beta_{1-989}$	Ap^R; ori-pBR322; *lacP-rpoB(1–989)*	***32***
$p\beta_{951-1342}$	Ap^R; ori-pBR322; *ϕ10P-rpoB(950–1342)*	***32***
$p\beta'_{1-580}$	Ap^R; ori-pBR322; ori-f1; *lacP-ϕ10P-rpoC(1–580)*	***33***
$p\beta'_{545-1407}$	Ap^R; ori-pBR322; ori-f1; *lacP-ϕ10P-rpoC(545–1407)*	***33***
$p\beta'_{1-878}$	Ap^R; ori-pBR322; ori-f1; *lacP-ϕ10P-rpoC(1–878)*	***33***
$p\beta'_{821-1407}$	Km^R; ori-pBR322; ori-f1; *ϕ10P-rpoC(821–1407)*	***33***

17. Buffer C: 40 m*M* Tris-HCl, pH 7.9, 300 m*M* KCl, 10 m*M* EDTA, 1 m*M* phenylmethylsulfonyl fluoride (PMSF), 1 m*M* dithiothreitol (DTT).
18. Buffer D: 50 m*M* Tris-HCl, pH 7.9, 6 *M* guanidine chloride, 10 m*M* $MgCl_2$, 0.01 m*M* $ZnCl_2$, 1 m*M* EDTA, 10 m*M* DTT, and 10% glycerol.
19. Buffer E: 50 m*M* Tris-HCl, pH 7.9, 200 m*M* KCl, 10 m*M* $MgCl_2$, 0.01 m*M* $ZnCl_2$, 1 m*M* EDTA, 5 m*M* β-mercaptoethanol, and 20% glycerol.
20. Buffer F: 50 m*M* Tris-HCl, pH 7.9, and 5% glycerol.
21. α Storage buffer: 50 m*M* Tris-HCl, pH 7.9, 200 m*M* KCl, 10 m*M* $MgCl_2$, 1 m*M* EDTA, 5 m*M* β-mercaptoethanol, and 20% glycerol.
22. 2X SDS loading buffer: 63 m*M* Tris-HCl, pH 8.3, 2% SDS, 5% β-mercapto-ethanol, 25% glycerol, and 0.3% bromophenol blue.
23. SDS running buffer: 25 m*M* Tris, 250 m*M* glycine pH 8.3, and 0.1% SDS.
24. Destaining solution: 10% acetic acid, 50% methanol, and 40% water.
25. 100 m*M* PMSF in ethanol (Sigma).
26. 2% lysozyme (Sigma, cat. no. L-6876) (approx 50,000 U/mg).
27. 10% sodium deoxycholate (Sigma).
28. 10% *n*-octyl-β-D-glucopyranoside (Sigma).
29. Triton X-100 (Sigma).

30. 2 *M* imidazole (pH adjusted to 8.0 with 10 *M* HCl) (Sigma).
31. Glycerol (Fisher).
32. Trichloroacetic acid (Aldrich).
33. Coomassie Brilliant Blue G-250 (Bio-Rad).
34. Acetone (Aldrich).
35. 10% polyacrylamide (37.5:1 acrylamide:*bis*-acrylamide), and 0.1% SDS, slab gel (10 × 7 × 0.075 cm).
36. Prestained protein molecular-weight markers (7-210 kDa) (Bio-Rad).
37. Bio-Rad Protein Assay Kit (cat. no. 500-0002).
38. Ni:NTA-agarose (Qiagen).
39. Dialysis membranes (10-kDa molecular-weight cutoff) (VWR, cat. no. 25223-821).
40. Dialysis-membrane closures (VWR).
41. Collodion dialysis bags (10-kDa molecular-weight cutoff) (Schleicher & Schuell).
42. Nanosep-30K centrifugal concentrators (VWR).
43. Econo-Pac 20-mL chromatography columns (Bio-Rad).
44. Chromatography column frits (1.5 × 0.3 cm) (Bio-Rad).
45. 15 mL culture tubes (autoclave-sterilized) (VWR).
46. Culture-tube stainless-steel closures (autoclave sterilized) (VWR).
47. 2.8-L triple-baffled Fernbach flask (autoclave sterilized) (Bellco Glass, cat. no. 2551-02800).
48. 30-mL polypropylene copolymer centrifuge tube with cap (VWR, cat. no. 21010-567).
49. 250-mL polypropylene copolymer centrifuge bottle with cap (VWR, cat. no. 21020-028).
50. 1-L polypropylene copolymer centrifuge bottle with cap (VWR, cat. no. 21020-061).
51. 200-mL steel beaker (VWR).
52. Branson 450 sonicator (VWR).
53. Sorvall RC-3B centrifuge (DuPont).
54. Sorvall RC-5B centrifuge (DuPont).

2.4. In-Gel Photocrosslinking

1. Cystamine dihydrocloride (Sigma).
2. Acryloyl chloride (Aldrich).
3. Acrylamide (Bio-Rad).
4. TEMED (Bio-Rad).
5. 10% ammonium persulfate (freshly made).
6. SurfaSil siliconizing agent (Pierce).
7. Derivatized promoter DNA fragment (**Subheading 3.2.**).
8. RNAP or RNAP derivative (**Subheading 3.3.**).
9. DNase I (126 units/μL) (Sigma, cat. no. D7291).
10. Micrococcal nuclease in nuclease dilution solution (50 U/μL) (Pharmacia, cat no. 27-0584).
11. Nuclease dilution solution: 5 m*M* $CaCl_2$, 0.1 m*M* PMSF, and 50% glycerol.

12. 2X DTT-free transcription buffer: 50 m*M* HEPES–HCl, pH 8.0, 200 m*M* KCl, 20 m*M* $MgCl_2$, and 10% glycerol.
13. Nondenaturing loading buffer: 0.3% bromophenol blue, 0.3% xylene cyanol, and 30% glycerol.
14. 5X SDS loading buffer: 300 m*M* Tris-HCl, pH 8.3, 10% SDS, 20 m*M* EDTA, 25% β-mercaptoethanol, 0.1% bromophenol blue, 50% glycerol.
15. 0.5X TBE: 45 m*M* Tris-borate, pH 8.0, and 1 m*M* EDTA.
16. SDS running buffer: 25 m*M* Tris, 250 m*M* glycine, pH 8.3, and 0.1% SDS.
17. 10% SDS.
18. 1 *M* DTT (freshly made).
19. 0.2 m*M* PMSF (Sigma).
20. 0.22 mg/mL heparin (Sigma, cat. no. H-3393) (grade I-A, from porcine intestinal mucosa, approx 170 USP units/mg).
21. 4–15% gradient polyacrylamide (37.5:1 acrylamide:*bis*-acrylamide) Tris–HCl slab gel (Bio-Rad, cat. no. 161-1176).
22. Prestained protein molecular-weight markers (7–210 kDa) (Bio-Rad).
23. Silicone rubber heating mat (200 W, 120 V AC; 25 × 10 cm) (Cole-Parmer, cat. no. P-03125-40).
24. Variable voltage controller (Cole-Parmer, cat. no. P-01575-10).
25. Digital thermometer (Cole-Parmer, cat. no. P-91000-00).
26. Thermocouple probe (needle, 0.7 mm in diameter) (Cole-Parmer, cat. no. P-91000-00).
27. Large binder clips (5-cm width) (Staples).
28. Filter unit (22-μm pore size, 250 mL) (Millipore).
29. 50-mL Büchner funnel with glass frit (10 μm pore size) (Fisher).
30. 500-mL separating funnel (Fisher).
31. Disposable scalpels (VWR).
32. X-ray exposure holder with intensifying screen (Kodak).
33. Light box (VWR).
34. Rayonet RPR-100 photochemical reactor equipped with 16 RPR-3500 Å tubes (Southern New England Ultraviolet).
35. Speedvac evaporator (Savant).

3. Methods

3.1. Preparation of Derivatized DNA Fragment, Chemical Reactions

3.1.1. Preparation of Phosphorothioate Oligodeoxyribonucleotide

1. Perform 24 standard cycles of solid-phase β-cyanoethylphosphoramidite oligodeoxyribonucleotide synthesis to prepare CPG-linked precursor containing residues 3–26 of desired oligodeoxyribonucleotide. Use the following settings: cycle, 1.0 μ*M* CE; DMT, on; end procedure, manual.
2. Replace iodine/water/pyridine/tetrahydrofuran solution (bottle 15) by tetraethylthiuram disulfide/acetonitrile solution. Perform one modified cycle of solid-

phase β-cyanoethylphosphoramidite oligodeoxyribonucleotide synthesis to add residue 2 and phosphorothioate linkage. Use the following settings: cycle, 1.0 m*M* sulfur; DMT, on; end procedure, manual.

3. Replace tetraethylthiuram disulfide/acetonitrile solution (bottle 15) by iodine/water/pyridine/tetrahydrofuran solution. Place collecting vial on the DNA synthesizer. Perform one standard cycle of solid-phase β-cyanoethylphosphoramidite oligodeoxyribonucleotide synthesis to add residue 1. Use the following settings: cycle, 1.0 μ*M* CE; DMT, on; end procedure, CE.
4. Remove collecting vial, screw cap tightly, and deblock by incubating 8 h at 55°C. Transfer sample to 6-mL polypropylene round-bottomed tube, place tube in Speedvac, and spin 20 min with Speedvac lid ajar and with no vacuum (allowing evaporation of ammonia). Close Speedvac lid, apply vacuum, and dry.
5. Detritylate and purify approx 0.075 μmol on OPC according to supplier's protocol.
6. Dry in Speedvac.
7. Resuspend in 100 μL TE. Remove 2-μL aliquot, dilute with 748 μL TE, and determine concentration from UV absorbance at 260 nm (molar extinction coefficient = 240,000 AU/*M*/cm).
8. To confirm purity of oligodeoxyribonucleotide, mix aliquot containing 1 nmol oligodeoxyribonucleotide with equal volume of formamide. Apply to 12% polyacrylamide (29:1 acrylamide:*bis*-acrylamide), 8 *M* urea, 0.5X TBE slab gel (10 × 7 × 0.075 cm). As marker, load in adjacent lane 5 mL denaturing loading buffer. Electrophorese 30 min at 25 V/cm. Disassemble gel, place on intensifying screen, and view in dark using 254 nm germicidal lamp. Oligodeoxyribonucleotide should appear as dark shadow against green background and should migrate more slowly than bromophenol blue. If purity is ≥95%, proceed to next step.
9. Divide remainder of sample into 50-nmol aliquots, transfer to 1.5-mL siliconized polypropylene microcentrifuge tubes, dry in Speedvac, and store at –20°C (stable for at least 2 yr).

*3.1.2. Derivatization of Oligodeoxyribonucleotide (All Steps Carried Out Under Subdued Lighting [see **Note 4**])*

1. Dissolve 10 mg (42 μmol) azidophenacyl bromide in 1 mL chloroform. Transfer 100-μL aliquots (4.2 μmol) to 1.5-mL siliconized polypropylene microcentrifuge tubes, and dry in Speedvac. Wrap tubes with aluminum foil, and store desiccated at 4°C (stable indefinitely).
2. Resuspend 50-nmol aliquot of phosphorothioate oligodeoxyribonucleotide (**Subheading 3.1.1.**) in 50 μL water, and resuspend 4.2-μmol aliquot of azidophenacyl bromide in 220 μL methanol.
3. Mix 50 μL (50 nmol) phosphorothioate oligodeoxyribonucleotide solution, 5 μL 1 *M* potassium phosphate (pH 7.0), and 55 μL (1 μmol) azidophenacyl bromide solution in a 1.5-mL siliconized polypropylene microcentrifuge tube. Incubate 3 h at 37°C in the dark.
4. Precipitate derivatized oligodeoxyribonucleotide by adding 11 μL of 3 *M* sodium acetate (pH 5.2) and 275 μL ice-cold 100% ethanol. Invert tube several times,

and place at –80°C for 30 min. Centrifuge 5 min at 13,000*g* at 4°C. Remove supernatant, and wash pellet with ice cold 70% ethanol. Air-dry 15 min at room temperature (RT). Store at –20°C (stable for at least 1 yr).

3.1.3. Purification of Derivatized Oligodeoxyribonucleotide (All Steps Carried Out Under Subdued Lighting [see **Note 4**])

1. Resuspend derivatized oligodeoxyribonucleotide in 100 µL of 50 m*M* triethylammonium acetate (pH 7.0).
2. Analyze 5-µL aliquot by C_{18} reversed-phase HPLC to confirm efficiency of derivatization reaction. Use LiChrospher 100 RP-18 C_{18} reversed-phase HPLC column (5 µm), with solvent A = 50 m*M* triethylammonium acetate (pH 7.0) and 5% acetonitrile; solvent B = 100% acetonitrile; and flow rate = 1 mL/min. Equilibrate column with 10 column volumes of solvent A before loading sample. After loading sample, wash column with 6 column volumes of solvent A and elute with a 50 min gradient of 0–70% solvent B in solvent A. Derivatized and underivatized oligodeoxyribonucleotides elute at approx 25% solvent B and approx 16% solvent B, respectively (*see* **Notes 5** and **6**).
3. If derivatization efficiency is ≥80%, purify remainder of sample using procedure in **step 2**, collecting peak fractions (*see* **Notes 5** and **6**).
4. Pool peak fractions, divide into 1-mL aliqouts, and dry in Speedvac. Store desiccated at –20°C in the dark (stable for at least 1 yr).
5. Resuspend one aliquot in 100 µL TE. Remove 5 µL, dilute with 495 µL water, and determine concentration from UV absorbance at 260 nm (molar extinction coefficient = 242,000 AU/*M*/cm).
6. Divide remainder of derivatized-oligodeoxyribonucleotide/TE solution from **step 5** into 20, 5-pmol aliquots and one larger aliquot, dry in Speedvac, and store desiccated at –20°C in the dark (stable for at least 1 yr).

3.2. Preparation of Derivatized DNA Fragment, Enzymatic Reactions

3.2.1. Radiophosphorylation of Derivatized Oligodeoxyribonucleotide (All Steps Carried Out Under Subdued Lighting [see **Note 4**])

1. Resuspend 5 pmol derivatized oligodeoxyribonucleotide in 12 µL water. Add 2 µL of 10X phosphorylation buffer, 5 µL of [γ-^{32}P]ATP (50 µCi) and 1 µL (10 U) T_4 polynucleotide kinase. Incubate 15 min at 37°C. Terminate reaction by heating 5 min at 65°C (*see* **Note 7**).
2. Add 15 µL water.
3. Desalt radiophosphorylated derivatized oligodeoxyribonucleotide into TE using CHROMA SPIN+TE–10 spin column according to supplier's protocol.
4. Immediately proceed to next step, or, if necessary, store radiophosphorylated derivatized oligodeoxyribonucleotide solution at –20°C in the dark (stable for up to 24 h).

3.2.2. Annealing, Extension, and Ligation of Radiophosphorylated Derivatized Oligodeoxyribonucleotide (All Steps Carried Out Under Subdued Lighting [see **Note 4**])

1. In 1.5-mL siliconized polypropylene microcentrifuge tube, mix 34 µL radiophosphorylated derivatized oligodeoxyribonucleotide, 1 µL of 10 µ*M* upstream

primer, 1 μL of 1 μ*M* M13mp2*(ICAP-UV5)* ssDNA (for analysis of crosslinks to template DNA strand) or M13mp2*(ICAP-UV5)*-rev ssDNA (for analysis of crosslinks to nontemplate DNA strand), and 4 μL of 10X annealing buffer.

2. Heat 5 min at 65°C (*see* **Note 7**). Transfer to 500-mL beaker containing 200 mL water at 65°C, and place beaker at room temperature to permit slow cooling (65°C to 25°C in approx 60 min).
3. Add 2 μL of 25 m*M* dNTPs, 1 μL of 100 m*M* ATP, 3 μL (9 units) T4 DNA polymerase, and 1 μL (5 units) T4 DNA ligase. Perform parallel reaction without ligase as "no-ligase" control.
4. Incubate 15 min at room temperature, followed by 3 h at 37°C. Terminate reaction by adding 1 μL of 10% SDS.
5. Desalt into TE using CHROMA SPIN+TE-100 spin column according to supplier's protocol. Immediately proceed to next step.

*3.2.3. Digestion and Purification of Derivatized DNA Fragment (All Steps Carried Out Under Subdued Lighting [*see **Note 4**])

1. In 1.5-mL siliconized polypropylene microcentrifuge tube, mix 40 μL product from **Subheading 3.2.2.**, 4.5 μL of 10X digestion buffer, 0.25 μL (10 units) *Hae*III or 0.25 μL (10 units) *Pvu*II (*see* **Note 8**). Incubate 1 h at 37°C.
2. Perform parallel reaction using 40 μL "no-ligase" control from **step 3** of **Subheading 3.2.2.**
3. Mix 3 μL aliquots of reaction of **step 1** and of "no-ligase" control reaction of **step 2**, each with 7 μL denaturing loading buffer. Heat 5 min at 65°C, and then apply to 12% polyacrylamide (29:1 acrylamide:bis-acrylamide), 8 *M* urea, 0.5X TBE slab gel (10 × 7 × 0.075 cm). As a marker, load 5 μL denaturing loading buffer in the adjacent lane. Electrophorese 30 min at 25 V/cm. Dry gel, expose to X-ray film 1 h at room temperature, and process film. Estimate ligation efficiency by comparing reaction and "no-ligase" control lanes. If the ligation efficiency is ≥80%, proceed to the next step. If not, repeat the steps of **Subheadings 3.2.1.** and **3.2.2.**
4. Mix remainder of reaction of **step 1** (42 μL) with 10 μL 50% glycerol. Apply to *nondenaturing* 7.5% polyacrylamide (29:1 acrylamide:bis-acrylamide), 0.5X TBE slab gel (10 × 7 × 0.15 cm). As a marker, load 5 μL *nondenaturing* loading buffer in the adjacent lane. Electrophorese at 25 V/cm until the bromophenol blue reaches the bottom of the gel.
5. Remove one glass plate, and cover the gel with plastic wrap. Attach two autoradiography markers to the gel. Expose to X-ray film for 60 s at room temperature and process the film. Cut out the portion of the film corresponding to the derivatized DNA fragment. Using a light box, superimpose the cut-out film on the gel, using autorad markers as the alignment reference points. Using disposable scalpel, excise portion of gel corresponding to derivatized DNA fragment.
6. Place the excised gel slice in a 1.5-mL siliconized polypropylene microcentrifuge tube, and crush with a 1-mL pipet tip. Add 300 μL elution buffer, centrifuge 5 s at 5000*g*, and incubate 12 h at 37°C.
7. Transfer supernatant to Spin-X centrifuge filter and centrifuge 1 min at 13,000*g* at room temperature in fixed-angle microcentrifuge.

8. Transfer filtrate to a 1.5-mL siliconized polypropylene microcentrifuge tube. Precipitate the derivatized DNA fragment by the addition of 1 mL ice-cold 100% ethanol. Invert the tube several times and place at –20°C for 30 min. Centrifuge 5 min at 13,000*g* at 4°C in a fixed-angle microcentrifuge. Remove and dispose of supernatant, wash pellet with 500 µL ice cold 70% ethanol, and air-dry for 15 min at room temperature.
9. Resuspend in 30 µL Low-EDTA TE. Determine radioactivity by Cerenkov counting. Remove 1 µL aliquot, and determine DNA concentration using PicoGreen dsDNA quantitation kit according to supplier's protocol. Calculate specific activity (expected specific activity ≈5,000 Ci/mmol).
10. Store derivatized DNA fragment at 4°C in the dark (stable for ≈1 wk).

3.3. Preparation of RNAP and RNAP Derivatives

3.3.1. Preparation of Hexahistidine-Tagged Recombinant α-Subunit

1. Transform *E. coli* strain BL21(DE3) pLysS with plasmid pHTT7f1-NHα. Plate to TYE agar containing 200 µg/mL ampicillin and 35 µg/mL chloramphenicol, and incubate 12 h at 37°C.
2. Inoculate single colony into 5 mL LB containing 200 µg/mL ampicillin and 35 µg/mL chloramphenicol in a 15-mL culture tube with a culture tube stainless-steel closure, and shake vigorously for 12 h at 37°C. Transfer to a 15 mL polypropylene centrifuge tube, and centrifuge 5 min at 3000*g* at room temperature. Discard supernatant, wash cell pellet twice with 5 mL LB, and resuspend cell pellet in 5 mL LB in a new 15-mL polypropylene centrifuge tube.
3. Inoculate into 1 L LB containing 200 µg/mL ampicillin and 35 µg/mL chloramphenicol in a 2.8-L Fernbach flask, and shake vigorously at 37°C until OD_{600} = 0.6. Add 1 mL of 1 M IPTG, and shake vigorously for an additional 3 h at 37°C.
4. Transfer culture to a 1-L polypropylene copolymer centrifuge bottle. Harvest cells by centrifugation 20 min at 5000*g* at 4°C.
5. Resuspend cell pellet in 100 mL buffer A at 4°C. Transfer into a 200-mL steel beaker and place beaker on ice. Lyse cells with four 40-s sonication pulses at 25% maximum sonicator output (2-min pause between each pulse).
6. Transfer lysate to a 250-mL polypropylene copolymer centrifuge bottle. Centrifuge for 15 min at 15,000*g* at 4°C. Collect supernatant.
7. Transfer supernatant to a 250-mL glass beaker. Add 35 g ammonium sulfate and stir for 20 min on ice.
8. Transfer suspension to a 250-mL polypropylene copolymer centrifuge bottle. Centrifuge for 20 min at 15,000*g* at 4°C. Discard the supernatant.
9. Resuspend the pellet in 28 mL buffer B containing 5 m*M* imidazole. Transfer to a 30-mL polypropylene copolymer centrifuge tube and rock gently for 30 min at 4°C. Centrifuge for 15 min at 15,000*g* at 4°C.
10. Load the supernatant onto a 5-mL Ni:NTA–agarose column pre-equilibrated with 25 mL buffer B containing 5 m*M* imidazole (*see* **Note 9**). Collect flowthrough and reload onto column. Wash column with 50 mL buffer B containing 5 m*M* imidazole, and 25 mL buffer B containing 10 m*M* imidazole. Elute column with

15 mL buffer B containing 20 m*M* imidazole, 15 mL buffer B containing 30 m*M* imidazole, 15 mL buffer B containing 40 m*M* imidazole, and 15 mL buffer B containing 150 m*M* imidazole. Collect 5-mL fractions.

11. Transfer a 10-µL aliquot of each fraction to a 1.5-mL siliconized polypropylene microcentrifuge tube, add 90 µL water and 100 µL of 10% trichloroacetic acid. Place on ice for 20 min. Centrifuge for 5 min at 13,000*g* at room temperature. Discard the supernatant. Wash the pellet with 500 µL acetone and air-dry for 15 min. Dissolve the pellet in 5 µL water, add 5 µL of 2X SDS loading buffer, heat for 3 min at 100°C, and apply to 10% polyacrylamide (37.5:1 acrylamide:bis-acrylamide) and 0.1% SDS slab gel (10 × 7 × 0.075 cm). As a marker, load 5 µL prestained protein molecular weight markers into the adjacent lane. Electrophorese in SDS running buffer at 25 V/cm until the bromophenol blue reaches the bottom of gel. Stain the gel by gently shaking for 5 min in 50 mL of 0.2% Coomassie Brilliant Blue G–250 in the destaining solution. Destain by gently shaking for 10 h in 100 mL destaining solution.
12. Pool fractions containing homogenous α (typically fractions with buffer B containing 40–150 m*M* imidazole). Dialyze using a 10-kDa molecular-weight-cutoff dialysis membrane against two 1-L changes of α storage buffer for 16 h at 4°C.
13. Determine protein concentration and total protein amount using Bio-Rad Protein Assay according to the supplier's protocol.
14. After dialysis, measure the volume and transfer to a 30-mL polypropylene copolymer centrifuge tube. Add 3 g ammonium sulfate per 10 mL and rock gently for 20 min at 4°C. Centrifuge for 20 min at 15,000*g* at 4°C.
15. Remove and discard 10 mL of supernatant. Resuspend pellet in the remaining supernatant. Divide into 50-µL aliquots and transfer to 1.5-mL siliconized polypropylene microcentrifuge tubes. Centrifuge aliquots for 5 min at 13,000*g* at 4°C. Store at –80°C (stable for at least 1 yr). Expected yield: 20–30 mg (250–500 µg/aliquot). Expected purity: >99%.

3.3.2. Preparation of Crude Recombinant RNAP Subunits and Subunit Fragments

1. Transform plasmid encoding RNAP subunit or subunit fragment into *E. coli* strain BL21(DE3) pLysS (for plasmids with *ϕ10P*- or *lacP*-*ϕ10P*-based expression; **Tables 1** and **2**) or *E. coli* strain XL1-blue (for plasmids with *lacP*-based expression; **Tables 1** and **2**). Plate transformants of BL21(DE3) pLysS to TYE agar containing 200 µg/mL ampicillin (40 µg/mL kanamycin for plasmid $p\beta'_{821-1407}$) and 35 µg/mL chloramphenicol, and incubate for 12 h at 37°C. Plate transformants of XL1-blue to TYE agar containing 200 µg/mL ampicillin (40 µg/mL kanamycin for plasmid $p\beta_{1-235}$) and 20 µg/mL tetracycline, and incubate for 16 h at 37°C.
2. Inoculate a single colony into 5 mL LB containing antibiotics at concentrations specified in **step 1** in a 15-mL culture tube with stainless-steel closure, and shake vigorously for 12 h at 37°C. Transfer to a 15-mL polypropylene centrifuge tube and centrifuge for 5 min at 3000*g* at room temperature. Discard the supernatant,

wash the cell pellet twice with 5 mL LB, and resuspend the cell pellet in 5 mL LB.

3. Inoculate into 1 L LB containing antibiotics at concentrations specified in **step 1** in a 2.8-L Fernbach flask, and shake vigorously at 37°C until $OD_{600} = 0.6$. Add 1 mL of 1 *M* IPTG and shake vigorously for an additional 3 h [transformants of BL21(DE3) pLysS] or 5 h (transformants of XL1-blue) at 37°C.
4. Transfer culture to a 1 L polypropylene copolymer centrifuge bottle. Harvest the cells by centrifugation for 20 min at 5000*g* at 4°C.
5. Resuspend the cell pellet in 10 mL buffer C containing 0.2% sodium deoxycholate and 0.02% lysozyme in a 30-mL polypropylene copolymer centrifuge tube at 4°C. Place the tube on ice. Lyse cells with five 30-s sonication pulses at 25% maximum sonicator output (2-min pause between each pulse).
6. Centrifuge 20 min at 15,000*g* at 4°C. Discard the supernatant.
7. Resuspend the pellet in 10 mL buffer C containing 0.2% *n*-octyl-β-D-glucopyranoside (0.5% Triton X–100 for preparation of σ^{70}) and 0.02% lysozyme at 4°C. Sonicate as in **step 5**. Centrifuge for 20 min at 15,000*g* at 4°C. Discard the supernatant.
8. Resuspend pellet in 10 mL buffer C containing 0.2% *n*-octyl-β-D-glucopyranoside (0.5% Triton X–100 for preparation of σ^{70}) at 4°C. Sonicate as in **step 5**. Centrifuge for 20 min at 15,000*g* at 4°C. Discard the supernatant.
9. Resuspend pellet in 10 mL buffer C at 4°C. Place tube on ice and sonicate for 10 s at 25% maximum sonicator output. Divide into 500-μL aliquots and transfer to 1.5-mL siliconized polypropylene microcentrifuge tubes. Centrifuge for 5 min at 13,000*g* at 4°C. Discard the supernatant.
10. Add 100 μL ice-cold buffer C containing 10% glycerol to each aliquot. Store at –80°C (stable for at least 2 yr). Expected yield: 50–100 mg (1.5–3 mg/aliquot). Expected purity: 50–90%.

3.3.3. Reconstitution of RNAP and RNAP Derivatives

1. Thaw aliquots containing purified subunit (from **Subheading 3.3.1., step 15**) and crude recombinant RNAP subunits and subunit fragments (from **Subheading 3.3.2., step 10**) by placing on ice for 10 min. Centrifuge for 30 s at 13,000*g* at 4°C. Discard supernatants.
2. Resuspend each pellet in 500 μL buffer D. Incubate for 30 min at 4°C, rocking gently. Centrifuge for 5 min at 13,000*g* at 4°C.
3. Transfer supernatants to new 1.5-mL siliconized polypropylene microcentrifuge tubes at 4°C. Determine the protein concentrations using Bio-Rad Protein Assay according to the supplier's protocol (expected concentrations: 3–6 mg/mL).
4. Prepare core reconstitution mixture by combining in a 1.5-mL siliconized polypropylene microcentrifuge tube the following: 30 μg N-terminally hexahistidine-tagged α, 300 μg β (or 170 μg β_{1-235} and 800 μg $\beta_{235-1342}$; or 700 μg β_{1-989} and 300 μg $\beta_{951-1342}$) and 500 μg β' (or 400 μg β'_{1-581} and 500 μg $\beta'_{545-1407}$; or 700 μg β'_{1-877} and 330 μg $\beta'_{821-1407}$), and diluting with buffer D to a total protein concentration of 450 μg/mL.

5. Prepare σ^{70} reconstitution mixture by adding 250 µg σ^{70} to a 1.5-mL siliconized polypropylene microcentrifuge tube and diluting with buffer D to a total protein concentration of 1500 µg/mL.
6. Dialyze core and σ^{70} reconstitution mixtures separately in collodion dialysis bags against two 1-L changes of buffer E for 16 h at 4°C.
7. Transfer core and σ^{70} reconstitution mixtures to separate 2.0-mL siliconized polypropylene microcentrifuge tubes. Centrifuge for 5 min at 13,000*g* at 4°C. Combine supernatants in a single, new 2.0-mL siliconized polypropylene microcentrifuge tube.
8. Incubate 45 min at 30°C. Centrifuge for 10 min at 13,000*g* at 4°C.

3.3.4. Purification of RNAP and RNAP Derivatives

1. During incubation of **step 8** of **Subheading 3.3.3.**, place 200 µL Ni:NTA–agarose in a 2.0-mL siliconized polypropylene microcentrifuge tube and centrifuge for 2 min at 13,000*g* at 4°C. Remove supernatant.
2. Resuspend Ni:NTA–agarose in 1 mL buffer F containing 5 m*M* imidazole at 4°C. Centrifuge 2 min at 13,000 × g at 4°C. Remove supernatant. Repeat two times.
3. Add supernatant of **step 8** of **Subheading 3.3.3.** to Ni:NTA–agarose from **step 2**. Incubate 45 min at 4°C, rocking gently. Centrifuge for 2 min at 13,000*g* at 4°C. Discard the supernatant.
4. Resuspend in 1.5 mL buffer F containing 5 m*M* imidazole at 4°C. Rock gently for 15 s at 4°C. Centrifuge 2 min at 13,000*g* at 4°C. Discard the supernatant. Repeat two times.
5. Resuspend in 250 µL buffer F containing 150 m*M* imidazole. Rock gently for 2 min at 4°C. Centrifuge for 2 min at 13,000*g* at 4°C.
6. Transfer supernatant to Nanosep-30K centrifugal concentrator. Centrifuge at 13,000*g* at 4°C until sample volume is reduced to approx 50 mL (approx 15 min).
7. Transfer the sample to a 1.5-mL siliconized polypropylene microcentrifuge tube. Add, in order, 1 µL of 0.1 *M* β-mercaptoethanol and 50 µL glycerol, mix well, and store at –20°C (stable for at least 1 mo).
8. Determine the protein concentration using the Bio-Rad Protein Assay according to the supplier's protocol. Expected yield: 100 µg. Expected purity: >90%.

3.4. In-Gel Photocrosslinking

3.4.1. Synthesis of N,N'*-Bisacryloylcystamine (BAC) (*see **Note 10***)*

1. Acryloyl chloride is highly toxic. Therefore, all manipulations in this section must be performed in a fume hood.
2. Dissolve 4.0 g (18 mmol) cystamine dihydrochloride in 40 mL of 3 *M* NaOH (120 mmol). Dissolve 4.3 mL (54 mmol) acryloyl chloride in 40 mL chloroform. Mix solutions in 500 mL flask (*see* **Note 11**). (Two phases will form: an upper, aqueous phase; and a lower, organic phase.) Place flask on a plate stirrer and stir 3 min at room temperature, followed by 15 min at 50°C.
3. Discontinue stirring. Immediately transfer reaction mixture to a 500-mL separating funnel, allow phases to separate (approx 2 min), and transfer lower, organic phase to a 250-mL beaker.

4. Place on ice for 10 min. Collect precipitate by filtration in Büchner funnel.
5. Transfer precipitate to a 250-mL beaker with 30 mL chloroform at room temperature. Place beaker on plate stirrer and stir 1 min at room temperature, followed by 5 min at 50°C. Place on ice for 10 min and collect precipitate (BAC) by filtration in Büchner funnel.
6. Transfer precipitate to a 50-mL polypropylene centrifuge tube. Seal tube with Parafilm, pierce seal several times with a syringe needle, place tube in vacuum desiccator, and dry under a vacuum for 16 h at room temperature. Expected yield: 1.5–1.9 g.

3.4.2. Preparation of Polyacrylamide:BAC Gel

1. Prepare 20% acrylamide:BAC (19:1) stock solution by dissolving, in order, 19 g acrylamide and 1 g BAC in 80 mL water in a 200-mL beaker at room temperature. Place on a plate stirrer and stir for 10 min at 60°C (*see* **Note 12**). Adjust volume to 100 mL with water. Allow solution to cool to room temperature. Filter stock solution using 0.22-μm filter unit and store at room temperature in the dark (stable for at least 2 mo).
2. Mix 9 mL of 20% acrylamide:BAC (19:1) stock solution, 1.8 mL of 10X TBE, and 25.2 mL water. Add 180 μL TEMED and 90 μL freshly prepared 10% ammonium persulfate (*see* **Note 13**). Immediately pour into slab gel assembly with siliconized notched glass plate (27 × 16 × 0.1 cm) (*see* **Note 14**). Insert comb and heat slab gel assembly to approx 60°C by positioning a bench lamp with a 60-W tungsten bulb 2 cm from the outer glass plate (*see* **Note 15**). Allow 10–20 min for polymerization. (The polyacrylamide:BAC gel is stable for up to 72 h at 4°C.)

3.4.3. Formation and Isolation of RNAP–Promoter Complexes

3.4.3.1. Formation and Isolation of RNAP–Promoter Intermediate Complex (All Steps Carried Out Under Subdued Lighting [*see* **Note 4**])

1. Place polyacrylamide:BAC slab gel in electrophoresis apparatus and pour 0.5X TBE into upper and lower reservoirs.
2. Prerun gel for 2 h at 20 V/cm.
3. Prechill electrophoresis unit by placing in 15°C cabinet for 3 h.
4. During 15°C prechilling of **step 3**, dilute RNAP or RNAP derivative to 180 μg/mL (400 nM) in buffer F containing 1 m*M* β-mercaptoethanol and 50% glycerol, and place tubes containing diluted RNAP, derivatized DNA fragment, 2X DTT-free transcription buffer, and water, at 15°C.
5. Immediately after 15°C gel prechilling of **step 3**, add the following, in order, to a 1.5-mL siliconized polypropylene microcentrifuge tube: 2 μL of 5 nM derivatized DNA fragment (approx 5000 Ci/mmol), 5 μL 2X DTT-free transcription buffer, 2 μL water, and 1 μL of 180 μg/mL (400 nM) RNAP or RNAP derivative (all at 15°C).
6. Incubate 20 min at 15°C in the dark.
7. During incubation of **step 6**, apply voltage to gel in a 15°C cabinet: 16 V/cm. Wash wells of gel carefully with 0.5X TBE to remove unpolymerized acrylamide and BAC. (**Caution:** care must be exercised to avoid electrocution.)

8. After completing incubation of **step 6**, immediately add 1 μL of 0.22 mg/mL heparin (prechilled to 15°C), mix, and immediately apply sample to gel in 15°C cabinet (*see* **Note 16**). Load 5 μL nondenaturing loading buffer into adjacent lane. (**Caution:** care must be exercised to avoid electrocution.) Continue electrophoresis in a 15°C cabinet for 20 min at 16 V/cm. Monitor gel temperature at 5-min intervals by inserting the thermocouple probe into the gel for 5 s (and removing immediately thereafter). Maintain the gel temperature at 15°C. If the gel temperature rises above 15°C, temporarily reduce electrophoresis voltage to 10 V/cm.
9. Immediately proceed to the next step (**Subheading 3.4.4.**).

3.4.3.2. Formation and Isolation of RNAP–Promoter Open Complex (All Steps Carried Out Under Subdued Lighting [*see* Note 4])

1. Place polyacrylamide:BAC slab gel in electrophoresis apparatus, clip 10 × 25-cm silicone heating mat directly to outer glass plate of the slab gel assembly with four large binder clips, and pour 0.5X TBE buffer in upper and lower reservoirs.
2. Prerun gel for 2 h at 20 V/cm.
3. Prewarm electrophoresis unit by placing in 37°C cabinet for 3 h.
4. During 37°C prewarming of **step 3**, dilute RNAP or RNAP derivative to 180 μg/mL (400 n*M*) in buffer F containing 1 m*M* β-mercaptoethanol and 50% glycerol.
5. Immediately after 37°C prewarming of **step 3**, add the following, in order, to a 1.5-mL siliconized polypropylene microcentrifuge tube: 2 μL of 5 n*M* derivatized DNA fragment (approx 5000 Ci/mmol), 5 μL of 2X DTT-free transcription buffer, 2 μL water, and 1 μL of 180 μg/mL (400 n*M*) RNAP or RNAP derivative (all at room temperature).
6. Incubate for 20 min at 37°C in the dark.
7. During incubation of **step 6** apply voltage to the gel: 16 V/cm. Wash wells of gel carefully with 0.5X TBE to remove unpolymerized acrylamide and BAC. (**Caution:** Care must be exercised to avoid electrocution.) Connect heating mat to variable-voltage controller. Monitor the gel temperature at 5-min intervals by inserting thermocouple probe into the gel for 5 s (and removing immediately thereafter). Maintain gel temperature at 37°C, adjusting heater voltage as necessary (typically 12–14V).
8. After completing incubation of **step 6**, immediately add 1 μL of 0.22 mg/mL heparin (prewarmed to 37°C), mix, and immediately apply sample to gel (*see* **Note 16**). Load 5 μL nondenaturing loading buffer into adjacent lane. (**Caution:** care must be exercised to avoid electrocution.) Continue electrophoresis 20 min at 16 V/cm. Monitor the gel temperature at 5-min intervals by inserting the thermocouple probe into the gel for 5 s (and removing immediately thereafter). Maintain the gel temperature at 37°C, adjusting heater voltage as necessary (typically 12–14V).
9. Immediately proceed to next step (**Subheading 3.4.4.**).

3.4.4. In-Gel UV Irradiation of RNAP–Promoter Complex

1. Remove gel with both glass plates in place (*see* **Note 17**) and mount vertically in a Rayonet RPR-100 photochemical reactor equipped with 16 RPR-3500 Å tubes.

2. Immediately UV irradiate for 3 min (17 mJ/mm^2 at 350 nm) (*see* **Note 18**).
3. Immediately proceed to the next step (**Subheading 3.4.5.**).

3.4.5. Identification, Excision, and Solubilization of Portion of Gel Containing RNAP–Promoter Complex

1. Remove one glass plate, and cover gel with plastic wrap (leaving the other glass plate in place). Attach two autorad markers. Expose to X-ray film for 1.5 h at room temperature (*see* **Note 19**). Process the film.
2. Cut out the portion of film corresponding to the RNAP–promoter complex of interest. Using a light box, superimpose the cut-out film on gel, using autorad markers as reference points. Using disposable scalpel, excise the portion of gel corresponding to the RNAP–promoter complex. Transfer excised gel slice to a 1.5-mL siliconized microcentrifuge tube.
3. Solubilize the gel slice by adding 10 µL of 1 *M* DTT (approx 0.4 *M* final) and heating for 5 min at 37°C (*see* **Note 20**).
4. Immediately proceed to the next step (**Subheading 3.4.6.**).

3.4.6. Nuclease Digestion

1. During X-ray film exposure of **step 1** of **Subheading 3.4.5.**, dilute DNase I and micrococcal nuclease with ice cold nuclease dilution solution to a final concentration of 10 U/µL.
2. Transfer 10 µL of the solubilized gel slice (**Subheading 3.4.5., step 3**) to a new 1.5-mL siliconized polypropylene microcentrifuge tube and add 1 µL of 200 m*M* $CaCl_2$, 1 µL of 0.2 m*M* PMSF, 0.5 µL (5 U) micrococcal nuclease, and 0.5 µL (5 U) DNase I. Incubate for 20 min at 37°C. Terminate reaction by adding 3 µL of 5X SDS loading buffer and heating for 5 min at 100°C.
3. Immediately proceed to next step (**Subheading 3.4.7.**).

3.4.7. Analysis

1. Apply the entire sample (16 µL) to a 4–15% gradient polyacrylamide (37.5:1 acrylamide:*bis*-acrylamide) slab gel. As a marker, load 5 µL prestained protein molecular-weight markers into the adjacent lane. Electrophorese in SDS running buffer at 25 V/cm until the bromophenol blue reaches the bottom of the gel.
2. Dry gel, and autoradiograph or phosphorimage.

4. Notes

1. M13mp2*(ICAP-UV5)* carries the *lacP(ICAP-UV5)* promoter, a derivative of the *lacP* promoter having a consensus DNA site for CAP *(37)* and a consensus –10 element *(38)*. M13mp2*(ICAP-UV5)* was prepared from M13mp2-*lacP1(ICAP)* *(39)* by use of site-directed mutagenesis to introduce a consensus –10 element *(40)*. M13mp2*(ICAP-UV5)*-rev carries the *lacP(ICAP-UV5)* promoter in an orientation opposite to that in M13mp2*(ICAP-UV5)*. M13mp2*(ICAP-UV5)*-rev was prepared from M13mp2*(ICAP-UV5)* by excising the *Pvu*II-*Pvu*II segment corre-

sponding to positions –217 to -125 of *lacP(ICAP-UV5)* and inverting the *Pvu*II–*Pvu*II segment corresponding to positions –124 to +145 of *lacP(ICAP-UV5)*.

2. M13mp2*(ICAP-UV5)* and M13mp2*(ICAP-UV5)*-rev ssDNAs carry respectively the nontemplate and template strands of *lacP(ICAP-UV5)*. M13mp2*(ICAP-UV5)* and M13mp2*(ICAP-UV5)*-rev ssDNAs are prepared as in **ref. *7***.
3. The specified 10X digestion buffer is for *Pvu*II and *Hae*III. Use 10X digestion buffer recommended by supplier—omitting DTT (*see* **ref. *41***)—for other restriction enzymes.
4. Fluorescent light and daylight must be excluded. Low to moderate levels of incandescent light (e.g., from a single bench lamp with a 60-W tungsten bulb) are acceptable.
5. The derivatized oligodeoxyribonucleotide tolerates exposure to the Hitachi L-3000 diode-array HPLC UV detector. The derivatized oligodeoxyribonucleotide can be identified unambiguously by monitoring the UV-absorbance spectrum from 200 nm to 350 nm. The derivatized oligodeoxyribonucleotide exhibits an absorbance peak at 260 nm, attributable to DNA, and a shoulder at 300–310 nm, attributable to the azidophenacyl group.
6. The derivatization procedure yields two diastereomers in an approximately one-to-one ratio: one in which azidophenacyl is incorporated at the sulfur atom corresponding to the phosphate O1P, and one in which azidophenacyl is incorporated at the sulfur atom corresponding to the phosphate O2P (*see* **refs. *4*** and ***6***). Depending on oligodeoxyribonucleotide sequence and HPLC conditions, the two diastereomers may elute as a single peak, or as two peaks (e.g., at 24% and 25% solution B). In most cases, no effort is made to resolve the two diastereomers, and experiments are performed using the unresolved diastereomeric mixture. This permits simultaneous probing of protein–DNA interactions in the DNA minor groove (probed by the O1P-derivatized diastereomer) and the DNA major groove (probed by the O2P-derivatized diastereomer).
7. Phenyl azides are unstable at temperatures above 70°C. Avoid heating above 70°C.
8. *Hae*III digestion, which yields a DNA fragment corresponding to positions –141 to +63 of *lacP(ICAP-UV5)*, is used for preparation of DNA fragments derivatized between positions –80 and –1, inclusive. *Pvu*II digestion, which yields a DNA fragment corresponding to positions –124 to +145 of *lacP(ICAP-UV5)*, is used for preparation of DNA fragments derivatized between positions +1 and +80 inclusive. (Use of DNA fragments with >60 bp between the site of derivatization and the nearest DNA fragment end eliminates "nonspecific" crosslinking from the subpopulation of complexes having RNAP bound at a DNA-fragment end *[42]* rather than at the promoter.)
9. Pour 10 mL Ni:NTA–agarose suspension into a 20-mL Econo-Pac column. Remove snap-off tip at bottom and allow liquid to drain. Place the frit on the top of the column bed.
10. BAC is a disulfide-containing analog of bis-acrylamide *(43–45)* Polyacrylamide:BAC gels can be solubilized by addition of reducing agents ***(43–45)***. The synthesis of BAC in this chapter is adapted from **ref. *43***.

11. Acryloyl chloride reacts violently with water. Add acryloyl chloride in 0.5-mL portions, waiting 30 s between successive additions.
12. BAC is substituted for bis-acrylamide on a mole-equivalent, not mass-equivalent, basis ***(43–45)***. The solubility of BAC in water is increased by adding acrylamide before adding BAC and by performing additions at 60°C.
13. TEMED and ammonium persulfate concentrations are critical variables in the preparation of polyacrylamide:BAC gels ***(44,45)***. (Use of nonoptimal TEMED and ammonium persulfate concentrations in preparing of polyacrylamide:BAC results in difficulties in subsequently solubilizing gels.)
14. Siliconize notched glass plate by applying 30 µL SurfaSil siliconizing agent and spreading evenly with a Kimwipe.
15. Heating during polymerization yields polyacrylamide:BAC gels that are maximally solubilizable upon the addition of reducing agents ***(44,45)***. Heat the glass plates of the gel assembly evenly. (If necessary, use two task lamps.) Avoid heating above 70°C, as this can result in the formation of bubbles and/or detachment of gels from the glass plates.
16. Do not add loading buffer to the reaction mixture. The reaction mixture is sufficiently dense for loading (because of the presence of glycerol).
17. Ultraviolet irradiation is performed with both glass plates in place. The glass plates exclude wavelengths <300 nm, minimizing photodamage to the protein and DNA. It is important to verify that the plates exhibit absorbances of ≤1.5 AU at 320 nm (e.g., by sacrificing a glass plate and placing a piece in the cuvet holder of a UV/Vis spectrophotometer). Glass plates purchased from Aladin (San Francisco, CA, USA) have performed satisfactorily.
18. For in-gel UV irradiation of the RNAP–promoter intermediate complex, prechill photochemical reactor for 15 min in a 15°C cabinet.
19. Do not use tight-fitting X-ray autoradiography cassettes, which can squeeze and distort the gel on the glass plate during exposure. The Kodak X-ray exposure holder with intensifying screen has performed satisfactorily.
20. 2–4 *M* β-mercaptoethanol can be substituted for 1 *M* DTT.

Acknowledgments

The basic protocol for preparation of derivatized DNA fragments was developed by T. Lagrange ***(1)***, the basic protocol for preparation of RNAP was developed by H. Tang and K. Severinov ***(34,46)***, and the basic protocol for in-gel UV irradiation was developed by T.-K. Kim ***(2)***.

We thank K. Severinov for plasmids and T.-K. Kim, T. Lagrange, D. Reinberg, and K. Severinov for discussions; and we are grateful for the Howard Hughes Medical Institute Investigatorship and National Institutes of Health grant GM41376 to R.H.E. for financial support.

References

1. Lagrange, T., Kim, T. K., Orphanides, G., Ebright, Y., Ebright, R., and Reinberg, D. (1996) High-resolution mapping of nucleoprotein complexes by site-specific

protein–DNA photocrosslinking: organization of the human TBP–TFIIA–TFIIB–DNA quaternary complex. *Proc. Natl. Acad. Sci. USA* **93,** 10,620–10,625.

2. Kim, T.-K., Lagrange, T., Wang, Y.-H., Griffith, J., Reinberg, D., and Ebright, R. (1997) Trajectory of DNA in the RNA polymerase II transcription preinitiation complex. *Proc. Natl. Acad. Sci. USA* **94,** 12,268–12,273.

2a. Kim, T.-K., Ebright, R., and Reinberg, D. (2000) Mechanisms of ATP-dependent promoter melting by transcription factor IIH. *Science* **288,** 1418–1421.

3. Lagrange, T., Kapanidis, A., Tang, H., Reinberg, D., and Ebright, R. (1998) New core promoter element in RNA polymerase II-dependent transcription: sequence-specific DNA binding by transcription factor IIB. *Genes Dev.* **12,** 34–44.

4. Fidanza, J., Ozaki, H., and McLaughlin, L. (1992) Site-specific labeling of DNA sequences containing phosphorothioate diesters. *J. Am. Chem. Soc.* **114,** 5509–5517.

5. Yang, S.-W. and Nash, H. (1994) Specific photocrosslinking of DNA–protein complexes: identification of contacts between integration host factor and its target DNA. *Proc. Natl. Acad. Sci. USA* **91,** 12,183–12,187.

6. Mayer, A. and Barany, F. (1995) Photoaffinity cross-linking of TaqI restriction endonuclease using an aryl azide linked to the phosphate backbone. *Gene* **153,** 1–8.

7. Sambrook, J., Fritsch, E., and Maniatis, T. (1989) *Molecular Cloning: A Laboratory Manual.* Cold Spring Harbor Laboratory, Cold Spring Harbor, NY.

8. Wang, Y. and Stumph, W. (1998) Identification and topological arrangement of *Drosophila* proximal sequence element (PSE)-binding protein subunits that contact the PSEs of U1 and U6 small nuclear RNA genes. *Mol. Cell. Biol.* **18,** 1570–1579.

9. Kim, Y., Geiger, J., Hahn, S., and Sigler, P. (1993) Crystal structure of a yeast TBP/TATA-box complex. *Nature* **365,** 512–520.

10. Kim, J., Nikolov, D., and Burley, S. (1993) Co-crystal structure of TBP recognizing the minor groove of a TATA element. *Nature* **365,** 520–527.

11. Geiger, J., Hahn, S., Lee, S., and Sigler, P. (1996) Crystal structure of the yeast TFIIA/TBP/DNA complex. *Science* **272,** 830–836.

12. Tan, S., Hunziker, Y., Sargent, D., and Richmond, T. (1996) Crystal structure of a yeast TFIIA/TBP/DNA complex. *Nature* **381,** 127–134.

13. Nikolov, D., Chen, H., Halay, E., Usheva, A., Hisatake, K., Lee, D. K., et al. (1995) Crystal structure of a TFIIB–TBP–TATA-element ternary complex. *Nature* **377,** 119–128.

14. Bartholomew, B., Kassavetis, G., Braun, B., and Geiduschek, E. P. (1990) The subunit structure of *Saccharomyces cerevisiae* transcription factor IIIC probed with a novel photocrosslinking reagent. *EMBO J.* **9,** 2197–2205.

15. Bartholomew, B., Kassavetis, G., and Geiduschek, E. P. (1991) Two components of *Saccharomyces cerevisiae* transcription factor IIIB (TFIIIB) are stereospecifically located upstream of a tRNA gene and interact with the second-largest subunit of TFIIIC. *Mol. Cell. Biol.* **11,** 5181–5189.

16. Braun, B., Bartholomew, B., Kassavetis, G., and Geiduschek, E. P. (1992) Topography of transcription factor complexes on the *Saccharomyces cerevisiae* 5 S RNA gene. *J. Mol. Biol.* **228,** 1063–1077.

17. Kassavetis, G., Kumar, A., Ramirez, E., and Geiduschek, E. P. (1998) Functional and structural organization of Brf, the TFIIB-related component of the RNA polymerase III transcription initiation complex. *Mol. Cell. Biol.* **18,** 5587–5599.
18. Bell, S., and Stillman, B. (1992) ATP-dependent recognition of eukaryotic origins of DNA replication by a multiprotein complex. *Nature* **357,** 128–134.
19. Coulombe, B., Li, J., and Greenblatt, J. (1994) Topological localization of the human transcription factors IIA, IIB, TATA Box-binding protein, and RNA polymerase II-associated protein 30 on a class II promoter. *J. Biol. Chem.* **269,** 19,962–19,967.
20. Gong, X., Radebaugh, C., Geiss, G., Simon, S., and Paule, M. (1995) Site-directed photo-crosslinking of rRNA transcription initiation complexes. *Mol. Cell. Biol.* **15,** 4956–4963.
21. Pruss, D., Bartholomew, B., Persinger, J., Hayes, J., Arents, G., Moudrianakis, E., and Wolffe, A. (1996) An asymmetric model for the nucleosome: a binding site for linker histones inside the DNA gyres. *Science* **274,** 614–617.
22. Robert, F., Forget, D., Li, J., Greenblatt, J., and Coulombe, B. (1996) Localization of subunits of transcription factors IIE and IIF immediately upstream of the transcriptional initiation site of the adenovirus major late promoter. *J. Biol. Chem.* **271,** 8517–8520.
23. Forget, D., Robert, F., Grondin, G., Burton, Z., Greenblatt, J., and Coulombe, B. (1997) RAP74 induces promoter contacts by RNA polymerase II upstream and downstream of a DNA bend centered on the TATA box. *Proc. Natl. Acad. Sci. USA* **94,** 7150–7155.
24. Record, M. T., Reznikoff, W., Craig, M., McQuade, K., and Schlax, P. (1996) *Escherichia coli* RNA polymerase (Eσ70), promoters, and the kinetics of the steps of transcription initiation, in *Escherichia coli and Salmonella*, vol. 1 (Neidhart, F., ed.), ASM, Washington, DC, pp. 792-820.
25. deHaseth, P., Zupancic, M., and Record, M. (1998) RNA polymerase–promoter interactions. *J. Bact.* **180,** 3019–3025.
26. Ebright, R. (1998) RNA polymerase–DNA interaction: structures of intermediate, open, and elongation complexes. *Cold Spring Harbor Symp. Quant. Biol.* **63,** 11–20.
27. Buc, H. and McClure, W. (1985) Kinetics of open complex formation between *Escherichia coli* RNA polymerase and the *lac* UV5 promoter. Evidence for a sequential mechanism involving three steps. *Biochemistry* **24,** 2712–2723.
28. Spassky, A., Kirkegaard, K., and Buc, H. (1985) Changes in the DNA structure of the *lac* UV5 promoter during formation of an open complex with *Escherichia coli* RNA polymerase. *Biochemistry* **24,** 2723–2731.
29. Kirkegaard, K., Buc, H., Spassky, A., and Wang, J. (1983) Mapping of single-stranded regions in duplex DNA at the sequence level: single-strand-specific cytosine methylation in RNA polymerase–promoter complexes. *Proc. Natl. Acad. Sci. USA* **80,** 2544–2548.
30. Becker, M. and Wang, J. (1984) Use of light for footprinting DNA *in vivo. Nature* **309,** 682–687.
31. Straney, D. and Crothers, D. (1985) Intermediates in transcription initiation from the E. coli lac UV5 promoter. *Cell* **43,** 449–459.

32. Severinov, K., Mustaev, A., Severinova, E., Bass, I., Kashlev, M., Landick, R., et al. (1995) Assembly of functional *Escherichia coli* RNA polymerase containing β subunit fragments. *Proc. Natl. Acad. Sci. USA* **92,** 4591–4595.
33. Severinov, K., Mustaev, A., Kukarin, A., Muzzin, O., Bass, I., Darst, S., et al. (1996) Structural modules of the large subunits of RNA polymerase. Introducing archaebacterial and chloroplast split sites in the β and β' subunits of *Escherichia coli* RNA polymerase. *J. Biol. Chem.* **271,** 27,969–27,974.
34. Tang, H., Severinov, K., Goldfarb, A., and Elbright, E. (1995) Rapid RNA polymerase genetics: one-day, no-column preparation of reconstituted recombinant Escherichia coli RNA polymerase. *Proc. Natl. Acad. Sci. USA* **92,** 4902–4906.
35. Martin, E., Sagitov, V., Burova, E., Nikiforov, V., and Goldfarb, A. (1992) Genetic dissection of the transcription cycle. A mutant RNA polymerase that cannot hold onto a promoter. *J. Biol. Chem.* **267,** 20,175–20,180.
36. Zalenskaya, K., Lee, J., Chandrasekhar, N. G., Shin, Y. K., Slutsky, M., and Goldfarb, A. (1990) Recombinant RNA polymerase: inducible overexpression, purification and assembly of *Escherichia coli rpo* gene products. *Gene* **89,** 7–12.
37. Ebright, R., Ebright, Y., and Gunasekera, A. (1989) Consensus DNA site for the *Escherichia coli* catabolite gene activator protein (CAP): CAP exhibits a 450-fold higher affinity for the consensus DNA site than for the *E. coli lac* DNA site. *Nucleic Acids Res.* **17,** 10,295–10,305.
38. Gilbert, W. (1976) Starting and stopping sequences for the RNA polymerase, in *RNA Polymerase* (Losick, R., and Chamberlin, M., eds.), Cold Spring Harbor Laboratory, Cold Spring Harbor, NY, pp. 193–206.
39. Zhang, X., Zhou, Y., Ebright, Y., and Ebright, R. (1992) Catabolite gene activator protein (CAP) is not an acidic-activating-region transcription activator protein: negatively charged amino acids of CAP that are solvent-accessible in the CAP–DNA complex play no role in transcription activation at the *lac* promoter. *J. Biol. Chem.* **267,** 8136–8139.
40. Kunkel, T. (1985) Rapid and efficient site-specific mutagenesis without phenotypic selection. *Proc. Natl. Acad. Sci. USA* **82,** 488–492.
41. Staros, J., Bayley, H., Standring, D., and Knowles, J. (1978) Reduction of aryl azides by thiols: implications for the use of photoaffinity reagents. *Biochem. Biophys. Res. Commun.* **80,** 568–572.
42. Melancon, P., Burgess, R., and Record, M. (1983) Direct evidence for the preferential binding of *Escherichia coli* RNA polymerase holoenzyme to the ends of deoxyribonucleic acid restriction fragments. *Biochemistry* **22,** 5169–5176.
43. Hansen, J. N. (1976) Electrophoresis of ribonucleic acid on a polyacrylamide gel which contains disulfide cross-linkages. *Anal. Biochem.* **76,** 37–44.
44. Hansen, J. N. (1980) Chemical and electrophoretic properties of solubilizable disulfide gels. *Anal. Biochem.* **105,** 192–201.
45. Hansen, J. N. (1981) Use of solubilizable acrylamide disulfide gels for isolation of DNA fragments suitable for sequence analysis. *Anal. Biochem.* **116,** 146–151.
46. Tang, H., Kim, Y., Severinov, K., Goldfarb, A., and Ebright, R. (1996) *Escherichia coli* RNA polymerase holoenzyme: rapid reconstitution from recombinant α, β, β', and σ subunits. *Methods Enzymol.* **273,** 130–134.

25

Site-Directed DNA Photoaffinity Labeling of RNA Polymerase III Transcription Complexes

Jim Persinger and Blaine Bartholomew

1. Introduction

Site-specific DNA photoaffinity labeling is a useful technique for mapping interactions of proteins with DNA in complex systems such as the yeast RNA polymerase III (Pol III) transcription complex, which consists of at least 25 different proteins *(1,2)*. This technique allows probing of protein–DNA interactions across large stretches of DNA and can be done in relatively crude extracts. The regions or domains of the protein contacting DNA can be identified by peptide mapping of the photoaffinity labeled protein. Our discussion of DNA photoaffinity labeling will focus on (1) the synthesis of photoreactive nucleotide analogs, (2) the manner in which the photoreactive nucleotide is incorporated into DNA, and (3) experimental details of DNA photoaffinity labeling.

Our group has used this technique to map the locations of many of the proteins of the Pol III transcription complex to sites within the *SUP4* tRNA^{Tyr} gene *(3–5)*. Some of the advantages of this approach are (1) detailed mapping of protein interactions with DNA in large multisubunit protein–DNA complexes and (2) the ability to use crude protein extracts potentially containing important auxiliary factors that may be lost upon purification. Solid-phase DNA probe synthesis allows for the synthesis of multiple probes in a single day, whereas in the past, this process would have taken several days. The synthesis of modified analogs for dATP, dCTP, and dTTP allow for the incorporation at nearly all positions in DNA. The photoreactive moiety can be changed on these nucleotides to place the more photoreactive phenyl diazirine into DNA to better target all potential protein surfaces. These photoreactive groups have short half-lives of less than 1 ns to approx 5 μs and can be used for kinetic

From: *Methods in Molecular Biology, vol. 148: DNA–Protein Interactions: Principles and Protocols, 2nd ed*
Edited by: T. Moss © Humana Press Inc., Totowa, NJ

analysis of changes in specific protein–DNA contacts. The 4-thiothymidine nucleotide has also been used for zero-distance crosslinking of protein to DNA and may be use useful for probing very close protein–DNA contacts *(6)*.

2. Materials

2.1. Synthesis of Modified Nucleotides

1. *Para*-azidobenzoic acid (4-ABA) (Molecular Probes).
2. Succinimidyl esters of 4-azidobenzoic acid, 4-azido-2,3,5,6-tetrafluorobenzoic acid, and 4-benzoylbenzoic acid (Molecular Probes).
3. 5-[*N*-(3-Aminoallyl)]-deoxyuridine triphosphate (5-aa-dUTP) (Sigma).
4. dCTP (Sigma).
5. Ethylene diamine, dicyclohexylcarbodiimide, anhydrous dioxane (99+%), ethyl ether, and sodium metabisulfite (Aldrich).
6. DEAE–Sephadex A-25 resin (Pharmacia).
7. Glycine, glycyl glycine, and glycylglycyl glycine (Sigma).
8. pH indicator strip (Panpeha, Schleicher & Schull).
9. Polyethyleneimine (PEI)–cellulose thin-layer chromatography (TLC) plates (J. T. Baker, with fluorescence indicator).
10. TE: 10 m*M* Tris-HCl, pH 8.0, 1 m*M* EDTA.

2.2. Immobilized DNA Templates

1. Buffer A: 10 m*M* Tris-HCl (pH 8.0), 10 m*M* $MgCl_2$, 50 m*M* NaCl, and 1 m*M* dithiothreitol (DTT).
2. Buffer B: 1 *M* LiCl, 10 m*M* Tris-HCl (pH 8.0), 1 m*M* EDTA, and 0.1% sodium dodecyl sulfate (SDS).
3. Plasmid DNA pTZ1 containing the *SUP4* $tRNA^{Tyr}$ gene with promoter-up mutation inserted into pGEM1 *(7)*.
4. Magnetic separation stand for DNA bead isolation (Promega).
5. M-280 Streptavidin Dynabeads (Dynal).
6. Buffer C: 2 *M* NaCl, 10 m*M* Tris-HCl (pH 7.5), and 1 m*M* EDTA (pH 8.0).
7. Buffer D: 30 m*M* Tris-HCl (pH. 8.0), 50 m*M* KCl, 7 m*M* $MgCl_2$, 1 m*M* of 2-mercaptoethanol, and 0.05% Tween-20.
8. Polystrene chromatography columns (5 in. [12.5 cm]) with a 45- to 90-µm filter (Evergreen Scientific). These disposable columns are ideal for the 2.5-mL spin columns.
9. Bio-11–dUTP and Bio-14–dATP (Sigma).

2.3. DNA Probe Synthesis

1. Buffer E: 150 m*M* Tris-HCl (pH 8.0), 250 m*M* KCl, 35 m*M* $MgCl_2$, 5 m*M* of 2-mercaptoethanol, and 0.25% Tween-20.
2. Storage buffer F: 50 m*M* potassium phosphate (pH 7.0), 5 m*M* of 2-mercaptoethanol, and 50% glycerol.

3. Storage buffer G: 50 m*M* KCl, 10 m*M* Tris (pH 7.5), 0.1 m*M* EDTA, 5 m*M* of 2-mercaptoethanol, 200 mg/mL bovine serum albumin (BSA), and 50% glycerol.
4. Site-specific oligonucleotides and upstream oligonucleotide (50-nmol-scale synthesis).
5. 4-Thiothymidine triphosphate (Amersham/Pharmacia).
6. Exonuclease-free version of the Klenow fragment of DNA Polymerase I (Amersham/Pharmacia, 5 U/µL) diluted to 0.25 U/µL with storage buffer F.
7. T_4 DNA ligase (New England Biolabs, high concentration form, 2000 U/µL) diluted to approx 300 U/µL with storage buffer G containing 5 m*M* of 2-mercaptoethanol.
8. TE: 10 m*M* Tris-HCl (pH 8.0) and 1 m*M* EDTA.
9. PBS: phosphate-buffered saline solution (pH 7.4).
10. T_4 DNA polymerase from New England Biolabs, comes stored in 100 m*M* potassium phosphate (pH 6.5), 10 m*M* of 2-mercaptoethanol, and 50% glycerol.

2.4. Photoaffinity Labeling

1. Buffer H: 100 m*M* Tris-HCl (pH 8.0), 25 m*M* $MgCl_2$ and 250 m*M* NaCl.
2. Buffer I: 100 m*M* NaCl, 40 m*M* Tris-HCl (pH 8.0), 5 m*M* $MgCl_2$, 1 m*M* EDTA, 20% glycerol, 10 m*M* of 2-mercaptoethanol, 0.5 m*M* phenylmethylsulfonyl fluoride (PMSF), 1 µg/mL pepstatin, and 1 µg/mL leupeptin.
3. Zinc acetate solution: 0.5 *M* glacial acetic acid and 12.5 m*M* zinc acetate.
4. 5X DB: 10% SDS, 25% 2-mercaptoethanol, 0.3 *M* Tris-HCl (pH 6.8), 0.4% bromophenol blue.
5. 3 NTP mix: 2 µL of 100 m*M* ATP, UTP, and CTP (Boehringer Mannheim), 20 µL buffer H, 35 µL Buffer I, and 43 µL sterile deionized water.

2.5. Peptide Mapping

1. Formic acid (99%, Sigma).
2. Diphenylamine (ACS 99+%, Aldrich).
2. Cyanogen bromide (97%, Aldrich.
3. Centricon 30 (Millipore).

3. Methods

Synthesis of modified nucleotides and DNA photoaffinity probes is done with indirect lighting conditions using 40-W incandescent lamps.

3.1. Synthesis of Modified Nucleotides

We have used a variety of modified nucleotides to probe the RNA polymerase III transcription complex. This section contains procedures for the synthesis of some of the commonly used nucleotides.

3.1.1. AB–dUTP

AB–dUTP (**Fig. 1**) is synthesized as follows:

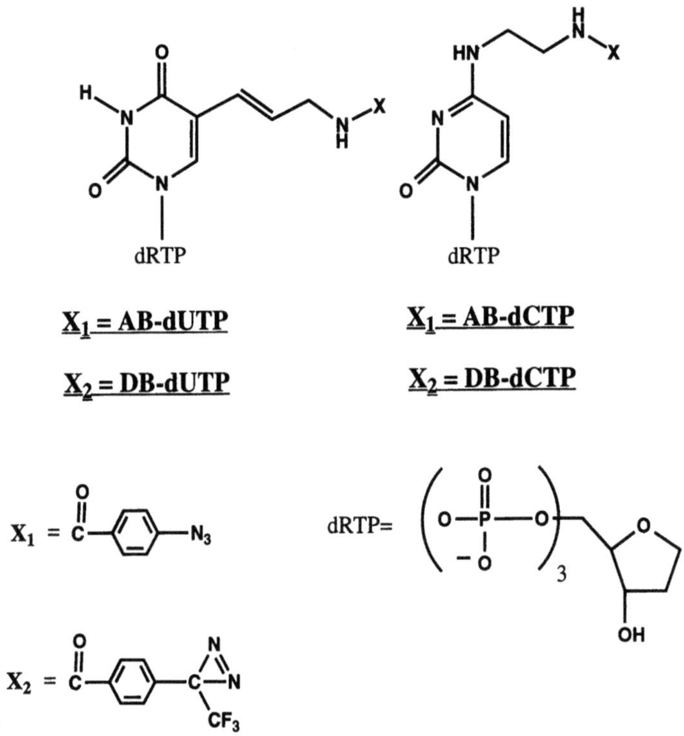

Variable Chain Length

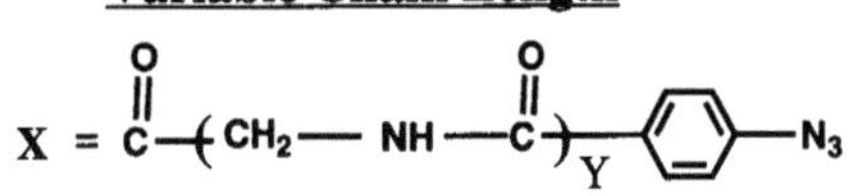

	dUTP analogs	dCTP analogs	Chain Length
Y= 0	AB-dUTP	AB-dCTP	[~10.0 Å]
Y= 1	ABG-dUTP	ABG-dCTP	[~14.3 Å]
Y= 2	ABG_2-dUTP	ABG_2-dCTP	[~18.6 Å]
Y= 3	ABG_3-dUTP	ABG_3-dCTP	[~22.9 Å]

Fig. 1. Structures of each of the photoreactive nucleotide analogs are shown. The complete IUPAC names for each nucleotide is given in the text.

1. Add 100 µL of 100 m*M* 4-azidobenzoic acid *N*-hydroxysuccinimide solution (ABA-NHS) in dimethylformamide (DMF) to 100 µL of 20 m*M* 5-aa-dUTP in 100 m*M* sodium borate (pH 8.5) *(8)*. The reaction is incubated at 25°C for 4 h,

and the pH of the reaction is checked with pH indicator strips (pH 8.5). Any excess precipitation that forms can be eliminated by addition of DMF.
2. The coupling reaction is stopped by the addition of 200 µL sterile deionized water and the product applied to a 0.7 × 8-cm (1.6-mL) DEAE–Sephadex A-25 column equilibrated in 100 m*M* TEAB (pH 8.0). The column is washed with 5 mL of the same buffer at a flow rate of 5 mL/h, and eluted with a 30-mL linear gradient of 0.1 to 1.5 *M* TEAB (pH 8.0), and 500 µL fractions are collected.
3. Every third fraction from the column is evaporated to dryness by vacuum centrifugation and resuspended in 250 µL of sterile deionized water. The samples are dried down and this process repeated a further time.
4. The fractions are resuspended in 50 µL of sterile deionized water and 2 µL of each analyzed on PEI-cellulose TLC plates (*see* **Note 1**). The plates were developed with 1 *M* LiCl and visualized with ultraviolet (UV) light source (254 n*M*). The reported R_f values for 5-aa-dUTP and AB–dUTP are 0.54 and 0.098, respectively *(8)*.
5. All of the fractions containing AB–dUTP are combined and TEAB is removed by repeated drying and resuspension in sterile deionized water. The final product is resuspended in 200 µL of TE. An estimated extinction coefficient of AB–dUTP at 270 nm is 10.3×10^3/M/cm at pH 8.0 (based on the sum of the extinction coefficients of ABA and 5-aa-dUTP at the indicated pH and wavelength). A concentrated stock of AB–dUTP is stable at –80°C for several years, and a 0.2-m*M* working stock can be stored at –20°C wrapped in foil (*see* **Notes 2–4**).

3.1.2. Varied Tether-Length Nucleotides

The tether of AB–dUTP is 9–10 Å in length and places the photoreactive group near the edge of the major groove of DNA. We have synthesized different dUTP and dCTP analogs with varying tether lengths by the addition of glycine residues into the tether *(4)*. Synthesis of these nucleotide analogs is similar to that of AB–dUTP and is as follows (**Fig. 1**).

Para-azidobenzoic acid (4-ABA) was esterified with *N*-hydroxysuccinimide (NHS) using the coupling reagent dicyclohexylcarbodiimide (DCI) and the product was recrystallized from anhydrous dioxane and ethyl ether (1:1) *(9)*.

1. A typical reaction contained 28 mmol ABA and 28 mmol NHS in 50 mL of anhydrous dioxane (99+%).
2. The solution is cooled in ice, and DCI in 15 mL dioxane is added and stirred for approx 24 h at room temperature.
3. Dicyclohexyl urea is removed by centrifugation and the supernatant was evaporated to dryness by vacuum centrifugation.
   The ABA-NHS is coupled to glycine, (Gly-Gly), or (GlyGlyGly) to make ABA derivatives with glycine, Gly-Gly, or Gly-Gly-Gly, respectively, attached to the carboxylic group of 4-ABA.
4. The reaction is started on ice and contains 2 mmol of glycine, the Gly-Gly, or Gly-Gly-Gly and 4 mmol of sodium bicarbonate in 4 mL of deionized

water to which is added 2 mmol of ABA-NHS in 8 mL of dioxane with constant stirring.

5. The reaction is allowed to proceed for 10–15 min on ice and is then transferred to room temperature and left with stirring for an additional 24 h.
6. Any insoluble material is removed from the reaction by centrifugation.
7. The pH of the reaction is lowered to 2 with concentrated HCl to precipitate the product.
8. Products are washed with deionized water.
9. The products are esterified with *N*-hydroxysuccinimide as described for ABA (**Subheading 3.1.1.**) except that dimethyl sulfoxide is used instead of dioxane for the ABG_3-NHS because of the limited solubility of this compound.
10. The dimethyl sulfoxide (DMSO) or dioxane is removed by vacuum centrifugation and the product is recrystallized from dioxane/isopropyl alcohol (1:1). Any residual solvent is removed by vacuum centrifugation. These products are coupled to 5-aa-dUTP in the same fashion as described for AB–dUTP (**Subheading 3.1.1.**).

3.1.3. Varied Photochemistry Nucleotides

We have also varied the photoreactive group attached to 5-aa-dUTP to contain either a phenyldiazirine, tetraflouro aryl azide, or a benzophenone group to optimize for nonselective crosslinking *(5)*. The coupling reactions of 5-aa-dUTP to the NHS esters of 4-azido-2,3,5,6-tetrafluorobenzoic acid, 4-benzoylbenzoic acid (commercially available from Molecular Probes), and 4-[3-9trifluoromethyl)diazirin-3-yl]benzoic acid (synthesized as described in **ref. *10***) are similar to that for the synthesis of AB–dUTP (**Fig. 1**).

3.1.4. Synthesis of dCTP Analogs

The synthesis of dCTP nucleotides begins with the synthesis of N^4-aminoethyl deoxycytidine triphosphate (daeCTP) by a bisulfite-catalyzed transamination reaction.

1. A bisulfite-amine solution is made by adding dropwise 2 mL of freshly distilled ethylene diamine to 4.0 mL of concentrated HCl and 3.5 mL of deionized water on ice.
2. Next, sodium meta-bisulfite (1.895 g) is added and the pH is adjusted to 5.0 with concentrated HCl.
3. Then, 100 µL of 1 mg/mL hydroquinone in ethanol is added to the reaction to scavenge free radicals. Bisulfite-amine solutions are always made up fresh.
4. The transamination reaction is initiated by adding 9 vol of the bisulfite-amine solution to 1 vol of 100 m*M* dCTP in 50 m*M* TEAB (pH 8.0).
5. The sample is incubated with constant vortexing at 42°C for 4 h.
6. The reaction is stopped by adjusting the pH to 8.2 with 5 *M* KOH.
7. The product is purified by DEAE–Sephadex A-25 chromatography as described for the purification of AB–dUTP (**Subheading 3.1.1.**) and dae-dCTP eluted from 0.84 *M* to 1.0 *M* TEAB.

8. Column fractions are analyzed by TLC as described in **Subheading 3.1.1.** The R_f of daeCTP is 0.221 *(3)*.
9. Fractions containing product are pooled and concentrated to 10–12 m*M*. The aryl azido or phenyl diazirine (**Fig. 1**) are coupled to daeCTP as discussed in **Subheading 3.1.1.** for 5-aa-dUTP. Tether length versions of AB–dCTP have also been synthesized with similar lengths of tether to those for the other tether-length nucleotides discussed (**Fig. 1**), and the R_f values of these range from 0.067 for ABG–dCTP to 0.078 for ABG_3–dCTP.

3.2. Immobilized DNA Templates

pTZ1 plasmid DNA is used for the synthesis of the *SUP4* $tRNA^{Tyr}$ DNA photoaffinity probes.

1. DNA is biotinylated by initially digesting 200 pmol of pTZ1 plasmid with either HindIII for nontranscribed strand templates or *Eco*RI for transcribed strand templates (*see* **Note 6**) (**Fig. 2**, step 1).
2. The 5' overhangs are biotinylated by the incorporation of Bio-11–dUTP and Bio-14–dATP (Sigma Chemical Co.) using the exonuclease-free version of the Klenow fragment of DNA Polymerase I (Amersham/Pharmacia). The 200-µL reaction contains 200 pmol of linearized pTZ1, 20 µ*M* Bio-14–dATP, dCTP, and dGTP, 25 µ*M* Bio-11–dUTP, and 150 U of Klenow fragment in buffer A (**Fig. 2**, step 2).
3. Unincorporated dNTPs are removed by spin-column chromatography *(11)* with a 2.5-mL Sephacryl S-200 spin column (Pharmacia) equilibrated in buffer B (**Fig. 2**, step 3). Aliquots of the samples are removed before and after the spin column to quantitate recovery.
4. Biotinylated DNA is precipitated by the addition of 2.5 vol of ethanol and placing the sample at –20°C overnight.
5. The sample is then resuspended and digested with *Eco*RI and *Rsa*I for nontranscribed strand templates or *Hind*III and *Pvu*2 for transcribed strand templates to generate a 315-base-pair biotinylated DNA fragment containing the *SUP4* $tRNA^{Tyr}$ gene (**Fig. 2**, step 4).

Biotinylated DNA (40 pmol) is bound to Dynabeads M-280 Streptavidin (Dynal) in the following procedure:

1. Washing Dynabeads
   a. The supernatant from 200 µL Streptavidin Dynabeads (10 mg/mL) is removed using a MagneSphere® Technology magnetic separation stand (Promega) and the beads are resuspended in 200 µL PBS + 0.1 mg/mL BSA.
   b. The beads are washed one time with 200 µL buffer C, and resuspended in 400 µL 2X B/W at 5 mg/mL (*see* **Note 7**).
2. Binding Reaction
   a. A reaction is assembled consisting of 400 µL of the 5-mg/mL washed Dynabeads and 400 µL of the 0.4-pmol/µL biotinylated DNA (*see* **Note 8**).

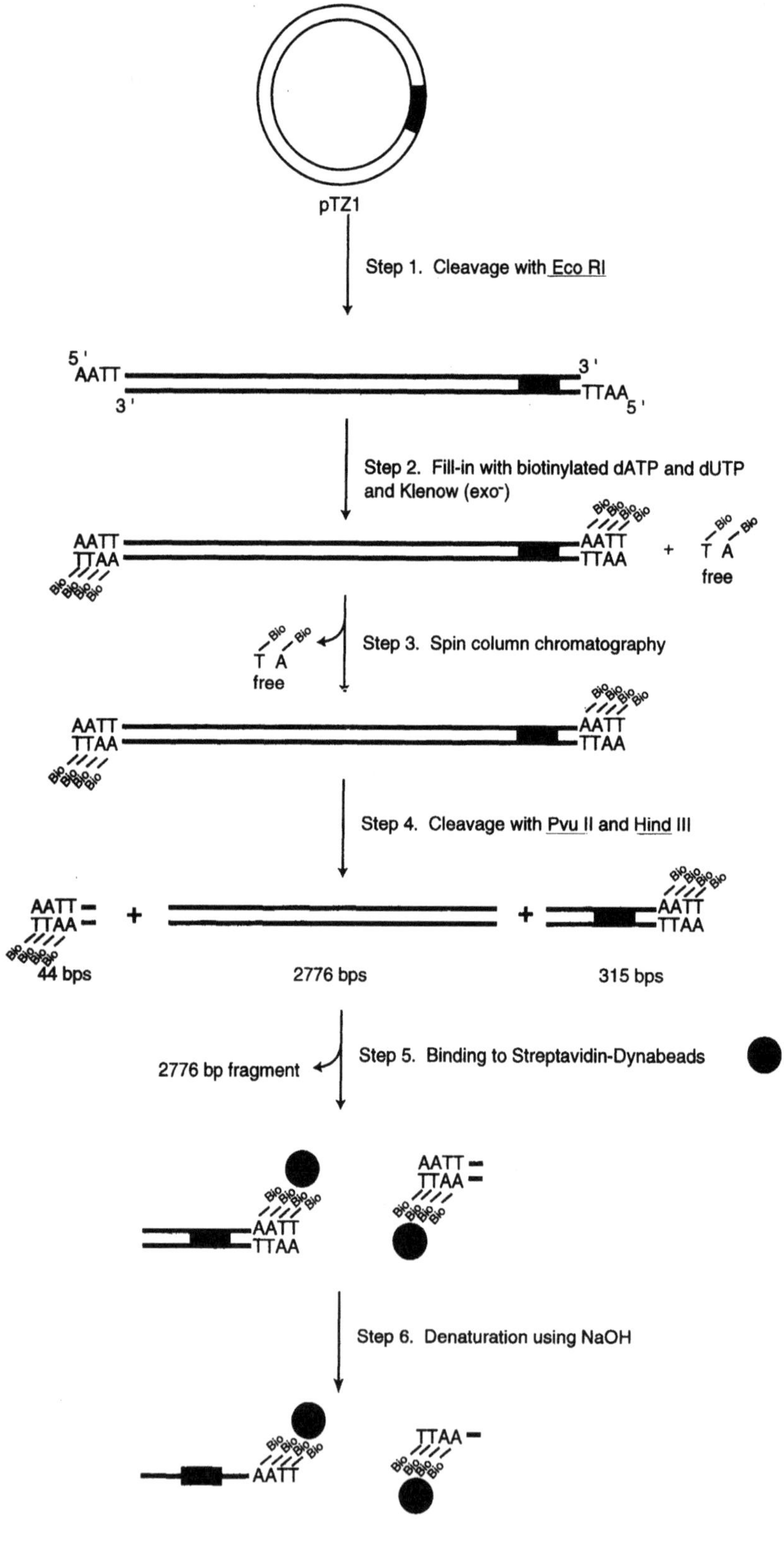
pTZ1
Step 1. Cleavage with Eco RI
5'
AATT
3'
3'
TTAA
5'
Step 2. Fill-in with biotinylated dATP and dUTP and Klenow (exo⁻)
AATT
TTAA
Bio
+
T A
free
Step 3. Spin column chromatography
Step 4. Cleavage with Pvu II and Hind III
44 bps
2776 bps
315 bps
Step 5. Binding to Streptavidin-Dynabeads
2776 bp fragment
Step 6. Denaturation using NaOH

   b. The reaction is mixed by gentle vortexing every 5 min during a 45-min incubation at 37°C (*see* **Note 9**).
   c. Buffer is removed using a magnetic stand and analyzed for binding efficiency by agarose gel electrophoresis (**Fig. 3**, lanes 2 and 5).
   d. Beads are washed with 200 µL of 0.5X buffer C and the washes saved for further analysis (**Fig. 3**, lanes 3 and 6).
   e. The beads are resuspended in 100 µL of 0.5X buffer C and stored at 4°C (**Fig. 2**, step 5) for extended periods of time.
3. Stripping off nonbiotinylated DNA strand
   a. Supernatant is removed from the beads and the beads are resuspended in 20 µL of 0.1 *M* NaOH.
   b. The sample is incubated with occasional vortexing for 10 min at room temperature.
   c. The supernatant is removed and beads are washed one time with 50 µL of 0.1 *M* NaOH and washed one time with sterile deionized water (**Fig. 2**, step 6).
4. Dephosphorylating DNA Beads
   a. DNA beads are washed three times with 50 µL buffer D and resuspended in 100 µL buffer D.
   b. Five units of shrimp alkaline phosphatase (Amersham/Pharmacia) is added to the reaction and incubated at 37°C for 60 min.
   c. Another 5 units of enzyme is added and incubated for an additional 60 min.
   d. Beads are washed three times with 50 µL of TE (pH 8.0) + 0.1% SDS.
   e. The beads are washed three times with 50 µL PBS + 0.1 mg/mL BSA and resuspended in 100 µL PBS + 0.1 mg/mL BSA (final concentration, 0.2 pmol/µL).
   f. The remaining enzyme is heat inactivated by incubating the reaction at 65°C for 15 min and gently vortexing every 5 min. The single-stranded DNA beads can be stored for several months up to 1 yr at 4°C.

3.3. DNA Probe Synthesis

Immobilized DNA templates made by the previous procedure (**Subheading 3.2.**) are used to construct DNA photoaffinity probes on either the transcribed or nontranscribed strand of the gene in the following procedure.

1. First primer extension. Remove 1 pmol of immobilized template and place in a 1.5-mL microcentrifuge tube.
2. Wash the beads three times with buffer D and resuspend in 10 µL of buffer D.
3. To each reaction add 5 µL of sterile deionized water, 1 µL of 2 mg/mL BSA, 2 µL buffer E, and 2 µL of 2 pmol/µL site-specific oligonucleotide (*see* **Note 10**).
4. Reactions are vortexed and heated at 70°C for 3 min and vortexed gently again before placing at 37°C for 30 min to allow the oligonucleotide primer to anneal

---

Fig. 2. *(previous page)* Diagrammatic representation DNA template preparation for solid-phase DNA probe synthesis.

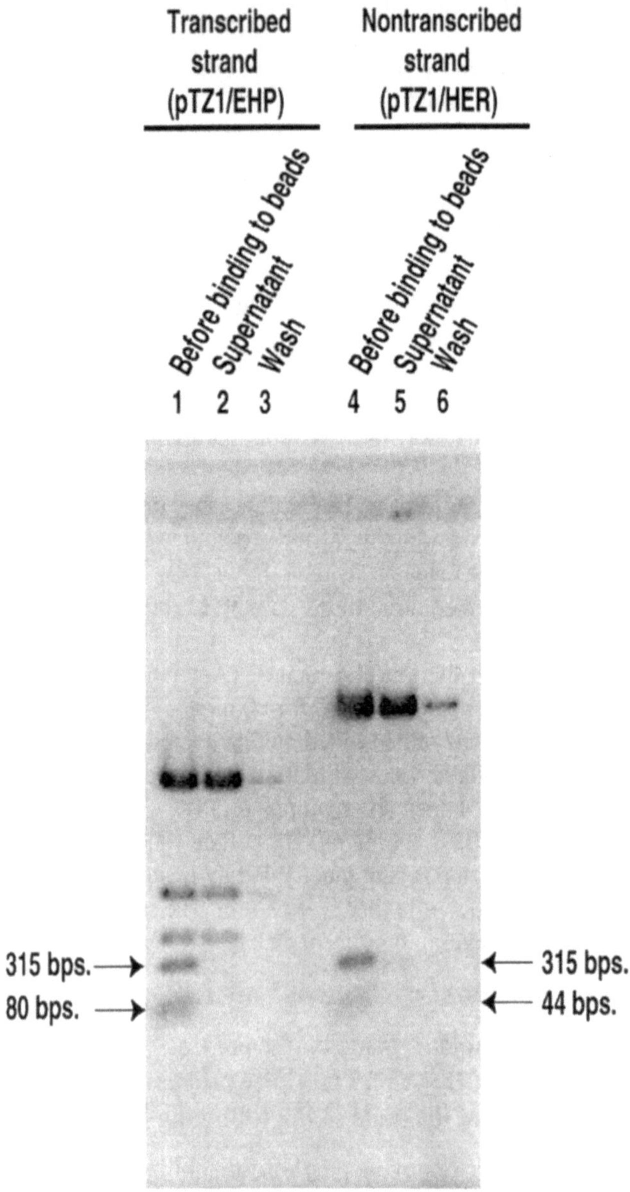

Fig. 3. Agarose gel analysis of DNA bead preparation. Lanes 1 and 4 are biotinylated pTZ1 DNA cut with *Eco*RI, *Hind*III, and *Pvu*II (pTZ1/EHP) used for the modification of the transcribed strand, or *Hind*III, *Eco*RI, and *Rsa*I (pTZ/HER) used for the modification of the nontranscribed strand before binding to Streptavidin Dynabeads. The biotinylated 315-base-pair fragments contain the *SUP4* tRNA^{Tyr} gene. The 80- and 40-base-pair fragments are also biotinylated, but are not used as a DNA template. Lanes 2 and 5 are taken from the supernatant of the binding reaction of DNA with Streptavidin Dynabeads after incubation. The abseπnce of the 315-base-pair fragments in these samples shows efficient binding of probe DNA to Dynabeads. Lanes 3 and 6 are samples from a final wash of the beads after binding.

(**Fig. 4**, step 1). During all incubations, samples must be vortexed gently every 5 min to keep the beads resuspended.

5. To the reaction add 1 µL of 100 µ*M* modified nucleotide described in **Subheading 3,** an α-[P^{32}]-labeled nucleotide, and 0.25 U of the exonuclease-free version of the Klenow fragment of DNA polymerase I (Amersham/Pharmacia) (*see* **Note 11**) and incubate at 37°C for 5 min (*see* **Note 12**) (**Fig. 4**, step 2 and **Fig. 5**, lanes 1–4).
6. Next, full-length extension is done by the addition of dNTPs to a final concentration of 0.5 m*M* and incubation for 5 min at 37°C (**Fig. 4**, step 3 and **Fig. 5**, lanes 5–8).
7. Klenow fragment and dNTPs are removed by washing the beads three times with 50 µL of TE (pH. 8.0) + 0.1% SDS and two times each with 50 µL buffer D.
8. Beads are resuspended in 20 µL of buffer D.
9. Three microliters of 3 pmol/µL upstream oligonucleotide, 1 µL of 2 mg/mL BSA, and 1 µL of buffer E are added to a 20 µL reaction containing the DNA beads.
10. The sample is vortexed gently every 5 min and incubated at 37°C for 30 min (**Fig. 4**, step 4).
11. The upstream oligonucleotide is extended by the addition of 3 µL 5 m*M* dNTPs and 3 units of T4 DNA polymerase (*see* **Note 11**) and incubation for 10 min at 37°C (**Fig. 4**, step 5 and **Fig. 5**, lanes 9–12).
12. The upstream strand is ligated to the site-specific primer by the addition of 1 µL of 10 m*M* ATP and 5 U of T4 DNA ligase and incubated at 37°C for 60 min (**Fig. 4**, step 6 and **Fig. 5**, lanes 13–16).
13. Beads are washed two times with 50 µL TE (pH 8.0) + 0.1% SDS, two times with 50 µL of buffer D, and resuspended in 20 µL of buffer D.
14. Next, the reaction is heated at 65°C for 15 min to inactivate any residual enzyme.
15. The DNA probe is released from the bead by the addition of 12–20 U of specific restriction enzyme (*see* **Notes 11** and **13**) to cut at a site between the *SUP4* $tRNA^{Tyr}$ gene and the attachment site (**Fig. 4**, step 7 and **Fig. 5**, lanes 17–20).
16. After restriction enzyme digestion, the probe is washed from the beads in a series of three washes of 50 µL each with buffer D, which are pooled in a fresh microcentrifuge tube.
17. The sample is extracted with phenol:chloroform (1:1) followed by extraction with chloroform.
18. The DNA probe is ethanol precipitated by the addition of 1/10 vol of 10 *M* lithium chloride and 2.5 vol of ethanol.
19. Samples are placed at –20°C overnight.
20. Samples are spun down at maximum speed in microfuge at 4°C for 30 min.
21. The supernatant is decanted and the pellet is allowed to dry.
22. Samples are resuspended in TE (pH 8.0) + 0.05% Tween-20 at a final concentration of 2–10 fmol/µL and stored at 4°C.
23. After resuspension of the probes, 1 µL is removed for analysis on a 4% native acrylamide gel, 20 cm × 20 cm × 0.8 mm (**Fig. 6**).

3.4. DNA Photoaffinity Labeling

Transcription complexes were formed on probe DNA using the 500-m*M* KCl fraction from Bio-Rex 70 chromatography of the S-100 extract, made from

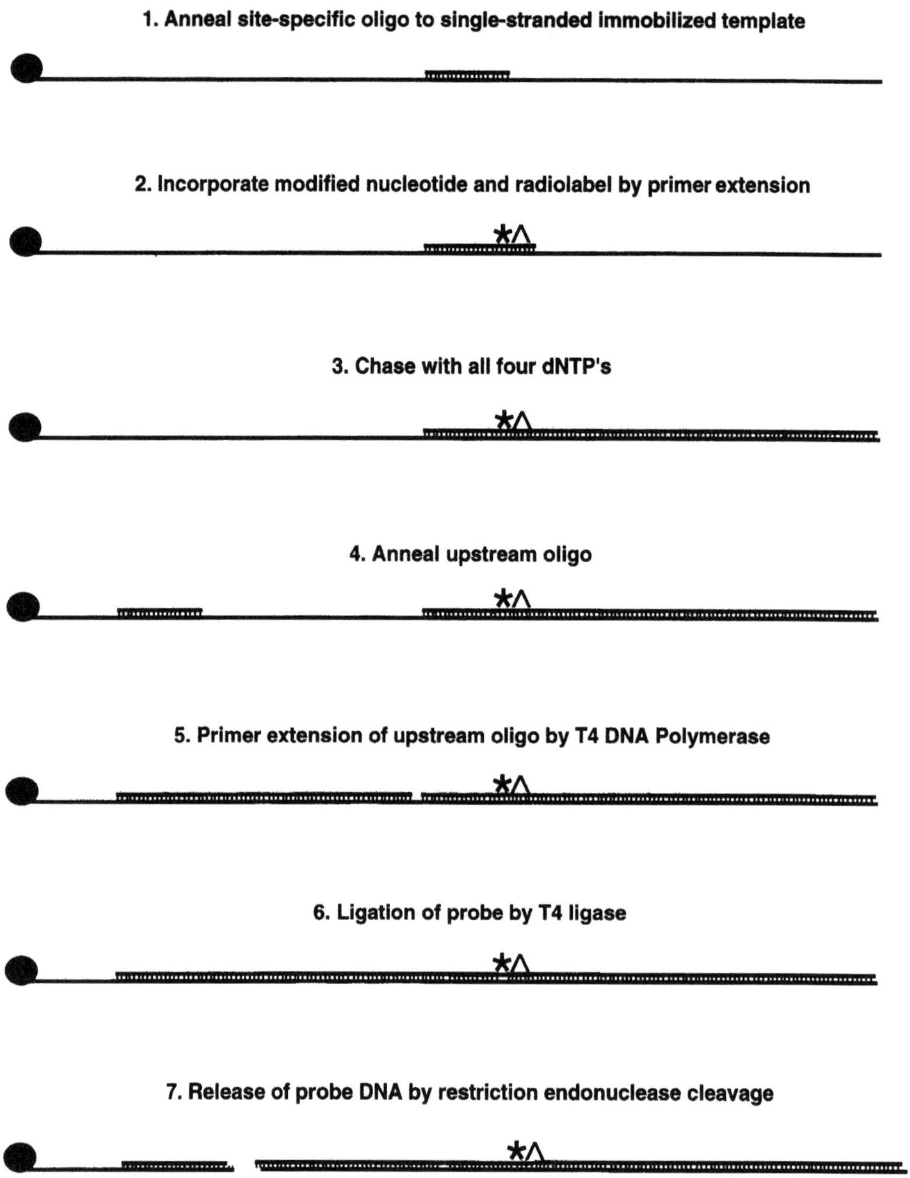

Fig. 4. Schematic representation of solid-phase DNA probe synthesis.

Fig. 5. *(opposite page)* Analysis of DNA probe synthesis modified at bps +11 on 10% urea PAGE. Incorporation of BP–dUMP, FAB–dUMP, DB–dUMP, and AB–dUMP and [α-^{32}P] dATP (lanes 1–4). Full-length extension of oligonucleotide primers in the presence of all four dNTPs (lanes 5–8). Upstream oligo annealed to template and extended to site-specific primer by T4 DNA polymerase in the presence

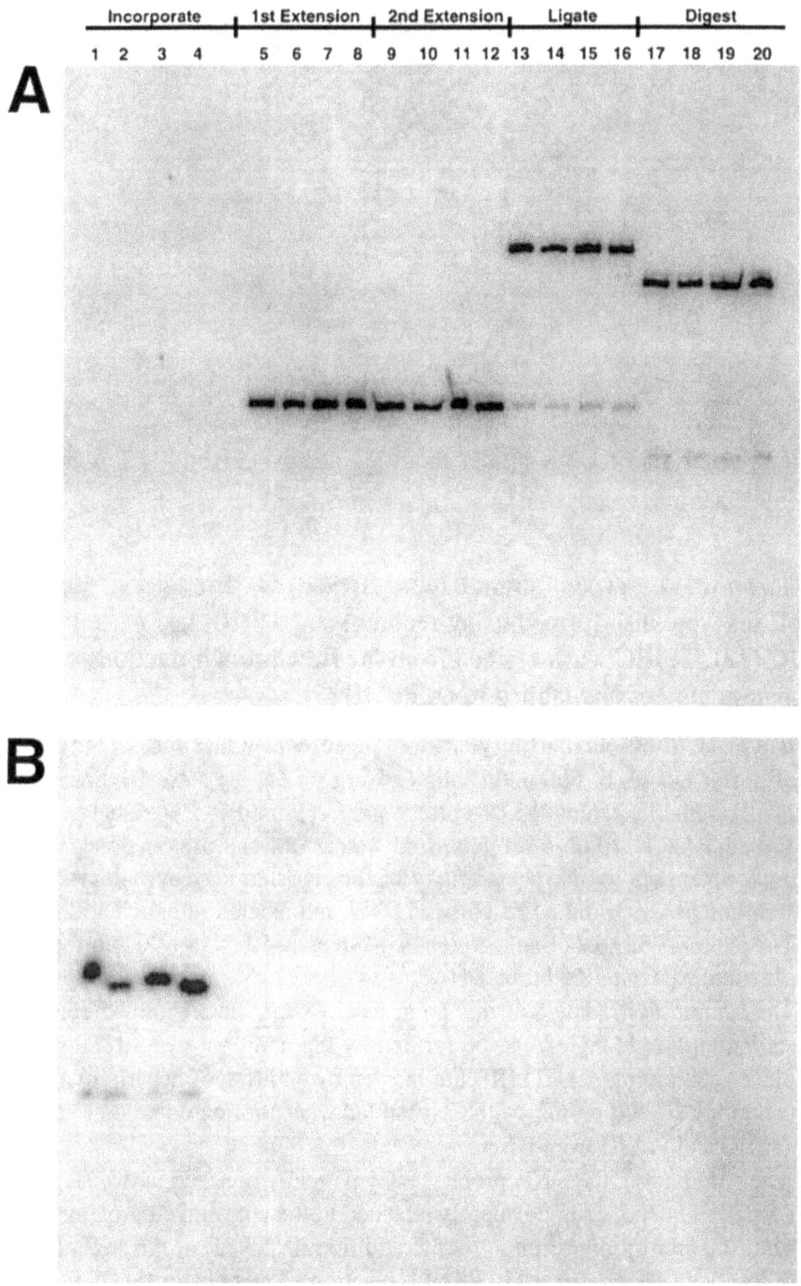

of all four dNTPs (lanes 9–12). Ligation of upstream extension product to site-specific oligonucleotide by T4 DNA ligase (lanes 13–16). Digestion of the probe DNA to release it from the Streptavidin Dynabeads by BamHI (lanes 17–20).

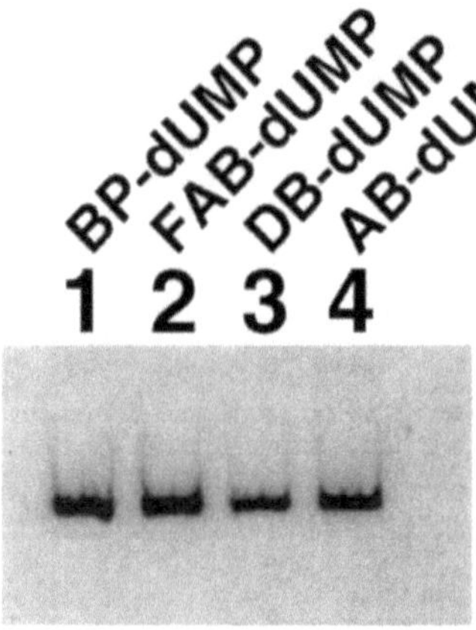

Fig. 6. Analysis of DNA probes modified at bps +11 on a 4% nondenaturing acrylamide gel.

Saccharomyces cerevisiae strain BJ926 (BR500) *(3)*. In addition, transcription complexes were also formed using recombinant TFIIIB and partially purified TFIIIC *(12)*. TFIIIC was obtained from the flowthrough fractions of Ni-NTA chromatography of His-tagged RNA Pol III *(3)*.

1. A typical 20-µL photoaffinity-labeling reaction contains 4 µL of buffer H, 9 µL of buffer I, 1 µL of 500 ng/µL pLNG-56 or pTZ1, see **Note 14**, linearized with *Eco*RI, 1 µL of a 2 fmol/µL DNA probe, and 1–4 µL of BR-500 extract (*see* **Note 15**) and adjusted to 20 µL with deionized water. Optimal protein concentration for transcription activity and photoaffinity labeling is determined by multiple round transcription assays using pTZ1 plasmid DNA and labeled ribonucleotides *(3)*.
2. The photoaffinity-labeling reaction is incubated at 25°C for 30 min for assembly of complexes onto the probe DNA.
3. The sample is irradiated at this point to crosslink the assembled complex or a stalled ternary complex can be formed by the addition of a 3 NTP mix. Complexes containing only TFIIIB are formed by addition of heparin to a final concentration of 100 µg/mL to the assembled transcription complex for release of TFIIIC and Pol III from DNA.
4. After irradiation, the DNA probe is enzymatically digested in two steps to leave a small radioactive tag covalently attached to the crosslinked protein.
5. The first step of digestion is by the addition of 2.3 µL of 0.5 mg/mL DNase I (Gibco Life Technologies) to a 21-µL reaction and incubation at 25°C for 10 min (*see* **Note 16**).
6. Immediately add 1 µL of 10% SDS to each sample and incubate at 90°C for 3 min and then place on ice for 5 min.
7. Next, 2 µL of the zinc acetate solution and 1 µL of 20 U/µL S1 nuclease (Gibco Life Technologies) is added, and samples are incubated at 37°C for 10 min.

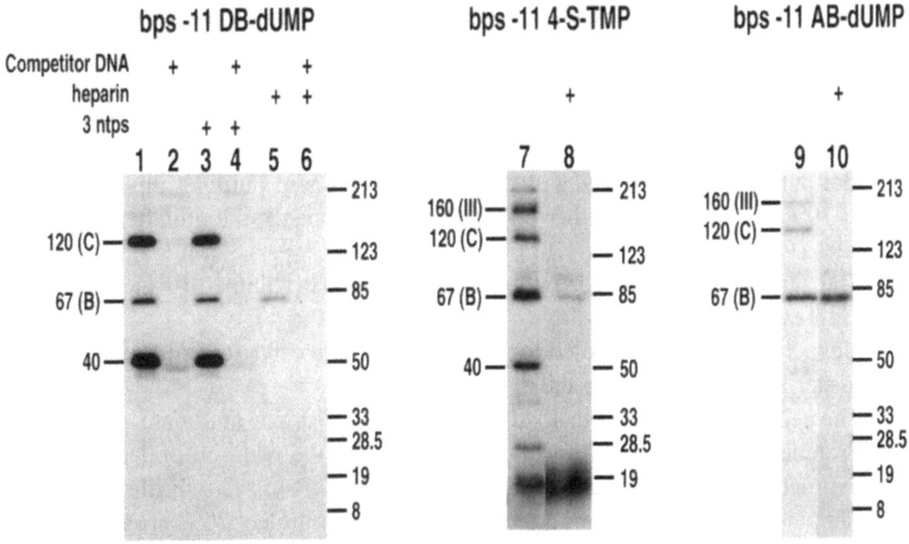

Fig. 7. Comparison of three photoreactive moieties incorporated at bps –11. Subunits labeled in both preinitiation (lanes 1, 7, and 9) and heparin-stripped complexes (lanes 5, 8, and 10) differ from one photoreactive group to the next. Examples of TFIIIC-specific competition (lanes 2, 4, and 6).

8. The reaction is stopped by the addition of 1 μL of 0.5 *M* Tris base to adjust the pH to approx 7.0 and 7 μL of 5X DB buffer.
9. The sample is heated at 90°C for 3 min and cooled on ice for 5 min prior to being loaded on to a 4–20% SDS-PAGE that is 20 × 20 cm × 0.8 mm.
10. After electrophoresis, the gel is dried with a slab gel dryer with heating at 80°C and vacuum for 2 h, and it is exposed to film for autoradiography or to a phosphorimager screen (*see* **Note 17**) (**Fig. 7**).

3.5. Peptide Mapping

1. Large-scale photoaffinity-labeling reactions for peptide mapping experiments are as described in **Subheading 3.4.**, except everything is scaled up to a final reaction volume of 2 mL (*see* **Note 18**).
2. Samples are irradiated in a multichannel pipettor tray instead of the original sample tube in order to keep the depth of the sample the same as a standard 20-μL labeling reaction.
3. DNase I and S1 nuclease digestion is done as described in **Subheading 3.4.** with volumes scaled up to the appropriate amounts for the increased reaction size.
4. Next, the samples are concentrated by ultrafiltration using a Centricon 30 (Millipore) to lower the sample volume and allow for loading on a 0.8-mm-thick 8% SDS polyacrylamide gel with a 1.4-cm well.

5. Photoaffinity-labeled BRF was excised from the gel and electrophoretically eluted using a Bio-Rad Model 422 Electro-Eluter for 4 h at 10 mA per gel slice into a volatile buffer consisting of 50 m*M* ammonium bicarbonate and 0.1% SDS.
6. The eluate is dried down by vacuum centrifugation, resuspended in 200 μL sterile deionized water, and dried down again.
7. Gel-purified BRF is treated with 25 μL 70% formic acid and 1.4% diphenylamine at 70°C for 20 min to further digest DNA and to cleave the protein at Asp-Pro sites (**Fig. 8**).
8. The sample was extracted five times with an equal volume of water-saturated ethyl ether (fresh).
9. Next, samples are evaporated to dryness by vacuum centrifugation, resuspended in sterile deionized water, and dried down again.
10. The pellet is resuspended in 40 μL of 2% SDS and 0.1 m*M* of 2-mercaptoethanol.
11. Proteins are cleaved with cyanogen bromide by the addition of 1 μL of 1 *M* hydrochloric acid and 1 μL of 1 *M* cyanogen bromide in acetronitrile to a 15-μL sample or addition of formic acid to a final concentration of 70% and 1 μL of 1 *M* cyanogen bromide in acetonitrile for a more complete digestion.
12. Samples are incubated at 25°C for 10 min or 2 h.
13. Samples are resolved on a 10–20% tricine gel *(13)*. After electrophoresis the gel is stained by Coomassie R-250 staining, dried, and placed on autoradiography film to visualize radiolabeled fragments (*see* **Note 19**) (**Fig. 8**).

4. Notes

1. We have found that differing brands of TLC plates cause products to migrate somewhat differently and may result in different observed *Rf* values.
2. We store concentrated stocks of modified nucleotides wrapped in foil at –80°C. Only small diluted stocks are stored at –20°C wrapped in foil to help protect the major stock from inadvertent photolysis.
3. Photoreactive nucleotides are tested to determine the range of nucleotide concentrations that can be used to incorporate the nucleotides by primer extension and can be compared to the unmodified nucleotide. An aliquot of each sample can be irradiated and the DNA will crosslink the Klenow fragment of DNA polymerase I and cause the labeled oligonucleotide to have a much slower electrophoretic mobility. Some of the oligonucleotide does not get crosslinked to DNA polymerase, but is visibly photolyzed as evident by smearing of the free oligonucleotide band.
4. Another question we have addressed is whether a modified nucleotide at a given position affects normal DNA–protein interactions in that region. This can be done for transcription complexes by gel shift analysis or performing transcription assays on wild type DNA and probe DNA to determine if DNA modification affects the level of transcription.
5. Our lab has now prepared and is characterizing a modified dATP analog to increase the number of DNA sites that can be modified (M. Zofall, manuscript in preparation).

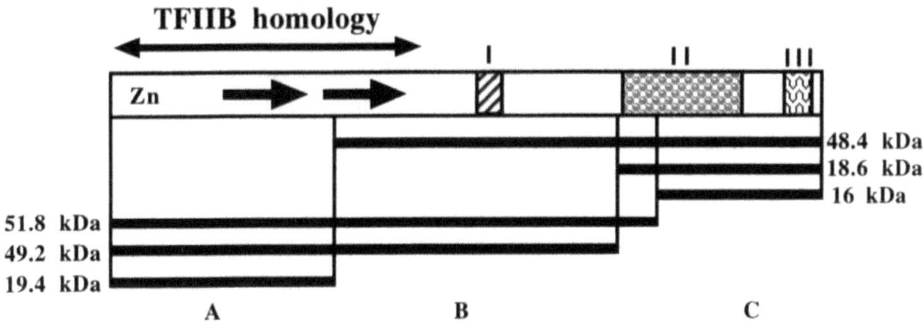

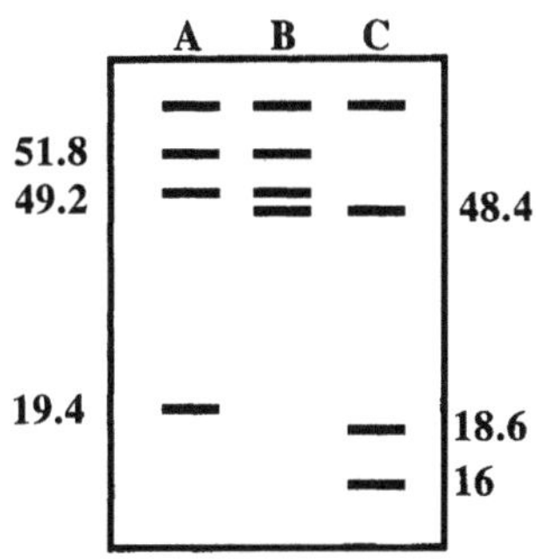

Fig. 8. Representation of possible results of peptide mapping of the BRF subunit of TFIIIB. **(Top)** Linear representation of the protein with Asp-Pro linkages labeled below, along with the fragments generated by single-hit digestion. **(Bottom)** A representation of protein gel analysis of labeled proteolytic products from crosslinking occurring in region A, B, or C.

6. It is recommended to have at least 50 base pairs of DNA between the biotinylation site or attachment site to the bead and the restriction endonuclease cut site, because of potential steric hindrance of the restriction endonuclease if the site is too close to the bead.
7. We have found the DNA beads to be sensitive to certain buffers containing high salt in combination with SDS. These conditions lead to the degradation of the magnetic particles, so it is important to avoid washing with buffers containing both SDS and high concentrations of salt.
8. The DNA beads have a preference for binding shorter DNAs with a size restriction of less than approx 2000 bp in length. It is critical to determine the extent of binding the biotinylated DNA by agarose gel electrophoresis.
9. Beads should always be mixed with gentle vortexing action in about 5-min intervals to ensure that they remain in solution; avoid vigorous vortexing because it will scatter the beads and lead to sample loss.

10. Oligonucleotides for probe synthesis are usually between 18- and 20-mers in length with a GC rich 5' end.
11. Enzymes containing DTT need to be avoided because DTT can reduce aryl azides. Some enzymes with DTT can be sufficiently diluted in buffers containing 2-mercaptoethanol (5 m*M*) in place of DTT.
12. If the DNA probe is to be used for peptide mapping, the modified nucleotide will have to be incorporated first, followed by dATP or dGTP, to leave a nucleotide tag covalently attached to the crosslinked protein upon chemical degradation of the DNA probe.
13. Choosing restriction enzymes with good cutting efficiency to remove probes from the DNA beads can be very beneficial to overall probe yields.
14. Proper controls such as DNA competition with specific-pTZ1 DNA (containing an up mutation in the box B region of DNA) vs nonspecific-pLNG56 DNA (containing a down mutation in the Box B region of the DNA) and heparin striping of samples are necessary to ensure labeling specificity.
15. We have found the crude extract BR-500 to be more efficient for photoaffinity labeling, probably because of higher activities of the proteins and factors that may be lost during extensive purification.
16. Because of the short sequential incubation times, it is best to cycle samples into and out of incubation baths to ensure equal incubation times for all samples. This means the addition of the enzyme or solution to the first sample and immediately placing it in the temperature bath, followed by processing the second sample in the same way and so on. The samples are removed in the same order so that they enter the bath with a 15-s delay between each sample.
17. Crosslinked proteins have a slightly greater electrophoretic mobility than unmodified protein because of the DNA tag left behind, and it is more noticeable for smaller proteins.
18. Labeling reactions used for peptide mapping are scaled up to ensure adequate signal. Sufficient amounts of labeled protein is necessary because of losses during the extensive purification and the partial proteolytic conditions creating only a small percentage of labeled proteolytic fragments.
19. Care should be taken in constructing a map of potential proteolytic fragments created by single-hit digests for peptide mapping. Chemical cleavage of protein is preferable to enzymatic cleavage, because chemical cleavage has no apparent specificity or site preference that can make it difficult to interpret the peptide mapping results.

References

1. Huet, J., Manaud, N., Dieci, G., Peyroche, G., Conesa, C., Lefebvre, O., et al. (1996) RNA polymerase III and class III transcription factors from *Saccharomyces cerevisiae. Methods Enzymol.* **273,** 249.
2. Geiduschek, E. P. and Kassavetis, G. A. (1992) *RNA polymerase III transcription complexes.*, in *Transcriptional Regulation* (McKnight, S. L. and Yamamoto, K. R., eds.) Cold Spring Harbor Laboratory, Cold Spring Harbor, NY. p. 247.

3. Lannutti, B. J., Persinger, J., and Bartholomew, B. (1996) Probing the protein-DNA contacts of a yeast RNA polymerase III transcription complex in a crude extract: solid phase synthesis of DNA photoaffinity probes containing a novel photoreactive deoxycytidine analog. *Biochemistry* **35,** 9821.
4. Persinger, J. and Bartholomew, B. (1996) Mapping the contacts of yeast TFIIIB and RNA polymerase III at various distances from the major groove of DNA by DNA photoaffinity labeling. *J. Biol. Chem.* **271,** 33,039.
5. Tate, J. J., Persinger, J., and Bartholomew, B. (1998) Survey of four different photoreactive moieties for DNA photoaffinity labeling of yeast RNA polymerase III transcription complexes. *Nucleic Acids Res.* **26,** 1421.
6. Bartholomew, B., Braun, B. R., Kassavetis, G. A., and Geiduschek, E. P. (1994) Probing close DNA contacts of RNA polymerase III transcription complexes with the photoactive nucleoside 4-thiodeoxythymidine. *J. Biol. Chem.* **269,** 18,090.
7. Kassavetis, G. A., Riggs, D. L., Negri, R., Nguyen, L. H., and Geiduschek, E. P. (1989) Transcription factor IIIB generates extended DNA interactions in RNA polymerase III transcription complexes on tRNA genes. *Mol. Cell. Biol.* **9,** 2551.
8. Bartholomew, B., Meares, C. F., and Dahmus, M. E. (1990) Photoaffinity labeling of RNA polymerase III transcription complexes by nascent RNA. *J. Biol. Chem.* **265,** 3731.
9. Galardy, R. E., Craig, L. C., Jamieson, J. D., and Printz, M. P. (1974) Photoaffinity labeling of peptide hormone binding sites. *J. Biol. Chem.* **249,** 3510.
10. Nassal, M. (1983) 4-(1-Azi-2,2,2-trifluroethyl)benzoic scid, a highly photolabile carbine generating label Readily fixable to biochemical agents. *Liebigs Ann. Chem.* 1510.
11. Bartholomew, B., Tinker, R. L., Kassavetis, G. A., and Geiduschek, E. P. (1995) Photochemical cross-linking assay for DNA tracking by replication proteins. *Methods Enzymol.* **262,** 476.
12. Bartholomew, B., Durkovich, D., Kassavetis, G. A., and Geiduschek, E. P. (1993) Orientation and topography of RNA polymerase III in transcription complexes. *Mol. Cell. Biol.* **13,** 942.
13. Schagger, H. and von Jagow, G. (1987) Tricine-sodium dodecyl sulfate-polyacrylamide gel electrophoresis for the separation of proteins in the range from 1 to 100 kDa. *Anal. Biochem.* **166,** 368.

26

Use of Site-Specific Protein–DNA Photocrosslinking to Analyze the Molecular Organization of the RNA Polymerase II Initiation Complex

François Robert and Benoit Coulombe

1. Introduction

Site-specific protein–DNA photocrosslinking has proved to be the method of choice for analysis of the formation of nucleoprotein complexes such as those involved in transcription by mammalian RNA polymerase II (RNA Pol II). The method has two principal advantages. First, it yields structural information on large, multisubunit complexes that in general cannot be analyzed using standard high-resolution techniques such as X-ray crystallography or nuclear magnetic resonance (NMR). For example, site-specific protein–DNA photocrosslinking, in conjunction with complementary methods such as protein-affinity chromatography and electron microscopy, has produced information on both the molecular organization and the composition of the RNA Pol II pre-initiation complex on promoter DNA *(1–4)*. This complex contains RNA Pol II and the general transcription factors TBP, TFIIA, TFIIB, TFIIE, TFIIF (RAP74 and RAP30), and TFIIH, and is composed of more than 25 polypeptides ranging in M_r from 10 to 220 kDa *(5)*. Neither X-ray crystallography nor NMR, which can only resolve the structure of complexes containing short protein fragments bound to small pieces of promoter DNA, could provide any detailed structural information on this complex. Second, the method has sufficient technical flexibility so as to allow the rapid analysis of complexes assembled under various conditions. Over the past few years, we have analyzed a large collection of complexes assembled in the presence of various combinations of the general transcription factors (wild-type or different deletion

From: *Methods in Molecular Biology, vol. 148: DNA–Protein Interactions: Principles and Protocols, 2nd ed.*
Edited by: T. Moss © Humana Press Inc., Totowa, NJ

mutants) and RNA Pol II *(1–4)*. These experiments have enabled us to draw conclusions on the dynamics of RNA Pol II pre-initiation complex assembly and has led to the notion that isomerization of the RNA Pol II pre-initiation complex proceeds through wrapping of the promoter DNA around the enzyme *(4)*.

Site-specific protein–DNA photocrosslinking is a method composed of two successive steps. First, a number of photoprobes that place one (or a few) photoreactive nucleotide(s) into juxtaposition with one (or a few) radiolabeled nucleotide(s) at various specific positions along the promoter DNA are prepared. Second, transcription complexes are assembled onto the various photoprobes, irradiated with ultraviolet (UV) light so as to induce protein–DNA crosslinking, and the processed in order to identify the crosslinked polypeptides. Because the crosslinking of protein to DNA is site-specific, the use of a series of photoprobes that place the photonucleotide derivative at various positions along the promoter DNA provides information on the relative position of the various factors within the complex.

2. Materials

1. Buffer A (10X): 300 m*M* Tris-HCl, pH 8.0, 500 m*M* KCl, and 70 m*M* $MgCl_2$, freshly prepared.
2. Bovine serum albumin (BSA) solution: Prepare a 25-mg/mL solution of BSA in deionized distilled water. Store in aliquots at –20°C.
3. Dilute with water to 5 mg/mL prior to use.
4. dNTP mix: 20 m*M* each of dATP, dCTP, dGTP, and dTTP in buffer A (1X), freshly prepared.
5. ND buffer: 20 m*M* HEPES, pH 7.9, 100 m*M* KCl, 20% glycerol, 0.2 m*M* EDTA, 0.2 m*M* EGTA, and 10 m*M* of β-mercaptoethanol. Store in aliquots at –20°C.
6. TBE buffer (10X): Prepare 1 L by mixing 108 g Tris base, 55 g boric acid, and 40 mL EDTA (0.5 *M*, pH 8.0).
7. Gel loading solution (10X): For polyacrylamide native gels, use a solution containing 0.25% bromophenol blue, 0.25% xylene cyanol, and 25% Ficoll (type 400) in deionized distilled water.
8. MBS (5X): 40 m*M* $MgCl_2$, 100 m*M* HEPES, pH 7.9, 100 μg/mL BSA, and 5 m*M* ATP.
9. Store in aliquots at –20°C.
10. Complex mix: 50 μL MBS (5X) and 2 μL $(NH_4)_2SO_4$ (2 *M*), freshly prepared.
11. Poly(dI.dC–dI.dC) stock: Prepare a 25-mg/mL solution of poly(dI.dC–dI.dC) in deionized distilled water. Store in aliquots at –20°C.
12. DNase mix: A solution containing 200 units/mL DNase I and 32 m*M* $CaCl_2$, freshly prepared.
13. Acid mix: Mix equal volumes of 5% acetic acid and 30 m*M* $ZnCl_2$, freshly prepared.
14. S1 mix: 30,000 U/mL S1 nuclease in deionized distilled water, freshly prepared.

15. Dilution mix: 50 m*M* Tris-HCl, pH 8.0, 10 m*M* EDTA, 10 m*M* EGTA, 500 m*M* NaCl, 5 m*M* NaF, 1 m*M* phenymethylsulfonyl fluoride (PMSF), and 1% (v/v) Triton X-100, freshly prepared.
16. Protein A mix: 100 mg/mL Protein A in the dilution mix.
17. Wash 1 solution: 10 m*M* Tris-HCl, pH 8.0, 150 m*M* NaCl, 1% Triton X-100, 1% NP-40, 0.1% sodium dodecyl sulfate (SDS), 1 mg/mL BSA, 0.5% NaN_3, and 0.5% Na-deoxychlorate, freshly prepared.
18. Wash 2 solution: 10 m*M* Tris-HCl, pH 8.0, and 1.5 *M* NaCl.
19. Wash 3 solution: 10 m*M* Tris-HCl, pH 8.0, and 150 m*M* NaCl.

3. Methods

3.1. Day 1: Synthesis and Purification of the Photoprobes

The first step of the procedure is the synthesis of the photoprobes. This part is illustrated in **Fig. 1** which shows, as an example, a scheme for the synthesis of a photoprobe designed to place the photoreactive nucleotide at position –2 relative to the transcriptional initiation site of the adenovirus major late promoter. The site-specific incorporation of the photoreactive nucleotide (*see* **Note 1**) and the radiolabeled nucleotide is directed through the annealing of a primer, referred to as the specific primer, with a single-stranded DNA template containing the promoter DNA. The promoter is flanked by two restriction sites (in this example, *Dra*I and *Sac*I). A second primer, referred to as the upstream primer, is annealed a few base pairs upstream of the *Dra*I site. After annealing, the photoreactive and radiolabeled nucleotides are incorporated by primer extension using T_4 DNA polymerase with limiting amounts of dNTPs (*see* **Note 2**). After the labeling step, the extension reaction is completed by the addition of an excess of cold dNTPs (*see* **Note 3**). The photoprobe is generated by digestion with the restriction enzymes and gel purified (*see* **Fig. 2** for an example of a gel on which the products of a photoprobe synthesis reaction have been separated).

1. Mix 700 ng of single-stranded (ss) DNA (approx 0.5 pmol) with 40 ng (approx 5 pmol) of both the specific and upstream primers. Add 1 µL of Buffer A (10X) and complete to 10 µL with deionized distilled water.
2. Mix well and incubate for 3 min at 90°C.
3. Incubate for 30 min at room temperature.
4. From this point on, all manipulations must be carried out under reduced light conditions (*see* **Note 4**). Add 1 µL BSA (5 mg/mL), 1 µL N_3R–dUMP (*see* **Note 1**), 25 µCi of the appropriate [α-^{32}P]–dNTP ([α-^{32}P]–dCTP for the example shown in **Fig. 1**), 5–10 U T4 DNA polymerase, and 1 µL buffer A (10X). Complete to a final volume of 20 µL with deionized distilled water.
5. Incubate for 30 min at room temperature.
6. Add 5 µL of dNTP mix.
7. Incubate for 5 min at room temperature.

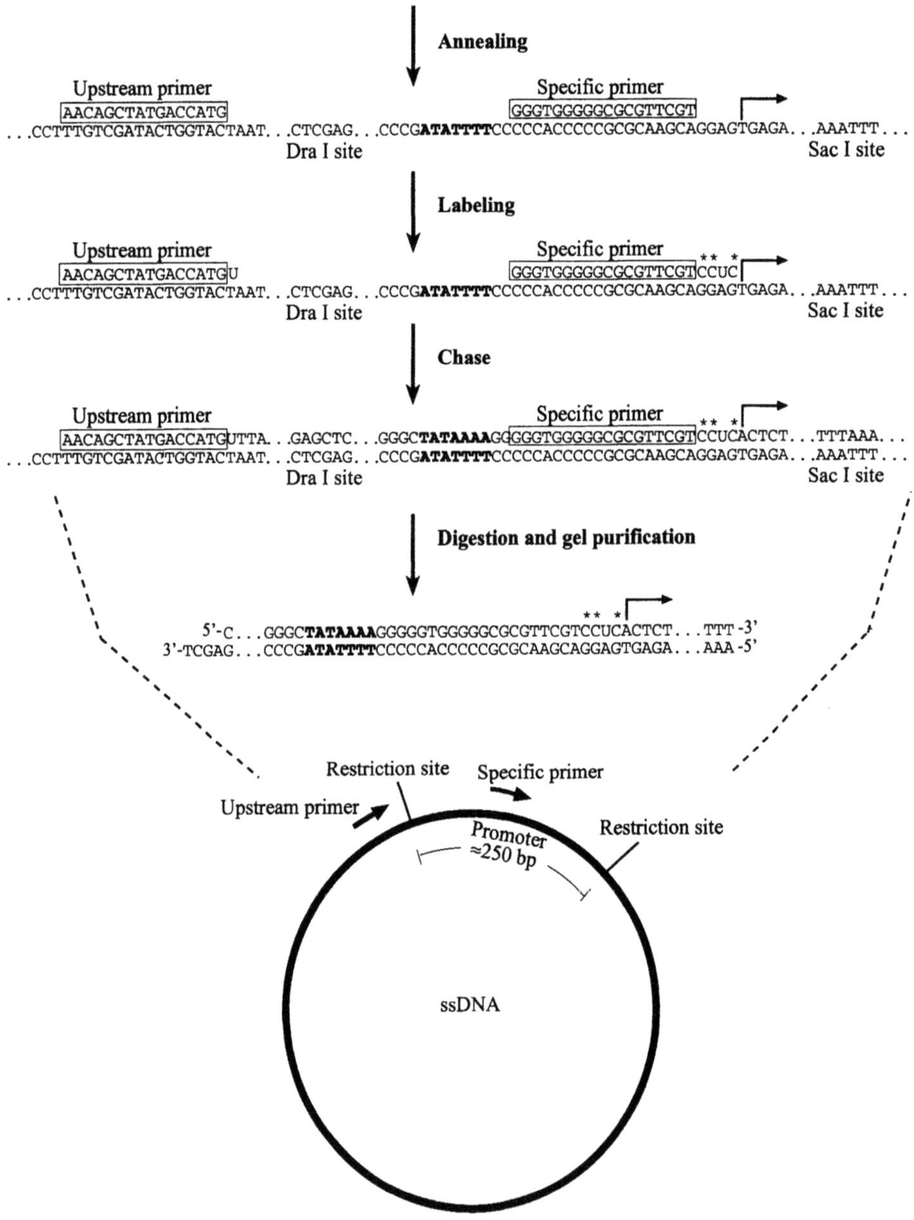

Fig. 1. Synthesis and purification of the photoprobes. Schematic representation of the synthesis of a photoprobe that places a photoreactive nucleotide at position –2 and three radiolabeled nucleotides at positions –4, –3, and –1 of the adenovirus major late promoter. The sequence of the adenovirus major late promoter flanked by plasmid DNA is shown. The primers used to direct the incorporation of N_3R–dUMP (U) and radiolabeled nucleotides (∗) are shown in boxes. The TATA element is in bold type and the transcriptional initiation site is indicated by an arrow. *Dra*I and *Sac*I restric-

8. Incubate for 20 min at 37°C.
9. Add 10–20 U of each restriction enzyme (*Sac*I and *Dra*I in the example shown in **Fig. 1**).
10. Incubate for 90 min at the temperature recommended by the supplier of restriction enzymes (37°C for *Sac*I and *Dra*I).
11. Add 3.5 µL of gel loading solution (10X).
12. Load on a native 8% polyacrylamide gel in TBE buffer (1X).
13. Run at 150 V for about 1 h in TBE buffer (1X).
14. Remove the glass plates containing the gel from the gel box.
15. Separate the glass plates and leave the gel on one of them.
16. Wrap the gel/glass plate in plastic wrap.
17. Wrap the entire package in aluminum foil.
18. Move to a dark room (*see* **Note 5**).
19. Place a Kodak X-OMAT AR film on a clean bench.
20. Remove the foil and place the gel on the film with the glass plate facing up (e.g., gel side down).
21. Expose 5 min.
22. During the exposition time, mark the film using a sharp tool by tracing the contour of the glass plate (this will be helpful later for the localization of the photoprobe in the gel).
23. Remove the gel and rewrap it with the foil.
24. Develop the film (*see* **Note 6**).
25. Using a scalpel, cut the film so that the square piece containing the band corresponding to the photoprobe is removed. This operation leaves the film with a window at the position of the photoprobe.
26. Superimpose the film on the gel by taking advantage of the marks made in **step 22**, and mark the square corresponding to the photoprobe on the saran wrap using a pen.
27. Cut out the gel slice containing the photoprobe using a clean scalpel.
28. Cut the gel slice in small pieces (six to eight fragments).
29. Place the gel fragments in an Eppendorf tube and add water in order to completely submerge the gel (usually 100–150 µL of water).
30. Incubate overnight at room temperature.
31. Collect the liquid containing the probe.
32. Purify the probe on a Micro-Spin S-200 HR column (Pharmacia Biotech) to remove any salts and other putative contaminants.

---

ion sites used for the excision of the photoprobe are indicated. In the first step, the primers are annealed to single-stranded DNA containing the promoter sequence (Annealing). In the second step, the photoreactive and the radiolabeled nucleotides are incorporated using T_4 DNA polymerase (Labeling). In the third step, the extension reaction is completed by a chase with excess of cold dNTPs (Chase). In the fourth step, the site-specifically labeled promoter fragment is excised using restriction enzymes and gel purified (Digestion and purification).

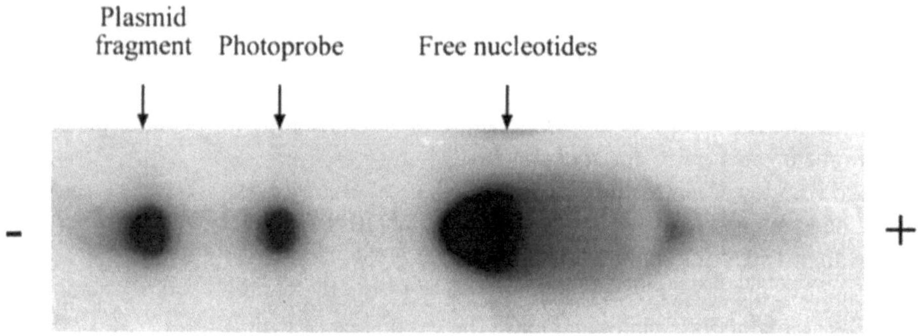

Fig. 2. Example of the autoradiogram of a gel after electrophoresis of the products of a synthesis reaction. The positions of the photoprobe, the free nucleotides, and the plasmid fragment are shown. The photoprobe is then purified from the gel.

33. Count 1 µL of the photoprobe solution by liquid scintillation, and dilute the probe to 1250 cpm/µL with deionized distilled water.
34. The probe is ready for use and can be stored in the dark at 4°C for 1–2 wk (*see* **Note 7**).

3.2. Day 2: Protein–DNA Photocrosslinking

The gel-purified photoprobe is used for pre-initiation complex assembly with the purified transcription factors (TBP, TFIIA, TFIIB, TFIIE, TFIIF, and TFIIH) and RNA Pol II. Reactions are irradiated with ultraviolet (UV) light to induce protein–DNA crosslinking and treated with DNase I and S1 nuclease in order to liberate polypeptides that are covalently attached to a very short piece of DNA carrying one to four radiolabeled nucleotides. After sodium dodecyl sulfate–polyacrylamide gel electrophoresis (SDS-PAGE) separation of the photocrosslinked polypeptides, the gel is dried and exposed to X-ray film. Examination of the ensuing autoradiogram permits identification of the protein(s) that interact with a particular site. For the RNA Pol II initiation complex, the specificity of the photocrosslinking signals can be assessed by comparing reactions performed with photoprobes containing either a wild-type or a mutated TATA element and/or by comparing reactions performed in either the presence or the absence of TBP (*see* **Note 8**). The photocrosslinked polypeptides can be identified according to their M_r (**Fig. 3A,B**) and to their immunoreactivity with specific antibodies (**Fig. 3C**). In the latter case, crosslinking reactions are submitted to immunoprecipitation with antibodies directed against specific factors prior to SDS-PAGE (*see* **Note 9**).

35. Mix the proteins (50–200 ng each) and complete the volume to 16 µL with ND buffer (*see* **Note 10**). Add 5.2 µL of the complex mix, 1 µL of diluted poly(dI.dC–

dI.dC) (*see* **Note 11**) and 5000 cpm of the photoprobe (4 µL of 1250 cpm/µL). The final volume is 26.2 µL.

36. Mix well and incubate for 30 min at 30°C.
37. Open the lids of the tubes and irradiate 10 min with UV light (*see* **Notes 12** and **13**).
38. Add 5 µL of DNase mix.
39. Incubate for 20 min at 37°C.
40. Add 1.5 µL of 10% SDS.
41. Incubate for 3 min at 90°C.
42. Spin for 10 s in a microfuge (16,000*g*)
43. Add 2 µL of Acid mix.
44. Add 1 µL of S1 mix.
45. Incubate for 20 min at 37°C.
46. For regular reactions, go directly to **step 62**.
47. For immunoprecipitation reactions, add 315 µL of the dilution mix.
48. Add 5 µL of antibody (*see* **Note 14**).
49. Incubate for 60 min at 4°C.
50. Add 80 µL of Protein A mix.
51. Agitate 60 min at 4°C using a rocker.
52. Spin down the Protein A beads (1 min in a microfuge at 16,000*g*).
53. Remove the supernatant by pipetting.
54. Wash with 400 µL of wash 1 solution.
55. Repeat **steps 52–54** twice.
56. Wash with 400 µL of wash 2 solution.
57. Spin down the Protein A beads (1 min in a microfuge at 16,000*g*).
58. Remove the supernatant by pipetting.
59. Wash with 400 µL of wash 3 solution.
60. Spin down the Protein A beads (1 min in a microfuge at 16,000*g*).
61. Remove the supernatant by pipetting.
62. Add SDS gel loading solution and boil 5 min (*see* **Note 15**).
63. Resolve the photocrosslinked polypeptides by SDS-PAGE (run at 30 mA in the stacking gel and 50 mA in the separating gel) (*see* **Note 15**).
64. Transfer the gel to Whatman paper and dry.
65. Expose the dried gel to X-ray film using an intensifying screen (*see* **Note 16**).

4. Notes

1. The nucleotide derivative we use, namely 5-[*N*-(azidobenzoyl)-3-aminoallyl]–dUMP (N_3R–dUMP or AB–dUMP) (*see* Chapter 25 for further details on the structure and the chemical synthesis of N_3R–dUMP), possesses a side chain that places a reactive nitrene 10 Å away from the DNA backbone in the major groove of the double helix. For this reason, the crosslinking of a polypeptide to the photoprobe does not require a direct interaction of the polypeptide with the DNA helix. The amount of N_3R–dUMP to be added to the reaction is determined empirically for each preparation of the photoreactive nucleotide and is generally between 0.5 and 2 µL (often 1 µL).

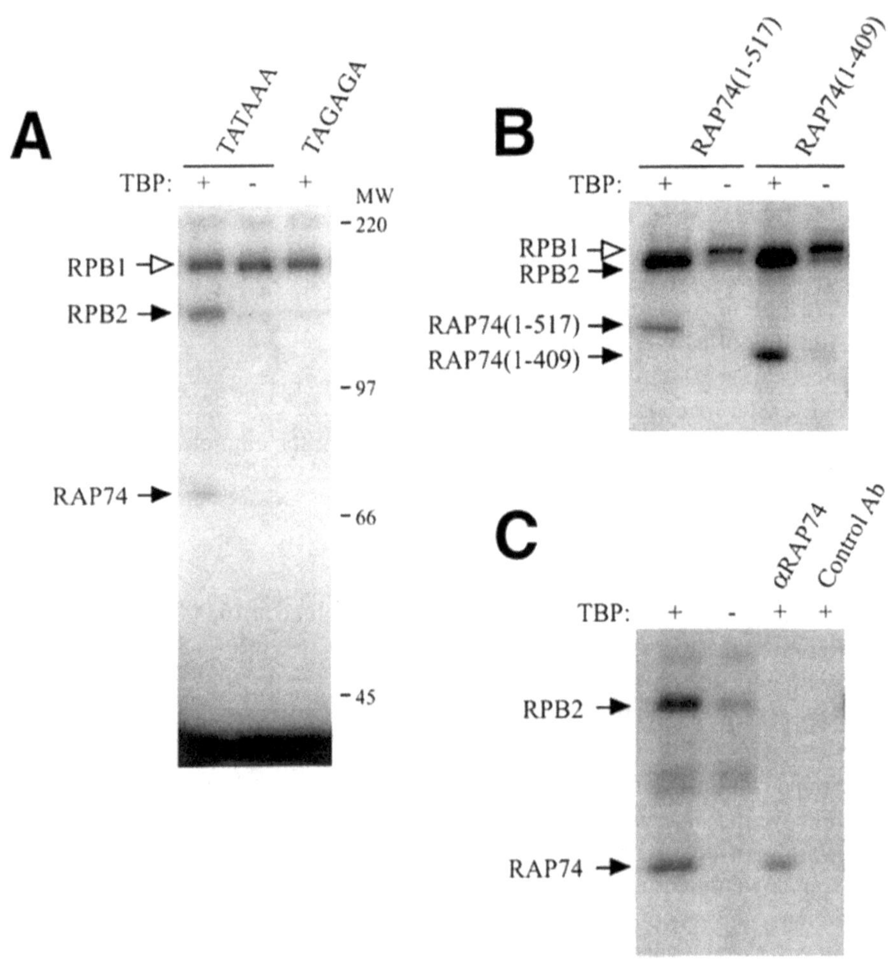

Fig. 3. Typical SDS-PAGE gels of photocrosslinked proteins. **(A)** Crosslinking reactions assembled either in the absence of TBP or using a probe with a mutated TATA box give identical results. Crosslinking reactions were performed with TFIIB, RAP30, RAP74, TFIIE34, TFIIE56, and RNA Pol II in the presence (+) or in the absence (–) of TBP using either a wild-type (TATAAA) or a mutated (TAGAGA) TATA element. In the example shown here (the photonucleotide derivative is placed at position –15 of the adenovirus major late promoter), the crosslinking of both the second largest subunit of RNA Pol II (RPB2) and RAP74 are considered to be specific (dark arrowhead). The crosslinking of RPB1 is considered as nonspecific (open arrowhead) because it is not affected by the omission of TBP or the use of a probe with a mutation in the TATA box. The positions of the molecular weight markers (MW) are indicated. **(B)** The use of truncated polypeptides as a tool to identify the photocrosslinked polypeptides. Crosslinking reactions were performed using TFIIB, RAP30, TFIIE34, TFIIE56, and RNA Pol II in the presence (+) or in the absence (–) of TBP,

2. The specific primer must be designed in such a manner that T_4 DNA polymerase only adds a few nucleotides. In the example shown in **Fig. 1**, the incorporation is restricted to positions –4 to –1 by omitting dATP from the reaction (*see* **refs. *1–4*** for additional examples). The success of this step can be monitored by analysis of the reaction products on a sequencing gel.
3. The addition of dNTPs in large excess is crucial because it is necessary to limit primer extension to the incorporation of standard dNTPs, preventing the addition of any more radiolabeled and photoreactive nucleotides.
4. The use of a standard dark room is not necessary at this stage. As a rule, we find that conditions that provide just enough light to be able to work are acceptable.
5. A conventional red light can be used.
6. An example of gel photoprobe purification is shown in **Fig. 2**. The position of the band corresponding to the photoprobe can be easily identified because the size of the DNA fragment generated by digestion with restriction enzymes is known.
7. Fresh probes (less than a week old) give the best results.
8. Because some of the general transcription factors and RNA Pol II bind nonspecifically to DNA, it is important to discriminate between specific and nonspecific crosslinking signals. For the adenovirus major late promoter, it is well-documented that mutations of two bases in the TATA box (TATAAA to TAGAGA) completely abolish pre-initiation complex formation. The comparison of crosslinking signals obtained with probes containing either a wild-type or a mutated TATA box permits differentiation between specific and nonspecific signals. However, this strategy is not simple, as it doubles the number of probes to be synthesized. Because the basic function of the TATA box is to bind TBP, the specificity can be assessed by comparing crosslinking reactions performed in the presence or the absence of TBP. The absence of TBP was found to give the same result as the use of a probe with a mutated TATA box *(1–4)*. An example is shown in **Fig. 3A**. The crosslinking signal at the top of the gel (RPB1) is considered as nonspecific because its intensity is not affected by either a mutation in the TATA box or the absence of TBP. The two additional crosslinking signals (RPB2 and RAP74) are specific because their intensities are significantly affected by both mutation of the TATA box and the omission of TBP in the crosslinking reaction.
9. Identification of the photocrosslinked polypeptides is a central issue in the method. The main difficulty comes from the fact that several factors have a M_r between 30 and 40 kDa. At least three different means can be used to identify the

---

using full-length (1–517) or truncated (1–409) RAP74. The different mobilities of RAP74 fragments are diagnostic for RAP74 contact at this promoter position. **(C)** Immunoprecipitation of photocrosslinked polypeptides. Crosslinking reactions were performed using TFIIB, RAP30, RAP74, TFIIE34, TFIIE56, and RNA Pol II in the presence (+) or in the absence (–) of TBP. The photocrosslinked polypeptides were either processed normally (first two lanes) or immunoprecipitated using an antibody directed against RAP74 or a control antibody.

crosslinking signals. First, SDS-PAGE analysis provides direct information on the size of photocrosslinked polypeptides (*see* **Fig. 3A**). Second, the use of truncated forms of a factor can be useful. An example is shown in **Fig. 3B** in which all factors included in the crosslinking reactions are the same, except for RAP74, which is either full-length (lanes 1 and 2) or truncated in its C-terminus (lanes 3 and 4). The different fragments of RAP74 migrate with different mobilities, allowing the identification of the photocrosslinked polypeptide. Third, the photocrosslinked polypeptides can be identified after immunoprecipitation with a specific antibody. Following nuclease treatment, the crosslinking products are immunoprecipitated and then submitted to SDS-PAGE analysis. An example is shown in **Fig. 3C**, where the photocrosslinked polypeptides shown in lane 1 have been immunoprecipitated using an antibody raised against RAP74 (lane 3) or a control antibody (lane 4). This shows that the photocrosslinked polypeptide is RAP74.

10. In the crosslinking reactions, we routinely use 200 ng of each recombinant human (rh) TFIIB, rhRAP30, rhRAP74, rhTFIIE34, and rhTFIIE56, 100 ng of calf thymus RNA Pol II, 50 ng of natural human TFIIH, 50–200 ng of rhTFIIA, and 200 ng of recombinant yeast TBP. The amounts of the different protein factors should be optimized for each different combination of proteins and for each protein preparation.
11. The poly(dI.dC–dI.dC) stock should be diluted just prior to use. The exact dilution should be determined experimentally in order to favor specific versus nonspecific signals without adversely affecting the intensity of the specific signals.
12. Irradiation time with UV light should be optimized by performing a time-course with the particular system to be used. We use a UV Stratalinker 2400 (Stratagene) with 254-nm bulbs.
13. From this point on, normal light conditions can be used.
14. The amounts of antibody to be used should be optimized. Purified antibodies provide the best results.
15. Detailed procedures for SDS-PAGE electrophoresis have been described (e.g., *see* **ref. *6***).
16. The use of BioMax (Kodak) screens and films is recommended.

Acknowledgments

We thank members of our laboratory for valuable discussions and Will Home for critical reading of the manuscript. B.C. is the recipient of funding from the Medical Research Council of Canada, the Cancer Research Society and the Fonds de la Recherche en Santé du Québec. F.R. holds a studentship from the FCAR.

References

1. Coulombe, B., Li, J., and Greenblatt, J. (1994) Topological localization of human transcription factors IIA, IIB, TATA box-binding protein and RNA polymerase II-associated protein 30 on a class II promoter. *J. Biol. Chem.* **269,** 19,962–19,967.

2. Robert, F., Forget, D., Li, J., Greenblatt, J., and Coulombe, B. (1996) Topological localization of transcription factors IIE and IIF immediately upstream the transcription start site of a class II promoter. *J. Biol. Chem.* **271,** 8517–8520.
3. Forget, D., Robert, F., Grondin, G., Burton, Z. F., Greenblatt, J., and Coulombe, B. (1997) RAP74 induces promoter contacts of RNA polymerase II upstream and downstream of a DNA bend at the TATA box. *Proc. Natl. Acad. Sci. USA* **94,** 7150–7155.
4. Robert, F., Douziech, M., Forget, D., Egly, J. M., Greenblatt, J., Burton, Z. F., et al. (1998) Wrapping of promoter DNA around the RNA polymerase II initiation complex induced by TFIIF. *Mol. Cell* **2,** 341–351.
5. Hampsey, M. (1998) Molecular genetics of the RNA polymerase II general transcription machinery. *Microbiol. Mol. Biol. Rev.* **62,** 465–503.
6. Sambrook, J., Fritsch, E. F., and Maniatis, T. (1989) *Molecular Cloning: A Laboratory Manual*, 2nd ed., Cold Spring Harbor Laboratory, Cold Spring Harbor, NY.

27

UV Laser-Induced Protein-DNA Crosslinking

Stefan I. Dimitrov and Tom Moss

1. Introduction

1.1. The Method

Photochemical crosslinking is a powerful method for studying all types of protein-nucleic acids interactions. In particular UV-induced crosslinking has been successfully applied to the study of protein–DNA interactions (e.g., **ref.** ***1***, *see* Chapters 23–26 and 43). Ultraviolet (UV) light is a zero-length crosslinking agent. It is therefore not subject to the steric problems that can be associated with chemical crosslinking agents and provides strong evidence for close protein–DNA interactions. However, crosslinking with conventional UV-light sources requires exposure times ranging from minutes to several hours (e.g., *see* **refs.** ***1–3***), permitting protein redistribution and the crosslinking of UV-damaged molecules. Because UV-laser irradiation is intense enough to induce crosslinking after very short exposure times, artifactual crosslinking can be avoided. The typically nanosecond or picosecond exposures times also allow UV-laser-induced crosslinking to be applied to study the intermediate states in rapid protein–DNA binding reactions ***(4,5)***.

Ultraviolet-crosslinking of protein to DNA occurs in two distinct steps. In the first step, the bases of DNA are excited by light absorption. This excitement rapidly gives rise to radicals of the bases, resulting in chemical crosslinking and macromolecular damage, mainly to the DNA. The time to excite the bases is simply the time of UV irradiation. Completion of the crosslinking reaction then occurs in less than a microsecond ***(6)***. Most micro-conformational transitions of macromolecules take more than 100 µs (*see* **ref.** ***7***). Nanoseconds or picosecond UV-laser irradiation therefore avoids almost all possibility of artifactual rearrangement during the crosslinking reaction. What is more, such a rapid crosslinking reaction can be used to freeze protein–

From: *Methods in Molecular Biology, vol. 148: DNA–Protein Interactions: Principles and Protocols, 2nd ed.*
Edited by: T. Moss © Humana Press Inc., Totowa, NJ

DNA interactions, providing "time-lapse views" of the assembly of protein–DNA complexes. Laser-induced reactions, as opposed to those generated by conventional UV sources, proceed via the higher (S_n and T_n) excited states of nucleotide bases, which are induced by the rapid sequential absorption of two photons. This leads to a higher quantum yield of cationic radicals and to higher crosslinking efficiencies ***(6,8–10)***. Although the photochemistry is still not understood, UV-laser irradiation probably also induces mechanisms of crosslinking that simply cannot occur when using conventional light sources ***(6,10)***. In a typical UV crosslinking experiment, 5–15% of UV-laser-irradiated protein–DNA complexes are crosslinked ***(6,10,11)***, nearly two orders of magnitude higher than with conventional UV sources ***(4,5,8,10,12)***. UV-laser-induced crosslinking produces exclusively protein–DNA adducts ***(4,6,8,9)*** and is applicable to a broad range of protein–DNA complexes. Even complexes too weak to be seen by methods such as gel shift (electrophoretic mobility shift assay [EMSA]) or footprinting can be crosslinked. In fact, only nonspecific protein–DNA interactions with association constants of less than $10^3/M$ are not crosslinked ***(4,6)***.

1.2. Practical Applications

Photocrosslinking induced by UV-laser irradiation has been applied to measure binding constants ***(6)***, to map the extent of protein–nucleic acid binding sites ***(6)***, to determine protein–DNA ***(5,6)*** and protein–RNA ***(9)*** interactions, and to identify protein–DNA contacts ***(13)***. It was found possible to study the weak DNA-ATPase complex from the T4 DNA replication system by UV-laser crosslinking despite this complex being invisible to DNase I footprinting ***(6)***. UV-laser crosslinking has also provided data on the distribution of chromosomal proteins in vivo ***(8)***. The presence of the histones and the high-mobility group 1 proteins on the *Xenopus* ribosomal DNA was determined and shown to be regulated ***(8,14,15)***. Only the N-terminal domains of the histones were found to be crosslinked to DNA, and hyperacetylation of these domains did not affect their interaction with the DNA ***(14–16)***. An excellent guide to the practical application of in vitro laser protein–DNA crosslinking can be found in the methodological review by von Hippel and co-authors ***(6)***. It is likely that new developments in the use of femtosecond lasers will provide important improvements to the use of UV-laser crosslinking (***17*** and *see* Chapter 63).

1.3. The Experimental Approach

Here, we describe a procedure used to induce histone–DNA crosslinking in cell nuclei and to determine the DNA sequence distribution of the various histone fractions. The procedures used are quite general and could well be applied to study any type of protein–DNA complex. Indeed, we have essentially used a similar approach to determine the kinetics of the TBP–polymerase II promoter interaction ***(11)***.

1. UV-laser irradiation of nuclei.
2. Isolation of the crosslinked protein–DNA complexes (*see* **Note 1**).
3. Detection of specific proteins crosslinked to bulk DNA using immunochemical techniques.
4. Immunoprecipitation of the crosslinked protein–DNA complex.
5. Identification and quantitation of the DNA sequences covalently attached to a given protein using hybridization with specific DNA probes.

2. Materials

2.1. Laser Irradiation

When using a passively mode-locked picosecond neodymium–yttrium–aluminum–garnet (Nd:YAG) laser ***(8)***, the parameters of the laser radiation at 266 nm were as follows: pulse duration, 30 ps (in a Gaussian pulse shape assumption), pulse energy 4 mJ; diameter of the beam, 0.5 cm; repetition rate 0.5 Hz. The intensity of irradiation was controlled by focusing and defocusing using fused silica lenses. The energy of radiation was measured with pyroelectrical detectors calibrated with a Model Rj7200 energy meter (Laser Precision Corp.). (The Nd:YAG lasers produce pulsed radiation at 1064 nm and the conversion to 266 nm [wavelength at which the samples are irradiated] was performed by quadrupling the main frequency by means of angle-matched KDP crystals.) A nanosecond Nd:YAG laser (Model DCR-3J, Spectra-Physics Inc. [Mountain View, CA] or YAGMaster YM1000/1200, Lumonics [Canada]) can also be used ***(6,11)***. Pulses in the UV are about 5 ns in duration, and the energy per pulse at 266 nm was typically 80 mJ (*see* **Note 2**).

2.2. Reagents and Solutions

2.2.1. Procedures Described in **Subheading 3.1.**

1. 8 *M* urea.
2. 1% sodium dodecyl sulfate (SDS).

2.2.2. Procedures Described in **Subheading 3.2.**

1. 10 m*M* Tris-HCl, pH 7.5, and 1 m*M* $CaCl_2$.
2. 0.48 *M* Na phosphate buffer, pH 6.8.
3. 0.12 *M* Na phosphate buffer, pH 6.8.
4. 0.12 *M* Na phosphate buffer (pH 6.8) 2 *M* NaCl, and 5 *M* urea.
5. CsCl.
6. Micrococcal nuclease.
7. Hydroxyapatite (Bio-Rad).

2.2.3. Procedures Described in **Subheading 3.3.**

1. Phosphate-buffered saline (PBS).
2. PBS–0.05% Triton X-100 (PBS-T).
3. 1% bovine serum albumin (BSA) in PBS-T.
4. PBS, 0.4% Triton X-100.
5. 0.3% 4-chloronaphtol, 0.03% H_2O_2 in 50 m*M* Tris-HCl, pH 7.5, and 150 m*M* NaCl.
6. Nitrocellulose filter (Schleicher & Schuell).

2.2.4. Procedures Described in **Subheading 3.4.**

1. IgGsorb (The Enzyme Center, Malden, MA).
2. Antibody buffer: 50 m*M* HEPES, pH 7.5, 2 *M* NaCl, 0.1% SDS, 1% Triton X-100, 1% Na-deoxycholate, 5 m*M* EDTA, and 0.1% BSA.
3. Rinse buffer: 50 m*M* HEPES, pH 7.5, 0.15 *M* NaCl, 5 m*M* EDTA.
4. 3.5 *M* KSCN and 20 m*M* Tris-HCl, pH 7.5.
5. RNase A (1 mg/mL).
6. Pronase (1 mg/mL).
7. Ethanol.
8. 10 m*M* Tris-HCl, pH 7.5, and 0.25 m*M* EDTA.

2.2.5. Procedures Described in **Subheading 3.5.**

1. Zeta Probe blotting membranes (Bio-Rad).
2. Pre-hyb buffer: 6X SSC, 10X Denhardt's solution, 0.1 mg/mL denatured *Escherichia coli* DNA, 1% SDS, 0.2% Na-pyrophosphate, and 50% formamide.
3. 0.5X SSC and 0.5% SDS.
4. 0.1% SSC.

3. Methods

3.1. Irradiation Techniques

One to two milliliters of the nuclei/cell suspension are placed in a standard rectangular fused silica cuvet, thermostated at 4°C and the sample constantly stirred (*see* **Note 3**). The optical density of the solution should be kept in the range of $2 < A_{260} < 5$ (i.e., optically thick samples (*see* **Note 4**). In the case of a picosecond laser, the conditions of irradiation should be such that 10–20 photons are absorbed per nucleotide (about 500 mJ of incident light per 1 OD_{260} of optically thick sample) at a constant laser intensity 0.7 GW/cm^2. In the case of the nanosecond UV laser use 250 mJ of incident light per 1 A_{260} of optically thick sample. Dependent on the power of the laser used, it may be necessary to irradiate with multiple pulses.

3.2. Separation of Covalently Crosslinked Histone–DNA Complexes

1. Digest the irradiated nuclei with microccocal nuclease (5 U/A_{260} unit, 15 min, 37°C) in 10 m*M* Tris-HCl, pH 7.5, 1 m*M* $CaCl_2$ This is simply to reduce the DNA size (*see* **Notes 1** and **5**).

2. Add 0.12 *M* phosphate buffer (pH 6.8) 2 *M* NaCl, 5 *M* urea to stop the reaction
   a. Load the material on a hydroxyapatite column (1 g hydroxyapatite/mg DNA) equilibrated with 0.12 *M* phosphate buffer (pH 6.8) 2 *M* NaCl , 5 *M* urea.
   b. Wash the column with 5 vol of the same buffer, then with 0.12 *M* phosphate buffer.
   c. Elute the free DNA and the crosslinked complex with 0.48 *M* phosphate buffer, pH 6.8.
4. Apply the eluted material on a preformed CsCl gradient (four layers, 2.2 mL each, density (ρ) = 1.76, 1.57, 1.54, and 1.32 g/mL) and run in a SW 41 Beckman rotor at 15°C for 35–40 h at 35,000 rpm.
5. Collect 250-µL fractions and monitor optical density at 260 nm.
6. The gradient profile should show a clear shoulder to the light side of the major (free-DNA) peak. This shoulder is a highly enriched fraction of protein–DNA crosslinked complexes.
7. Collect the material from the peak (or, better, from the shoulder only) and dialyze extensively against 10 m*M* Tris-HCl, pH 7.5, and 0.25 m*M* EDTA.

3.3. Dot Immunoassay for Abundant Crosslinked Protein–DNA Complexes

1. Dot the crosslinked material (about 0.5 µg DNA) on nitrocellulose filters.
2. Wash filters twice for 5 min in PBS-T with gentle shaking to remove unbound antigen.
3. Repeat **step 2** with PBS only.
4. Block the filters in BSA (1% in PBS-T) for 1 h at 37°C.
5. Incubate overnight at 4°C with a suitable dilution of specific antibody in PBS-T, and 1% BSA.
6. Wash three times with PBS-T then three times with PBS only with gentle shaking.
7. Incubate with peroxidase-conjugated goat anti-rabbit IgG (Sigma, 1:1000 dilution) in PBS containing 1% BSA for 4 h at 37°C.
8. Repeat **step 6**.
9. Develop filters in 0.3% 4-chloro-1-naphtol, 0.03% H_2O_2 in 50 m*M* Tris-HCl, pH 7.5, and 150 m*M* NaCl.

3.4. Immunoprecipitation of Crosslinked Protein–DNA Complex

1. Suspend 0.05 mL IgGsorb in 0.5 mL of 1% BSA in PBS and shake for 30 min at room temperature to block the sites of nonspecific absorption.
2. Centrifuge for 30 s in a microcentrifuge (*see* **Note 8**) and suspend IgGsorb directly in 0.5 mL of a mixture of the specific antibody and the crosslinked DNA–protein complexes (w:w ratio 1:2.5) in antibody buffer (20–50 µg of crosslinked material, *see* **Note 9**).
3. Shake for 2 h at room temperature.
4. Recover IgGsorb by microcentrifugation and resuspend in 0.5 mL antibody buffer. Repeat this procedure five times.

5. Recover IgGsorb by microcentrifugation and resuspend in 0.5 mL of rinse buffer. Repeat this procedure three times.
6. Release the immunoprecipitate in 0.1 mL of 3.5 *M* KSCN, and 20 m*M* Tris-HCl, pH 8.2, and remove IgGsorb by microcentrifugation.
5. Treat eluate with RNase A (0.015 mg per sample, 30 min, room temperature [RT]) and then with Pronase (1 mg/mL for at least 4 h, RT).
6. Precipitate released DNA with ethanol (3 vol, leave overnight at –20°C).
7. Recover precipitate by 10-min microcentrifugation and resuspend in 10 m*M* Tris-HCl, pH 7.5, and 0.25 m*M* EDTA.

3.5. Identification of Crosslinked DNA Sequences by DNA Hybridization

1. The DNA samples from **Subheading 3.4.** are alkali denatured and loaded onto Zeta-Probe Blotting membranes (Bio-Rad); *see* **Notes 9** and **10**.
2. Prehybridize the membranes in Pre-hyb Buffer for 3–5 h at 42°C.
3. Hybridize at 42°C for 16–20 h with 50–100 ng of DNA probe, ^{32}P-labeled by random priming.
4. Wash the filters extensively with 0.5X SSC, 0.5% SDS at 42°C and, finally, with 0.1X SSC at 65°C and autoradiograph at –70°C using Cronex Lightning-plus intensifying screens (Dupont).

4. Notes

1. When working with the laser crosslinked protein–DNA complexes, never use acids: The crosslinked adducts are unstable under acidic conditions.
2. To date we have found that Nd:YAG lasers are most suitable for protein–DNA crosslinking. The description of the lasers and the irradiation techniques may sound obscure to biologists, but most laser spectroscopy laboratories will have the equipment necessary.
3. 2-Mercaptoethanol or dithiothreitol in solutions will reduce the efficiency of crosslinking by 50–60%.
4. The optical density of the nuclei can be determined in 8 *M* urea or 1% SDS.
5. Micrococcal nuclease was used to reduce the molecular weight of DNA in irradiated samples. Alternatively, sonication could be used. Recover the irradiated nuclei by centrifugation and resuspend them in 300–400 µL of 1% sarkosyl in an Eppendorf tube (A_{260} = 10–12). Sonicate the sample with a Model W-35 sonicator (Heat Systems Ultrasonics Inc.) or equivalent, using a microtip at a power setting of 5 for ten 30-s bursts in an ice bath. Under these conditions, the size of DNA is reduced to about 150–200 bp.
6. Hydroxyapatite is used to remove excess protein and so forth from the crude nuclear lysate. To avoid large losses, the molecular weight of the DNA must be reduced to 200–300 bp before it is applied to the hydroxyapatite column. Even when purifying low molecular material on this column, we sometimes lose 20–25% of the loaded material. This step is not essential and hence is best omitted if only small quantities of the irradiated samples are available.

7. Although shoulder in the CsCl sedimentation profile contains an enriched fraction of the protein–DNA complexes, some crosslinked material is present in other regions of the DNA peak. Thus, if it is essential to recover all crosslinked material, the complete DNA peak must be pooled. Free DNA does not significantly interfere with the immunoprecipitation procedure. Overexposure of DNA to UV-laser irradiation can induce such severe damage that this of itself will reduce the DNA density and give rise to aberrant peaks. This situation can be recognized because usually a large proportion of the DNA will be shifted toward the light end of the gradient.
8. When washing the IgGsorb, microcentrifuge for 30 s only. Longer centrifugation causes the pellet to become very compact and difficult to resuspend.
9. Usually, 20–50 µg of crosslinked material (from the whole CsCl peak) were taken for the immunoprecipitations. The results with smaller samples were found to be irreproducible.
10. In our studies, different membranes were used for the hybridization. Best results were obtained with the Zeta-Probe membrane (Bio-Rad).
11. If a quantitative estimation of the protein(s) present on a specific DNA sequence(s) is required, as for example was made for the histones on the *Xenopus* rDNA ***(16)***, the following procedure is recommended. Dot aliquots from the antibody-precipitated DNA preparations on a Zeta-Probe filter. Apply increasing amounts of genomic DNA (in the range 50–1000 ng) to the same filter to produce a calibration curve and repeat exactly the same set of dots on a second filter. Hybridize one filter with the specific DNA probe and the second with labeled total genomic DNA. Exposure of the autoradiogram must be made within the linear range of the film or using a phosphorimager. Scan the hybridization signals and estimate (in nanograms of bulk DNA) the amount of DNA precipitated by the antibody, using the respective calibration curve (genomic hybridization signal versus nanogram of total DNA or sequence-specific signal, versus nanogram of total DNA). If the crosslinked protein does not exhibit sequence-specific DNA binding, the signals obtained with the two hybridization probes, although differing in magnitude, should indicate the same amount of bulk DNA. If, however, the protein is a sequence-specific one, the two results will differ.

Acknowledgments

This work was supported by project grant from the Medical Research Council (MRC) of Canada. T.M. is an MRC of Canada Scientist and a member of the Centre de Recherche en Cancérologie de l'Université Laval which is supported by the FRSQ of Québec.

References

1. Welsh, J. and Cantor, C.R. (1984) Protein–DNA crosslinking. *Trends Biochem. Sci.* **9,** 505–507.
2. Markovitz, A. (1972) Ultraviolet light-induced stable complexes of DNA and DNA polymerase. *Biochim. Biophys. Acta* **281,** 522–534.

3. Labbé, S., Prévost, J., Remondelli, P., Leone, A., and Séguin, C. (1991) A nuclear factor binds to the metal regulatory elements of the mouse gene encoding metallothionein-I. *Nucleic Acids Res.* **19,** 4225–4231.
4. Hockensmith, J. W., Kubasek, W. L., Vorachek, W. R., and Von Hippel, P. H. (1986) Laser cross linking of nucleic acids to proteins. Methodology and first applications to the phage T4 DNA replication system. *J. Biol. Chem.* **261,** 3512–3518.
5. Harrison, C. A., Turner, D. H., and Hinkle, D. C. (1982) Laser crosslinking of *E. coli* RNA polymerase and T7 DNA. *Nucleic Acids Res.* **10,** 2399–2414.
6. Hockensmith, J. W., Kubasek, W. L., Vorachek, W. R., Evertsz, E. M., and Von Hippel, P. H. (1991) *Methods Enzymol.* **288,** 211–236.
7. Careri, G., Fasella, P., and Gratton, E. (1975) Statistical time events in enzymes: a physical assessment. *CRC Crit. Rev. Biochem.* **3,** 141–164.
8. Angelov, D., Stefanovsky, V. Y., Dimitrov, S. I., Russanova, V. R., Keskinova, E., and Pashev, I. G. (1988) Protein–DNA crosslinking in reconstituted nucleohistone, nuclei and whole cells by picosecond UV laser irradiation. *Nucleic Acids Res.* **16,** 4525–4538.
9. Budowsky, E. I., Axentyeva, M. S., Abdurashidova, G. G., Simukova, N. A., and Rubin, L. B. (1986) Induction of polynucleotide–protein crosslinkages by ultraviolet irradiation. Peculiarities of the high-intensity laser pulse irradiation. *Eur. J. Biochem.* **159,** 95–101.
10. Pashev, I. G., Dimitrov, S. I., and Angelov, D. (1991) Crosslinking proteins to nucleic acids by ultraviolet laser irradiation. *Trends Biochem. Sci.* **16,** 323–326.
11. Moss, T., Dimitrov, S. I., and Houde, D. (1997) UV-laser crosslinking of proteins to DNA. *Methods* **11,** 225–234.
12. Dobrov, E. N., Arbieva, Z. K., Timofeeva, E. K., Esenaliev, R. O., Oraevsky, A. A., and Nikogosyan, D. N. (1989) UV-laser-induced RNA–protein crosslinks and RNA chain breaks in tobacco mosaic virus RNA *in situ. Photochem. Photobiol.* **49,** 595–598.
13. Buckle, M., Geiselmann, J., Kolb, A., and Buc, H. (1991) Protein–DNA crosslinking at the lac promoter. *Nucleic Acids Res.* **19,** 833–840.
14. Stefanovsky, V. Y., Dimitrov, S. I., Russanova, V. R., Angelov, D., and Pashev, I. G. (1989) Laser-induced crosslinking of histones to DNA in chromatin and core particles: implications in studying histone–DNA interactions. *Nucleic Acids Res.* **17,** 10,069–10,081.
15. Stefanovsky, V. Y., Dimitrov, S. I., Angelov, D., and Pashev, I. G. (1989) Interactions of acetylated histones with DNA as revealed by UV laser induced histone-DNA crosslinking. *Biochem. Biophys. Res. Commun.* **164,** 304–310.
16. Dimitrov, S. I., Stefanovsky, V. Y., Karagyozov, L., Angelov, D., and Pashev, I. G. (1990) The enhancers and promoters of the *Xenopus laevis* ribosomal spacer are associated with histones upon active transcription of the ribosomal genes. *Nucleic Acids Res.* **18,** 6393–6397.
17. Russmann, C., Stollhof, J., Weiss, C., Beigang, R., and Beato, M. (1998) Two wavelength femtosecond laser-induced DNA–protein crosslinking. *Nucleic Acids Res.* **26,** 3967–3970.

28

Plasmid Vectors for the Analysis of Protein-Induced DNA Bending

Christian Zwieb and Sankar Adhya

1. Introduction

Bending of DNA by proteins plays an important role in transcription initiation, DNA replication, and recombination. The degree of protein-induced DNA-bending can be simply and conveniently determined by combining gel electrophoresis of DNA–protein complexes with the use of special plasmid vectors carrying the bendable DNA sequence *(1,2)*. The vectors contain duplicate sets of restriction sites in a direct repeat order and cloning sites for insertion of protein-binding sequences between the two sets. Restriction enzyme digestion readily generates DNA fragments, which are identical in size but differ in the location of the protein-binding site (**Fig. 1**).

The mobility of a DNA fragment is less when a bend is located at its center than when the same bend is located toward one or the other end *(3,4)*. The bending angle α is defined as the angle by which a segment of the rod-like DNA duplex bends away from linearity. Thus, the bending angle is 0° for a straight DNA-fragment. α can be estimated by measuring μ_M (mobility of the complex with the protein bound in the *middle* of the fragment) and μ_E (mobility of the complex with the protein bound near the *end* of a DNA fragment) using the empirical relationship $\mu_M/\mu_E = \cos(1/2\alpha)$ *(4,5)*.

To carry out a bending experiment, the protein-binding site is inserted into the bending vector. Next, DNA fragments with the binding site located at a variety of positions are generated by digestion of the vector with different restriction enzymes. Finally, DNA–protein complexes formed with these DNA fragments are analyzed by gel electrophoresis to determine the bending angle α. As an example, we illustrate bending at the *lac* promoter by cyclic AMP receptor protein (CRP) *(4,6)*. However, the method is easily adopted to other

From: *Methods in Molecular Biology, vol. 148: DNA–Protein Interactions: Principles and Protocols, 2nd ed.*
Edited by: T. Moss © Humana Press Inc., Totowa, NJ

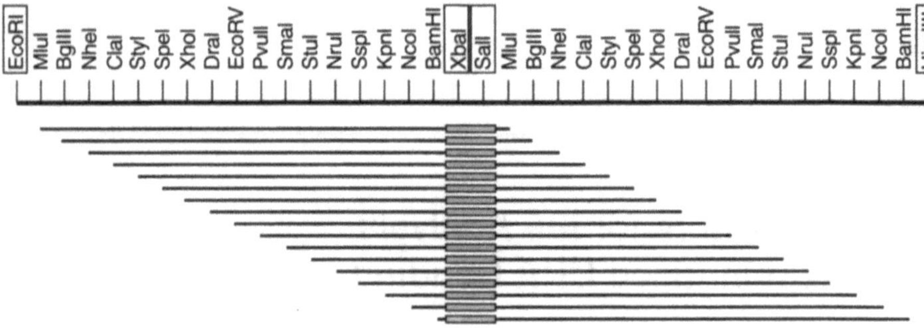


```
EcoRI      MluI  BglII     NheI       ClaI  StyI    SpeI  XhoI     DraI     EcoRV     PvuII SmaI    StuI
GAATTCACGC GTAGATCTGC TAGCATCGAT CCATGGACTA GTCTCGAGTT TAAAGATATC CAGCTGCCCG GGAGGCCTTC

 NruI SspI   RsaI    NcoI            BamHI XbaI   SalI    MluI   BglII    NheI    ClaI   StyI SpeI
GCGAAATATT GGTACCCCAT GGAATCGAGG GATCCTCTAG AGTCGACACG CGTAGATCTG CTAGCATCGA TCCATGGACT

  XhoI      DraI    EcoRV   PvuII  SmaI   StuI    NruI SspI   RsaI NcoI        BamHI     HindIII
AGTCTCGAGT TTAAAGATAT CCAGCTGCCC GGGAGGCCTT CGCGAAATAT TGGTACCCCA TGGAATCGAG GGATCCAAGC TT
```

Fig. 1. Schematic representation of the pBend3 insert located between the *Eco*RI and *Hin*dIII sites. pBend3 was constructed by cloning of the 236-base-pair *Eco*RI–*Hin*dIII fragment of pBend2 *(4)* into pBluescript SK (Stratagene). pBluescript is a high-copy-number plasmid and generates a large amount of DNA upon plasmid extraction. The *Eco*RI–*Hin*dIII fragment contains 17 duplicated restriction sites. The duplicated sites can be used to generate DNA fragments of identical length, but in which the protein-binding sequence (gray rectangle) is shifted. The sites *Xba*I and *Sal*I (in boxes) are unique and suitable for cloning of the protein binding sequence. Restriction sites are not drawn to scale. The sequence of the insert is shown in the lower part of the figure.

protein–DNA complexes if they can be separated from free DNA by gel electrophoresis. General considerations for successful complex formation and for the selection of restriction sites are discussed in the Notes section.

pBend3, 4, and 5 (**Figs. 1** and **2**) are different versions of the bending vector pBend2 *(4)*. pBend2 and pBend3 contain the same 236-base-pair *Eco*RI–*Hind*III fragment with 17 duplicated restriction sites. Because of the higher copy number of pBend3, preparation of plasmid DNA is more efficient. The remainder of the digested vector DNA is less likely to comigrate with protein–DNA complexes; therefore, larger DNA-binding proteins with a high degree of bending can be analyzed without tedious purification of individual fragments containing the protein-binding site. pBend4 and pBend5 contain an additional *Hpa*I cloning site to facilitate the insertion of DNAs with blunt ends. The promoters for T_3 and T_7 polymerase have potential use for analyzing the

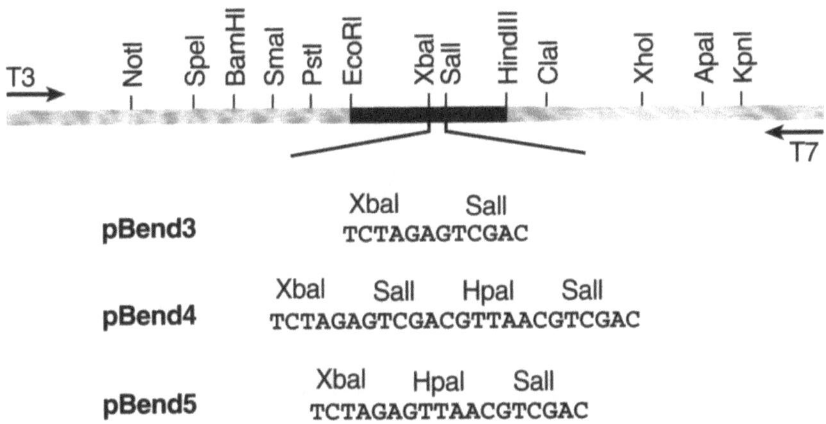

Fig. 2. Restriction and cloning sites of pBend3, 4, and 5. Portions of pBluescript SK (Stratagene) are shown in gray; the *Xba*I and *Sal*I sites of pBluescript are abolished by partial digestion with *Xba*I or *Sal*I, filling in with DNA polymerase (Klenow) and ligation with T4 DNA ligase. Promoters for T_3 and T_7 polymerase are indicated by the arrows. They can be used to study the bending by proteins that bind to double-stranded RNA or to RNA–DNA hybrids. The region between the *Eco*RI and *Hin*dIII sites (indicated in black) is identical to the one shown in **Fig. 1**. pBend4 and pBend5 contain additional *Sal*I and *Hpa*I sites as shown.

bending of double-stranded RNA or RNA/DNA hybrids by proteins such as transcription factor TFIIIA *(7)*. Details of the construction of pBend3, 4, and 5 are described in the legends to **Figs. 1** and **2**.

For unidirectional cloning of the protein-binding site into any of the pBend vectors, one has to use two cloning sites (e.g., *Xba*I and *Sal*I). Because the efficiency of cleavage by both enzymes is somewhat impaired when the two target sites are close to each other, incomplete cutting results in a large amount of recircularized vector DNA without insert after ligation. This may make cloning of a binding site difficult. To make detection of cloned inserts easier, another pBend-derivative pBendBlue (**Fig. 3**) has been made *(8)*. pBendblue, available from the authors in this reference, allows a color (blue/white) screening of the transformants after cloning of the inserts. The construction of the pBendBlue plasmid is described in the legend to **Fig. 3**.

2. Materials

2.1. Insertion of the Protein-Binding Site into a pBend Plasmid

1. pBend vector (provided by the authors upon request).
2. Oligonucleotides or DNA fragments containing the binding site of the protein being investigated with ends compatible with the cloning sites of pBend.

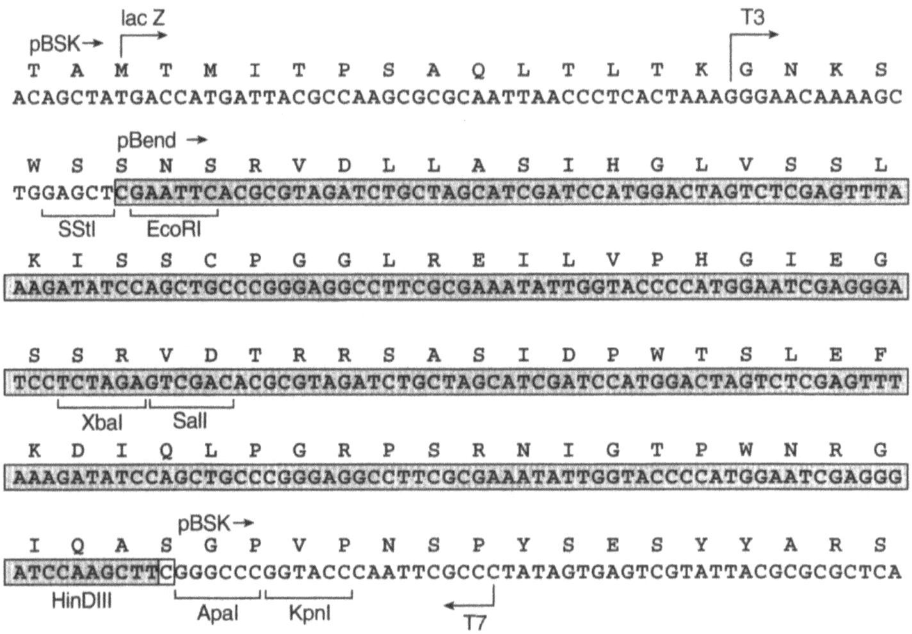

Fig. 3. Restriction and cloning sites of pBendBlue along with the ORF of the *lacZ*(α) gene passing through the entire permuted sequence of pBend3, including the two cloning sites (*Xba*I and *Sal*I). The 241-bp permuted sequence of pBend, including the *Xba*I and *Sal*I cloning sites (shown in shade) was inserted between the *Sst*I (*Sac*I) and ApaI sites of plasmid pBSK (strategene). The *Sst*I and ApaI sites are two of the cloning sites within the *lacZ* gene of pBSK plasmid. The 241-bp pBend2 segmented is cloned in such a way that the *lacZ* ORF remain uninterrupted when passing through the entire *Eco*RI and *Hin*dIII segment of pBend3. The amino acid sequence of the *lacZ*(α) is shown above the DNA sequence. Note that pBSK carries only the α-complementing portion of *lacZ*, necessitating the use of Δ*lacZ*α host (strain *Epicuran coli*, Strategene), which is an w donor. A complementation between α and ω makes a cell β-galactosidase proficient. The inserted segment is also bounded by T_3 and T_7 promoter sequence, respectively, both within the ORF.

3. Restriction enzymes: *Hpa*I, *Eco*RI, *Hin*dIII, and *Mlu*I at about 10 U/μL with 10 times concentrated digestion buffers as specified or provided by the vendors.
4. 500 m*M* EDTA pH 8: Dissolve 93 g of disodium ethylene diamine tetraacetate. $2H_2O$ in 400 mL water, adjust pH to 8.0 by adding about 20 g of NaOH pellets (only then will EDTA dissolve completely), adjust volume to 500 mL, and autoclave.
5. TE: 10 m*M* Tris-HCl, pH 7.5, 1 m*M* EDTA. Mix 1 mL of 1 *M* Tris-HCl, pH 7.5, and 200 μL of 500 m*M* EDTA, pH 8.0 in a total volume of 100 mL water. Store at 4°C.

6. 7.5 *M* ammonium acetate: Dissolve 57.8 g ammonium acetate in a total volume of 100 mL water. Store at 4°C.
7. 80% Ethanol: Mix 80 mL ethanol with 20 mL water. Store at 4°C.
8. Oligonucleotide annealing buffer: Mix 1 mL of 1 *M* Tris-HCl, pH 7.5, 200 µL of 500 m*M* EDTA pH 8.0, and 10 mL of 1 *M* NaCl in a total volume of 100 mL.
9. Ligation buffer, five times concentrated: 500 m*M* Tris-HCl, pH 7.6, and 100 m*M* $MgCl_2$, 100 m*M* dithiothreitol (DTT). Store at –20°C
10. 5 m*M* ATP: Dissolve 3 mg of ATP (disodium salt) in 1 mL of TE. Store at –20°C.
11. T_4 polynucleotide kinase: 10 U/µL.
12. T_4 DNA ligase: 1 U/µL.
13. Competent *Escherichia coli* cells (e.g., strain DH5α.
14. LB-amp plates: Suspend 10 g of LB powder and 7.5 g of Bacto-agar in 500 mL of water; autoclave and dissolve agar by swirling, cool solution to 55°C in a water bath, add 50 mg of ampicillin (for final concentration of 100 µg/mL), dissolve by swirling, and pour plates and store in a plastic bag at 4°C.
15. LB media: Suspend 10 g of LB powder in 500 mL water and autoclave. Store at room temperature. For the ampicillin containing LB, add and dissolve the specified amount to the media at room temperature.
16. Tris–sucrose: 5 m*M* Tris-HCl (pH 7.8) containing 10% (w/v) sucrose (prepare fresh).
17. Lysozyme solution: 10 mg/mL in 250 m*M* Tris-HCl, pH 8.0. Keep frozen aliquots at mL 20°C, thaw once immediately before use, and discard unused portions.
18. 200 m*M* EDTA, pH 8.0: Prepare by dilution of 500 m*M* EDTA (**item 4**).
19. TLM: Mix 3 mL of 10% Triton X100, 75 mL of 250 m*M* EDTA pH 8.0, 15 mL of 1 *M* Tris-HCl, pH 8.0, and 7 mL water.
20. Phenol: Add 62.5 mL water to bottle with 250 g phenol, mix, warm as little as possible, add 300 mg of 8 mL hydroxychinoline, and fill 20-mL aliquots in 30-mL Falcon tubes. Upon use, thaw, add 1 mL of 1 *M* Tris base to one aliquot. Keep refrigerated not longer than 1 mo.
21. RNase A: 250 µg/mL in 10 m*M* Tris-HCl, pH 7.5. Store at mL –20°C.
22. 20X Tris–acetate: Dissolve 96.8 g of Tris base, 22.84 mL glacial acetic acid, and 40 mL EDTA (500 m*M*, pH 8.0), and add water to a final volume of 1 L.
23. 2% Agarose gel, about 5 mm thick, 7 cm long, and 10 cm wide: Mix 0.6 g agarose (electrophoresis grade), 1.5 mL of 20X concentrated Tris–acetate and 30 mL water in a 100-mL reagent bottle. Be sure that the cap of the bottle is loose before melting agarose in a microwave oven. Swirl the mixture occasionally to dissolve the agarose completely. Adjust the volume to 30 mL with water and pour the gel. Insert a comb for about 3-mm-wide slots and let the agarose solidify. Cover the gel with Tris–acetate electrophoresis buffer containing 1 µg/mL ethidium bromide (**Caution:** Ethidium bromide is mutagenic).
24. 10 mg/mL Ethidium bromide solution: Dissolve 1 g of ethidium bromide in 100 mL of water by stirring for several hours. Store in the dark at 4°C.
25. Agarose loading buffer: Mix 1 vol of 50% glycerol, 1 vol of Tris–acetate electrophoresis buffer, and 1/10 vol of a 2.5% (w/v) bromophenol blue solution in TE.

26. DNA molecular-weight markers in the range of 100–1000 base pairs (e.g., *Hae*III digest of bacteriophage ΦC174 DNA).
27. Ultraviolet (UV) transilluminator.
28. X-ray film for autoradiography.
29. Horizontal electrophoresis apparatus for agarose gel, approx dimensions: 7 × 10 cm.

2.2. Detection of Plasmid Clones by Blue/White Color

LB-amp–XG plates: LB-amp plates (*see* **Subheading 2.1., item 14**) containing 0.05 mg/mL of X-gal and 5×10^{-5} *M* of IPTG. Stocks of 2% X-gal in dimethylsulfoxide and 0.1 *M* IPTG in water can be made and appropriate amounts can be spread onto LB-amp plates the day before use.

2.3. Purification of Plasmid DNA

1. *E. coli* suspension buffer: 50 m*M* Tris-HCl, pH 8.0, and 100 m*M* NaCl; prepare by mixing 50 mL of 1 M Tris-HCl, pH 8.0, and 100 mL of 1 *M* NaCl in a total volume of 1 L. Store at 4°C.
2. Tris–glucose–EDTA: 25 m*M* Tris-HCl, pH 8.0, 50 m*M* glucose, and 10 m*M* EDTA, prepare freshly by mixing and dissolving 2.5 mL of 1 *M* Tris-HCl, pH 8.0, 1 g glucose, and 2 mL of 500 m*M* EDTA pH 8.0 in a total volume of 100 mL of water.
3. NaOH–SDS (sodium dodecyl sulfate): 200 m*M* NaOH and 1% SDS, prepare by mixing 10 mL of 10% SDS and 2 mL of 10 *N* NaOH in a total volume of 100 mL of water.
4. 3 *M* sodium acetate pH 4.8. Adjust pH with glacial acetic acid.
5. 2 *M* ammonium acetate: dissolve 154.2 g ammonium acetate in a total volume of 1 L of water. Store at 4°C.
6. Isopropanol.
7. Cheese cloth.
8. Quick-Seal polyallomer centrifuge tubes (e.g., Beckman No. 342413).
9. Cesium chloride.
10. 1 *M* NaCl: Dissolve 58.44 g of NaCl in 1 L of water, autoclave, and store at room temperature.
11. CsCl mix: Dissolve 122.1 g of cesium chloride (DNA grade) in 128 mL TE; add 4.13 mL of ethidium bromide (10 mg/mL). The final volume is 165 mL.
12. *n*-butanol: Mix *n*-butanol with an equal volume of water and an amount of cesium chloride that leaves some undissolved. Use only the upper (*n*-butanol) phase.

2.4. Analysis of DNA–Protein Complexes

1. Vertical electrophoresis apparatus for polyacrylamide gel, approx dimensions 15 × 15 cm.
2. Acrylamide/*bis*-acrylamide (30/0.8, w/w): Add 30 g acrylamide and 0.80 g *bis*-acrylamide dissolved in 100 mL water.
3. 10X TBE: Dissolve 108 g Tris-base, 55 g boric acid, and 9.3 g EDTA (ethylenediaminetetraacetic acid, disodium salt) in water. Fill up to a total volume of 1 L.

4. 8% Polyacrylamide slab gel (about 15 × 15 cm, 1 mm thick): Mix 31 mL of water, 13.3 mL of 30% acrylamide/*bis*-acrylamide, 5 mL of 10X TBE, 600 µL of 10% APS (dissolve 1 g of ammonium persulfate in 10 mL of water), and 60 µL of TEMED (*N,N,N',N'*-tetramethylethylene diamide). Pour solution between the glass plates of the assembled vertical electrophoresis apparatus. Insert a comb for about 1-cm-wide slots. Let the gel polymerize for several hours, preferably overnight.
5. 5X DNA–protein binding buffer for Gal repressor: 50 m*M* KCl, 10 m*M* Tris-HCl, pH 7.5, 50 m*M* KCl, 1.0 m*M* EDTA, 20 µ*M* cyclic AMP, 50 µg/mL bovine serum albumin (BSA), 10% glycerol.
6. Ethidium bromide stain: Add 100 µL of ethidium bromide solution (10 mg/mL) to 1 L of TE.

2.5. E. coli *Host Strains*

For pBendBlue, Epicuran coli SURE2 competent cells (Strategene) are recommended. For other pBend plasmids, any transformation proficient *E. coli* strains may be used.

3. Methods

3.1. Insertion of the Protein-Binding Site into pBend

1. Add 25 µL of pBend DNA (1 mg/mL), 10 µL of 10X *Hpa*I digestion buffer, 55 µL of water, and 10 µL (100 U) of *Hpa*I restriction endonuclease to a 1.5-mL Eppendorf tube. Mix and incubate at 37°C for 2 h or overnight.
2. To precipitate the restriction enzyme, place the digest on ice, add 3 µL of 500 m*M* EDTA pH 8, 100 µL of ice-cold TE, and 100 µL of ice-cold 7.5 *M* ammonium acetate. Keep the sample at 4°C for 10 min. Centrifuge in a tabletop centrifuge for 10 min. Remove the supernatant (containing the DNA) and add it to 600 µL ice-cold ethanol. Incubate at mL –70°C for 20 min, and centrifuge in a tabletop centrifuge for 10 min. Discard the supernatant, add 500 µL of ice-cold 80% ethanol, centrifuge for 5 min, and discard the supernatant. Dry the DNA pellet in a vacuum centrifuge and dissolve the sample in 250 µL of TE. Store the linearized pBend DNA at mL –20°C.
3. Synthesize two complementary oligonucleotides that, when annealed to each other, form the protein-binding site. (*See* **Note 1** for the design of DNA inserts.) The purification of the oligonucleotides is likely to be unnecessary if they are shorter than 30 nucleotides. Dissolve each oligonucleotide in autoclaved distilled water at a concentration of 200 µg/mL.
4. Add 10 µL of each oligonucleotide and 180 µL of oligonucleotide annealing buffer to a 1.5-mL Eppendorf tube. Incubate the sample for 3 min in a 300-mL beaker with about 150 mL of boiling water. Place the beaker with the tubes in the coldroom at 4°C to allow for annealing of the oligonucleotides over a period of several hours. Store the DNA at mL –20°C.
5. Mix in a 1.5-mL Eppendorf tube, 2 µL of annealed oligonucleotides, 3 µL of five times concentrated ligation buffer, 1 µL of 5 m*M* ATP , 8.5 µL of water,

and 0.5 µL of T_4 polynucleotide kinase. Incubate for 10 min at 37°C. Place the sample on ice and add 3 µL of five times concentrated ligation buffer, 1 µL of linearized vector DNA (from **step 2**), 1 µL of 5 m*M* ATP, 9 µL of water, and 1 µL of T_4 DNA ligase. Incubate at 15°C for several hours or overnight. The samples can be stored in a refrigerator for several days and aliquots can be used for several transformations.

6. Transform competent *E. coli* cells according to the protocol provided by the vendor and plate on LB-amp plates. Incubate the plates at 37°C overnight or until the colonies appear.
7. For preparation of the plasmid DNA on a small scale, use sterile toothpicks to transfer individual colonies to 15 mL tubes containing 5 mL of LB with 200 µg/mL ampicillin; also, streak cells from each transformant onto a LB-amp plate. Incubate this master plate at 37°C and shake the liquid cultures at 37°C overnight.
8. Pellet the cells by centrifugation for 15 min at about 700*g* at 4°C (e.g., at 3000 rpm in a Sorvall RT6000B refrigerated centrifuge with a H1000B rotor). Decant the supernatant, add 200 µL Tris–sucrose and transfer to 1.5-mL Eppendorf tubes. Add 25 µL of lysozyme solution. Mix and add 130 µL of 200 m*M* EDTA pH 8.0, and 130 µL of TLM. Mix and place at 65°C until lysis occurs (which usually takes a few min). Vortex briefly and centrifuge for 15 min in a tabletop centrifuge. Remove the pellet with a sterile toothpick; add half a volume of the prepared phenol and half a volume of chloroform. Vortex for 10 s, centrifuge for 10 min and carefully remove about 200 µL of the aqueous (upper) phase while staying clear of the interface. Add 400 µL of ice-cold ethanol, mix and centrifuge for 5 min, decant the supernatant, add 1 mL of 80% ethanol, centrifuge for 2 min, carefully decant the supernatant and dry the pellet in a vacuum centrifuge. Dissolve the pellet in 30 µL of TE with occasional mixing. Store the samples at –20°C.
9. To verify successful insertion of the protein-binding site, digest an aliquot of the DNA with *Eco*RI and *Hin*dIII. To a 5-µL aliquot of the plasmid preparation, add 2 µL of water, 1 µL of 10 times concentrated *Eco*RI digestion buffer, 1 µL of *Eco*RI, 1 µL of *Hin*dIII, and 1 µL of RNase. As a control, digest 1 µg of pBend DNA. Incubate all samples at 37°C for several hours or overnight. Place digests on ice, add 1 µL of 200 m*M* EDTA pH 8.0, 90 µL of ice-cold TE, and 50 µL of ice-cold 7.5 *M* ammonium acetate. Keep on ice for 10 min. Centrifuge in a tabletop centrifuge for 10 min. Collect the supernatant and add it to an Eppendorf tube containing 300 µL ice-cold ethanol. Mix and incubate at mL 70°C for 20 min. Centrifuge in a tabletop centrifuge for 10 min. Remove supernatant, add 300 µL ice-cold 80% ethanol to the pellet, centrifuge for 5 min, and discard supernatant. Carefully dry the pelleted DNA in a vacuum centrifuge and dissolve it in 5 µL of TE. Add 5 µL of Tris–acetate loading buffer and mix briefly.
10. Prepare a 2% agarose gel. Load the samples from **step 9** in parallel with DNA molecular-weight markers. Electrophorese at 80 V until the bromophenol blue has migrated about 4 cm. Examine the DNA under a UV transilluminator and take a picture with a Polaroid camera (film type 57 or 55). Successful insertion is

indicated by an *Eco*RI–*Hin*dIII fragment of the expected mobility (242 base pairs plus insert). Electrophoresis can be continued to discover minor mobility differences, but then fresh electrophoresis buffer should be used. Eventually, the nature of the positive clone must be verified by DNA sequencing, which also reveals the orientation of the inserted binding site (*see* **Note 2** for the selection of suitable sequencing primers).

3.2. Detection of the Protein-Binding Site Cloned into pBend Blue by Blue/White Color Screening

After ligation of the DNA segment corresponding to the protein-binding site into pBendBlue DNA, as described in **steps 1–5** of **Subheading 3.1.**, transform competent *E. coli* cells (strain Epicuvan) and plate on LB-amp–XG plates. Incubate the plates at 37°C overnight or until the colonies are large enough to distinguish their blue/white color phenotype. By this procedure, usually 1–10% of the transformed colonies on LB-amp–XG plates are white. Verify the white colonies by purifying on LB-amp–XG plates. Sequence verification shows that almost all of the white colonies contain the desired insert. Thus, when the cloning efficiency is poor, the color screening allows successful use of pBendBlue in cloning short DNA sequences for studying DNA binding.

3.3. Purification of pBend DNA

1. To obtain pure DNA of the positive pBend derivative, set up a 5-mL culture of the positive clone in LB with 500 µg/mL of ampicillin starting from an individual colony of the master plate (**step 7** of **Subheading 3.1.**). Shake at 37°C for several hours until the culture becomes turbid. Transfer the cells to a 2-L sterile Erlenmeyer containing 400 mL of LB with ampicillin. Shake overnight at 37°C.
2. Place the culture on ice and transfer the cells into centrifuge bottles. Pellet the cells by centrifugation at 4°C at about 1600*g* (e.g., 3500 rpm in a H6000A rotor of a Sorvall RC3C centrifuge). Decant the supernatant, resuspend the pellet in 20 mL *E. coli* suspension buffer and transfer the cells to a 50-mL centrifuge tube (preferably Nalgene, cat. no. 3131-0024). Centrifuge at 4°C for 10 min at about 5000*g* (e.g., in Sorvall SS34 rotor at 10,000 rpm). Freeze the pellet completely by placing the sample on dry ice or in a –80°C freezer.
3. Thaw the pellet and resuspend the cells in 6 mL of Tris–glucose–EDTA. Add 12 mg of lysozyme powder, mix and keep on ice for 30 min. Bring to room temperature and add 12 mL of NaOH–SDS. Mix and place on ice for 5 min.
4. Add 9 mL of 3 *M* sodium-acetate pH 4.8, shake, and leave on ice for 30 min. Centrifuge at about 12,000*g* at 4°C for 15 min (e.g., in an SS34 rotor at 15,000 rpm).
5. Transfer the supernatant to a new centrifuge tube by filtering through a cheesecloth. Add half a volume of isopropanol to the transferred solution, leave 5 min at room temperature and centrifuge at 4°C for 10 min at 5000 g (e.g., in the SS34 rotor at 10,000 rpm).

6. Discard the supernatant and add 6 mL of ice-cold 2 *M* ammonium-acetate to the pellet. Vortex repeatedly to dissolve the plasmid DNA until only small particles are visible. Centrifuge at 4°C for 10 min at 5000*g*.
7. Transfer the supernatant (containing the plasmid DNA) to a new centrifuge tube, add 4 mL of isopropanol, mix, and centrifuge at 4°C for 10 min at 5000*g*.
8. Discard the supernatant and completely dissolve the pellet in 7 mL of TE with occasional shaking. Add 8 g of cesium chloride, 400 µL of 1 *M* NaCl, 400 µL of 1 *M* Tris-HCl, pH 8.0, 160 µL of ethidium bromide (10 mg/mL) and 2.5 mL of CsCl mix.
9. After the CsCl is dissolved, draw the solution into a 20-mL syringe and transfer it into a Quick-Seal centrifuge tube. Fill a second tube with CsCl mix and make sure that the two tubes are balanced. Seal the tubes and centrifuge overnight at about 20,000*g* at 20°C (e.g., at 50,000 rpm in a Beckman NTV65 rotor).
10. Remove the tubes from the rotor, puncture the top of the tube, then collect the lower of the two visible bands with a syringe by puncturing the side of the tube. Transfer the DNA into a 15-mL Corex glass centrifuge tube.
11. Extract the ethidium bromide by adding 1 mL of *n*-butanol that has been saturated with water and cesium chloride. Vortex and remove the upper phase with a glass pipet. Repeat this process beyond the point where the color becomes invisible (usually about six times).
12. Add 2.5 mL of water and 7 mL of ethanol. Mix and incubate at mL –70°C for 15 min. (Do not leave too long, otherwise CsCl will precipitate). Centrifuge for 15 min at 4°C at about 7000*g* preferably in a swinging-bucket rotor (e.g., Sorvall HB4 rotor at 10,000 rpm).
13. Pour off the supernatant, add 5 mL of ice-cold 80% ethanol to the pellet, repeat the centrifugation, discard the supernatant and evaporate excess ethanol under vacuum.
14. Dissolve the DNA in 500 µL of TE and determine the absorbance at 260 nm. Add the appropriate amount of TE to adjust the concentration of the plasmid DNA to 1 mg/mL (1 A_{260} is equivalent to 50 µg/mL). Store the DNA at 4°C.

3.4. Analysis of DNA–Protein Complexes

1. To generate restriction fragments with the protein-binding site located at the end or in the middle, digest the pBend construct with *Mlu*I (end) and *Eco*RV (middle) separately or with the entire set of restriction enzymes with duplicated targets individually. One digestion contains 100 µL (100 µg) of plasmid DNA from **step 14** of **Subheading 3.3.**, 30 µL of 10 times concentrated *Mlu*I (or *Eco*RV) digestion buffer, 160 µL of water, and 10 µL of *Mlu*I or *Eco*RV restriction enzyme (100 U). Incubate the samples at 37°C for several hours or overnight. Place the samples on ice, add 15 µL of 500 m*M* EDTA, pH 8.0, 150 µL of ice-cold 7.5 *M* ammonium acetate and leave at 4°C for 10 min. Centrifuge in a table-top machine for 10 min. Remove and add the supernatant (containing the DNA) to a new Eppendorf tube filled with 900 µL of ice-cold ethanol. Mix and incubate at –70°C for 20 min. Centrifuge for 10 min. Discard the supernatant, add 500 µL of ice-cold 80% ethanol, centrifuge again for 5 min and decant the super-

natant. Carefully dry the pellet in a vacuum centrifuge and dissolve the DNA in 50 µL of water. Verify the success of the digestion by electrophoresis of an aliquot on a 2% agarose gel (described in **item 23** of **Subheading 2.1.**).

2. Pour a vertical 8% polyacrylamide slab gel. Assemble the electrophoresis apparatus and pre-electrophorese at room temperature for 1 h at 100 V with TBE buffer containing 1.0 µ*M* cyclic AMP in the reservoirs.
3. Isolate the different DNA fragments containing the protein-binding site after gel electrophoresis for end labeling, if necessary (*see* **Note 6**). Label the ends of the DNA fragments by T_4 polynucleotide kinase and [γ-^{32}P]ATP as recommended by the supplier of the enzyme.
4. Mix at room temperature 2 µL of digested DNA fragments (from **step 1** or **step 3**), 1.6 µL of five times concentrate solution of CRP binding buffer, 4.4 µL of water, and 2 µL of diluted CRP. Keep the time between diluting the protein and addition to the DNA as short as possible. Do not vortex; mix gently with the tip of the pipet. Prepare a control without added protein. Incubate all samples for 10 min at room temperature.
5. Flush the wells of the polyacrylamide gel with reservoir buffer and load samples without the addition of loading buffer and tracking dyes. The glycerol in the binding buffer gives the sample sufficient density. A long plastic microcapillary tip is helpful for delivering the sample to the bottom of the slot. Glass capillaries should be avoided because proteins tend to stick to glass. In a separate slot, load DNA molecular-weight markers with bromophenol blue. Electrophorese at room temperature for 5 h at 200 V. Separate the glass plates and immerse the gel in ethidium bromide stain for visualization of the DNA under UV light. Take a picture with a Polaroid camera and subsequently autoradiograph gel (consult **Notes 3** and **4** if complexes cannot be detected).
6. Measure the distances between the gel loading slot and the position of the protein–DNA complex of the *Mlu*I (μ_E) and the *Eco*RV digest (μ_M). Also, examine the mobilities of the free DNA (**Fig. 4**) to make sure that the DNA fragments contain no intrinsic bending. Calculate the bending angle α using the empirical formula $\mu_M/\mu_E = \cos(1/2\alpha)$.

4. Notes

1. Protein-binding sites can be inserted into the pBend vectors using restriction fragments or synthetic oligonucleotides. Restriction fragments should not be considerably larger than the protein-binding site to be tested. Oligonucleotides are normally available with blunt ends and can be cloned into the *Hpa*I sites of pBend4 or pBend5 (**Fig. 2**). Newly synthesized oligonucleotides can be designed with "sticky" ends such that they are compatible with the *Xba*I and the *Sal*I sites; they can be cloned more efficiently and inserted in a single orientation.
2. Insertion of the protein-binding site may not occur if the oligonucleotides are of poor quality. In this case, they should be purified and checked by polyacrylamide gel electrophoresis. If the transformation efficiency with supercoil control DNA is high, yet very few transformants are obtained with the annealed oligonucle-

otides, reduce the amount of insert DNA. Multiple insertion of the binding site can occur and is detected by gel electrophoresis and sequencing. For sequencing, use primers named T_3, T_7, M13-20, or reverse primer (Stratagene). Do not use the SK and KS primers (Stratagene) because they are not fully complementary to pBend3, 4, and 5.

3. Bending experiments with radioactively labeled DNA are particularly useful if the protein has not been purified or if its availability is limited. Labeling can be accomplished with T_4 polynucleotide kinase and [γ–^{32}P]ATP. Often, the DNA ends generated by the various restriction enzymes are labeled to different degrees. This problem can be overcome by loading the gel with aliquots of the binding reaction adjusted for the efficiency of fragment labeling. It is best to purify and isolate the protein-binding fragments because the radioactively labeled plasmid–DNA might obscure the region where the complexes are located. Another potential problem (which is also the case with unlabeled DNA) might be that bands appear which represent minor digestion products. In order to identify these, make sure to include controls without added protein.
4. One of the frustrating aspects of conducting a bending experiment can be the inability to detect a complex on the polyacrylamide gel. Even if the binding and electrophoresis conditions are known one should be careful to avoid solutions and equipment that have been in contact with SDS. If possible, dedicate one electrophoresis setup to "gel-shift" experiments. Many DNA-binding proteins are insoluble in the low salt concentration of the electrophoresis buffer and must be stored at high ionic strength. Limit the time between dilution and addition to the DNA. Avoid vortexing during complex formation and do not add tracking dyes because they interact with the complex and might change its mobility. Larger protein–DNA complexes (e.g., Lac repressor *[6]*) behave better in low percentage polyacrylamide gels (e.g., 4%) with a high acrylamide/bis-acrylamide ratio (80:1). The protein concentration for obtaining about equal amounts of free and complexed DNA should always be determined in a preliminary experiment.
5. If many transformants are obtained, but none contain the protein-binding site, the pBend DNA might not have been fully linearized. Alter the DNA–enzyme ratio in favor of the enzyme and confirm complete digestion of an aliquot by electrophoresis on an agarose gel.
6. In the initial bending experiment, it is advisable to restrict only with *Mlu*I or *Bam*HI (to place the binding site close to the ends) and *Eco*RV or *Pvu*II (to place the binding site in the middle). Do not select restriction sites that also occur in the protein-binding sequence. When exploiting the 17 circular permutated restriction sites, attention must be paid to the property of some of the restriction enzymes as follows: ClaI sites are methylated in most *E. coli* strains; its use is therefore limited to prior growth of the plasmid in a methylation-defective (*dam*-) host. *Sty*I will also cut at *Nco*I of the repeat; for the purpose of a bending experiment, it can therefore only be used under partial digestion conditions. An additional *Spe*I site is present in the vicinity of the single *Eco*RI site as part of pBluescript SK (*see* **Fig. 2**). *Spe*I digestion generates an additional small fragment, which contains no

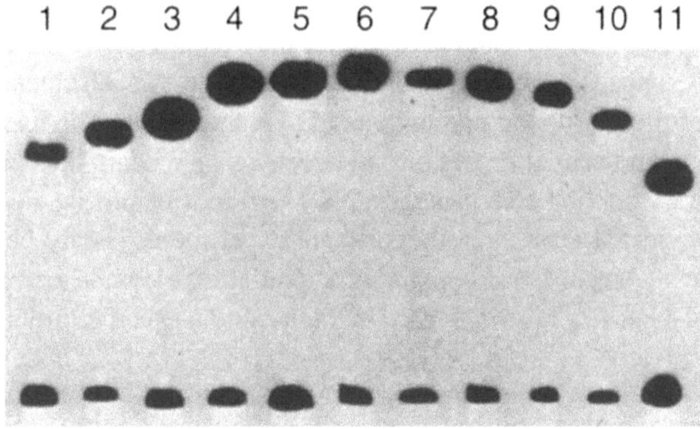

Fig. 4. Gel electrophoresis of permuted fragments of *lac* CRP sites. CRP was mixed separately with 11 different 5'-end ^{32}P-labeled DNA fragments. In a volume of 20 µL, a sample of each labeled fragment was mixed with 10 m*M* Tris-HCl, pH 7.5, 1 m*M* EDTA, 50 m*M* KC1, 20 µ*M* cAMP, 50 µg/mL BSA, 10% glycerol, and 1 n*M* CRP. Polyacrylamide concentration was 10%. The protein and DNA concentrations were such that 50% of the DNA was engaged in complexes with CRP. The DNA fragments used were generated by restriction enzymes, which, from left to right, are *Mlu*I, *Bgl*II, *Nhe*I, *Spe*I, *Xho*I, *Eco*RV, *Pvu*II, *Stu*I, *Nru*I, *Kpn*I, and *Bam*HI. The fragments at the bottom of the gel are free DNA and those at the upper part are bound to the cAMP–CRP complex.

protein-binding site and does not interfere with the bending assay. Three additional *Dra*I sites are located in the plasmid corresponding to pBluescript coordinates 1912, 1931, and 2623. Depending on their electrophoretic property, some of the vector-derived DNA fragments might comigrate with certain protein–DNA complexes. Make sure to include a control without added protein. Likewise, two additional *Pvu*II sites correspond to pBluescript coordinates 529 and 977. A *Sma*I site is present close to *Eco*RI (*see* **Fig. 2**). Two additional SspI sites correspond to pBluescript coordinates 442 and 2850, and two additional RsaI sites correspond to pBluescript coordinates 665 and 2526. *Nco*I will also cut at *Sty*I of the repeat and can only be used under partial digestion conditions. An additional *Bam*HI-site is present close to the *Eco*RI site (*see* **Fig. 2**).

7. The bending angle a assumes a value of 0° for a straight duplex. Since the mobility of a rigid DNA fragment is related to its end-to-end distance, the latter equals $L \cos(^1/_2\alpha)$, with L being the length of the unbent DNA. The end-to-end distance of a fragment bent at the end will be virtually the same as L. Thus, $\mu_M/\mu_E = L \cos(^1/_2\alpha)/L = \cos(^1/_2\alpha)$, where μ_M is the mobility of the complex with the protein bound centrally and μ_E the mobility of the complex with the protein bound at the end of the DNA fragment. The apparent bending angle for the *lac* promoter, induced by CRP is 96° (*see* **Fig. 4** of **ref. 2**). We measure the distance between the top of the gel and the front of the band representing the protein–DNA complex. Whatever method is used, one must be consistent. Possible intrinsic bend-

ing in the free DNA must be considered in the calculation of the bending angle. It should be noted that the calculated values may be different from absolute bending angles, because factors other than the end-to-end distance influence the mobility of protein-bound and unbound DNA fragments. The method measures the net bend and cannot distinguish between a single sharp bend at one position and a smooth curving over a larger DNA region. For precise determination of bending angles, at least three independent experiments should be conducted. If possible, use control lanes containing a similar size complex in which a DNA fragment is bent to a known degree by binding to a specific protein.

References

1. Crothers, D. M. and Fried, M. G. (1983) Transmission of long-range effects in DNA. *Cold Spring Harbor Symp. Quant. Biol.* **47,** 263–269.
2. Zwieb, C., Kim, J., and Adhya, S. (1989) DNA bending by negative regulatory proteins: Gal and Lac repressors. *Genes Dev.* **3,** 606–611.
3. Wu, H.-M. and Crothers, D. M. (1986) The locus of sequence-directed and protein-induced DNA bending. *Nature (London)* **308,** 509–513.
4. Kim, J., Zwieb, C., Wu, C.. and Adhya, S. (1989) Bending of DNA by gene-regulatory proteins: construction and use of a DNA bending vector. *Gene* **85,** 15–23.
5. Thompson, J. F. and Landy, A. (1988) Empirical estimation of protein-induced DNA bending angles: application to site-specific recombination complexes. *Nucleic Acids Res.* **20,** 9687–9705.
6. Fried, M. G. and Crothers, D. M. (1983) CAP and RNA polymerase interaction with the lac promoter: binding stoichiometry and longe-range effects. *Nucleic Acids Res.* **11,** 141–185.
7. Zwieb, C. and Brown, R. S. (1990) Absence of substantial bending in the *Xenopus laevis* transcription factor IIIA–DNA complex. *Nucleic Acids Res* **18,** 583–587.
8. Sperbeck, S. J. and Wistow, G. J. (1998) pBendBlue: modification of the pBend system for color selectability. *BioTechniques* **24,** 66–68.

29

Engineering Nucleic Acid-Binding Proteins by Phage Display

Mark Isalan and Yen Choo

1. Introduction

In the phage display method, peptides *(1)* or protein domains *(2,3)* cloned as fusions to the coat proteins of filamentous bacteriophage are displayed on the capsid, which encloses the viral genome. Proteins of interest and their associated phage can be selected from a large pool of variants (a library) by affinity purification using an appropriate ligand bound to a solid support. Thus, while weakly interacting phage are removed by washing, strongly bound phage are retained and can be subsequently amplified by passage through a bacterial host. Sequential rounds of selection and amplification lead to enrichment of those clones with the highest affinity for the target ligand. The identities of these clones can then be deduced by sequencing part of the phage genome.

Protein–nucleic acid interactions *(4,5)*, which often involve complicated networks of intermolecular contacts, can be investigated expeditiously using phage display because large numbers of protein variants can be screened simultaneously. We and others have used this powerful technique to study DNA binding by the TFIIIA-type zinc finger motif *(6–11)* and, more recently, by RNA binding by zinc fingers *(12,13)*, the HIV-1 Tat protein *(14)*, and an RNP domain from the U1A spliceosomal protein *(15)*. In this chapter, we describe the steps involved in cloning phage display libraries of nucleic-acid-binding proteins (1) preparation of high-quality vector for cloning; (2) preparation of a cassette coding for the protein library to be expressed on phage; and (3) ligation of these two components and transformation of competent bacteria. Also described are the protocols we use to (1) perform selections of nucleic-acid-binding proteins displayed on phage and (2) assay the binding affinity and specificity of selected clones using phage enzyme-linked immunosorbent assay

From: *Methods in Molecular Biology, vol. 148: DNA–Protein Interactions: Principles and Protocols, 2nd ed.*
Edited by: T. Moss © Humana Press Inc., Totowa, NJ

(ELISA). Moreover, since we have found that the success of such phage-display experiments is as dependent on experimental strategy as it is on technique, we have commented on the rationale underlying our published phage-display experiments *(8,11,16)* in **Subheading 4.** We anticipate that these general principles and methods can be applied to the phage display of many other nucleic-acid-binding motifs.

2. Materials

2.1. Preparation of Phage Vector

1. 2X TY medium: 16 g/L Bactotryptone, 10 g/L Bactoyeast extract, and 5 g/L NaCl).
2. Tetracycline.
3. *Escherichia coli* strain expressing the F pilus (e.g., *E. coli* TG1 [F' *traD36 lacIq* Δ*(lacZ)M15 proA+B+/supE* Δ*(hsdM-mcrB)5(*r_K*-*m_K*-McrB-) thi* Δ*(lac-proAB*]) grown on minimal medium.
4. Phage vector suitable for phage display, (e.g., Fd-TET-SN *[8]*).
5. Plasmid purification system (e.g., Wizard Maxiprep kit [Promega]).
6. TE buffer: 10 m*M* Tris, pH 7.4, and 1 m*M* EDTA, pH 8.0.
7. Cesium chloride.
8. Water-saturated butan–2-ol.
9. *Sfi*I 20,000 U/mL, and NEBuffer 2 (New England Biolabs).
10. *Not*I 10,000 U/mL, and NEBuffer 3 (New England Biolabs).
11. Mineral oil.
12. TAE, 50X stock solution: 242 g Tris base, 57.1 mL glacial acetic acid, 37.2 g $Na_2EDTA \cdot 2H_2O$, and H_2O to 1 L.
13. Ethidium bromide, 10 mg/mL.
14. AgaraseI (Sigma).

2.2. Construction of a Gene Cassette Coding for a Protein Library

1. T_4 polynucleotide kinase (10 U/µL) and buffer (New England Biolabs).
2. T_4 DNA ligase (400 U/µL) and buffer (New England Biolabs).

2.3. Cloning of Library DNA Cassette into Phage Vector

1. Electrocompetent *E. coli (*e.g., strain TG1).
2. Electroporation cuvets (2-mm path, Equibio).
3. SOC medium: 0.5% (w/v) yeast extract, 2% (w/v) tryptone, 10 m*M* NaCl, 2.5 m*M* KCl, 10 m*M* $MgCl_2$, 10 m*M* $MgSO_4$, and 20 m*M* glucose.
4. TYE medium: 1.5% (w/v) agar, 1% (w/v) Bactotryptone, 0.5% (w/v) Bactoyeast extract, and 0.8% (w/v) NaCl.

2.4. Phage Selection Against Nucleic Acid Targets

1. 2X nucleic acid annealing buffer: 40 m*M* Tris buffer, pH 8.0, and 200 m*M* NaCl).
2. Streptavidin-coated paramagnetic beads (Dynal AS).

3. Phosphate-buffered saline (PBS): 10X stock: 80 g NaCl, 2 g KCl, 11.5 g $Na_2HPO_4 \cdot 7H_2O$, 2 g KH_2PO_4, water to 1 L.
4. Fat-free freeze-dried milk (Marvel).
5. Tween-20.
6. Sonicated salmon sperm DNA (10 mg/mL).
7. 0.1 *M* triethanolamine.
8. 1 *M* Tris-HCl, pH 7.4.

2.5. Assaying Binding Properties of Selected Clones by Phage ELISA

1. Streptavidin-coated microtiter wells (Roche).
2. Horseradish peroxidase-conjugated anti-M13 IgG (Pharmacia Biotech).
3. 3,3',5,5'-tetramethyl-benzidine (TMB, Sigma).
4. Dimethyl sulfoxide (DMSO).
5. 3 *M* Sodium acetate (pH 5.5).
6. 30% (v/v) hydrogen peroxide.
7. 1 *M* sulfuric acid.

3. Method

3.1. Preparation of Phage Vector

1. Prepare vector DNA (*see* **Note 1**) from a 1-L bacterial culture by using a large-scale plasmid preparation kit (e.g., Wizard Maxipreps, Promega) followed by additional purification on a cesium chloride gradient (*see* **Note 2**). We have found that only cesium-chloride-pure phage DNA is suitable for library construction.
2. Resuspend 40 µg pure vector in 460 µL of 1X NEBuffer 2 containing 100 µg/mL bovine serum albumin (BSA). Add 10 µL (200 U) of *Sfi*I, overlay with mineral oil, and incubate at 50°C. Supplement the reaction with 10 µL (200 U) of *Sfi*I every 2 h to a total incubation time of 8 h.
3. Purify DNA by extracting once with phenol and once with chloroform, followed by ethanol precipitation.
4. Resuspend DNA in 460 µL of 1X NEBuffer 3 containing 100 µg/mL BSA. Add 10 µL (100 U) of *Not*I and incubate at 37°C. Supplement the reaction with 10 µL (200 U) of *Not*I every 2 h to a total incubation time of 8 h.
5. Purify the cut vector by agarose gel electrophoresis prior to cloning. Run the cut vector DNA on a 1% low-melting-point agarose gel, made up in 1X TAE containing 0.5 µg/mL ethidium bromide.
6. Excise the vector DNA band under ultraviolet (UV) light.
7. Extract the vector DNA from the gel slice by digestion using AgaraseI (Sigma). Purify DNA by extracting once with phenol and once with chloroform, followed by ethanol precipitation.
8. Resuspend DNA in sterile water and quantitate by spectrophotometry. Vector may be stored in aliquots at –20°C.

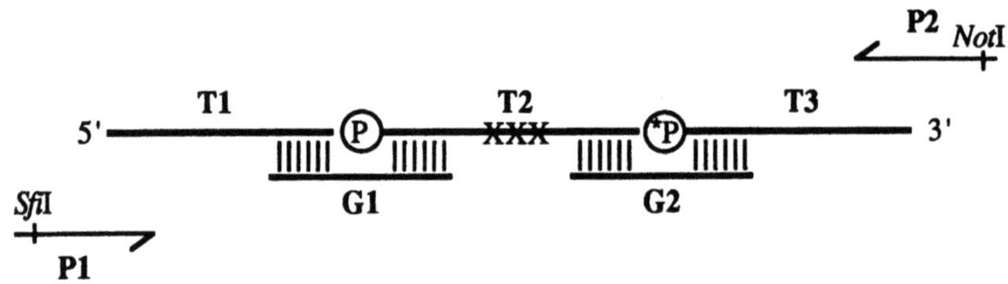

Key:

T1 - 3	Template oligos (coding sequence)
G1 - 2	Guide oligos (non-coding sequence)
P1 - 2	PCR primers
(P)	5' phosphate
(*P)	^{32}P-labelled 5' phosphate
XXX	Nucleotide randomisations
||||||	Base pairs

Fig 1. Construction of a gene cassette coding for the protein library to be displayed on phage. The pool of genes is constructed by end-to-end ligation of synthetic "template" oligonucleotides (T1–T3), containing nucleotide randomizations (marked X), which is directed using "guide" oligonucleotides (G1, G2) that bridge the junctions by sequence-specific annealing. After the ligation reaction, a full-length, double-stranded gene cassette is amplified by the PCR using primers that contain restriction enzyme sites for cloning.

3.2. Construction of a Gene Cassette Coding for a Protein Library

1. Design and synthesize the pool of related genes using end-to-end ligation of synthetic "template" oligonucleotides, directed by annealing "guide" oligonucleotides, according to the schematic in **Fig. 1**. The DNA cassette must code for the genes in frame with the leader sequence and geneIII. **Note 3** provides guidelines regarding the design of randomizations in the protein library.
2. Use polyacrylamide gel electrophoresis to purify about 100 mg of each oligonucleotide. Adjust purified oligonucleotide concentration to 10 pmol/µL in distilled water.
3. 5'-Phosphorylate the "template" oligonucleotides prior to ligation (**Fig. 1**). Take 10 µL of each oligonucleotide at 10 pmol/µL and add 1.2-µL of 10X T_4 ligase buffer (this is essentially T_4 polynucleotide kinase buffer containing 1 m*M* ATP). Add 1 µL (10 U) of T_4 polynucleotide kinase and incubate at 37°C for 1 h. Heat inactivate the enzyme by incubating at 65°C for 10 min.
4. Take 100 pmol of each phosphorylated template oligonucleotide in (12-µL volumes from kinase reactions) and mix with 100 pmol of each guide oligonucleotide

(10 µL of 10 pmol/µL in 1X T_4 DNA ligase buffer). Place in a boiling waterbath for 3 min, then turn off the heat source and leave to cool to room temperature to allow annealing of the template and guide oligonucleotides. Finally, place on ice.

5. Add an equal volume of 1X T_4 DNA ligase buffer to the oligonucleotide mixture. Supplement with 8 µL (3200 U [DNA end ligation units]) of T_4 DNA ligase per 100 µL. Incubate at 16°C for 8–24 h.
6. Carry out a small-scale (25-µL) trial polymerase chain reaction (PCR) reaction, with appropriate primers (as shown in **Fig. 1**), to amplify full-length double-stranded DNA cassette (*see* **Note 4**). Typically, 1 µL of the ligation reaction contains sufficient template. The correct size of the PCR product should be verified by agarose gel electrophoresis. When PCR conditions are optimized, scale up the PCR reaction 10-fold. If the PCR product is not of the predicted size, then gel purification of the ligated single-stranded template may be required. Purification is facilitated by ^{32}P end labeling of the 3'-terminal oligonucleotide used in constructing the gene cassette.
7. Prepare the PCR product for restriction digestion by extracting once with phenol and once with chloroform, followed by ethanol precipitation.
8. Resuspend DNA in 460 µL 1X NEBuffer 2 containing 100 µg/mL BSA. Add 10 µL (200 U) of *Sfi*I, overlay with mineral oil, and incubate at 50°C. Supplement the reaction with 10 µL (200 U) of *Sfi*I every 2 h to a total incubation time of 8 h.
9. Prepare DNA for the second digestion by extracting once with phenol and once with chloroform, followed by ethanol precipitation.
10. Resuspend DNA in 460 µL 1X NEBuffer 3 containing 100 µg/mL BSA. Add 10 µL (100 U) of *Not*I and incubate at 37°C. Supplement the reaction with 10 µL (200 U) of *Not*I every 2 h to a total incubation time of 8 h.
11. Remove small restriction fragments and prepare the cassette for cloning using agarose gel purification: run the cut cassette on 2% low-melting-point agarose gel in 1X TAE containing 0.5 µg/mL ethidium bromide. Excise the band of interest from the gel and extract DNA by any suitable method such as AgaraseI digestion.
12. The DNA yield should be quantitated and the cassette stored in distilled water at –20°C. This protocol gives 1–3 µg of library cassette, of which less than 0.5 µg is typically required to clone a library of 5 million transformants.

3.3. Cloning of the Library DNA Cassette into Phage Vector

1. Ligate DNA insert into a 1 µg (approx 0.2 pmol) phage vector. This is carried out in a total volume of 30 µL in 1X T_4 DNA ligase buffer supplemented with 3 µL (1200 U) of T_4 DNA ligase. Incubate at 16°C for 16 h. A 5:1 molar ratio of insert to the vector should give good yields, but results can be improved by optimizing this ratio after each new vector and insert preparation.
2. Purify ligation products by phenol/chloroform extraction followed by ethanol precipitation. Wash the precipitate thoroughly with 70% ethanol and dry for 5 min in a lyophilizer. Resuspend pellet in 10 µL of distilled water. It is important to minimize the amount of salt carried through because this can cause arcing during electroporation.

3. Electroporate 2-μL (approx 200 ng) samples of ligated vector into 50–70 μL of electrocompetent *E. coli* (*see* **Note 5**) in a 2-mm-path electroporation cuvet (Equibio).
4. Pulse cells in electroporation apparatus set to 2.5 kV and 25 μF, with a pulse controller set to 200 Ω.
5. Immediately resuspend cells in 1 mL room-temperature SOC. Incubate for 1 h at 37°C with shaking.
6. Plate cells out on TYE medium containing 15 μg/mL tetracycline and grow 16 h at 30°C. Large 24-cm × 24-cm plates are convenient for plating out approx 10^6 transformant colonies. Plate out a dilution series of the transformation on smaller plates to estimate the total number of colonies obtained (i.e., the library size).
7. The efficiency of the cloning can be verified by PCR screening 20 randomly selected colonies using internal primers that amplify the cloned cassette (*see* **Note 4**). Individual colonies can be cultured in liquid medium to produce phage (*see* **Note 6**) for DNA sequencing to confirm that the library contains unbiased randomizations at the expected positions.
8. Harvest the library by gently scraping colonies into 1–3 mL of 2X TY. Bacteria may be stored at –70°C after the addition of 50% (v/v) glycerol.
9. To obtain phage particles, grow phage-infected bacteria in 2X TY containing 15 μg/mL tetracycline. Use 50 mL of culture medium/mL of harvested bacteria and incubate for 8–24 h at 30°C.
10. Centrifuge cultures at 500*g* for 20 min to obtain clear phage-containing supernatant. This may be filtered through a 2-μm filter to remove bacteria, stored at 4°C or –20°C, or alternatively used for selection experiments.

3.4. Phage Selection Against Nucleic Acid Targets

1. Nucleic acid targets used for selection are prepared by annealing complementary DNA oligonucleotides together, or refolding an RNA hairpin structure. At least one of the oligonucleotides must be biotinylated. Mix 10 μL of (each) oligonucleotide at 10 pmol/μL with 20 μL of 2X nucleic acid annealing buffer. Place in a boiling waterbath for 3 min, then turn off the heat source and allow to cool to room temperature. Finally, place on ice. Annealed binding sites may be diluted in water to a 1 pmol/μL stock solution stored at –20°C.
2. Fresh phage are prepared for selection by growing phage-infected bacteria in 2X TY containing 15 μg/mL tetracycline for 8–24 h at 30°C. Grow 1 mL of culture for each selection experiment (*see* **Note 7**).
3. Take 50 μL of dynabeads solution and separate the beads from the preservative buffer using a magnet. Then, add 1 pmol of nucleic acid target site to the blocked beads in 50 μL PBS and allow to bind for 30 min (*see* **Note 8**).
4. Start growing 1 mL of fresh TG1 bacteria per selection experiment, in 2X TY at 37°C. The bacteria will be ready when an OD_{600} of 0.6 is reached (this takes about 3 h if 2–5 mL of 2X TY is inoculated with a single colony). In the meantime, **steps 5–10** of the selection should be performed.
5. Block the nucleic-acid-coated beads for 1 h at 20°C by adding 1 mL of PBS containing 4% (w/v) fat-free freeze-dried milk (Marvel).

6. Centrifuge bacterial cultures from **step 2** on a bench-top microfuge at top speed for 10 min to obtain clear phage-containing supernatants.
7. Prepare 1 mL phage binding mixture for each selection (*see* **Note 9**):
   897 μL of PBS containing 2% (w/v) fat-free dried milk (Marvel) and 1% (v/v) Tween-20
   100 μL phage-containing supernatant (*see* **Note 10**)
   2 μL (10 mg/mL) of sonicated salmon sperm DNA (or tRNA) competitor
   1 mL (10 pmol) unbiotinylated competitor nucleic acid
8. Separate the nucleic-acid -coated beads from the blocking mixture and add 1 mL of phage binding mixture. Incubate for up to 1 h at 20°C. RNA selections may require shorter incubations at lower temperatures. We have selected variants of the U1A RNA-binding motif by incubating for 5 min on ice. In this case, the addition of ribonuclease inhibitors (e.g., RNasin [Promega]), is optional.
9. Wash away unbound phage from the beads with 15–20 washes of 1 mL PBS containing 2% (w/v) fat-free freeze-dried milk (Marvel) and 1% (v/v) Tween-20. Vortex the beads thoroughly during washes. The number of washes may need to be optimized for a particular selection. If too few washes are carried out, then selection is poor as many nonspecific binding phage are carried through. Too many washes can reduce phage yield to the point where specific binders are lost. Carry out one final wash with 1 mL PBS.
10. Remove PBS and elute phage from beads by adding 100 μL of 0.1 *M* triethanolamine. Mix on vortex for 1 min, then collect beads and remove the supernatant that should be immediately neutralized with 100 μL of 1 *M* Tris buffer, pH 7.4.
11. Infect 500 μL of fresh logarithmic phase TG1 (from **step 4**) with 50 μL of eluted, neutralized phage solution. Incubate without shaking at 37°C for 1 h. Dilutions of the infected *E. coli* culture may be plated out on TYE containing tetracycline to estimate phage yields in order to follow the progress of the selection over several rounds.
12. Centrifuge the infected culture for 5 min at top speed in a bench-top microfuge to harvest the bacteria. Resuspend bacterial pellet in 2 mL 2X TY supplemented with 15 μg/mL tetracycline, and grow for 16 h at 30°C to obtain phage for further rounds of selection (*see* **Note 11**).
13. After the final round of selection, infected bacteria are plated out and individual colonies are cultured to produce phage for binding assays and DNA sequencing.

3.5. Assaying Binding Properties of Selected Clones by Phage ELISA

1. Prepare biotinylated nucleic acid binding sites as in **step 1** of **Subheading 3.4.**
2. Prepare a fresh phage culture for ELISA (*see* **Note 12**) by inoculating 2 mL of 2X TY containing 15 μg/mL tetracycline with a single bacterial colony and incubating for 8–24 h at 30°C.
3. Add between 0 and 100 pmol of biotinylated nucleic acid target site (in 50 μL of PBS) to each streptavidin-coated microtiter well (250 μL capacity, Boehringer-Mannheim) and allow to bind for 30 min. Initially, it may be worthwhile to carry out binding assays using a range of nucleic acid target concentrations. Bear in

mind the biotin binding capacity of the streptavidin-coated microtiter plate.
4. Block microtiter-plate wells for 1 h at 20°C by adding 150 µL PBS containing 4% (w/v) fat-free freeze-dried milk (Marvel).
5. Centrifuge phage cultures from **step 2** on a bench-top microfuge for 10 min at top speed to obtain clear phage-containing culture supernatant.
6. Prepare 50 µL phage-binding mixture for each ELISA well by mixing 5 µL phage supernatant (*see* **Note 10**) with 45 µL of PBS containing 2% (w/v) fat-free freeze-dried milk (Marvel), 1% (v/v) Tween-20, and 1-µg competitor nucleic acid (e.g., calf liver tRNA or sonicated salmon sperm DNA, depending on the application).
7. Discard blocking mixture from microtiter plate wells and add 50 µL of phage-binding mixture per well. Incubate for up to 1 h at 20°C. Assays using RNA targets may require shorter incubations at lower temperatures. We have assayed variants of the U1A RNA-binding motif by incubating for 5 min on ice. The addition of ribonuclease inhibitors (e.g., RNasin [Promega]) is optional.
8. Remove unbound phage by washing microtiter-plate wells seven times with PBS containing 1% (v/v) Tween-20, followed by three washes with PBS.
9. Remove all PBS and add 100 µL of PBS containing 2% (w/v) fat-free freeze-dried milk (Marvel) and 0.02% (v/v) peroxidase-conjugated anti-M13 IgG (Pharmacia Biotech). Incubate for 1 h at 20°C.
10. About 10 min before the incubation from **step 9** is over, prepare the reagents for the ELISA developing solution. Dissolve a 1-mg TMB tablet in 100 µL of DMSO. Make up a 10-mL solution of 0.1 *M* sodium acetate (pH 5.5) and add 2 µL of 30% (v/v) hydrogen peroxide.
11. Remove unbound antibody by washing microtiter-plate wells three times with PBS containing 0.05% (v/v) Tween-20, followed by three washes with PBS. Discard all traces of PBS from the final wash.
12. Mix the solutions from **step 10** to make developing solution and immediately add 100 µL to each microtiter-plate well.
13. Allow the colorimetric reaction to develop between 1–10 min. The time required for developing depends on factors such as the concentration of nucleic acid target bound to the plate, the amount of phage added to the reaction, and the quality of the antibody used. A blue color develops in microtiter-plate wells that contain phage, whereas empty control wells and wells to which phage have not bound remain colorless.
14. Stop the reaction with 100 µL of 1 *M* sulfuric acid per well, which converts any blue signal to yellow. Absorbance may be quantitated in a plate-reader spectrophotometer at 450 nm.

4. Notes

1. Nucleic-acid-binding motifs such as the zinc finger *(17)* and the ribonucleoprotein motif *(18)* have been displayed on the surface of filamentous bacteriophage as fusions to the minor coat protein encoded by gene III. This type of display can be achieved using either phage vectors (which encode gIII and all functions required for phage replication, packaging and infection, as well as antibiotic

resistance) or phagemid vectors (which comprise only plasmid-encoded gIII and require "rescue" with "helper phage" in order to replicate and package the single-stranded genome) ***(19)***. Although either vector type is suitable for the experiments described, each system has its own particular advantage. Phagemid vectors are advantageous because they can be used to transform *E. coli* with greater efficiency (10^7–10^8 clones/µg DNA) relative to phage vectors (10^5–10^6 clones/µg DNA), hence facilitating cloning of larger combinatorial libraries. On the other hand, we have found that phage vectors are more suitable for phage ELISA, which can be used to study protein–nucleic acid interactions without the need to subclone the selected genes into protein expression vectors. This may be because the phage vectors permit polyvalent display of proteins, allowing for stronger binding to ligands attached onto a solid support. Because the choice between phage and phagemid depends on the application, we have almost always used a phage vector, as we consider this vector system easier to work with and phage ELISA to be particularly useful in our experiments.

2. Cesium chloride gradient purification of the phage vector is carried out as follows. Dissolve 10–100 µg of DNA in 4 mL TE buffer containing 4 g of cesium chloride and 0.1 mL of ethidium bromide solution (10 mg/mL). Mix well to dissolve and dispense into a 5-mL ultracentifuge tube. Centrifuge at 370,000*g* for 20 h at 20°C using a Beckman Vti 65.2 rotor. Two bands of DNA should be visible under UV light. The lower (supercoiled plasmid) band should be collected with a syringe and washed four times with water-saturated butanol to remove all traces of ethidium bromide. The resulting solution is diluted threefold in TE buffer and then ethanol precipitated to recover pure phage vector.
3. A key step in designing the randomized gene cassette that will code for proteins to be displayed on phage is to clearly define the purpose of the phage-display experiment in order to decide the nature and extent of combinatorial randomizations. These decisions concern mainly (1) the part of the protein structure that is to be varied, (2) the number of positions that should be randomized, and (3) the amino acids to be represented at each varied position.

   The process can be illustrated by analyzing the strategy behind our initial experiments using the phage display of zinc fingers, which were designed to study DNA recognition by the three-finger DNA-binding domain from transcription factor Zif268. In the earliest experiment ***(8)***, our goal was to determine which residue positions of the zinc-finger structure were responsible for DNA base recognition, and the amino acids at those positions which would effect nucleotide discrimination. We therefore chose to randomize residues in only one zinc finger of the three-finger DNA-binding domain such that the register of the protein–DNA interaction would be defined by the interaction of the other two wild-type fingers with the DNA-binding site.

   The X-ray crystal structure of the Zif268 DNA-binding domain in complex with DNA ***(20)*** showed that only three or four amino acids on the α-helix of each zinc finger were responsible for base contacts. To determine whether it was indeed these α-helical positions that were responsible for DNA-binding specificity, we decided to randomize several additional residues (in fact a total of seven residues) in the zinc-finger α-helix in order to compare biases in the identity of selected amino acids throughout the helix.

As we did not want to bias the selection to particular residues, we decided that each randomization would represent most of the 20 amino acids, but we reduced the theoretical library size by designing the randomized codons such that the first base in the triplet was not thymine, thus precluding all stop codons and the codons for Phe, Tyr, Trp, and Cys, which occur very rarely in the α-helices of zinc fingers. The theoretical library size for the combinatorial variation of seven positions, each with 16 possible amino acids, was approx 5×10^8; however, because of the practical limitations (essentially the poor efficiency of transformation with phage vector) we were only able to clone approx 2×10^6 members of the library. Thus, the randomizations to be designed into a phage display library depend partly on the aims of the experiment and partly on the practical limitations of the technique, particularly the library size that can be cloned in practice.

Another important consideration in the phage display of proteins is that the assembly of phage particles requires the gIII fusion protein to be translocated to the periplasmic space of *E. coli* to await packaging (*see* **ref. *15***). Normally, this can be achieved by the presence of a cleavable signal peptide sequence that is coded by most phage-display vectors. However, in the event that phage yield is poor or that phage particles are not assembled at all, particular attention must be paid to the amino acid sequence directly C-terminal of the signal peptide. In many cases, it is worthwhile randomizing two to three amino acids of this sequence in a pilot experiment, to select phage that are capable of displaying the relevant nucleic-acid-binding protein. It should be noted, however, that certain nucleic acid binding motifs will prove very difficult or indeed impossible to display on phage. In this case, the failure will become apparent when phage particles fail to be produced on growing a culture of *E. coli* that have been transformed with the recombinant phage vector.

4. Polymerase chain reactions typically contain 10 µL 10X *Taq* reaction buffer, 4 µL of each primer (stock conc., 10 pmol/µL), 1 µL of 20 m*M* dNTPs, 1.4 µL *Taq* polymerase, and 1 µL ligation mixture, in a final volume of 100 mL.
5. Assuming all prior steps have proceeded smoothly, the efficiency of transformation will dictate the practical limit of the library size: 10^5–10^6 transformants per microgram of well-prepared phage vector is a typical yield. For best results, transformations are carried out by electroporation of fresh (unfrozen) electrocompetent cells prepared to a competency of at least 10^9 transformants per microgram supercoiled pUC19 plasmid. As this technique requires practice and is also unpredictable, it may be prudent to purchase such cells from a commercial supplier.
6. At this stage, it is worthwhile growing phage from individual colonies that contain the recombinant phage vector in order to check that the construct is competent for phage particle assembly. The phage titer can be estimated by reinfecting fresh *E. coli* from a liquid culture grown to mid-log phase, and plating on TYE containing tetracycline.

7. Certain nucleic-acid-binding motifs may require supplements to the growing medium to ensure biological activity. For example, during phage display of zinc fingers, we recommend supplementing all solutions with 50 μ*M* zinc chloride.
8. The 1 pmol of target site will be used to select phage from 1 mL of binding mixture. The concentration of binding site is therefore roughly in the nanomolar range; phage that can bind with a K_d below this range will be retained. For nucleic-acid-binding motifs that bind their targets more weakly, the amount of target site may have to be increased. Streptavidin-coated tubes or plates can also be used for selection.
9. Phage displaying nucleic-acid-binding motifs of interest are selected from a large library by affinity purification using biotinylated nucleic acid targets that have been immobilized on a streptavidin-coated matrix. Unbound phage are washed away and the phage that remain are eluted and amplified by passaging through bacteria. After several (three to six) rounds of selection, the phage library becomes enriched in those members that display proteins with the highest affinities for the target sites.

   Conditions of binding, washing, and elution can influence the results of a phage-display experiment. We have found that selections using streptavidin-coated paramagnetic beads (e.g., Dynabeads, Dynal) are more reliable than selections using streptavidin-coated microtiter plates or tubes. The amount of biotinylated nucleic acid target used for selection will determine the stringency of the selection (*see* **Note 8**).

   The amount and type of nonspecific competitor nucleic acid can also have an effect. In early rounds of selection, nonspecific competitor such as sonicated salmon sperm DNA, or tRNA, may be used. In later rounds, the selection may be refined by adding competitor sequences that are closely related to the desired target site but containing systematic base variations from the target sequence.

   The composition of the binding and washing buffers will also affect the stringency of selection. High-ionic-strength buffers tend to increase the stringency, whereas relatively high concentrations of detergents help to reduce background. It is important to wash away nonspecific binding phage thoroughly, and for this purpose, we recommend at least 15 changes of buffer.

   Finally, the method of elution can affect the results of the selection experiments and can be manipulated accordingly; for example to yield all retained phage (elution with alkali or acid) or only those bound phage with the highest specificity (elution with nucleic competitors).
10. This protocol suggests a 1:10 dilution of the phage-containing supernatant, which we find typically contains about 10^9–10^{10} colony-forming units/mL, but this is only intended as a guideline. Depending on the application, we have also previously used 1:1 dilutions of the phage-containing supernatant, or indeed 50-fold concentrates of the phage-containing supernatant. Higher concentrations of phage are particularly useful during the early rounds of selection, when specifically binding phage are rare in the library. Precipitation of the phage (e.g., by centrifugation at 60,000 rpm for 20 min) is also useful prior to binding reactions involv-

ing RNA, in order to minimize exposure to ribonucleases present in the bacterial culture supernatant.

11. It is convenient to do one round of selection per day so that transfected phage can be grown overnight for another round the next day. In practice, three to six rounds of selection are usually sufficient to obtain clones of interest. Depending on the nature of the library, however, the number of rounds required may vary greatly. An alternative strategy is to plate out the bacteria infected with selected phage after each round and to start a culture using a pool of the bacteria the next day. This increases the number of phage that are produced in the overnight culture, and reduces the selection advantage of fast-growing clones.
12. Phage clones isolated by selection against a nucleic acid target site must be assayed to characterize their range of binding affinities and specificities. This is conveniently done by phage ELISA, in which phage-displaying nucleic-acid-binding proteins bind to microtiter-plate wells coated with appropriate nucleic-acid-binding sites and are detected using an antiphage antibody. We routinely perform phage ELISA using phage derived from Fd-TET-SN and displaying either zinc-finger domains or the RNA-binding domain from U1A protein. However, we do not have any data on phage ELISA carried out with phagemid-derived phage particles.

References

1. Smith, G. P. (1985) Filamentous fusion phage: novel expression vectors that display cloned antigens on the virion surface. *Science* **228,** 1315–1317.
2. McCafferty, J., Griffiths, A. D., Winter, G., and Chiswell, D. J. (1990) Phage antibodies: filamentous phage displaying antibody variable domains. *Nature* **348,** 552–554.
3. Bass, S., Greene, R., and Wells, J. A. (1990) Hormone phage: an enrichment method for variant proteins with altered binding properties. *Proteins* **8,** 309–314.
4. Pabo, C. O. and Sauer, R. T. (1992) Transcription factors: structural families and principles of DNA recognition. *Annu. Rev. Biochem.* **61,** 1053–1095.
5. Klug, A. (1993) Protein designs for the specific recognition of DNA. *Gene* **135,** 83–92.
6. Rebar, E. J. and Pabo, C. O. (1994) Zinc finger phage: affinity selection of fingers with new DNA-binding specificities. *Science* **263,** 671–673.
7. Jamieson, A. C., Kim, S.-H., and Wells, J. A. (1994) In vitro selection of zinc fingers with altered DNA-binding specificity. *Biochemistry* **33,** 5689–5695.
8. Choo, Y. and Klug, A. (1994) Toward a code for the interactions of zinc fingers with DNA: Selection of randomised zinc fingers displayed on phage. *Proc. Natl. Acad. Sci. USA* **91,** 11,163–11,167.
9. Wu, H., Yang, W.-P., and Barbas III, C. F. (1995) Building zinc fingers by selection: Toward a therapeutic application. *Proc. Natl. Acad. Sci. USA* **92,** 344–348.
10. Rebar, E. J., Greisman, H. A., and Pabo, C. O. (1996) Phage display methods for selecting zinc finger proteins with novel DNA-binding specificities. *Methods Enzymol.* **267,** 129–149.

11. Choo, Y. and Klug, A. (1995) Designing DNA-binding proteins on the surface of filamentous phage. *Curr. Opin. Biotech.* **6,** 431–436.
12. Friesen, W. J. and Darby, M. K. (1997) Phage display of RNA binding zinc fingers from transcription factor IIIA. *J. Biol. Chem.* **272,** 10,994–10,997.
13. Friesen, W. J. and Darby, M. K. (1998) Specific RNA binding proteins constructed from zinc fingers. *Nat. Struct. Biol.* **5,** 543–546.
14. Hoffmann, S. and Willbold, D. (1997) A selection system to study protein–RNA interactions: functional display of HIV-1 Tat protein on filamentous bacteriophage M13. *Biochem. Biophys. Res. Com.* **235,** 806–811.
15. Laird-Offringa, I. A. and Belasco, J. G. (1996) In vitro genetic analysis of RNA-binding proteins using phage display libraries. *Methods Enzymol.* **267,** 149–168.
16. Choo, Y. and Klug, A. (1994) Selection of DNA binding sites for zinc fingers using rationally randomized DNA reveals coded interactions. *Proc. Natl. Acad. Sci. USA* **91,** 11,168–11,172.
17. Klug, A. and Rhodes, D. (1987) "Zinc fingers": a novel protein motif for nucleic acid recognition. *Trends Biochem. Sci.* **12,** 464–469.
18. Nagai, K., Oubridge, C., Ito N, Avis J., and Evans P. (1995) The RNP domain: a sequence-specific RNA-binding domain involved in processing and transport of RNA. *Trends Biochem. Sci.* **20,** 235–240.
19. Harrison, J. L., Williams, S. C., Winter, G., and Nissim, A. (1996) Screening of phage antibody libraries. *Methods Enzymol.* **267,** 83–109.
20. Pavletich, N. P. and Pabo, C. O. (1991) Zinc finger-DNA recognition: crystal structure of a Zif268-DNA complex at 2.1 Å. *Science* **252,** 809–817.

30

Genetic Analysis of DNA–Protein Interactions Using a Reporter Gene Assay in Yeast

David R. Setzer, Deborah B. Schulman, and Michael J. Bumbulis

1. Introduction

Understanding the underlying structural and physico-chemical basis for the recognition of specific DNA sequences by regulatory proteins is a central goal of modern biochemical genetics. A method for the rapid identification of mutant molecules altered in the affinity and/or specificity of such interactions could be a powerful tool in the hands of those studying this difficult problem. Conventional genetic approaches for obtaining and analyzing interesting mutant forms of specific DNA-binding proteins are often infeasible because of the genetic intractability of the species being studied or as a result of difficulties in identifying relevant and specific phenotypes associated with alterations in the interaction under investigation. High-resolution genetic analysis of DNA–protein interactions is particularly problematic in metazoans. We have devised an approach that makes use of the modularity in structure and function of eukaryotic transcription factors *(1)*, the power of the polymerase chain reaction (PCR) to generate specific DNA fragments with defined levels of mutagenesis in vitro *(2,3)*, and the recombinogenic potential of *S. cerevisiae* *(2,4)* to carry out a high-resolution genetic analysis of the sequence-specific DNA-binding properties of *Xenopus* transcription factor IIIA (TFIIIA) *(5)*. It seems likely that this approach will be generally applicable to the study of many DNA–protein interactions.

1.1. Outline of the Approach

In its simplest form, the method we describe here involves the construction and introduction of two plasmids into an appropriate yeast strain (**Fig. 1**). The

From: *Methods in Molecular Biology, vol. 148: DNA–Protein Interactions: Principles and Protocols, 2nd ed.*
Edited by: T. Moss © Humana Press Inc., Totowa, NJ

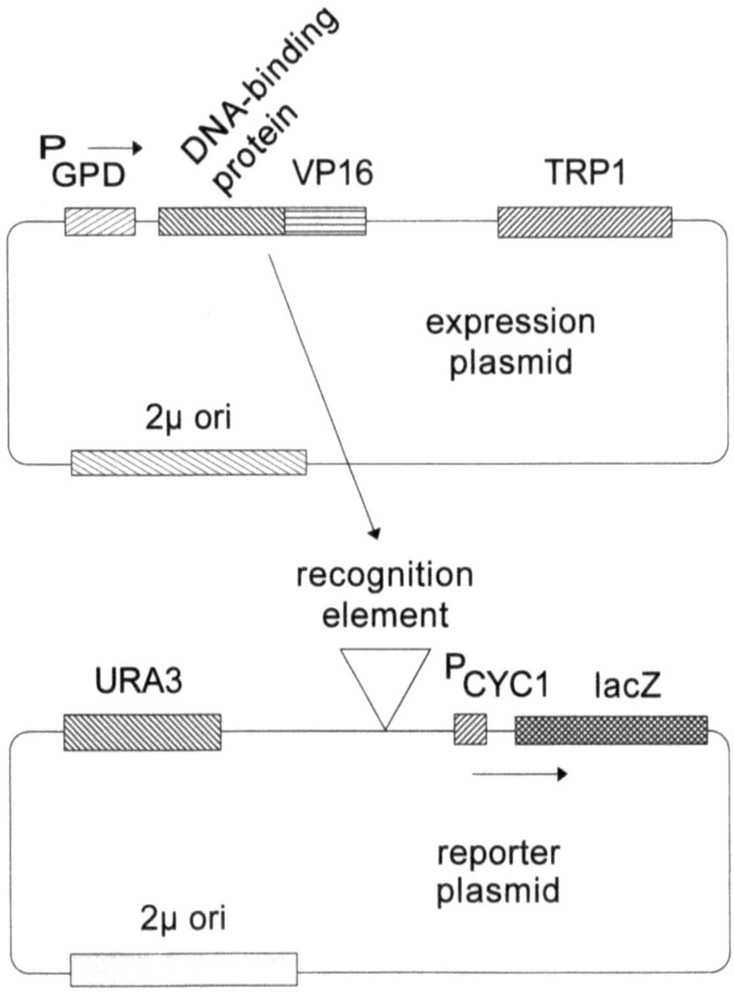

Fig. 1. Schematic representation of generic expression and reporter plasmids derived from pG1 and pΔSS, respectively. Only the plasmid components functional in yeast cells are shown; the plasmids also contain colE1 origins and β-lactamase genes for replication and selection in *E. coli*.

first, called the reporter plasmid, contains a reporter gene (*Escherichia coli* β-galactosidase) under control of the core promoter of the *S. cerevisiae* iso-1-cytochrome-*c* (*CYC1*) gene. In the parent plasmid, the upstream activator sequence (UAS) normally required for expression from the *CYC1* promoter has been deleted so that β-galactosidase is expressed at very low levels, and yeast strains containing this plasmid are white on X-gal indicator plates. A DNA fragment containing the cognate recognition site for the DNA-binding protein to be analyzed is substituted for the normal UAS. In the second plasmid, called the expression construct, the VP16 activation domain from *Herpes simplex* is fused in-frame to a sequence encoding the DNA-binding domain of the protein of interest. The DNA-binding-domain–VP16 fusion protein is expressed under control of the constitutive glyceraldehyde-3-phosphate dehydrogenase (GPD) promoter of *S. cerevisiae*. The reporter and expression plasmids carry different selectable markers (*URA3* and *TRP1*, for example) so that both can be selected and maintained in an appropriate yeast strain (ura3⁻trp1⁻, for example). When both plasmids are introduced into a single yeast cell and if the DNA-binding domain of the protein of interest binds with sufficiently high affinity and specificity to its recognition site in the reporter construct, the VP16 activation domain will be displayed in the vicinity of the core *CYC1* promoter and result in activation of transcription of the β-galactosidase reporter gene. On X-gal indicator plates, such a strain will be blue. Thus, this blue phenotype can be used as a marker for high-affinity interaction of the DNA-binding domain of interest with its recognition sequence. Mutations in either the DNA-binding protein or the DNA sequence to which it binds may adversely affect binding, resulting in white or light blue colonies, or may increase the affinity of binding, resulting in dark blue colonies (*see* **Notes 1–5**).

Generation of randomly mutated sequences encoding either the protein of interest or its cognate recognition site and the introduction of these mutated sequences into their appropriate contexts in either the expression or reporter plasmids is achieved with technical ease and high efficiency using a combination of error-prone PCR in vitro and homologous recombination in vivo following transformation (**Fig. 2**). Unusually long oligonucleotide primers are used for error-prone PCR (about 60–70 nucleotides, but *see* **Note 6**). The 3' portion of these primers anneals to a substrate plasmid at sites flanking the sequence to be mutagenized. The primers also contain 5' sequences identical to those which flank the ends of a linear version of the plasmid construct into which the mutated sequences are to be introduced. Error-prone PCR is used to synthesize a population of mutant DNA fragments containing the sequence of interest flanked by the sequences defined by the long amplification primers. In parallel, the plasmid into which the mutagenized PCR product is to be inserted is linearized or gapped by digestion at one or two sites, respectively, such that the unique ends of the PCR product correspond to sequences at the ends of the linear or gapped plasmid. When cotransformed into competent yeast cells, the linear plasmid and mutant PCR products undergo homologous recombination in vivo to produce a circular plasmid product in which the mutagenized fragment is integrated into the target plasmid at or between the restriction sites used in linearization/gapping. Successful recombination events can be clonally selected using the marker (*URA3* or *TRP1*, for example) on the

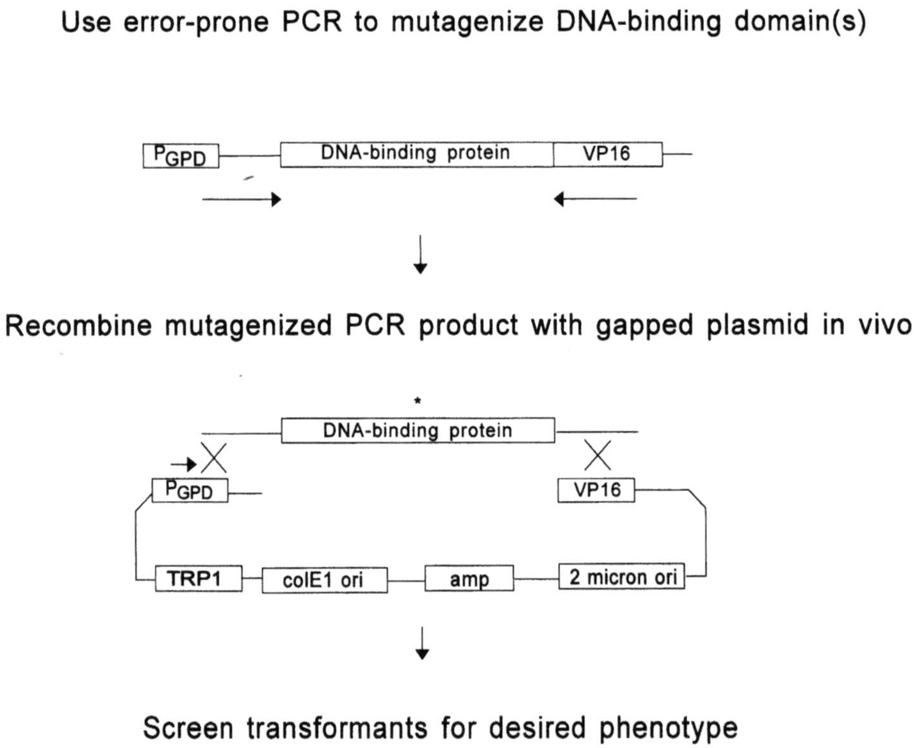

Fig. 2. Error-prone PCR in vitro and homologous recombination in vivo to mutagenize the DNA-binding protein of interest and introduce the mutagenized fragment into the expression vector. The asterisk represents a mutation introduced during error-prone PCR. In the example shown here, the substrate used for error-prone PCR is the expression plasmid itself, derived in this case from pG1, and the entire DNA-binding protein is subjected to mutagenesis. It is also possible to target only a portion of the DNA-binding protein for mutagenesis or to use a different plasmid as a substrate for PCR, provided the expression plasmid and amplification primers are appropriately designed.

target plasmid. If the yeast strain used for transformation already contains the second plasmid component of the system, then resulting colonies containing both reporter and expression plasmids can be selected and screened subsequently by replica-plating on indicator plates to identify mutants resulting in altered phenotypes (white, light blue, or dark blue). After further tests to ensure the mutant phenotype is authentic and the mutant protein or DNA sequence is likely to be of interest, the mutant plasmid is recovered in *E. coli* and the mutation identified by DNA sequence analysis, If desired, the mutant DNA–protein interaction can be subjected to detailed biochemical or further genetic analysis.

2. Materials

2.1. Initial Design and Testing of the System

1. *S. cerevisiae* strain BJ2168 *(6)* or other haploid strain with appropriate genotype (stable mutant alleles of genes used as selectable markers in expression and reporter plasmids) (*see* **Note 1**).
2. pΔSS *(7)* or other plasmid to be used as parent for reporter plasmid construction (*see* **Note 2**).
3. pG1 *(7)* or other plasmid to be used as parent for expression plasmid construction (*see* **Note 3**).
4. pSJT-1193-CRF1 *(8)* or other source of DNA encoding the VP16 activation domain (*see* **Note 4**)
5. Source of DNA encoding the DNA-binding protein or DNA-binding domain of interest.
6. Source of DNA including the recognition sequence for the DNA-binding protein or DNA-binding domain of interest.
7. Complete medium (C) agar plates lacking appropriate nutrients to permit selection of yeast strains containing reporter and expression plasmids. These will include C-uracil, C-tryptophan, and C-uracil–tryptophan for systems making use of derivatives of pΔSS and pG1. Procedures for preparation of liquid C medium and C agar are described by Rose et al. *(9)*. Our specific procedures are as follows:
   a. Dissolve in 1 L water the following: 20 g dextrose, 20 g Bactoagar, 1.7 g yeast nitrogen base without amino acids and without ammonium sulfate, 5 g ammonium sulfate, and 0.5 g amino acid mixture (*see* **step 7b**). Autoclave to sterilize and use to pour approx forty 100-mm plates.
   b. The amino acid mixture used to prepare C agar plates contains 0.2 g arginine, 0.2 g histidine, 0.5 g lysine, 0.4 g methionine, 0.2 g phenylalanine, 0.4 g tryptophan, 2.0 g threonine, 0.4 g tyrosine, 0.5 g serine, 0.2 g adenine, and 0.1 g uracil. For selective plates, the appropriate combination of nutrients (uracil and tryptophan [e.g., to select for plasmids containing *URA3* and *TRP1* markers]) should be omitted from the mixture.
8. SSX agar plates lacking appropriate nutrients as described in **step 7**, but also containing 40 μg/mL X-gal, prepared as follows:
   a. Dissolve the following in 900 mL water: 1.7 g yeast nitrogen base without amino acids and without ammonium sulfate, and 5 g ammonium sulfate, 20 g dextrose, 14 g Sigma agar, 0.5 g appropriate amino acid mixture (*see* **step 7b**). Autoclave to sterilize.
   b. Cool to 48°C and add aseptically: 1 mL of 40 mg/mL X-gal prepared in *N,N*-dimethylformamide and 100 mL 10X phosphate buffer (*see* **step 8c**).
   c. 10X phosphate buffer is prepared by mixing the following in 1 L water: 136.1 g KH_2PO_4 (1 *M*), 19.8 g $(NH_4)_2SO_4$ (0.15 *M*), and 42.1 g KOH (0.75 *N*). Adjust the pH to 7.0 and autoclave to sterilize.
9. Standard reagents and methods for subcloning DNA fragments into plasmids.

2.2. Error-Prone PCR

1. Plasmid(s) or other source(s) of DNA containing the sequence encoding the protein of interest and/or the DNA sequence recognized by the protein of interest.
2. Oligonucleotide primers that contain, at their 5' ends, approx 50 nucleotides of sequence identity to the site immediately adjacent to the end of the linear plasmid into which the PCR product is to be inserted. The 3' 15–20 nucleotides of these primers should have sequence identity with the parts of the substrate plasmid that define the DNA sequence to be amplified and mutagenized (*see* **Fig. 2** and **Note 6**).
2. *Taq* DNA polymerase.
3. Stock solutions of 100 m*M* $MgCl_2$ and 100 m*M* $MnCl_2$.
4. 10X stock solution of *Taq* PCR buffer, lacking $MgCl_2$: 100 m*M* Tris, pH 9.0, 500 m*M* KCl, and 1% Triton X-100 (Promega).
5. Individual stock solutions of dATP, dGTP, dCTP, and dTTP, each at a concentration of 10 m*M*. In addition, individual stock solutions of the same, each at a concentration of 2 m*M*.
6. Thermal cycler.

2.3. Yeast Transformation and Homologous Recombination

1. *S. cerevisiae* strain BJ2168 or other appropriate strain (*see* **Note 1**).
2. If the DNA-binding protein is to be mutagenized, BJ2168 containing the reporter plasmid and BJ2168 containing the parent of the reporter plasmid (pΔSS, for example).
3. If the DNA recognition site is to be mutagenized, BJ2168 containing the expression plasmid and BJ2168 containing the parent of the expression plasmid (pG1, for example).
4. Linearized or gapped plasmid to be used as the target for integration of the PCR-generated DNA fragment.
5. Crude product of the error-prone PCR
6. C-agar plates lacking the relevant nutrients for selection and maintenance of both reporter and expression plasmids (uracil and tryptophan for pΔSS- and pG1-derived plasmids).
7. Sterile stock solution of 100 m*M* lithium acetate, 1 m*M* EDTA, 10 m*M* Tris–Cl, pH 8.0.
8. Sonicated salmon sperm DNA of about 10,000 bp average length, denatured by heating to 100°C for 5 min at a concentration of about 5–10 mg/mL. Commercially available salmon sperm DNA should be extracted with phenol/chloroform and precipitated prior to use.
9. 40% (w/v) Polyethylene glycol (PEG) (average molecular weight of 3350). To prepare this solution, autoclave 2 g solid PEG in a sealable tube. The PEG will melt during sterilization and resolidify at room temperature. Many sterile PEG aliquots can be prepared simultaneously. On the day of use, add 3.5 mL sterile solution from **step 7** to one of these aliquots, heat to 65°C, and mix vigorously to

dissolve (final concentration of PEG is 40%). This quantity of solution is sufficient for approx 10 transformations.

10. Sterile SOS solution: 2 mL 2 *M* sorbitol, 1.3 mL YEPD medium (*9*), 0.26 mL 100 m*M* $CaCl_2$, 0.4 mL H_2O. YEPD medium: per liter of water, add 10 g Bactoyeast extract, 20 g Bactopeptone, and 20 g glucose.

2.4. Screening for Mutants

1. C-agar plates lacking both uracil and tryptophan, and containing 40 mg/mL X-gal.
2. Liquid C-trp or C-ura medium for selection of only one of the two plasmids in BJ2168.
3. C-agar plates lacking uracil and tryptophan individually, as well as plates lacking both.
4. Standard *E. coli* strain for plasmid transformation, propagation, and isolation, along with reagents for distinguishing reporter and expression plasmids by restriction endonuclease analysis.
5. 2% Sodium dodecyl sulfate (SDS)
6. Acid-washed glass beads (0.45 mm in diameter, from Sigma) prepared by washing overnight in 3 *N* HCl and then rinsing repeatedly in water.
7. Reagents for protein concentration determination using the BCA method (Pierce).
8. If possible, antibodies to the DNA-binding protein of interest and/or the activation domain used in the construction of the expression plasmid; reagents for Western blotting.

2.5. Analysis of Mutants

1. Z buffer for determination of β-galactosidase activity: 60 m*M* Na_2HPO_4, 40 m*M* NaH_2PO_4, 10 m*M* KCl, 1 m*M* $MgSO_4$, and 40 m*M* β-mercaptoethanol, pH 7.0.
2. Other reagents for determination of β-galactosidase activity: chloroform, 0.1% SDS, 4 mg/mL *o*-nitrophenol-β-D-galactoside (ONPG) prepared in Z buffer, and 1 *M* Na_2CO_3.
3. Reagents for DNA sequence determination.

3. Methods

3.1. Initial Design and Testing of the System

Details of the construction of appropriate reporter and expression constructs will depend on the specific features of the plasmids and clones to be used. It is therefore impossible to describe a step-by-step protocol for use in every case, but standard recombinant DNA methods should suffice for preparation of the desired plasmids. We will briefly outline the steps necessary for construction and testing of reporter and expression constructs prepared in pΔSS and pG1, respectively.

3.1.1. Construction of the Reporter Plasmid

It is necessary that a DNA fragment containing one or more copies of the DNA sequence recognized by the protein of interest be subcloned upstream of the *CYC1* core promoter in pΔSS. The only unique restriction site in pΔSS that is suitable for insertion of such a fragment is an *Xho*I site. A DNA fragment with *Xho*I-compatible ends and containing one or more copies of the relevant DNA sequence should be subcloned into the *Xho*I site of pΔSS. Most typically, this fragment would be either a restriction fragment from another plasmid or a PCR product digested to produce *Xho*I-compatible ends. It is possible, and probably desirable, to obtain pΔSS derivatives with multiple inserts of the DNA sequence of interest, and with single inserts in either orientation. The number and orientation of insert fragments must be diagnosed by some means, typically including restriction endonuclease mapping using enzymes that cut asymmetrically within the insert fragment to determine orientation, PCR with primers flanking the insert site to determine number of inserts, or DNA sequence analysis to determine either orientation or number of inserts if the insert fragment is not too long. An alternative to subcloning the insert fragment into the *Xho*I site of pΔSS is to use homologous recombination as described in **Subheading 3.3.** to integrate a PCR-generated DNA fragment into *Xho*I-digested pΔSS. In this case, the PCR fragment should be produced under high-fidelity conditions; even so, we recommend sequencing of the inserted fragment in the resulting plasmid to ensure that no mutations were introduced during amplification. If homologous recombination is used to generate the reporter plasmid, judicious choice of sequences in the long primers used for PCR can be used to regenerate either or both of the *Xho*I sites at the end of the insert, or even to introduce novel restriction endonuclease recognition elements. This may facilitate the introduction of multiple copies of the DNA-binding site into pΔSS.

3.1.2. Construction of the Expression Plasmid

One or more DNA fragments encoding an in-frame fusion of the DNA-binding protein of interest and a transcriptional activation domain must be introduced into pG1 downstream of the GPD promoter. The unique *Sal*I site in pG1 is probably the most convenient site for doing this. Fusion of the DNA-binding protein of interest and the VP16 activation domain can be done directly in pG1 or, perhaps more conveniently, in a smaller, simpler plasmid vector and then subcloned as a unit into pG1. Subsequent mutagenesis (**Subheading 3.2.**) can be more directly targeted to the DNA-binding protein rather than to the activation domain if a unique restriction site (more precisely, one that does not occur elsewhere in the plasmid outside of the sequence encoding the DNA-binding protein) can be engineered at the junction of the DNA-binding protein and the activation domain. It is also important to note that some DNA-binding proteins, and particularly transcriptional activator proteins acting through the RNA polymerase II core machinery, may contain endogenous transcriptional activation domains that will function in *S. cerevisiae*; in that event, transcriptional activity in the absence of the VP16 activation domain may be observed. The existence of an intrinsic activation domain in the protein of interest might obviate the need to prepare a fusion construct, but one must be careful in the subsequent analysis to distinguish mutations affecting DNA-binding affinity from those affecting transcriptional activation directly. The VP16 activation domain coding sequence followed by a polyadenylation signal from the *H. simplex* thymidine kinase gene can be excised on a *Kpn*I–*Hin*dIII fragment of approx 760 bp from the plasmid pSJT-1193-CRF1 *(8)*. At the *Kpn*I cleavage site, the reading frame for fusion to VP16 is XXG-GTA-CCX, but other plasmids in which the VP16 reading frame is shifted relative to the *Kpn*I cleavage site have also been constructed *(8)*. The *Kpn*I–*Hin*dIII fragment from this family of constructs is suitable for preparing a fusion protein in which the VP16 domain is at the C-terminus. Depending on what is known about the polarity of DNA binding by the protein of interest, this may or may not be desirable. Construction of N-terminal fusions may be preferable in some cases (*see* **Note 7**); these can be made by taking advantage of one of a number of vectors intended for the construction of libraries for use in two-hybrid screens (e.g., pACT-II, pGAD-GH, and pB42-AD from Clontech). One must be cautious in the choice of vector, however (*see* **Note 3**). In the case of *Xenopus* TFIIIA binding to the *Xenopus* 5S rRNA gene, some of these vectors (including pGAD10) result in very low levels of protein expression and no detectable transcription activation. Also, *see* **Note 4** concerning choice of activation domains vis-a-vis sensitivity of the genetic assay.

3.1.3. In Vivo Assay of the Reporter and Expression Constructs

For each reporter plasmid constructed and the parent vector (pΔSS), as control, two strains derived from BJ2168 should be prepared, one containing the expression plasmid in addition to the reporter, and one containing the parent plasmid from which the expression plasmid was derived (pG1) in combination with the reporter. The different selectable markers on these two plasmids (URA3 and TRP1) allow their simultaneous maintenance in BJ2168 (ura3$^-$ trp1$^-$) by selecting for growth in medium lacking uracil and tryptophan. The requisite strains should be constructed by sequentially transforming BJ2168 with the reporter and expression plasmids. Methods for transformation of BJ2168 (or other yeast strains) are described in detail in **Subheading 3.3.** One need only adjust the protocol to reflect the nutritional requirements of the strain being transformed and the selectable marker on the plasmid being introduced. Thus, the doubly transformed strain would be selected on C-ura-trp plates.

Colonies of strains containing both reporter and expression plasmids can be replica plated, spotted, or streaked onto C-ura-trp plates containing 40 μg/mL

X-gal. Colony color is assessed at an empirically determined time after robust colony growth has occurred. For analysis of the *Xenopus* TFIIIA–5S rRNA gene interaction, this was done typically after 2–3 d of growth at 30°C and an additional 2–3 d at room temperature. For the system to be exploited successfully, one must be able to distinguish reproducibly the color of strains containing both the expression and reporter plasmids from that of all the other control strains (lacking either expression of the fusion protein containing the DNA-binding domain[s] of interest or the cognate recognition site in the reporter construct, or both). If this is not the case, it may be possible to correct the problem by manipulation of parameters as described in **Note 5**. Of course, it is also possible that the particular interaction being studied will not be amenable to analysis with this method; among other reasons, this could result from a low-affinity/specificity interaction or from the existence of endogenous yeast factors that interact with the binding site introduced into the reporter plasmid, resulting in high levels of transcriptional activity in the absence of the interaction being targeted for study.

3.2. Error-Prone PCR

The DNA-binding protein or its recognition site can be subjected to random mutagenesis using error-prone PCR. In the following protocol, we assume that the DNA-binding protein is targeted for mutagenesis, but the procedure can be adapted readily for mutagenesis of the recognition site.

1. Set up a 50-μL polymerase chain reaction mixture containing 10–50 ng of plasmid DNA containing the sequence encoding the region to be mutagenized. This can be the expression plasmid itself (*see* **Note 8**) or another plasmid containing the sequence of interest. In addition, add 5 mL of 10X PCR buffer lacking $MgCl_2$, long amplification primers (*see* **Subheading 2.2.**, **Fig. 2**, and **Note 6**) to a final concentration of 0.3 μ*M* each, three deoxynucleoside triphosphates to a final concentration of 1 m*M* each, the fourth deoxynucleoside triphosphate to a final concentration of 0.2 m*M*, $MgCl_2$ to a final concentration of 3 m*M*, $MnCl_2$ to a final concentration of 0.05 m*M*, and 1 unit *Taq* DNA polymerase (*see* **Note 9**).
2. Amplify using a thermal cycler for 25 cycles, with each cycle being 94°C for 1 min, 42°C for 2 min, and 72°C for 1 min. After 25 cycles, use a final extension step of 72°C for 7 min (*see* **Note 9**).
3. Use the crude PCR product (without purification) in a yeast transformation with linearized/gapped target plasmid as described in **Subheading 3.3.**

3.3. Yeast Transformation and Homologous Recombination

1. Prepare a stock of linearized or gapped target plasmid by digesting to completion with one or two restriction endoncleases that result in ends corresponding to the site at which integration of the mutagenized DNA fragment is to occur. As an example, with an expression plasmid derived from pG1, this might be a double

digest with *Sal*I and an enzyme recognizing the fusion junction between the DNA-binding protein and the VP16 activation domain (*see* **Note 10**). A stock of this gapped plasmid can be prepared in 10 m*M* Tris–Cl, pH 8.0, and 1 m*M* EDTA at a concentration of 10–100 ng/µL and stored at –20°C.

2. Prepare competent yeast cells:
   a. Grow a 50-mL culture of the yeast strain containing the reporter plasmid (if mutations in the DNA-binding protein are to be analyzed) or expression plasmid (if mutations in the DNA recognition site are to be analyzed) overnight at 30°C until the OD_{600} = 0.5–1.0.
   b. Pellet the yeast cells by spinning for 5 min in a clinical centrifuge (1000*g*) at room temperature.
   c. Resuspend the cells in 10 mL of 100 m*M* lithium acetate, 1 m*M* EDTA, and 10 m*M* Tris–Cl, pH 8.0. Pellet again as in **step 2b**.
   d. Resuspend again in 10 mL of 100 m*M* lithium acetate, 1 m*M* EDTA, and 10 m*M* Tris–Cl, pH 8.0.
   e. Incubate at 30°C for 30 min without agitation and pellet again as in **step 2b**.
   f. Resuspend in 500 mL of 100 m*M* lithium acetate, 1 m*M* EDTA, and 10 m*M* Tris–Cl, pH 8.0. Place on ice until used for transformation.
3. Combine 100 µL competent yeast cells with 40 µL of the error-prone PCR mixture (**Subheading 3.2.**), 50 µg sonicated, denatured salmon sperm DNA, and 150 ng gapped plasmid (*see* **Notes 11** and **12**).
4. Incubate at 30°C for 30 min without agitation.
5. Add 0.5 mL of 40% PEG (3350 mol. wt.) in 100 m*M* lithium acetate, 1 m*M* EDTA, and 10 m*M* Tris–Cl, pH 8.0, and incubate further at 30°C for 60 min.
6. Incubate at 37°C for 5 min.
7. Centrifuge for 10 s at room temperature in a microcentrifuge at 14,000 rpm (16,000*g*).
8. Remove the supernatant and resuspend in 100 µL SOS solution.
9. Plate half of the resuspended transformed cells on each of the two C-agar plates lacking the appropriate nutrients for selection of both expression and reporter plasmids (e.g., C-ura-trp).
10. Incubate at 30°C for 2–3 d, until robust colony growth is obtained.

3.4. Screening for Mutants

1. Initial screen.
   a. Replica plate the colonies obtained in **step 10** of **Subheading 3.3.** onto selective C agar plates (C-ura-trp, for example) containing 40 µg/mL X-gal.
   b. Place replica plates at 30°C for 2–3 d and then at room temperature for an additional 2–3 d.
   c. Identify potentially interesting mutants by color. These could include light blue, white, or dark blue colonies; in each case, the color is in comparison to that exhibited by colonies containing wild-type versions of both the expression and reporter plasmids.
2. Pick candidates for further study and respot on selective C-agar plates. Grow for about 2 d at 30°C. Replica plate onto indicator media and score colony color

again after growth as in **step 1**. If the phenotype is consistent with that seen initially, proceed to **step 3**.

3. Enrich for cells containing only the mutagenized plasmid.
   a. Pick one or more colonies with an interesting phenotype from selective C-agar plates without X-gal indicator. Grow in 2 mL liquid C-trp medium (if the expression plasmid has been mutagenized; grow in C-ura medium if the reporter plasmid has been mutagenized) overnight at 30°C.
   b. Prepare a 1:200 dilution of this overnight culture in liquid medium and plate 10 µL of this diluted culture on a C-trp plate (assuming throughout that the expression plasmid has been mutagenized).
   c. After 3 d of growth at 30°C, replica plate onto C-ura-trp as well as onto C-trp.
   d. Grow overnight at 30°C and compare the colonies that grew on the two replica plates. Choose several that grew on C-trp but not on C-ura-trp. These are likely to have lost the reporter plasmid but not the expression plasmid. Use these to inoculate 2-mL liquid cultures of C-trp and grow overnight at 30°C.
4. Isolate the mutated plasmid using Method 1 as described by Strathern and Higgins ***(10)*** (*see* **Note 13**).
5. Use the isolated plasmid to transform *E. coli*. Pick individual colonies and perform standard plasmid minipreps. Analyze the plasmids thus obtained using appropriate restriction digests to ensure that the plasmid isolated is the mutagenized plasmid (e.g., the expression plasmid) and not the other plasmid component of the system (e.g., the reporter plasmid).
6. Analyze isolated plasmids by retransformation.
   a. Use the purified plasmid to retransform the appropriate yeast strain (already containing the reporter plasmid) and obtain colonies on selective C-ura-trp plates of cells with both reporter and expression plasmids.
   b. Analyze transformants by replica plating onto indicator plates as described in **step 1**. If the phenotype observed is consistent with that seen initially, then one can conclude that the phenotype is plasmid dependent. If the expression plasmid was mutagenized and it is the DNA-binding protein that is under analysis, continue with **step 7**. If, instead, the reporter plasmid was mutagenized and is being studied, one can move directly to **Subheading 3.5.**
7. Western blot screen to ensure the expressed fusion protein is full length and expressed at normal levels
   a. Inoculate 30 mL of C-ura-trp medium with a mutant strain and grow overnight at 30°C.
   b. Pellet the cells in a clinical centrifuge at (1000*g*) for 5 min. Resuspend the pellet in 1 mL of 2% SDS.
   c. Add 1 g glass beads and vortex vigorously for 15 min.
   d. Transfer the mixture to another tube and spin in a microcentrifuge at top speed (16,000*g*) for 15 s. Transfer the supernatant to another tube.
   e. Determine protein concentration in the extract using the BCA method (reagents from Pierce) ***(11)*** with bovine serum albumin as the standard.

f. Using standard Western blot methods, analyze 20 µg total protein from each mutant and compare to equivalent amounts of protein from a control strain expressing wild-type DNA-binding protein as well as a second, negative control strain containing the parent plasmid from which the expression construct was derived, but which lacks a coding sequence for the DNA-binding protein. The primary antibody used in the Western blot should be specific for the DNA-binding protein of interest; if such antibodies are unavailable, it may be possible to obtain, from commercial sources, antibodies to the activation domain used in the fusion. The point of the Western blot exercise is to exclude from further consideration those mutants whose steady-state level of expression is different from that of wild-type protein or which are truncated as a result of chain-termination or frame-shift mutations. Some truncation mutants may be interesting, but recall that the activation domain will be completely absent from truncated proteins if the activation domain was fused to the C-terminus of the DNA-binding protein. We have found that a substantial fraction of potential loss-of-function mutants (white phenotype) of TFIIIA are truncation mutants and a Western blot screen was important in removing them from further, more laborious analysis *(5)*.

3.5. Analysis of Mutants

3.5.1. Quantitative Determination of β-Galactosidase Activity

We have found that relative DNA-binding affinities of various mutant forms of TFIIIA can be predicted with reasonable precision from measurements of β-galactosidase activity in vivo in yeast reporter strains expressing the mutant protein fused to the activation domain of VP16. Whether this will prove to be generally true remains to be seen, but β-galactosidase activities can be readily measured using standard methods and may prove informative.

1. Inoculate 12.5 mL liquid C-ura-trp medium with the yeast strain to be analyzed. In addition, prepare a similar culture of a control strain containing the same reporter plasmid but with the parent vector (e.g., pG1) of the expression construct, rather than the expression construct itself. This control will permit determination of β-galactosidase activity in the absence of the DNA–protein interaction of interest. Grow the cultures overnight at 30°C until the OD_{600} = 0.3–1.0. Note the actual optical density.
2. Pellet cells by spinning at 1000*g* for 5 min in a clinical centrifuge. Resuspend in 3 mL of Z buffer.
3. Divide into three 1-mL fractions. In addition, process a tube containing 1 mL Z buffer in parallel. This will serve as the blank for spectrophotometric determination of β-galactosidase activity. Add 30 µL chloroform and 20 µL of 0.1% SDS to each and vortex for 10 s.
4. Incubate at 30°C for 5 min and add 200 µL ONPG at a concentration of 4 mg/mL in Z buffer. Incubate at 30°C until a yellow color becomes apparent.
5. Add 500 mL Na_2CO_3 and note the elapsed time since ONPG was added.

6. Transfer 800 µL of the reaction mixture to a microfuge tube and centrifuge at 16,000*g* for 2 min.
7. Measure the absorbance at 420 nm relative to that of the control processed in parallel.
8. Calculate the number of units of β-galactosidase activity according to the following formula:

$$\frac{\text{Absorbance at 420 nm} \times 100}{(\text{Assay duration [min]})\ (\text{culture volume analyzed [mL]})\ (\text{OD}_{600}\ \text{of culture})}$$

10. Average the three different determinations of activity to obtain a single value (*see* **Note 14**).
11. Normalize the final average β-galactosidase activity relative to that of the negative control culture (*see* **Note 14**).

3.5.2. Sequence Determination

Ultimately, one must determine the sequence of the mutant DNA-binding protein or recognition site in the expression or reporter plasmid, respectively. The details of the sequencing strategy to be used will depend on the specific system under analysis and the technology available to the investigator. We do recommend that the complete set of screens described in **Subheading 3.4.** be completed prior to sequence analysis, as the latter is typically laborious, expensive, or both.

3.5.3. Biochemical Analysis of Mutants

Until correlations between in vivo phenotypes and DNA-binding affinity and/or specificity can be verified in a particular system, the approach described here can be applied best as a method for generating interesting mutants that can be subjected to further biochemical analysis. In most cases, this will require expression and purification of the mutant DNA-binding protein under analysis, preferably without a fusion to the artificial activation domain. Again, the details of how this can be done will depend on the specific interaction being studied. We recommend, however, that the necessity of further sequence manipulation, including subcloning into other plasmid vectors, be taken into account in the initial design and construction of the yeast expression vector.

4. Notes

1. The genotype of strain BJ2168 is reported to be *MATa ura3-52 trp1 leu2 pep4-3 prc1-407 prb1-1122 gal2* (***6***). A variety of alternative strains should also be suitable for use. The only relevant genotypes for the approach used here pertain to the nutritional markers used for plasmid selection. If *HIS3* were to be used as a reporter, then it would also be important that the parent strain be *his3-* or that the chromosomal *HIS3* locus be placed under control of the hybrid promoter being used in the analysis. Clearly, the mutant chromosomal alleles of the nutritional

markers should be stable and not give rise to revertants at measurable frequencies. BJ2168 also contains mutations in three vacuolar proteases (the *PEP4*, *PRC1*, and *PRB1* genes), but it is unlikely that these are relevant to the approach described here. In fact, we have made use of other strains that presumably contain wild-type alleles at these loci without detectably altering the results obtained using the *Xenopus* TFIIIA/5S rRNA gene interaction.

2. A variety of reporter gene constructs probably can be used successfully. Plasmid-based reporters could be constructed using various parent plasmids containing different selectable markers, and alternative reporter genes might also be chosen. In addition, the use of a chromosomally integrated reporter gene is possible; in fact, we have successfully used *HIS3* as a chromosomal reporter gene in this fashion to study the *Xenopus* TFIIIA–5S rRNA gene interaction. Advantages of β-galactosidase as a reporter include the fact that it permits a range of phenotypes to be scored and the ease with which quantitative measurements of activity can be made using a simple spectrophotometric assay. *HIS3*, on the other hand, is probably much more sensitive, permitting detection of weaker DNA–protein interactions or use of a single-copy reporter when multicopy β-galactosidase reporters are required. It also allows selection of rare gain-of-function mutants that might occur at such a low frequency that identifying them would be difficult or impossible using a screening protocol like that required with β-galactosidase as a reporter. *HIS3* expression does not provide a wide dynamic range of phenotypes in a single-plate assay; however, this limitation can be overcome partially by analyzing growth on plates containing variable 3-amino-triazole (3-AT) concentrations. 3-AT is an inhibitor of the *HIS3* gene product and can, therefore, be used to adjust the level of *HIS3* expression required to permit growth.

We have not investigated the use of different core promoter elements other than that derived from *CYC1*. Even for *CYC1*, however, levels of reporter gene expression potentially can be manipulated by adjusting the number of copies and orientation of the DNA target sequence for the binding protein under study relative to the core promoter. We have found that both orientation and number of TFIIIA-binding sites in the promoter can have substantial effects on reporter gene activity when a TFIIIA-VP16 fusion protein is expressed. In our case, the orientation dependence could be explained readily on the basis of the known polarity of TFIIIA binding to the 5S rRNA gene ***(12)***. If similar information is available for the DNA-binding protein of interest, it may be possible to design reporter constructs rationally. Otherwise, it is probably advisable to generate multiple constructs for testing. It is possible that homodimers binding to symmetrical sites may exhibit little or no orientation dependence. With TFIIIA, we found that increasing the number of binding sites from one to two in the reporter gene promoter resulted in a doubling of β-galactosidase activity ***(5)***. For many natural Pol II activators, the effect of having multiple binding sites is synergistic ***(13)***, so it is possible that even larger effects will be observed when the number of binding sites for other DNA-binding proteins is increased. We have not investigated the effects of varying spacing between the artificial upstream activator sequence and the core promoter, but others have shown that transcriptional activity is only moderately affected by alterations in the wild-type spacing between these two elements *(14–16)*. Nonetheless, it is possible that attaining maximal activity in any particular system may require optimizing the spatial relationship between the DNA recognition site of interest and the core promoter. In the end, it may be necessary or desirable to adjust several of these parameters to obtain a level of reporter gene activity that is phenotypically detectable but that also provides for reasonable sensitivity in the detection of mutants weakening the DNA–protein interaction under analysis.

3. A variety of alternative expression plasmids are available, including several from commercial sources that may be preferable to pG1 in some respects (*see* **Subheading 3.1.2.**). We have used vectors other than pG1 and, in some cases, these vectors are better engineered for the easy introduction of DNA fragments encoding the DNA-binding protein of interest while producing an in-frame fusion to an N-terminal activation domain. Choice of expression vector may be affected by the selectable marker used, the presence of an activation domain in the vector, whether the domain is to be N- or C-terminally fused to the DNA-binding domain, and the strength of the promoter driving expression of the fusion protein. It is also possible that regulated expression of the fusion protein will be desirable, in which case a regulated promoter, like that of the *GAL1* gene, might be chosen over the constitutive GPD promoter of pG1.
4. The VP16 activation domain we have used *(8)* is probably the best studied activation domain acting to stimulate transcription by RNA polymerase II, and is probably also the strongest one in common use. Depending on the system being studied (*see* **Note 5**) and the choice of expression plasmid, an alternative activation domain might be chosen. We have used both the GAL4 activation domain and the synthetic domain B42 in addition to that of VP16 to study the *Xenopus* TFIIIA–5S rRNA gene interaction.
5. It may be necessary to adjust the system's parameters to achieve a level of transcriptional activity that affords a reasonable sensitivity of the assay to variations in binding affinity. Potentially, activity may be affected not only by the affinity of the interaction but also by the number, spacing, functional synergy between, and orientation of the binding sites in the reporter construct (*see* **Note 2**), by the level of expression of the fusion protein (*see* **Note 3**) and the fraction of protein that is available for binding (effective in vivo concentration), by the copy number of the reporter gene (*see* **Note 2**), by the choice of reporter gene (*see* **Note 2**), and by the choice of activation domain (*see* **Note 4**). Our experience in analyzing the *Xenopus* TFIIIA–5S rRNA gene interaction suggests that, for TFIIIA–VP16 under control of the GPD promoter on a 2 μ-containing plasmid, the effective in vivo concentration of protein was much lower than the total in vivo concentration as determined by Western blot *(5)*. Transcriptional activation in this system could not be detected on X-gal plates when the K_d was above 2 n*M*, using an *E. coli* β-galactosidase reporter gene containing two TFIIIA-binding sites upstream of the *CYC1* promoter on a plasmid with a 2 μ origin of replication.

6. The length of oligonucleotide primers we recommend for error-prone PCR and recombination in vivo is based on our success with primers of this length for mutagenizing TFIIIA and introducing the resulting mutant DNA sequences into an appropriate target vector. In reality, it is likely that shorter primers would work equally well ***(17,18)***, but we have not empirically demonstrated that to be the case.
7. If the DNA-binding domain of the protein of interest interacts with its binding site in an asymmetric manner, then the fused activation domain may be presented in two different ways, depending on the orientation of the binding site in the reporter gene promoter and whether it is fused to the N- or C-terminus of the DNA-binding protein. In the case of the *Xenopus* TFIIIA–5S rRNA gene interaction, it was important that the activation domain be promoter-proximal in order to obtain high-level transcriptional activation. Either N- or C-terminal activation domain fusions could be made active, but only in combination with the appropriate orientation of binding site in the reporter construct. It is difficult to know how common this phenomenon will prove to be, but it clearly has the potential to impact on the successful use of a particular set of reporter and expression constructs.
8. If the wild-type expression plasmid is used as the template for error-prone PCR, one should keep in mind that this wild-type plasmid will be carried over into the subsequent transformation reaction. This should not generally be a problem unless, for some reason, a low background of wild-type plasmid cannot be tolerated. One circumstance in which this might prove problematic is if very low efficiencies of recombination are obtained, so that most transformants recovered are derived from the template plasmid originally used for PCR and not from integration of PCR products into the gapped target vector. If necessary, the product of the error-prone PCR can be purified prior to transformation.
9. The extent of mutagenesis can be adjusted by manipulating the concentrations of $MgCl_2$ and $MnCl_2$ in the polymerase chain reaction. The concentrations given resulted in about 1–2 amino acid substitutions per 344 amino acids when used to mutagenize *Xenopus* TFIIIA. In general, the mutation frequency was higher at greater concentrations of $MgCl_2$ and $MnCl_2$, with the highest mutation rate being observed at the highest concentrations tested (4.5 m*M* $MgCl_2$, 0.5 m*M* $MnCl_2$). The lowest mutation rate was observed at "standard" PCR conditions (1.5 m*M* $MgCl_2$, no $MnCl_2$). Our evidence would also suggest that there is a substantially greater rate of mutation at positions where the low-concentration deoxynucleoside triphosphate should be incorporated. For this reason, we recommend performing a series of reactions in which each of the four deoxynucleoside triphosphates is represented in turn as the low-concentration nucleotide. One should also note that the conditions suggested for thermal cycling are intended only as a starting point; it may be necessary or desirable to adjust these conditions, particularly the annealing temperature, to reflect differences in primer T_m and specificity.
10. A potential problem in this protocol is the recircularization of the target plasmid without integration of the PCR-generated fragment. This can result in a high background of apparent loss-of-function mutants because no protein is being pro-

duced from the expression plasmid. To minimize this problem, we recommend several steps. If possible, use two different restriction endonucleases to generate the linear, gapped plasmid. Second, treat the cut plasmid with phosphatase to dephosphorylate the 5' termini generated by restriction digestion. Third, gel-purify the gapped plasmid to eliminate any small fragment that was removed by the restriction digestion as well as to eliminate any uncut molecules that might remain as a result of incomplete cleavage.

11. The quantities of gapped plasmid and error-prone PCR product that are optimal for recovery of the maximal number of colonies containing plasmids with integrated PCR products have not been carefully optimized, and probably should be. We have some evidence that suggests that transformation efficiencies are improved by reducing the amount of PCR product relative to the amount of gapped plasmid recommended here.
12. We have routinely used the crude, unpurified products of the error-prone PCR reaction in cotransformations with the gapped expression plasmid to generate and study mutant forms of TFIIIA without encountering significant problems *(5)*. In some related applications, however, we have noted that a high background of colonies can be produced as a consequence of poorly characterized recombination events that we believe derive from the production and integration of aberrant PCR products into the gapped target plasmid. It is most likely that these are small DNA fragments, perhaps "primer dimers." If these are produced at anything other than a very low frequency and are capable of recombining with the target plasmid, they can complicate the identification of loss-of-function mutants in the DNA-binding protein of interest, because they result in the production of white colonies on X-gal plates. If initial studies in a new system suggest that a similar problem occurs with the use of the crude PCR product, then we recommend purifying the mutagenized PCR product prior to recombination/transformation. Simply gel-isolating a product of the correct length should eliminate any problems derived from the presence of unincorporated primers, primer dimers, or other aberrant products of the PCR.
13. As minor modifications to the procedure of Strathern and Higgins ***(10)***, we have isolated plasmids directly from yeast colonies in lieu of growing liquid cultures, have used Prep-a-Gene (Bio-Rad) rather than Geneclean reagents, do not precipitate DNA after elution from Prep-a-Gene, and transform *E. coli* with 10% of the DNA sample recovered from the Prep-a-Gene reagent (2 µL of 20 µL total).
14. We recommend repeating the triplicate assays described here for each mutant at least three times with independent cultures to obtain an overall average value. Normalization of activity to a standard seems to be important in correcting day-to-day variability in the determination of β-galactosidase activity.

References

1. Brent, R. and Ptashne, M. (1985) A eukaryotic transcriptional activator bearing the DNA specificity of a prokaryotic repressor. *Cell* **43,** 729–736.

2. Muhlrad, D., Hunter, R., and Parker, R. (1992) A rapid method for localized mutagenesis of yeast genes. *Yeast* **8,** 79–82.
3. Kuipers, O. P. (1996) Random mutagenesis by using mixtures of dNTP and dITP in PCR. *Methods Mol. Biol.* **57,** 351–356.
4. Ma, H., Kunes, S., Schatz, P. J., and Botstein, D. (1987) Plasmid construction by homologous recombination in yeast. *Gene* **58,** 201–216.
5. Bumbulis, M. J., Wroblewski, G., McKean, D., and Setzer, D. R. (1998) Genetic analysis of *Xenopus* transcription factor IIIA. *J. Mol. Biol.* **284,** 1307–1322.
6. Jones, E. W. (1990) Vacuolar proteases in yeast *Saccharomyces cerevisiae. Methods Enzymol.* **185,** 372–386.
7. Schena, M. and Yamamoto, K. R. (1988) Mammalian glucocorticoid receptor derivatives enhance transcription in yeast. *Science* **241,** 965–967.
8. Triezenberg, S. J., Kingsbury, R. C., and McKnight, S. L. (1988) Functional dissection of VP16, the trans-activator of *Herpes simplex* virus immediate early gene expression. *Genes Dev.* **2,** 718–729.
9. Rose, M. D., Winston, F., and Hieter, P. (1990) *Methods in Yeast Genetics: A Laboratory Course Manual.* Cold Spring Harbor Laboratory, Cold Spring Harbor, NY.
10. Strathern, J. N. and Higgins, D. R. (1991) Recovery of plasmids from yeast into *Escherichia coli*: shuttle vectors. *Methods Enzymol.* **194,** 319–329.
11. Smith, P. K. (1985) Measurement of protein using bicinchoninic acid. *Anal. Biochem.* **150,** 76–85.
12. Smith, D. R., Jackson, I. J., and Brown, D. D. (1984) Domains of the positive transcription factor specific for the *Xenopus* 5S RNA gene. *Cell* **37,** 645–652.
13. Carey, M., Lin, Y.-S., Green, M. R., and Ptashne, M. (1990) A mechanism for synergistic activation of a mammalian gene by GAL4 derivatives. *Nature* **345,** 361–364.
14. Struhl, K. (1982) The yeast his3 promoter contains at least two distinct elements. *Proc. Natl. Acad. Sci. USA* **79,** 7385–7389.
15. Brent, R. and Ptashne, M. (1984) A bacterial repressor protein or a yeast transcriptional terminator can block upstream activation of a yeast gene. *Nature* **312,** 612–665.
16. Guarente, L. and Hoar, E. (1984) Upstream activation sites of the CYC1 gene of *Saccharomyces cerevisiae* are active when inverted but not when placed downstream of the "TATA" box. *Proc. Natl. Acad. Sci. USA* **81,** 7860–7864.
17. Baudin, A., Ozier-Kalogeropoulos, O., Denouel, A., Lacroute, F., and Cullin, C. (1993) A simple and efficient method for direct gene deletion in *Saccharomyces cerevisiae. Nuceic Acids Res.* **21,** 3329–3330.
18. Oldenburg, K. R., Vo, K. T., Michaelis, S., and Paddon, C. (1997) Recombination-mediated PCR-directed plasmid construction in vivo in yeast. *Nucleic Acids Res.* **25,** 451–452.

31

Assays for Transcription Factor Activity

Virgil Rhodius, Nigel Savery, Annie Kolb, and Stephen Busby

1. Introduction

Most transcription activator proteins have three important features that can be probed at the molecular level: they bind to specific sequences near promoters, they can be interconverted between active and inactive forms by covalent or noncovalent modification, and, when bound at target promoters, they can stimulate the initiation of transcription by RNA polymerase *(1)*. This chapter is concerned with in vitro methods for measuring the transcription activation function of this important class of proteins. In most cases, these methods are applied to activators that have been substantially purified, and for which the target sequences are known and the binding sites characterized (other chapters in this volume cover methods for locating and investigating the binding sites for such activators). Here we are concerned with the measurement of the products of activation. Because *Escherichia coli* transcription activators have been studied more than any others, we will take these as the paradigm, though, in principle, the techniques can be applied to any organism for which in vitro systems have been developed.

The starting point of the methodology was the observation, made in the early 1970s, that purified *E. coli* RNA polymerase could initiate transcription at promoters in purified DNA *(2)*. With improvements in RNA methodology, it was found that the transcription start site in vitro, in many cases, was the same as in vivo. Further, at a number of promoters, interactions with specific transcription activators could be demonstrated and factor-dependent transcription in vitro occurred. The literature is now full of instances of factor-dependent transcription initiation with purified proteins and DNA, setting the scene for studies on the mechanism of transcription activation. However, as with all in vitro techniques, it is worth noting that the conditions found in the plastic tube differ

From: *Methods in Molecular Biology, vol. 148: DNA–Protein Interactions: Principles and Protocols, 2nd ed.*
Edited by: T. Moss © Humana Press Inc., Totowa, NJ

greatly from those in vivo and that, for some instances, the reconstituted transcription system simply may not work.

Two principal methods can be used to monitor transcription activation: transcript assays and abortive initiation (**Fig. 1**). In transcript assays, RNA polymerase is incubated with a DNA template carrying the promoter of interest either with or without the transcription activator protein. Radioactively labeled nucleotides are then added and RNA polymerase molecules that have formed transcriptionally competent complexes start to make RNA. RNA polymerase molecules then "run" to a suitably placed downstream terminator (or to the end of the fragment), thus making labeled RNA of a discrete length that can be monitored and quantified *(3)*. In abortive initiation assays, nucleotide precursors corresponding just to the start of the message are added. RNA polymerase forms transcriptionally competent complexes at the chosen promoter, but elongation is prevented because some of the four nucleoside triphosphates are withheld *(4)*. The result is that RNA polymerase is "trapped" at the promoter and can only synthesize a short oligonucleotide, which is released as the polymerase cycles between a number of conformations in the abortive complex (it is termed abortive because a longer transcript cannot be made). The consequence of the cycling is that each polymerase molecule, trapped at a single promoter, synthesizes the oligonucleotide continuously, and the appearance of the product can be measured directly. The rate of product formation will be dependent on the number of promoters that are occupied by polymerase in an open complex. This, in turn, will usually depend on the activity of the transcription factor under study. In most cases, transcript assays are used to locate transcription starts and to monitor factor activity qualitatively, whereas abortive initiation assays are exploited for quantitative and kinetic work.

2. Materials

1. *E. coli* RNA polymerase. This enzyme can be purified *(5,6)* or can be purchased from Boehringer (Mannheim, Germany), Pharmacia (Piscataway, NJ), or other molecular biology companies. The enzyme can also be reconstituted from individual subunits that have been overproduced using overexpression vectors *(7,8)*. Purity can be easily checked by denaturing polyacrylamide gel electrophoresis and activity can be verified using standard templates. Preparations of RNA polymerase are usually stored at –20°C in buffer containing 50% glycerol.
2. Transcription factors. These must be purified, at least partially, away from any nuclease activities. Transcription assays can be used to monitor the purification.
3. Plasmids and DNA fragments. The DNA used for transcription assays should be free from nicks and RNase activity. Short DNA sequences carrying different promoters can be constructed using standard recombinant DNA methodology and cloned into plasmid vectors such that a strong transcription terminator is located downstream of the promoter *(9)*. Transcript assays can be performed directly on

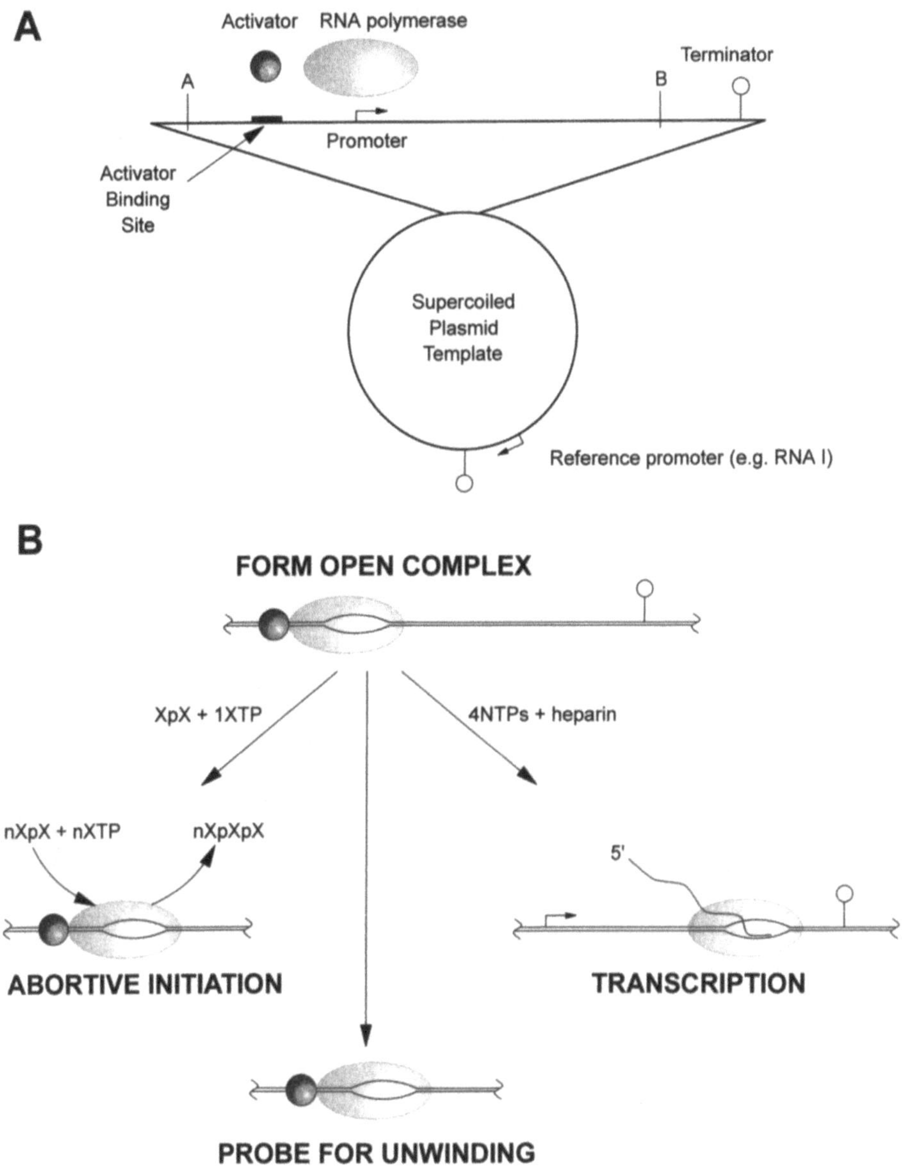

Fig. 1. Overview of techniques discussed in this chapter. Section A illustrates a plasmid carrying a test promoter, cloned on a restriction fragment, upstream of a terminator. The location of the reference RNA I promoter is shown. Section B illustrates the various techniques described: see **Subheading 3.1.** for transcript assays, **Subheading 3.2.** for abortive initiation assays, and **Note 10** for probes to detect unwinding.

such recombinant plasmids, providing they have been purified, for example by using cesium chloride gradient centrifugation *(10)*. Alternatively, linear, promoter-containing fragments can be purified from restricted plasmid DNA by polyacrylamide or agarose gels by a variety of methods *(10)*. Although any DNA fragments can be used (*see* **Note 1**), fragments around 200–1000 bp are most desirable. Stock solutions of most short fragments will be adjusted to around 20 mg/mL, the concentration being checked after gel electrophoresis.

4. Transcription buffer: 20 m*M* Tris-HCl, pH 8.0, 100 m*M* NaCl, 5 m*M* $MgCl_2$, 0.1 m*M* EDTA, 1 m*M* dithiothreitol (DTT), 50 µg/mL nuclease-free bovine serum albumin (BSA), and 5% glycerol. This is a standard 1X buffer for many in vitro transcription assays and can be prepared as a 10X stock. There are many variations of this and the literature must be checked for any particular instance.
5. Heparin, cat. no. H6279 from Sigma (St. Louis, MO), made up as a 10 mg/mL stock solution in water.
6. Nucleotides. [α-^{32}P] UTP from NEN (Boston, MA) or Amersham (Arlington Heights, IL) can be used in conjunction with the four nucleoside triphosphates from Boehringer or Pharmacia. For transcription assays, most workers use final concentrations of 200 µ*M* ATP/CTP/GTP, 10 µ*M* UTP, 0.5–5.0 µCi of [α-^{32}P] UTP per reaction, and 100 µg/mL heparin to prevent reinitiation. Typically, an 8X stock NTP + heparin solution is prepared containing 80 µ*M* UTP, 1.6 m*M* of the three other NTPs, and 800 µg/mL heparin in 2X transcription buffer. A "hot" NTP + heparin mix is then made by diluting this 1:1 with [α-^{32}P] UTP in water.
7. Dinucleotides. These can be bought from Sigma and used without further purification as 10-m*M* stock solutions in water. The choice of the appropriate dinucleotide for priming abortive initiation assays is discussed in **Subheading 3.2.1.**
8. Transcription stop solution: 80% deionized formamide, 0.1% xylene cyanol FF, 0.1% bromophenol blue, 20 m*M* EDTA in the standard gel running buffer, and 1X TBE.
9. RNA gels. Standard 6% polyacrylamide sequencing gels containing 6 *M* to 8 *M* urea and run in TBE *(10)* can be used to separate runoff transcripts. Autoradiography or a phosphor screen are used to detect the products.
10. Gel running buffer, TBE. This is usually made up in large volumes and kept as a 5X stock solution. To make up 1 L of 5X stock use 54 g Tris base, 27.5 g boric acid, and 20 mL of 0.5 *M* EDTA.
11. Whatman 3MM paper. Cut into strips 20 cm in length for chromatography of abortive products.
12. Chromatography buffer: 18:80:2 (v/v/v) water/saturated ammonium sulfate/isopropanol.
13. RNA size markers. These are usually generated from runoff transcripts of well-characterized DNA fragments. Alternatively, sequence ladders can be used.
14. 0.1 *M* EDTA.

3. Methods

3.1. Transcript Assays

1. The first step is the binding reaction. Purified transcription factor and supercoiled plasmid or linear template DNA (*see* **Note 1**) are mixed gently in 1X transcrip-

tion buffer (*see* **Note 2**) in a final volume of 8 µL and incubated for 5 min at 37°C (*see* **Note 3**). It is important to include any cofactor required by the transcriptional activator. For example, the cyclic AMP receptor protein (CRP) requires cAMP in the transcription buffer for activity. Some other transcription activators require covalent modification, such as phosphorylation. The active form must be used.

2. Next, 4 µL of RNA polymerase, diluted in 1X transcription buffer, is added and mixed in gently. The binding reaction is then incubated for a further 5–20 min at 37°C. Typically, the incubations are performed in a final volume of 12 µL with a template concentration of 0.5–5 n*M*, a range of transcription factor concentrations from 5 to 50 times the promoter concentration and up to 100 n*M* RNA polymerase.
3. The second step is the transcription reaction. Add 4 mL of "hot" NTP + heparin mix (*see* **Note 4**) to each 12 µL binding reaction, mix gently, and incubate at 37°C for 5 min (*see* **Note 5**). For each individual DNA molecule, the RNA polymerase may or may not have reached an open complex depending on the activity of the transcription factor. At molecules where an open complex has formed, the polymerase will then "run" from the promoter to the downstream terminator (or to the end of the fragment) making a discrete-sized RNA product. Because only one molecule of polymerase can occupy a promoter at any time and because the inclusion of heparin prevents further initiation, the amount of any particular run-off transcript will be directly proportional to the amount of open-complex formation (*see* **Notes 6** and **7**).
4. Terminate the reactions by adding 12 µL of transcription stop solution. The samples can be stored for short periods on ice, or for longer periods at –20°C, until ready for loading on a sequencing gel.
5. Heat the samples for 2 min at 90°C and load 8 µL on a sequencing gel, together with size markers, and perform the electrophoresis. We routinely use the S2 model from Gibco-BRL (Gaithersburg, MD), running the gel for 2–3 h at 60 W constant power. After running, the gel is dried, an autoradiograph is exposed, and the film is developed. From the sequence marker it is possible to identify bands caused by transcription initiation at the promoter under study and to determine the effects of the transcription activator on the appearance of these bands. An example is shown in **Fig. 2** (taken from **ref. *11***) (*see* **Notes 8–10**).

3.2. Abortive Initiation Assays

1. Choose the nucleotides to be employed in the assay. Typically, this is done by selecting a dinucleotide appearing in the sequence anywhere from position –4 to +2, and using the next nucleotide as the labeled precursor. For example, at the *E. coli galP1* promoter ***(12)***, the sequence at the transcription start is 5'-TCATA-3' with the central A as +1. The dinucleotide CpA and [α-^{32}P] UTP can be used to give the product, ^{32}P-labeled CpApU. It is important to ensure that no extended products can form (*see* **Note 11**).
2. Before the assay is performed, set up a series of Whatman 3MM paper chromatograms (typically 20 cm long). Spot the origins with 20 µL of 0.1 *M* EDTA to

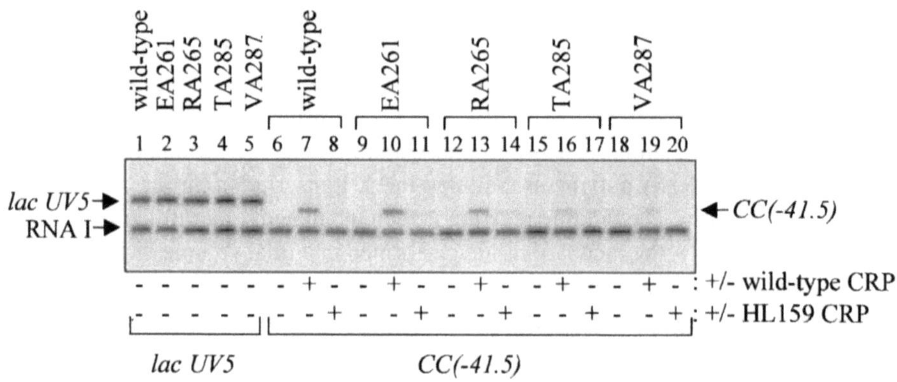

Fig. 2. In vitro transcription from plasmid carrying a CRP-independent promoter, *lacUV5* (lanes 1–5) or a CRP-dependent promoter, *CC(–41.5)* (lanes 6–20), cloned upstream of a transcription terminator. The figure shows the transcripts produced using purified RNA polymerase containing wild-type or mutant α-subunits (EA261, RA265, TA285, and VA287 as indicated). Prior to the addition of RNA polymerase, CRP or CRP carrying the HL159 substitution (that interferes with the CRP–RNA polymerase interaction) was added to the reaction mixtures as indicated. The position of transcripts initiated at the *lacUV5* or *CC(–41.5)* promoters, and the position of the plasmid-encoded *RNA I* transcript are indicated. (From **ref. *11***.)

ensure that product formation ceases the moment that the samples are loaded on the chromatogram.

3. Set up the standard assay, using concentrations of reagents as for the transcript analysis experiment. In a typical starting experiment, excess RNA polymerase and transcription factor will be premixed with DNA and incubated long enough to reach complete open complex formation. The experiment will be started by the addition of nucleotides. The final reaction mix will contain, for example, 0.5–5 n*M* promoter DNA, 100 n*M* RNA polymerase, 0.5 m*M* dinucleotide, and 0.05 m*M* UTP with 2.5 mCi [α-^{32}P] UTP in 100 µL. Run experiments both with and without the transcription factor and perform a control with no DNA.
4. At different times after addition of the [α-^{32}P] UTP, remove 15-µL aliquots and spot at the origin of the chromatogram. Six aliquots taken every 5 min will suffice.
5. Develop the chromatogram using chromatography buffer. After the solvent front has progressed 20 cm, remove the chromatogram and dry. Cut the paper into 5-mm slices and count each slice for Cerenkov radiation to locate the bands resulting from product and unincorporated UTP. For each time point, determine the number of counts incorporated into the product ($CPM_{product}$), and the number of counts in the unincorporated UTP (CPM_u). From the ratio of counts in the product to the total counts ($CPM_{product}$+CPM_u), the amount of product at each time-point can be deduced. Alternatively, the products can be analyzed and quantified using a phosphorimager. In this case, it is sufficient to spot 2-µL aliquots onto the chro-

matogram, and the transcription reactions can be scaled down threefold to fivefold (*see* **Note 12**).

A plot of product formed versus time should be linear, and from the slope, the rate of product formation can be deduced (*see* **Note 13**). The rate of product formation per promoter (TON, turnover number) can then be calculated from the molar amount of DNA fragment that was used in the experiment. A control run without the transcription factor will give factor-independent activity and will allow the effect of the activator to be quantified. Assuming that the rate of product formation in the presence of the activator reflects 100% occupancy at the promoter, the occupancy in the absence of the activator can be calculated (from the ratio of the TON values in the absence and the presence of the activator) (*see* **Note 14**).

6. The analysis can then be taken a stage further (e.g., *see* **refs. *13–16***). In the experiment above, RNA polymerase is preincubated with the DNA template prior to the addition of substrate, so product formation is linear from zero time. However, if the reaction is started by the addition of polymerase, the plot of product formation versus time shows a lag where the RNA polymerase "installs" itself at the promoter. This lag time (τ) can easily be measured and is a function of the initial binding of polymerase to the promoter and subsequent isomerizations to the open complex (*see* **Fig. 3**).

   The interaction of holoenzyme with promoters involves at least two steps: a rapid and reversible binding to promoter DNA, characterized by an association constant K_B, which leads to the "closed" inactive complex, followed by a conformation change to the "open" complex characterized by the rate constant k_f. For most promoters, the reverse of open-complex formation is extremely slow. Thus, according to McClure *(4)*, the measured lag time (τ) is related to the enzyme concentration [RNP] by the relation

$$\tau = 1/k_f + 1/(K_B k_f[\mathrm{RNP}])$$

   To make a kinetic analysis, it is necessary to perform the assays with a range of different polymerase concentrations (typically 5–200 n*M*: for kinetic analysis, RNA polymerase should always be present in significant excess over promoter DNA). The lag time (τ) is measured in each case and is plotted as function of the reciprocal of the RNA polymerase concentration (**Fig. 4**). This plot can be extrapolated to infinite RNA polymerase concentrations (the intersect with the *y*-axis) to give the reciprocal of k_f, and K_B can be deduced from the intercept of the τ plot with the *x*-axis. Alternatively, K_B can be calculated from the ratio of the lag time at infinite enzyme concentration and the slope of the straight line. Data are normally fitted using a computer program, such as Enzfitter or Fig-P (*see* **Notes 15** and **16**).

4. Notes

1. Transcript analysis assays are generally performed on templates with the transcription start of interest positioned 50–150 bp upstream from a transcription

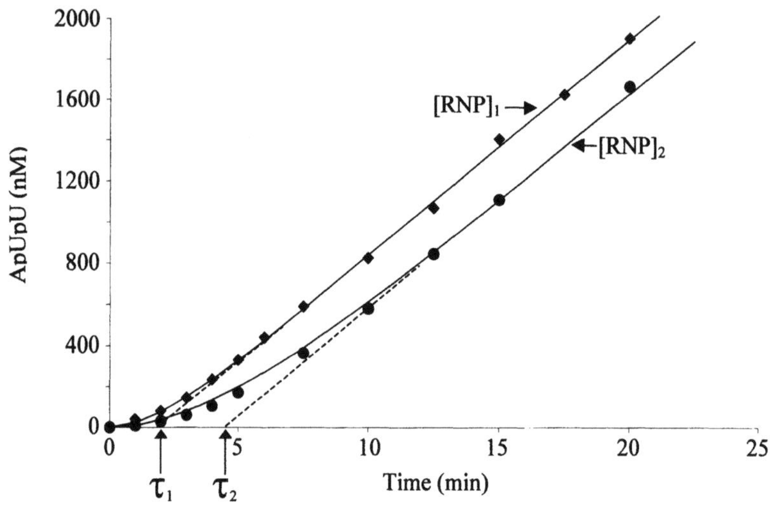

Fig. 3. Lag plots of a CRP-dependent promoter in the presence of different concentrations of RNA polymerase (50 n*M* $[RNP]_1$ and 16.7 n*M* $[RNP]_2$). A 15-µL sample of 4 n*M* DNA fragment in standard buffer containing CRP HL159 and cAMP was preincubated for 10 min at 37°C with 5 µL of a mixture containing 3 m*M* ApU and 300 m*M* UTP with 0.75 µCi [α-^{32}P]UTP. At time 0, 10 µL of a prewarmed RNA polymerase solution was added (150 n*M* or 50 n*M*) and the reaction carefully mixed. At the indicated times, 2-µL portions of the reactions were removed for product quantification. The normalized quantity of ApUpU product was fitted using the Fig-P program according to the equation $Y = V_t - V_\tau(1 - e^{-t/\tau})$, where *V* is the final steady-state velocity (moles of ApUpU per mole of promoter per minute). Care was taken to run the reaction until $t = 5\tau$ and to check that the final slope V was in agreement ±15% with the value of the TON, determined after preincubation of promoter and holoenzyme as described in **step 5** of **Subheading 3.2.**

terminator or the end of the fragment. Often, a longer fragment that carries more than one promoter will be chosen; longer transcripts can be sized by running the sequence gels further. Individual transcripts can be identified by using families of fragments that are truncated from one end. Transcription assays can be performed using both relaxed and supercoiled DNA. Reference promoters can be used to aid in the quantification of transcripts (e.g., if *colE1* plasmid derivatives are used as vectors for the promoter under study, the 107 nucleotide transcript from the RNA I promoter can be used; *see* **Figs. 1** and **2**).

2. A number of alternative buffer systems can be used and the final choice is largely a matter of trial and error. An alternative system is 40 m*M* Tris, pH 8.0, 100 m*M* KCl, 10 m*M* $MgCl_2$, 1 m*M* DTT, and 100 µg/mL acetylated bovine serum albumin. In some cases, the effects of substituting different anions or cations may be significant *(17)*. Many recent studies have used glutamate-containing buffers to enhance DNA binding of different factors.

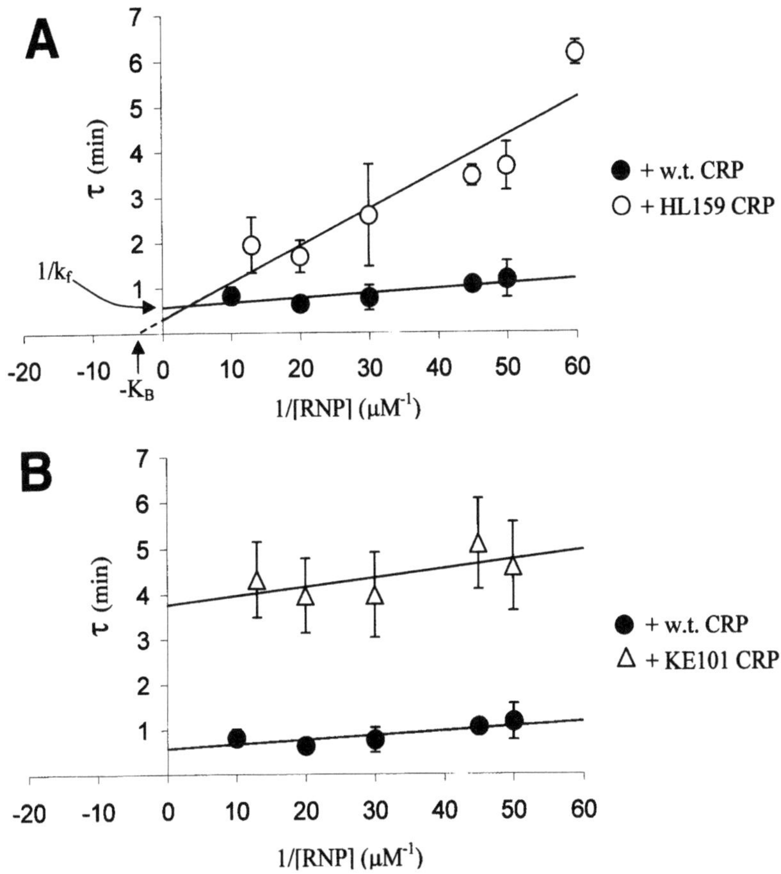

Fig. 4. Tau plots comparing the effects of different substitutions in CRP on transcription activation at a CRP-dependent promoter. The lag time (τ) before linear production of ApUpU is plotted against the reciprocal of RNA polymerase concentration. Plot A compares CRP carrying the HL159 substitution (which inactivates Activating Region 1 and decreases K_B) with wild-type CRP. Plot B compares CRP carrying the KE101 substitution (which inactivates Activating Region 2 and decreases k_f) with wild-type CRP. K_Bk_f and k_f can be calculated from the slope and intercept of each plot, respectively. Each data point represents the average of three independent assays and the error bars show one standard deviation on either side of the mean (data taken from **ref.** ***16***).

3. Care should be taken to avoid introducing RNase contamination during protein and DNA purifications, in the preparation of solutions and in the handling of plasticware. If necessary, commercially available RNase inhibitors can be added to the transcription reactions to counteract low levels of nuclease contamination.
4. The inclusion of heparin in assays ensures a single round of transcript formation. However, multiround assays can be performed by omitting the heparin. This

can be useful when working with promoters where the open complex is sensitive to heparin.

5. The transcription step of the protocol is very fast and complete in min. Sometimes a doublet band is seen corresponding to a particular transcript. This is often due to "hesitation" by the polymerase at the end of the transcript. The relative intensities of the doublet can depend on temperature or the length of time of the elongation step. In some cases, multiple bands are caused by ambiguity in the starting base of the transcript. This can be resolved by working with [γ-^{32}P]-labeled initiating nucleotide (e.g., *see* **ref. *18***).
6. The kinetics of open-complex formation can be monitored using transcript assays. After the addition of polymerase, take aliquots at different times and add to the "hot" NTP + heparin mix. Because heparin blocks reinitiation, the amount of transcript from that sample will be proportional to the amount of open complex formed at that time (for an example, *see* **ref. *19***). Typically, for these conditions, the half-time for open-complex formation ranges from 20 s to 30 min. In principle, it is possible to make these measurements at different polymerase concentrations and make the τ plot analysis, as for abortive initiation; in practice, this is extremely difficult and the abortive initiation assay is preferable.
7. Elongation can be studied by preforming open complexes and then adding nucleotide precursors one by one. 3' *O*-methyl ***(20)*** or dideoxy ***(21)*** derivatives of nucleotides can be used to freeze elongation complexes at particular lengths.
8. Transcript analysis assays provide a simple method for monitoring the effects of transcription factors and their cofactors. However, it can also be exploited to investigate effects of conditions (e.g., temperature, salt, etc.) on open-complex formation. It is important to note that changes may affect elongation rather than transcription initiation. This can be checked simply by preforming open complexes and then altering the conditions. In our experience, the elongation step is usually unaltered by changes in the assay conditions, and differences reflect changes at one or other step in the formation of the open complex ***(22)***.
9. Many, but not all, promoters are active in transcript assays and there is no way of predicting whether a particular activator will or will not work in vitro. Many workers find that such experiments produce more bands than "ought" to be seen. In particular, some runoff transcripts made with purified fragments as templates exceed the size of the template fragment ***(3)***. This results from RNA polymerase molecules failing to stop when reaching the end of the fragment, turning around, and continuing to transcribe the opposite strand. This effect can be partially circumvented by lowering NTP concentrations or decreasing the temperature. Another problem may arise because any DNA sequence will contain a number of potential transcription starts that are normally not used in vivo, because the competition for polymerase in vivo favors stronger promoters. In vitro conditions are such that there is little discrimination against weak promoters (e.g., *see* **ref. *23***). If the appearance of bands from these weak promoters "spoils" the results, they can be reduced by using higher salt concentrations or lower concentrations of polymerase to increase specificity.

10. Although the appearance of a transcript makes a good assay for the activity of a transcription factor, there may be situations in which the transcript cannot be detected. In this case, the best strategy is to attempt to monitor open-complex binding directly by the opening of the strands. In most cases, the activity of a transcription factor will cause a measurable unwinding of the DNA duplex around the –10 sequence and transcription start. Many chemical reagents can be used to monitor unwinding, but one of the simplest is potassium permanganate, that preferentially attacks nonbase-paired thymines. To measure the unwinding, start with end-labeled DNA and make open complexes with RNA polymerase and activator proteins, as in the runoff assays. Typically, then add 1 µL of fresh 200 m*M* potassium permanganate per 20-µL sample and incubate for 1 min (still at 37°C). After the addition of 50 µL stop buffer (3 *M* ammonium acetate, 0.1 m*M* EDTA, 1.5 *M* β-mercaptoethanol), phenol-extract the sample and alcohol-precipitate the DNA. The labeled DNA can then be cleaved at the sites of permanganate modification by using the Maxam–Gilbert piperidine protocol. The resulting fragments are run on a sequence gel to find the sites of unwinding. A typical experiment will include runs with DNA alone, DNA plus polymerase, and DNA plus polymerase plus activator. This provides a simple method for checking that the activator is functional and provides information on the size of the region of unwinding in the open complex ***(24,25)***.
11. A great feature of the abortive initiation assay is that it can be performed on promoters carried by both circular DNA and linear fragments: the dinucleotide primer picks out one promoter from others. Obviously, there is more chance of interference from other promoters with longer DNAs. Thus, if working with circular plasmid, it is prudent to test the reaction using plasmid either with or without the insertion carrying the promoter under study. It may be possible to reduce interfering signals from the vector by altering the dinucleotide used. Some primers can be used without being completely specific for the promoter tested. For instance, CpA and UTP gives the trinucleotide CpApU at *gal*P1, but also the longer oligonucleotides CpApUpU and CpApUpUpU starting from *gal*P2, which can be separated on the chromatogram ***(28)***.
12. The abortive initiation assay is tedious because of the chromatographic analysis of the products, which takes 2–3 h. One way to accelerate the procedure is to replace radioactive UTP with a fluorescent analog, UTP-γ-ANS (1-naphtylamine-5-sulfonic acid UTP). The assay can then be measured fluorometrically by following the increase in light emission caused by the release of the pyrophosphate–ANS moiety each time a unit is incorporated ***(27)***. A considerable advantage of this method is that it allows the continuous monitoring of product formation. A disadvantage is that the fluorescent label may alter the kinetics, although, to date, this has not been reported.
13. In some cases, product formation may never become linear with respect to time. Assuming that there are no contaminating nucleases, this is likely to be a result of the consumption of nucleoside triphosphates, which reduces the reaction velocity. This can be overcome by lowering the dinucleotide concentration. Ideally, any time-course needs to be run for at least five times τ.

14. Before starting any kinetics, it is advisable to check chosen combinations of primer and nucleotide for specificity and for product formation: a TON value of <10/min is useless for kinetic studies. Some promoters give no abortive cycling reaction, whereas others may give homopolymer synthesis caused by slippage in the enzyme's active site *(28)*, rendering the abortive initiation assay useless.
15. The most powerful use of abortive initiation is to determine the microscopic rate constants of individual steps during transcription initiation. Measurements of these rates in the absence or presence of a transcription factor can provide mechanistic information about the enzymology of activation. However, the method relies on a number of assumptions that are true for most, but not all, promoters *(4)*. First, active RNA polymerase must be present in significant (i.e., >5X) excess over the promoter DNA; second, the isomerization from the closed to open complex must be essentially irreversible over the time-course of the experiment; and third, in order for the equation in **step 6** of **Subheading 3.2.** to hold true, the closed complex must be in rapid equilibrium with free polymerase and DNA.
16. In different situations, transcription activators can affect K_B *(14)*, k_f*(13)*, or TON *(14)*. In a small number of cases, transcription factors have no effect on abortive initiation parameters. In such instances, the activator cannot be intervening at the level of open complex formation, but must be affecting later steps of the transcription process (e.g., *see* **ref. *29***). Such situations can be analyzed by single or multiple rounds of transcript assays. Note that in some complex cases (e.g., overlapping promoters), microscopic rate parameters cannot be deduced from abortive initiation assays *(30)*.

References

1. Raibaud, O. and Schwartz, M. (1984) Positive control of transcription initiation in bacteria. *Annu. Rev. Genet.* **18,** 173–206.
2. Losick, R. and Chamberlin, M. (eds.) (1976) *RNA Polymerase.* Cold Spring Harbor Laboratory, Cold Spring Harbor, NY.
3. Zubay, G. (1980) The isolation and properties of CAP, the catabolite gene activator. *Methods Enzymol.* **65,** 856–877.
4. McClure, W. (1980) Rate-limiting steps in RNA chain initiation. *Proc. Natl. Acad. Sci. USA* **77,** 5634–5638.
5. Burgess, R. and Jendrisak, J. (1975) A procedure for the rapid, large-scale purification of *Escherichia coli* DNA-dependent RNA polymerase involving Polymin P precipitation and DNA-cellulose chromatography. *Biochemistry* **14,** 4634–4638.
6. Hager, D., Jun Jin, D., and Burgess, R. (1990) Use of mono Q high resolution ionic exchange chromotography to obtain highly pure and active *Escherichia coli* RNA polymerase. *Biochemistry* **29,** 7890–7894.
7. Tang, H., Severinov, K., Goldfarb, A., and Ebright, R. (1995) Rapid RNA polymerase genetics: one-day, no-column preparation of reconstituted recombinant *Escherichia coli* RNA polymerase. *Proc. Natl. Acad. Sci. USA* **92,** 4902–4906.
8. Fujita, N. and Ishihama, A. (1996) Reconstitution of RNA polymerase. *Methods Enzymol.* **273,** 121–130.

9. Kolb, A., Kotlarz, D., Kusano, S., and Ishihama, A. (1995) Selectivity of the *Escherichia coli* RNA polymerase $E\sigma^{38}$ for overlapping promoters and ability to support CRP activation. *Nucleic Acids Res.* **23,** 819–826.
10. Sambrook, J., Fritsch, E., and Maniatis, T. (1989) *Molecular Cloning. A Laboratory Manual, 2nd ed.* Cold Spring Harbor Laboratory, Cold Spring Harbor, NY.
11. Savery, N., Lloyd, G., Kainz, M., Gaal, T., Ross, W., Ebright, R., et al. (1998) Transcription activation at Class II CRP-dependent promoters: identification of determinants in the C-terminal domain of the RNA polymerase alpha subunit. *EMBO J.* **17,** 3439–3447.
12. Herbert, M., Kolb, A., and Buc, H. (1986) Overlapping promoters and their control in *Escherichia coli*: the *gal* case. *Proc. Natl. Acad Sci. USA* **83,** 2807–2811.
13. Hawley, D. and McClure, W. (1982) Mechanism of activation of transcription initiation from the λ P_{RM} promoter. *J. Mol. Biol* **157,** 493–525.
14. Malan, T., Kolb, A., Buc, H., and McClure. W. (1984) Mechanism of CRP-cAMP activation of *lac* operon transcription initiation: activation of the P1 promoter. *J. Mol. Biol.* **180,** 881–909.
15. Leirmo, S. and Gourse, R. (1991) Factor independent activation of *E. coli* rRNA transcription (I) Kinetic analysis of the role of the upstream activator region and supercoiling on transcription of the *rrnB* P1 promoter *in vitro. J. Mol. Biol.* **220,** 555–568.
16. Rhodius, V., West, D., Webster, C., Busby, S., and Savery, N. (1997) Transcription activation at Class II CRP-dependent promoters: the role of different activating regions. *Nucleic Acids Res.* **25,** 326–333.
17. Leirmo, S., Harrison, S., Cayley, D., and Burgess, R. (1987) Replacement of potassium chloride by potassium glutamate dramatically enhances protein DNA interactions *in vitro. Biochemistry* **26,** 2095–2101.
18. Hsu, L. (1996) Quantitative parameters for promoter clearance. *Methods Enzymol.* **273,** 59–71.
19. Chan, B. and Busby, S. (1989) Recognition of nucleotide sequences at the *Escherichia coli* galactose operon *P1* promoter by RNA polymerase. *Gene* **84,** 227–236.
20. Straney, D. and Crothers, D. (1985) Intermediates in transcription initiation from the *E. coli lac* UV5 promoter. *Cell* **43,** 449–459.
21. Krummel, B. and Chamberlin, M. (1992) Structural analysis of ternary complexes of *E. coli* RNA polymerase: individual complexes halted along different transcription units have distinct and unexpected biochemical properties. *J. Mol. Biol.* **225,** 221–237.
22. Grimes, E., Busby, S., and Minchin, S. (1991) Different thermal energy requirement for open complex formation by *Escherichia coll* RNA polymerase at two related promoters. *Nucleic Acids Res.* **19,** 6113–6118.
23. Ponnambalam, S.. Spassky, A., and Busby, S. (1987) Studies with the *Escherichia coli* galactose operon regulatory region carrying a point mutation that simultaneously inactivates the two overlapping promoters. Interactions with RNA polymerase and the cyclic AMP receptor protein. *FEBS Lett.* **219,** 189–196.

24. Gaston, K., Bell, A., Kolb, A., Buc, H., and Busby, S. (1990) Stringent spacing requirements for transcription activation by CRP. *Cell* **62,** 733–743.
25. Chan, B., Minchin, S., and Busby, S. (1990) Unwinding of the duplex DNA during transcription initiation at the *Escherichia coli* galactose operon overlapping promoters. *FEBS Lett.* **267,** 46–50.
26. Goodrich, J. and McClure, W. (1992) Regulation of open complex formation at the galactose operon promoters. Simultaneous interaction of RNA polymerase, *gal* repressor and CAP/cyclic AMP. *J. Mol Biol* **224,** 15–29.
27. Bertrand-Burggraf, E., Lefevre. J. F., and Daune, M. (1984) A new experimental approach for studying the association between RNA polymerase and the *tet* promoter of pBR322. *Nucleic Acids Res.* **12,** 1697–1706.
28. Qi, F., Liu, C., Heath, L., and Turnbough, C. (1996) In vitro assay for reiterative transcription during transcriptional initiation by *Escherichia coli* RNA polymerase. *Methods Enzymol.* **273,** 71–85.
29. Menendez, M., Kolb, A., and Buc, H. (1987) A new target for CRP action at the *malT* promoter. *EMBO J.* **6,** 4227–4234.
30. Gussin, G. (1996) Kinetic analysis of RNA polymerase–promoter interactions. *Methods in Enzymology* **273,** 45–59.

32

Assay of Restriction Endonucleases Using Oligonucleotides

Bernard A. Connolly, Hsiao-Hui Liu, Damian Parry, Lisa E. Engler, Michael R. Kurpiewski, and Linda Jen-Jacobson

1. Introduction

Type II restriction endonucleases are familiar to most investigators in the biological sciences as indispensable reagents for a variety of molecular biology techniques. Usually, these dimeric enzymes cut double-stranded DNA at defined, palindromic, sequences 4, 6, or 8 base pairs in length *(1)*. Restriction endonucleases have extremely high specificities, a key property that underlies their applications in genetic engineering and which has attracted the attention of scientists interested in sequence-specific DNA discrimination. Considerable effort has been expended in trying to elucidate the mechanisms that underlie specificity, often using the "structural perturbation" approach *(2,3)*. Here, an alteration is made to the protein by site-directed mutagenesis, or to the nucleic acid by substituting with natural base pairs, or with synthetic base, sugar and phosphate analogs and the effects on binding and cleavage observed. A useful measure of the perturbation can be obtained by evaluation of the parameter

$$(k_{st}/K_D)_{modified}/(k_{st}/K_D)_{unmodified}$$

where "unmodified" refers to the native endonuclease and the natural DNA target sequence and "modified" indicates that either the protein or the nucleic acid has been changed in order to delete a selected interaction. The parameter k_{st} is the first-order rate constant obtained under single-turnover conditions. Provided that the association of DNA and the essential cofactor Mg^{2+} with the protein are rapid (*see* **Note 1**), k_{st} measures the slowest step following the assembly of the endonuclease–DNA–Mg^{2+} complex up to and

From: *Methods in Molecular Biology, vol. 148: DNA–Protein Interactions: Principles and Protocols, 2nd ed.*
Edited by: T. Moss © Humana Press Inc., Totowa, NJ

including the hydrolysis step. K_D (the equilibrium dissociation constant) = $[E]_f[D]_f/[ED]$.*

The evaluation of $(k_{st}/K_D)_{modified}/(k_{st}/K_D)_{unmodified}$ can be highly informative, revealing how much a particular interaction contributes to the energetics of DNA discrimination and whether the effect arises at substrate or transition state binding *(2,3)*. Examples include the use of oligonucleotides containing base and phosphate analogs to probe DNA recognition by the *Eco*RI restriction endonuclease *(4,5)*. The underlying rationale of the "structural perturbation" method have been treated comprehensively elsewhere *(2)* and are beyond the scope of this chapter. Similarly, methodology for the alteration of proteins by site-directed mutagenesis and the preparation and uses of oligonucleotides containing modified bases, sugars, and phosphates will not be discussed. Rather, this chapter will concentrate on the practical aspects of k_{st} and K_D measurement.

In order to measure k_{st}, the oligonucleotide is labeled with ^{32}P, most commonly at the 5' terminus using γ-[^{32}P]-ATP and polynucleotide kinase. Alternatively, labeling at the 3' end with α-[^{32}P]-ddNTP and terminal transferase can be used. Restriction endonucleases have an absolute requirement for Mg^{2+}, enabling the reaction to be initiated in a number of ways: usually the addition of Mg^{2+} to a premixed solution of an enzyme plus oligonucleotide or the addition of an enzyme to a solution containing the oligonucleotide and Mg^{2+} (*see* **Note 1**). At various times aliquots are removed and the reaction quenched. The substrate and product DNA are separated using denaturing polyacrylamide gel electrophoresis (PAGE) and quantitated by phosphorimaging. The data obtained are fitted to appropriate equations to give k_{st} values. Several considerations need to be taken into account:

- Measurement of a rate constant under single-turnover conditions requires all the substrate to be bound to the enzyme at the start of the reaction. This necessitates (a) $[E]_t > [D]_t$ and (b) $[D]_t >> K_D$ (*see* **Note 2**).
- When a wild type restriction endonuclease is used to cut its natural DNA target, both strands of the duplex are usually cut at the same rate. However, introducing a modification into only one of the DNA strands often results in the individual strands being cut at different rates. In these cases cleavage is most simply and accurately measured if the two substrate strands and the two labeled product strands can be separated, as shown in **Fig. 1**.
- Measurement of DNA cleavage by restriction endonucleases, under single-turnover conditions, approaches the limit of manual manipulation. In some cases, it has been possible to mix the enzyme with its substrate and withdraw aliquots by

*Throughout this chapter, E = endonuclease; D = DNA; ED = enzyme–DNA complex. The subscripts t and f denote the total and free amounts of E and D present (i.e., $[E]_f = [E]_t - [ED]$; $[D]_f = [D]_t - [ED]$). Several investigators use K_A (= $1/K_D$) rather then K_D. Both terms are acceptable, and in some instances in this chapter K_A has been used.

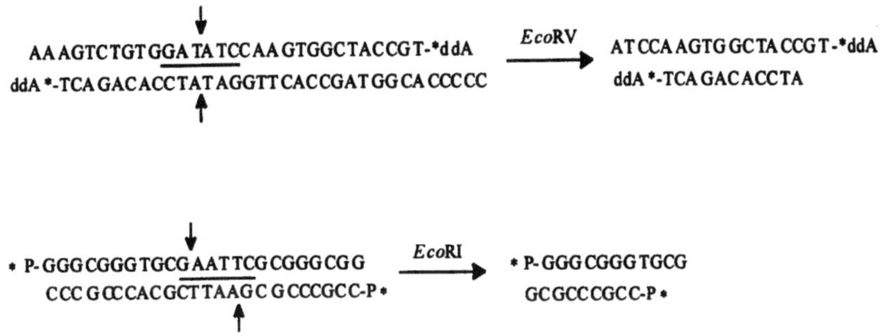

Fig. 1. Oligonucleotides that have been used to measure single turnovers with the *Eco*RV and *Eco*RI restriction endonucleases (recognition sequence shown with line, cleavage sites with arrows). With *Eco*RV, both the substrate strands are different lengths. In the case of *Eco*RI, the two substrate strands can be separated by denaturing gel electrophoresis, despite having the same length, as the top strand is dG rich and the bottom dC rich. dC-rich strands migrate faster than dG (dA and T base-pairs affect migration much less). With *Eco*RV 3' labeling (*ddA) was used, and with *Eco*RI, 5' labeling (*P) was used. The offset recognition sites gives products of different sizes. The two arrangements ensure the separation of both substrate and both labelled product strands and so allow independent measure of the rate of cleavage of the two strands. Only the labeled products are shown. For both enzymes the "top" strands are written in the 5' → 3' direction and the "bottom" are written in the 3' → 5' direction.

hand. In others, the reaction is too fast to be evaluated in this manner and a rapid-mixing quenched flow apparatus must be used. Alteration to the endonuclease or the DNA often cause a reduction in the single-turnover rate constant and it is often possible to use manual methods in these cases.

The most suitable method for K_D determination depends critically on its magnitude and hence the stability of the protein–DNA complex (*see* **Note 3**). Several approaches for the measurement of K_D are presented, enabling equilibrium constant evaluation under conditions that range from very tight ($K_D \approx$ p*M*) to very weak ($K_D \approx$ m*M*) binding. When commencing experiments, the K_D will not be known and good practice requires an initial estimate, followed by an accurate evaluation using the optimal approach. Modification to the endonuclease or the DNA often weakens their interaction, sometimes by a considerable amount. Therefore, the method used in the "natural" case may not be the best when alterations to the macromolecules are present. Whatever method is used, it is important to prevent DNA hydrolysis and this is most simply achieved by omission of Mg^{2+} and addition of EDTA.

Two commonly used methods for K_D determination are filter binding and gel retardation (electrophoretic mobility shift assay [EMSA]). Both are discussed elsewhere in this volume (*see* Chapters 1 and 2) and this chapter emphasizes their application to restriction endonucleases and also the use of competition titration. Most often, a ^{32}P-labeled oligonucleotide is incubated with increasing amounts of restriction enzyme to produce a protein–DNA complex. The free and protein-bound DNA are then separated and the relative amounts in each pool determined. This allows the construction of a binding isotherm and K_D evaluation. Data analysis is simplified when $[E]_t \approx [E]_f$ and experimentation is usually carried out under these conditions (*see* **Note 4**). It is also useful to carry out a "reverse" titration: incubating a fixed amount of endonuclease with increasing quantities of ^{32}P-labeled oligonucleotide. Titrations with a fixed [DNA] may give rise to shifts in the oligomeric state of the protein because of its increasing concentration, which can produce biphasic isotherms or curves that do not show asymptotic behavour. This arises due to the coupled reactions $E_2 + E_2 \rightarrow E_4 \rightarrow E_n$ (where E_2 is the active form) and is particularly relevant to modified substrates that show weak binding; requiring titration with concentrations of protein that may exceed the K_D for the coupled dimer–tetramer equilibrium. If both procedures give identical K_D values, one has confidence that the coupled reaction can be ruled out as an obfuscating factor. In addition, the combination of the two approaches permits precise determination of the number of active protein molecules, providing that the DNA concentration is accurately known.

Filter binding and gel retardation differ in the manner used to separate the free DNA and the protein–DNA complex. In filter binding, separation is achieved using a nitrocellulose filter that retains proteins, and hence protein–DNA complexes, but allows the passage of free DNA. Some proteins are poorly bound by nitrocellulose filters although this difficulty can sometimes be circumvented by using a different filter material (e.g., pure nitrocellulose rather than mixed-ester filters). Gel retardation uses non denaturing PAGE; free DNA migrates more rapidly than the protein–DNA complex giving, under ideal conditions, two well-resolved bands. Both approaches are sensitive (can be used at very low [DNA] to measure tight binding), experimentally very simple to carry out, and do not require specialized equipment.

Whether filter-binding or gel-shift methods are chosen, the direct titration protocol is not recommended for modified substrates that show very weak binding (i.e., $K_D > 10^{-7}$ *M*), because the lifetime of the complexes (*see* **Note 3**) is always shorter than the time required either for filtration or for entry into the gel (*see* **Note 5**). This may result in partial dissociation of the endonuclease–DNA complex during the measurement (i.e., a "nonequilibrium" situation). For weak binding, a competition titration ***(6)*** protocol should be used. Here, a fixed quantity of an endonuclease and a ^{32}P-labeled oligonucleotide (most often containing an unmodified recognition sequence for the endonuclease under study) are allowed to form a complex. Progressively increasing amounts of a

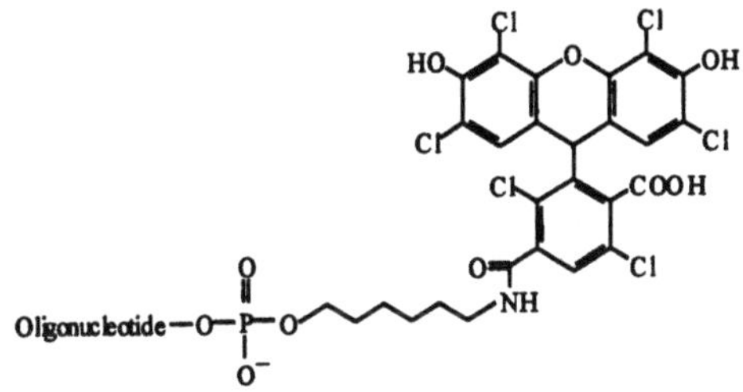

Fig. 2. The structure of hexachlorofluorescein and its linkage to the 5'-phosphate of an oligonucleotide.

nonradioactive competitor DNA (e.g., a variant oligonucleotide in which there has been a base-analog or natural base-pair substitution) are added. The unlabeled DNA molecules compete with the ^{32}P-labeled oligonucleotide for the DNA-binding site on the protein and so leads to a displacement of the radioactive probe. It is good practice to compare some of the K_D values obtained by the direct method with those obtained by competition titration, both to test whether equilibrium is significantly perturbed in the direct method and to provide confidence that all binding reactions have achieved equilibrium in the more indirect competition method. Thus, one should include in the experimental set of modified oligonucleotides at least one "internal control" for which both direct and competition titrations can be performed (*see* **Note 6**).

Equilibrium constants can also be determined using fluorescence anisotropy *(7,8)* with oligonucleotides labeled with hexachlorofluorescein. When a fluorophore is excited with plane polarized light, the extent to which the emitted light becomes depolarized depends on the rate at which the fluorophore tumbles (which, in turn, depends on molecular weight) and also the lifetime of the fluorescence excited state (*see* **Note 7**). When a protein binds to an fluorescent labeled oligonucleotide, the mass associated with the fluorophore, and therefore the fluorescence anisotropy, is increased, allowing K_D determination. Hexachlorofluorescein is commercially available as a phosphoramidite suitable for automated DNA synthesis, enabling its attachment to the 5' terminus of an oligonucleotide via a six-carbon-chain linker (**Fig. 2**). The flexible linker gives the probe a degree of motion independent of the dynamics of the DNA or the protein–DNA complex. However, although the probe is not rigidly attached to the DNA, it has enough movement coupled to the DNA, together with an appropriate fluorescence lifetime of about 3 ns for the probe attached to a 21-mer *(9)*, to make it sensitive to protein binding (*see* **Note 7**).

The minimum concentration of hexachlorofluorescein that can be easily measured is about 1 n*M* and this limits the technique to measuring K_D values of approx 1 n*M* and above. It has been used to measure K_D values that approach

1 µ*M*. Thus, fluorescence anisotropy is much less sensitive than filter binding or gel retardation, which rely on the detection of ^{32}P. The biggest advantage of this method is that it is a strictly equilibrium approach and does not suffer from problems caused by protein–DNA dissociation. Drawbacks include the requirement for expensive equipment and inability to measure tight binding (K_D < 1 n*M*).

2. Materials

2.1. Oligodeoxynucleotides and Restriction Endonucleases

1. [^{32}P]-labeled oligodeoxynucleotides. These should be of high specific activity, prepared using either γ-[^{32}P]-ATP (3000–6000 Ci/mmol)/polynucleotide kinase (5' labeling) or α-[^{32}P]-ddATP(>5000 Ci/mmol)/terminal transferase (3'-labeling) and purified by standard procedures ***(10)***. Duplex oligodeoxynucleotides are prepared by heating equimolar amounts of each strand to 95°C in buffer with the same pH value and salt concentration to be used in k_{st} or K_D measurement and cooling slowly to room temperature. For k_{st} determination, both strands must be labeled; for K_D evaluation, only one strand needs to be labeled.
2. Oligodeoxynucleotides containing hexachlorofluorescein at their 5' termini. These can be prepared by chemical synthesis using commercially available hexachlorofluorescein–phosphoramidite (Glen Research, Sterling, VA; Cruachem Ltd., Glasgow, Scotland) and purified by reverse-phase high-performance liquid chromatography (HPLC) ***(9)***. Duplexes, only one strand of which needs to contain the fluorophore, are prepared as in **item 1**.
3. Restriction endonucleases under study. In this case, *Eco*RI, *Eco*RV, and *Bam*HI purified from overproducing *Escherichia coli* strains.

2.2. k_{st} Determination

1. Vertical slab gel apparatus (140 × 160 k 0.75 mm) (e.g., Hoefer SE600 or equivalent) and power pack with ≈200 V direct current output.
2. Denaturing polyacrylamide gel. Prepared from 16% acrylamide (acrylamide/*bis*-acrylamide, 19/1) in 0.089 *M* Tris–borate, pH 8.0, containing 8 *M* urea and 1 m*M* EDTA and polymerized with 0.05% ammonium persulfate (added from a freshly prepared 10% aqueous solution) and 1% TEMED (*N,N,N'N'*-tetramethylethylene diammine). The gel should be prerun at a constant power of 30 W, for 1 h, prior to use.
3. Gel running buffer: 0.089 *M* Tris–borate, pH 8.0, containing 1 m*M* EDTA.
4. Hydrolysis buffer (made up at 2X the final concentration, freshly prepared and stored at 4°C): in the examples given using *Eco*RV endonuclease; 20 m*M* HEPES, pH 7.5, 200 m*M* NaCl, and 20 m*M* $MgCl_2$ (*see* **Note 8**).
5. Enzyme dilution buffer: Incubation buffer plus 5% (v/v) glycerol (*see* **Note 8**).
6. Stop solution: 0.1 *M* Tris-HCl, pH 8.0, 0.1 *M* EDTA, 2.5 *M* urea, 10% (w/v) sucrose, and 125 mg/mL (each) bromophenol blue and xylene cyanol FF.
7. Vacuum bag sealer and plastic sheets.
8. Phosphorimager (e.g., Fuji BAS-150) and phosphorimager screen (e.g., Fuji BAS-MP) (or equivalent).

9. Quenched flow apparatus (e.g., Hi-Tech RQF–63 [Hi-Tech Scientific, Salisbury, UK]) or equivalent.

2.3. K_D Determination (Filter Binding)

1. Vacuum filtration manifold (e.g., Millipore 1225 Sampling Vacuum Manifold, Bedford, MA).
2. Mixed ester nitrocellulose membrane filters (e.g., Gelman GN6, Millipore HAWP02500, Schleicher & Schuell ME25) or pure nitrocellulose membrane filters (e.g., Schleicher & Schuell, type BA85), 25 mm disks, 0.45-µm pore size.
3. Enzyme dilution buffer: Binding buffer plus 5% glycerol and 0.2 *M* or 0.6 *M* NaCl (final concentrations), pH 7.4 (*see* **Note 8**).
4. Filter buffer (stored at 4°C): 10 m*M* BTP (*bis*-Tris propane), 1 m*M* EDTA, 0.02% NaN_3 at desired pH (e.g., pH 7.4) and desired salt concentration (e.g., 0.2 *M* NaCl) (*see* **Note 8**).
5. Binding buffer (stored at 4°C): Filter buffer plus 0.1 mg/mL bovine serum albumin and 50 µ*M* dithiothreitol (DTT).
6. Polyethylene liquid-scintillation vials (6 mL capacity).
7. Liquid-scintillation fluid (e.g., Scintisafe 30%, Fisher Biotech, Pittsburgh, PA).
8. Liquid-scintillation counter (e.g., Packard model 1600; Meridan, CT).

2.4. K_D Determination (Gel Retardation)

1. Vertical slab gel apparatus (140 × 160 × 0.75 mm) (e.g., Hoefer SE600 or equivalent) and power pack with ≈200 V direct current output.
2. 10% Nondenaturing polyacrylamide gel (acrylamide/*bis*-acrylamide, 37.5/1) in 0.089 *M* Tris–borate, pH 8.0, and 1 m*M* EDTA and polymerized with 0.05% ammonium persulfate (added from a freshly prepared 10% aqueous solution) and 1% TEMED. The gel should be prerun at a constant power of 20 W with gel running buffer and the wells rinsed with this buffer, prior to use.
3. Gel running buffer: Usually 0.089 *M* Tris–borate, pH 8.0, containing 1 m*M* EDTA. However, the running buffer can, to a limited extent, be varied to produce a better match with the binding buffer (*see* **Note 9**).
4. Binding buffers and enzyme dilution buffers: As *in* **Subheading 2.3., items 3** and **5** except that the binding buffer should additionally contain 3% (v/v) glycerol (*see* **Note 10**).
5. Bromophenol blue solution: 0.25% (w/v) in 30% (v/v) glycerol.
6. Vacuum bag sealer and plastic sheets.
7. Phosphorimager e.g., Fuji BAS–150 and phosphorimager screen (e.g., Fuji BAS-MP) (or equivalent).

2.5. K_D Determination (Fluorescence Anisotropy)

1. Fluorimeter, with thermostated cuvet compartment, capable of measuring fluorescence anisotropy e.g., Aminco SLM-8100.
2. 3-mm-thick 570-nm longpass filter, (OG 570, Schott Glaswerke).
3. 0.5 mL semimicro quartz fluorescence cuvets (excitation and emission pathlengths 5 mm).

4. Fluorescence binding buffer and enzyme dilution buffer: In the examples given with the *Eco*RV endonuclease, 50 m*M* Tris-HCl, pH 7.5, 100 m*M* NaCl, 1 m*M* DTT, 1 m*M* EDTA, and 0.1 mg/mL acetylated bovine serum albumin. The fluorescence binding buffer should be prepared using the best quality reagents and HPLC-grade water and both degassed and passed through 0.22-μm filters (e.g., Millex-GV_{13}, Millipore) prior to use.

2.6. Data Analysis

Software capable of fitting reaction data to single and multiple exponential decay(s) and binding data to single-site binding isotherms using nonlinear regression analysis. The software used for binding analysis should be able to deal with direct titrations when the simplifying assumption $[E]_f = [E]_t$ cannot be made (*see* **Note 4**) and competitive titrations. Three packages described in this chapter are GraFit *(11)*, Scientist *(12)*, and SigmaPlot *(13)*, but any software with a similar capability is perfectly acceptable.

3. Methods

3.1. Determination of k_{st}

3.1.1. Slow Hydrolysis ($t_{1/2}$ 15 s or More ≈k_{st} Values Slower Than 3 min^{-1}; K_D Between 2 and 40 nM) (e.g., with Mutant EcoRV or Altered Oligodeoxynucleotides)

1. In a plastic microcentrifuge tube, prepare 11 μL of a solution containing 1.5 μ*M* of radiolabeled DNA duplex in hydrolysis buffer (in this case for the *Eco*RV endonuclease, 10 m*M* HEPES, pH 7.5, 100 m*M* NaCl, and 10 m*M* $MgCl_2$) (*see* **Note 8**). Keep on ice.
2. Withdraw 1 μL of the above solution and add to 10 μL of stop solution (**item 6** of **Subheading 2.2.**). This serves as a zero time-point.
3. Prepare 5 μL of a solution containing 30 μ*M* of *Eco*RV endonuclease in 10 m*M* HEPES, pH 7.5, 100 m*M* NaCl, and 10 m*M* $MgCl_2$. Keep on ice.
4. Incubate the solutions prepared in **steps 1** and **3** at 25°C for 10 min.
5. Mix the two solutions to give a final oligonucleotide concentration of 1 μ*M* and a final endonuclease concentration of 10 μ*M* (*see* **Notes 1** and **2**). Vortex briefly to ensure thorough mixing and incubate at 25°C.
6. Withdraw 1-μL aliquots at times up to 25 h (the exact time range to be used must be found by trial and error and will depend on the particular mutant enzyme and/or altered oligodeoxynucleotide under study) and quench the reaction by the addition to 10 μL of stop solution (**item 6** of **Subheading 2.2.**). Store the quenched samples on ice.
7. Run the samples on 16% denaturing polyacrylamide gels (**item 2** of **Subheading 2.2.**) until the bromophenol blue dye marker reaches the bottom of the gel.
8. Remove the gel from the electrophoresis apparatus and seal in a plastic bag using a vacuum bag sealer.

9. Determine the amount of radioactivity present in each substrate and product band using a phosphorimager (*see* **Note 11**).
10. From the data obtained from the phosphorimager, determine the percentage of the two substrate and the two product strands present at each time-point.

3.1.2. Rapid Hydrolysis ($t_{1/2}$ 15 s or Less ≈k_{st} Values Faster Than 3 min^{-1}; K_D Between 2 and 40 nM) (e.g., with Wild-Type EcoRV and Cognate GATATC Recognition Sites

1. Fill the three syringes of the quenched flow apparatus with the following:
   a. Reaction syringe 1: 0.1 mL of 2 µ*M* radiolabeled DNA in 10 m*M* HEPES, pH 7.5, 100 m*M* NaCl and 10 m*M* $MgCl_2$.
   b. Reaction syringe 2; 0.1 mL of 20 µ*M* *Eco*RV endonuclease in 10 m*M* HEPES, pH 7.5, 100 m*M* NaCl, and 10 m*M* $MgCl_2$.
   c. Quench syringe 3; 0.1 mL of 0.3 *M* EDTA.
2. Set the apparatus to mix the 0.1 mL of the oligonucleotide and enzyme solution (final concentrations of each 1 µ*M* and 10 µ*M*, respectively) and to quench the reaction at the first time-point (0.051 s) by the addition of the 0.1 mL of the EDTA solution.
3. Keep the quenched sample on ice.
4. Repeat for each subsequent time-point (in this case, 19 further points between 0.094 and 20 s) (*see* **Fig. 4**).
5. As a zero time-point, 0.1 mL of the oligonucleotide solution manually mixed with 0.1 mL of 10 m*M* HEPES, pH 7.5, 100 m*M* NaCl, and 10 m*M* $MgCl_2$, and 0.1 mL of 0.3 *M* EDTA can be used.
6. Add 5 µL of each of the quenched samples to 5 µL of stop solution (**Subheading 2.2., item 6**).
7. Proceed with the analysis by gel electrophoresis and phosphoroimaging (**Subheading 3.1.1., step 6–9**).

3.1.3. Data Analysis; k_{st} Determination

1. Cleavage of a duplex oligonucleotide by a restriction endonuclease involves parallel sequential reactions (**Fig. 3**), giving intermediates in which one strand is nicked, and described by four rate constants *(14,15)*. In some instances, the reaction scheme can be simplified. This is usually true for wild-type endonucleases acting on their natural, unmodified, target sequences. Here, both strands are efficiently cut in a concerted reaction and nicked intermediates do not accumulate. In this case, the cutting of each of the two substrate strands (or the accumulation of the two radiolabeled products) can be described by a single rate constant with an acceptable degree of accuracy.
2. If simplification is possible, fit the data to an equation describing a single exponential using GraFit *(11)*. This equation is supplied within GraFit and almost all biological kinetic software packages. The design of the oligonucleotides used (**Fig. 1**) means that the cutting of each strand can be evaluated individually. GraFit requires time and the percentage of substrate or product present at these times to

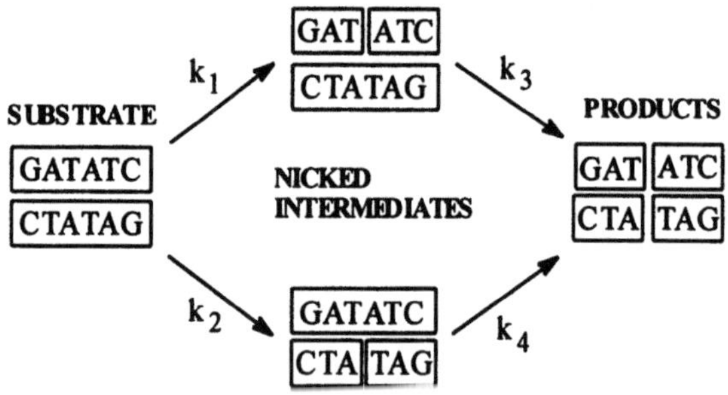

Fig. 3. Parallel sequential cutting of a double-stranded oligodeoxynucleotide by the *Eco*RV endonuclease.

be entered into a spreadsheet and to use these data to evaluate k_{st}. Concerted cutting of both strands implies that the k_{st} for each of them should be, within the limits of experimental error, identical. An example is shown in **Fig. 4**.

3. With modified oligonucleotides or enzymes altered by mutagenesis, cutting often becomes inefficient and nicked intermediates accumulate. Altering any one point in a palindromic recognition site produces structural asymmetry, leading to different values of k_1 and k_2. In such cases, the following equations are used to obtain values of the four rate constants that describe the unsimplified reaction scheme (**Fig. 3**):

$$P_1 = S_1[1 - \bar{e}^{\lambda t} - \frac{k_2}{k_4 - \lambda}(e^{-\lambda t} - e^{k4t})]$$

$$P_2 = S_2[1 - \bar{e}^{\lambda t} - \frac{k_1}{k_3 - \lambda}(e^{-\lambda t} - e^{k3t})]$$

where $\lambda = k_1 + k_2$ and S_1 and S_2 are the percentages of total radioactivity in each DNA substrate strand at $t = 0$. If end labeling of the two strands is equal, then $S_1 = S_2 = 50\%$. The ability to separate the two DNA substrate strands permits normalizing for "differential labeling" in the event that the two strands have been radiolabeled with slightly different efficiencies (e.g., if $S_1 = 40\%$ and $S_2 = 60\%$, these values should be entered into the equations). If the radioactivity of the two strands is markedly different, the program has more difficulty finding the best fit. This is especially the case when a strand that has an intrinsically faster cleavage rate constant has lower specific radioactivity than the other strand. Therefore, it is preferable to attempt to label both strands to the same specific activity.

Reaction times (t) and the amount of products (P_1 and P_2, expressed as percentage of total radioactivity for each time point) are entered into the Scientist ***(12)*** software spreadsheet. A plot of the experimental and "fitted" values of P_1 and P_2 is generated along with numerical values of the rate constants k_1 to k_4. Such fits are most statistically robust when the uncertainties in the independent

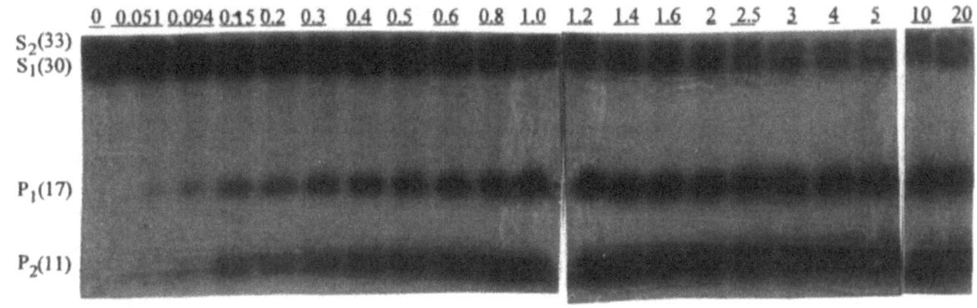

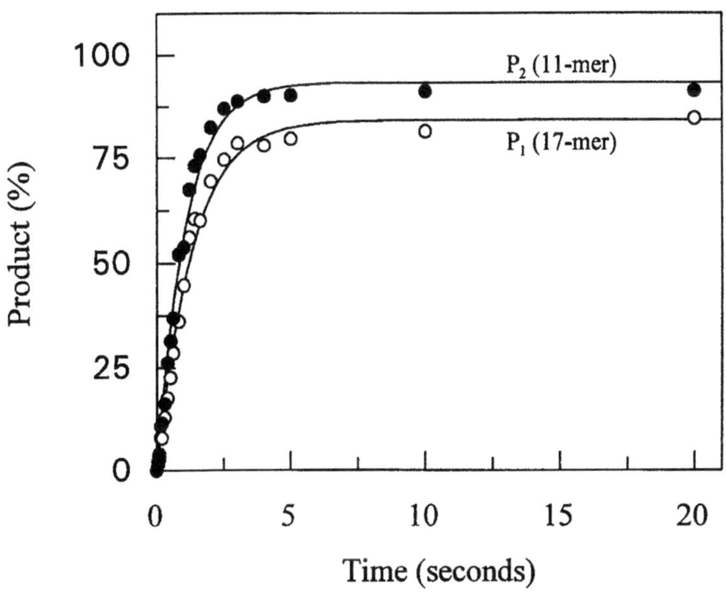

Fig. 4. Cleavage of the duplex oligonucleotide (1 µ*M*) produced by mixing 5'-AAAGTCTGTG**GATATC**CAAGTGGCTACCGT-*ddA and 5'-CCCCCACGG-TAGCCACTTG**GATATC**CACAGACT-*ddA (**Fig. 1**) with *Eco*RV endonuclease (10 µ*M*) using a quench-flow apparatus. The top part shows an autoradiograph of denaturing PAGE analysis of the two substrates and the two labeled products present after various mixing times (given on top of the gel lanes in s). The bottom part shows fits for the formation of each product using the simplified model (i.e., formation of each product is fitted using a single exponential with Grafit *[11]*). It should be noted that data were obtained by phosphorimaging of the gel, and the autoradiograph is presented for illustrative purposes only. The k_{st} values found were 0.94 s^{-1} and 0.77 s^{-1} for the production of P_2 and P_1 respectively. Fits to the full parallel sequential model (figure 3) using Scientist *(12)* (not shown) gave a k_2 (for P_2 and production) of 0.65 s^{-1} and a k_1 (for P_1 production) of 0.51 s^{-1}. Thus, in the case of wild-type *Eco*RV with its cognate GATATC sequence, the simplified model gives rates that are very similar to the rigorously correct full model.

variable (time) are much smaller than the uncertainties in the dependent variable (here, each radioactivity measurement in a product) and when the number of data points is sufficiently high to provide a good sampling of experimental uncertainties. The "goodness of fit" of the experimental data to the fitted curve is judged by the individual reaction point residuals, which should vary no more than 10%. We have found that a rigorous test of the robustness of the fit to the above equations is to measure the second strand cleavage rate constants for "nicked" constructs, assembled from three individual fragments. When using the full parallel-sequential model, it is good practice to apply the same calculation method to reference reactions with the normal enzyme–substrate pair. A comparison of the rate constants obtained from the "full" and "simplified" calculation methods then tests the validity of using the simplified-model fit for the cleavage rate constants k_1 and k_2 for the reference pair (*see* **Note 12**).

3.2. Filter Binding

3.2.1. K_D Determination: Direct Binding–Titration of DNA with Protein. Example: $K_D = 0.66$ nM $K_A = 1/K_D = 1.5 \times 10^9$ M^{-1}

1. Presoak nitrocellulose membrane filters in filter buffer at the same salt concentration and pH used in the binding reaction. Place in the vacuum manifold.
2. Prepare a solution of 0.5 n*M* radiolabeled duplex DNA by diluting the radiolabeled stock DNA solution (100 n*M*) with binding buffer at the desired salt concentration and pH. Add 10 μL of radiolabeled duplex oligonucleotide (0.5 n*M*) to each of 10 microcentrifuge tubes; the final concentration of radiolabeled duplex DNA should be 0.05 n*M* after **step 3** (*see* **Note 4**). Add the appropriate volume of binding buffer such that the total volume of the reaction after step 3 would be 100 μL. Equilibrate on ice for 5 min.
3. To each of the above microcentrifuge tubes add restriction endonuclease to give final concentrations ranging from 0.05 n*M* to 5 n*M* (i.e., the midpoint protein concentration should be approx equal to the K_D; *see* **Note 4**); equilibrate reactions on ice for 5 min more. During an experimental series, the protein stock should always be kept on ice and any intermediate dilutions of the stock that are not used directly in the final reactions are made using enzyme dilution buffer (salt type and pH always match the experimental conditions). The final dilution to give the sample used in the reaction should be made with binding buffer (no glycerol) at the appropriate salt concentration (*see* **Note 8**). The salt derived from the diluted enzyme stock and the DNA source is always accounted for in designing the experiment.
4. Set up two "blank" tubes containing DNA, binding buffer, but no enzyme. One blank tube will be filtered to obtain R_B (background counts) and the other used to determine R_T (the input counts).
5. Also set up a tube containing DNA, binding buffer, and enzyme at a concentration 100-fold the K_D (here, 50 n*M* enzyme). This tube will be filtered to obtain R_{max}, which represents the total available DNA for binding.

6. Transfer the reaction tubes, the two blank tube, and the R_{max} tubes to the desired temperature (e.g., 25°C) and incubate for 30 min.
7. Pipet 85 µL from each reaction tube, the R_B blank, and the R_{max} tube onto a presoaked 25 mm nitrocellulose filter in the vacuum filtration manifold with the vacuum applied. The vacuum (flow rate about 0.5 mL/10 s) is applied continuously throughout the experiment to pull the reaction aliquot through the filter.
8. As quickly as possible, wash each filter with 350 µL of filter buffer to remove trapped free DNA.
9. Place each filter in a liquid-scintillation vial and add 2.5 mL of liquid-scintillation fluid. Count in a liquid scintillation counter.
10. Pipet 85 µL from the R_T blank tube onto a nitrocellulose filter. Do not wash, but place the filter directly into a liquid-scintillation vial, add 2.5 mL of liquid-scintillation fluid, and count.

3.2.2. K_D Determination: Direct Binding–Titration of Protein with DNA Example: $K_D = 40$ pM, $K_A = 1/K_D = 2.5 \times 10^{10}$ M^{-1}

1. Presoak nitrocellulose membrane filters (**Subheading 3.2.1., step 1**).
2. Prepare a 4 n*M* stock solution of radiolabeled DNA. For a 4 n*M* stock DNA solution, the ratio of radiolabeled DNA to unlabeled DNA is typically 1:3 or 1:4, depending on the specific activity of the radiolabeled DNA.
3. To each of 10 microcentrifuge tubes, add radiolabeled DNA to give final concentrations ranging from 10 p*M* to 600 p*M*. (the midpoint DNA concentration should be approx equal to the K_D, *see* **Note 4**).
4. Add the appropriate volume of binding buffer plus salt at the desired pH such that the total volume of the reaction after **step 6** will be 100 µL. Transfer to ice and equilibrate for 5 min.
5. Prepare a solution (40 p*M*) of restriction endonuclease, at the desired salt concentration and pH, by making intermediate dilutions of the stock enzyme solution with enzyme dilution buffer and the final dilution (to give the 40 p*M* solution) with binding buffer as described in **Subheading 3.2.1., step 3**.
6. To the 10 reaction tubes prepared above, add 10 µL enzyme to give a final concentration of 4 p*M*. Keep on ice for an additional 5 min.
7. Set up 10 "blank" reaction tubes (each with total volume 100 µL) that correspond to the reaction tubes prepared in **steps 3** and **4** (i.e., 100 µL total volume; radiolabeled DNA concentrations ranging from 10 p*M* to 600 p*M*) except that no enzyme is added.
8. Transfer the reaction and blank tubes to the desired temperature (e.g., 25°C) and incubate for 30 min.
9. Remove 85 µL from each tube and carry out filter binding as described in **Subheading 3.2.1., steps 5–7**. As the DNA concentration varies in this experiment, a different blank (with a DNA concentration that corresponds to its reaction partner) is required for each reaction point.

3.2.3. Data Analysis: Direct Binding

1. For "normal titrations," where DNA is titrated with protein, values of K_{obs} are obtained by nonlinear least-squares fits to a single-site binding isotherm:

$$\frac{[ED]}{[D]_t} = \frac{K_A[E]_f}{1 + K_A[E]_f}$$

Note that here K_{obs} is K_A and not K_D (*see* the footnote in **Subheading 1.**). The actual equation entered into SigmaPlot is (this titration is carried out under conditions $[E]_f \approx [E]_t$, *see* **Note 4**):

$$R_F - R_B = MAX\frac{K_A[E]_t}{1 + K_A[E]_t}$$

The experimental data entered into the spreadsheet are the enzyme concentration, $[E]_t$, for each titration point and R_F (counts representing complex retained on the filter) for each titration point; R_B (counts obtained when radiolabeled DNA is filtered without enzyme). A plot of the experimental counts (circles) and idealized counts (line) versus $[E]_t$ is generated as shown in **Fig. 5**. In this figure, $[ED]/[ED]_{max}$ i.e., $(R_F - R_B)/MAX$ is plotted against $[E]_t$. The nonlinear least squares best fit to the equation will give both K_A and MAX. The parameter MAX is the asymptote of the binding isotherm and represents the theoretical maximum counts retained by the filter. This theoretical MAX can be checked against the experimentally determined R_{max} (*see* **Subheading 3.2.1., step 5**). The retention efficiency, $MAX/(R_T - R_B)$ or $R_{max}/(R_T - R_B)$, is a good index of reproducibility for each set of conditions (enzyme, DNA, salt, pH, temperature). The three enzymes (*Eco*RI, *Bam*HI, and *Eco*RV) have different retention efficiencies.
2. In the case of "reverse titration" (i.e., when protein is titrated with DNA), values for K_A are obtained essentially as above, except that here the equation for a single-site binding isotherm is

$$\frac{[ED]}{[E]_t} = \frac{K_A[D]_f}{1 + K_A[D]_f} \text{ and } R_F - R_B = MAX\left(\frac{K_A[D]_t}{1 + K_A[D]_t}\right)$$

Because the concentration of enzyme is held constant and titrated with DNA, $[E]_t$ should be 5–10% of the K_D value (*see* **Note 4**). This is in contrast to a "normal titration" (DNA held constant and titrated with protein), where $[D]_t$ is 5–10% of the K_D.

3.2.4. K_D Determination: Competitive Equilibrium Binding. Examples: K_D of Reference DNA = 0.25 nM; K_D of Competitor DNA = 0.69 nM K_D of reference DNA = 0.25 nM; K_D of competitor DNA = 0.81 µM

1. Presoak nitrocellulose membrane filters as in **Subheading 3.2.1., step 1.**
2. To each of 10 microcentrifuge tubes, add 10 µL of radiolabeled reference specific DNA from an 10 n*M* stock solution to give a final concentration of 1 n*M* (*see* **Note 12**).

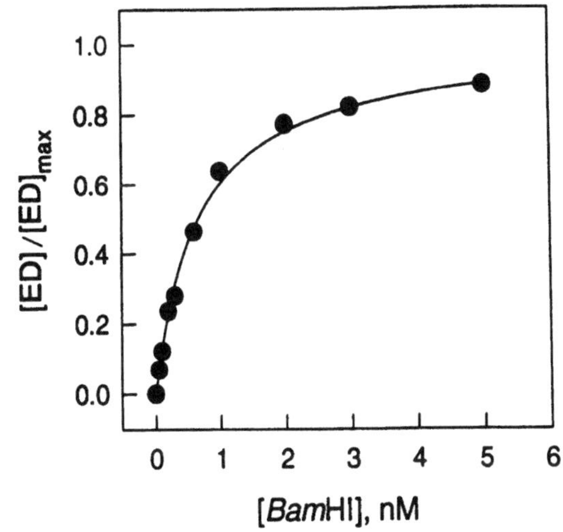

Fig. 5. Representative binding isotherm determined by the direct equilibrium filter-binding assay. The specific *Bam*HI substrate 5'-CGCGGGCGGC**GGATCC**GGGCGGGC was titrated with *Bam*HI endonuclease. The binding buffer contained 0.14 *M* potassium acetate at pH 7.3 and the experiment was carried out at 25°C. Solid circles are experimental points. The fitted Sigmaplot curve gives a K_A value of $1.51 \times 10^9/M$.

3. Add the appropriate volume of unlabeled competitor DNA such that the final concentrations of competitor DNA will vary between 0 and 40 n*M* for an expected K_D of competitor DNA ≈1 n*M* (*see* **Fig. 6A**). If the K_D expected for the competitor DNA is much higher (e.g., approx 1 µ*M*), then the final concentrations of unlabeled competitor DNA should vary in the range from 0 to 20 µ*M* (*see* **Fig. 6B**). Add the appropriate volume of binding buffer (at desired salt concentration and pH) such that the total volume of the reaction after **step 4** will be 100 µL. Place tubes on ice for at least 10 min.
4. To each of the above 10 microcentrifuge tubes, add restriction endonuclease to give a final concentration of 0.8 n*M*, taking into account the considerations detailed in **Subheading 3.2.1., step 3** for enzyme dilution (*see* **Notes 8** *and* **12**). Keep on ice for 5 min or more.
5. Set up two additional tubes: a "blank" tube containing radiolabeled DNA, binding buffer, but no competitive DNA and no enzyme to give R_B and a tube to obtain the R_{max} (i.e., maximum counts retainable by the filter) containing radiolabeled DNA, binding buffer, and a final concentration of 80 n*M* enzyme (no competitive DNA). The R_{max} tube should also be chilled on ice.
6. Transfer all reactions to the desired temperature (e.g., 25°C) and equilibrate for 30 min.
7. Remove 85 µL from each tube and carry out filter binding as described in **Subheading 3.2.1., steps 5–7**. Background counts (R_B) are obtained by filtering the tube containing radiolabeled DNA in the absence of protein and competitor DNA.

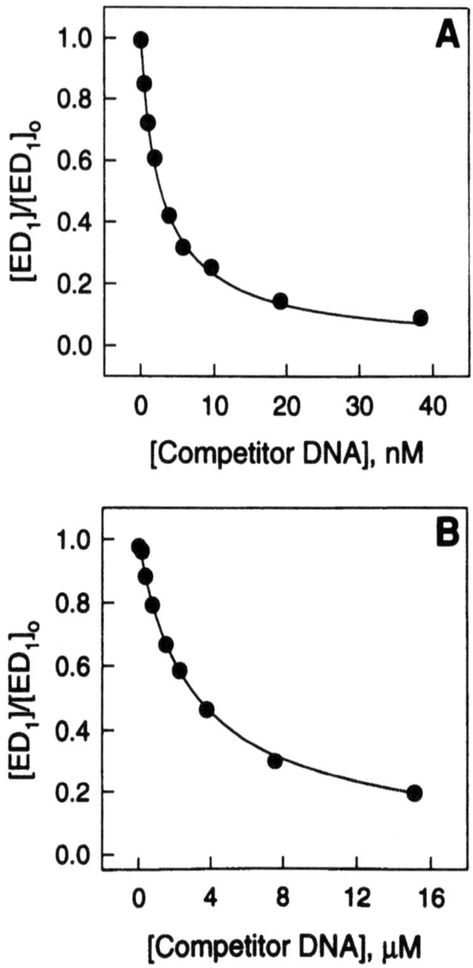

Fig. 6. Representative equilibrium-competition curves for the interaction of *Bam*HI endonuclease with specific and nonspecific sites. A 40-base-pair specific "snapback" substrate 5'-TGGGTG**GGATCC**CACCCACCCCCTGGGTG**GGATTC**CACCC was used as a radiolabeled probe for both competition curves. The binding buffer contained 0.14 *M* potassium acetate at pH 7.3 and the experiment was carried out at 25°C. Solid circles are experimental points. Test unlabeled specific competitors were (a) the same specific substrate used for the direct binding assay in **Fig. 4** and (b) a nonspecific site (**CCTAGG**) embedded in the same flanking context. The K_D value determined from the Sigmaplot fit to curve A is 0.69 n*M* ($K_A = 1.45 \times 10^9\ M^{-1}$) and to curve B is 0.81 m*M* ($K_A = 1.23 \times 10^6/M^{-1}$).

Corrected reaction counts are obtained by subtracting the background counts from each reaction. The tube containing radiolabeled DNA and 80 n*M* enzyme is used to determine the R_{max} value (*see* **step 5**).

3.2.5. Data Analysis: Competitive Equilibrium Titration

Results are fitted, using SigmaPlot, to a binding isotherm as previously described ***(6)*** using the equation

$$[ED_1] = \frac{[E]_t([D_1]_t - [ED_1])}{K_1\left(1 + \frac{[D_2]_t}{K_2}\right) + ([D_1]_t - [ED_1])}$$

where $[E]_t$ = total enzyme concentration, $[D_1]_t$ = total radiolabeled reference DNA concentration, $[D_2]_t$ = total unlabeled competitor DNA concentration, K_1 = dissociation constant for the radiolabeled reference DNA; K_2 = dissociation constant for the unlabeled competitor DNA. Solving the equation for $[ED_1]$ yields

$$[ED_1] = \frac{1}{2}\left\{K_1 + \frac{K_1}{K_2}[D_2]_t + [E]_t + [D_1]_t - \sqrt{\left[(K_1 + \frac{K_1}{K_2}[D_2]_t + [E_1]_t + [D_1]_t)^2 - 4[D_1]_t[E_1]_t\right]}\right\}$$

where

$$[ED_1] + [D_1]_t \frac{R_F - R_B}{R_{max} - R_B}$$

The dissociation constant K_1 for the reference DNA is always determined by direct equilibrium binding at the start of a competition experiment. R_{max}, obtained at saturating concentrations of protein and without competitor DNA (*see* **Subheading 3.2.4., step 5**) represents the maximum available $[D_1]_t$. The known values for K_1, $[D_1]_t$, [E], R_{max}, and R_B (*see* **Subheading 3.2.4., step 5**), are entered into the SigmaPlot software at the start of each calculation. The experimental data entered into spreadsheet columns are the corrected counts (counts retained on the filter for each titration point minus background counts; $R_F - R_B$) and the competitor DNA concentration $[D_2]_t$ for each point. The curve generated by plotting the corrected counts as a function of increasing concentrations of competitor DNA ($[D_2]_t$) is fitted to the best value for the equilibrium dissociation constant K_2 using SigmaPlot nonlinear regression analysis. **Figure 6** presents the results as a plot of the ratio of $[ED_1]/[ED_1]_0$ where $[ED_1]_0$ is the concentration of enzyme–DNA complex obtained in the absence of competitor. The $[ED_1]_0$ found in this experiment should not be significantly different from the $[ED_1]_0$ calculated for a direct binding experiment.

3.3. Gel Retardation

3.3.1. K_D Determination: Direct Binding–Titration of DNA with Protein

1. Set up and equilibrate binding reactions as described in **Subheading 3.2.1., steps 2–4**. The final reaction tubes should additionally contain 3% (v/v) glycerol (*see* **Note 10**). A "blank" should contain all the components except endonuclease. Smaller samples (typically 10 μL) are used in gel retardation than are used for

filter binding (85 µL in the examples given above). Therefore, if only gel retardation is being carried out, the final volume should be 10 µL. Alternatively, it can be very useful to make up a binding reaction and split into aliquots for simultaneous analysis by both filter binding and gel retardation (either direct or competition titration).

2. Following incubation, load the aliquots of the samples, typically 10 µL, into the wells of 10% nondenaturing polyacryalmide gels (*see* **Subheading 2.4., items 1** and **2** and **Note 9**).
3. Run the gels at a constant power of 20–25 W, with cooling, until a bromophenol blue marker (*see* **Subheading 2.4., item 5** and **Note 10**) in an adjacent lane, not containing the experimental samples, reaches the gel front (about 1 h).
4. Remove the gel from the electrophoresis apparatus and seal in plastic using a vacuum bag sealer.
5. Determine the amount of radioactivity present in the bands that correspond to the free and bound DNA using a phosphorimager (*see* **Note 11**).

3.3.2. K_D Determination: Competitive Equilibrium Binding

1. Set up equilibrium competition reactions according to the steps outlined in **Subheading 3.2.4., steps 2–4**, except that the binding buffer for all reactions should include 3% (v/v) glycerol (*see* **Note 10**). If only gel retardation is being carried out smaller volumes (*see* **Subheading 3.3.1., step 1**) (e.g., 10 µL should be used. A "blank" should contain all the components except endonuclease and competitor DNA.
2. Perform gel retardation analysis and phosphorimaging as described in **Subheading 3.3.1., steps 1–5**.

3.3.3. Data Analysis: Gel Retardation

1. For direct titration it is possible to determine K_{obs} in an analogous manner to that described in **Subheading 3.2.3.** In this case, however,

   $$[ED]/[D]_t = Counts_{complex}/Counts_{free} + Counts_{complex} = \text{fraction of } counts_{complex}$$

   where $counts_{complex}$ and $counts_{free}$ are the counts in the shifted (enzyme-DNA) and unshifted (free DNA) bands, respectively. The experimental data entered into the spreadsheet are fraction of $counts_{complex}$ and the enzyme concentration $[E]_t$ for each titration point.
2. For competition binding the equations given in **Subheading 3.2.5.** are applicable. In this case,

   $$[ED_1] = [D_1]_t \frac{counts_{complex}}{counts_{max}}$$

   where $counts_{complex}$ and $counts_{max}$ are, respectively, the counts in the shifted band for each titration point at a particular competitor DNA concentration and for the control using large amounts of enzyme and no competitor DNA (*see* **Subheading 3.2.4., step 5**), where all the DNA is shifted into the complex. The known values for K_1, $[D_1]_t$, $[E]_t$, and $counts_{max}$ are entered into SigmaPlot at the start of each calculation. The experimental data entered into spreadsheet columns are $counts_{complex}$ and the competitor DNA concentration $[D_2]_t$ for each titration point.

3.4. Fluorescence Anisotropy

3.4.1. K_D Determination. Example: $K_D \approx 40$ nM

1. Set the fluorimeter to measure anisotropy in the "single-point polarization" mode. Each anisotropy value should be measured 10 times (measurement time, 5 s) and averaged (automatically carried out by the fluorimeter). Set the G-factor to "per measurement" (*see* **Note 14**).
2. An excitation wavelength of 530 nm is used with all the slits on the excitation monochromator set to 8.
3. Anisotropy is measured in the "L" format through the right photomultiplier with a 570-nm long-pass filter between the sample and the photomultiplier. Using a filter rather than a monochromator on the emission side greatly increases the light intensity at the photomultiplier and so allows measurement of lower concentrations of fluorophore.
4. Place a 0.5-mL fluorescence cuvet in the fluorimeter. The cuvet should not be removed from the fluorimeter or otherwise moved during the experiment. Add fluorescence binding buffer such that the final volume after **step 3** will be 0.5 mL. A measurement temperature of 25°C is used.
5. Add duplex DNA, one strand of which is labeled with hexachlorofluorescein (**Subheading 2.1., item 2**) to give a final concentration of 10 n*M* (*see* **Note 15**). Note the anisotropy.
6. Add *Eco*RV endonuclease in small aliquots (*see* **Note 16**), using a microliter Hamilton syringe, to cover the range 0–400 n*M*. Mix thoroughly by gently withdrawing the contents with a plastic pipet tip and readding to the cuvet. After each addition, measure the anisotropy.
7. At the beginning (all oligonucleotide free in solution) and end (all oligonucleotide bound to endonuclease) of the titration, the fluorescence emission intensity should be noted (*see* **Note 14**).

3.4.2. Data Analysis

1. The fluorescence anisotropy measurements were carried out at $[D]_t \approx K_D$ (*see* **Note 15**). Under these circumstances, $[E]_f \neq [E]_t$ and so the simplified, hyperbolic, form of the binding equation cannot be used. Therefore, data must be fitted using the full quadratic binding equation (*see* **Note 4**). A solution, in terms of anisotropy, is

$$A = A_{min} + \left(\frac{A_{max} - A_{min}}{[D]_t}\right)\left[\times\frac{1}{2}\left(([D]_t + [E]_t + K_D) - \sqrt{\left[([D]_t + [E]_t + K_D)^2 - (4\,[D]_t[E]_t)\right]}\right.\right]$$

where A is the measured anisotropy; A_{min} is the anisotropy of free DNA, and is the A_{max} anisotropy of DNA when fully bound to the endonuclease.

2. The data should be fitted to the above equation using GraFit *(11)*. This requires entry of A and $[E]_t$ into the spreadsheet provided by the software. $[D]_t$ is a con-

stant (i.e., the concentration of oligonucleotide present at the start of the experiment). The titration consists of the addition of multiple aliquots of endonuclease to the cuvet, resulting in a dilution of the components. Correction should be made for the dilution of the endonuclease as aliquots are progressively added. Anisotropy, a ratio of fluorescence intensities, does not depend on concentration and is, therefore, unaffected by dilution. However, as $[D]_t$ is treated as a constant by this software it is not possible to compensate for its dilution. Therefore, the final dilution should not exceed about 10%. This may require endonuclease addition from several stocks of different concentration. GraFit also requires an estimate of A_{min}, A_{max}, and K_D; suitable values for the first two can be obtained from the measured anisotropy values at the start (before endonuclease addition) and (after the final addition of endonuclease) at the end of the experiment. The software uses nonlinear regression analysis to calculate best-fit values of these three parameters. Note that the equation given above only holds if the fluorescence emission intensity of the free and protein-bound DNA are identical (*see* **Note 14**). This is determined in **Subheading 3.4.1., step 6**.

3. A representative titration curve is shown in **Fig. 7**.

4. Notes

1. With restriction endonucleases, it is often found that the binding of DNA and Mg^{2+} are rapid relative to subsequent events such as conformational changes, hydrolysis, and product release. This means that rate constants measured under single-turnover conditions are the same whether the reaction is initiated by adding Mg^{2+} to an endonuclease–DNA solution or adding the enzyme to a solution containing DNA and Mg^{2+}. This should, however, be checked by trying both initiating methods.
2. The K_D values for the *Eco*RV endonuclease and the oligonucleotides used in **Subheadings 3.1.1.** and **3.1.2.** vary between 2 and 40 n*M*. The DNA concentration used in **Subheadings 3.1.1.** and **3.1.2.** ($25 \times K_D$ for a K_D of 40 n*M*) and an endonuclease level 10 times higher ensures that ≥95.5% of the nucleic acid is bound to the protein. Many other concentrations of the two macromolecules formally meet this requirement. A useful combination, especially for proteins that are unstable, poorly soluble, and prone to form inactive tetramers and insoluble higher aggregates when free in solution is $[D]_t = 50\ K_D$ to100 K_D and $[E]_t = 2 \times [D]_t$. This results in at least 98% of the DNA being bound and avoids a large excess of the labile free protein. When bound to DNA, many endonucleases are both stabilized and protected from aggregation. For substrates with poor K_D and k_{st} (especially so-called "star" sites, where one of the bases in the recognition sequence is replaced by another natural base), dissociation rate constants are much greater than the cleavage rate constants. Therefore, multiple protein dissociations and associations occur prior to a productive cleavage event and so very poor substrates are never really cleaved under true single-turnover conditions. However, high concentrations of protein and nucleic acid ensure that the rate ($k_a \times$ [DNA][E]) of association is as fast as possible and minimizes the

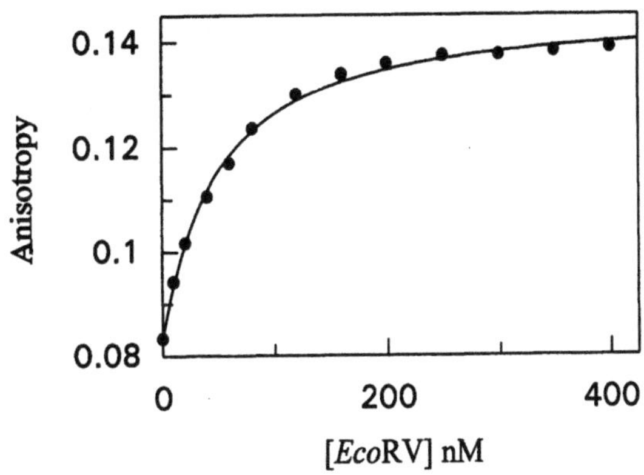

Fig. 7. Representative binding isotherm determined by fluorescence anisotropy assay. An oligonucleotide composed of the complementary 14-mers 5'-Hex-TCCG**GATATC**ACCT and 5'-AGGT**GATATC**CGGA was titrated with the *Eco*RV endonuclease. The binding buffer contained 50 m*M* Tris-HCl and 100 m*M* NaCl and the experiment was carried out at 25°C. The solid circles are experimental points. The K_D value determined from the GraFit fit to the curve is 42 n*M*.

chances that it contributes to the rate-determining step. Whatever levels of protein and DNA are selected, the rate constants should be independent of protein concentration if true single-turnover conditions apply. A useful empirical test is to measure rates at several protein concentrations to confirm this assumption.

3. For the equilibrium,

$$\text{Endonuclease + DNA} \underset{k_{off}}{\overset{k_{on}}{\rightleftharpoons}} \text{Endonuclease–DNA} \quad K_D = k_{off}/k_{on}$$

the stability of a protein–DNA complex depends on k_{off}. Most proteins have k_{on} values of between 10^6 and $10^8\ M^{-1}s^{-1}$ enabling an estimate of the half-life (0.693/k_{off}) expected for an endonuclease–DNA complex at different K_D values:

$K_D = 1\ \mu M$ ($K_A = 1 \times 10^6/M^{-1}$); half-life between 0.01 and 1 s
$K_D = 1$ nM ($K_A = 1 \times 10^9/M^{-1}$); half-life between 10 s and 10 m
$K_D = 1$ pM ($K_A = 1 \times 10^{12}/M^{-1}$); half-life between 100 m and 150 h

4.

$$K_D = [E]_f\,[D]_f/[ED] = ([E]_t - [ED]) \times ([D]_t - [ED])/[ED]$$

Therefore, the fractional saturation of the DNA with protein, $[ED]/[D]_t$ (the parameter measured using filter binding, gel shift, or fluorescence anisotropy), has a quadratic relationship to $[E]_t$. However, under the condition $[E]_f = [E]_t$ (which occurs if $[E]_t >> [ED]$),

$$K_D = [E]_t \times ([D]_t - [ED])/[ED]$$

Here, $[ED]/[D]_t$ has a hyperbolic relationship to $[E]_t$, simplifying data analysis. In order to achieve the condition $[E]_t >> [ED]$ and to obtain a satisfactory titration it is also necessary to have $[D]_t < K_D$. DNA concentrations ≤15% K_D are sufficient to satisfy the assumption $[D]_t < K_D$ and give insignificantly different K_D values whether the hyperbolic or quadratic binding equation is used. However, very low $[D]_t$ often gives binding curves of very poor quality. This may result from insufficient ligand stabilization of the protein by the nucleic acid. Therefore, a $[D]_t$ in the 5–15% K_D range is recommended. Concentrations of protein are chosen so that the midpoint of the titration curve yields a protein concentration approx equal to the expected K_D. Clearly, the K_D will not be known prior to experimentation, emphasizing the need for an initial estimation, followed by a second, more accurate, determination.

5. Filter binding requires passage of the mixture through a nitrocellulose filter and a brief washing of the filter with buffer to remove excess unbound DNA. The entire process can be carried out in seconds. Gel retardation takes longer; about 1–2 min to load the samples and the same time to run them into the gel. Running the gel takes 30–60 min. Therefore, gel retardation can suffer, to a greater extent than filter binding, from dissociation of the protein–DNA complex. To some degree the problem is self-diagnostic, because it results in a smeared protein–DNA band rather than the tightly focused one which arises when no dissociation takes place. In these cases, K_D determination may be achieved by measurement of the free-DNA band ***(16)***, but if this approach is used, an alternative method is best used to check the validity of the result.
6. As an example, consider an unmodified DNA with a K_D of 0.1 n*M* and four modified DNAs with K_D values that range from 1 n*M* to 1 m*M*. Direct equilibrium analyses are performed for the unmodified substrate and the modified site with K_D = 1 n*M*. Equilibrium competition analyses are also performed for the four modified sites. If identical values of K_D are obtained from both direct and competition analyses for the modified site with K_D = 1 n*M*, the modified site serves as an internal control for the three modified sites which bind more weakly.
7. The relationship between fluorescence anisotropy, lifetime, and rotation is given by the Perrin equation ***(7,8)***:

$$A = A_0/(1 + \tau/\phi)$$

where A is the measured anisotropy, A_0 = intrinsic anisotropy (i.e., the anisotropy measured under conditions where no depolarization takes place), τ is the lifetime of the fluorescence excited state, and ϕ is the rotational correlation time, a measure of rotational diffusion that depends on the molar volume and, hence, the molecular mass attached to the fluorophore. The values of τ and ϕ must be balanced: a very short lifetime, relative to the correlation time, would lead to $A = A_0$; a very long lifetime to $A = 0$. Use of this method requires $0 < A < A_0$. The τ value of approx 3 ns for hexachlorofluorescein attached to oligonucleotides is suitable for distinguishing between free and protein-bound oligonucleotides. The measured anisotropy is given by:

$$A = (I_{\parallel} - I_{\perp})/(_{\perp} I_{\parallel} + 2I_{\perp})$$

$I_\parallel$ and $I_\perp$ are the intensities of the parallel and perpendicular components of the emitted light when parallel excitation is used.

8. The buffer components used for k_{st} and K_D determination will obviously vary somewhat depending on the enzyme under study and the particular experiment. However, for most DNA-binding proteins, including *Eco*RI, *Eco*RV, and *Bam*HI restriction endonucleases there is pronounced nonlinearity in the dependence of log K_D versus salt concentration ***(17)*** at salt concentrations ≤0.1 *M*, presumably because at these low salt concentrations, coupled protein–protein equilibria leading to aggregation ***(18)*** become significant. This phenomenon can produce misleading quantitative comparisons and thus salt concentrations below 0.1 *M* should be avoided. Bis-Tris propane is useful as it has two pK values (6.8 and 9), allowing pH-dependence studies with a single buffer. However, other buffers, of appropriate pK can equally well be used. Bovine serum albumin and DTT can help to stabilize certain restriction endonucleases, especially at low concentrations. With several DNA-binding proteins, including *Eco*RI endonuclease, aggregation and ultimately precipitation, in the absence of the stabilizing DNA substrate, is quite common at salt concentrations below 0.4 *M*. In these cases, intermediate dilutions should be made with buffers containing 0.6 *M* NaCl. Five percent glycerol, in the dilution buffers, also aids stability.
9. The "classic" gel-shift buffer, used to prepare and run nondenaturing polyacrylamide gel is 0.089 *M* Tris-borate, pH 8.0, containing 1 m*M* EDTA ***(19)***. Many protein–DNA complexes, particularly those of high affinity, are stable in this buffer. Therefore, it is often possible to incubate enzyme and DNA in a selected binding buffer and then carry out gel retardation in Tris-borate electrophoresis buffer without disturbing the equilibrium initially set up. However, the binding of restriction endonucleases to DNA depends on pH, salt concentration (NaCl or KCl), and divalent cations (Ca^{2+}, a Mg^{2+} mimic that does not allow hydrolysis, often considerably strengthens binding) ***(18,20,21)***. Ideally it would be best to use identical incubation and electrophoresis buffers. We have obtained satisfactory gel retardation data using Tris-borate at pH 7.5. Replacing the EDTA with up to 5 m*M* $CaCl_2$ is also not detrimental. However, many buffers are not suitable for electrophoresis, giving poorly resolved or smeared bands, especially if NaCl or KCl is added.
10. Gel retardation requires that the sample applied to the gel be denser than the electrophoresis buffer and so sinks into the well. The presence of 3% glycerol in the binding reactions ensures efficient sinking of the sample ***(18)***. If this method is used, it should be checked that glycerol does not perturb K_D values (e.g., by carrying out filter binding ± glycerol. An alternative involves addition of half a volume (i.e., in this case, 5 µL to the 10 µL binding reaction) of 25% (w/v) sucrose solution following incubation and immediately prior to loading on the gel. This dilutes the sample, and in cases of weak binding, the associated fast k_{off} rates may lead to some dissociation. However, if the binding is strong, the small dilution factor is unlikely to be a problem. It is best not to add the bromophenol blue marker to the solution of enzyme and DNA, as the dye can diminish ***(18)*** binding.

11. Phosphorimaging is the best way to quantitate the amount of radioactivity in bands on gels. Phosphorimaging systems are now fairly standard in most laboratories. As an alternative autoradiography followed by scanning of the bands may be used, but it is more cumbersome and less accurate.
12. Fitting to the full kinetic model is actually the least robust for the unmodified enzyme-oligonucleotide pair because the rate constants are so similar numerically that the program has difficulty partitioning among them. Thus, many closely spaced, and accurate, time-points should be taken and this will usually require rapid quench.
13. The total concentration of endonuclease and radiolabeled reference DNA (D_1) are chosen to be at least fourfold over the predicted equilibrium dissociation constant (K_1) for their interaction. $[E]_t$ and $[D_1]_t$ in the range K_1 to 30 K_1 have been tested; the K_2 (equilibrium dissociation constant for competitor DNA) is not affected significantly.
14. The change in anisotropy is only strictly related to the molecular mass of the fluorophore and hence the amount of oligonucleotide complexed with endonuclease, if the following two criteria are met. First, the detector response to the parallel and perpendicular polarized light must be identical. The G-factor measures this response, which is automatically corrected for, at each reading, with the fluorimeter used here. Second, the quantum yield of the free oligonucleotide and the enzyme-bound oligonucleotide must be the same (i.e., under identical conditions, the free and bound oligonucleotide should emit the same amount of light). If the quantum yield of the free and bound oligonucleotide varies the equation used to fit the data will not be valid *(7,22,23)*. In the example shown in **Fig. 7**, the emission intensity of the free and bound oligonucleotide varies less than 10%. However, the use of other hexachlorofluorescein oligonucleotides and different buffers with the *Eco*RV endonuclease often results in large intensity changes (>10%). In these cases, fluorescence anisotropy cannot be used unless corrections are made *(7,22,23)*.
15. Good results have been obtained using a DNA concentration about equal to the K_D and varying protein concentrations from 0 to (20–30) K_D. However, lower concentrations of DNA can also be used, although 1 n*M* represents the detection limit. DNA levels much higher than the K_D should be avoided, as this promotes stoichiometric titration (where each protein molecule added binds to the nucleic acid), conditions under which accurate K_D evaluation is not possible.
16. Purified endonucleases are often stored in buffers containing glycerol. Fluorescence anisotropy is very sensitive to viscosity and the addition of large amounts of glycerol during the titration adversely effects the measurement. Therefore, intermediate dilutions should be made with buffers lacking glycerol. This usually ensures that the amount of glycerol added to the fluorescence cuvet is negligible (≤2%). In the case of weak binding, when an enzyme may have to be added directly from the storage buffer, dialysis is recommended to remove the glycerol.

Acknowledgments

The Newcastle group would like to thank G. Baldwin and S. Halford (Bristol) for the use of their quench-flow apparatus and help with the data shown in **Fig. 4**. B.A.C. is supported by the UK BBSRC and MRC. L.J.J. is supported by a grant (GM-29207) from the National Institutes of Health (USA). B.A.C. and L.J.J. have a Wellcome Trust biomedical research collaboration grant.

References

1. Roberts, R. J. and Macelis, D. Rebase; the restriction enzyme database. *http://www. neb. com/rebase/rebase. html*
2. Jen-Jacobson, L. (1995) Structural-perturbation approaches to thermodynamics of site-specific protein-DNA interactions. *Methods Enzymol.* **259,** 305–344.
3. Jen-Jacobson, L. (1997) Protein-DNA recognition complexes: conservation of structure and binding energy in the transition state. *Biopolymers* **44,** 153–180.
4. Lesser, D. R., Kurpiewski, M. R., Waters, T., Connolly, B. A., and Jen-Jacobson, L. (1993) Facilitated distortion of the DNA site enhances *Eco*RI endonuclease DNA interactions. *Proc. Natl. Acad. Sci. USA* **90,** 7546–7552.
5. Kurpiewski, M. R., Koziolkiewicz, M., Wilk, A., Stec, W. J., and Jen-Jacobson, L. (1996) Chiral phosphorothioates as probes of protein interactions with individual DNA phosphoryl oxygens: essential interactions of the *Eco*RI endonuclease with the phosphate at pGAATTC. *Biochemistry* **35,** 8846–8854.
6. Lin, S. Y. and Riggs, A. D. (1972) lac Repressor binding to nonoperator DNA: detailed studies and a comparison of equilibrium and rate competition methods. *J. Mol. Biol.* **72,** 671–690.
7. James, D. M and Sawyer, W. H. (1995) Fluorescence anisotropy applied to biomolecular interactions. *Methods Enzymol.* **246,** 283–300.
8. Hill, J. J. and Roger, C. A. (1997) Fluorescence approaches to the study of protein-nucleic acid complexes. *Methods Enzymol.* **278,** 390–416.
9. Powell, L. M, Connolly, B. A., and Drained, D. T. F. (1998) The DNA binding characteristics of trimeric *EcoKI* methyltransferase and its partially assembled dimeric form detected by fluorescence polarization and DNA footprinting. *J. Mol. Biol.* **283,** 947–961.
10. Sambrook, J., Fritsch, E. F., and Maniatis, T. (1989) *Molecular Cloning. A Laboratory Manual,* 2nd ed., Cold Spring Harbor Laboratory, Cold Spring Harbor, NY, pp. 5.56–5.87 and 11.31–11.39.
11. GraFit (1998) Version 4.05, Erithacus Software, Staines, UK.
12. Scientist (1995) Version 2.0, MicroMath Software Inc., Salt Lake City, UT.
13. Sigma Plot (1996) Version 4.0 SPSS Inc, Chicago, IL.
14. Lesser, D. R., Kurpiewski, M. R., and Jen-Jacobson, L. (1990) The energetic basis of specificity in the *Eco*RI endonuclease–DNA interaction. *Science* **250,** 776–786.
15. Jen-Jacobson, L., Lesser, D. R., and Kurpiewski, M. R. (1991) DNA sequence discrimination by the *Eco*RI endonuclease, in *Nucleic Acids and Molecular Biology, Vol. 5* (Eckstein, F. and Lilley, D. M. J., eds.), Springer-Verlag, NY, pp. 141–170.
16. Carey, J. (1991) Gel retardation. *Methods Enzymol.* **208,** 103–118.
17. Record, M. T., Jr., Anderson, C. F., and Lohman, T. M. (1978) Thermodynamic analysis of ion effects on the binding and conformational equilibria of proteins

and nucleic acids: the roles of ion association or release, screening, and ion effects on water activity. *Quart. Rev. Biophys.* **11,** 103–178.
18. Engler, L. E., Welch, K. K., and Jen-Jacobson, L. (1997) Specific binding by the *Eco*RV restriction endonuclease to its recognition site GATATC. *J. Mol. Biol.* **269,** 82–101.
19. Revzin, A. (1987) Gel electrophoresis assays for DNA–protein interactions *BioTechniques* **7,** 346–355.
20. Taylor, J. D. and Halford, S. E. (1989) Discrimination between DNA sequences by the *Eco*RV restriction endonuclease. *Biochemistry* **28,** 6198–6207.
21. Vipond, I. B. and Halford, S. E. (1995) Specific DNA recognition by *Eco*RV restriction endonuclease induced by Ca^{2+} ions. *Biochemistry* **34,** 1113–1119.
22. Eftink, M. R. (1997) Fluorescence methods for studying equilibrium macromolecule-ligand interactions. *Methods Enzymol.* **278,** 221–257.
23. Gutfreund, H. (1995) The role of light in kinetic investigations, in *Kinetics for the Life Sciences Receptors, Transmitters and Catalysts.* Cambridge University Press, Cambridge, pp. 291–295.

33

Analysis of DNA–Protein Interactions by Intrinsic Fluorescence

Mark L. Carpenter, Anthony W. Oliver, and G. Geoff Kneale

1. Introduction

Changes in the fluorescence emission spectrum of a protein upon binding to DNA can often be used to determine the stoichiometry of binding and equilibrium binding constants; in some cases the data can also give an indication of the location of particular residues within the protein. The experiments are generally quick and easy to perform, requiring only small quantities of material *(1)*. Spectroscopic techniques allow one to measure binding at equilibrium (unlike, for example, gel retardation assays and other separation techniques that are strictly nonequilibrium methods). Fluorescence is one of the most sensitive of spectroscopic techniques, allowing the low concentrations (typically in the nanomolar to micromolar range) required for estimation of binding constants for many protein–DNA interactions. Considerable care, however, needs to be exercised in the experiment itself and in the interpretation of results. The fundamental principles of fluorescence are discussed briefly in the remainder of the Introduction.

A molecule that has been electronically excited with ultraviolet (UV)/visible light can lose some of the excess energy gained and return to its ground state by a number of processes. In two of these, fluorescence and phosphorescence, this is achieved by emission of light. Phosphorescence is rarely observed from molecules at room temperature and will not be considered further. Although electrons can be excited to a number of higher-energy states, fluorescence emission in most cases only occurs from the first vibrational level of the first excited state. This has two implications for the measurement of fluorescence emission spectra. First, some of the energy initially absorbed is lost prior to emission, which means that the light emitted will be of longer wavelength

From: *Methods in Molecular Biology, vol. 148: DNA–Protein Interactions: Principles and Protocols, 2nd ed.*
Edited by: T. Moss © Humana Press Inc., Totowa, NJ

(i.e., lower energy) than that absorbed. This is known as the Stoke's shift. Second, the emission spectrum and therefore the wavelength of maximum fluorescence will be independent of the precise wavelength used to excite the molecule. Thus, for tyrosine, the wavelength of the fluorescence maximum is observed around 305 nm, regardless of whether excitation is at the absorption maximum (approx 278 nm) or elsewhere in the absorption band. Of course, the fluorescence intensity will change as a consequence of the difference in the amount of light absorbed at these two wavelengths.

The fraction of light emitted as fluorescence compared to that initially absorbed is termed the quantum yield. The value of the quantum yield for a particular fluorophore will depend on a number of environmental factors such as temperature, solvent, and the presence of other molecules that may enhance or diminish the probability of other processes deactivating the excited state. The deactivation or quenching of fluorescence by another molecule, either through collisional encounters or the formation of excited-state complexes, forms the basis of many of the fluorescence studies on protein–DNA interactions.

The study of protein–nucleic acid interactions is greatly simplified by the fact that all detectable fluorescence arises from the protein, all four of the naturally occurring DNA bases being nonfluorescent by comparison. Tyrosine and tryptophan residues account for almost all the fluorescence found in proteins. As a general rule, when both residues are present, the emission spectrum will be dominated by tryptophan, unless the ratio of tyrosines to tryptophans is very high. The quantum yield of a tyrosine residue in a protein compared to that observed in free solution is generally very low, illustrating the susceptibility of tyrosine to quenching. Tryptophan residues are highly sensitive to the polarity of the surrounding solvent, which affects the energy levels of the first excited state with the result that the emission maximum for tryptophan can range from 330 nm in a hydrophobic environment to 355 nm in water. Thus, in proteins containing only one tryptophan, the general environment surrounding the residue can be ascertained. Tryptophan fluorescence, like that of tyrosine, can be quenched by a number of molecules, including DNA. Unlike tyrosine, the emission maximum can also change if tryptophan is involved in the interaction and this can also be used to monitor DNA binding *(2)*.

The extent to which the fluorescence of a protein is quenched by DNA is proportional to the concentration of quencher. As quenching is the result of the formation of a complex between the protein and the DNA, the extent of quenching is proportional to the amount of bound protein. Thus by determining the extent to which the protein fluorescence is quenched when fully bound to DNA (i.e., at saturation), the fraction of bound and free protein at any point in a titration can be determined. From these data, the stoichiometry and binding constant of the interaction can often be obtained. Note that to establish an accurate stoichiometry, a high concentration of protein is preferred when titrating with DNA (i.e., well above the K_d of the complex) to ensure stoichiometric binding. To establish the binding constant itself, one should be working at much lower concentrations of protein so that at the stoichiometric point, there is a measurable concentration of unbound protein. In the case of protein–DNA interactions having a low K_d, this may not be possible.

If fluorescence quenching is being used to follow DNA binding, it is vital to take account of sample dilution, as well as the increased absorption of the sample as DNA is titrated in. The latter effect is known as the inner filter effect and arises from the absorption of the excitation beam (and generally to a lesser extent, the emission beam) on passing through the sample (*see* **Fig. 1**). One should aim to keep the absorption of the sample (at the excitation wavelength) as low as possible, although absorbances up to 0.2 can normally be corrected without too much difficulty. A small-pathlength cell will also help (if rectangular, the excitation beam should pass through the smallest path). Ideally, the absorption of the sample at the excitation and emission wavelengths (A_{ex} and A_{em}) should be measured for each point in the titration (if not, one can calculate these values from the known concentrations of protein and nucleic acid at each point). For normal right-angled geometry of observation, the corrected fluorescence F_{corr} can be obtained from the observed fluorescence F_{obs} by the formula

$$F_{corr} = F_{obs} \times 10^{(Aex/2 + Aem/2)} \quad (1)$$

Often, the value of A_{em} is small enough to ignore (for a detailed treatment of the inner filter correction, *see* **ref. 3**). Note that it is equally important to correct for the inner filter effect whether titrating protein into DNA or vice versa.

The following method deals only with the determination of DNA binding curves by intrinsic fluorescence quenching. However, fluorescence anisotropy can also be used if the molecular size of the complex is sufficiently different from the free protein; for example, to investigate proteins that bind cooperatively to DNA *(4)*. Time-resolved fluorescence techniques are also advantageous (for measurements of fluorescent lifetimes or rotational correlation times) but require sophisticated instrumentation *(5)*.

The use of intrinsic fluorescence, as a method for investigating protein–DNA interactions, is widespread. For example, both binding parameters (K_{obs}) and stoichiometric ratios have been derived for the interaction of the HIV-1 nucleocapsid protein NCp7 with the natural primer $tRNA_3^{Lys}$ and other related RNA molecules *(6)*. Similarly, estimates of binding constants have been determined for the interaction of human replication protein A (hRPA) with single-stranded homopolynucleotides (e.g., poly[dT] and poly[dA]) *(7)*. Furthermore, both steady-state and time-resolved fluorescence have been used in a binding study of the single-stranded DNA-binding protein of phage ϕ29 *(8)*.

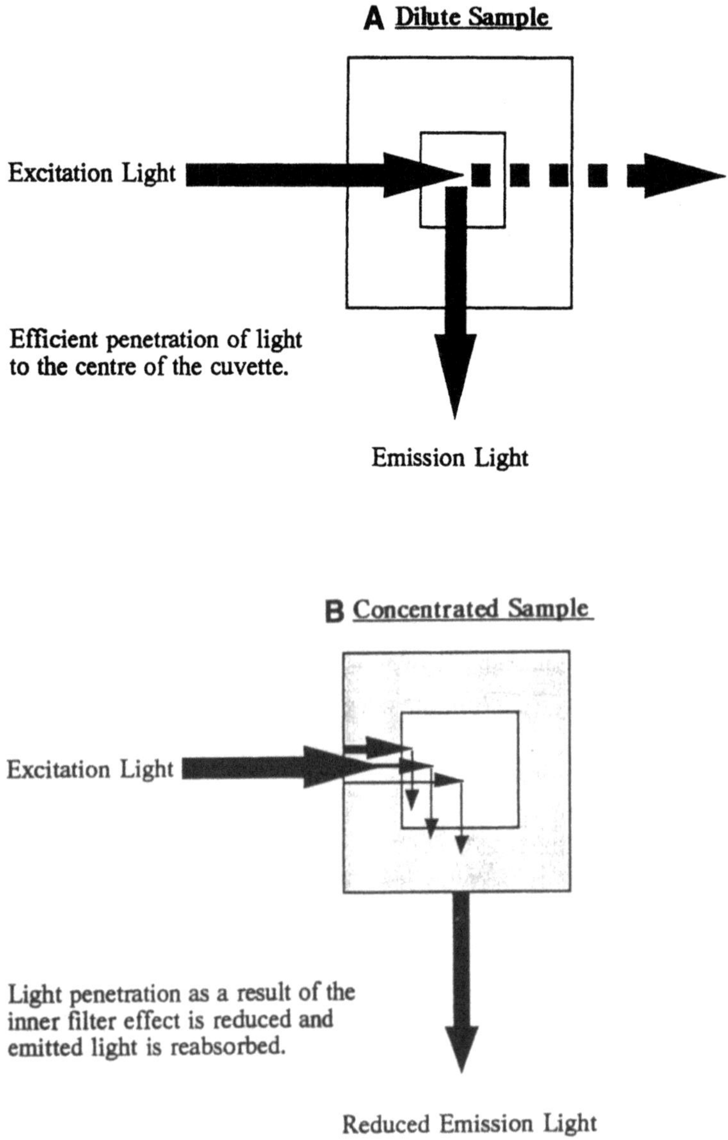

Fig.1. Schematic representation of the inner filter effect in fluorescence, showing the effect of high concentration on the absorbance of the excitation beam.

2. Materials

1. Reagents used in buffer solutions should be of the highest purity available and the solutions prepared in doubly-distilled water. The buffer should have negligible absorbance in the 260- to 300-nm wavelength range and should not be used if it shows any fluorescence in the region 290–400 nm. High-molecular-weight

ions should be avoided. Phosphate buffer should not be used if tyrosine fluorescence is being monitored.

2. Stock solutions of protein and DNA should be divided into small aliquots and stored at –20°C, assuming this is not detrimental to the protein.
3. High-quality quartz cuvets with all four faces polished (*see* **Notes 1** and **2**).
4. Most commercially available fluorimeters allow scanning by both the excitation and emission monochromators and are suitable for use in these studies. Cell compartments that can be thermostatically controlled are preferable (*see* **Notes 3** and **4**). We routinely use a Perkin-Elmer LS 50B fluorimeter.
5. An on-line computer is normally linked to the fluorimeter, which allows fluorescence spectra to be recorded and analyzed. We routinely use the software provided with the Perkin-Elmer LS 50B fluorimeter "FL WinLab."

3. Method

The method described in this section assumes that nothing is known concerning the fluorescence properties of the protein or its complex with DNA. Consequently, the initial steps described in **Subheading 3.1.** are concerned with characterizing some of the fluorescence properties of the two species such that the optimal conditions for obtaining accurate and reliable data can be obtained. **Subheading 3.2.** describes the procedure for obtaining data for a protein that is quenched by DNA, which results only in a decrease in fluorescence intensity, and how these data can be used to obtain information on binding. Several variations of the method are mentioned in the **Subheading 4.**

3.1. Preliminary Experiments

1. Switch on the fluorimeter and allow 10 min for the components to stabilize. Set the excitation and emission slits to intermediate values (e.g., 10 nm bandpass).
2. Fill the cuvet with protein solution (*see* **Note 5**). Allow time for the solution to equilibrate to the temperature of the compartment. To prevent local heating of the solution or possible photodecomposition, the excitation shutter should be kept closed, except when taking measurements.
3. If the absorption spectrum of the protein is known, set the excitation wavelength to that corresponding to the absorption maximum between 265 and 285 nm; if no peak exists, the protein does not contain tyrosine or tryptophan residues and will not fluoresce. If the absorption spectrum is unknown, set the excitation wavelength to 280 nm.
4. Open the excitation shutter and quickly scan the emission between 285 and 400 nm, looking for the wavelength at which a maximum value for the intensity is given on the readout. Return the emission monochromator to this wavelength.
5. Find the excitation wavelength maximum between 265 and 285 nm in the same manner, with the emission monochromator set at the wavelength of maximum fluorescence. **Note:** The aforementioned "FL WinLab" software allows simultaneous scanning of both the excitation and emission wavelengths, in what is

termed a "3D" scan. This allows both the excitation and emission maxima to be determined in one experiment. However, for proteins, this is not normally necessary.

6. With both the excitation and emission wavelengths set at their peak values, adjust the instrument to give a reading corresponding to about 90% of the full scale. Narrow slit widths and a lower amplification (expansion factor, gain) are preferred and a compromise between the two may have to be found (*see* **Note 6**).
7. Determine the emission spectrum by scanning the emission monochromator over the entire wavelength range over which fluorescence occurs. A scan speed of 60 nm/min is generally suitable.
8. Add a small aliquot of a concentrated DNA solution to the cuvet such that the concentration of DNA is in excess. Mix and immediately check the fluorescence emission at the emission maximum of the protein. Check several times over the next few min until a consistent reading is obtained. Allow this time for equilibration in subsequent experiments. Do not adjust slit widths or the amplication.
9. Obtain an emission spectrum and compare with that obtained for the protein only. If fluorescence quenching is suspected, make sure that allowance for sample dilution has been made. If an inner filter correction is required, measure the absorbance of the sample in a spectrophotometer (in the same cuvet) and correct the observed fluorescence as discussed in **Subheading 1.**
10. Add aliquots of DNA until there is no further change in fluorescence intensity in the emission spectrum (*see* **Notes 7** and **8**).

3.2. Protein-DNA Titrations

1. Examine the emission spectrum of the free protein. If it is characteristic of tyrosine fluorescence, check for interference from the Raman band (*see* **Note 9**). If it is characteristic of tryptophanlike, check for tyrosine contributions that may be masked (*see* **Note 10**).
2. Examine the emission spectrum of the protein bound to DNA. The titration method described in the following passage is particularly applicable when the only change in the spectrum is a change in fluorescence intensity. Several variations of this method are described briefly in **Notes 11** and **12** including an example where the emission spectrum of the protein shifts on binding DNA.
3. Accurately determine the concentration of protein and DNA solutions by UV spectroscopy. Because we are titrating DNA into protein, try to use a stock concentration of DNA, which is at least 20 times the concentration of protein used in the experiment multiplied by the estimated stoichiometric ratio; for example, if the protein concentration used is 10 μM and the estimated stoichiometry is 5 bases per protein, then the DNA concentration should be at least $20 \times 10\ \mu M \times 5 = 1000\ \mu M$ (1 μM). This would mean that the dilution of the original protein solution will be only 5% at the stoichiometric point.
4. Using the protein solution set up the instrument as described in **steps 1–6** of **Subheading 3.1.** If measuring tyrosine fluorescence, use an excitation wavelength near the maximum. This wavelength can also be used for tryptophan excitation if tyrosine fluorescence is insignificant, otherwise use an excitation wavelength of 295 nm.

5. Run a buffer blank and check that the profile of the emission spectrum is consistent with that previously obtained. Subtract this spectrum from subsequent spectra if this can be done automatically.
6. Set the emission monochromator to the emission wavelength maximum and ensure that the readout is about 90% of its maximum value. Note the value.
7. To begin the titration, add a small aliquot from the stock DNA solution to the protein in the cuvet. Mix and allow the sample to equilibrate (use the time period determined earlier) before taking a reading. The aliquots should be sufficiently small such that the protein is still greatly in excess and a linear change is observed as more DNA is added.
8. Continue to add the same quantity of DNA for 8–10 points. If changes are still approximately linear at this stage, gradually increase the volume of the DNA added, noting the total amount added at each point.
9. When the change in intensity begins to deviate significantly from linearity, decrease the size of the aliquot so that more data points are obtained in this region.
10. As quenching approaches the maximum, larger aliquots of DNA can be added. Continue until no change in quenching is observed for several points.
11. After the last point, check that the emission spectrum of the complex is consistent with that previously obtained for the bound protein.
12. Remove the sample, wash the cuvet thoroughly, and run a blank spectrum consisting of cell plus buffer. This should have negligible or no fluorescence. Subtract any value at the emission wavelength maximum from the data points, if not already done automatically (*see* **Note 9**).
13. For each data point, calculate the fluorescence quenching ($Q = 100\,[F_0 - F]/F_0$, where F is the measured fluorescence and F_0 is the fluorescence in the absence of DNA), having made any corrections for inner filter effects and dilution of the sample. Also calculate the nucleotide concentration at each point ($N = ny/[n + x]$, where n is the total volume of DNA added up until that point, x is the initial volume of sample in the cuvet and y is the molar concentration of DNA stock solution). From the DNA concentration, calculate R, which is the ratio of the concentration of DNA to that of protein. (For polynucleotides, it is usual to express N as the concentration of nucleotides; for short synthetic duplexes, it is more usual to use the molar concentration of the duplex).
14. Plot a graph of Q against R (or N). If only one mode of binding is occurring, the graph should look like one of the curves shown in **Fig. 2**. If the "break point" in the titration is sharp, it indicates a high value for the binding constant (i.e., a small dissociation constant compared with the protein concentration used). Conversely, too weak a binding constant (or too dilute a protein solution) will give rise to a smoothly rising curve with no apparent break point.
15. The stoichiometry of binding is the value of R at which the slope obtained from the initial linear range of the titration crosses the horizontal line defined by Q_{max}, at which no further change in intensity occurs.
16. Further information can be extracted from the binding curve by fitting it to an appropriate model. In the simplest case of a bimolecular interaction ($P + N =$

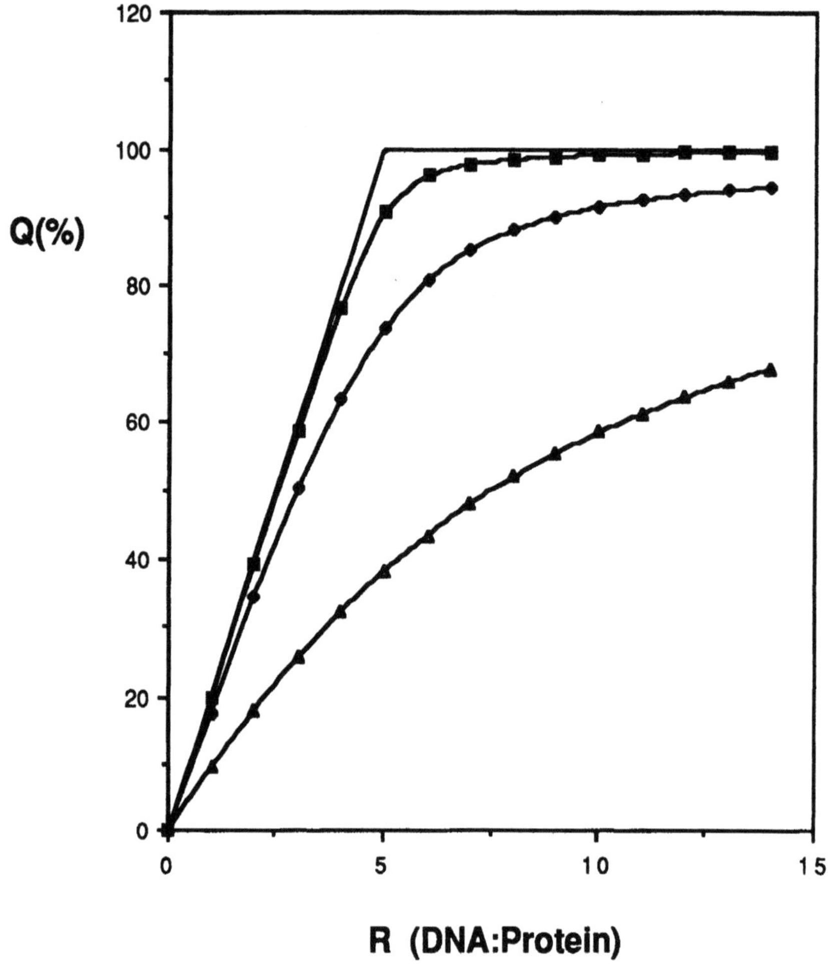

Fig. 2. A graph of fluorescence quenching against DNA:protein ratio (R). The curves illustrate the addition of DNA (in this case, a polynucleotide) to protein (10 μ*M*). The theoretical binding curve for infinitely strong binding (upper curve) shows a stoichiometry of five nucleotides bound per protein. Typical curves are shown for binding constants of 10^7 (■), 10^6 (◆), and 10^5 (▲) per molar.

PN), then a useful expression to estimate the binding constant is $K = [P_0]\theta/(1-\theta)^2$ where θ is the fraction of bound protein at the stoichiometric point and $[P_0]$ is the total protein concentration in the cuvet. This expression also applies to more complex cooperative binding along a linear DNA lattice ***(9)***, assuming the cooperativity is sufficiently high, when K becomes equal to the apparent binding constant (and approx the product of the cooperativity factor and the intrinsic binding constant for one site). For a more extensive discussion of complex DNA binding equilibria, *see* **ref. *10***.

4. Notes

1. Most fluorimeter cell compartments are designed to take cuvets with a 1-cm pathlength (distance between opposite faces) and usually require 2.5–3.0 cm^3 of sample for measurements. If smaller sample volumes are required, then reduced volume cuvets similar to those used in absorption studies but suitable for fluorescence work can usually be obtained from most cuvet suppliers. Also, most suppliers are prepared to construct cuvets to your own specifications (at a cost). The major requirement is that the sample be located in the center of the cell (assuming standard right-angle observation). Cuvets holding as little as 300 µL have been successfully used by the authors.
2. Care should be taken when handling fluorescence cuvets, as both fingerprints and scratches can introduce significant artifacts into the experiment. After use, cuvets should be thoroughly washed with distilled water and a mild detergent, if necessary. If greatly contaminated, then immerse the cells in a 50:50 mix of ethanol with sulfuric acid (4 *M*) for several hours and then rinse thoroughly with water.
3. If the fluorimeter is not equipped with a magnetic stirrer unit, adequate mixing can usually be achieved by gently drawing the solution in and out through a plastic pipet tip. Avoid introducing bubbles into the sample, as this can both denature the protein and cause light scattering. For accurate measurements, temperature control is essential, as fluorescence is highly sensitive to temperature.
4. If the use of a reduced-volume cell prevents the use of commercially available magnetic fleas, then substitutes can be made as follows: (1) Seal the narrow end of a Pasteur pipet or micropipet by heating it in the flame of a Bunsen burner. (2) Insert a small length of iron wire (cut up a paperclip) and shake it down to the sealed end. (3) Cut the pipet just above the wire using a glass cutter and seal the open end in the flame. These fleas should only be used once, as some rusting occurs with time.
5. Fluorescence intensity is only proportional to concentration when the absorbance is no greater than 0.1 absorbance units at the excitation wavelength selected. If a molar extinction coefficient for the protein is known, use this to calculate a suitable protein concentration. Remember that the absorption bands for proteins and nucleic acids overlap, and in titrations, the contribution of the nucleic acid to the overall absorption must be considered.
6. Although it is preferable to have narrow excitation and emission slit widths and a low amplification factor, there may be a need to compromise in order to obtain a stable reading. For proteins displaying tyrosine fluorescence, the small wavelength difference between the excitation and emission maxima suggests that it would be better to maintain narrow slits and increase the signal amplification. For proteins dominated by tryptophan fluorescence, the greater the difference between the excitation and emission wavelengths, the greater the feasibility of increasing the slit widths and maintaining a lower amplification. In general, when measuring emission spectra, it is better to use a narrow emission slit width and widen the excitation slit width. For broad-banded spectra such as that seen with tryptophan, both slits can be widened.

7. If no changes in the emission spectrum of the protein are observed when DNA is added (after inner filter and dilution corrections if necessary), then either the protein is not binding to DNA or binding cannot be detected by this procedure and will need to be assessed by another method, such as fluorescence anisotropy or the use of an extrinsic probe (*see* Chapter 18).
8. Note the molar ratio of DNA: protein at which no further changes occur. This will provide a rough guide for future experiments.
9. Tyrosine emission can often be confused with Raman scattering of light, which occurs around 305 nm when an excitation wavelength of 280 nm is used. The presence of the Raman band can be assessed by measuring the emission spectrum using a different excitation wavelength. The fluorescence emission spectrum is independent of excitation wavelength, whereas Raman scattering occurs at a constant wave number (= $1/\lambda$, in cm^{-1}) from that used for excitation and will shift in the same direction as the change in excitation wavelength. The contribution of the Raman band to the overall intensity of the signal can be assessed by running an emission spectrum of a buffer blank. Automatically subtract out this spectrum from subsequent spectra where possible. **Note:** The "FL WinLab" software offers a prescan mode, in which the Raman peak can be identified automatically. Other fluorescence software packages may offer a similar facility.
10. To check the contribution tyrosine may make to a fluorescence emission spectrum dominated by tryptophan, run an emission spectrum using an excitation wavelength of 295 nm. At this wavelength, only tryptophan emission will be observed. If the emission spectrum is unchanged, then it can be concluded that the contribution from tyrosine residues is negligible. (Of course, the intensity will be lower, as tryptophan absorption is greater at 280 nm than it is at 295 nm.)
11. In cases where both the emission maximum shifts and the fluorescence intensity is quenched, the method described can be used provided that an emission wavelength is chosen outside the wavelength region overlapped by the emission spectra of the free and bound protein. Alternatively, the ratio of the intensity of the emission maxima of the free and bound proteins can be followed (for an example, *see* **ref. *2***). The use of a ratio method means that the dilution factor and inner filter correction can usually be ignored, although, strictly speaking, the ratio is not a linear function of degree of binding.
12. In some cases it may be preferable to titrate DNA with protein (for an example method, *see* **ref. *11***). The procedure is similar to that given here, but in this case the experiment should be repeated by adding protein to the buffer in the absence of DNA as a reference. Subtraction of the two curves should yield a clear binding curve (*see* **Fig. 3**). (For a discussion of the merits of whether to titrate protein with DNA or vice versa, *see* **ref. *4***). We have found that in some cases, different results can be found dependent on the direction of the titration; this can occur when the fluorescence changes observed include contributions from protein–protein interactions accompanying DNA binding in addition to (or instead of) contributions from the interaction with the DNA itself.

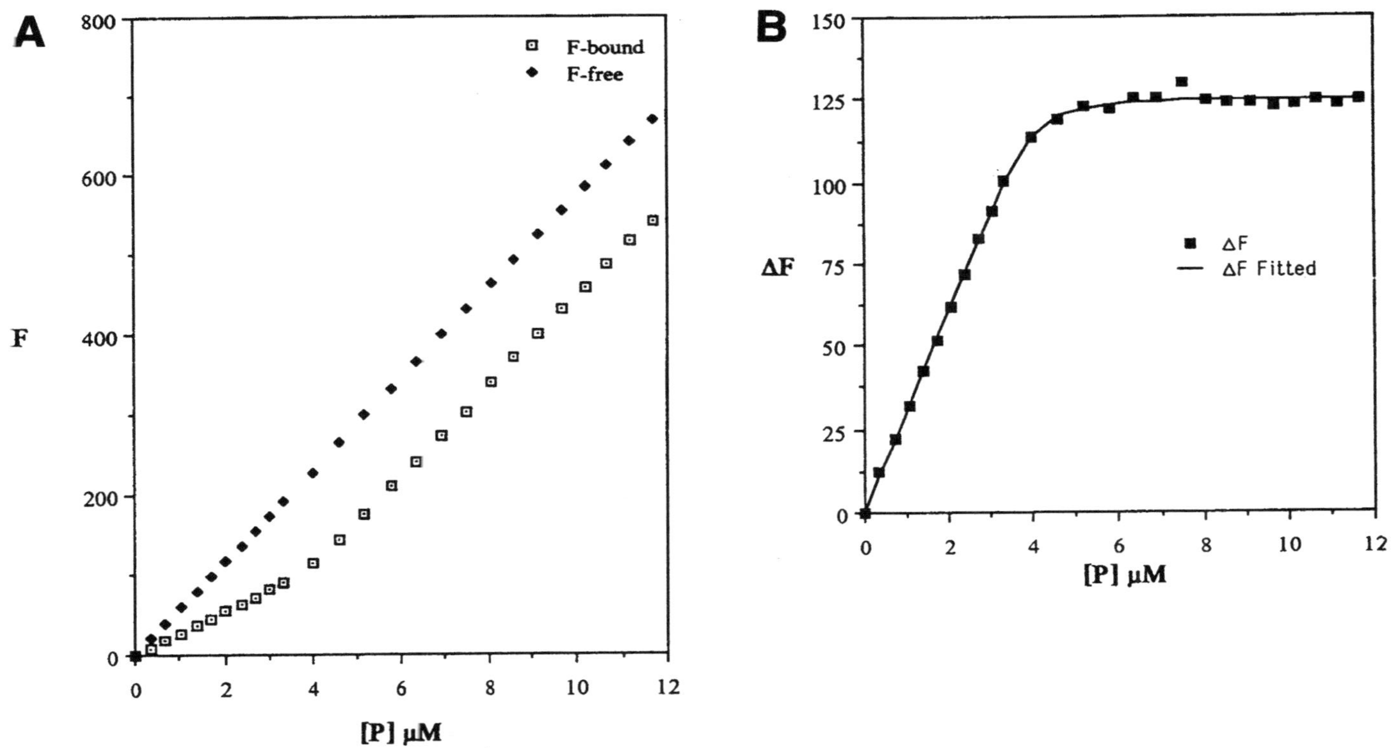

Fig. 3. **(A)** Titration of an oligonucleotide with fd gene 5 protein. Fluorescence (305 nm) of increasing concentrations of protein (P) is measured in the presence (lower curve) and absence (upper curve) of DNA. The difference between these two curves is plotted in **(B)** along with the theoretical binding curve. In this experiment, the starting concentration of DNA was 16.7 µ*M*.

References

1. Harris, D. A. and Bashford, C. L. (eds.) (1987) *Spectrophotometry and Spectrofluorimetry: A Practical Approach.* IRL Press, Oxford, UK. (Chapters 1 and 4 are particularly relevant.)
2. Kneale G. G. and Wijnaendts van Resandt, R. W. (1985) Time resolved fluorescence of the Pf1 bacteriophage DNA-binding protein: determination of oligo- and polynucleotide binding parameters. *Eur. J. Biochem.* **149,** 85–93.
3. Birdsall, B., King, R. W., Wheeler, M. R., Lewis, C. A., Goode, S. R., Dunlap, R. B. et al. (1983) *Anal. Biochem.* **132,** 353–361.
4. Carpenter, M. L. and Kneale, G. G. (1991) Circular dichroism and fluorescence analysis of the interaction of Pf1 gene 5 protein with poly(dT). *J. Mol. Biol.* **217,** 681–689.
5. Greulich, K. O., Wijnaendts van Resandt, R. W., and Kneale G. G (1985) Time resolved fluorescence of bacteriophage Pf1 DNA-binding protein and its complex with DNA. *Eur. Biophys. J.* **11,** 195–201.
6. Mély, Y., de Rocquigny, H., Sorinas-Jimeno, M., Keith, G., Roques, B. P., Marquet, R., et al. (1995) Binding of the HIV-1 nucleocapsid protein to the primer $tRNA_3^{Lys}$ in vitro, is essentially not specific. *J. Biol. Chem.* **270,** 1650–1656.
7. Kim, C. and Wold, M. S. (1995) Recombinant human replication protein A binds to polynucleotides with low cooperativity. *Biochemistry* **34,** 2058–2064.
8. Soengas, M. S., Mateo, C. R., Salas, M., Acuña, A. U., and Gutiérrez, C. (1997) Structural features of ϕ29 single-stranded DNA-binding protein. *J. Biol. Chem.* **272,** 295–302.
9. Kelly, R. C., Jensen, D. E., and von Hippel, P. H. (1976) Fluorescence measurements of binding parameters for bacteriophage T4 gene 32 protein to mono-, oligo-, and polynucleotides. *J. Biol. Chem.* **251,** 7240–7250.
10. McGhee, J. D. and von Hippel, P. H. (1974) Theoretical aspects of DNA–protein interactions: cooperative and non-cooperative binding of large ligands to a one-dimensional homogeneous lattice. *J. Mol. Biol.* **86,** 469–489.
11. Alma, N. C. M., Harmsen, B. J. M., de Jong, E. A. M., Ven, J. V. D., and Hilbers, C. W. (1983) Fluorescence studies of the complex formation between the gene 5 protein of bacteriophage M13 and polynucleotides. *J. Mol. Biol.* **163,** 47–62.

34

Circular Dichroism for the Analysis of Protein–DNA Interactions

Mark L. Carpenter, Anthony W. Oliver, and G. Geoff Kneale

1. Introduction

The asymmetric carbon atoms present in the sugars of nucleotides and in all the amino acids (with the exception of glycine) results in nucleic acids and proteins displaying optical activity. Further contributions to the optical activity of the polymers result from their ability to form well-defined secondary structures, in particular helices, which themselves possess asymmetry. As a consequence, circular dichroism (CD) has found widespread use in secondary structure prediction of proteins *(1)*. Similar studies, though less widespread, have sought to correlate structural parameters of DNA with their CD spectrum *(2)*, with some success particularly in assigning quaternary structures to nucleic acids (e.g., in the case of DNA triplexes and G-quartet mediated structures) *(3,4)*. It follows that the disruption of secondary structure by, for example, denaturation or ligand binding can be usefully followed by circular dichroism.

Plane polarized light can be resolved into left- and right-handed circularly polarized components. Circular dichroism measures the difference in the absorption of these two components,

$$\Delta\varepsilon = \varepsilon_L - \varepsilon_R \qquad (1)$$

where ε is the molar extinction coefficient (M^{-1}/cm^{-1}) for the left (*L*) and right (*R*) components (*see* **Note 1**). When passing through an optically active sample, the plane of polarized light is also rotated, which means that the emerging beam is elliptically polarized. Thus, CD is often expressed in terms of ellipticity (θ_λ, in degrees) or molar ellipticity ($[\theta]_\lambda$, in degrees cm^2/dmol).

$$[\theta]_\lambda = 100\ \theta_\lambda/cl \qquad (2)$$

From: *Methods in Molecular Biology, vol. 148: DNA–Protein Interactions: Principles and Protocols, 2nd ed.*
Edited by: T. Moss © Humana Press Inc., Totowa, NJ

where c is the molar concentration and l is the pathlength (in cm). The two expressions are interconvertible via the expression $[\theta]_\lambda = 3300\ \Delta\varepsilon$.

The overlap of the ultraviolet (UV) absorption bands of nucleic acids and proteins means that CD studies of protein–DNA interactions can be complicated by the contributions observed from both components. This is particularly true for wavelengths less than 250 nm. In practice, CD spectra between 250 and 300 nm are dominated by that of the nucleic acid, the contribution arising from the aromatic chromophores of the protein being weak by comparison. Changes in conformation can usually be attributed to the polynucleotide, as the random distribution of aromatic amino acids means a large conformational change throughout the protein would be required to cause a significant change in the CD spectrum.

The low-molar ellipticity of polynucleotides means that for accurate CD measurements, high concentrations (10^{-4} *M* to 10 *M*) of nucleotide are required. For this reason, circular dichroism is not generally used to determine binding constants of protein–DNA interactions. However, circular dichroism can be used to obtain accurate values for the stoichiometry of protein–nucleic acid interactions *(5)*, and in the case of the bacteriophage, the fd gene 5 protein was used to show the existence of two distinct binding modes *(6)*. Circular dichroism has also been used to show that conformational changes induced by the bound *Lac* repressor are different for operator DNA and for random sequence DNA *(7)*. Similar studies on the *Gal* repressor demonstrated the involvement of the central G-C base pairs of the operator sequence in repressor-induced conformational changes *(8)*. Studies on the interaction of the *cro* protein of bacteriophage λ have also revealed different conformational changes for specific and nonspecific DNA binding *(9)*. Despite the apparent lack of any direct interaction of the central base pair of the operator sequence with the *cro* protein, base substitution at this site was shown to affect the CD spectrum considerably.

Some additional examples in which CD has been used to examine protein–DNA interactions include the SRY-related protein Sox-5 *(10)*, the type IC DNA methyltransferase M. *Eco*R124I *(11)*, and the bacteriophage Pf3 single-stranded DNA-binding protein (ssDBP) *(12)*.

It should be emphasized that circular dichroism provides complementary data to other spectroscopic techniques such as fluorescence, because each technique can monitor different components of the interaction. For this reason, even the apparent stoichiometry of binding estimated by each technique could be significantly different despite it being measured under the same solution conditions *(5)*.

2. Materials

1. A high-quality quartz cell with low strain is required for accurate measurements. The cell pathlength will depend on the absorption properties and concentration of

the sample. Cells with pathlengths between 0.05 cm and 1 cm are often used, depending on the CD signal to be measured and the absorption of the sample (including the buffer). A pathlength of 1 cm is usually recommended for measurement of the DNA signal in the vicinity of 275 nm, where the signal is weak and buffer absorption is negligible.
2. Buffers should be prepared using high-quality reagents and water. Use buffers that have low absorbance in the wavelength region of interest. Tris-HCl, perchlorate, and phosphate are routinely used.
3. Stock solutions of appropriate protein and nucleic acid solutions in the same buffer. The protein should be as concentrated as possible to minimize dilution during the titration. If a synthetic DNA fragment containing the recognition sequence is to be used, it should be close to the minimum size required for binding to maximize the change in CD signal. To avoid denaturation and degradation, keep concentrated solutions of protein and DNA frozen in small aliquots, assuming it has been established that this procedure does not damage the protein.
4. A supply of dry nitrogen (oxygen free).
5. (+)10-Camphor–sulfonic acid at a concentration of 0.5 mg/mL, used as a calibration standard. Check the accurate concentration by UV spectroscopy using a molar extinction coefficient of 34.5 at 285 nm ***(1)***.
6. A circular dichroism spectrometer (spectropolarimeter). We currently make use of a Jasco J720 spectrometer.

3. Methods

For most proteins, there is only a weak signal from aromatic amino acids in the region of the CD spectrum between 250 and 300 nm compared to that seen for nucleic acids. Experiments involving the addition of protein can thus be conveniently carried out in this wavelength range, as described in the following. Below 250 nm, both proteins and DNA have optical activity and any experiments here may require resolution of the spectrum into protein and DNA components (*see* **ref. *13*** for an example of where a mutant protein has been used to assign and interpret overlapping CD bands).

1. To prevent damage to the optics, flush the instrument with nitrogen for 10–15 min before switching on the lamp. Continue to purge the instrument for the duration of the experiment (*see* **Note 2**).
2. Switch on the lamp and allow the instrument to stabilize for 30 min before making any measurements.
3. While waiting for the instrument to warm up, measure the UV spectrum of both the DNA and protein and calculate the concentration of the stock solutions from their extinction coefficients. The stock solution of protein should be as high a concentration as possible, to minimize corrections for dilution in subsequent titrations.
4. Measure the UV absorbance of the cell to be used in the CD experiment against an air blank (*see* **Note 3**).

5. Using the data from **steps 3** and **4**, determine the concentrations of DNA and protein that can be used in the experiment such that the total absorbance of all components including the cell is <1.0.
6. Once the CD instrument has warmed up, use a 1-cm cell filled with water to determine the baseline spectrum between 250 and 320 nm (*see* **Note 4**). This should be flat.
7. Calibrate the instrument by replacing the water with the previously prepared solution of camphor sulfonic acid and measure the CD between 250 and 320 nm. A 0.5-mg/mL solution in a 1-cm pathlength cell has ellipticity of 168 mdeg at 290.5 nm.
8. Take the cell set aside for the experiment and fill with buffer. Place the cell in the instrument, taking care to note the orientation of its faces in the beam. If using a cylindrical cell, place it so that the neck of the cell rests against the side of the cell holder. Run a baseline spectrum between 250 and 320 nm.
9. Replace the buffer with the DNA solution and run the spectrum under the same conditions. If you remove the cell to do this, remember to place the cell back in the holder with the same face toward the light source.
10. Accurately pipet a small aliquot of the stock protein solution into the DNA in the cell and mix. Allow time for equilibration and measure the CD spectrum. The volume of the protein solution added will be determined by its concentration, the DNA concentration in the cuvet, and a rough idea of the expected stoichiometry. For initial experiments, the molar quantity of protein added at each step should be perhaps 10% of that of the DNA.
11. Repeat the addition of protein to the DNA until no further changes in the CD are observed.
12. Thoroughly wash the cell and refill with buffer. Rerun the baseline spectrum. If the baseline has drifted from its initial value, this will have to be considered when analyzing your results. If drifting is severe, *see* **Note 5**.
13. The collected data will need to be processed (i.e., to remove the CD signal due to buffer components and/or to correct baseline values) (*see* **Note 6**.)
14. Plot the measured CD parameter at a particular wavelength against the concentration of protein added. The stoichiometry can be determined from the point at which a line drawn along the initial slope intersects that of the titration end point, which should be horizontal assuming dilution (if significant) has been corrected for (*see* **Fig. 1**).
15. Once the spectral changes and stoichiometry have been established, it is often useful to repeat the experiment with rather more titration points, using smaller

---

Fig. 1. *(opposite page)* Circular dichroism titrations of Pf1 gene 5 protein to poly dT. In this example, the protein is a single-stranded DNA-binding protein that binds cooperatively to DNA with little sequence preference. **(A)** Successive CD spectra during the titration of poly dT with Pf1 gene 5 protein. The top curve corresponds to free DNA; below are increasing concentrations of protein corresponding to molar ratios of nucleotide to protein of 19.2, 6.7 and 4.0 (which represents saturation).

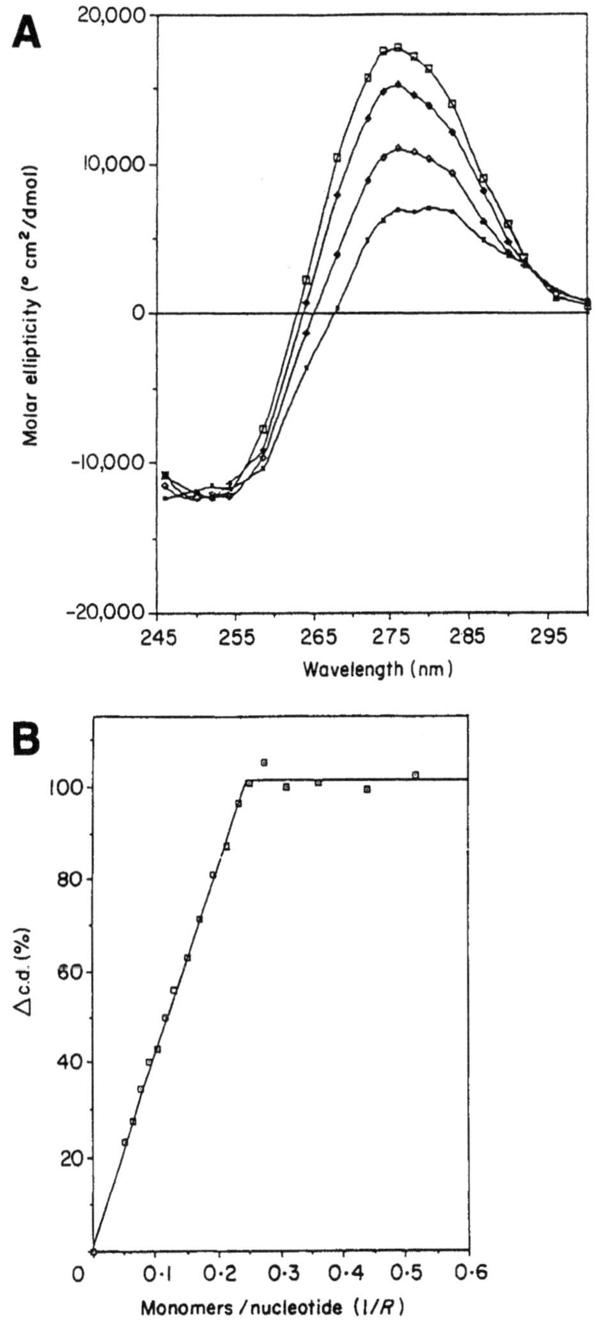

(**B**) The normalized change in CD signal at 276 nm is plotted against the ratio of gene 5 protein:poly(dT) (expressed as subunit concentration:nucleotide). The data show clear stoichiometric binding with an end point of 0.25 (i.e., 4.0 nucleotides bound per protein subunit).

aliquots of protein. This can be done at a single wavelength, without the need for scanning. Although the maximum CD signal from DNA is normally obtained around 275 nm, one should work at the wavelength that corresponds to the largest difference between free and bound DNA, which may well be different.

The CD spectrum of the DNA at saturation can be used to assess conformational changes that result from protein binding to a given sequence. Although analysis of the spectrum in terms of molecular structure is not straightforward *(2)*, it may be possible to interpret CD spectra of double-stranded DNA in terms of changes in helical twist angle (underwinding or overwinding of the helix). The spectral changes that accompany protein binding to different DNA sequences or with a variety of cofactors can also be informative.

4. Notes

1. For proteins and polynucleotides, molarity is sometimes expressed in terms of moles of amino acid or nucleotide residue respectively for such calculations.
2. High-intensity UV radiation converts oxygen to ozone, which damages the optics. Failure to purge will lead to deterioration in instrument performance.
3. For most experiments at these wavelengths, cells with pathlengths of 0.5 cm or 1 cm can usually be used. Longer-pathlength cells for work with more dilute solutions are available at a price.
4. A number of instrument parameters will normally need to be set before spectra can be recorded. For the Jasco J-700 series, these typically include slit width, CD scale, wavelength range, step resolution, scan speed, response time and number of accumulations. Typically, the slit width is set to 1 nm, although for high-sensitivity (i.e., low CD signal) measurements, it can be increased to 2 nm (thus improving the signal-to-noise ratio). The CD scale is set to reflect the expected CD of the sample being measured (e.g., 0–50 mdeg). Similarly, the wavelength range is set to cover the region required for the experiment. We typically use a step resolution of 0.2 nm (which allows a maximum wavelength range of 400 nm to be covered) and a scan speed of between 20 and 100 nm/min. The response time is set in accordance with the scan speed (e.g., for a scan speed of 20 nm/min, the response time is set to 2 s [for 100 nm/min an appropriate response time would be 0.5 s]). It is advised that the manufacturer's recommendations be followed when setting both scan speeds and response times, as the quality of the data collected (in terms of the signal-to-noise ratio) can be adversely affected if incorrect or incompatible settings are used. Between three and five accumulations are generally recorded (although more can be collected), the resulting averaged data offering an improvement in the CD signal-to-noise ratio.
5. If significant drifting of the baseline has occurred, the experiment may have to be repeated. To minimize the effect of drifting, use two cells, one as a baseline reference and one containing sample. The baseline can then be standardized at each point in the titration. Remember that when swapping cells, it is vitally important to present the same section of the cell to the beam each time.

6. The CD software supplied by Jasco allows simple calculations/derivations to be performed with collected data (e.g., the subtraction of CD signal as a result of the sample buffer, subtraction of a constant factor, or conversion between standard units (millidegrees to molar ellipticity). If the signal-to-noise ratio is particularly problematic, data can also be treated with a number of smoothing or noise reduction functions; although if the data collection parameters are correctly set up and the sample is at a sufficient concentration, these procedures should not be necessary. A further useful function of the Jasco CD software allows the collected data to be "dumped" and imported into other software packages (e.g., Microsoft Excel or other spreadsheet/ graphing packages, which may be more familiar to the user).

References

1. Johnson, W. C., Jr. (1990) Protein secondary structure and circular dichroism: a practical guide. *PROTEINS: Struct. Funct. Genet.* **7,** 205–214.
2. Johnson, B. B., Dakl, K. S., Tinoco, I., Jr., Ivanov, V. I., and Zhurkin, V. B. (1981) Correlations between deoxyribonucleic acid structural parameters and calculated circular dichroism spectra. *Biochemistry* **20,** 73–78.
3. Gray, D. M., Hung, S. H., and Johnson, K. H. (1995) Absorption and circular dichroism spectroscopy of nucleic acid duplexes and triplexes. *Methods Enzymol.* **246,** 19–34.
4. Hardin, C. C., Henderson, E., Watson, T., and Prosser, J. K. (1991) Monovalent cation induced structural transitions in telomeric DNAs: G-DNA folding intermediates. *Biochemistry* **30,** 4460–4472.
5. Carpenter, M. L. and Kneale, G. G. (1991) Circular dichroism and fluorescence analysis of the interaction of Pf1 gene 5 protein with poly(dT). *J. Mol. Biol.* **27,** 681–689 .
6. Kansy, J. W., Cluck, B. A., and Gray, D. M. (1986) The binding of fd gene 5 protein to polydeoxyribonucleotides: evidence from CD measurements for two binding modes. *J. Biomol. Struct. Dynam.* **3,** 1079–1110.
7. Culard, F. and Maurizot, J. C. (1981) Lac repressor–lac operator interaction. Circular dichroism study. *Nucleic Acids Res.* **9,** 5175–5184.
8. Wartell, R. M. and Adhya, S. (1988) DNA conformational change in Gal repressor–operator complex: involvement of central G-C base pair(s) of dyad symmetry. *Nucleic Acids Res.* **16,** 11,531–11,541.
9. Torigoe, C., Kidokoro, S., Takimoto, M., Kyoyoku, Y., and Wada, A. (1991) Spectroscopic studies on lambda cro protein–DNA interactions. *J. Mol. Biol.* **219,** 733–746.
10. Connor, F., Cary, D., Read, C., Preston, N. S., Driscoll, P. C., Denny, P., et al. (1994) DNA binding and bending properties of the post-meiotically expressed Sry-related protein Sox–5. *Nucleic Acids Res.* **22,** 3339–3346.
11. Taylor, I. A., Davis, K. G., Watts, D., and Kneale, G. G. (1994) DNA binding induces a major structural transition in a type I methyltransferase. *EMBO J.* **13,** 5772–5778.

12. Powell, M. D. and Gray, D. M. (1995) Role of Tyr-22 in the binding of Pf3 ssDNA binding proteins to nucleic acids. *Biochemistry* **34,** 5635–5643.
13. Mark, B. L. and Gray, D. M. (1997) Tyrosine mutant helps define overlapping CD bands from fd gene 5 protein–nucleic acid complexes. *Biopolymers* **42,** 337–348.

General Reading

1. Bayley, P. M. (1973) The analysis of circular dichroism of biomolecules. *Prog. Biophys.* **27**, 1–76.
2. Bush, C. A. (1974) Ultraviolet spectroscopy, circular dichroism and optical rotatory dispersion, in *Basic Principles in Nucleic Acid Chemistry,* vol. 2 (Ts'O, P.O.P., ed.), Academic, New York, pp. 91–169.
3. Drake, A. F. (1994) Circular dichroism. *Methods Mol. Biol.* **22**, 219–44.
4. Gratzer, W. B. (1971) Optical rotatory dispersion and circular dichroism of nucleic acids, in *Procedures in Nucleic Acid Research*, vol. 2 (Cantoni G. L. and Davies D. R., eds.), Harper and Row, New York, pp. 3–30.
5. Woody, R. W. (1995) Circular dichroism. *Methods Enzymol.* **246**, 34–71.

35

Calorimetry of Protein-DNA Complexes and Their Components

Christopher M. Read and Ilian Jelesarov

1. Introduction

The many detailed, though static, structures of protein–nucleic acid complexes have revealed the underlying structural determinants of the binding process: a high surface complementarity and a precise orientation of interacting groups. However, another important question not yet fully answered is *why* the molecules interact with each other. To answer such a question means that we have to rationalize a structure energetically. This requires a deconvolution of the relatively large negative value of the free energy of association (ΔG) into its enthalpic (ΔH) and entropic (ΔS) contributions. Only then can one appreciate the true nature of the forces that drive the interaction of protein and nucleic acid. Despite some promising theoretical developments and the steady accumulation of experimental data, the deconvolution of ΔG into its components remains a difficult task

Proteins and nucleic acids behave cooperatively and often undergo structural rearrangements on associating. These changes range from subtle adjustments of dihedral angles to rearrangements in the orientation of domains and even the refolding of entire binding domains. Many DNA-binding domains are either very flexible or partly unstructured or even fully unfolded in isolation and become folded only when bound to the specific DNA target site *(1)*. The DNA itself sometimes undergoes large structural deformation upon complex formation *(2–5)*. In energetic terms, the view that binding specificity is simply the accumulation of specific favorable interactions between rigid binding partners no longer holds and conformational changes in the two components as well as solvent rearrangement in the binding site play an important role. Thus,

From: *Methods in Molecular Biology, vol. 148: DNA–Protein Interactions: Principles and Protocols, 2nd ed.*
Edited by: T. Moss © Humana Press Inc., Totowa, NJ

there is a complicated energetic profile involving changes in ΔG, ΔH, and ΔS on going from the free components to the final complex.

This chapter focuses on the practical aspects of measuring the energetics of binding of DNA-binding domains to short DNA duplexes. As an example, we will consider the HMG domain of mouse Sox-5, hereafter referred to as just Sox-5, as it illustrates the importance of structural changes that accompany complex formation *(6)*. HMG domains contain three α-helices arranged into an L-shaped fold. Nuclear magnetic resonance (NMR) derived structures of the HMG domain of human SRY bound to an 8-bp DNA duplex *(4)* as well as the HMG domain of mouse LEF-1 complexed to 15 bp DNA *(5)* show that the domain binds almost exclusively into a widened minor DNA groove. A large bend with considerable base unstacking is introduced into the DNA, in part as a result of the partial intercalation of an amino acid side chain (Ile in SRY and Met in LEF-1).

The interacting components, the free Sox-5 protein and the DNA duplexes have been examined and both structures are seen to depend significantly on temperature *(7,8)*. A practical problem then arises because the temperature dependence of the enthalpy of binding, as measured by isothermal titration calorimetry (ITC), can no longer be interpreted as being solely the result of a unique heat capacity increment, ΔC_p. As a result, the energetics of complex formation require a thorough understanding of the energetics of the free components (i.e., the study of the heat capacity C_p, of the complex, and its free components over a broad temperature range) as can be directly obtained by differential scanning calorimetry (DSC). We thus wish to emphasize that for the study of the energetics of protein–nucleic acid complexes the combined use of ITC and DSC is necessary, because the DSC data are essential to correct the data obtained by ITC.

1.1. The Energetics of Biomolecular Interaction

To characterize the thermodynamics of a binding reaction, it is necessary to determine the association free energy ΔG and its enthalpic and entropic components, ΔH and ΔS, at a given reference temperature. The heat capacity increment on binding, ΔC_p, is required to predict the change of the above three quantities with temperature, according to the general thermodynamic relationships. Values for ΔG, ΔH, and ΔS may then be obtained at temperatures that are experimentally inaccessible. Furthermore, ΔC_p is a fundamental thermodynamic quantity and contains important structural information. For example, it has been shown that ΔC_p changes with the proportion and chemical properties of the molecular surface buried at the binding site *(9,10)*.

ΔG is accessible through different types of binding experiments because it is related to the association constant, K_A, by $\Delta G = -RT\ln K_A$, where R is the gas constant and T is the absolute temperature. For the simple equilibrium describing the interaction of protein P with DNA, D:

$$\mathrm{P} + \mathrm{D} \Leftrightarrow \mathrm{PD} \text{ then } K_A = \frac{[\mathrm{PD}]}{[\mathrm{P}][\mathrm{D}]} \quad (1)$$

Techniques such as equilibrium dialysis, ultracentrifugation, the gel-shift, or radio-ligand binding assays yield direct values for [P], [D], and [PD] and, therefore, values for K_A and ΔG at any fixed temperature. More indirect, mostly spectroscopic, methods allow the construction of a binding isotherm in terms of the degree of saturation, Y:

$$Y = \frac{\Delta p_i}{\Delta p_{max}} = \frac{[\mathrm{PD}]}{[\mathrm{D}] + [\mathrm{PD}]} = \frac{K_A[\mathrm{P}]}{1 + K_A[\mathrm{P}]} \quad (2)$$

where Δp_i is the signal change obtained at some particular addition of P to a constant amount of D and Δp_{max} is the maximal signal change at full saturation ($Y = 1$). If the experiment is repeated at different temperatures, $K_A(T)$ and $\Delta G(T)$ are obtained. $\Delta H(T)$ and $\Delta S(T)$ can be derived from the integrated form of the van Hoff equation. Further, the s derivative of $\Delta G(T)$ with respect to temperature yields ΔC_p. This noncalorimetric analysis of binding data (sometimes referred to as a van Hoff analysis) can be flawed. First, the analysis assumes a model, which may or may not be correct, to describe the interaction of the components. Second, for technical reasons, experiments can be performed only over a limited temperature range and the experimental errors propagate into large errors in ΔH, ΔS, and especially ΔC_p when the extrapolation from the reference temperature to any other temperature is large.

1.2. The ITC Experiment

Modern mixing microcalorimeters allow precise measurements of the enthalpy of binding, ΔH_{app}, between 5°C and 80°C. The basic principle is that one binding partner (protein, P) is titrated at a constant temperature into a known amount of the other binding partner (DNA, D) placed in the sample cell of the calorimeter. Heat is produced or released when binding occurs. The instrument measures the electrical power (in units of J/s) required to maintain a small temperature difference between the sample cell and the reference cell, which is filled with buffer. The contents of the sample cell are stirred to allow rapid mixing and effective heat transfer over the surface of the cell. If K_A is high and the degree of saturation is still low, electrical power peaks of similar magnitude appear in the thermogram at each addition. Integration with respect to time yields the apparent heat change, $\Delta q_{i,app}$, between additions i and i–1 as $\Delta q_{i,app} = q_i - q_i{-}1$.

As the fractional saturation increases, $\Delta q_{i,app}$ gradually decreases and eventually all binding sites become saturated. Small nonspecific heat effects, $\Delta q_{i,ns}$

registered after complete saturation may be caused by the heat of protein dilution, an imperfect match between the buffer composition of the protein solution and the DNA solution, or by other nonspecific effects. $\Delta q_{i,app}$ is proportional to the volume of the calorimetric cell, V_{cell}, to the change in concentration of the bound protein $\Delta[P]_{i,bound} = [P]_{i,bound} = [P]_{i-1,bound}$ and to the apparent molar enthalpy of association ΔH_{app}. Thus,

$$\begin{aligned}\Delta q_{i,app} &= \Delta q_i + \Delta q_{i,ns} \\ &= \Delta[P]_{i,bound}\, V_{cell}\, \Delta H_{app}\end{aligned} \tag{3}$$

ΔH_{app} at temperature T may therefore be calculated because V_{cell} is known and $\Delta q_{i,ns}$ may be obtained from a blank titration of protein into buffer.

For the binding of protein, P, with n identical and noninteracting sites to DNA, D,

$$\begin{aligned}\Delta q_i &= \Delta q_{i,app} + \Delta q_{i,ns} \\ &= nR\, V_{cell}\, \Delta H_{app}\end{aligned} \tag{4}$$

Δq_i is the effective heat change caused by the formation of complex, PD, at the ith step of the titration and R is the root of the quadratic equation

$$Y_i^2 - Y_i\left(1 + \frac{1}{nK_A[D]_{total}} + \frac{[P]_{i,total}}{n[D]_{total}}\right) + n[P]_{i,total}[D]_{total} = 0 \tag{5}$$

where the fractional saturation is defined by $Y_i = \Delta[P]_{i,bound} / [D]_{total}$. $[P]_{i,total}$ is the total concentration of protein, P, added until injection, i. A nonlinear regression procedure based on **Eq. 4** yields n, K_A, and ΔH_{app} at temperature T, from a single titration experiment. More complicated systems with multiple and interacting sites require a statistical mechanical treatment of the data *(11,12)*. $\Delta C_{p,app}$ the apparent heat capacity change of association can be calculated from a plot of ΔH_{app} versus temperature, because in general, $\Delta C_p = \Delta H / \Delta T$.

As in other binding experiments, the dissociation constant, K_D (equal to $1/K_A$) can only be reliably measured over a narrow concentration range. If the concentration of binding sites is much larger than K_D, the binding isotherm is of rectangular shape with a sharp step at saturation. In the case of the binding-site concentration being much smaller than K_D, the binding isotherm is very shallow and is of limited use for calculating the dissociation constant. In practice, values of the ratio of the binding-site concentration to K_D that range between 10 and 100 represent an optimal experimental window to obtain precise values of the dissociation constant. Unfortunately, this window is not always accessible to ITC. For very small K_D values, the optimal concentrations are so low that the released heat is below the sensitivity of the instrument, the specific enthalpy of biomolecular interactions never being very large. For this reason, there is an effective lower limit to the ITC measurement of K_D to values of approx 10^{-9} M (ΔG about –50 kJ/mol). Alternatively, if K_D is high, then very high concentrations are required and the heat of binding may be obscured by aggregation effects.

The two types of ITC measurement are normally obtained under different conditions. The procedure to obtain n, K_A, and ΔH_{app} is best carried out as a full titration at concentrations near the stoichiometric point of the interaction (i.e., at a DNA concentration 10–100 times the estimated value for K_D and with the injection of up to a 2 M to 4 M excess of protein. However, the direct measurement of ΔH_{app} is best carried out in conditions where the system is fully associated and the degree of saturation is low (i.e., at high DNA concentrations (>1000K_D) and with the injection of low amounts of protein). The total number of moles of added protein is then much less than the number of moles of binding sites available and virtually all protein molecules are bound to DNA, such that $\Delta[P]_{i,bound}$ approximately equals $\Delta[P]_{i,total}$. Each injection of protein, $\Delta[P]_{i,total}$ thus produces an approximately equal change in heat, $\Delta q_{i,app}$. If the overall fractional saturation is still low after completion of a series of injections, then the average may be taken and **Eq. 3** becomes

$$\frac{1}{m}\sum_{i=1}^{m}\Delta qi_{,app} = \frac{1}{m}\sum_{i=1}^{m}\left(\Delta qi + \Delta qi_{,ns}\right) \tag{6}$$

$$= \frac{1}{m}\sum_{i=1}^{m}\left(\Delta[P]_{i,bound} + V_{cell}\,\Delta H_{app}\right)$$

where m is the number of injections. If low amounts of protein are injected, aggregation effects and nonspecific binding are also minimized.

1.3. The DSC Experiment

In a DSC experiment, the heat capacity of a macromolecule is measured as a function of temperature. Typically, two thermally insulated cells, a sample cell containing the macromolecule of interest in buffer and a reference cell containing buffer, are electrically heated at a known rate. At a temperature-induced transition, which is typically endothermic, the temperature of the sample cell will lag behind that of the reference cell. An electrical feedback mechanism is used to maintain the reference cell at the same temperature as the sample cell. This amount of compensatory electrical power (in units of J/s or Watts) at temperature T divided by the heating rate is the apparent difference in heat capacity between the cell containing the sample and the reference cell, $\Delta C_{p,app}(T)$ (in units of J/K). Because the sample containing cell has a smaller volume fraction of buffer as compared to that in the reference cell, the partial molar heat capacity of the dissolved macromolecule $C_{p,f}(T)$ at temperature T (J/mol/K) is given by

$$CT_{p,\phi}(T) = \frac{C_{p,buffer}(T)V_{\phi}^{\circ} \quad M\Delta C_{p,app}(T)}{V_{buffer} \qquad\qquad m} \tag{7}$$

where $C_{p,buffer}(T)$ and V_{buffer} are the partial molar heat capacity and molar volume of buffer, respectively. V_f° and M are the partial molar volume and the molar mass of the macromolecule, respectively, and m is the mass of macromolecule in the sample cell.

The excess molar heat capacity function is the heat absorbed in the melting transition above that of the intrinsic heat capacity of the macromolecule. Integration of the excess molar heat capacity function with respect to temperature yields the enthalpy of the melting transition:

$$\Delta H = \int_{T_1}^{T_2} \langle C_p(T) \rangle dT \tag{8}$$

The intrinsic heat capacity of the initial (folded) and final (unfolded) states of the macromolecule must be estimated within the melting transition. For monomeric proteins, the heat capacity of the folded state is an approximate linear function of temperature ***(13,14)*** and the unfolded state is a shallow parabolic function of temperature ***(13)***.

The relationship of the partial molar heat capacity to the excess molar heat capacity functions are schematically shown in **Fig. 1**.

1.4. Preliminary Characterization of the Sox-5–DNA Interaction

It must be emphasized that the in vitro interaction of the protein with DNA must be characterized prior to the calorimetry experiments. Preliminary knowledge of the number of (independent) binding sites, the stoichiometry of the interaction, and the magnitude of the association constant are all required for the optimal design of the ITC experiments (*see* **Subheading 1.2.**). For DSC, the experiments must be performed with a fully associated complex (at low temperature) and therefore at a concentration well above the estimated K_D. Furthermore, the protein–DNA complex and its components need to be soluble at this concentration, even at raised temperatures, and all temperature-induced conformational transitions must be fully reversible. Knowledge of the mode of interaction is also required to obtain van Hoff enthalpies from the DSC data.

Site-selection and DNAse I footprinting assays (Chapter 3) defined the Sox-5 target sequence as 5'-AACAAT-3' within a DNAse I protected region of approx 14 nucleotides on both DNA strands ***(15,16)***. This determined the sequence and size of the DNA duplexes that were to be used. A circular permutation assay showed that the binding of Sox-5 to these duplexes was able to introduce a DNA bend of similar magnitude to that obtained with the complete

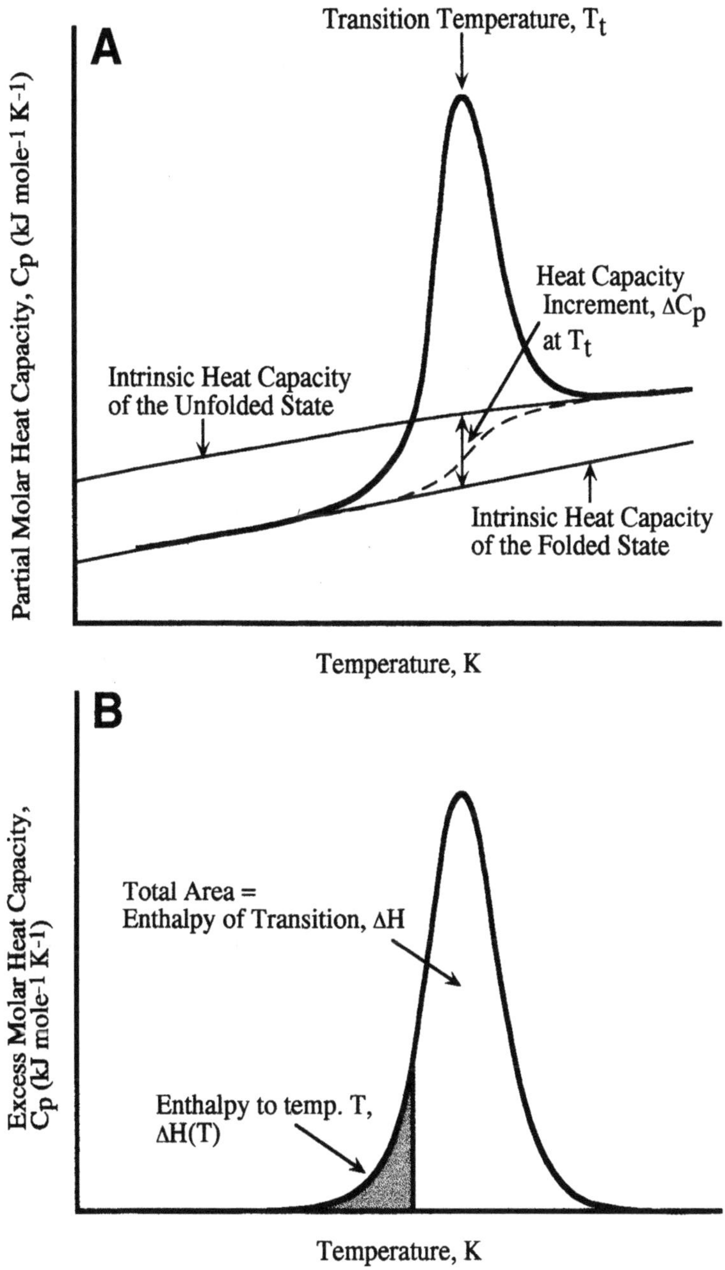

Fig. 1. Schematic representation of (**A**) the partial molar heat capacity function as would be observed for the melting of a single domain monomeric protein. Over the transition region the intrinsic heat capacity function is an interpolation of the folded and unfolded states, weighted in proportion to their relative contributions (dashed line). (**B**) replots the C_p/T function above that of the intrinsic heat capacity of the system. This is known as the molar excess heat capacity function.

Sox-5 protein *(6,16)*. The dissociation constant and the stoichiometry of the Sox-5–DNA interaction were determined from circular dichroism (CD) and gel-shift assays (Chapters 2 and 34):

1. The binding of Sox-5 to a 12-bp DNA duplex (10 μ*M*) was followed by CD, monitoring the ellipticity of the positive DNA peak at 280 nm upon successive additions of the protein. The CD data were fitted to a 1:1 binding model with an estimated K_D of lower than 100 n*M* *(6)*. This same estimate was obtained at temperatures between 10°C and 37°C, suggesting no significant variation of binding constant with temperature.
2. A gel-shift assay using a radioactively labeled 12-bp DNA duplex (at 1 n*M*) was titrated in molar excess with Sox-5 *(6)*. This indicated that the primary binding site is characterized by a K_D of about 35 n*M* at 4°C. No secondary protein binding was observed with <1 μ*M* Sox-5.

These assays thus indicate a bimolecular interaction of Sox-5 with the DNA, with a single primary binding site characterized by a K_D in the low nanomolar range. Furthermore any secondary binding of Sox-5 to DNA, if present, must have a K_D of >10 μ*M* (i.e., at least two orders greater than the primary binding site). Thus, for the DSC measurements, typically at a concentration of 100–500 μ*M* complex, the preformed Sox-5–DNA complex is fully associated and the effect of secondary binding is negligible.

Given the estimated K_D value, the ITC measurement of a full binding isotherm under optimal conditions requires a DNA concentration of between approx 0.35 and 3.5 μ*M* (*see* **Subheading 1.2.**). Preliminary titration experiments however showed only a small heat effect of association (a result of the temperature dependency of ΔH_{app} which passes through zero at 17°C) and the presence of a secondary binding event, with a large exothermic effect, after saturation of the primary binding site (**Fig. 2**). It was concluded that measurement of the entire binding isotherm was not possible by ITC.

However, conditions could be chosen that were optimal for the direct determination of ΔH_{app} by ITC (i.e., total association at partial saturation (*see* **Subheading 1.2.**). At the chosen DNA concentration of 60 μ*M*, total association of injected Sox-5 in an ITC experiment is achieved (the ratio of the DNA concentration to the estimated K_D being approx 1700). The fractional molar saturation was always kept lower than 0.5 to avoid any secondary binding effects. Because high concentrations were to be used, the total heat effect was greater and more easily detectable, which meant that ΔH_{app} values could be accurately obtained at temperatures close to 17°C.

Additional experiments included low-speed analytical ultracentrifugation at different temperatures to demonstrate that the free Sox-5 is monomeric. For the free DNA, UV melting was performed as an initial check on its melting temperature and for two-state melting of the duplexes. UV melting of the

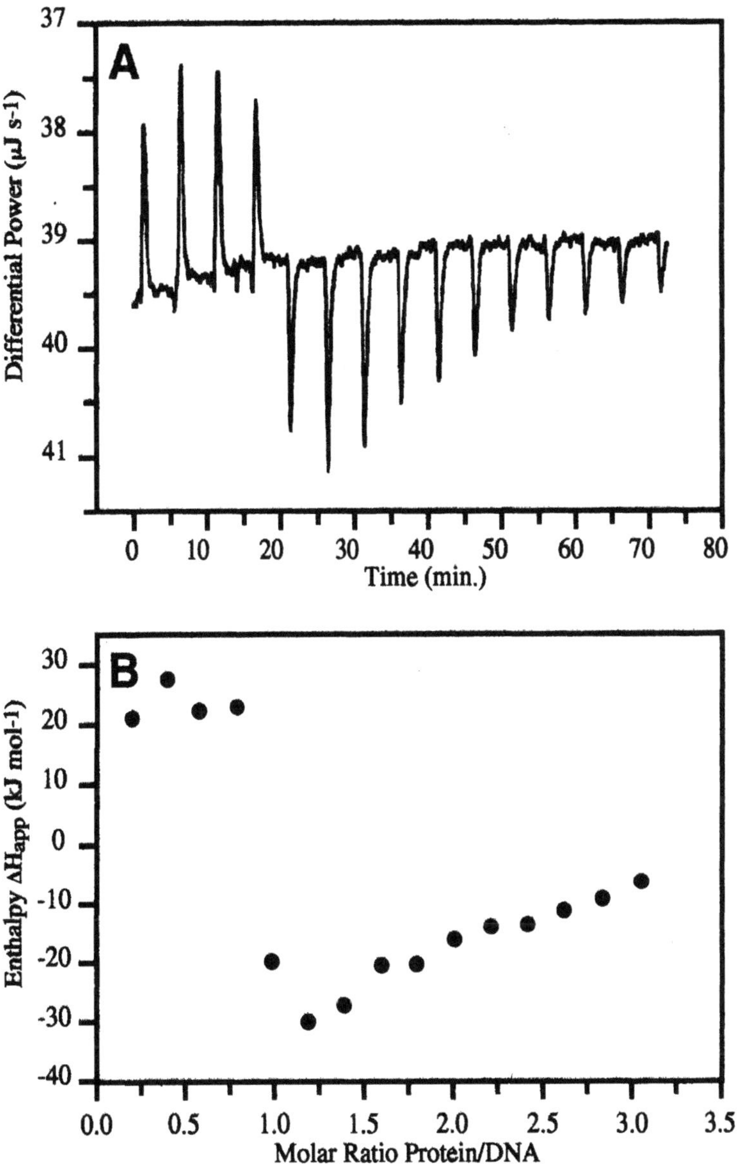

Fig. 2. ITC titration experiment of Sox-5 into a 12-bp DNA duplex. (**A**) Fifteen injections of 540 µ*M* Sox-5 into 10 µ*M* of 12-bp DNA at 9°C. The injection rate was 5 µL in 8 s, with a time interval of 5 min between injections. (**B**) a plot of ΔH_{app} against ratio of Sox-5 to DNA showing the primary endothermic and secondary exothermic reactions that occur on either side of the stoichiometric point.

Sox-5–DNA complex showed a single cooperative transition with a melting temperature above that of the free DNA.

2. Materials

2.1. Calorimeters

Isothermal titration calorimetry was performed on an OMEGA or MCS-ITC titration calorimeter (MicroCal Inc., Northampton, MA). Technical details on the construction of mixing microcalorimeters, their performance and sensitivity, and on the theory of data analysis have been described elsewhere ***(17–20)***.

Differential scanning calorimetric experiments were usually carried out on a Nano-DSC calorimeter (Calorimetric Science Corp., Utah). The instrument's performance and data acquisition are detailed in **ref.** ***21***. In brief, the instrument operates over the temperature range –20°C to 130°C at any set heating/cooling rate. The sample and reference calorimetric cells are of approx 900 µL volume. Control of the calorimeter and scan acquisition is achieved via the supplied DSC_Acquisition program running on a PC computer connected to the calorimeter. In experiments in which the amount of sample was limiting, a new version of the Nano-DSC calorimeter having cells of approx 300 µL volume was used.

2.2. Reagents and Solutions

2.2.1. Calorimetry

1. For calorimetric experiments, it is important to ensure that the protein and DNA samples are homogeneous, because even a few percent of contaminating species might cause significant heat effects. Homogeneity of the samples is most reliably verified by mass spectrometry.
2. Concentrations must be accurately determined because this affects the DSC and ITC data (*see* **Subheading 3.4.**). The preparation of equimolar mixtures (of complementary oligonucleotides or Sox-5-DNA complex) can then be made precisely, without further purification. Dilutions are made by weight on a precision balance.
3. The DSC and ITC experiments with the Sox-5–DNA complex used a working buffer of 100 m*M* KCl, 10 m*M* potassium phosphate, and 1 m*M* EDTA (pH 6.0) (*see* **Note 1**). All buffer solutions were prepared from the highest-quality reagents and ultrapure water. For ITC experiments, the solutions were filtered through a 0.45-µm membrane and thoroughly degassed prior to use.

2.2.2. Preparation of Sox-5 HMG Box–DNA Complexes and Components

1. DNA oligonucleotides were synthesized using phosphoramidite chemistry and purified on a Mono Q HR16/10 column fitted to a Pharmacia FPLC system, eluting with a linear 0.1 *M* to 1.0 *M* NaCl gradient in 10 m*M* Tris-HCl, 1 m*M* EDTA, and 20% (v/v) acetonitrile (pH 7.0). Fractions containing oligonucleotide were precipitated with 3 vol of ethanol at –20°C overnight, then centrifuged down and

redissolved in water. Oligonucleotides were extensively dialyzed using Spectrapor tubing (molecular-weight cutoff 500 Daltons) against three changes of 1 L working buffer at 4°C.

2. DNA duplexes were prepared by mixing equimolar amounts of the complementary oligonucleotides in working buffer and annealed by heating to 95°C in a water-bath followed by slow cooling to 4°C over a period of approx 4 h. DNA duplexes were then extensively dialyzed using 3000-Dalton molecular-weight cutoff tubing against three changes of 1 L working buffer at 4°C.
3. The HMG box of mouse Sox-5 (amino acids 182 to 260, *[15]*) was expressed as a fusion protein in pGEX-2T, using *E. coli* BL21 (DE3) plysS cells. After affinity purification with glutathione-agarose and thrombin cleavage while still attached to the column *(22)*, reverse-phase high-performance liquid chromatography (HPLC) was used to purify the protein. Protein was redissolved in water and refolded by extensive dialysis against three changes of 1 L working buffer at 4°C.
4. The Sox-5–DNA complex was prepared by the mixing together of equal volumes of the components (at equimolar concentration) in working buffer at 4°C. Sox-5 was added in 10% aliquots, at 5-min intervals, to the DNA. The complex was then extensively dialyzed using 3000-Dalton molecular-weight cutoff tubing against three changes of 1 L working buffer at 4°C. The accuracy of the concentrations of protein and DNA used to form the 1:1 complex may be checked by trial additions of Sox-5 to DNA at various protein:DNA ratios followed by electrophoresis on a nondenaturing polyacrylamide gel.

2.2.3. DNA Concentrations

1. 100 m*M* Tris-HCl (pH 8.0).
2. Snake venom phosphodiesterase I (PDE1, from *Crotalus durissus terrificus*, Sigma).

3. Methods

3.1. Isothermal Titration Calorimetry

Before a series of experiments, the calorimeter was calibrated either by applying electrically generated heat pulses or by standardized chemical reactions (e.g., the protonation of tris[hydroxymethyl] aminomethane or the binding of Ba^{2+} to 18-crown-6 ether). It is recommended to equilibrate the jacket with a circulating water-bath for about 10 h. at a temperature lower than the temperature of the experiment by 3–5°C. Although such equilibration is not strictly necessary when operating at above room temperature, it substantially improves the baseline stability. The stirring speed during both the equilibration and the experiment was 350 rpm (*see* **Note 2**).

1. Sample and reference cells are first rinsed with dialysis buffer. The reference cell is then filled with buffer and the sample cell filled with the DNA solution. The system is heated to the working temperature and equilibrated until the differen-

tial power signal levels off. The injection syringe containing Sox-5 is inserted into the sample cell, stirring is initiated, and the baseline is established over a period of 30–60 min. Typically, a baseline drift (differential power signal drift) of less than 20–30 nJ/min and an rms noise of less than 15–20 nJ/s (or nW) indicates complete thermal equilibration of the system under stirring (*see* **Note 3**).

2. The experiment is started with a small injection of 1–2 μL. The reason for this is that during the long equilibration period, diffusion through the injection ports occurs, thus causing a change in protein concentration near the syringe needle tip. The actual injection schedule is then executed. To measure the enthalpy of Sox-5 binding to DNA, six to eight injections each of 8–12 μL and of 12–15 s duration were performed, with a 5-min interval between injections (*see* **Note 4**). Typical thermograms are depicted in **Fig. 3A**, traces a and c.
3. After completion of the experiment, the cells are thoroughly cleaned. Cleaning of the calorimetric cells and filling syringes follows a standard laboratory protocol (*see* **Note 5**).
4. The cells may then be filled with buffer and **step 2** repeated. Injections of protein into buffer will yield the heat associated with protein dilution and other nonspecific effects. Typical control thermograms are depicted in **Fig. 3A**, traces b and d. Because the heats obtained in **steps 2** and **4** are directly compared in the data analysis, it is crucial that (1) the blank titration is performed at exactly the same temperature as the main experiment, (2) the same protein solution and dialysis buffer are used, and (3) an identical injection scheme is executed. If any of the listed requirements is not fulfilled, the results of the experiment could be misleading.
5. The titrations are performed at different temperatures to collect data on the temperature dependence of the binding enthalpy.

3.1.1. ITC Data Analysis

If a complete binding isotherm has been recorded over an optimal concentration range, the data can be subjected to a nonlinear least-squares analysis to obtain a full set of parameters (ΔH_{app}, K_A, and n) according to **Eq. 4**. In the case of Sox-5 binding to DNA, the experiments were designed to measure only the enthalpy and heat capacity changes. Thus integration of the differential power peaks (*see* **Note 6**) collected as in **step 2** of **Subheading 3.1.** yields $\Delta q_{i,app}$ and the nonspecific heats $\Delta q_{i,ns}$ are obtained by integration of the peaks collected as in **step 4** of **Subheading 3.1.** Under conditions of total association $\Delta[P]_{i,bound}$ equals $\Delta[P]_{i,total}$ and is easily calculated from the known concentration of protein in the injection syringe, the volume of each injection and the

---

Fig. 3. *(opposite page)* Representative records of a Sox-5 titration into DNA duplex (**A**) and enthalpies of binding of Sox-5 to a 12-bp DNA duplex as measured by ITC (**B**). (**A**) Trace a: six injections of 550 μ*M* Sox-5 into 58 μ*M* of 16-bp DNA at 9°C (i.e., up to a fractional saturation of 0.45); trace b: control titration of Sox-5 into buffer at

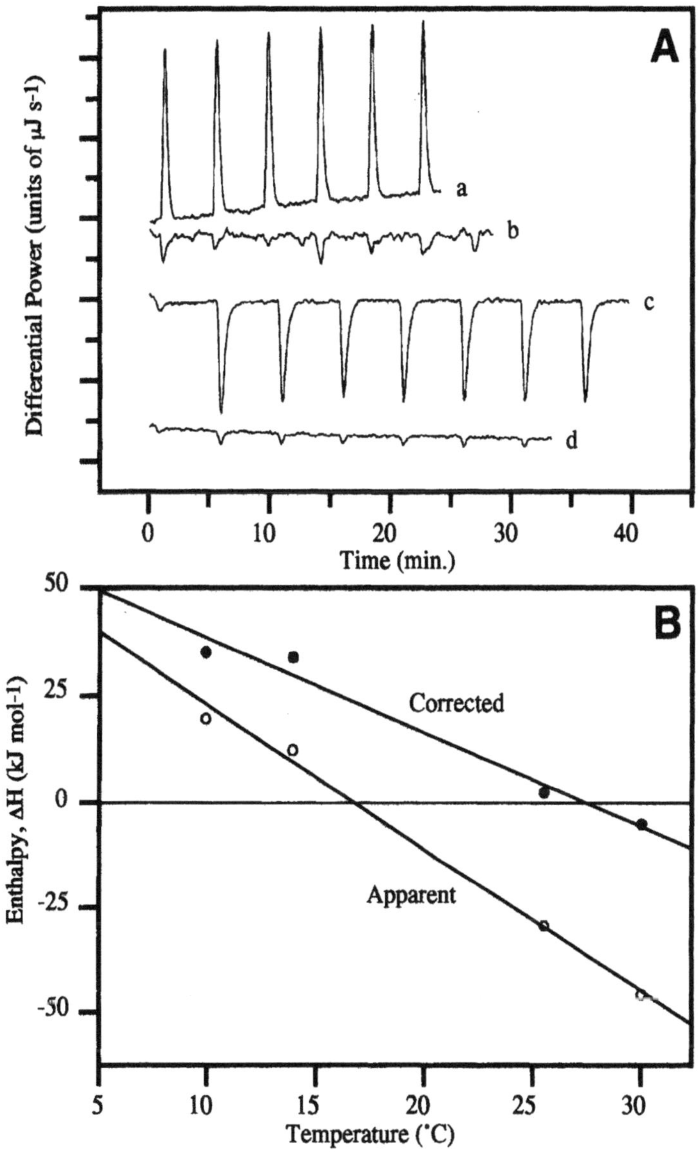

9°C; trace c: eight injections of 490 μ*M* Sox-5 into 60 μ*M* of 12-bp DNA at 30°C; trace d: control titration of Sox-5 into buffer at 30°C. The injection rate was 8 μL in 12 s. The time intervals between injections were 4 min (traces a and b) and 5 min (traces c and d). **(B)** The open circles represent ΔH_{app} measured between 9 and 30°C and calculated according to **Eq. 6** (mean values of six to eight injections). Note that ΔH_{app} changes sign at about 17°C. The standard deviations from the mean values shown are of the order of ±5–10%. Filled circles represent the binding enthalpy obtained after correction of ΔH_{app} for protein and DNA refolding and complex unfolding. The lines correspond to linear least-squares fits to the data, the slope of which is equal to ΔC_p.

volume of the cell, V_{cell}. The enthalpy of association, ΔH_{app}, at temperature T may then be calculated from **Eq. 6**. This procedure is repeated for the pairs of titrations at different temperatures. The experimentally observed enthalpies (ΔH_{app}) obtained at a number of temperatures can be plotted as illustrated in **Fig. 3B**.

3.2. Differential Scanning Calorimetry

1. Prior to starting experiments ensure that the Nano-DSC calorimeter sample and reference cells are at room temperature, thoroughly cleaned, and thermally balanced.
2. Add working buffer (100 m*M* KCl, 10 m*M* potassium phosphate, and 1 m*M* EDTA (pH 6.0)) which is at room temperature to both sample and reference cells. It is most important that no air bubbles are present in the calorimeter cells (*see* **Note 7**). The presence of air bubbles will yield incorrect values for the partial molar heat capacity of the macromolecule, because the volume of solution in the sample and reference cells is not identical (assumed in **Eq. 7**). Furthermore expansion of the minutes of air bubbles with heating results in significant heat output (and *vice versa* on cooling).
3. Seal the top of the chamber with the screw-threaded piston and let the calorimeter settle to thermal equilibrium. Apply approx 2 atm of overpressure to the cells by screwing down the piston. As pressure is applied note the compensatory power reading—it should not change by more than a few microwatts (*see* **Note 8**).
4. Scan to obtain a buffer–buffer baseline. Initially, a number of scans are recorded in order to obtain a number of reproducible baselines (to within 0.5 µW, in the linear region). A complete cycle of heating and cooling at a rate of 1 K/min actually takes about 4 h because at the end of each heating or cooling scan, there is a period of thermal equilibration.
5. Remove the buffer from the sample cell. This is most easily accomplished with a vacuum line attached to a water pump.
6. Slowly and carefully apply a 1.8-mL dialyzed sample to the cell, using the method outlined in **Note 7**. For concentrated protein samples, care is needed to remove all air bubbles. The excess sample is put to good use: first to determine sample concentration and, second, it is diluted with the final dialysate for use in further calorimetric runs. Again, apply a pressure of approx 2 atm to the cells.
7. Perform one heating and cooling scan to obtain the denaturation and renaturation curves of the sample-buffer. Usually samples were heated from 0°C to 80°C and then cooled to –8°C, both at a rate of 1 K/min. The heat effects of unfolding and refolding should appear as mirror images, slightly shifted in temperature because the slower kinetics of refolding. Reproducibility shows that the heat-induced unfolding/refolding is reversible. However, this may not be the case (*see* **Note 9**).
8. Remove the sample from the cell. Thoroughly wash out the cell with buffer and then perform another scan of buffer. A heat-asorption peak may be observable if the previous sample is not completely removed (*see* **Note 10**).

3.2.1. DSC Data Analysis

The subtraction of buffer scans to obtain an accurate baseline, the conversion to partial molar heat capacities, and the deconvolution of the excess molar heat capacity function into separate transitions were all performed using the CPCalc program. CPCalc provides a simple interactive mechanism, based on Data Exchange Ports, for the parsing of data from one step to another.

1. The functions of differential power and temperature vs time are extracted from the DSC data acquisition files. Each file may contain more than one scan, but further analysis is on a one-scan basis.
2. A matching buffer–buffer scan is subtracted from a sample-buffer scan. Both must be from a heating (or cooling) scan. Since the absolute values of the molar heat capacities depend on good baseline subtraction and the buffer-buffer scans vary to a small extent, it is important that a number of subtractions are tried using different scans from the acquisition file.
3. The subtracted compensatory power curve is converted to the partial molar heat capacity function by use of **Eq. 7**. Accurate values for the concentration (**Subheading 3.4.**), molecular mass, and partial specific volume of the macromolecule are required (*see* **Note 11**).
4. Appropriate functions for the intrinsic heat capacity of the initial and final states may then be put onto the partial molar heat capacity function (**Fig. 4**). The calorimetric ΔH, ΔS, and melting temperature of the peak are obtained. Enthalpies and entropies over a restricted temperature interval, $\Delta H(T)$ and $\Delta S(T)$, are obtained by viewing the partial molar heat capacity function between the two temperatures.
5. The excess molar heat capacity function is deconvoluted into separate transitions.

3.3. Correction of ITC Derived Enthalpies

The value for $\Delta C_{p,app}$ calculated from the slope of the ΔH_{app} vs T plot (**Fig. 3**) is about –3.3 kJ/K/mol, implying that a considerable amount of hydrophobic surface is buried at the complex interface. However, over the temperature range (10–30°C) used for the ITC measurements, the partial molar heat capacity functions for both free Sox-5 and DNA (DSC data, **Fig. 4A**) are greater than the calculated intrinsic heat capacity functions for the fully folded molecules. Thus, the DSC experiments show significant heat absorption resulting from a temperature-dependent unfolding of the free Sox-5 and DNA. Significant refolding of both components must therefore take place when they associate at these temperatures: a simple interaction between rigid-body molecules is not observed. If one wishes to relate observed energetic parameters to structural features of a rigid fully-folded complex, then the enthalpic contributions arising from refolding of the free Sox-5 and DNA, as determined from DSC, must be added to the ITC-measured enthalpies of association to obtain the enthalpies applicable to a rigid-body interaction. Furthermore, because the

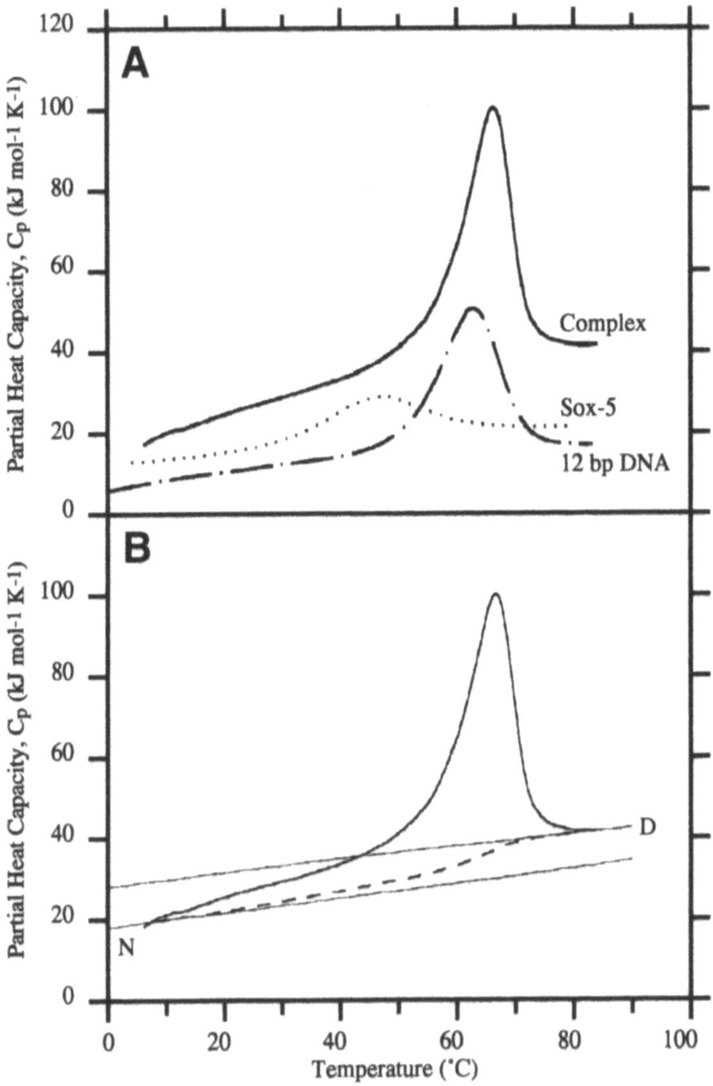

Fig. 4. **(A)** Partial molar heat capacity functions obtained for the free Sox-5 protein (dotted curve), the 12-bp DNA duplex that includes the AACAAT motif (dot–dash curve) and the 1:1 Sox-5–12-bp DNA complex (solid curve). Note that the complex dissociates in a single cooperative trimolecular transition with a melting temperature only slightly above that of the 12-bp DNA duplex. The free Sox-5 protein and 12-bp DNA duplex dissociate in a monomolecular and bimolecular manner, respectively. **(B)** Partial molar heat capacity function for the Sox-5–12-bp DNA complex showing the intrinsic heat capacity functions for the fully folded (N) and unfolded (D) complex (thin lines). The area above the dashed line corresponds to the total enthalpy of all the temperature-induced changes in the complex.

complex also absorbs heat over the 10–30°C temperature range (i.e., becomes partially unfolded [**Fig. 4A**]), such heats must be subtracted from the ITC-measured enthalpies of association to generate the net enthalpies applicable to the formation of a fully folded complex. These corrections considerably affect the enthalpy profile of binding (*see* **Fig. 3B**). The corrected plots of ΔH_{app} vs T are now linear, with an average slope of about –1.5 kJ/K/mol and this corrected value of ΔC_p is in very good agreement with the amount and type of molecular surface screened from the solvent upon association, calculated assuming a rigid-body interaction.

3.4. Determination of Concentrations

The protein and DNA samples must be highly homogenous and their concentrations must be known to the highest possible precision. In the first place this enables an accurate preparation of DNA duplex from its complementary oligonucleotides and of the Sox-5–DNA complex from protein and DNA duplex. In DSC, the accuracy with which the sample concentration can be determined directly affects the molar heat capacity values obtained. In ITC, unlike other binding assays, errors in the concentration of protein are directly reflected in the values of ΔH_{app} (**Eqs. 3** and **4**). Thus, in both ITC and DSC, concentrations are determined after dialysis. In the DSC experiments, it is also possible that a small amount of buffer from a previous scan remains in the gold capillary cell. To eliminate this potential source of error, concentrations were determined using the excess sample that remains after filling the calorimeter cell.

3.4.1. DNA

For oligonucleotides and DNA duplexes, concentrations were determined from their UV absorption at 260 nm, after digestion to nucleotides with snake venom phosphodiesterase I (PDE1).

1. Accurately dilute the solution of DNA to about 1.0 mL with 100 m*M* Tris-HCl (pH 8.0) so as to give an absorbance at 260 nm of about 0.5 in a 1-cm pathlength cell. To overcome any reliance on the presumed accuracy of the pipets being used, the dilutions are performed on a precision balance.
2. Record the UV absorption spectrum of the diluted DNA solution, using as reference a blank cell containing buffer. Note that for small oligonucleotides and DNA duplexes with a biased nucleotide composition, the absorption maximum is not necessarily at 260 nm. This is in fact an optional step, but the recording of UV spectra both before and after the addition of PDE1 indicates that digestion has occurred and enables a determination of the hypochromicity of the DNA solution.
3. Remove the diluted DNA solution from the cell and place into a screw-capped 1.5-mL tube. Add either 0.008 U (for oligonucleotides) or 0.08 U (for DNA

duplexes) of PDE1 to the solution, mix well, and incubate at 37°C overnight. If necessary, a time-course of the digestion may be followed (at room temperature) in the UV spectrophotometer by direct addition of PDE1 to the DNA solution in the quartz cell.

4. Accurately record the UV absorption spectrum of the DNA solution after PDE1 digestion, using a black cell containing buffer as a reference. The contribution to the absorbance at 260 nm arising from overlap of the PDE1 280-nm peak may be neglected
5. Optionally, calculate the percent hypochromicity (%*H*) from the equation

$$\%H = \frac{(A_a - A_b)}{A_a} \times 100$$

where A_b and A_a are the absorbances at 260 nm before and after PDE1 digestion respectively, with dilution the result of the addition of PDE1 taken into account. Values of %*H* are usually approx 20% for oligonucleotides and approx 35–40% for duplex DNA. Lower values indicate that digestion may not be complete, that the original DNA is degraded, or that there is a failure to form duplex DNA.

6. Calculate the molar nucleotide concentration and thus the molar DNA concentration from the following equations

$$[\text{Nucleotide}] = \frac{\text{Aa(Dilution)N}}{(12{,}010G) + (15{,}200A) + (8400T) + 7050C)}$$

and

$$[\text{DNA}] = \frac{[\text{Nucleotide}]}{N}$$

where *N* is the total number of nucleotides in the DNA and *G*, *A*, *T*, and *C* are respectively the number of dG, dA, dT, and dC nucleotides. The extinction coefficients at 260 nm for the four nucleotides are from **ref. *23***.

3.4.2. Protein

The concentration of Sox-5 (in working buffer) was determined from its UV absorption at 280 nm using an extinction coefficient of 17,460/*M*/cm. This extinction coefficient is based on the addition of the extinction coefficients of tryptophan and tyrosine (of which there are two and five in Sox-5, respectively). The accuracy of the concentration determination is within 5%.

3.4.3. Sox-5–DNA Complex

The concentration of complex was determined from its UV absorbance at 260 nm. At this wavelength, the absorption is mainly the result of the DNA, but there is a significant absorption from the protein that must be taken into account. There may also be additional hyperchomic effects because the DNA of the Sox-5–DNA complex is highly bent with considerable base unstacking.

Thus, the approach taken was to measure the UV absorbance at 260 nm for a solution of DNA duplex and then again after the addition of an equimolar amount of Sox-5 protein to form the complex. The measurement was performed at approx 60 μ*M* DNA in a 1-mm pathlength cell, such that the complex was fully associated. For the Sox-5–12-bp DNA complex, an extinction coefficient at 260 nm of 158.6 × 10^3/Molar complex/cm was obtained.

4. Notes

1. The choice of assay buffer depends on the system under study and on the requirements of other assays performed in the context of a particular investigation. Buffers with a different heat of protonation may be used in an ITC experiment to check whether the protonation state of the system changes upon association.
2. In setting the stirring speed, the particular system under study must be considered. Some proteins do not resist the forces arising from rapid stirring of the solution placed in the narrow calorimetric cell and aggregate or unfold. On the other hand, very slow stirring may result in low rates of heat transfer, thus causing broadening of the peaks observed in the thermogram and a decrease in sensitivity.
3. When working below room temperature, it is recommended to fill the cells with cold solutions, otherwise the equilibration time may be very long. Any particles in the stirred sample cell as well as the formation of bubbles during filling may cause problems in establishing the baseline.
4. There is no general rule about the number, volume, and duration of injections. To obtain an entire binding isotherm a 2–4 molar excess of ligand over the concentration of receptor binding sites should be injected by at least 10–15 additions. Five to 10 injections suffice to reliably measure the enthalpy of reaction. In this case, the degree of saturation must be less than 0.5 at the end of experiment. The volume and duration of injections should be chosen in such a way that the released heat is well above the threshold of sensitivity and sharp peaks appear in the thermogram. Typically, 5-min intervals between injection are sufficiently long times for the signal to return to the baseline. However, for very slow reactions, this interval should be prolonged. Because the heat of reaction is a temperature-dependent quantity, the injection scheme might be changed for experiments carried out at different temperatures.
5. Alternating cycles of washing with 0.5 *M* NaOH and isopropanol are routinely used. To remove heavily precipitated proteins, the cell can be rinsed with a hot solution of 20% sodium dodecyl sulfate (SDS). The protocol for cleaning will depend on the particular system under study and the material of the cell.
6. Integration requires construction of a proper baseline. Some software products support automatic procedures for baseline determination. However, it is often found that manual adjustment of the baseline is a better practice, particularly in cases when the signal-to-noise ratio is low or when the instrument baseline drift is substantial. In constructing the baseline manually, it is important to use the

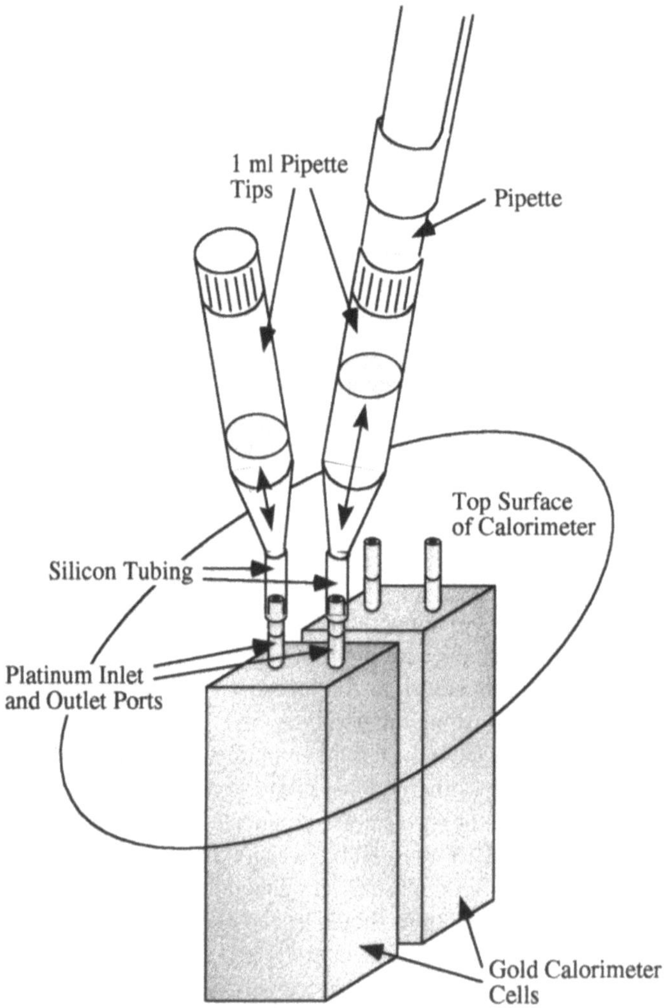

Fig. 5. Arrangement for filling the capillary cells in the Nano-DSC calorimeter.

same integration window for all the peaks observed, both in the specific binding and in the blank titration experiments. This will avoid the introduction of nonrandom bias in the data.

7. We would advocate filling each cell using a minimum of approx 1.8 mL of solution by the following method (*see also* **Fig. 5**). First attach a 1-mL pipet tip using a short length of silicon tubing to one port of the cell. Draw up 0.9 mL of solution using a pipettor fitted with a 1-mL tip and silicon tubing. Connect this to the other side of the cell and slowly introduce the solution into the cell. Disconnect and apply a second 0.9 mL of solution, without introducing any air bubbles. As the cell is only about 0.9 mL in volume, excess solution rises up into the 1-mL

pipet tip attached to the other side. The solution is then pipetted up and down, at first slowly and then more rapidly, for sufficient time to expel minute air bubbles from the cell. Eventually, excess solution may be removed by carefully withdrawing it into the 1.0-mL tip connected to the pipettor and then disconnecting by simultaneously pulling off both 1-mL tips with their connecting tubing.

8. If the microwatt reading rapidly increases or decreases on applying pressure, then air bubbles are present in either of the cells. The cells must then be emptied and refilled.
9. One reason may be temperature-induced aggregation, manifesting itself as a large peak of heat evolution that typically commences close to the peak of maximum heat absorption of the cooperative transition (approximately at the melting temperature, T_m). A s possible reason for irreversibility is the presence of disulfide bonds in the protein, which may disrupt on heating but, on cooling form, mixed disulfide bonds with incorrect Cys residues. This may be overcome by using proteins (or domains) with no Cys residues (as in Sox-5) or by using engineered versions in which the Cys residues are mutated to Ser. This strategy will, however, depend on the contribution of the disulfide bonds in stabilizing the overall protein fold.
10. The calorimetric cells may be thoroughly cleaned by filling them with 50% (v/v) formic acid, followed by heating from 25°C to 65°C (at 1 K/min). The cells are then thoroughly rinsed with water and buffer. Multiple baseline scans of the buffer are recommended before proceeding with the next sample.
11. The partial specific volumes used for oligonucleotides and DNA duplexes were 0.53 and 0.54 mL/g. The partial specific volume of Sox-5 was calculated from its composition as 0.723 mL/g, using the known partial specific volumes of the amino acids. The Sox-5–12-bp DNA complex partial specific volume of 0.650 mL/g was calculated as the weight average of the partial specific volumes of Sox-5 and the DNA duplex.

Acknowledgments

We would like to thank Peter Privalov (Johns Hopkins University, Baltimore), in whose laboratory the calorimetry was performed. We thank Colyn Crane-Robinson for critical reading of the manuscript. Financial support from an NIH grant to the Baltimore laboratory (GM48036-06), a Wellcome Trust grant to the Portsmouth laboratory, and NATO Collaborative Research Grants are gratefully acknowledged.

References

1. Patikoglou, G. and Burley, S. K. (1997) Eukaryotic transcription factor–DNA complexes. *Ann. Rev. Biophys. Biomol. Struct.* **26,** 289–325.
2. Schultz, S. C., Shields, G. C., and Steitz, T. A. (1991) Crystal structure of a CAP–DNA complex: the DNA is bent by 90°. *Science* **253,** 1001–1007.
3. Kim, J. L., Nikolov, D. B., and Burley, S. K. (1993) Co-crystal structure of TBP recognising the minor groove of a TATA element. *Nature* **365,** 520–527.

4. Werner, M. H., Huth, J. R., Gronenborn, A. M., and Clore, G. M. (1995) Molecular basis of human 46X,Y sex reversal revealed from the three-dimensional solution structure of the human SRY–DNA complex. *Cell* **81,** 705–714.
5. Love, J. J., Li, X., Case, D. A., Giese, K., Grosschedl, R., and Wright, P. E. (1995) Structural basis for DNA bending by the architectural transcription factor LEF-1. *Nature* **376,** 791–795.
6. Privalov, P. L., Jelesarov, I., Read, C. M., Crane-Robinson, C., et al. (1999) The energetics of HMG box interactions with DNA. Thermodynamics of the DNA binding of the HMG box from mouse Sox-5. *J. Mol. Biol.* **294,** 997–1013.
7. Crane-Robinson, C., Read, C. M., Cary, P. D., Driscoll, P. C., Dragan, A. I., and Privalov, P. L. (1998) The energetics of HMG box interactions with DNA. Thermodynamic description of the box from mouse Sox-5. *J. Mol. Biol.* **281,** 705–717.
8. Jelesarov, I., Crane-Robinson, C., and Privalov, P. L. (1990) The energetics of HMG box interactions with DNA. Thermodynamic description of DNA duplexes the target. *J. Mol. Biol.* **294,** 981–995.
9. Murphy, K. P. and Freire, E. (1992) Thermodynamics of structural stability and co-operative folding behaviour in proteins. *Adv. Protein Chem.* **43,** 313–361.
10. Spolar, R. S., Livingstone, J. R., and Record, M. T., Jr. (1992) Use of liquid hydrocarbon and amide transfer data to estimate contributions to thermodynamic functions of protein folding from the removal of nonpolar and polar surface from water. *Biochemistry* **31,** 3947–3955.
11. Ferrari, M. E. and Lohman, T. M. (1994) Apparent heat capacity change accompanying a nonspecific protein-DNA interaction. *Escherichia coli* SSB tetramer binding to oligodeoxyadenylates. *Biochemistry* **33,** 12,896–12,910.
12. Bruzzese, F. J. and Connelly, P. R. (1997) Allosteric properties of inosine monophosphate dehydrogenase revealed through the thermodynamics of binding of inosine 5'-monophosphate and mycophenolic acid. Temperature dependent heat capacity of binding as a signature of ligand-coupled conformational equilibria. *Biochemistry* **36,** 10,428–10,438.
13. Privalov, P. L. and Makhatadze, G. I. (1990) Heat capacity of proteins. *J. Mol. Biol.* **213,** 385–391.
14. Makhatadze, G. I. and Privalov, P. L. (1995) Energetics of protein structure. *Adv. Protein Chem.* **47,** 307–325.
15. Denny, P., Swift, S., Connor, F., and Ashworth, A. (1992) A Sry-related gene expressed during spermatogenesis in the mouse encodes a sequence-specific DNA-binding protein. *EMBO J.* **11,** 3705–3712.
16. Connor, F., Cary, P. D., Read, C. M., Preston, N. S., Driscoll, P. C., Denny, P., et al. (1994) DNA binding and bending properties of the post-meiotically expressed Sry-related protein Sox-5. *Nucleic Acids Res.* **22,** 3339–3346.
17. McKinnon, I. R., Fall, L., Parody-Morreale, A., and Gill, S. J. (1984) A twin titration microcalorimeter for the study of biochemical reactions. *Anal. Biochem.* **139,** 134–139.
18. Wiseman, T., Williston, S., Brandts, J. F., and Lin, L. N. (1989) Rapid measurement of binding constants and heats of binding using a new titration calorimeter. *Anal. Biochem.* **179,** 131–137.

19. Freire, E., Mayorga, O. L., and Straume, M. (1990) Isothermal titration. *Anal. Chem.* **62,** 950A–959A.
20. Breslauer, K. J., Freire, E., and Straume, M. (1992) Calorimetry: a tool for DNA and ligand–DNA studies. *Methods Enzymol.* **211,** 533–567.
21. Privalov, G., Kavina, V., Freire, E., and Privalov, P. L. (1995) Precise scanning calorimeter for studying thermal properties of biological macromolecules in dilute solution. *Anal. Biochem.* **232,** 79–85.
22. Read, C. M., Cary, P. D., Preston, N. S., Lnenicek-Allen, M., and Crane-Robinson, C. (1994) The DNA sequence specificity of HMG boxes lies in the minor wing of the structure. *EMBO J.* **13,** 5639–5646.
23. Wallace, R. B. and Miyada, C. G. (1987) Oligonucleotide probes for the screening of recombinant DNA libraries. *Methods Enzymol.* **152,** 432–442.

36

Surface Plasmon Resonance Applied to DNA–Protein Complexes

Malcolm Buckle

1. Introduction

1.1. The Relationship Between Refractive Index and Mass

Surface plasmon resonance (SPR) measures refractive index changes (Δn) at or near a surface and relates these to changes in mass at the surface (**Fig 1**). This relationship is given by the Clausius Mossotti form (**Eq. 2**) of the Debye equation (**Eq. 1**):

$$\frac{\varepsilon - 1}{\varepsilon + 2} = \frac{N}{3\varepsilon_0}\left(\frac{\alpha + \mu}{kT}\right) \quad (1)$$

$$\frac{\varepsilon - 1}{\varepsilon + 2} = \frac{N\alpha}{3\varepsilon_0} \quad (2)$$

where ε is the real part of the dielectric constant or permittivity constant related to the refractive index by $\varepsilon = n^2$, N is the number density given by $N_A\rho/M_a$ (N_A is Avogadro's number, ρ is the density and M_a is the molecular mass). It is assumed that $\Delta n/\Delta C$ is a constant.

1.2. SPR Using the BIACORE Instrument

In physical terms, the detection system of this SPR machine consists of a monochromatic, plane polarized light source, and a photodetector that are connected optically through a glass prism (**Fig. 1**). A thin gold film (50 nm thick), deposited on one side of the prism, is in contact with the sample solution. This gold film is, in turn, covered with a long-chain hydroxyalkanethiol, which forms a monolayer (approx 100 nm thick) at the surface. This layer essentially serves as an attachment point for carboxymethylated dextran chains that create a hydrophilic surface to which ligands can be covalently coupled. Light

From: *Methods in Molecular Biology, vol. 148: DNA–Protein Interactions: Principles and Protocols, 2nd ed.*
Edited by: T. Moss © Humana Press Inc., Totowa, NJ

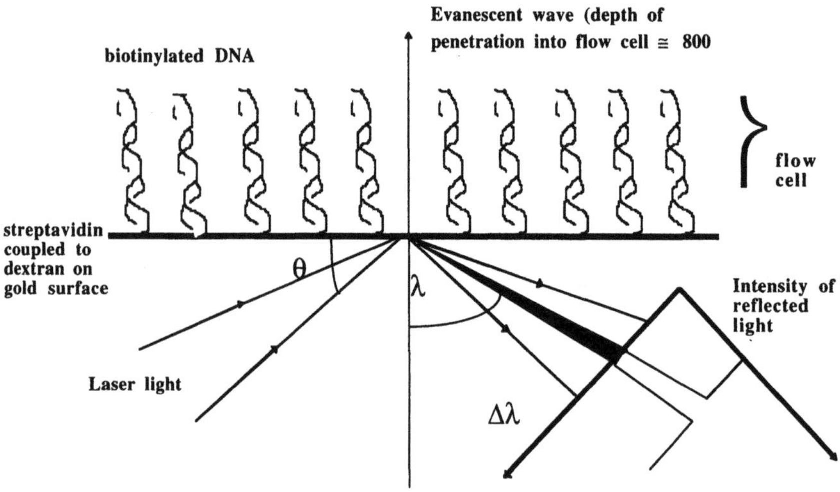

Fig. 1. Schema showing the principle of surface plasmon resonance. Light from a laser source arriving through a prism at a gold surface at the angle of total internal reflection (θ) induces a nonpropagative evanescent wave that penetrates into the flow cell opposite the prism. The intensity of the reflected light is continuously monitored. At a given angle (λ) dependent on the refractive index of the solution in the flow cell, resonance between the evanescent wave and free electrons in the gold layer results in a reduction in the intensity of reflected light. The change in angle of reduced intensity (Δλ) reflects changes in the refractive index (n) of the solution in the flow cell immediately adjacent to the gold layer. A dextran surface coupled to the gold layer allows immobilization of ligands (e.g., DNA) within the evanescent field.

incident to the back side of the metal film is totally internally reflected onto the diode-array detector. A property of this situation is that a nonpropagative evanescent wave penetrates into the solution side of the prism away from the light source. Free electrons in the gold layer enter into resonance with the evanescent wave. In fact, such resonance implies that the amplitude vector characterizing a transversal wave propagating along the gold surface ($\overrightarrow{ksp}$) is equal to the component ($\overrightarrow{kx}$) of the evanescent wave. Because $\varepsilon = n^2$, if ω is the frequency of the wave and c the speed of light, then

$$|\mathrm{ksp}| = \frac{\omega}{\mathrm{c}}\sqrt{\frac{\varepsilon_1 - \varepsilon_2}{\varepsilon_1 + \varepsilon_2}} \tag{3}$$

furthermore, given that for the evanescent wave,

$$|\mathrm{kx}| = \frac{\omega}{\mathrm{c}}\sin\theta\sqrt{\varepsilon_g} \tag{4}$$

when resonance occurs, $|ksp| = |kx|$ and the intensity of the reflected light decreases at a sharply defined angle of incidence, the SPR angle, given by the simple expression

$$\sin\theta_0 = \frac{\varepsilon_1\varepsilon_2}{\varepsilon_g(\varepsilon_1 + \varepsilon_2)} \quad (5)$$

Thus, θ_0, the SPR angle at which a decrease in the intensity of reflected light occurs, measures the refractive index of the solution in contact with the gold surface and is dependent on several instrumental parameters (e.g., the wavelength of the light source and the metal of the film). When these parameters are kept constant, the SPR angle shifts are dependent only on changes in refractive index of a thin layer adjacent to the metal surface. Any increase of material at the surface will cause a successive increase of the SPR angle, which is detected as a shift of the position of the light intensity minimum on the diode array. This change can be monitored over time, thus allowing changes in local concentration to be accurately followed. The SPR angle shifts obtained from different proteins in solution have been correlated to surface concentrations determined from radio-labeling techniques and found to be linear over a wide range of surface concentration. The instrument output, the resonance signal, is indicated in resonance units (RU); 1000 RU correspond to a 0.1° shift in the SPR angle, and for an average protein, this corresponds to a surface concentration change of about 1 ng/mm^2 (for nucleic acids, *see* **Note 1**). It is remarkable that the present instrument (Biacore 2000) can measure a deviation of 10^{-3o}, in other words, a variation of 10^{-5} in the refractive index .

1.3. Immobilization of DNA to a Surface

Although a variety of techniques exist for the immobilization of DNA on the dextran surface, the most efficient for the majority of protein–DNA interactions is the use of immobilized streptavidin that can then interact with a suitably end-labeled DNA molecule. The streptavidin is immobilized via a carbodiimide–*N*-hydroxyl succinimide coupling reaction to the carboxyl groups of the dextran (**Fig. 2**). DNA is easily obtained either by direct purchase of oligomers end-labeled with biotin, or, for larger fragments, direct polymerase chain reaction (PCR) from biotinylated oligomers. Unless a particularly unusual configuration is required, biotin is generally present at one end of the DNA molecule and on one strand if the DNA is double stranded. The end biotinylated DNA is then flowed across the surface and allowed to bind to the desired final concentration (**Fig. 3**).

1.4. Protein Binding to Immobilized DNA

The protein is flowed across the immobilized DNA in a buffer and at a temperature suitable for the interaction being studied (**Fig. 4**). A range of concen-

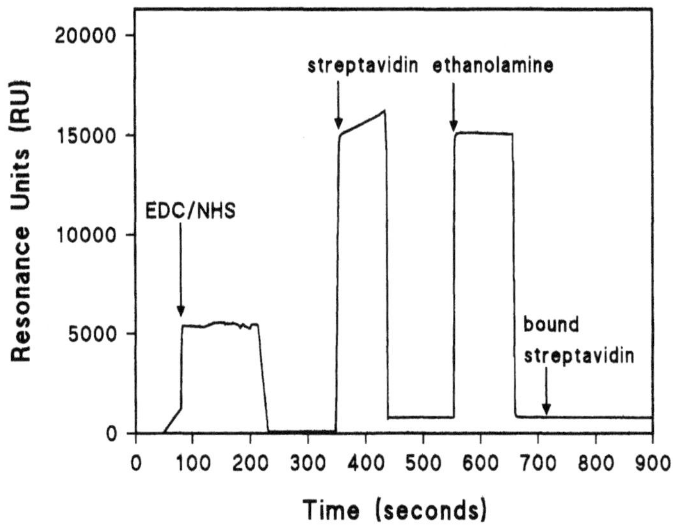

Fig. 2. Coupling of streptavidin to a flow cell activated by carbodiimide and hydroxyl succinimide. The figure shows a sensorgram of a carboxymethylated dextran surface (CM5) activated by carbodiimide and *N*-hydroxylsuccinimide prior to coupling with streptavidin. Sharp changes in the resonance units (RU) reflect bulk refractive index changes as a result of differences in the buffer. Ethanolamine is used to block all unreacted activated carboxyl groups. The difference between the final RU value and the initial value presents an accurate measure of the amount of streptavidin covalently coupled to the surface.

trations should be investigated. A flow rate in excess of 10 μL/min is advised and a sufficient contact time during the association phase to saturate the immobilized DNA (as seen by a steady-state plateau for the RU values) at protein concentrations in excess of the anticipated K_d for the interaction. The dissociation phase should also be allowed to continue for at least sufficient time to allow over a third of the complex to dissociate.

1.5. Binding Curve Analysis

1.5.1. Stoichiometry and Equilibrium Analysis

For an immobilized DNA fragment (D), the interaction with a mobile protein (P) can be written as

$$\mathrm{D} + \mathrm{P} \underset{k_d}{\overset{k_a}{\longleftrightarrow}} \mathrm{DP} \qquad \text{(Scheme 1)}$$

A classical Langmuir adsorption isotherm requires that the fraction of available sites on the DNA occupied by the protein (θ_D) be given by

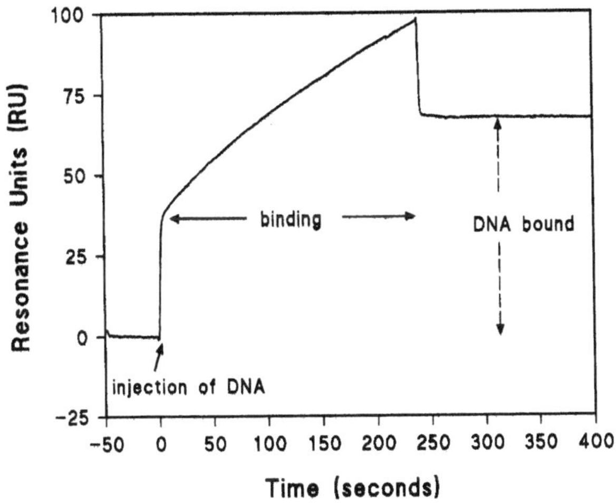

Fig. 3. Immobilization of a biotinylated DNA fragment to a streptavidin surface. In this example, a 200-bp fragment of DNA (10 µg/mL) containing a single biotin label at one 5' end was flowed at 20 µL/min in HBS buffer over the streptavidin surface. The initial bulk refractive index change was followed by a gradual increase reflecting DNA binding to the streptavidin. At the end of the DNA injection phase, the bulk refractive index change was recovered and the difference in absolute RU values compared to the initial value reflects the number of molecules of DNA now bound to the surface.

$$\theta_D = \frac{\mathrm{DP}}{\mathrm{D}_t} \tag{6}$$

Furthermore, in such a simple case, the equilibrium association constant K_a is given by the expression

$$\theta_D = \frac{\mathrm{KaP}}{1 + \mathrm{KaP}} \tag{7}$$

Thus, in an SPR experiment, the steady-state level of bound protein at a given concentration of total protein should be calculated from the asymptote of the sensorgram and the RU values converted into moles of bound protein. Assuming that in the continuous-flow system typical of Biacore SPR machines, $[P]_T = [P]$, a plot of θ_D against $[P]_T$ should allow a direct fit by **Eq. 7** to give an estimation of K_a, from which we obtain $K_d = 1/K_a$.

1.5.2. Kinetic Analysis

The protein that is injected across the surface should after an infinite time arrive at an association equilibrium giving a signal R_{eq}, and the resonance signal R at time t during this process following injection at $t = 0$ when $R = R_0$, should, in simple instances, obey the expression

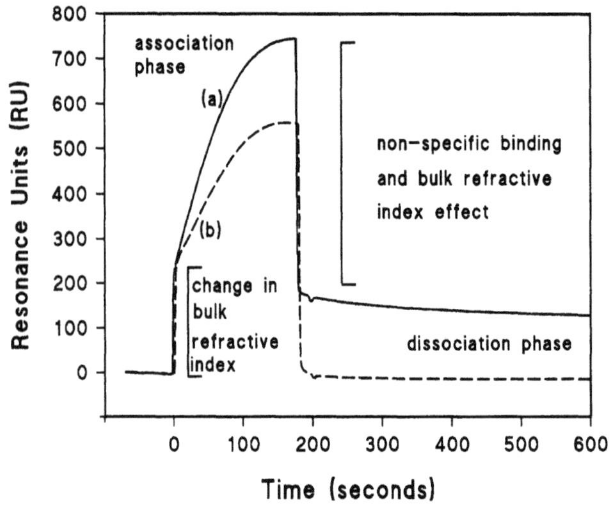

Fig. 4. Protein binding to immobilized DNA. In this example, purified RNA polymerase (120 n*M*) from *Eschericia coli* was injected at 20 μL/min across an immobilized 203-bp DNA fragment containing a promoter sequence (continuous line [a]). The dotted line (b) shows the same protein flowing across a streptavidin surface without immobilized DNA. Note the large bulk refractive index effects resulting principally from the presence of glycerol in the protein solution and nonspecific binding. This is a complex phenomenon composed not only of electrostatic interactions with the dextran but also necessary transient interactions with nonpromoter DNA. The dissociation phase is characterized by a steady decrease in signal lending itself to the type of analysis described in the text. The association phase is more complicated. In this instance an example of how the association phase may be dealt with is given in **ref. *7***.

$$R_t = R_0 - (R_{eq} - R_0)\,(1 - e^{-k_{obs}t}) \tag{8}$$

Similarly, for the dissociation of the bound protein,

$$R_t = R_0 + (R_{eq} - R_0)\,(e^{-k_{off}t}) \tag{9}$$

assuming that the bound molecule completely dissociates from the immobilized ligand. Consequently, the observed reaction rate k_{obs} for the interaction is given by

$$k_{obs} + k_{on}[P] + k_{off} \tag{10}$$

There is thus a linear relationship between the value for k_{obs} and the total concentration of protein [P]. The value for k_{obs} can be obtained from a direct fit of the association phase using **Eq. 8**, or by linear regression of a semi-log plot. It thus follows that linear regression analysis of the dependence of k_{obs} on [P] allows the calculation of k_{on} and k_{off} using **Eq. 7**. If we assume that the reaction is in fact activation controlled (were it otherwise, then the association rate would be of the order of 10^9/M/s, which is well beyond the range of current SPR devices), then

$$k_{\text{off}} = k_{\text{on}} K_d \quad (11)$$

Thus, the equilibrium dissociation constant (K_d) can be obtained from the ratio of the off and on rates.

There are many pitfalls to using SPR, which are covered in several fairly recent reviews ***(2,3)***. In the case of the Biacore instrumentation, the new data evaluation software deals with certain situations. What it cannot do is to determine the best strategy for setting up an experiment. In summary, certain important points must be taken into account even in the simple analysis given above. Several of these are covered in **Notes 2–6**.

2. Materials

1. An SPR device. In this chapter, a Biacore instrument is referred to either as the classic Biacore or the Biacore 2000. It is recommended (but not essential) that the machine be modified such that the two racks into which samples are placed in the machine are separately thermostated. Rack 1 should be thermostated to 4°C; rack 2 should be thermostated to the temperature at which the interaction is to be measured. The protocols illustrated here require rack D in the first position and rack A in the second position.
2. Streptavidin from Pierce resuspended in 0.22 µm filtered distilled water to a final concentration of 5 mg/mL. This preparation may be stored at 4°C for up to 3 mo.
3. HBS buffer: 10 m*M* HEPES, pH 7.4, 150 m*M* NaCl, 3.4 m*M* EDTA, and 0.005% Biacore surfactant.
4. *N*-Ethyl-*N*'-(diethylaminopropyl) carbodiimide (EDC) and *N*-hydroxyl succinimide (NHS) purchased from Biacore as lyophilized powders are resuspended in 0.22 µm filtered distilled water to a final concentration of 100 m*M* each.
5. 1 *M* ethanolamine hydrochloride (pH 8.5), purchased from Biacore, stored at 4°C.
6. HBS buffer: 10 m*M* HEPES, pH 7.4, 150 m*M* NaCl, 3.4 m*M* EDTA, and 0.005% surfactant P20.
7. Sensor chip surface CM5 research grade installed in the Biacore apparatus and preprimed with HBS buffer.
8. Reaction vials for the Biacore (small, plastic = 7 mm; medium, glass = 16 mm; large, glass = 2 mL) purchased from Biacore.
9. End biotinylated DNA suspended in HBS buffer to 10 µg/mL. This DNA can either be purchased directly or constructed by polymerase chain reaction (PCR) using templates and an oligomer primer carrying a biotin group (purchased from Genset for example) as one of the primers. It is advisable to gel purify or high-performance liquid chromatography (HPLC) purify the DNA prior to immobilization.

3. Methods

3.1. Immobilization of the Ligand on the Surface

3.1.1. Coupling of Streptavidin

1. Prime the apparatus with HBS buffer.
2. The thawed EDC solution, in an Eppendorf tube with the top removed, is placed in rack 1 position a1 (r1a1).
3. The thawed NHS solution in an Eppendorf tube with the top removed, is placed in r1a2.
4. Streptavidin (5 mg/mL, 50 µL), in an Eppendorf tube with the top removed, is placed in r1a3; 2 mL of filtered (0.2 µm) distilled water is placed in a large glass vial in r2f7.
5. 1 *M*, sodium acetate buffer (1 mL, pH 4.5) is placed in a large glass vial in r2f3.
6. 1 *M* ethanolamine (200 µL) is placed in a large tube in r2f4.
7. Two small clean plastic vials are placed in r2a1 and r2a2.
8. An empty large glass vial is placed in r2f5.
9. The following method is programmed into the Biacore or Biacore 2000, checked for errors, and run.

```
DEFINE APROG mixing
            FLOW                         20
            TRANSFER     r1a1 r2a1       50   !rack1a1 = EDC
            TRANSFER     r1a2 r2a1       50   !rack1a2 = NHS
            MIX          r2a1            50   !rack2a1 = EDC/NHS mix
            TRANSFER     r2f7 r2a2      200   !rack2f7 = distilled water
            TRANSFER     r2f7 r2a2      290   !rack2f7 = distilled water
            TRANSFER     r2f3 r2a2        5   !rack2f3 = 1 M acetate pH 4.5
            TRANSFER     r1a3 r2a2        5   !rack1a3 = streptavidin (5 µg/mL)
            MIX          r2a2            50
END
DEFINE APROG bind
   CAPTION activation
            FLOW                         20
       *    INJECT       r2a1            50
   -0:20    RPOINT       EDC/NHS -b
       *    INJECT       r2a2            30
   -0:20    RPOINT       streptavidin
       *    INJECT       r2f4            35   !Ethanolamine (1 M)
   -0:20    RPOINT       ethanolamine
   15:00    RPOINT       bound
            END
            MAIN
            FLOWCELL 1
            APROG        mixing
            FLOWCELL 1
            APROG        bind
END
```

(*See* **Note 3**.)

3.1.2. Immobilization of the DNA

1. Place the streptavidin-activated sensor chip surface CM5 research grade in the Biacore apparatus and preprime with HBS buffer.
2. Select a surface pretreated with streptavidin.
3. Flow HBS buffer at 20 µL/min across the surface.
4. Inject the DNA solution across the surface, set the baseline to the point of injection, and monitor the change in RU during the injection phase. Ideally, between 20 and 100 RU of DNA should be immobilized (**Fig. 3**).
5. Wash the surface with a 50-µL injection of 1 *M* NaCl in filtered (0.2 µm) distilled water.
6. Allow the surface to equilibrate in HBS buffer to a stable baseline, the difference in RU between the beginning of the injection phase and the end of the wash period reflects the amount of DNA bound. For stoichiometry, and availability of sites, *see* **Note 6**.
7. Note that values in excess of 100 RU for DNA molecules of appro 100–1000 bp are to be avoided for a number of reasons (*see* **Notes 2–6**).

3.2. Protein Binding to the Immobilized DNA

1. The protein should be prepared in the required buffer over a range of concentrations, at least two orders of magnitude on either side of the suspected K_d. The detergent P20 should be present at concentrations of around 0.005% unless it has been shown to have a deleterious effect on the interaction with the DNA.
2. Samples should be injected over the both the immobilized DNA surface and a surface that has been treated with streptavidin and ethanolamine but no DNA as a blank (**Fig. 4**).
3. The baseline should be stable with a slope inferior to 10 RU/min. If this is not the case, then check the temperature of both the apparatus and the continuous-flow buffer. If the problem persists, replace the sensor surface with an old used chip and carry out a desorb and sanitize. Re-equilibrate the immobilized DNA chip in new filtered (0.22-µm filters) buffer by running the prime command. If the problem persists, a potential reason may be degradation of the integrated fluid cartridge necessitating its replacement (see **Note 8**).
4. A typical sensorgram is shown in **Fig. 4**. Note that at low protein concentrations it is very difficult to obtain steady-state saturation levels. Note also that there is a nonspecific interaction with the control surface (curve b) and also a contribution from the bulk refractive index effect and that both of these must be taken into consideration when deducing kinetic or equilibrium values.

4. Notes

1. The relationship between mass and refractive index changes as measured by changes in the angle at which resonance occurs in this system, although theoretically available through the additive properties of molar refractivity, has been empirically established such that $10^{-1\circ}$ is equivalent to 1000 resonance units (RU)

and relates to a change in mass of 1 ng of a globular protein at the surface. Using these values, a similar relationship has been demonstrated where 0.78 ng of a nucleic acid gives the same response (1000 RU) *(4,5)*.

2. Mass transport: In instances where the association rate is particularly elevated and the diffusion rate of the nonimmobilized molecules is not especially fast, then the interaction of the free molecules with the immobilized ligand may deplete a layer of solvent immediately surrounding the immobilized ligand such that the rate-limiting step for association now becomes the rate of repletion of this layer from the bulk solvent. Two practical solutions to this problem are first to use a low immobilization density and, second, to use relatively elevated flow rates (>20 µL/min).
3. Estimation of the amplitude of the signal allowing calculation of the final equilibrium or steady-state level is often hampered by bulk refractive index changes that mask the initial and final phases of the injection period. Blank runs over free surfaces may be used to correct for this provided that the free surface reflects as close as possible the ligand surface. For example, if streptavidin is being used to immobilize the DNA, then a surface containing a comparable quantity of streptavidin to the sample surface should be used as a blank (**Fig. 4**). Incidentally, this is again an argument in favor of low levels of DNA immobilization so that the control surface is very similar in refractive properties to the surface under study. Alternatively, and indeed if possible, preferably samples being injected across the surface should be desalted into the injection buffer so as to minimize bulk refractive index effects. An ideal method is to use fast desalting columns on HPLC/FPLC systems such as the SMART (Amersham Pharmacia Biotech) which produce little dilution, are rapid and allow automatic quantification of the protein.
4. The asymptote of the binding curve should provide the stoichiometry of the reaction assuming that all the DNA molecules are available for binding. If this is uncertain, one way of ensuring that a double-stranded target DNA is accessible would be to hybridize the protein binding site *in situ* on the surface by immobilizing a single strand and then hybridizing a second homologous strand to constitute the double stranded site. The mass increase in the second hybridization step would give the number of accessible DNA molecules because it results from a successful hybridization at the surface. It goes without saying however that this requires great care in eliminating nonspecific adsorption of DNA onto the surface.
5. Steric hindrance: Let us imagine for the sake of argument that we have immobilized 1000 RU of a 100-bp double-stranded DNA molecule via biotin/streptavidin to a surface. This constitutes 1.2×10^{-14} mole of DNA or 2×10^{8} molecules. Let us further assume that these molecules are evenly and randomly distributed across the 1-mm^2 surface so a simple calculation shows that each molecule is separated from its neighbor by 70 nm. The DNA molecules are around 30 nm long, so if the dextran is itself flexible, then these DNA molecules are going to be in contact. This may produce problems such as occlusion of sites, creation of new potential binding sites, or unusual DNA structures possessing aberrant binding modes for

target proteins. Even if a relatively large molecule such as a prokaryotic RNA polymerase (500 kDa, approximate diameter-10 nm) were to bind to a single site on the DNA, then there would be every possibility that adjacent bound molecules are going to interact with complicated consequences upon binding kinetics.

6. The amplitude of each signal should be estimated manually from the steady-state signal at the end of the injection period or, if this is difficult because of a slow overall relaxation time or because of a perturbing bulk refractive index effect, from the projected R_{eq} value calculated from direct fits. Note that the projected R_{eq} values should be identical for fits of both the association and dissociation phases. It is essential that in the equilibrium analysis of these curves, the stoichiometry reflects the biology of the situation. An immobilized DNA ligand containing a single site for a protein should bind only one protein per site at saturation. If this is not the case, then no single- or double-binding site analysis is going to provide meaningful affinity values. It will be clear from careful inspection of the dissociation phase, for example, that several interactions are involved and a simple analysis cannot be made.
7. This procedure will activate surface 1. Other surfaces can be activated simply by changing the designated flow cell.
8. The Biacore machines are robust but susceptible to poorly prepared solutions. All solutions entering the fluid system must be filtered through 0.2-μm membranes and preferentially be sterile. The machine should be cleaned periodically with the desorb and sanitize protocols, and when not in use, it is recommended to maintain a continuous flow of distilled water containing 0.005% P20 at 20°C.
9. As pointed out in **ref. *6***, there are at least three consistency tests that should be applied to a given analysis: First, the equilibrium or steady-state analysis provides a dissociation constant K_d from the Langmuir isotherm derived from **Eq. 6**:

$$R_{eq}\,[\mathrm{P}] = R_0 + (R_{sat} - R_0)\,\frac{[\mathrm{P}]}{K_d + [\mathrm{P}]} \tag{12}$$

where R_{sat} corresponds to the asymptote of binding curves of the type shown in **Fig. 4**. The value obtained for K_d here should be equal to that obtained from the thermodynamic relationship described in **Eq. 11**. In cases where this is not the case neither treatment may reflect the correct situation. Second, the use of **Eq. 10** in a linear least squares analysis provides a value for k_{off}. This value should agree with that obtained from direct analysis of the dissociation phase of all sensorgrams at all concentrations of P (**Eq. 9**). Third, it should be obvious from **Eq. 10** that the value for k_{off} should be inferior to that of k_{obs} over all values of [P] because k_{off} must always be in excess of zero. Any values that do not comply to these simple tests are thus derived from an erroneous treatment of the interaction.
10. Recapture: At high levels of immobilization, when a captured ligand dissociates from the immobilized surface, it may subsequently be recruited to an adjacent molecule. The effect of this will be to decrease the numerical values ascribed to derived dissociation constants. Thus, in practice, the density of immobilization

must be adjusted so that the dissociation rate is independent of immobilized ligand density. If this proves difficult, then one can, with considerable error, extrapolate to infinite dilution. Finally, a free ligand may be included during the dissociation phase in order to calculate an affinity for the competitor and thus allow an estimation of a true dissociation constant.

References

1. Buc, H. (1998) L'utilisation de la résonance de plasmon de surface pour la détermination des constantes d'equilibre). *Regard Biochim.* **2,** 21–26.
2. Schuck, P. (1997) Reliable determination of binding affinity and kinetics using surface plasmon resonance biosensors. *Curr. Opin. Biotechnol.* **8(4),** 498–502.
3. Schuck, P. and Minton, A. P. (1996) Analysis of mass transport-limited binding kinetics in evanescent wave biosensors. *Analy. Biochem.* **240(2),** 262–272.
4. Buckle, M., et al. (1996) Real time measurements of elongation by a reverse transcriptase using surface plasmon resonance. *Proc. Natl. Acad. Sci. USA* **93(2),** 889–894.
5. Fisher, R. J., et al. (1994) Real-time DNA binding measurements of the ETS1 recombinant oncoproteins reveal significant kinetic differences between the p42 and p51 isoforms. *Protein Sci.* **3(2),** 257–266.
6. Schuck, P. and Minton, A. P. (1996) Kinetic analysis of biosensor data: elementary tests for self-consistency [see comments]. *Trends Biochem. Sci.* **21(12),** 458–460.
7. Adelman, K., et al. (1998) Stimulation of bacteriophage T4 middle transcription by the T4 proteins MotA and AsiA occurs at two distinct steps in the transcription cycle. *Proc. Natl. Acad. Sci. USA* **95,** 15,247–15,252.

37

Reconstitution of Protein–DNA Complexes for Crystallization

Rachel M. Conlin and Raymond S. Brown

1. Introduction

An increasing number of structural studies are aimed at identifying the principles that govern protein-DNA recognition in gene regulation *(1)*. This work depends on the successful reconstitution of protein–DNA complexes from their purified components. X-ray crystallography and two-dimensional nuclear magnetic resonance (NMR) techniques require large amounts of pure protein and DNA. These can be supplied through expression in bacteria of the cDNA coding for intact proteins or their smaller DNA-binding domains and the automated chemical synthesis of DNA in the laboratory.

Expression of the protein of interest is usually achieved at high levels in bacteria. An increasingly popular system is the combination of *Escherichia coli* strain BL21(DE3) transformed with a pRSET expression vector containing a strong phage promoter adjacent to the cloning site *(2)*. This particular bacterial strain has an integrated T7 RNA polymerase gene that can be induced with IPTG *(3)*.

Target DNA sequences are easily identified by DNase I footprinting of a radioactively labeled DNA restriction fragment to which the protein is bound *(4)*. Little technical difficulty is experienced in the chemical synthesis of these DNA sequences, up to about 40 base pairs in length, in amounts necessary to perform structural studies.

Considerable progress has been made in solving three-dimensional structures of protein–DNA complexes *(1)*, largely because proteins and their isolated DNA-binding domains are able to recognize and form stable complexes with quite short duplexes containing the binding sequence. Indeed, protein–DNA complexes have been crystallized that contain duplexes that are as short as 8 base pairs in length *(5)*.

From: *Methods in Molecular Biology, vol. 148: DNA–Protein Interactions: Principles and Protocols, 2nd ed.*
Edited by: T. Moss © Humana Press Inc., Totowa, NJ

In this chapter, we describe the preparation of a protein–DNA complex composed of a six-zinc-finger fragment (22 kDa) of *Xenopus laevis* transcription factor TFIIIA and a synthetic DNA duplex. The experimental strategy, reconstitution conditions, and technical problems to be discussed are probably quite similar to those encountered with most other protein–DNA complexes. In this example, the solution properties of the protein are known and the limits of the target DNA sequence are precisely defined. It has generally been assumed that association of the components takes place spontaneously to produce protein–DNA complexes of the desired molar composition. We have chosen to perform biochemical analysis in order to optimize authentic binding and ensure the efficient scaling up of reconstitution conditions.

2. Materials

1. Expression plasmid pRSET B (Invitrogen, Carlsbad, CA).
2. Buffer A: 500 m*M* NaCl, 2 m*M* benzamidine–HCl, 1 m*M* dithiothreitol (DTT), 1 m*M* NaN_3, 50% (v/v) glycerol, and 50 m*M* Tris-HCl, pH 7.5.
3. Standard protein solution: 1 mg/mL bovine serum albumin (BSA).
4. Protein assay concentrate (Bio-Rad, Richmond, CA). Dilute 1:5 with water and filter through Whatman 2^v paper.
5. 0.01% (w/v) Coomassie brilliant blue R250 in 20% (v/v) ethanol and 10% (v/v) acetic acid.
6. 1 *M* Tris-HCl, pH 8.0.
7. DEAE–Sephacel (Pharmacia, Piscataway, NJ).
8. 2 *M* sodium acetate.
9. Repelcote solution (Hopkin and Williams, Chadwell Heath, Essex, UK).
10. Millex HA and Millex HV4 (0.45 µm) filter units (Millipore, Bedford, MA).
11. Buffer B: 100 m*M* NaCl and 10 m*M* Tris-HCl, pH 8.0.
12. Amberlite MB-150 resin (Sigma, St. Louis, MO).
13. 150-mL (0.2-µm) NYL/50 filter unit (Sybron Corp., Rochester, NY).
14. Thermal cycler for polymerase chain reaction (PCR).
15. 5X binding buffer: 250 m*M* NaCl, 5 µ*M* $MgCl_2$, 5 m*M* DTT, 50 µ*M* zinc acetate, 50% (v/v) glycerol and 100 m*M* Tris-HCl, pH 7.5.
16. 1 *M* HEPES–NaOH, pH 7.5.
17. 5 *M* NaCl.
18. Buffer C: 100 m*M* NaCl, 1 m*M* DTT, 1 m*M* NaN_3, and 20 m*M* Tris-HCl, pH 7.5.
19. Collodion bags and 300-mL glass vacuum dialysis flask (Sartorius AG, Göttingen, Germany).
20. Magnetic micro flea 5-mm × 2-mm spinbar (Bel-Art products, Pequannock, NJ).
21. Microcon-10 microconcentrators (Amicon, Beverly, MA).
22. Natrix nucleic acid sparse matrix kit (Hampton Research, Lagnua Niguel, CA).
23. Linbro tissue culture multiwell plate with cover. Twenty-four flat-bottomed wells (1.7 cm × 1.6 cm) (Flow Laboratories, McLean, VA).
24. AquaSil water-soluble siliconizing fluid (Pierce, Rockford, IL).

25. DiSPo plastic cover slips M6100 (American Scientific Products, McGaw Park, IL).
26. High-vacuum grease (Dow Corning, Midland, MI).
27. A 10cc plastic B-D syringe (Becton-Dickinson, Rutherford, NJ).

3. Methods

3.1. Purification of a Recombinant DNA-Binding Protein

Standard laboratory techniques for protein purification *(6)* will not be described in detail in this chapter. Advantage is usually taken of the rather basic nature of DNA-binding proteins to isolate them from the bacterial cell extract. Few of the bacterial proteins bind so strongly to chromatography matrices, such as CM–Sepharose, Bio-Rex 70, S-Sepharose, phosphocellulose, and hydroxyapatite–Ultrogel. The protein of interest can usually be eluted with a NaCl gradient. Affinity chromatography with heparin–Sepharose, an immobilized dye–Sepharose, DNA–agarose or phenyl–Sepharose often provides an adequate final purification step.

3.1.1. Preparation of Protein

The cDNA for the 22-kDa fragment of TFIIIA (amino acids 1–190) was cloned into the vector pRSET B and expressed in strain BL21 (DE3) (*see* **Note 1**). After sonication, the cell pellet is stirred in 0.5 *M* NaCl and 7 *M* urea for 48 h at 4°C. Purification of the protein (**Fig. 1**) is carried out in 7 *M* urea on columns of Bio-Rex 70 and heparin–Sepharose (*see* **Notes 2–4**). Workup of the protein for use in DNA binding is described in detail as follows:

1. Pool those fractions that contain protein after heparin–Sepharose column chromatography.
2. Concentrate the protein to 5 mg/mL by vacuum dialysis at 4°C in a collodion bag (*see* **Note 5**).
3. Dialyze against ice-cold buffer A.
4. Store the protein at –20°C in a 1.5-mL Eppendorf tube.

3.1.2. Measurement of Protein Concentration

1. Construct a calibration curve of protein concentration versus absorbance from samples (in triplicate) containing 15, 25, 35, and 45 µL of the standard protein solution and water added to 100 µL in glass tubes (*see* **Note 6**).
2. Add 5 mL of protein assay reagent and read the absorbance at 595 nm with a spectrophotometer after the blue color has developed for 10 min.
3. Calculate the relative concentration from comparison of the values for the protein sample with the calibration curve (*see* **Notes 7** and **8**).
4. Examine the purity of the protein sample by standard sodium dodecyl sulfate (SDS)/polyacrylamide gel electrophoresis. A 5% stacking/13.5% resolving slab gel (30:0.8 acrylamide:bis) is suitable for this purpose (*see* **Note 9**). At least 0.1 µg

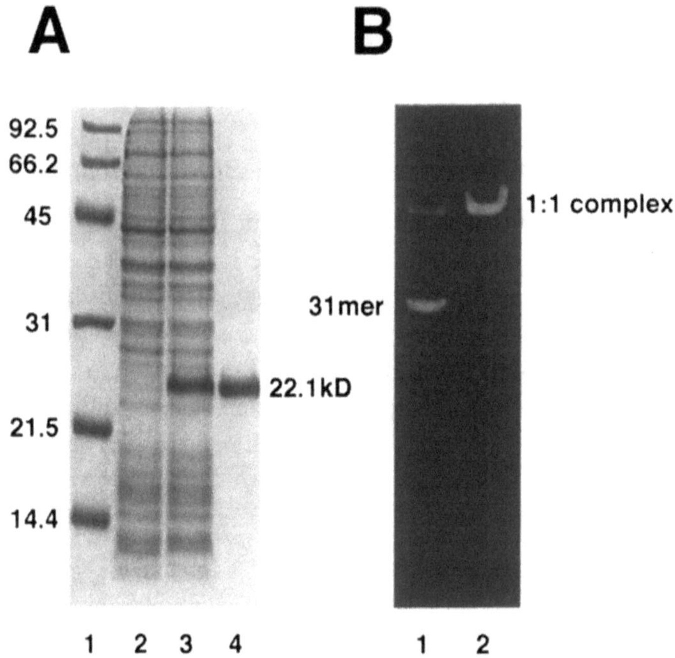

Fig. 1. **(A)** 13.5% SDS-PAGE gel showing the expression of TFIIIA (residues 1–190) in *E. coli* BL21(DE3). Dalton markers are in lane 1. Whole-cell extracts are run in lanes 2 and 3 before and after induction with 1 m*M* IPTG. The TFIIIA protein after purification is shown in lane 4. **(B)** 6.5% PAGE mobility shift gel for analysis of protein-DNA binding. One microgram of the synthetic 31-mer DNA duplex was mixed with 0.5 μg (lane 1) or 1 μg (lane 2) of the 22.1-kDa TFIIIA fragment. Bands that contain DNA are visualized by ethidium bromide stain and UV illumination.

of protein is detected in a band with Coomassie blue stain. If necessary, adjust the protein concentration by an estimate of its gel purity.

3.2. Preparation of Synthetic Oligomers

Typically a 1-μmol scale synthesis with a commercially available machine provides 1–2 mg of a purified 31-mer. The standard laboratory methods for isolation of the 5'-dimethoxytrityl full-length oligomer and subsequent chemical removal of base protecting groups will not be described here (*see* **Note 10**). Oligomers pure enough for this work can be obtained with a fast protein liquid chromatography system (FPLC) and suitable columns *(7)*.

3.2.1. Recovery and Concentration of Oligomers

1. Adjust the pH to 8.0 and apply the pooled Mono Q column fractions containing the oligomer onto a 0.5-mL DEAE-Sephacel column equilibrated with 50 m*M* Tris-HCl (pH 8.0) at room temperature.

2. Elute the bound oligomer with 5X 1 mL of 2 *M* sodium acetate into a silanized 30-mL Corex tube.
3. Add 20 mL of 95% ethanol and precipitate the oligomer overnight at –20°C.
4. Centrifuge (45 min at 7500*g*, 4°C) to recover the oligomer.
5. Wash the pellet with 25 mL of 75% ethanol to remove excess sodium acetate and centrifuge as in **step 4**. Pour off the supernatant and dry pellet under vacuum.
6. Redissolve the oligomer in 1 mL of water (*see* **Note 11**) and then filter through a Millex HV4 unit (0.45 µm).
7. Measure the absorbance at 260 nm with a spectrophotometer (dilute the oligomer as necessary with 100 m*M* NaCl). The concentration is calculated with a conversion factor of 25 A_{260} units = 1 mg. Store the oligomer at 1 mg/mL at –20°C.

3.2.2. Annealing the Duplex

1. Mix equal amounts of complementary DNA strands in 100 µL of buffer B at 1 mg/mL in 0.25-mL PCR tubes.
2. Heat at 95°C for 5 min and then slowly cool to 4°C (*see* **Note 12**).
3. Examine duplex formation by electrophoresis at 50 V in a nondenaturing 6.5% polyacrylamide slab gel at room temperature. One-tenth microgram of DNA in a gel band is easily visible after staining in ethidium bromide (1 mg/L) and illumination with ultraviolet (UV) light.

3.3. Testing the Reconstitution Conditions

The duplex is titrated with increasing amounts of protein in order to measure DNA-binding activity (**Fig. 1**). The resulting complexes are monitored by mobility shift on a nondenaturing 6.5% polyacrylamide gel (*see* **Notes 13–15**). This method is used to systematically optimize 1:1 molar complex formation with respect to incubation time, temperature, concentration of monovalent and divalent ions, pH, and the presence of glycerol and nonionic detergents like Nonidct P-40.

1. Mix 1 µg of DNA, 2 µL of 5X binding buffer, and 5 *M* NaCl (to make 0.5 *M* NaCl after **step 2**) in a 1.5-mL Eppendorf tube.
2. Add 0.5, 1, 1.5, and 2X molar excess of protein and stir in gently with the Pipetman tip (*see* **Note 16**).
3. Dilute with 10 µL of 1X binding buffer and incubate at room temperature for 15 min.
4. Apply the 20-µL samples, without tracking dyes, to a nondenaturing 6.5% polyacrylamide gel (*see* **Note 17**).
5. Load dyes (bromophenol blue and xylene cyanol FF) in adjacent tracks to indicate progress of the electrophoresis. The protein–DNA complex migrates between the dyes.
6. Stain the gel first with ethidium bromide (1 mg/L) for DNA and then with 0.01% Coomassie blue for protein.

3.3.1. Scaling-Up Reconstitution of the Complex

In low-salt conditions, the protein shows a strong tendency to aggregate. This is detected as material trapped at the top of the nondenaturing polyacrylamide gel. An excess of protein results in formation of some protein–DNA complexes with higher molar stoichiometries. We have discovered that these unwanted effects can be substantially eliminated by dilution of NaCl in the scaled-up reconstitution mix in steps; from 0.75 *M* to 0.225 *M* NaCl (*see* **Note 18**).

1. Mix together 500 µg of DNA, 200 µL of 5X binding buffer, and 5 *M* NaCl water in a 50-mL polypropylene Falcon tube (the final reconstitution mix, including the protein, is 0.75 *M* NaCl).
2. Add 625 µg of protein and mix together by gentle swirling.
3. Dilute with 1 mL of 1X binding buffer and incubate for 5 min at room temperature.
4. Add 1 mL of 1X binding buffer and incubate for 5 min.
5. Add 1 mL of 1X binding buffer and incubate for 5 min.
6. Concentrate the complex to 10 mg/mL by centrifugation 4°C with Microcon-10 microconcentrator units (Amicon).

3.3.2. Crystallization Trials

Crystallization conditions are often screened according to an incomplete factorial design *(8)*. Supplies and kits are available from Hampton Research. Some typical crystallization results, mainly employing the hanging drop/vapor diffusion method, are shown in **Table 1**. Conditions are screened at 4°C and at room temperature with small droplets (1–2 µL) of the protein–DNA complex in which the concentrations of salts, buffers, and precipitants are systematically varied by small amounts. Crystallization often occurs at or close to the point of precipitation of the protein–DNA complex.

1. Pass all solutions of salts, buffers and precipitants through 0.45 µ filters (Millex HA units) before use.
2. Blow dust from plastic pipet tips and then place 1-µL aliquots of the protein–DNA complex onto silanized plastic cover slips (*see* **Note 19**).
3. Mix with 1 µL of appropriate salts, buffer and precipitant or the well solution (*see* **Note 20**).
4. Apply vacuum grease to the upper rim of the wells of the tissue culture plate (*see* **Note 21**).
5. Into each well put 1 mL of suitable precipitant.
6. Carefully invert a cover slip without disturbing the droplet and place onto the greased rim so as to seal each well.
7. Store the tissue culture plates in closed polystyrene boxes to avoid excessive vibration and changes in temperature.
8. After a week inspect the droplets for signs of crystallization with a stereo microscope at ×50 magnification.

4. Notes

1. Bacteria grown in LB broth are induced by addition of 1 m*M* IPTG and 100 µ*M* zinc acetate for 2 h at 37°C.
2. Losses of protein during column chromatography may be decreased by addition of 10% (v/v) glycerol.

Table 1
Crystallization Conditions for Protein–DNA Complexes

Protein–DNA Complex	Method	Well solution
Sso7d: 8-mer, Gao et al. ***(5)***	DNA and protein in 2 m*M* Tris-HCl, pH 6.5, and 2.6% PEG 400 at room temperature	15% PEG 400
NF-κB p50/p65: 12-mer, Chen et al. ***(9)***	Final conditions are 50 m*M* sodium acetate (pH 5.5) 100 m*M* $CaCl_2$, 0.125% β-octyl-glucoside, 1 m*M* spermine, 10 m*M* DTT, and 8% PEG 3350 at 18°C	
Antp homeodomain: 15-mer, Fraenkel and Pabo ***(10)***	Mix protein and DNA in 5 m*M bis*-tris–propane (pH 7.0) with 1 vol of well solution at room temperature	1–5% MPD 20 m*M bis*-tris–propane (pH 7.0), and10 m*M* $NiCl_2$
Stat3beta: 18-mer glycerol, Becker et al. ***(11)***	DNA and protein in 20 m*M* HEPES (pH 7.0) 200 m*M* NaCl, 10 m*M* $MgCl_2$, 5 m*M* DTT, and 0.5 m*M* PMSF. Mix with 1 volume of well solution at room temperature	10% v/v glycerol, 0.1–0.4 *M* NaCl. 5 m*M* $MgSO_4$, 50 m*M* MES, pH 5.6–6.0, 0.1 *M* ammonium aceta te
NFAT/AP–1: 20-mer, Chen et al. ***(12)***	Final conditions are 10 m*M* HEPES (pH 7.5), 1 m*M* DTT, 100 m*M* NaCl, 20% (v/v) glycerol, 500 m*M* ammonium acetate	300 m*M* ammonium acetate
GABPalpha/beta: 21-mer, Batchelor et al. ***(13)***	Mix protein and DNA in 1 m*M* EDTA, 1 m*M* DTT, 20 m*M* Tris-HCl, pH 8.0, and 0.001% NaN_3 with 1 vol of the well solution at 20°C	100 m*M bis*-tris–propane, pH9.0, 5 m*M* cobaltic hexammine chloride, and 9% PEG 1000
Topo I: 22-mer Redinbo et al. ***(14)***	Mix 1μL of DNA in 6 m*M* NaCl with 2 μL of protein in 10 m*M* Tris-HCl (pH 7.5), 1 m*M* EDTA, 5 m*M* DTT, and 3 μL H_2O. Add to 3 μL of well solution at 22°C	100 m*M* $MgCl_2$, 10 m*M* DTT, 100 m*M* Tris-HCl (pH 7.7), and 24% PEG 400
T7 DNA polymerase: 21-mer + 26-mer Doublie et al. ***(15)***	Microseeding	100 m*M* ACES, pH 7.5, 30 m*M* $MgCl_2$, 120 m*M* ammonium sulfate, 5 m*M* DTT, and 12% PEG 8000
TFIIIA (residues 1–190): 31-mer, Nolte et al. ***(16)***	Final conditions are 165 m*M* NaCl, 35 m*M* sodium acetate, 3.2 m*M* DTT, 9.2% (v/v) glycerol, 1.8 m*M* NaN_3, 1.8 m*M* cadaverine–2HCl, 5.5 m*M* Tris-HCl (pH 8.0), and 22.5% PEG 4000 at 18°C	
DtxR repressor: 33-mer, White et al. ***(17)***	Mix 2μL of protein and DNA in 1 m*M* $NiCl_2$, and 100 m*M* Tris-HCl, pH 7.5 with 1 μL well solution at room temperature	6–12% PEG 4000, 10 m*M* $MgCl_2$, and 100 m*M* MES (pH 6.0)

Abbreviations: ACES: 2-[(2-Amino–2-oxoethly)amino]ethanesulfonic acid; DTT: dithiothreitol; EDTA: disodium ethylenediaminetetraacetic acid; HEPES: (*N*-[2-hydroxyethyl]piperazine-*N*′-[2-ethanesulfonic acid]); MES: 2-[*N*-morpholino]ethane-sulfonic acid; MPD: 2-methyl-2, 4-pentanediol; PEG: polyethylene glycol; PMSF: phenylmethylsulfonyl flluoride.

3. Customized gradients can be supplied to standard laboratory chromatography columns using a programmed FPLC control unit and pumps. Flow rates of 1–50 mL/h are possible without significant back pressure.
4. The protein binds to heparin–Sepharose in 0.25 *M* NaCl and 7 *M* urea. A reverse gradient is applied to remove urea. The protein is eluted in a 0.25 *M* NaCl to 2 *M* NaCl gradient and 10 µ*M* zinc acetate.
5. The collodion bag is pre-equilibrated in buffer and contains a small magnetic stirrer (5 mm × 2 mm) to aid dialysis and recovery of the protein.
6. This assay can be used to measure protein concentrations between 50 µg/mL and 50 mg/mL according to the manufacturer's directions.
7. Data may be conveniently analyzed by linear regression using commercial software like KaleidaGraph.
8. The protein concentration obtained by the Bradford colorimetric assay is relative. An absolute value requires multiplication by a conversion factor. This factor can be derived from simultaneous amino acid analysis or microKjeldahl nitrogen determination ***(18)***. In the case of the TFIIIA 22-kDa fragment, the deduced concentration of BSA is multiplied by 0.39.
9. TFIIIA and its fragments have anomalous electrophoretic mobilities. Reduction and carboxymethylation with 50 m*M* iodoacetamide restores the predicted gel mobility to the intact protein. Fragments containing zinc fingers generally run slower than expected after this treatment.
10. Chemical detritylation can be efficiently performed by passing 0.5% trifluoroacetic acid for 3 min at room temperature through a reverse-phase ProRPC HR column (Pharmacia). Following deprotection of the bases (30% [w/v] NH_4OH, 16 h at 65°C), the oligomer is purified with a Mono Q HR column (Pharmacia) and eluted in a NaCl gradient with 7 *M* urea according to the manufacturer's instructions.
11. After the addition of water, the Corex tube is sealed with parafilm to avoid spillage. The water is rolled around the walls of the silanized tube as well as on the pellet.
12. This is conveniently done in a thermal cycler PCR machine at a linear cooling rate of 1°C/3 min.
13. A stock acrylamide solution (30:0.8 acrylamide:*bis*) is deionized with Amberlite MB–150 resin (10 g/100 mL) for 1 h, filtered through a NYL/50 filter unit (0.2 µm), and stored at 4°C. This treatment minimizes the sequestering of the protein from protein–DNA complexes during gel electrophoresis.
14. Glass plates, combs, spacers and the gel apparatus must be cleaned and washed completely free of detergent to avoid disruption of the protein–DNA complex.
15. The polyacrylamide gel is 1.5 mm thick and contains 50 m*M* NaCl, 40 m*M* HEPES–NaOH (pH 7.5), and 5% (v/v) glycerol. The running buffer consists of 50 m*M* NaCl, 20 m*M* HEPES–NaOH (pH 7.5).
16. A duplex with an unrelated sequence of the same length may be used to monitor the level of nonspecific protein binding. The affinity of the protein for each of the single strands of the duplex can also be tested.
17. Samples can be applied smoothly to the gel by slowly winding the Pipetman volume control back to zero.

18. At high salt concentration, above 0.65 *M* NaCl, the protein–DNA complex is dissociated and both components are soluble. The salt concentration is lowered stepwise to 0.225 *M* NaCl to reconstitute the complex ***(19)***. At the same time, the protein is diluted and becomes less likely to aggregate or bind nonspecifically to the duplex.
19. Wash dust off the cover slips with deionized water and dry in a suitable rack at 50°C in an oven. Immerse cover slips for 15 min in dilute AquaSil (1:40) and leave to dry overnight. Wash in deionized water and dry at 50°C. Keep in a closed container to avoid contact with dust.
20. It is customary to equilibrate against a well solution that contains double the concentration of the droplet components. One volume of the well solution containing appropriate salts, buffers, and precipitant is added to the droplet of the protein–DNA complex.
21. The grease can be applied accurately to the rim with a 10cc plastic syringe filled with high-vacuum grease.

References

1. Steitz, T. A. (1990) Structural studies of protein-nucleic acid interaction: the sources of sequence-specific binding. *Quart. Rev. Biophys.* **23,** 205–280.
2. Dubendorff, J. W., and Studier, F. W. (1991) Controlling basal expression in an inducible T7 expression system by blocking the target T7 promoter with lac repressor. *J. Mol. Biol.* **219,** 45–59.
3. Studier, F. W., Rosenberg, A. G., Dunn, J. J., and Dubendorff, J. W. (1990) Use of T7 RNA polymerase to direct the expression of cloned genes. *Methods Enzymol.* **85,** 60–89.
4. Engelke, D. R., Ng, S.-Y., Shastry, B. S., and Roeder, R. G. (1980) Specific interaction of a purified transcription factor with an internal control regionof 5S RNA genes. *Cell* **19,** 717–728.
5. Gao, Y. G., Su, S. Y., Robinson, H., Padmanabhan, S., Lim, L., MacCrary, B. S., et al. (1998) The crystal structure of the hyperthermophile chromosomal protein Sso7d bound to DNA. *Nature Struct. Biol.* **5,** 782–786.
6. Deutscher, M. P. (ed.) (1990) *Guide to Protein Purification*, Methods in Enzymology, *Academic, New York, p.* 182.
7. Oliver, R. W. A. (ed.) (1989) *HPLC of macromolecules: a practical approach.* IRL, Oxford.
8. Carter, C. W.,Jr. (1992) *Crystallization of Nucleic Acids and Proteins: A Practical Approach* (Ducruix, A. and Giege, R., eds.), IRL, NY, pp. 47–71.
9. Chen, F. E., Huang, D.-B., Chen, Y.-Q., and Ghosh, G. (1998) Crystal structure of p50/p65 heterodimer. *Nature* **391,** 410–413.
10. Fraenkel, E. and Pabo, C. O. (1998) Comparison of X-ray and NMR structure for the Antennapedia homodomain–DNA complex. *Nature Struct. Biol.* **5,** 692–696.
11. Becker, S., Groner, B., and Muller, C. W. (1998) Three-dimensional structure of the Stat3beta homodimer bound to DNA. *Nature* **394,** 145–151.

12. Chen, L., Glover, J. N. M., Hogan, P. G., Rao, A., and Harrison, S. C. (1998) Structure of the DNA-binding domains from NFAT, Fos and Jun bound specifically to DNA. *Nature* **392,** 42–48.
13. Batchelor, A. H., Piper, D. E., de la Brousse, F. C., McKnight, S. L., and Wolberger, C. (1998) The structure of GABPalpha/beta: an ETS domain-ankyrin repeat heterodimer bound to DNA. *Science* **279,** 1037–1041.
14. Redinbo, M. R., Stewart, L., Kuhn, P., Champoux, J. J., and Hol, W. G. J. (1998) Crystal structures of human topoisomerase I in covalent and noncovalent complexes with DNA. *Science* **279,** 1504–1513.
15. Doublie, S., Tabor, S., Long, A. M., Richardson, C. C., and Ellenberger, T. (1998) Crystal structure of a bacteriophage T7 DNA replication complex at 2. 2Å resolution. *Nature* **391,** 251–258.
16. Nolte, R. T, Conlin, R. M., Harrison, S. C., and Brown, R. S. (1998) Differing roles for zinc fingers in DNA recognition: structure of a six-finger transcription factor IIIA complex. *Proc. Natl. Acad. Sci. USA* **95,** 2938–2943.
17. White, A., Ding, X., vanderSpek, J. C., Murphy, J. R., and Ringe, D. (1998) Structure of the metal-ion-activated diphtheria toxin repressor/tox operator complex. *Nature* **394**, 502–506.
18. Jaenicke, L. (1974) A rapid micromethod for the determination of nitrogen and phosphate in biological material. *Anal. Biochem.* **61,** 623–627.
19. Zwieb, C. and Brown, R. S. (1990) Absence of substantial bending in *Xenopus laevis* transcription factor IIIA-DNA complexes. *Nucleic Acids Res.* **18,** 583–587.

38

Two-Dimensional Crystallization of Soluble Protein Complexes

Patrick Schultz, Nicolas Bischler, and Luc Lebeau

1. Introduction

Structural data on biological macromolecules provide invaluable insights into the interactions of proteins with nucleic acids. Data obtained at atomic resolution by X-ray diffraction or nuclear magnetic resonance (NMR) studies ultimately describes the exact folding of polypeptide chains and the contacts between proteins and DNA. However, many complexes are difficult to analyze at atomic resolution because either they are too large or too difficult to crystallize in three dimensions (3-D). Recent progress in electron microscopy, specimen preservation, and image processing provides the possibility to calculate detailed molecular envelopes that are complementary to X-ray crystallography with little theoretical limit on specimen size. In the case of membrane proteins organized into one-molecule-thick two-dimensional (2-D) arrays, atomic models could be elaborated from electron microscopy data *(1,2)*. It is beyond the scope of this chapter to describe in detail the electron crystallographic methods and computer image analysis required to calculate a 3-D model of a crystallized protein complex; these aspects are described elsewhere *(3)*. Here, we will focus on the formation of 2-D crystals of soluble proteins, an essential preliminary step in high-resolution structure determination by electron microscopy, and provide the necessary information to set up, screen, and evaluate crystallization experiments.

The remarkable achievements in the field of membrane protein 2-D crystals prompted the pioneering work of Kornberg and collaborators aimed at transposing the crystallization mechanisms occurring in a lipid bilayer to soluble proteins *(4)*. The method consists in targeting the protein of interest to a lipid surface self-assembled as a monomolecular film at an air–water interface.

From: *Methods in Molecular Biology, vol. 148: DNA–Protein Interactions: Principles and Protocols, 2nd ed.*
Edited by: T. Moss © Humana Press Inc., Totowa, NJ

The buffer-exposed hydrophilic part of the lipid molecule may carry charged groups *(5,6)* or be chemically modified *(7,8)* in order to interact with the protein. Consequently, the protein is concentrated at the lipid plane and adopts only a few orientations relatively to it. If the lipid molecules are free to diffuse in the monolayer plane, the system permits the 2-D crystallization of the macromolecule. This achieved, the lipid–protein film is transferred onto a solid support and processed for electron microscopy observations. As was demonstrated by the study of streptavidin 2-D crystals *(9)*, which revealed structural information down to 3 Å in projection, such an approach can yield high-resolution structural data.

Two categories of lipid derivatives can be used according to the specificity of the interaction to be established with the protein complex *(10)*. On the one hand, a charged surface can be created by using lipids with positive or negative charges that will interact with the surface potential of the protein to be crystallized *(5)*. The characteristic behavior of the protein in ion-exchange chromatography may give a hint as to which type of charged lipids should be used. Typically used lipids are phosphatidyl serine which carries a negative charge, and stearyl amine or alkyl trimethyl ammonium, both of which are positively charged. On the other hand the lipid film may be derivatized by a specific ligand recognized by the protein of interest. The ligand is chemically grafted to the lipid moiety through a linker whose length modulates the accessibility for the protein. The grafted molecule can be a natural ligand of the protein such as dinitrophenol for specific anti-DNP antibodies *(4)*, novobiocin for gyrase B subunit *(8)* (**Fig. 1**), or biotin for streptavidin *(11)*. More recently lipid molecules have been designed to interact with specific tags (such as polyhistidines) introduced genetically into the sequence of the protein of interest *(12,13)*. The polar head group of the lipid molecule carries a nitrilotriacetate moiety, which chelates nickel ions and interacts with the histidine tag.

Stability and fluidity of the spread lipid layer are mainly provided by the hydrophobic part of the lipid (**Fig. 1**). The lipid layer has to be in the fluid phase at the incubation temperature because crystallization of the lipid chains was shown to prevent protein organization probably by lowering its 2-D mobility *(11,13)*. In most cases, a cis double bond in the alkyl chains provides sufficient fluidity. In addition, the lateral cohesion of the lipid molecules has to be strong enough to allow the spreading of a stable monolayer and prevent the solubilization of the lipid as micelles or liposomes. A double alkyl chain containing 18 carbon atoms (dioleyl) generally fulfills these requirements.

When protein–lipid interactions occur, the proteins are rapidly concentrated close to the lipid layer and are likely to be partly oriented. In the case of a specific interaction with a functionalized lipid, the macromolecules are tethered to the lipid film by a unique site and the extent of orientation is deter-

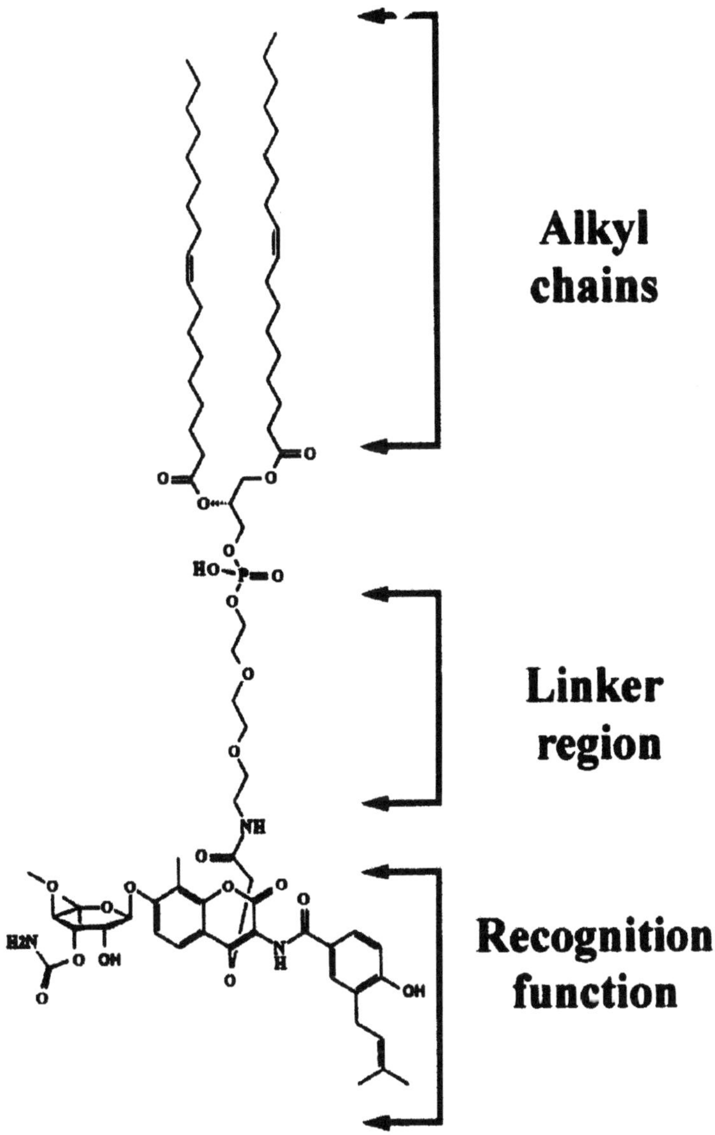

Fig. 1. Schematic representation of a lipid molecule used for 2-D crystallization of the b-subunit of DNA gyrase. The hydrophobic alkyl chain with a cis double bond confers fluidity and stability to the spread lipid layer. The linker region provides accessibility of the ligand for the protein of interest. The recognition function, here a novobiocin molecule, determines the interaction properties between the lipid and the protein.

mined by the length and flexibility of the linker region *(8)* (**Fig. 1**). Increased concentration, possible preferential orientation, and in-plane mobility facilitate contacts between macromolecules, which results in their increased organization when an ordered network of interactions is established. However, a dead end may be reached if the macromolecules interact too strongly or too rapidly with each other. As a consequence, 2-D aggregates can form, which appear as close-packed, noncrystalline assemblies.

The method requires limited amounts of protein because a single experiment may only need 300 ng of protein. However, a quantity of 1 mg is more realistic because any new project implies systematic trials starting without *a priori* knowledge of the factors affecting the lipid–protein interactions. When a specific interaction of the protein with the lipid is involved, the degree of purity of the biological sample appears to be of less crucial importance than when charged lipid are used ***(13,14)***.

2. Materials

1. A suitable lipid to interact with the protein of interest.
2. Teflon® or Nylon® blocks in which cylindrical wells 4 mm in diameter and 1 mm deep have been milled such that each well can contain about 15 µL of aqueous solution (**Fig. 2**).
3. Standard Cu or Cu/Rh 300 mesh electron microscopy grids.
4. Mica sheets 2.5 × 9 cm in size.
5. A carbon evaporator.
6. A 2% uranyl acetate solution.
7. A control electron microscope.
8. An optical diffraction bench.

3. Methods

3.1. Preparation of Electron Microscopy Supports

The lipid–protein assemblies will have to be transferred onto a standard electron microscopy grid. The grids need to be coated with a thin hydrophobic carbon film to support the assemblies and to allow the adsorption of the hydrophobic lipid alkyl chains.

1. The mica sheets are freshly cleaved to create a clean and flat surface.
2. The mica sheets are placed into a carbon evaporator and a 10 to 50-nm thick carbon film is evaporated under vacuum onto the cleaved face of the sheets.
3. Electron microscopy grids are placed on a supporting filter paper below the surface of a water bath. The size of the filter paper matches that of the mica sheet and will hold a total of 75–100 grids.
4. The carbon foil is floated at the clean air–water interface by dipping the mica sheet in the water bath with an angle of about 30°.
5. Finally, the carbon foil is gently lowered onto the grids by removing the water using a vacuum pump.

It was observed with some systems, such as streptavidin, that the contact of the lipid layer with the carbon foil interferes with the quality of protein arrays

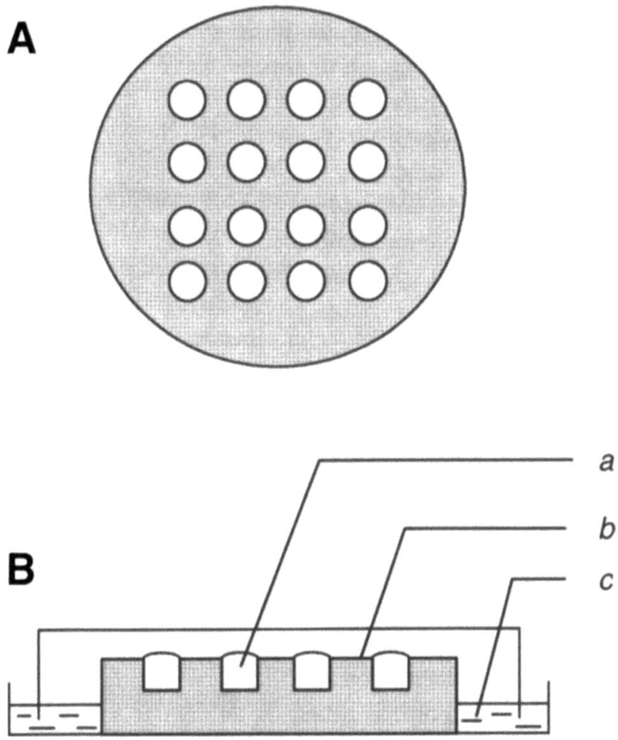

Fig. 2. Design of a Teflon® block for 2-D crystallization experiments. **(A)** A Teflon® cylinder 4 cm in diameter truncated into 1 cm-thick slices and 16 wells, 4 mm in diameter and 1 mm deep, are milled into the block such that each well can contain about 15 µL of solution. **(B)** During the crystallization experiments performed in the wells (a), the Teflon® block (b) is placed into a humid chamber consisting of a reverted Petri dish containing some buffer (c).

(15). "Holey" carbon grids can then be used to transfer the crystals without interactions with a carbon surface, the lipid–protein layer being spread over the holes. A protocol to prepare "holey" carbon films is as follows ***(16)***:

1. The surface of an optical microscope glass slide is extensively cleaned by boiling in an aqueous detergent solution and extensive rinsing with demineralized water (H_2Od).
2. The slide is immersed in a 0.1% Triton X405 (Sigma) solution for 30 min, briefly rinsed with H_2Od to remove the excess of detergent and left to dry. This will result in a clean hydrophobic surface.
3. The slide is placed on a precooled aluminum block to allow minute water droplets to form on the surface by condensation. The size of the droplets depends on the humidity of the room and on the condensation time.

4. One mL of a cellulose acetate or cellulose butyrate solution (0.4% [w/v] in ethyl acetate) is poured with a pipet over the surface, excess solution is removed from one end of the slide by touching a filter paper and left to dry. Upon drying, the cellulose forms a thin film around the water droplets, thus forming holes.
5. At this point, an optical microscope can be used to check the size of the holes and their distribution.
6. The slide is then immersed 30 min in a 0.5% (w/v) sodium dodecyl sulfosuccinate solution to peal off the cellulose film.
7. Electron microscopy grids are deposited on a supporting filter paper just below the surface of a water bath.
8. The holey cellulose film is floated on the clean air–water interface and deposited onto the grids.
9. A 50-nm-thick carbon film is evaporated onto the cellulose-coated grids.
10. The cellulose film is finally dissolved by placing the grids on a ethyl acetate-soaked filter paper.

3.2. Crystallization Experiments

1. The lipids are best stored as a dry powder under an argon atmosphere at –80°C. A mother solution at a concentration of 10 mg/mL is produced by solubilizing the lipids in an organic solvent such as a 1:1 mixture of chloroform:hexane. This solution can be stored under argon up to 1 yr at –20°C. The working lipid solution is at a concentration of 0.5–1 mg/L in an organic solvent. All solution are stored in 2 mL glass vials with Teflon® caps to prevent solvent evaporation.
2. The Teflon® wells have to be cleaned prior to use for crystallization experiments to remove residues of proteins or lipids. The Teflon® support should be dipped into a sulfochromic acid solution for 1 h, rinsed 10 times with H_2Od, dipped for 1 h into H_2Od, and rinsed again three times with H_2Od. Alternatively, the support can be rinsed 10 times with methanol to eliminate proteins, 10 times with a chloroform:methanol 2:1 or hexane:methanol 9:1 solution to remove lipids and rinsed 10 times with hexane to remove fatty acids. Finally, the support is brought into contact with a filter paper to remove the excess of H_2Od or organic solvent, without wiping to avoid electrostatic charging, and is allowed to dry in a dust-free chamber.
3. Incubations are performed in a humid chamber to prevent buffer evaporation (**Fig. 2B**). The Teflon® block is placed in a reverted Petri dish containing some buffer with an opening in the top to let air in and out during removal of the lid.
4. In each well, 10 µL of buffer are added (**Fig. 3B**).
5. To spread the lipid at the air-buffer interface, 1 µL of the lipid solution at a concentration of 0.5–1 mg/mL is placed on the top of the drop of buffer with a micropipet. At this moment it can be observed that the surface tension of the drop is released (**Fig. 3C**).
6. The organic solvent is allowed to evaporate for 5 min.
7. The protein solution (5 µL) is injected into the aqueous phase (**Fig. 3D**). The final protein concentration is generally set between 20 and 200 µg/mL.

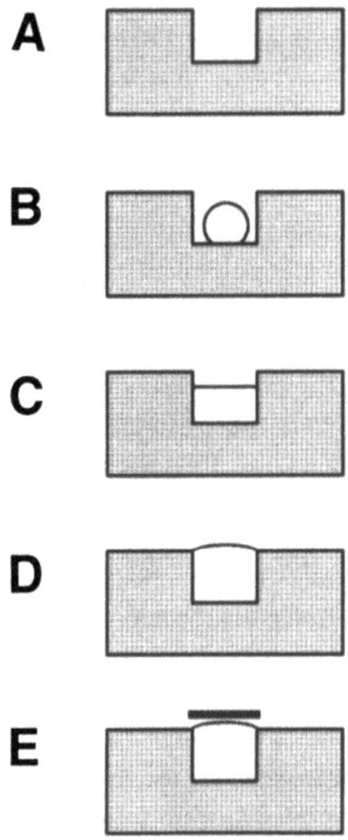

Fig. 3. Setup of a crystallization experiment. In each well (**A**), a 10-µL drop of buffer is placed (**B**). Because the Teflon well is hydrophobic, the drop does not wet the surface. Upon the addition of 1 µL of the lipid solution at a concentration of 0.5–1 mg/mL, the surface tension of the drop is released (**C**). After evaporation of the organic solvent, 5 µL of the protein solution is injected into the well (**D**) and is allowed to interact with the lipid layer. The resulting lipid–protein assembly is transferred to a carbon-coated electron microscopy grid placed on the top of the drop (**E**).

8. The incubation chamber is closed, and if oxidation is a problem air is replaced by argon.
9. The incubation time will vary from one system to another but is generally in the range of 1–36 h. Most experiments can be performed at room temperature, but longer incubation times at 4°C may improve crystal quality in some cases.

3.3. Electron Microscopy

1. The 2-D crystal is transferred to the electron microscopy grid through hydrophobic contacts between the lipid chains and the carbon foil (**Fig. 3E**). This is simply done by placing the grid over the well for 1–2 min. The grid is then withdrawn and prepared for observation (*see* **Notes 1** and **2**).
2. To be visualized by electron microscopy, the specimen has to be contrasted by creating a mould of heavy atoms around the proteins, a process named negative

staining. The transferred specimen placed on the grid held by forceps is washed with a drop of buffer that is quickly removed. The buffer is then replaced by a drop of a 1–2% aqueous solution of uranyl acetate and the grid is dried by touching a piece of filter paper with the edge of the grid (*see* **Notes 3** and **4**).

3. The crystallization experiments have then to be evaluated in terms of protein concentration on the lipid layer and degree of organization (*see* **Notes 11** and **12**). When the specimen is large enough (> 50 kDa), individual molecules can be identified visually during electron microscopy inspection. To ascertain that the specimen is specifically bound to the lipid layer and not in a nonspecific way to the carbon film, it is useful to locate breaks in the lipid layer in order to observe a difference in binding efficiency with the underlying carbon film. A frequently observed intermediate step in specimen ordering is the formation of symmetry-related oligomers that arise when a particular set of protein–protein interactions is energetically favored. The formation of oligomers probably favors further organization because interactions between symmetry related surfaces will propagate forming linear polymers or 2-D crystals. Once larger crystalline areas are obtained, electron micrographs are recorded and the extent of order is evaluated by optical diffraction.

3.4. Feedback Loops

1. If the protein is not concentrated on the lipid film, it is advisable to act on the lipid region involved in protein recognition, on its environment, or on the buffer composition. In the case of a specific lipid, the linker may be too short to allow the ligand to be recognized by the protein. Alternatively, the surface potential created by the lipid layer may have a repulsive effect on the protein and it may be of importance to modify the environment by the addition of a dilution lipid. Finally, the ionic strength of the buffer may be too high and screen electrostatic interactions between the protein and the charged lipid (*see* **Note 5**).
2. When the protein tends to form close-packed arrays, which do not evolve toward organized protein patches, it is advisable to reduce the kinetics of protein concentration either by increasing the viscosity of the medium by adding glycerol (up to 40%) or by reducing the temperature or the protein concentration. The specific or charged lipid can also be diluted with a neutral lipid to reduce the local concentration of ligand or the charge density of the surface (*see* **Note 10**).
3. The experiment has to be evaluated further in terms of macromolecular organization. Higher degrees of order are recognized visually during the electron microscopy inspection of the specimen by the appearance of patches of ordered arrays (**Fig. 4A**). Once larger crystalline areas are obtained, electron micrographs are recorded and the extent of order is evaluated by optical diffraction. A large number of parameters such as the pH, the ionic strength, the buffer composition, the presence of divalent cations, the protein concentration, the presence of glycerol, the incubation temperature, or the incubation time can be modified to improve crystal order (*see* **Note 6**). At this stage, the homogeneity of the specimen suspension may be crucial.

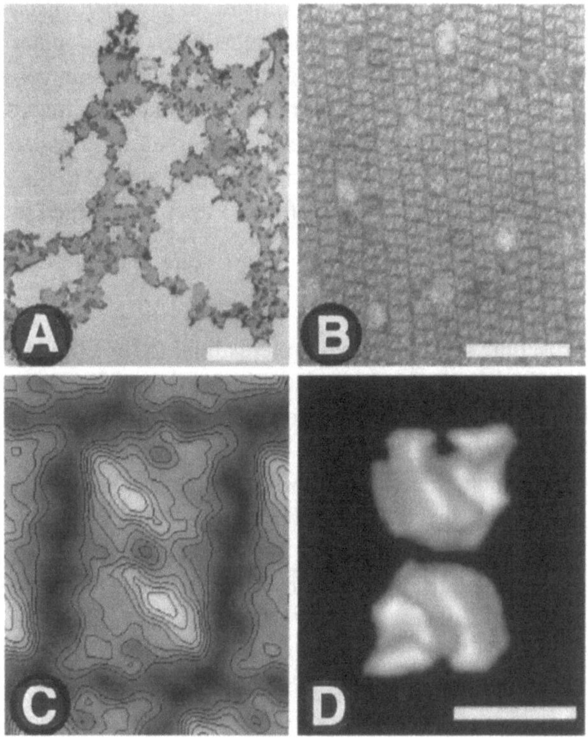

Fig. 4. Evaluation and exploitation of a 2-D crystallization experiment. Histidine-tagged yeast RNA polymerase I was incubated with nickel chelating lipids. **(A)** Low-magnification electron microscopy image showing the organization of the protein complex into domains. The bar represents 5 µm. **(B)** A higher magnification reveals ordered RNA polymerase arrays. The bar represents 50 nm. **(C)** A noise-free image is obtained by averaging multiple molecular images. The stain excluding protein densities are in white and represented as lines of equal densities. **(D)** A 3-D model of the protein complex can be calculated by combining several views of the macromolecule obtained by tilting the crystals in the microscope. The bar represents 10 nm in **(C)** and **(D)**.

4. In the case of streptavidin, the method for preparing the sample for electron microscopy and, in particular, the transfer mode proved to be essential to recover a large number of highly ordered crystals *(15)*. More generally, the manipulation of one-molecule-thick assemblies during transfer to the electron microscope is likely to introduce at least some of the defects observed in 2-D crystals such as rotational and translational distortions, fragmentation, and other forms of disorders (*see* **Notes 7–9**).
5. An improvement of the interpretable resolution once the specimen diffracts to about 0.5 nm^{-1} will probably need a change in the method of specimen preservation from negatively stained to frozen hydrated samples *(17)*.

6. To calculate a 3-D model, astigmatism-free and well-focused images of the oriented or crystallized macromolecules must be recorded under minimal exposure conditions and generally at low temperature. These images are analyzed to calculate a noise-free image representing a projection of the macromolecular densities (**Fig. 4C**). Because the particles are adsorbed on a planar surface, tilted views are then recorded to recover the information normal to the lipid plane. The images are then processed and the different views are merged into a 3-D model *(3)* (**Fig. 4D**).

4. Notes

1. A good macroscopic indication that proteins bind to the lipid layers and that the transfer is efficient is obtained by visual inspection of the carbon surface after transfer. The original hydrophobic grid becomes hydrophilic, as assessed by the change in its wetting properties.
2. Storage of the carbon-coated grids in hexane atmosphere may provide higher reproducibility in the specimen transfer step by preventing adsorption of contaminating material.
3. Do not use phosphate buffers or buffers with high ionic strength, which precipitate uranyl salts.
4. Other heavy metal solutions can be used for negative staining such as sodium phosphotungstate or ammonium molybdate.
5. It is useful to check the specificity of the protein–lipid interactions. In the case of charged lipids, the protein binding should be reduced by increasing the ionic strength. In the case of functionalized lipids, the amount of transferred protein should diminish by adding some competing ligand in solution. Note that in the case of nickel-chelating lipids, it was observed that addition of small amounts of imidazole prevented the nonspecific aggregation of the protein and allowed the selection of the specific interaction with the polyhistidine tag *(13)*.
6. Detergents should be avoided in the incubation buffer because they may solubilize the lipid layer.
7. In some cases, it was observed that the grid side on which the carbon foil was deposited affected crystal transfer *(13)*. This effect may be related to the surface roughness of the carbon foil and of the grid *(18)*.
8. To strengthen the crystals, 1 μL of a 0.5% glutaraldehyde solution can be added to the incubation drop before placing the electron microscopy grid in order to crosslink the specimen.
9. Another method of specimen transfer is the loop method *(19)*. A loop is formed with a thin Pt/Pd wire (0.075 mm in diameter). The inside diameter must be slightly larger than the outside diameter of the electron microscopy grid. The loop is then lowered onto the drop, the entire loop makes contact with the drop surface at the same time. This can be observed through drop deformation. So that no excess subphase is picked up, the loop should not go through the monolayer and into the subphase. The loop is then gently and carefully raised and lowered onto a glow-discharged grid. The grid is held with forceps and is parallel to the

film in the loop. The transfer is made by hydrophilic contacts between the carbon foil and the crystal. The film is then broken by tilting the loop to increase the angle between the film and the grid.

10. In order to better evaluate the organizational state of the molecule in the crystallization experiment, it is useful to control its shape and size by direct adsorption of the sample on a carbon film and negative staining. Such an experiment will also give an insight into the aggregation state of the protein in solution.
11. The appearance of vesicular structures is often an indication for a too large excess of lipids. The working lipid solution should then be diluted.
12. To remove excess lipids, a detergent solution at low concentration can be used ***(19)***. Care must be taken during this step because the drop might migrate to both sides of the grid and interfere with the staining process.

References

1. Henderson, R., Baldwin, J. M., Ceska, T. A., Zemlin, F., Beckmann, E., and Downing, K. H. (1990) Model for the structure of bacteriorhodopsin based on high-resolution electron cryo-microscopy. *J Mol. Biol.* **213,** 899–929.
2. Kuhlbrandt, W., Wang, D. N., and Fujiyoshi, Y. (1994) Atomic model of plant light-harvesting complex by electron crystallography. *Nature* **367,** 614–621.
3. Amos, L. A., Henderson, R., and Unwin, P. N. (1982) Three-dimensional structure determination by electron microscopy of two-dimensional crystals. *Prog. Biophys. Mol. Biol.* **39** , 183–231.
4. Uzgiris, E. E. and Kornberg, R. D. (1983) Two-dimensional crystallization technique for imaging macromolecules, with application to antigen-antibody-complement complexes. *Nature* **301,** 125–129.
5. Darst, S. A., Ribi, H. O., Pierce, D. W., and Kornberg, R. D. (1988) Two-dimensional crystals of *Escherichia coli* RNA polymerase holoenzyme on positively charged lipid layers. *J. Mol. Biol.* **203,** 269–273.
6. Schultz, P., Celia, H., Riva, M., Darst, S. A., Colin, P., Kornberg, R. D., et al. (1990) Structural study of the yeast RNA polymerase A. Electron microscopy of lipid-bound molecules and two-dimensional crystals. *J. Mol. Biol.* **216,** 353–362.
7. Ribi, H. O., Reichard, P., and Kornberg, R. D. (1987) Two-dimensional crystals of enzyme-effector complexes: ribonucleotide reductase at 18-A resolution. *Biochemistry* **26,** 7974–7979.
8. Lebeau, L., Regnier, E., Schultz, P., Wang, J. C., Mioskowski, C., and Oudet, P. (1990) Two-dimensional crystallization of DNA gyrase B subunit on specifically designed lipid monolayers. *FEBS Lett.* **267,** 38–42.
9. Avila-Sakar, A. J., and Chiu, W. (1996) Visualization of beta-sheets and side-chain clusters in two-dimensional periodic arrays of streptavidin on phospholipid monolayers by electron crystallography. *Biophys. J.* **70,** 57–68.
10. Lebeau, L., Schultz, P., Celia, H., Mesini, P., Nuss, S., Klinger, C., et al. (1996) Specifically designed lipid assemblies as tools for two-dimensional crystallization of soluble biological macromolecules, in *Handbook of Nonmedical Applica-*

tions of Liposomes, vol. II (Barenholz Y. and Lasic D. D., eds.), CRC, Boca Raton, FL, pp. 155–188.
11. Darst, S. A., Ahlers, M., Meller, P. H., Kubalek, E. W., Blankenburg, R., Ribi, H. O., et al. (1991) Two-dimensional crystals of streptavidin on biotinylated lipid layers and their interactions with biotinylated macromolecules. *Biophys. J.* **59,** 387–396.
12. Kubalek, E. W., Le Grice, S. F., and Brown, P. O. (1994) Two-dimensional crystallization of histidine-tagged, HIV-1 reverse transcriptase promoted by a novel nickel-chelating lipid. *J. Struct. Biol.* **113,** 117–123.
13. Bischler, N., Balavoine, F., Milkereit, P., Tschochner, H., Mioskowski, C., and Schultz, P. (1998) Specific interaction and two-dimensional crystallization of histidine tagged yeast RNA polymerase I on nickel-chelating lipids. *Biophys. J.* **74,** 1522–1532.
14. Mosser, G. and Brisson, A. (1991) Structural analysis of two-dimensional arrays of cholera toxin B- subunit. *J. Electron Microsc. Tech.* **18,** 387–394.
15. Kubalek, E. W., Kornberg, R. D., and Darst, S. A. (1991) Improved transfer of two-dimensional crystals from the air/water interface to specimen support grids for high-resolution analysis by electron microscopy. *Ultramicroscopy* **35,** 295–304.
16. Fukami, A. and Adachi, K. (1965) A new method of preparation of a self-perforated micro plastic grid and its application. *J. Electron Microscopy*, **14,** 112–118.
17. Dubochet, J., Adrian, M., Chang, J. J., Homo, J. C., Lepault, J., McDowall, A. W., et al. (1988) Cryo-electron microscopy of vitrified specimens. *Quart. Rev. Biophys.* **21,** 129–228.
18. Schmutz, M. and Brisson, A. (1996) Analysis of carbon film planarity by reflected light microscopy. *Ultramicroscopy* **63,** 263–272.
19. Asturias, F. J. and Kornberg, R. D. (1995) A novel method for transfer of two-dimensional crystals from the air/water interface to specimen grids. EM sample preparation/lipid-layer crystallization. *J. Struct. Biol.* **114,** 60–66.

39

Atomic Force Microscopy of DNA and Protein-DNA Complexes Using Functionalized Mica Substrates

Yuri L. Lyubchenko, Alexander A. Gall, and Luda S. Shlyakhtenko

1. Introduction

Atomic force microscopy (AFM; also called scanning force microscopy [SFM]) is a rather novel technique that offers unique advantages in the potential for the very high resolution of DNA and small ligands in the absence of stains, shadows, and labels *(1,2)*. Furthermore, the scanning can be performed in air or liquid. The latter is particularly important for resolving fully hydrated structures. The AFM is theoretically capable of resolving structural details at the level of atomic dimensions, provided that the specimen is not dynamic.

A serious practical limitation to the application of AFM to structural and conformational studies of DNA and its complexes with proteins and other biological macromolecules has been sample preparation. The macromolecules must be tethered to the substrate surface in order to avoid resolution-limiting motion caused by the sweeping tip during scanning. Progress in sample preparation for AFM studies of DNA has been achieved in a number of groups *(3–7)* and some of these approaches have been applied to studies of a number of protein–DNA complexes *(3,4,8)*.

A versatile approach based on functionalization of surfaces with silanes was suggested in **refs.** *9–11*. A weak cationic surface is obtained if aminopropyltrietoxy silane (APTES) is used to functionalize the mica surface with amine groups (AP-mica). This technique in addition to imaging nucleic acids under different conditions *(10,12–14)* was applied to imaging of a number of nucleoprotein complexes *(9,11,15–17)*. Here, we describe a sample preparation procedure for AFM using AP-mica substrates.

From: *Methods in Molecular Biology, vol. 148: DNA–Protein Interactions: Principles and Protocols, 2nd ed.*
Edited by: T. Moss © Humana Press Inc., Totowa, NJ

The method of functionalization of mica is based on covalent attachment of 3-aminopropyltriethoxy silane to the surface of the mica, as shown schematically in **Fig. 1**. The amino groups of APTES are bound covalently to the freshly cleaved mica surface, giving it properties similar to an anion-exchange resin used in affinity chromatography. This group after being exposed to a water solution becomes positively charged in a rather broad range of pH (aliphatic amino groups have a p*K* of around 10.5). Therefore, DNA, which is a negatively charged polymer, should adhere to this surface strongly. The binding of DNA to AP-mica was monitored directly by the use of radiolabeled DNA. AFM imaging of AP-mica showed that very low concentration of APTES (less than 1 μ*M*) should be used to obtain smooth surface ***(9,10)***. Vapor deposition of APTES allowed one to obtain the surface with mean roughness of several angstroms ***(9–12)***, so the DNA and DNA–protein complexes can be visualized easily (*see* **Fig. 2A,B**, respectively).

The features of this procedure of sample preparation are as follows ***(9,11)***:

- DNA binding to AP-mica is insensitive to the type of buffer and presence of Mg^{2+} or other divalent and miltivalent cations; hence sample preparation can be done in a variety of conditions.
- Deposition can be done in a wide variety of pH and over a wide range of temperatures.
- Once prepared, samples are stable and do not absorb any contaminants for months with minimal precautions for storing.
- As low as 10 ng of DNA is sufficient for the preparation of one sample.

These characteristics of AP-mica were crucial for routine imaging nucleic acids (DNA, dsRNA, kinetoplast DNA) and nucleoprotein complexes of different type ***(9,16,17)***.

2. Materials.

1. Chemicals: commercially available 3-aminopropyltriethoxy silane (e.g., Fluka, Chemika-BioChemika (Switzerland), Aldrich (USA), United Chemical Technology (USA), and *N,N*-diisopropylethylamine (Aldrich, Sigma). It is recommended to redistill APTES and store under argon.
2. Mica substrate: any type of commercially available mica sheets (green or ruby mica). Asheville-Schoonmaker Mica Co. (Newport News, VA) supplies both thick and large (more than 5 × 7 cm) sheets suitable for making the substrates of different sizes.
3. Water: Double glass distilled or deionized water filtered through a 0.5-μm filter.
4. 2-L glass desiccators and vacuum line (50 mmHg is sufficient).
5. Plastic syringes (5–10 mL) with a plastic tip for rinsing the samples.
6. Plastic syringes (1 mL) for imaging in liquid.
7. Gas tank with clean argon gas.
8. Vacuum cabinet for storing the samples.

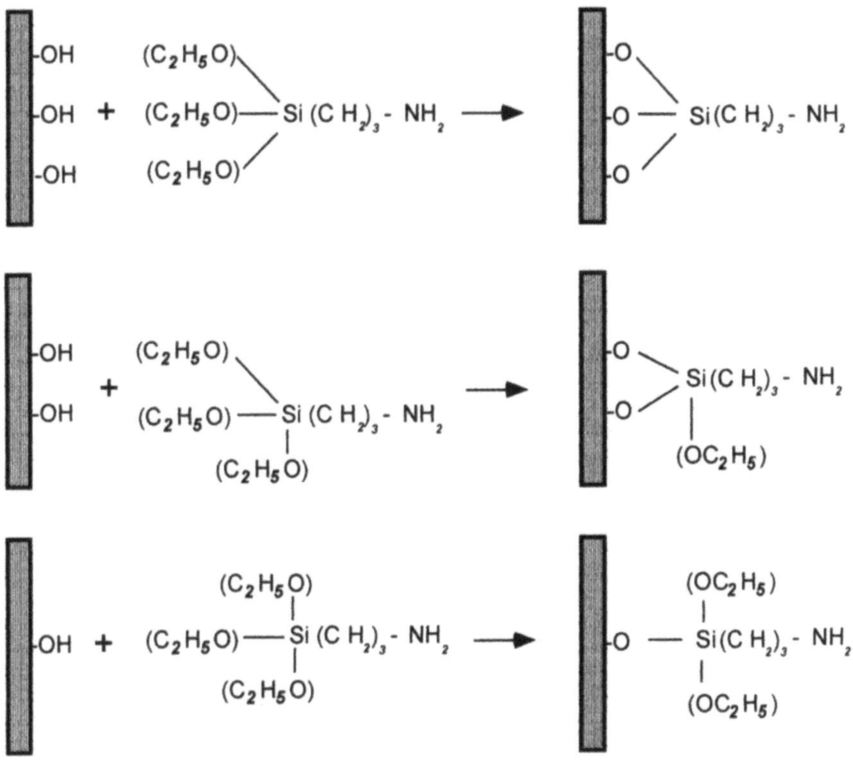

Fig. 1. The reaction of aminopropyltriethoxy silane (APTES) with mica. Three possible types of reaction are illustrated.

3. Methods

3.1. AP-Mica Preparation

1. Place two plastic caps (cut them from regular 1.5 mL plastic tubes) on the bottom of a 2-L desiccator, evacuate then purge with argon.
2. Cleave mica sheets (approx 5 × 5 cm) to make them as thin as 0.1–0.05 mm and mount at the top of the desiccator.
3. Put 30 µL of APTES into one plastic cap in the desiccator and 10 µL of *N,N*-diisopropylethylamine (Aldrich) into the other cap and allow the functionalization reaction to proceed for 1–2 h. Remove the cap with APTES and purge the desiccator with argon for 2 min.
4. Leave the sheets for 1–2 d in the desiccator to cure. The AP-mica is then ready for the sample deposition. (*See* **Note 1**.)

The procedure allows one to obtain a weak cationic surface with rather uniform charge distribution. This is illustrated in **Fig. 2A** by the uniform distribution of DNA fragments that can be obtained.

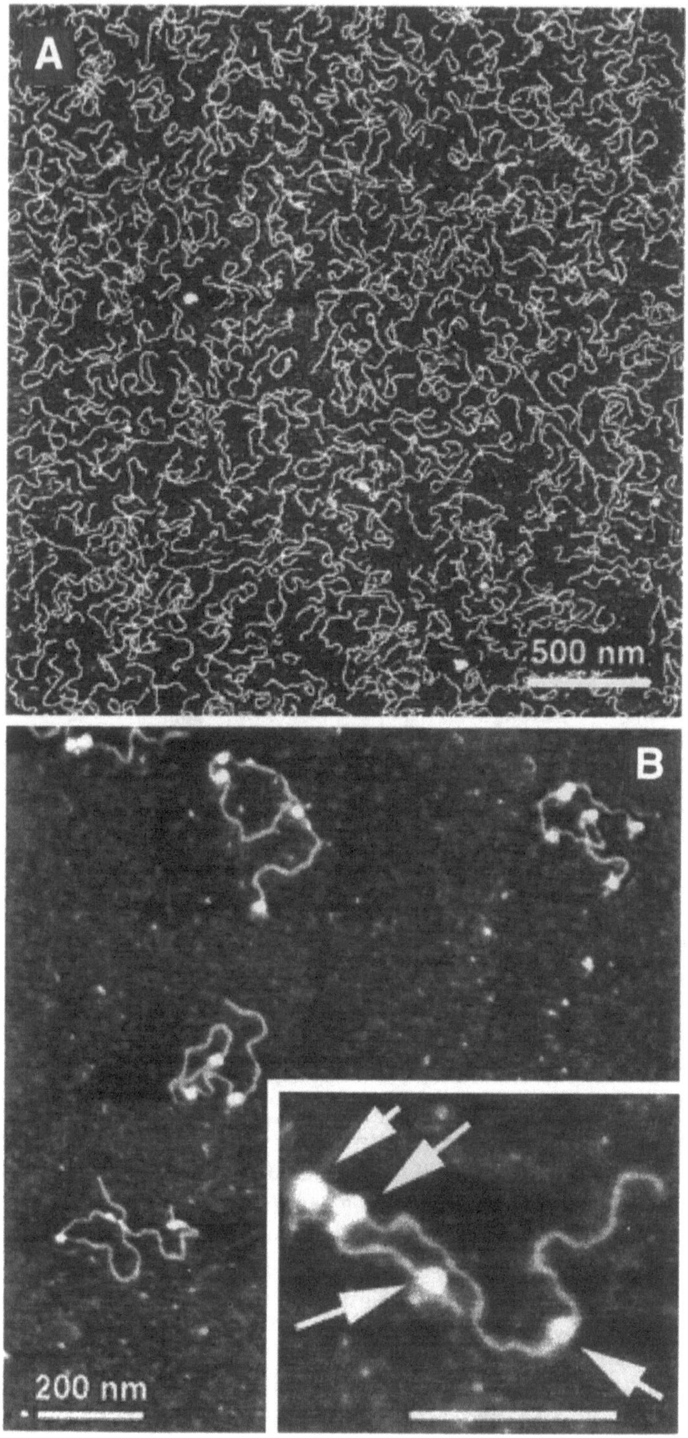
A
500 nm
B
200 nm

3.2. Sample Preparation for AFM Imaging in Air

3.2.1. The Droplet Procedure

1. Prepare the solution of the sample (DNA, RNA, protein–DNA complex) in appropriate buffer. DNA concentration should be between 0.1 and 0.01 μg/mL, depending on the size of the molecules (*see* **Notes 2** and **3**).
2. Place 5–10 μL of the solution in the middle of AP-mica substrate (usually 1 × 1-cm squares) for 2–3 min.
3. Rinse the surface thoroughly with water (2–3 mL per sample) to remove all buffer components. A 10-mL plastic syringe is very useful for rinsing, but attach an appropriate plastic tip instead of a metal needle.
4. Dry the sample by blowing with clean argon gas. The sample is ready for imaging. Store the samples in vacuum cabinets or desiccators filled with argon.

3.2.2. The Immersion Procedure

This procedure is recommended if the deposition should be performed at strictly controlled temperature conditions (0°C or elevated temperatures).

1. Prepare the solution (DNA, RNA, nucleoprotein complexes) and preincubate for 10–20 min to allow the temperature to equilibrate. Recommended concentration of DNA is 0.2 –0.01 μg/mL, depending on the size of the molecules (*see* **Notes 2** and **3**).
2. Immerse a piece of AP-mica into the vials and leave it for 10–20 min to allow the samples to adsorb to the surface.
3. Remove the specimen, rinse with water thoroughly, and dry under the argon flow. The sample is ready for imaging.
4. The samples can be stored in vacuum cabinet or under argon.

3.3. AFM Imaging in Air

1. Mount the sample and start approaching the probe.
2. Both the contact and intermittent (tapping) modes can be used, but the latter is preferable and allows one to obtain images of DNA and DNA–protein complexes routinely. Our experience is mostly limited to a NanoScope III microscope (MultiMode system, Digital Instruments, CA), but samples prepared on AP-mica were imaged on other commercially available instruments (e.g., the microscopes manufactured by Topometrix, Park Scientific Instruments, Molecular Imaging). With the MultiMode system, any type of probe designed for noncontact imaging can be used. NanoProbe TESP tips (Digital Instruments, Inc.) and conical sharp

---

Fig. 2. *(previous page)* AFM images of a 800-bp fragment **(A)** and reconstituted chromatin **(B)**. The concentration of DNA was 0.5 μg/mL in (A). Reconstituted chromatin was deposited onto the substrate after glutaraldehyde fixation. (The sample was from D. Lohr [Arizona State University] and the images were taken in air with TM AFM [NanoScope III]).

silicon probes from K-TEK International (Portland, OR) work well. Typically a tapping frequency of 240–380 kHz and a scanning rate 2–3 Hz allows one to obtain stable images.

3.4. Imaging in Solution

The capability of AFM to perform scanning in liquid is its most attractive feature for numerous biological applications, allowing imaging under near-physiological conditions. In addition, this mode of imaging permits one to eliminate the undesirable resolution-limiting capillary effect typical for imaging in air *(3,8,9)*. As a results, images of DNA filaments as thin as approx 3 nm were obtained in water solutions *(13)* and helical periodicity was observed when dried DNA samples were imaged in propanol *(18)*. AP-mica can be used as a substrate for imaging in liquid. Moreover, the first images of DNA in fully hydrated state were obtained by the use of AP-mica *(19)*. This section describes the procedures of two types of imaging in solution.

3.4.1. Imaging of Dried Sample in Solutions

This type of imaging was successfully applied for high-resolution imaging of DNA *(18,19)*.

1. Install an appropriate tip designed for imaging in liquid (fluid cell). Use stiff triangular Si_3Ni_4 cantilevers *(20)*.
2. Mount the sample on the stage of the microscope. Coating the stage of the scanner with a thin plastic film prevents it from being wetted because of accidental leakage of the fluid beneath the mica sheet.
3. Attach the head of the microscope with installed fluid cell and make appropriate adjustments to the microscope.
4. Approach the sample to the tip manually, leaving approx 20-µm gap between the tip and the surface.
5. Inject buffer solution or appropriate solvent with a 1-mL plastic syringe through the inlet hole in the fluid cell.
6. Change the position of the mirror to maximize the signal on the photodetector.
7. Find a resonance peak. Typically, it is quite broad peak around 8–9 kHz for the MultiMode system. Follow the recommendations given in the manual for the fluid cell on how to find the peak.
8. Minimize the drive amplitude. The numbers vary from tip to tip, but amplitudes as low as 10 nm or even less provide better quality pictures (*see* **Note 4**).
9. Allow the microscope to approach the sample and engage the surface.
10. Operate with the setpoint voltage and drive the amplitude parameters to improve the quality of images (*see* **Note 4**).

3.4.2. Imaging Without Drying of the Sample (*AFM* In Situ)

This type of imaging is recommended in cases in which dynamics are to be studied.

1. Prepare the solution of your sample in appropriate buffer. The concentration is the same as it is needed for imaging in air (**Subheading 3.2.**)
2. After installing the tip in the fluid cell, mount a piece of AP-mica on the sample stage. Mica pieces of 1 cm × 1 cm are sufficient for NanoScope design of the fluid cell. As earlier, coating the stage with a plastic film is recommended for secure operating of the microscope.
3. Take approx 50 µL of the sample with a 1-mL syringe. Use 200-µL plastic tips with capillary ends instead of a metal needle. Cut both ends of the tip to fit the syringe and the diameter of inlet hole of the fluid cell.
4. After necessary adjustments of the microscope (**Subheading 3.4.1.**) and manual approaching of the tip, inject the solution into the cell. The use of additional syringe attached to the outlet of fluid cell as a suction helps in manipulating the small volume of solution.
5. Start approaching and follow the steps described in **Subheading 3.4.1.**

3.5. Alternative Procedures for AFM Sample Preparation

Among other techniques applied to AFM studies of DNA, the method based on miltivalent cations *(3,5,8)* has permitted the imaging of a number of nucleoprotein complexes *(3,4,8)*. In this approach, the mica surface is simply treated with multivalent ions (e.g., Mg^{2+}) to increase its affinity for DNA, the DNA then being held in place strongly enough to permit reliable imaging by AFM. An alternative is to deposit the sample in the buffer containing a multivalent ion. This cation-assisted procedure of sample preparation was used for studies of the process of DNA degradation with nuclease *(21)* and interaction of DNA with photolyase *(22)*. The mechanism underlying this technique remains unclear and the protocol depends on the system studied and the type of the cation used, and the efficiency of DNA deposition is buffer sensitive *(23,24)*. In some cases, a special type of tips (electron-beam-deposited tips) is required for reliable imaging *(25)*. A protocol describing the use of Mg-assisted procedure has been published *(26)*.

4. Notes

1. A dry argon atmosphere is crucial for obtaining the substrates for AFM studies and for storage of the substrate. Allow the gas to flow while desiccator is opened. With these precautions, the AP-mica substrates retain their activity for several weeks.
2. DNA concentration. This parameter depends on the length of molecules. If the molecules are as small (e.g., several hundred base pairs), a concentration of approx 0.3 µg/mL is recommended to avoid intermolecular crossing. A lower DNA concentration is recommended for larger DNA molecules. For example, concentration of lambda DNA (approx 48 kb) of approx 0.01 µg/mL allows one to obtain images of individual DNA molecules *(9–11,19)*.
3. DNA preparation. Very little DNA is needed to prepare the samples by the droplet procedure. Typically, 10 ng of DNA is sufficient for the preparation of plas-

mid DNA (approx 3 kb long). Because a band of DNA from agarose gel usually contains 100 ng of DNA, DNA extracted from a single gel slot can be sufficient for preparation of a complete set of samples. The following procedures can be used for purification of DNA extracted from the gel:

- *Electrophoretic deposition of DNA bands onto DEAE paper.* Strips of DEAE paper are placed into a slot cut in the agarose gel 3–5 mm below the band to be recovered and the DNA is electrophoresed onto the paper for 5–10 min (the time can be determined by direct examination the gel under an ultraviolet [UV] source). The DNA is extracted from the paper by elution into 2 *M* NaCl followed by two rounds of spin-column desalting and extensive ethanol precipitation.
- *Extraction from the gel.* The procedure is based on the use of the extraction kit UltraClean15 (MoBio Laboratories, Solana Beach, CA). The purification consists of melting of the slice of agarose, immobilizing the DNA on the absorbent matrix, washing off all contaminants, and eluting DNA from the matrix with a low-salt buffer. At least one step of the ethanol precipitation is needed to remove UV absorbing low-molecular-weight materials. A similar procedure can be applied to purification of the sample eluted from polyacrylamide gel.

4. Imaging conditions. It was recommended to operate the instrument at the lowest possible drive amplitude. This recommendation is based on the following considerations; the oscillating tip provides rather large energy to the sample. According to **ref. *45***, a total energy provided to the sample by oscillating tip can be as high as 10^{-16}–10^{-17} J at 30 nm amplitude of oscillation. However, this value is almost three orders of magnitude lower if the microscope is operated at an amplitude as low as approx 3 nm. Such imaging conditions allow one to minimize the effect of the tip on the sample, to prevent damaging the tip, and to obtain images with high contrast. In addition, such conditions simplify considerably the study by AFM of dynamic processes such as segmental DNA mobility ***(13,14)*** or the process of protein–DNA interaction ***(27)***.

Acknowledgments

This work was supported by grant GM 54991 from the NIH.

References

1. Binnig, G., Quate, C. F., and Gerber, C. H. (1986) Atomic force microscope *Phys. Rev. Lett.* **56,** 930–933.
2. Hansma, P. K., Elings, V. B., Marti, O., and Bracker, C. E. (1988) Scanning tunneling microscopy and atomic force microscopy: some applications to biology and technology. *Science* **242,** 209–216.
3. Bustamante, C., Erie, D. A., and Keller, D. (1994) Biochemical and structural applications of scanning force microscopy. *Curr. Opin. Struct. Biol.* **3,** 750–760.
4. Bustamante, C. and Rivetti, C. (1996) Visualizing protein-nucleic acid interactions on a large scale with scanning force microscope. *Annu. Rev. Biophys. Biomol. Struct.* **25,** 395–429.

5. Vesenka, J., Guthold, M., Tang, C. L., Keller, D., and Bustamante, C. (1992) Substrate preparation for reliable imaging of DNA molecules with the scanning force microscopy. *Ultramicroscopy* **42–44,** 1243–1249.
6. Yang, J., Takeyasu, K., and Shao, Z. (1992) Atomic force microscopy of DNA molecules. *FEBS Lett.* **301,** 173–176.
7. Allen, M. J., Dong, X. F., O'Neil, T. E., Yau, P., Kowalczykowski, S. C., Gatewood, J., Balhorn, R., et al. (1993) Atomic force microscope measurements of nucleosome cores assembled along defined DNA sequences. *Biochemistry* **32,** 8390–8396.
8. Hansma, H., G., and Hoh, J. (1994) Biomolecular imaging with the atomic force microscopy. *Ann. Rev. Biophys. Biochem. Struct.* **23,** 115–139.
9. Lyubchenko Y. L., Jacobs B. L., Lindsay S. M., and Stasiak A. (1995) Atomic force microscopy of protein–DNA complexes [Review]. *Scanning Microsc.* **9,** 705–727.
10. Lyubchenko, Y. L., Gall, A. A., Shlyakhtenko, L. S., Harrington, R. E., Oden, P. I., Jacobs, B. L., et al. (1992) Atomic force microscopy imaging of double stranded DNA and RNA *J. Biomolec. Struct. Dynam.* **9,** 589–606.
11. Lyubchenko, Y. L., Blankenship, R. E., Lindsay, S. M., Simpson, L., Shlyakhtenko, L. S. (1996) AFM studies of nucleic acids, nucleoproteins and cellular complexes: The use of functionalized substrates. *Scanning Microsc.* **10(Suppl.),** 97–109.
12. Lyubchenko, Y. L., Jacobs, B. L., and Lindsay, S. M. (1992) Atomic force microscopy imaging of reovirus dsRNA: a routine technique for length measurements. *Nucleic Acids Res.* **20,** 3983–3986.
13. Lyubchenko Y. L. and Shlyakhtenko, L. S. (1997) Visualization of supercoiled DNA with atomic force microscopy*in situ. Proc. Natl. Acad. Sci. USA* **94,** 496–501.
14. Shlyakhtenko, L. S, Potaman, V. N., Sinden, R. R., and Lyubchenko, Y. L. (1998) Structure and dynamics of supercoil-stabilized DNA cruciform. *J. Mol. Biol.* **280,** 61–72.
15. Lyubchenko, Y. L., Oden, P. I., Lampner, D., Lindsay, S. M., and Dunker, K. (1993) Atomic force microscopy of DNA and bacteriophage in air, water and propanol: the role of adhesion forces, *Nucleic Acids Res.* **21,** 1117–1123.
16. Lyubchenko, Y. L., Shlyakhtenko, L. S., Aki, T., and Adhya, S. (1997) AFM visualization of GalR mediated DNA looping. *Nucleic Acids Res.* **25,** 873–876.
17. Herbert A., Schade, M., Lowenkaupt, K., Alfken, J., Schwartz, T., Shlyakhtenko, L. S., et al. (1998) The Za domain from human ADAR1 binds to the Z-DNA conformer of many different sequences. *Nucleic Acid Res.* **26,** 3486–3493.
18. Hansma, H. G., Laney, D. E., Bezanilla, M., Sinsheimer, R. L., and Hansma, P. K. (1995) Applications for atomic force microscopy of DNA. *Biophys. J.* **68,** 672–1677.
19. Lyubchenko, Y. L., Shlyakhtenko, L. S., Harrington, R. E., Oden, P. I., and Lindsay, S. M. (1993) AFM imaging of long DNA in air and under water. *Proc. Natl. Acad. Sci. USA* **90,** 2137–2140.
20. Hansma, P. K., Cleveland, J. P., Radmacher, M., Walters, D. A., and Hillner, P. (1994) Tapping mode atomic force microscopy in liquids. *Appl. Phys. Lett.* **64,** 1738–1740.

21. Bezanilla, M., Drake, B., Nudler, E., Kashlev, M., Hansma, P. K., and Hansma, H. G. (1994 Motion and enzymatic degradation of DNA in the atomic force microscope *Biophys. J.* **67,** 2454–2459.
22. Han, W. H., Lindsay, S. M., and Jing, T. W (1996) A magnetically-driven oscillating probe microscope for operation in liquids. *Appl. Phys. Lett.* **69,** 4111–4114.
23. Bezanilla, M., Manne, S., Laney, D. E., Lyubchenko, Y. L., and Hansma, H. G (1995) Adsorption of DNA to mica, silylated mica, and minerals: characterization by atomic force microscopy. *Langmuir* **11,** 655–659.
24. Hansma, H. G. (1996) A useful buffer for atomic force microscopy of DNA. *Sci. Tools Pharmacia Biotech.* **1(3),** 7.
25. Kasas, S., Thomson, N. H., Smith, B. L., Hansma, H. G., Zhu, X., Guthold, M., et al. (1997) *Escherichia coli* RNA polymerase activity observed using atomic force microscopy. *Biochemistry* **36,** 461–468.
26. Hansma, H. G. (1998) Atomic force microscopy of DNA on mica in air and fluid, in *Procedures for Scanning Probe Microscope* (Colton, R. J., et al., eds.), Wiley, Chichester, pp. 389–393.
27. van Noort, S. J. T. , van der Werf, K. O., Eker, A. P. M., Wyman, C., Grooth, B. G., van Hulst, N. F., et al. (1998) Direct visualization of dynamic protein–DNA interactions with a dedicated atomic force microscope. *Biophys. J.* **74,** 2840–2849.

Additional Reading

Colton et al., (eds.) (1998) *Procedures in Scanning Probe Microscopes.* Wiley, Chichester.

40

Electron Microscopy of Protein–Nucleic Acid Complexes

Uniform Spreading of Flexible Complexes, Staining with a Uniform Thin Layer of Uranyl Acetate, and Determining Helix Handedness

Carla W. Gray

1. Introduction

There are a number of proteins involved in DNA replication, recombination, or repair that bind stoichiometrically to single DNA strands irrespective of the nucleotide sequence, and some of these proteins also bind to single-stranded RNA. Some of the best known examples are the ssb protein of *Escherichia coli*, the gene 32 protein of phage T4, the gene 5 protein of the M13/fd/f1 filamentous bacterial viruses, and the more recently isolated human replication protein A *(1–4)*. Complexes formed by these proteins contain protein bound to the nucleic acid at defined ratios of the number of nucleotides per molecule of bound protein; the ratios are determined by the interactive properties of the protein. These ratios, and the structures of the complexes that are formed, may vary with factors such as changes in solution conditions that alter the binding properties of the proteins.

Stoichiometric, multiprotein complexes of proteins with nucleic acids will form structural repeats, arranged as discrete clusters of bound proteins or as a continuous nucleoprotein helix. Although the individual proteins may not be resolved, the structural repeats tend to be of a size that can be visualized by electron microscopy. "Negative" staining, in which protein masses are delineated by their exclusion of an electron-opaque stain, is a method of choice because negative staining provides a well-contrasted image at higher resolution than is attained with other techniques such as shadowing with refractory

From: *Methods in Molecular Biology, vol. 148: DNA–Protein Interactions: Principles and Protocols, 2nd ed.*
Edited by: I. Moss © Humana Press Inc., Totowa, NJ

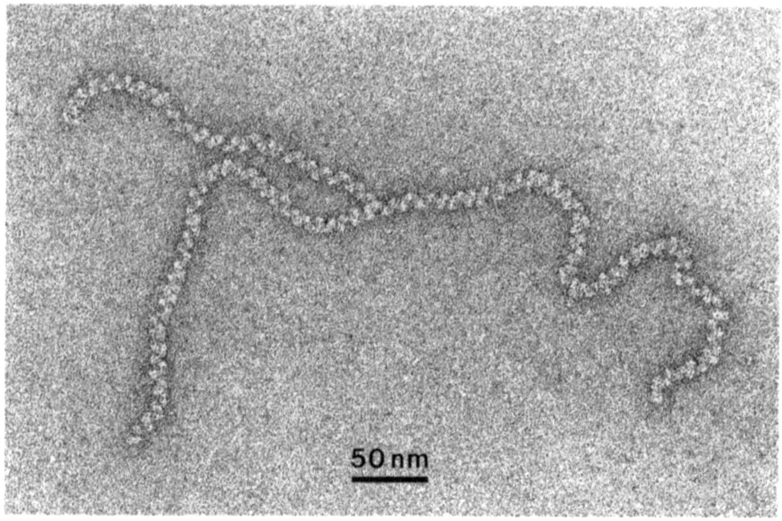

Fig. 1. A transmission electron micrograph of a helical complex of the fd gene 5 protein with circular, single-stranded fd viral DNA. The complex was formed in vitro and was spread and stained with uranyl acetate by the methods described in this chapter. The complex is not tilted; the plane of the support film is in the plane of the page.

metals. We have used negatively stained complexes with the M13/fd/f1 gene 5 protein to provide crucial information in conjuntion with crystallographic and low-angle X-ray scattering studies, making it possible to model the three-dimensional structures of the complexes *(5,6)*. Negatively stained nucleoprotein complexes can be prepared relatively quickly and examined immediately after preparation. This makes it advantageous to use negatively stained preparations prior to or in conjunction with cryo electron microscopy, a technique that, although it offers better preservation for detailed structural studies, is a more difficult and time-consuming method.

Nucleoprotein complexes are often highly flexible and the complexes are easily distorted, tangled, or partially dissociated during preparation for negative staining. The author has developed procedures to overcome these difficulties *(3,7)*, such that preparations consistently contain complexes with well-extended configurations free of any obvious distortions (**Fig. 1**). Complexes prepared in this manner are uniformly spread on a two-dimensional support film and can be used for quantitative analysis of such parameters as the number of protein clusters or helical turns in a complex, the axial length of the complex, and the extent of the local variations in interturn distances in a complex that forms a flexible helix. These preparations can also be used for analyses of three-dimensional structures using tilted specimens *(3)*. The most likely

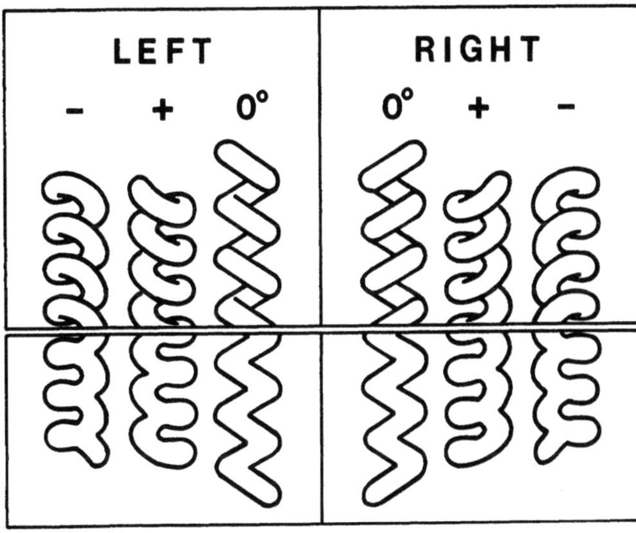

Fig. 2. Effects of tilting on the appearance of left- and right-handed helices. Three-dimensional images are drawn in the upper panel of the figure and two-dimensional parallel projections in the lower panel. A negatively stained complex will be seen as a two-dimensional projection. A helix at 0° tilt is parallel to the plane of the page; a "plus" (+) tilt brings the top of the helix nearer to the observer, whereas a "minus" (–) tilt brings the bottom of the helix closer to the observer. (Reprinted with permission from **ref.** ***3***.)

application is a determination of the handedness of helical nucleoprotein complexes, using the approach of Finch ***(8)***, which is illustrated for a general case in **Fig. 2**. Projections of left- or right-handed helices on a two-dimensional plane are identical, as shown in the lower part of the figure. However, a left-handed helix tilted with its top toward the viewer will show deeper indentations between the helical turns along its *right* side, whereas an identically tilted right-handed helix has the deeper indentations on its *left* side. Tilting the helices in the opposite direction, with the tops of the helices away from the viewer, produces an opposite set of left-hand and right-side indentations. In the description that follows, we describe a practical means of determining the absolute orientations of helices in images of tilted specimens.

2. Materials

1. Purified water, chemically softened and then predistilled in bulk, is twice redistilled in our laboratory using a series of two 24-in. borosilicate glass Vigreux columns. We do not find it necessary to use a quartz still. Alternatively, the bulk distilled water may be deionized to a resistivity of 18 mΩ·cm in a Millipore Milli-Q system consisting of one Milligard cellulose ester prefilter cartridge, two ion-

exchange cartridges, and a 0.22-µm filter, in series. No activated-charcoal filter is included in our system, because of the tendency of the charcoal filter to release minute charcoal particles, which are found in our preparations for electron microscopy. (*See* **Note 1**.)

2. Buffers in which the protein–DNA complexes are to be visualized. Concentrated buffer stocks are passed through a 0.22-µm fiberless polycarbonate filter (Nuclepore/Costar) to remove particulates and are then stored in borosilicate glass or polystyrene containers. The buffers should generally not contain high concentrations of salts or other nonvolatile components, as these components will tend to be retained on the carbon support film and can interfere with visualization of the protein–DNA complexes.
3. The protein–nucleic acid complexes are to be examined. These preparations should not contain significant excess quantities of noncomplexed or contaminating proteins, and they must generally be free of contaminating nonaqueous solvents as well as lipids, oils, salts, detergents, and other nonvolatile materials. We frequently repurify proteins and nucleic acids obtained commercially or from other laboratories, using ethanol precipitation of nucleic acids, molecular-sieve chromatography of proteins, or dialysis. About 0.2–2 nmol of DNA (measured as the concentration of nucleotides), together with protein added at an appropriate ratio, will be needed for a 50-µL incubation mixture from which the protein–DNA complexes are adsorbed to a single specimen grid.
4. Glutaraldehyde, purified for electron microscopy (distilled *in vacuo* and stored in sealed ampoules under inert gas). The contents of one ampoule are diluted to 8% (v/v) in purified water and are stored at –20°C in a tightly capped borosilicate glass tube with a Teflon-lined screw cap; this solution can be used for as long as 6 mo.
5. A suitable electron-opaque ("negative") stain, preferably analytical reagent grade. We generally use uranyl acetate, which provides good surface detail and yields satisfactory images with many proteins. However, one should always consider the possibility that other stains may be preferable for a particular application ***(9)***. A 2% (w/v) solution of a small amount of uranyl acetate in purified water is dissolved by stirring for 30 min in a borosilicate glass beaker. The beaker is sealed with a wax film and stored in the dark. The solution is used within a few days or weeks, but only if precipitates have not begun to form. Uranyl salts are weakly radioactive, and discarded solutions should be collected and properly disposed of as radioactive waste.
6. Carbon support films, 8–10 nm thick, on 500-mesh copper grids. We make our films in an Edwards E306A evaporator equipped with a liquid-nitrogen trap and a quartz crystal film thickness monitor (Edwards FTM5). The chamber is evacuated just to 1×10^{-4} mbar, contaminants are burnt off from the carbon rods (shutter closed, carbon rods brought to a red glow) and then evaporation is carried out over a period of several seconds. We use high-purity carbon rods (Bio-Rad/Polaron, <20 ppm impurities) that have been milled to form 1-mm tips. The carbon is evaporated onto a freshly cleaved mica film (Ladd, tested by flame spec-

troscopy). The carbon films are then floated from the mica onto purified water in a Teflon dish, picked up on copper grids, and dried for 30 min under a heat lamp.

3. Methods

3.1. Specimen Mounting and Negative Staining

1. Prepare stock solutions of DNA and protein, or of protein–DNA complexes, in high-quality, contamination-free, non-wetting polypropylene tubes. Care must be taken to avoid touching pipet tips or tube rims, to avoid contamination with oils and nucleases from the skin. (*See* **Note 2**.)
2. Prepare diluted DNA–protein mixtures for microscopy in a volume of 50 µL for each specimen grid. The final concentration of DNA will be approx 4–40 nmol of DNA nucleotides per milliliters; the optimal concentration must be determined experimentally, as it will vary with the adsorptive properties of the DNA and protein in a buffer of given composition, pH, and ionic strength.
3. Clean a Teflon surface with reagent-grade ethanol and rinse it with purified water. For each specimen grid, place a row of droplets on the Teflon surface: leave an empty space for one droplet, then deposit two 50-µL droplets of purified water, then one 50-µL droplet of 2% uranyl acetate.
4. Initiate glutaraldehyde fixation. This will be required in most cases to maintain noncovalent protein–DNA associations during adsorption to the charged (glow-discharge-activated) carbon film. (To confirm that fixation is effective and yields an unperturbed structure, stained preparations can be made by the method of Valentine et al. *[10]*, which does not require fixation even for some relatively labile complexes.) Add 0.5 µL of 8% glutaraldehyde to the bottom of a 0.5-mL conical polyethylene tube; immediately add 50 mL of the protein–DNA mixture and gently mix by pipetting up and down once. Incubate the reaction mixture at 20–25°C for 20 min.
5. Meanwhile, place carbon-coated grids, carbon side up, on a clean, inverted glass Petri dish and subject the grids to glow discharge. We use two parallel, L-shaped aluminum rods (6.5 mm in diameter) fitted to the high tension electrodes of the Edwards E306A evaporator. Glow discharge is carried out at 0.1–0.2 mbar (with only the rotary pump in operation), with the grids placed on the Petri dish at a distance of about 4 cm below the horizontal segments of the rods. The discharge is continued for about 50 s at 40% of maximum voltage (i.e., at approx 2 kV) and complexes are adsorbed to the grids within 10 min.
6. Place a 50-µL droplet of each protein–DNA mixture on the Teflon surface, at the beginning of a row of droplets prepared as described in **step 3**. Touch the grid, carbon side down, to the top of the droplet containing the protein–DNA mixture for 20–60 s, then wick off (remove) excess solution from the grid onto a filter paper, holding the grid perpendicular to the filter paper. Next, touch the grid to each of the two water droplets for 1 s each and, finally, to the uranyl acetate droplet for 20 s, wicking off excess liquid after each step.
7. Dry the grid for 10 s by holding it within 2–3 cm of a lamp bulb (we use an illuminator having a 30-W bulb and a polished metal reflector) and then dry it for

10 min, carbon side up, on a filter paper that is under the lamp and approx 12 cm from the bulb.

3.2. Specimen Tilting

Although specimen grids are readily tilted in an electron microscope fitted with a goniometer, when dealing with flexible helical nucleoprotein complexes one is confronted with two problems: first, to find a specimen that will show the helical asymmetries when it is tilted, and second, to determine which end of a tilted helix is closer to the viewer when the helix is seen on the fluorescent screen, in negatives, and in prints. We deal with these problems in the following:

1. Flexible nucleoprotein helices will generally not have straight helical axes and uniform turns as in **Fig. 2**, but a much less regular structure. It is essential to select a segment of a helical complex in which the turns are as regular as can be found, with a roughly linear (or only gently curved) helix axis extending for 5–10 helical turns. **Figure 3** contains a set of drawings taken from an actual tilt experiment *(3)*. The helix axis of a relatively linear and regular helical segment has been oriented approximately perpendicular to the tilt axis (which is horizontal in **Fig. 3**) to maximize the changes observed upon tilting. The characteristic left-side and right-side indentations are seen when the complex in **Fig. 3** is tilted in opposite directions (+55°, –55°), even though there is significant flexing in the helix axis. The change of indentations from one side to the other is faintly visible even in a diagonally oriented helical segment (*arrows*), but the effect is more convincing in a segment that is perpendicular to the tilt axis.
2. The complexes must also have been prepared under conditions that yield a minimum of flattening of the helical structure; the use of a relatively deep layer of negative stain will tend to help support the three-dimensional structure. Flattening of the helix makes the left-hand-side and right-hand-side indentations in tilted helices more difficult to observe, but they still can be seen when the helix is distorted. Note that in **Fig. 3** the width at *a* appears to be about half of the width at *b* in the helix that is tilted –55°, indicating that the helix is flattened so that the height of a helical coil above the support film is roughly half of the width of the coil measured in a direction parallel to the support film.
3. In order to determine the absolute orientations of helical specimens as they are seen on the fluorescent screen of an electron microscope, focusing effects are utilized. An asymmetric marker (such as a macroscopic letter R punched on a specimen grid) can be used to correlate what is seen on the screen with the known position of the tilted grid (oriented, for example, so that the top edge of the R is uppermost in the column at a tilt angle of +55°). We used such a device in the top-entry stage of our Zeiss EM10C to demonstrate that when the objective lens current is adjusted to focus on the central portion of a steeply tilted grid, then the edge of the grid that is uppermost in the microscope and furthest from the upper polepiece of the objective (imaging) lens will be *overfocused* (lens current too strong to focus on it), whereas the lower edge of the grid will be *underfocused*

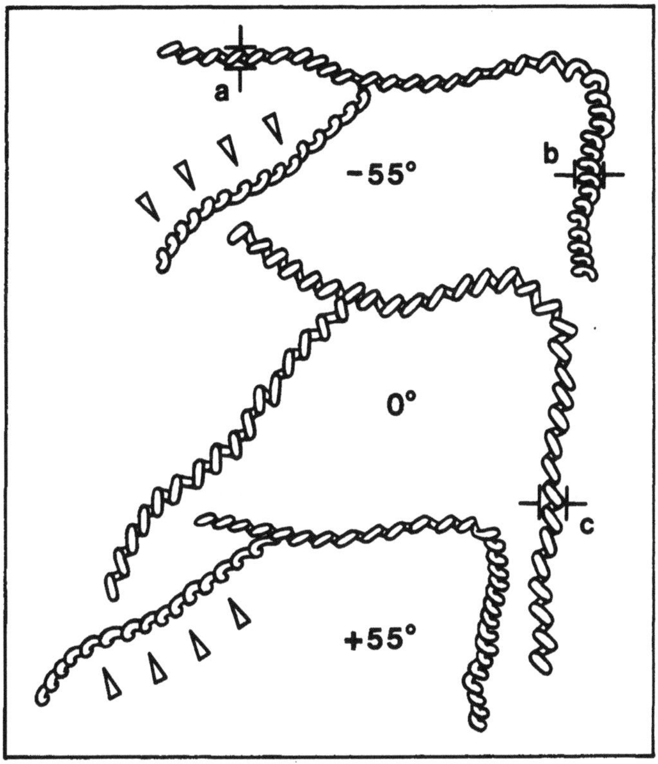

Fig. 3. Drawings made from electron micrographs of a left-handed helical complex of the IKe gene 5 protein with single-stranded fd viral DNA. The center image is of the nontilted complex that lies in the plane of the page. The upper image is of the same complex tilted –55° around a horizontal axis, so that the top of the complex is below the plane of the page and the bottom of the complex is closer to the observer. The lower image is of the same complex tilted +55°, so that the bottom of the complex is below the plane of the page and the top of the complex is nearer to the observer. The drawings show the helical coils as three-dimensional structures to show how they can account for the projection images seen in the original electron micrographs, which are shown in **ref. *3***. (Reprinted with permission from **ref. *3***.)

(lens current too weak). These focusing effects can be used to determine which end of any helical complex is uppermost in the column; that same upper end of the helix will be, in effect, the end that is nearest to the viewer who is looking at the helix on the fluorescent screen.

4. The differing Fresnel patterns due to underfocusing and overfocusing of a tilted specimen support film are visible in a negative bearing an image of the specimen. Hence, proof of the three-dimensional orientation of the specimens is contained in each negative. To correctly interpret the hand of a helix from an image, the negatives must be viewed from the emulsion side, which in the microscope faces

in the same direction (toward the electron beam and toward the viewer) as does the viewing surface of the fluorescent screen. It is, of course, essential that the operator correctly recognize the patterns corresponding to overfocusing as opposed to underfocusing; if in doubt, the patterns can be demonstrated by making exposures of deliberately underfocused and overfocused specimens at zero degrees of tilting.

5. The Fresnel patterns will also be visible in prints made at a suitable magnification. To present the helical asymmetries correctly, the prints must be made with the emulsion side of the negative facing the enlarger light source, so that the print will represent the same view of the object as that which is seen on the fluorescent screen.
6. Finally, we emphasize that it is essential to demonstrate that the indentations between helical turns switch from one side to the other of the same helical nucleoprotein segment as the helix is tilted in opposite orientations. We find that the flexing and partial compression of nucleoprotein helices can produce structures that show indentations on one side in one tilt orientation, but that do not show indentations on the other side when the helix is tilted in the opposite orientation. A satisfactory proof of the helical symmetry (hand of the helix) requires that *both* symmetrically related effects be demonstrated in the same helical segment.

4. Notes

1. The purity of the water used as solvent can be critical to the success of preparations for electron microscopy. The chemical content of the water supplied to a laboratory varies greatly with the locale, and it sometimes happens that procedures that worked in one location will fail in another, when the only reagent not carried to the new location is the water. If difficulties are encountered with experimental procedures, alternative sources of water and alternative water purification protocols should be explored.
2. The exercise of care to preclude the contamination of solutions is essential to the success of these procedures. Negative staining of specimens mounted on glow-discharge-activated grids is a widely used technique, but long and flexible nucleoprotein helices are particularly susceptible to the effects of contaminants that can interfere with the uniform adsorption of the complexes to an activated carbon film. The adsorption may fail to give satisfactory results if appropriate precautions are not taken.
3. We find that a more uniform layer of uranyl acetate stain can be obtained if the carbon-coated grids are hydrated prior to **step 5** of **Subheading 3.1.** This is done by placing the grids carbon side up in a small plastic Petri dish, 2 cm in diameter, and then floating the small Petri dish on distilled water that has been brought to 90°C and poured into a standard Petri dish bottom. Another standard Petri dish bottom is inverted over the first, creating a chamber in which steam is captured and will be exposed to the carbon coatings on the grids. After 10 min of hydration, the grids are dried under a lamp for 10 min and are immediately subjected to glow-discharge activation. The depth of the uranyl acetate stain layer can be con-

trolled by varying the times of hydration exposure and drying. In our hands, this procedure yields satisfactory staining on every grid, provided that there is no interference resulting from contaminants in the specimen.

Acknowledgments

The author gratefully acknowledges support of this work by National Institutes of Health Research grant 5-RO1-GM34293-03 (to C.W.G.), by NIH Biomedical Research Support grant 2S07-RR07133-21, by NIH Small Instrumentation grant 1-S15-NS25421-01, and by National Science Foundation Instrumentation grant PCM-8116109.

References

1. Bujalowski, W. and Lohman, T. M. (1991) Monomer–tetramer equilibrium of the *Escherichia coli ssb-1* single-strand binding protein. *J. Biol. Chem.* **266,** 1616–1626.
2. Kodadek, T. (1990) The role of the bacteriophage T4 *gene 32* protein in homologous pairing. *J. Biol. Chem.* **265,** 20,966–20,969.
3. Gray, C. W. (1989) Three-dimensional structure of complexes of single-stranded DNA binding proteins with DNA: IKe and fd *gene 5* proteins form left-handed helices with single-stranded DNA. *J. Mol. Biol.* **208,** 57–64.
4. Bochkarev, A., Pfuetzner, R. A., Edwards, A. M., and Frappier, L. (1007). Structure of the single-stranded-DNA-binding domain of replication protein A bound to DNA. *Nature* **385,** 176–181.
5. Skinner, M. M., Zhang, H., Leschnitzer, D. H., Guan, Y., Bellamy, H., Sweet, R. M., et al. (1994). Structure of the gene V protein of bacteriophge f1 determined by multiwavelength X-ray diffraction on the selenomethionyl protein. *Proc. Natl. Acad. Sci. USA* **91,** 2071–2075.
6. Olah, G. A., Gray, D. M., Gray, C. W., Kergil, D. L., Sosnick, T. R., Mark, B. L., et al. (1995) Structures of fd gene 5 protein-nucleic acid complexes: a combined solution scattering and electron microscopy study. *J. Mol. Biol.* **249,** 576–594.
7. Gray, C. W., Brown, R. S., and Marvin, D. A. (1981) Adsorption complex of filamentous fd virus. *J. Mol. Biol.* **146,** 621–627.
8. Finch, J. T. (1972). The hand of the helix of tobacco mosaic virus. *J. Mol. Biol.* **66,** 291–294.
9. Haschemeyer, R. H. and Myers, R. J. (1972). Negative staining, in *Principles and Techniques of Electron Microscopy* (Hayat, M. A., ed.), vol. 2, Van Nostrand Reinhold, New York, pp. 101–147.
10. Valentine, R. C., Shapiro, B. M., and Stadtman, E. R. (1968). Regulation of glutamine synthetase, XII. Electron microscopy of the enzyme from *Escherichia coli. Biochemistry* **7,** 2143–2152.

41

Scanning Transmission Electron Microscopy of DNA–Protein Complexes

Joseph S. Wall and Martha N. Simon

1. Introduction

The scanning transmission electron microscope (STEM) at Brookhaven National Laboratory* (BNL) is nearly unique in its ability not only to image isolated unstained and unshadowed biological molecules but also to obtain quantitative information about them such as their oligomeric state *(1)* (*see* **Note 1**). This makes it ideal, in principle, for looking at protein–DNA complexes and obtaining information about the masses bound. However, there are very stringent requirements on the purity and stability of the samples.

The quantitative STEM offers significant advantages over other techniques for characterizing biological complexes (*see* **Note 2**). The STEM operates in a dark-field imaging mode which gives high contrast for small objects. The STEM forms an image one point at a time by rastering a finely focused electron beam over a specimen in a TV-type scan. Typically 1000 electrons are used for the readout of each picture element (pixel) in the specimen. Most of these pass directly through the specimen without losing energy or changing direction. However, a few of these electrons interact with the specimen and are scattered into two annular detectors. The number of scattered electrons is directly proportional to the mass thickness at that point.

The digital STEM image is a two-dimensional projection map of the specimen's mass distribution. Adding up the mass in regions of the image containing a complex gives the molecular weight, which usually identifies the complex and determines its oligomeric state. Thus, a link is established between biochemistry and imaging.

*The BNL STEM is a NIH Supported Resource Center, NIH P41-RR01777, with additional support provided by the Department of Enrgy and Office of Biological and Environmental Research.

From: *Methods in Molecular Biology, vol. 148: DNA–Protein Interactions: Principles and Protocols, 2nd ed.*
Edited by: T. Moss © Humana Press Inc., Totowa, NJ

The high contrast and signal-to-noise-ratio (S/N) of a STEM dark-field image permits direct visualization of unstained double-stranded DNA (dsDNA) and can provide an accurate measurement of its total length and the position of a bound complex (*see* **Note 3**). The high contrast of the STEM also reveals any problems with the specimen such as denatured protein or residual salt. This forces rigorous attention to details of specimen preparation that, although sometimes annoying, leads to results having high credibility.

We distinguish three types of complex that place different constraints on specimen preparation, imaging, and analysis: (1) one-dimensional complexes *(2)* such as a protein binding to a specific sequence of nucleic acid, (2) two-dimensional complexes *(3)* such as chromatin, filamentous viruses and recA-type filaments, and (3) three-dimensional objects such as ribosomes *(4)*, "spherical" virus particles *(5)*,and intermediates in such viral assembly.

Objects of the last type tend to be the easiest to work with because the components are bound together in a relatively rigid and stable complex. Adsorbing such objects from solution onto a suitable EM substrate such as carbon film generally causes minimal perturbation. The particles tend to have characteristic shapes and relatively few distinct orientations, so it is easy to recognize broken or incompletely assembled particles. Compact shapes with distinct boundaries make it relatively easy to identify complete particles for mass measurements.

Two-dimensional complexes are usually curved and may also be tangled in solution, so depositing them on a planar substrate may cause distortions or topological problems. In this case also, the outside edges of the complex are usually composed of a relatively uniform compact protein shell, which is not significantly disrupted by contact with the substrate. Any areas that have unraveled are easily recognized as different from sharp, well-preserved sections for mass/length determinations.

One-dimensional complexes tend to be the most difficult with which to work. In solution, the dsDNA itself tends to be a random coil with low rigidity, so docking onto the substrate can give an uninterpretable tangle if the strands are long. Searching for bound proteins on this can be slow, so the best approach is to use the shortest strands compatible with the experiment. A more serious problem comes from the fact that such complexes contain relatively large stretches of bare nucleic acid that attach poorly to the carbon film substrate. The entire coil in solution is undergoing rapid translational and rotational Brownian motion. If the protein sticks first, as usually happens, the mechanical forces transmitted to the binding region are frequently greater than the specific binding forces and the complexes are disrupted. Shorter strands help, but covalent crosslinking is usually required to give reproducible complexes. Alternatively, poly-lysine pretreated grids will bind the dsDNA well, but, sometimes, this forces a weaker binding protein off so that the complexes are disrupted, and, again crosslinking may be required. However, if the strands are too short (e.g., made from oligomers of 20 bases) such that the protein complex covers them, it is not possible to determine the orientation of the complex on the DNA.

The use of general fixatives to crosslink, such as glutaraldehyde and formaldehyde, raises the question of generating artifacts such as nonspecific binding or binding of extraneous proteins and consequent increases in masses. The best scenario is to have is a specific crosslinker between the protein(s) and the dsDNA, but that is not always available.

Whatever is used should be tested on the complex, and the minimum conditions to maintain it should be determined. This can be done by footprinting, gel retardation, or some other assay of the crosslinking conditions. The crosslinking need not be 100% and that should actually probably be avoided, because anything that is *over*crosslinked will be uninterpretable (*See* **Note 4**.) If more than 50% of the molecules are in complexes and the concentration is good (roughly 10 complexes per square micron), searching in the microscope can be fairly rapid if the strands are reasonably short (<1000 bp). The histogram of mass bound as a function of distance from the end of the strand is the final test of specimen quality.

2. Materials

We will describe in some detail the materials we use, as the physical purity of most materials is critical to the success of a STEM project. The grids for the STEM are prepared by the STEM staff, but the specimens themselves come from the users. (*See* **Note 5**.) Knowing the details involved in specimen preparation enables users to understand the problems that can arise. One of the most important is the *physical purity* of all of the materials involving the sample, which may end up on the grid in the STEM.

2.1. Water, Buffers, and Salts

1. Of surprising importance is that the water used in all the steps of sample preparation must be physically clean. The water in some labs is very impure. STEM water is deionized and freshly distilled daily. It is used for all grid and sample preparation as well as for all buffers (*see* **Note 6**).
2. All solutions must be physically clean. Anything "dirty" involved in the purification such as fragments of column or gel material will show up on the grid and interfere with the observations and analysis. Any physical additives that scatter electrons will also interfere (*see* **Note 7**).
3. Most samples can be applied to the grid in the buffer which is necessary for their biological activity. Additives such as sucrose, glycerol, and dithiothreitol will usually wash off. If necessary, a sample can be applied to the grid in high salt. (A few washes with a high concentration of ammonium acetate will usually remove it.) (*See* **Note 8**.) However, some buffers and salts are known to cause

problems (*see* **Note 9**). HEPES and MOPS are good buffers. The final washes of the grid before freeze-drying must be with a volatile solution such as freshly prepared ammonium acetate or STEM water. Because ammonium acetate is a poor buffer, it can be used at any desired pH.

2.2. Grids and Films

1. We use titanium grids for the STEM. This is for ease of handling in the microscope and for their good thermal behavior. The supporting film (**step 3**) is thin carbon. The grids are at liquid-nitrogen temperature in the microscope to prevent contamination and reduce radiation damage to the specimens. Carbon and titanium have similar thermal properties, so there is no differential contraction or expansion that would result in tearing of the film. Also, compared to copper grids, titanium grids are flat and chemically inert. However, to mimic STEM conditions in a conventional TEM, there are many choices for grids.
2. The slots in our titanium grids are too large to support the thin carbon film necessary for STEM samples, so we put a "holey" film over the slots. This is a plastic film made on glass slides under controlled humidity conditions so little droplets of condensation determine the hole size, which is approx 5–10 mm in diameter. The plastic holey film is heavily shadowed with carbon for strength. Again, to mimic our conditions, there are many available choices of grids with support for small areas of thin film.
3. The thin carbon film is critical to simulating STEM conditions for samples. Because carbon film behaves like activated charcoal in adsorbing everything from the atmosphere, it is made so that the side to which the sample is adsorbed has never seen air. We make it by lightly shadowing carbon (for the STEM, it needs to be 2–3 nm thick) onto a piece of freshly cleaved rock salt in an ion-pumped vacuum system. (*See* **Note 10.**) We buy NaCl crystals from Bicron. We cleave them with a Weck safety razor blade, which is placed on a corner edge of the crystal (to give a piece several millimeters thick) and tapped with a hammer. The freshly cleaved side is lightly breathed on before putting it face down in the bell jar. The carbon rods (from Fullam) are located below the rock salt in the shadower so that cinders do not drop on it. (*See* **Note 11**.)
4. We often use poly-lysine pretreated grids for complexes on dsDNA, which does not stick well to carbon grids. However, if the grids are pretreated with poly-lysine, dsDNA binds very well. We use poly-lysine from Sigma, approx 3000 Daltons, and dilute and freeze away aliquots at 10 μg/mL.
5. We sometimes fix complexes with low concentrations (approx 0.1%) of glutaraldehyde for relatively short times (approx 15 min) before applying them to the grid. The sample washings will stop the fixation so that it does not have to be quenched. (*See* **Note 4.**) We start with a new ampoule of 8% EM grade from Polysciences. We usually open a new ampoule, but if the air is flushed out with dry nitrogen gas and it is resealed, it can be kept for a few days for further use.

2.3. Concentrations of Complexes and Components

1. Only small quantities of sample are needed for examination in the STEM, but it is useful if the concentration is relatively high. Because different samples absorb to the thin carbon film differently, a concentration series for a new specimen is made. Twenty to 50 µL of a sample at 100 or 200 µg/mL is adequate. (*See* **Note 12**.)
2. It is often a good idea to look at the components tht go into a complex separately. On poly-lysine pretreated grids, a dsDNA concentration of 1–5 µg/mL is plenty. Depending on its size, a protein concentration of 50–100 µg/mL is good. Again, only 10–50 µL is needed.
3. The ideal sample is frozen in aliquots to be thawed and used as needed. However, many samples cannot be frozen. They can be shipped overnight on wet ice and the grids can be made on the day of arrival (if it has been arranged). The grids are stored under liquid nitrogen until microscope time is available.
4. Occasionally, a complex is unstable, either over time or shipping conditions. Sometimes, frozen aliquots of the components can be shipped, but the complex itself cannot be frozen or it may not be stable overnight. Under those circumstances, we can assemble (and incubate) a complex just prior to putting it on the grids. We need a detailed protocol and all of the components.

3. Method

3.1. Specimen Preparation

Details of specimen preparation for grids for the STEM are given in this section. Specimens for the STEM are prepared by the STEM staff (*see* **Note 5**), but knowing the details enables users to understand the potential problems. A user with EM experience in other ways of visualizing proteins or DNA or complexes can mimic these conditions to screen their samples in a conventional TEM.

3.1.1. Wet Film, Hanging Drop Method

1. The thin (2–3 nm) carbon film, which has been shadowed onto freshly cleaved rock salt, is floated off the crystal onto a dish of STEM water (*see* **Note 6**).
2. The grids, covered with holey film (and for the STEM, in rings and caps) are placed face down on it for approx 1 h.
3. A grid is picked up from above and retains a droplet of water. It is washed and wicked with STEM water two or three times, never being allowed to dry. Water is applied from one side with an Eppendorf pipet and wicked from the other side with filter paper.
4. Three microliters of tobacco mosaic virus (TMV) at 100 µg/mL is injected into the drop and allowed to adsorb to the thin carbon film for 1 min.
5. The grid is then washed and wicked two or three times with injection buffer for the sample. Three microliters of the specimen is injected into the drop and allowed to adsorb for 1 min.

6. The grid is washed with sample buffer a couple of times, followed by washes (approx 10), ending with washes of 20–50 m*M* ammonium acetate (or, rarely, water). The intermediate washes depend on the sample buffer (e.g., high-salt buffers require washes with high ammonium acetate). (*See* **Notes 8** and **9**.).
7. For freeze-drying the grid, after the final wash the liquid is briefly blotted between two pieces of filter paper to a thickness of a few micrometers and rapidly plunged into liquid-nitrogen slush to freeze it.

3.1.2. Poly-lysine-Pretreated Grids

The dsDNA does not bind well to carbon films. However, if the grids are pretreated with poly-lysine, dsDNA binds very well to them. Poly-lysine grids that have been made by this method contribute no additional detectable background.

1. The thin carbon film is picked up as above and washed with water.
2. Three microliters of poly-lysine at 10 μg/mL is injected into the drop and allowed to adsorb to the carbon film for 1 min.
3. The grid is then washed with water, approx eight times, and allowed to air-dry.
4. Before use, 3 μL of water is put on its surface before the TMV is injected. Because TMV also binds very well, a solution of 10 μg/mL should be used for these grids. (*See* **Note 13**.)

3.1.3. Fixation

In any multicomponent system, one part may bind more strongly to the carbon film substrate. The rest of the complex may be coming apart through Brownian motion as described earlier. Frequently, some kind of crosslinking or fixation is required to keep the complex together.

The best results are when the users have been able to fix or crosslink their sample before sending it to us. Sometimes they have determined the minimal fixation or crosslinking conditions for their sample, but it needs to be done just prior to applying the sample to a grid. Unless something very unusual is called for, we can do that.

If the complex is falling apart and optimal fixation conditions to preserve it are not known, we will often try a brief glutaraldehyde fixation on the sample just prior to applying it to the grid. The washes involved in specimen preparation will stop the fixation without having to worry about quenching conditions. Sometimes, this will help preserve the complexes, but it may not.

3.2. STEM Operation

The STEM is usually run by a trained operator to assure optimum data quality and efficient use of microscope time. Frequent users can obtain training on request, but the alacrity of most STEM experiments makes this unnecessary.

The main factors considered by the STEM operator in taking data are described as follows.

All biological specimens are sensitive to electron irradiation (*see* **Note 14**). Because unstained, freeze-dried specimens have only unprotected biological material, radiation damage to them is particularly noticeable and limits quantitative interpretation. Therefore, searching for suitable areas must be done at the lowest possible dose (low magnification). Potential areas of interest (AOIs) are boxed on the display for later scanning. Near the AOIs, but not overlapping them, focusing areas are placed. The STEM beam is highly collimated with essentially no intensity outside the 0.3-nm-diameter focused spot, so scanning one area causes no damage to nearby areas. Once the microscope focus is verified, an AOI is scanned and displayed (8.5 s per scan). At that point, the scans are shut off (meaning there is no additional radiation damage) while the operator evaluates the quality of the data in the scan using the following criteria: background cleanliness, TMV profile, TMV mass, numbers, shapes, and masses of sample particles. If the area is suitable for more complete analysis, it is recorded digitally and processed with another computer program (*see* **Subheading 3.3.**).

The STEM control computer is a PC with a custom interface designed to take full advantage of STEM capabilities. The computer directs the focused beam, scanning the next AOI upon command and recording the electron counts striking the two annular detectors (large-angle and small-angle scattering) surrounding the beam. Most of the beam passes through the irradiated spot on the specimen and strikes a bright-field detector used for normalization. The large-angle annular detector provides the most useful dark-field signal, which directly maps the local mass thickness of the specimen. The computer display shows two 512 × 512-element images (large-angle and small-angle detectors) where the intensity is proportional to the detector signal. Objects as small as single heavy atoms or dsDNA are easily visible, especially with zoom or contrast enhancement. During this decision making, the beam is not scanning, so there is no additional damage.

Once the image is saved, it can be viewed by any networked computer. Image files are usually distributed by FTP transfer on the Internet (requiring 1–10 s per image) to users. For users without Ethernet, images can be put on a CD (1300 images/CD) and mailed.

3.3. Analysis

The STEM images stored on a CD or a hard drive can be viewed using commercial software or our mass-analysis program, PCMass, provided by the STEM Group. The format for the BNL STEM images is a header 4096 bytes long, followed by two 512 × 512, 8-bit images interleaved. Adobe PhotoShop can read these images directly using its "AS RAW" option with the above parameters. We recommend this program for producing publication-quality images with a high-quality printer.

The PCMass program provides for rapid viewing of STEM images, with the ability to perform accurate mass measurements on them. This program "reads" the image header and imports the image data with appropriate settings. Mass accuracy is most often limited by the cleanliness of the background carbon film between particles. The first step in analysis with PCMass is to mask out particles and dirt and measure the background mass per unit area (mass–thickness) in the cleanest areas. There are several diagnostic criteria to indicate the severity of background problems. The background determination is critical because the mass of the carbon substrate in the measuring area is usually equal to or greater than the particle mass and it must be subtracted to get the net particle mass. If the background quality is less than ideal, mass measurements can still be made, but their interpretation requires caution.

Manual mass measurements require the user to position a measuring circle or rectangle around particles of interest using a mouse and keystrokes to change size. A mouse click or '=' keystroke saves the measurement in a database and shows the running average and standard deviation for particles in the selected category. Automated mass measurements use a set of simple comparison models to align and categorize particles, providing size and shape information as well as mass. This offers the advantages of speed and reproducible particle selection, but it is not suitable for complicated specimens such as convoluted DNA–protein complexes.

Summarizing several thousand (automated) mass measurements for publication involves answering two basic questions: (1) Are the particle selections reliable and unambiguous? (2) Is the standard deviation of the mass measurements what would be expected for a homogeneous population of particles with the observed shape? Frequently, particles will break apart into subunits either during preparation or attachment to the grid. This may be obvious both by careful inspection of the images and by mass measurements of the various species observed. Knowing the molecular weights of the components facilitates this sorting. If the automatic particle selection with model fitting was used, one can view histograms for all the fitting parameters and choose those selection parameters that exclude clear outliers. However, one must be careful not to skew the "real" mass distribution.

The expected mass accuracy depends on particle size, shape, and measuring dose. A 100-kDa globular particle on a 2-nm carbon substrate measured with a dose of 10 el/$Å^2$ should give a standard deviation (SD) of 10%, whereas a 1-MDa particle should give 2% (*see* **Note 15**). If the particle is extended or the background is dirty, the SD will be worse. The TMV reference particles should give mass per unit length of 13.1 kDa/Å with a SD of 1%. With a large number of measurements, one may be tempted to quote the standard error of the mean, but this is overly optimistic. Instead, we find that the standard deviation of image averages (for all images having more than 10 particles) gives a truer idea of accuracy. One can also plot the total mass histogram and fit multiple Gaussians to it. If the spread in masses is larger than expected, caution is indicated.

A poor standard deviation, worse than that predicted, is most often caused by random background problems such as denatured protein, residual salt, and so forth. Unfortunately, these also give systematic errors that can shift the measurements in either direction. Residual salt tends to accumulate around the edges of particles (a meniscus effect), which raises the apparent mass. On the other hand, denatured protein may not be masked completely by the background program, giving too much subtraction for the background and lowering the apparent mass.

The example shown in **Fig. 1** is particularly complex because it has four dsDNA arms in a Holliday junction complex (*see* **Note 16**). In this example, the *M/L* of the dsDNA is greater than expected because of deviations from straightness and/or salt binding. The particle mass is also larger than the expected value (800 kDa + 2 × 2 kDa/nm × 26 nm = 900 kDa), presumably the result of binding of fixative or salt.

The TMV control particles are useful in identifying these problems. Residual salt tends to accumulate in their hollow central hole and along their edges, giving an increased *M/L* and an altered radial mass profile. Denatured protein from the sample tends to decrease the TMV *M/L*, because the TMV is added before the sample and tends to protect the substrate from deposition of this denatured protein. Consequently, one cannot assume that specimen problems observed on the TMV will be the same for the specimen of interest and some "fudge factor" will make everything come out right. Rather, it is well worth the extra effort to try to solve the specimen problems by using added purification, fixation, other buffers or stabilizing agents (*see* **Subheading 3.1.**).

The results from the automated PCMass program are tabulated as TMV *M/L*, TMV SD, and the number of particles passing the selection parameters vs the number measured for each image file. On the same line are the results for particles selected relative to one model, followed by apparent dose and background thickness. At the bottom of the summary, the global averages, standard deviations, and a mass histogram for the selected model are given. This provides a convenient format to identify images or blocks of images with problems (e.g., poor TMV). It is not unusual for a single grid to have significant variations in specimen quality from one area to another, so we strive to collect good image data in at least three widely separated areas on each specimen. What is meant by "good" data is then clear: Discarding data based on poor

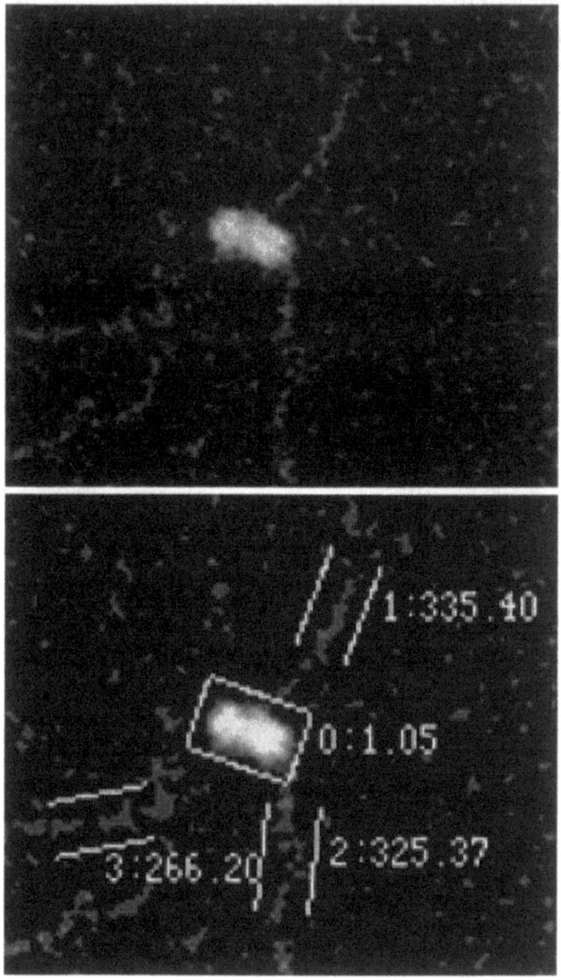

Fig. 1. A STEM micrograph **(top)** of a RuvAB-Holliday junction complex *(6)*, illustrating the mass measurement program **(bottom)**. The box encloses an area 26 nm wide and 15 nm long, which contains a mass of 1.05 MDa. The three sets of parallel lines (along three of the DNA branches) enclose areas 12 nm wide by 25 nm long with mass per length approx 300 Da/Å. Clean, straight dsDNA should have a $M/L = 200$ Da/Å.

TMV measurements is legitimate, whereas discarding data based on poor (unexpected) particle mass is suspect.

3.4. Conclusion

The STEM, with its ability to provide both mass and structural information, offers an excellent method for studying protein–DNA complexes. The samples do have to be very clean and the complexes have to be stable under conditions for applying them to our grids. All of the above information is meant to help address the potential problems that can arise. However, if these conditions are met, it is a relatively rapid and inexpensive method of obtaining a lot of information about the composition and location of proteins bound to dsDNA. The quantitative link between observed structure and biochemistry provided by mass mapping, makes image interpretation highly believable.

4. Notes

1. One cannot see bare dsDNA in a conventional TEM. It can be visualized if it is stained, shadowed, or imbedded in a spreading solution, but under those conditions, the mass of any protein bound to it cannot be determined.
2. The STEM provides both mass and structural information simultaneously, as does small-angle scattering. Although the STEM does not have the atomic resolution of X-ray crystallography, NMR, or cryoEM of 2-D arrays, it also does not depend on the ability of samples to form arrays. Comparatively, it is rapid and not very labor intensive. Scanning probe microscopies potentially can offer similar capabilities. Electron spectroscopic imaging techniques are quantitative and offer the possibility of distinguishing nucleic acids from proteins *(6)*.
3. Single-stranded DNA (ssDNA) has been visualized in the STEM but only under conditions where it has a lot of secondary structure. If it were fully extended or stretched, it would be barely detectable in the STEM above the thin-carbon-film background.
4. Crosslinking conditions can be quite different for different samples. Because conditions depend on concentration, temperature, and time, frequently the crosslinking has to be stopped by quenching. It is important that the quencher is either removed or is known not to interfere with visualization of the complex in the STEM. Overly crosslinked samples turn into large uninterpretable aggregates.
5. The STEM is a NIH Research Resource. As such, it is available to users with appropriate projects free of charge. A project is usually initiated by a discussion of it on the phone or by e-mail. A trial sample is sent by overnight mail to us. We can make the grids on the day it arrives, if it has been prearranged. The grids are stored in a grid fridge under liquid nitrogen (stable for years) until they can be freeze-dried and examined in the microscope. Additional details about the facility can be found on our web page, most easily accessed from www.bnl.gov, then click on Scientific Facilities, then on Scanning Transmission Electron Microscope. Our address is http://bnlstb.bio.bnl.gov/biodocs/stem/stem.htmlx
6. A water control grid can be made if it is likely that the water from a lab is a problem. A 5-µL drop of water is applied to and completely dried on a grid in a laminar-flow clean hood (which is where the samples are made) and examined in the STEM. If there is a problem, STEM water (deionized and freshly distilled) can be sent to the lab to make the buffers for the final steps of sample purification. Clean air during sample preparation is also critical, as "dust" particles can carry both physical and biological contaminants.

7. Most physical additives (which may improve the biological activity) such as bovine serum albumin, trypsin, and high concentrations of peptides, oligonucleotides, or PEG cannot be used. One way to remove these is by passing the sample of interest over an appropriate sizing column, if one can be found where it comes off in the void volume. This can also be a useful way of exchanging salts or buffers. If one is in a lab where bacteria are a problem, low levels (0.2%) of sodium azide can be tolerated in buffer solutions. It seems to wash off grids and not interfere with STEM samples.
8. Ca^{2+} often does not wash off well, but Mg^{2+} usually does. A buffer containing Ca^{2+} can be washed with Mg^{2+}, then ammonium acetate.
9. Some salts and buffers are likely to cause problems. Phosphate buffers often do not wash off well and leave bright spots, which interfere with the analyses. Sometimes, Tris buffers leave a bad background and they also interfere with glutaraldehyde fixation. A buffer control grid is made for a new sample (especially in an unusual buffer) to see how well the components of the buffer wash off.
10. Carbon films are often made by shadowing onto cleaved mica. We have found that films made on freshly cleaved NaCl crystals will float off onto a dish of water much easier than off of mica. Mass analysis of the two background films are indistinguishable, indicating that they are just as flat.
11. Many experimental details about the grids, films, and shadowing can be found in *Cells: A Laboratory Manual*, Vol. 3, pp. 125.2–125.7 (CSH Press, 1998). However, to simulate STEM conditions for a sample, the only critical step is the thin carbon film because it is the only surface that the biological materials will see. Some additional capabilities of the STEM can be found in the same volume, Chapter 124.
12. **Caution:** Steps taken to concentrate a sample, such as Centricon filters, may also concentrate contaminants of the same size.
13. Many, but not all, molecules adsorb well to poly-lysine-pretreated grids. It may be necessary to change the suggested concentrations to lower ones when using them.
14. Biological specimens are sensitive to radiation damage by the electron beam. This manifests itself in two ways, as a loss of mass and a loss of fine detail. For STEM mass measurements at 40 keV with the specimen at –150°C, the rate of mass loss is roughly 0.25% for every 1 el/Å^2. The normal STEM imaging dose is 10 el/Å^2, which results in 2.5% mass loss per scan. Measurements are usually done on first scan images, but it is instructive to do a dose-response curve using sequential scans of the same area for a new specimen type. Most protein specimens plateau at approx 50% mass loss even at high dose. DNA has much less mass loss, whereas carbohydrate has much more. Resolution loss is seen most easily by examining the ends of TMV rods. These have relatively sharp corners with one end concave and the other convex. After several scans, the corners become rounded, providing a visual monitor of the accumulated dose to that part of the specimen.
15. The expected error (standard deviation) for mass measurements arises from electron counting statistics and thickness variations in the thin carbon substrate. These

can be calculated from first principles, but a more straightforward approach is to use image simulation with the appropriate model shape and mass. The simulation program places particles in random positions and orientations and adds Poisson counting noise to produce an image that should closely resemble the specimen image, but with homogeneous mass. Mass measurements on the simulated image should reflect only the random variations inherent in the STEM technique.

16. This complex was particularly difficult because it contained four dsDNA branches of different lengths, ranging from 660 to 1940 base pairs, which are long enough for tangling. Also because of these long branches, the complex had to be exhaustively fixed to hold together on a STEM grid *(7)*.

References

1. Wall, J. S., Hainfeld, J. F., and Simon, M. N. (1998) Scanning transmission electron microscopy (STEM) of nuclear structures, in *Methods in Cell Biology*, vol. 53 (Berrios, M., ed.), Academic, New York, pp. 139–164.
2. Antoshechkin, I., Bogenhagen, D. F., and Mastrangelo, I. A. (1997) The HMG-box mitochondrial transcription factor xl-mtTFA binds DNA as a tetramer to activate bidirectional transcription. *EMBO J.* **16,** 3198–3206.
3. Citovsky, V., Guralnick, B., Simon, M. N., and Wall, J. S. (1997) The molecular structure of agrobacterium VirE2-single-stranded DNA complexes involved in nuclear import. *J. Mol. Biol.* **271,** 718–727.
4. Tumminia, S. J., Hellmann, W., Wall, J. S., and Boublik, M. (1994) Visualization of protein–nucleic acid interactions involved in the in vitro assembly of the *Escherichia coli* 50S ribosomal subunit. *J. Mol. Biol.* **235,** 1239–1250.
5. Caston, J. R., Trus, B. L., Booy, F. P., Wickner, R. B., Wall, J. S., and Steven, A. C. (1997) Structure of L-A virus: a specialized compartment for the transcription and replication of double-stranded RNA. *J. Cell Biol.* **138,** 975–985.
6. Bazett-Jones, D. P., Hendzel, M. J., and Kruhlak, M. J. (1999) Stoichiometric analysis of protein- and nucleic acid-based structures in the cell nucleus. *Micron* **30,** 151–157.
7. Yu, X., West, S. C., and Egelman, E. H. (1997) Structure and subunit composition of the RuvAB–Holliday junction complex. *J. Mol. Biol.* **266,** 217–222.

42

Determination of Nucleic Acid Recognition Sequences by SELEX

Philippe Bouvet

1. Introduction

Interactions of proteins with nucleic acids play important roles in biological phenomenon. Almost every stage in the regulation of gene expression involves the interaction of proteins with specific nucleic acids sequences. The identification of the nucleic acid recognition sequence of a given DNA-binding protein is therefore often the first step to be undertaken in the study of its biological function. Over the last 10 yr, the SELEX procedure (systematic evolution of ligands by exponential enrichment) has been used to identify high-affinity nucleic acids ligands for a large number of different proteins. The method was first described for the identification of the DNA and RNA target sequences of nucleic-acid-binding proteins *(1,2)* but has since been used for the selection of the nucleic acid sequence ligands for other kinds of molecules *(3)*. SELEX uses the power of genetic selection while taking advantage of in vitro biochemistry. It is a rapid technique that is relatively easy to implement and can accelerate and simplify nucleic-acid/protein interaction studies.

The SELEX procedure involves only the few simple steps described in **Fig. 1**. The procedure consists of the selection of a subset of oligonucleotides from a complex mixture of nucleic acid sequences by repeated rounds of binding to the protein of interest. First, ligand sequences that bind to the target protein are partitioned from the unbound sequences. The bound sequences are then amplified by polymerase chain reaction (PCR). This partitioning and amplification is repeated until a very significant enrichment of nucleic acid sequences that bind to the protein with high affinity is obtained. Finally, these sequences are cloned and analyzed.

From: *Methods in Molecular Biology, vol. 148: DNA–Protein Interactions: Principles and Protocols, 2nd ed.*
Edited by: T. Moss © Humana Press Inc., Totowa, NJ

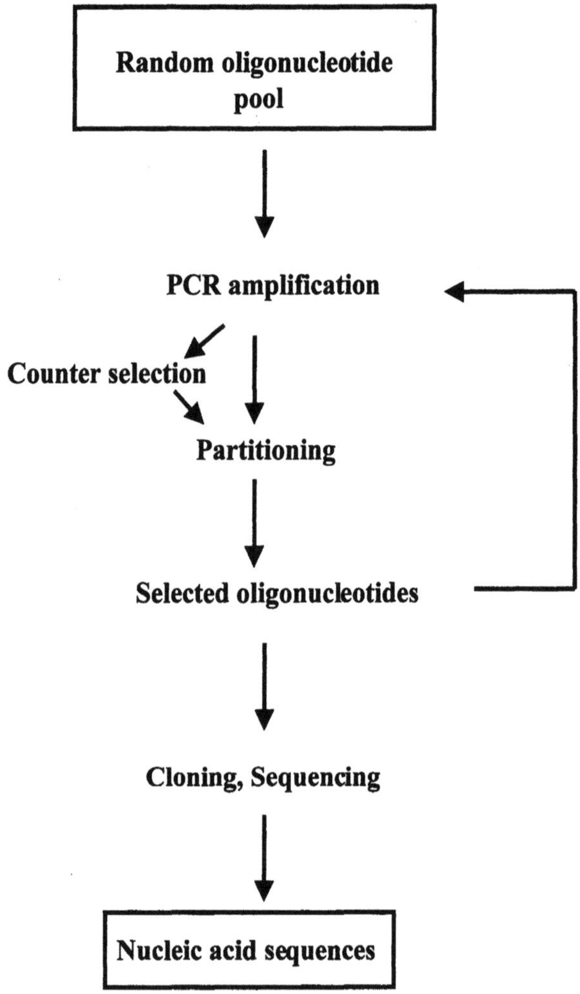

Fig. 1. Schematic representation of the different steps of SELEX.

The strategy is designed to determine an optimal nucleic-acid-binding sequence, also called an "aptamer" for proteins *(4)*. However, a high-affinity nucleic acid ligand can be isolated for apparently nonspecific partitioning agents (e.g., the matrices used for protein immobilization). Therefore, if the strategy is used to determine the nucleic acid recognition sequence of a protein, one must realize that the most difficult part of the study will be the analysis of isolated SELEX sequences and the demonstration that these sequences are relevant to the in vivo function of the protein studied.

Numerous protocols for the SELEX procedure have been used successfully by different laboratories. In fact, each step of this procedure can be optimized

(5) and modified as a function of the characteristics of the nucleic-acid-binding protein being studied.

We provide below a typical protocol that has been used successfully in several laboratories ***(6–9)*** to identify nucleic acid ligands for RNA-binding proteins. The same synthetic oligonucleotide DNA template can be used for the selection of DNA sequences by a DNA binding protein. In this case, **steps 6** and **18** of the method should be omitted.

2. Materials

1. The following synthetic DNA template (*see* **Note 7**) has been used with success by several laboratories: 5' TGGGCACTATTTATATCAAC (N25) AATGTCGTTG-GTGGCCC 3' with these flanking primers T_7 5'-CGCGGATCCTAATACG-ACTCACTATAGGGGCCACCAACGACATT–3' and Rev 5'-CCCGAC ACCCGCGGATCCATGGGCACTATTTATATCAAC–3'. The T_7-Xba primer 5' GGTCTAGATAATACGACTCACTATAGGGG 3' and Rev-HIII primer 5' ACCGCAAGCTTATGGGCACTATTTATAT 3' can be used for the final PCR amplification, and will allow an oriented cloning (*Xba*I and *Hin*dIII) of the PCR product in a cloning vector like pBluescript (Stratagene).
2. Thermocycler.
3. *Taq* polymerase.
4. TAE (Tris–acetate buffer); 40 m*M* Tris–acetate and 0.4 m*M* EDTA (pH 7.5).
5. Partitioning matrix (to be chosen as a function of the studied protein; *see* **Note 1**).
6. Nucleic acid electrophoresis system.
7. NT2 buffer: 50 m*M* Tris-HCl (pH 7.4), 150 m*M* NaCl, 0.05% NP 40, and 1 m*M* $MgCl_2$.
8. Binding buffer (BB): 50 m*M* Tris-HCl (pH 7.5), 150 m*M* NaCl, 20 m*M* KCl, 1 m*M* dithiothreitol (DTT), 0.05% NP 40, 1 m*M* $MgCl_2$, 2.5% polyvinyl alcohol (PVA), 1 m*M* EGTA, 50 µg/mL poly(A), 2 µL/mL vanadyl ribonucleoside complex (VRC), 0.5 µg/mL tRNA, and 125 µg/mL bovine serum albumin (BSA).
9. 5X reverse transcription buffer: 250 m*M* Tris-HCl (pH 8.5), 40 m*M* $MgCl_2$, 5 m*M* DTT, 250 µg/mL BSA, 150 m*M* KCl.
10. 1X transcription buffer: 40 m*M* Tris-HCl (pH 7.5), 6 m*M* $MgCl_2$, 2 m*M* spermidine, 10 m*M* NaCl, and 10 m*M* DTT.

3. Method

1. About 10 pmol of synthetic template DNA (N25) (*see* **Note 2**) is amplified by PCR in a standard 100-µL reaction, in a 500-µL test tube, using 2 µL of a mix of all four dNTPs (10 m*M* each) and 500 ng of each primer. One unit of *Taq* polymerase is added just before the start of the amplification procedure. If the thermocycler does not have a hot cover, the reaction mixture is overlayed with 2 drops of mineral oil.
2. Set up the thermocycler with the following cycle conditions: denaturation 1 min at 94°C, annealing 1 min at 50°C, and elongation 1 min at 72°C (*see* **Note 8**).

After 25 cycles of amplification, finish with an elongation of 10 min at 72°C. The PCR reaction can be stored at 4°C without further purification.

3. Analyze 5 µL of the PCR reaction on a 3% agarose gel in TAE. Run in parallel with a commercial DNA molecular-weight marker giving characteristic bands around 100 bp. The PCR reaction should produce a strongly staining product at 108 bp.
4. Add 100 µL of phenol:chloroform (1:1) to the PCR reaction and mix vigorously for 1 min. After a 5-min centrifugation at 15,000 rpm, the upper aqueous phase is extracted one more time with 1 volume of phenol:chloroform. Ten microliters of 3 *M* sodium acetate (NaOAc) (pH 5.0) and 300 µL of cold ethanol are added. Allow DNA precipitation for at least 15 min at –20°C.
5. The PCR product is recovered by centrifugation (15 min at 15,000 rpm), washed with 70% ethanol, dried, and resuspended in 10 µL of sterile water.
6. In vitro transcription. This step should be omitted for a SELEX with a DNA-binding protein. One microgram of PCR product of **step 5** is incubated in 1X transcription buffer containing 0.5 m*M* of each rNTP, 1 unit of RNasin, and 20 U of T_7 RNA polymerase. The reaction is incubated for 1 h at 37°C. The DNA template is eliminated by the addition of 1 unit of RNase-free DNase and incubation for an additional 10 min at 37°C. After two sequential phenol:chloroform extractions, the RNA is purified through a G50 column (to remove most of the unincorporated nucleotides) then precipitated with 0.1 vol of 3M NaAc (pH 5.0) and 2 vol of 100% EtOH for 15 min at –20 °C. The RNA is pelleted for 15 min at 15,000 rpm, washed with 70% EtOH, dried and resuspended in 20 µL of RNase-free water. One microliter of the transcription reaction is loaded on a 3% agarose gel to check the quality of the RNA. The RNA concentration is determined from its ultraviolet (UV) adsorption at 260 nm.
7. Preparation of the partitioning matrix. The nature of this matrix will depend on the protein to be used (*see* **Note 1**). We will provide here a detailed protocol for a selection procedure using a histidine-tagged protein. Other strategies are mentioned in **Note 1**. Take 2 µL of Ni–NTA agarose beads (Qiagen) and wash twice with 500 µL of sterile water to remove storage buffer (beads are recovered by a 15-s centrifugation in a microcentrifuge). Wash the beads twice with 500 µL of NT2 buffer. During the last wash, divide the suspension into two tubes (tubes A and B), centrifuge, and eliminate 150 µL of the supernatant (i.e., leave beads in 100 µL NT2).
8. Add about 1 pmol of purified histidine-tagged protein to tube A (*see* **Note 10**). Incubate 30 min at 4°C on a roller to allow binding of the protein on the Ni–NTA beads.
9. Centrifuges tube A for 15 s, remove supernatant, then wash the beads twice with 500 µL of NT2 buffer to remove all unbound protein.
10. Centrifuge tubes A and B, and remove most of the NT2 supernatant to leave about 10 µL of buffer above the beads. Ni–NTA beads must be visible at the bottom of the tube.
11. Add 100 µL of BB buffer in tube B (*see* **Note 9**).

12. Add 15 µg of nucleic acid from **step 5** (for a DNA-binding protein) or from **step 6** (for an RNA-binding protein) to tube B (*see* **Note 3**). Incubate 5 min at room temperature. Centrifuge tube B for 15 s. Remove and save supernatant (*see* **Note 4**).
13. Add supernatant from **step 12** to the tube containing the matrix-bound protein (tube A). Incubate 5 min at room temperature. Centrifuge tube A for 15 s, then remove and discard supernatant.
14. Add 1 mL of NT2 buffer (*see* **Note 5**) to tubes A and B. Mix well by inverting the tubes three times. Centrifuge for 15 s to pellet the Ni–NTA beads. Remove as much the supernatant as possible.
15. Repeat this wash four more times. At the last wash, transfer the matrix-bound nucleic acid–protein complex to a new test tube (*see* **Note 6**).
16. After the last wash, leave 100 µL of NT2 buffer in each tube. Add 100 µL of sterile water and 200 µL of phenol:chloroform (1:1). Vortex for 30 s and spin for 5 min at full speed in a microcentrifuge. Repeat the extraction one more time.
17. Recover the upper aqueous phase and add 2 µL of 1 *M* $MgCl_2$, 20 µL of 3 *M* NaOAc, and 700 µL of 100% EtOH. Precipitate for at least 30 min at –20°C, spin for 30 min at 15,000 rpm. Wash the pellet with 70% EtOH, dry and resuspend it in 13 µL of sterile water.
18. Reverse transcribe the bond RNA (should be omitted if the SELEX is performed with DNA). To each tube A and B, add 100 ng of Rev primer (in 1 µL), 2 µL of a dNTP mix (each dNTP at 10 m*M*), 4 µL of 5X reverse transcription buffer, 30 U of RNasin (Promega), and 25 U of AMV reverse transcriptase (Boehringer Mannheim). The reaction is incubated for 5 min at 55°C, then for 1 h at 42°C. Five microliters of this reverse transcription reaction is used directly, without further purification, for the next PCR amplification.
19. Add to the recovered nucleic acid of **step 17** or **18** the reagents necessary for the PCR reaction, as described in **step 1**. Include a control reaction without oligonucleotide template to assure that the PCR reaction is specific. No PCR product should be obtained from the products of tube B (*see* **Note 6**).
20. Purify the PCR product obtained with tube A as described in **steps 4** and **5**.
21. Repeat several rounds of partitioning on fresh binding protein matrix, **steps 6–18** (but *see* **Note 4**).
22. After several rounds of selection (usually between 4 and 10) check for an enrichment of the oligonucleotide pool in the high-affinity ligand for the target protein. This can be done by binding a labeled aliquot of the selected oligonucleotide pool of each cycle (either direct 5' [^{32}P] labeling of the PCR product with polynucleotide kinase for DNA-binding protein, or in vitro transcribed with an [α-^{32}P] rNTP for an RNA-binding protein) following **steps 7–16**. If a substantial enrichment is obtained, as determined by the fraction of labeled nucleic acid remaining on the matrix-bound protein and this fraction reaches a plateau after a given number of partitioning cycles, the selection can be stopped and one should proceed to the cloning step. If no enrichment is observed, more rounds of selection should be performed, either under the same experimental conditions or at higher stringency (*see* **Note 5**).

23. The final PCR can be performed with the primers T_7–Xba and Rev–HIII (*see* **Subheading 2.**). After gel purification, the PCR products are digested with the restriction enzymes *Xba*I and *Hin*dIII, and the oligonucleotides are cloned in a suitable vector (e.g., pBluescript). A number of individual clones are isolated and their inserts are sequenced using standard methodology.

4. Notes

1. Several methods of partitioning can be used as a function of what is available to the researcher. If the target is a tagged recombinant protein (GST, histidine, or any other tag), nucleic acid–protein complexes can be recovered by classical affinity chromatography as described in the present protocol. If the protein is pure but not tagged, filtration of the binding reaction mixture through nitrocellulose (*see* Chapter 1) may allow separation of the bound and unbound molecules ***(15)***. An alternative method of partitioning is to use gel-shift analysis or EMSA (Chapter 2). When using EMSA, labeled oligonucleotides are to be preferred in order to easily identify the nucleic acid–protein complex. Shifted oligonucleotides are eluted from the EMSA gel and used for PCR amplification.
2. A custom-made random oligonucleotide can be easily synthesized using standard chemistry. The variable sequence is flanked by fixed sequences that serve in the PCR amplification steps. Full-length oligonucleotides should be purified before their amplification by PCR. The length of the random sequence may vary from a few nucleotides to as many as a hundred. If a simple DNA-binding site is expected, a random sequence of as little as 20 nucleotides should be sufficient. A library with a short random sequence has also the advantage of being more likely to contain all possible random sequences and therefore to allow the selection of the best binding sequence. Sequencing of a few random sequences should be ideally performed to ensure that the synthesis of the random sequences has not been biased by a preferential incorporation of one deoxynucleotide. If the composition of the random sequence is severely biased, this should be corrected by modifying the percentage of the addition of nucleotides accordingly during synthesis. If a binding site is already known for the protein, the SELEX procedure can be used to determine nucleotides important for binding affinity and specificity. In this case, an oligonucleotide containing a degenerated sequence within the known binding site can be synthesized ***(10)***. Libraries that contain genomic sequences can also be used to identify potential natural binding sites ***(11,12)***.
3. The amount of oligonucleotide present in the binding reaction should be in large excess over the protein. This ensures an efficient competition between ligands for the protein. The ratio oligonucleotide/protein is often within the range of 10 and 1000. The volume of the binding reaction should also be determined as a function of the diversity of the library. A binding reaction of several milliliters might be required if one wished to test all possible sequences (4^n, *n*: number of random nucleotides) present in an initial library.
4. The interaction of the random oligonucleotide pool with the partitioning matrix alone (i.e., without bound protein [called counter selection or negative selec-

tion]) is important to remove oligonucleotide molecules with high affinity for the partitioning matrix itself. This counter selection is not necessary if the DNA bound to the protein is recovered instead using electrophoretic mobility shift assay (EMSA) gel shift (*see* Chapter 2) because only nucleic acid from the shifted band will be used for the next selection cycles. The counter selection should only be performed during the first and possibly the second cycles of selection and should subsequently be omitted in subsequent cycles.

5. The stringency of the binding and washing buffer can be increased if necessary. This could be done by increasing the salt concentration, or by adding 0.5–1.0 *M* urea. Usually, between 1% and 10% of the initial oligonucleotide pool binds to the target protein. Preliminary binding tests should determine the optimal buffer stringency to allow a binding that falls within this range. Buffer stringency can also be increased during the cycling process if no substantial enrichment is observed.
6. Although nucleic acids bind poorly to most plastic tubes, this binding is sometimes sufficient to give a PCR product after the selection reaction. This problem may be overcome by using siliconized tubes and/or by transferring the nucleic acid/protein complex to a new tubes during the washing procedure.
7. The random sequence is flanked by fixed sequences (17–20 nt) to allow PCR amplification with the corresponding primers (*see* **Subheading 2.** for an example of primer sequences). It is important to check that the flanking sequences are not a binding site for the protein. This can easily be done by performing a binding assay between the random pool and the studied protein.
8. The PCR reaction can be modified to allow the production of single-stranded oligonucleotides, the incorporation of modified nucleotides or random mutations, and so forth. If the PCR reaction produces aberrant products (higher-molecular-weight DNA products, smear, etc.), several tests reactions (with various amount of primers, number of cycle) should be made.
9. The composition of the binding buffer should be adapted to the protein that is being studied. The addition of nucleic acid competitors, such as tRNA and homopolymers like poly(A), may be necessary in some selection experiments in order to reduce nonspecific binding. Preliminary tests of interaction of the random pool with the protein can be performed with nucleic acid competitors to determine the best selection conditions.
10. In most published experiments, the SELEX procedure is performed with purified recombinant proteins. However, it can be performed with crude cell extracts containing the protein of interest or with multiprotein complexes if the partitioning procedure allows a specific recovering of the nucleic acid-protein target ***(13,14)***. Epitope-tagged protein can be expressed in cells, or added to cell extract and used for the SELEX. For some SELEX experiments, it could also be interesting to use truncated protein with only the nucleic-acid-binding domain. In some cases, this can significantly reduce nonspecific binding of the random oligonucleotide pool and therefore reduce the number of rounds necessary for the isolation of specific ligands.

References

1. Oliphant, A. R., Brandl, C. J., and Struhl, K. (1989) Defining the sequence specificity of DNA-binding proteins by selecting binding sites from random-sequence oligonucleotides: analysis of yeast GCN4 protein. *Mol. Cell. Biol.* **9,** 2944–2949.
2. Tuerk, C. and Gold, L. (1990) Systematic evolution of ligands by exponential enrichment: RNA ligands to bacteriophage T4 DNA polymerase. *Science* **249,** 505–510.
3. Gold, L., Polisky, B., Uhlenbeck, O., and Yarus, M. (1995) Diversity of oligonucleotide functions. *Annu. Rev. Biochem.* **64,** 763–797.
4. Ellington, A. D. and Szostak, J. W. (1990) In vitro selection of RNA molecules that bind specific ligands. *Nature* **346,** 818–822.
5. Irvine, D., Tuerk, C., and Gold, L. (1991) SELEXION. Systematic evolution of ligands by exponential enrichment with integrated optimization by nonlinear analysis. *J. Mol. Biol.* **222,** 739–761.
6. Bouvet, P., Matsumoto, K., and Wolffe, A. P. (1995) Sequence-specific RNA recognition by the *Xenopus* Y-box proteins. An essential role for the cold shock domain. *J. Biol. Chem.* **270,** 28,297–28,303.
7. Ghisolfi-Nieto, L., Joseph, G., Puvion-Dutilleul, F., Amalric, F., and Bouvet, P. (1996) Nucleolin is a sequence-specific RNA-binding protein: characterization of targets on pre-ribosomal RNA. *J. Mol. Biol.* **260,** 34–53.
8. Triqueneaux, G., Velten, M., Franzon, P., Dautry, F., and Jacquemin-Sablon, H. (1999) RNA binding specificity of Unr, a protein with five cold shock domains. *Nucleic Acids Res.* **27,** 1926–1934.
9. Tsai, D. E., Harper, D. S., and Keene, J. D. (1991) U1-snRNP-A protein selects a ten nucleotide consensus sequence from a degenerate RNA pool presented in various structural contexts. *Nucleic Acids Res.* **19,** 4931–4936.
10. Bartel, D. P., Zapp, M. L., Green, M. R., and Szostak, J. W. (1991) HIV-1 Rev regulation involves recognition of non-Watson–Crick base pairs in viral RNA. *Cell* **67,** 529–536.
11. Gao, F. B., Carson, C. C., Levine, T., and Keene, J. D. (1994) Selection of a subset of mRNAs from combinatorial 3' untranslated region libraries using neuronal RNA-binding protein Hel-N1. *Proc. Natl. Acad. Sci. USA* **91,** 11,207–11,211.
12. Singer, B. S., Shtatland, T., Brown, D., and Gold, L. (1997) Libraries for genomic SELEX. *Nucleic Acids Res.* **25,** 781–786.
13. Pollock, R. and Treisman, R. (1990) A sensitive method for the determination of protein–DNA binding specificities. *Nucleic Acids Res.* **18**, 6197–6204.
14. Ringquist, S., Jones, T., Snyder, E. E., Gibson, T., Boni, I., and Gold, L. (1995) High-affinity RNA ligands to *Escherichia coli* ribosomes and ribosomal protein S1: comparison of natural and unnatural binding sites. *Biochemistry* **34**, 3640–3648.
15. Tuerk, C., Eddy, S., Parma, D., and Gold, L. (1990) Autogenous translational operator recognized by bacteriophage T4 DNA polymerase. *J Mol Biol* **213**, 749–761.

43

High DNA–Protein Crosslinking Yield with Two-Wavelength Femtosecond Laser Irradiation

Christoph Russmann, Rene Beigang, and Miguel Beato

1. Introduction

DNA–protein interactions are instrumental in the replication, repair, recombination, and expression of genetic information. Their study under physiological conditions requires rapid methods that do not cause extensive cell damage. Among the available techniques, crosslinking by ultraviolet (UV) laser irradiation seems best placed to fulfill these requirements, but the crosslinking yield is still relatively low with the commonly used pulses in the nanosecond (ns) range (*see* Chapter 27). To attain higher crosslinking yields, laser parameters must be adjusted to the underlying two-photon process, which strongly depends on pulse intensity and pulse length *(1)*. The highest crosslink yields are achieved using femtosecond (fs) laser pulses, but these also result in higher DNA damage *(1)*. In order to reduce the UV-induced DNA damage, we have applied a combination of UV pulses for the first excitation step and blue pulses for the second step *(2)*. This strategy has the advantage that the intensity of the UV pulse in the first step can be kept low, thus reducing DNA damage caused by the UV photons. High crosslinking efficiency can, however, still be attained by applying the second pulse with a very short time delay and at a visible wavelength too long to excite DNA bases from the ground state. This second pulse cannot alone damage DNA, but it can provide enough additional energy for the UV excited bases to pass their ionization threshold, leading to crosslinking.

1.1. The Method

Two-wavelength femtosecond (TWF) laser-induced DNA–protein crosslinking provides a significant improvement to classical laser crosslinking.

From: *Methods in Molecular Biology, vol. 148: DNA–Protein Interactions: Principles and Protocols, 2nd ed.*
Edited by: T. Moss © Humana Press Inc., Totowa, NJ

Optimized laser parameters lead to an increased crosslinking efficiency while minimizing DNA damage.

1.2. Practical Applications

The practical applications of TWF-laser crosslinking are similar to the applications of classical laser crosslinking (*see* Chapter 27).

1.3. The Basic Approach

The basic approach of TWF-laser crosslinking differs from classical UV-laser crosslinking in the simultaneous irradiation of the sample with two precisely timed fs pulses of different wavelengths. As a consequence, the laser equipment and the irradiation procedure have been modified. In the following, we describe the method in detail.

2 Materials

2.1. Lasers

As a source for the fs laser pulses, we have used a continuous-wave (cw)-argon-ion-laser (514.5 nm)-pumped cw-mode-locked Ti:sapphire laser (Spectra Physics, model Tsunami), which generates 150-fs-long pulses at a repetition rate of 82 MHz with an average power of up to 2 W in a wavelength range from 720 to 850 nm *(2)*. Infrared pulses around 798 nm were frequency doubled and tripled in two β-barium borate crystals in a "flexible harmonic generator" (FHG), and pulses at the second harmonic frequency 2ω (λ = 399 nm) and the third harmonic frequency 3ω (λ = 266 nm) were used to irradiate the sample (**Fig. 1**). The pulse length in the ultraviolet was increased to about 200 fs as a result of the dispersion of the nonlinear crystals. We used the following physical parameters in our approach (**Table 1**).

Using a variable delay line with a temporal resolution of 10 fs, the sample was irradiated with UV and blue in a two-step excitation process with a well-defined time delay. The time delay between the UV and the blue pulses has to be chosen accurately, as the lifetime of the intermediate singlet level of the base pairs is of the order of the pulse length. It was determined with a delay line using a β-barium borate crystal and a photodiode for difference-frequency generation (DFG). If the two laser pulses arrive simultaneously in the crystal, the generated difference-frequency signal is maximized. This method is used to calibrate the delay line. For our approach, the optimum temporal delay between the UV and the blue pulse was determined to be 300 fs. The laser pulse was focused onto the sample with a quartz lens. The energy of the applied radiation was measured with a standard energy meter (*see* **Note 1**).

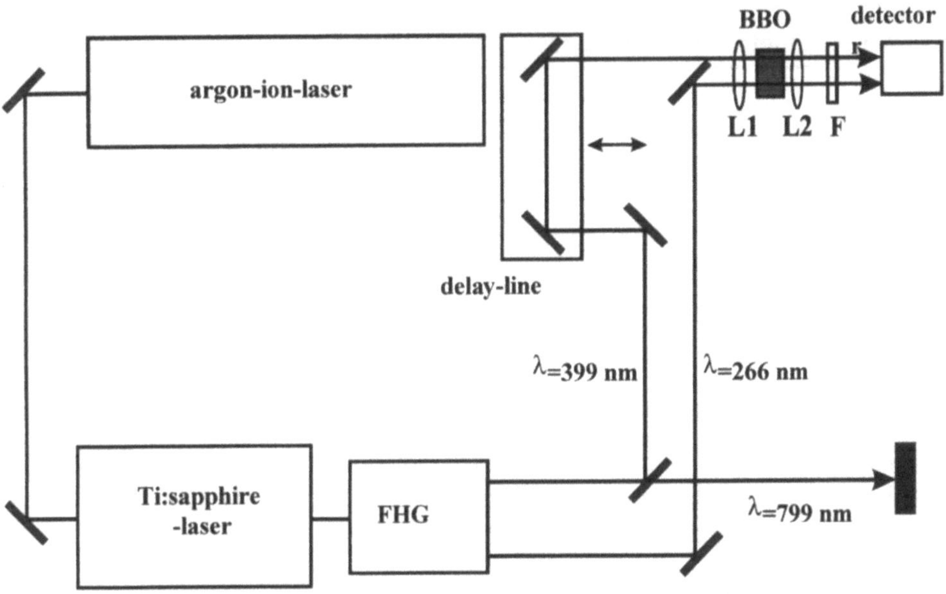

Fig. 1. The fs-laser equipment with delay line. The BBO crystal is used to determine the temporal delay between the blue and UV pulses. In the actual experiment, the combination of L1, L2, BBO, filter "F," and detector is replaced by the sample.

2.2. Reagents and Solutions for Irradiation and Analysis

1. The buffer should be optimal for DNA binding and exhibit minimal UV and visible absorbency. For our experiments we used the following buffer: 10 m*M* Tris-HCl, pH 7.5, 10% glycerol, 1 m*M* EDTA, and 300 m*M* NaCl.
2. Native polyacrylamide gel for binding studies: 5% acrylamide, acrylamide:*bis*-acrylamide ratio 40:1 in 0.05 *M* Tris base, 0.045 boric acid, and 0.5 m*M* EDTA.
3. Sodium dodecyl sulfate (SDS) sample buffer: 200 m*M* Tris-HCl, pH 6.8, 8% SDS, 0.4% bromophenol blue, 40% glycerol, 1% of 2-mercaptoethanol, and 1 *M* urea *(3)*.
4. SDS -polyacrylamide protein gel for crosslinking analysis: 8% acrylamide, (acrylamide:*bis*-acrylamide ratio 40:1).
5. Sequencing gel for analysis of DNA integrity: 6% acrylamide (acrylamide:bis-acrylamide ratio 40:1), 90 m*M* Tris-borate, pH 8.3, and 2 m*M* EDTA.

3. Methods

3.1. Irradiation Techniques

With the laser equipment described in **Subheading 2.1.**, it is possible to irradiate in vitro and in vivo. For comparison, the in vitro and in vivo samples should be irradiated under the following conditions. The molecular biological

Table 1
Laser Parameters for TWF-Laser Crosslinking

	Wavelength [J]	Pulse length Dt_p [fs]	Energy/pulse [nJ]	Total energy 2[J]
2. Harmonic	399	105	1.00	1.00
3. Harmonic	266	225	0.25	0.25

techniques used for treatment of the samples after the irradiation procedure are similar to the methods mentioned in Chapter 27.

3.1.1. In Vitro Crosslinking

The in vitro samples should be irradiated in microcentrifuge tubes (Eppendorf) with an unfocused beam (approx 5 mm in. diameter) of the laser *(1,2)*. The irradiation time should controlled with a shutter.

3.1.2. In Vivo Crosslinking

For in vivo crosslinking, it is recommended to irradiate cells in a culture dish. The blue and UV laser pulses should be focused with a broadband mirror perpendicular to the cell dishes. The cell dishes can be moved under the laser beams with an *xy*-positioning table (**Fig. 2**).

The medium should have no dominant absorption band in the UV and the visible-wavelength range. In principle, one can irradiate samples in Eppendorf tubes, but one should keep in mind that the maximal crosslink yield is a critical function of the optical density and the thickness of the sample (*see* **Note 2**).

3.2. Determination of the Optimal Irradiation Conditions

To optimize the irradiation conditions, it is necessary to determine the influence of various parameters on the "effective crosslinking yield" (e.g., the fraction of primer extendable or intact DNA multiplied by the crosslinking yield) *(1,2)*.

3.2.1. Determination of the Binding Efficiency by Electrophoretic Mobility Shift Assay

1. To determine the binding efficiency, different amounts of purified DNA-binding protein are incubated for 20 min under appropriate buffer conditions with a $[P^{32}]$-labeled DNA fragment that encompasses the binding site of the protein, bovine serum albumin and poly (dI–dC) to reduce unspecific binding (e.g., *see* Chapter 2).
2. The samples are separated on a native 5% native polyacrylamide gel and the gel dried onto Whatman paper.

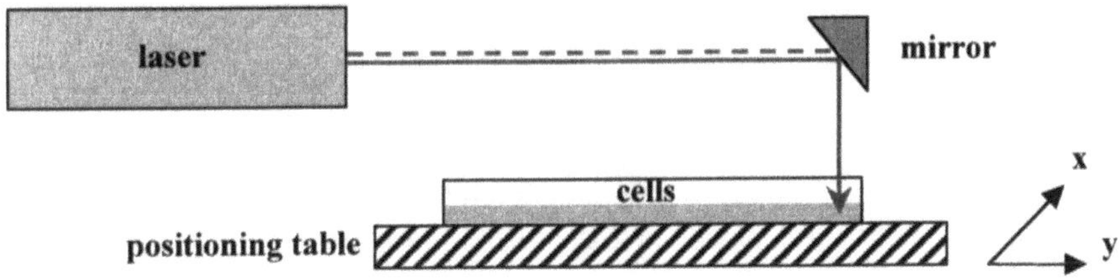

Fig. 2. Scanning cell dishes with the TWF-laser beam.

3. The dried gel is analyzed with a PhosphorImager™ (Molecular Dynamics). For optimal crosslinking, a binding efficiency of 50–70% should be reached.

3.2.2. Determination of the Crosslinking Yield

1. The DNA–protein complexes are irradiated as described in **Subheading 3.1.**.
2. The samples are heated for 5 min to 100°C with 1/3 vol SDS sample buffer and loaded onto a 8% SDS–polyacrylamide gel containing 1 *M* urea (*see* **Note 3**). After electrophoresis, the gel is dried and analyzed with a PhosphorImager.
3. The fraction of shifted DNA sample (i.e., crosslinked to protein, multiplied by the binding efficiency, as determined in the band-shift assay) is the "crosslinking yield."

3.2.3. Determination of the Integrity of DNA and Calculation of the Effective Crosslinking Yield

1. A plasmid containing the region of DNA to be analyzed is irradiated, restricted near one end of the insert, and analyzed by 30 cycles of primer extension across the insert toward the site of restriction using *Taq* polymerase and a suitable cycle of extension and denaturation.
2. The reaction products are visualized on a 6% sequencing gel followed by quantitative analysis of the dried gel on the PhosphorImager.
3. The amount of full-length extension product obtained from the irradiated DNA is divided by the amount of full-length extension products obtained from non-irradiated DNA to yield the fraction of intact or primer extendable DNA (*see* **Note 4**). This value does not represent a stringent measurement of DNA damage.
4. The fraction of intact DNA is multiplied by the crosslinking yield to obtain the "effective crosslinking yield."

3.3. Example

Using recombinant nuclear factor 1 (NF1) and the MMTV promoter region containing one degenerated NF1-binding site under conditions yielding 60% binding efficiency, the crosslinking yield was 30% with a single fs UV pulse and 30% with the TWF method *(2)*. However, the increased fraction of intact DNA using TWF laser crosslinking results in a threefold higher effective crosslinking yield with the TWF method (14%) as compared with a single fs UV pulse (4.8%).

4. Notes

1. The laser equipment used in the first experiments for TWF laser crosslinking is rather complex and sophisticated. Molecular biologists interested in using this method are best advised to contact a laser laboratory. However, the recent development of compact, diode-laser-pumped, all-solid-state femtosecond lasers and the use of new nonlinear materials for frequency conversion, like quasi-phase-matched crystals, will eventually result in reliable easy-to-operate femtosecond systems at a reasonable price. These systems are well-suited for routine applications of this new crosslinking method in molecular biology laboratories.
2. The irradiated samples should be as thin as possible to avoid dispersion effects, as these cause a change of the temporal distance between the blue and the UV pulses. The velocity of the positioning table should be precisely tunable. Changes in velocity can then be used to vary the applied energy per square millimeter of cell layer. This is an important parameter for increasing the in vivo crosslinking yield. However, irradiation should be fast enough to avoid drying of the cells.
3. Many crosslinked protein-DNA complexes do not enter a standard SDS–protein gel when solubilized in SDS–sample buffer without urea *(3)*. Probably, urea prevents the aggregation of crosslinked material.
4. The amount of full-length extension product is only an operational value, depending on the irradiation conditions and on the length of the template. One should keep in mind that this parameter is used to optimize the crosslink procedure and does not represent a stringent measurement of DNA damage.

References

1. Rubmann, C., Truss, M., Fix, A., Naumer, C., Herrmann, T., Schmitt, J., et al. (1997) Lasercrosslinking of progesterone receptor to DNA using tuneable nanosecond, picosecond and femtosecond laser pulses. *Nucleic Acids Res.* **25(12),** 2478–2484.
2. Rubmann, C., Stollhof, J., Weib, C., Beigang, R., and Beato, M. (1998) Femtosecond two wavelength crosslinking of DNA–protein-complexes. *Nucleic Acids Res.* **26(17),** 3967–3971.
3. Ho, D. T., Sauvé, D. M., and Roberge, M. (1994) Detection and isolation of DNA-binding proteins using single-pulse ultraviolet laser crosslinking. *Anal. Biochem.* **218,** 248–254.

Appendix I

EMSA/Gel Shift Conditions:

Buffer Conditions	Acrylamide: Bisacrylamide Ratio	Reference Chapter
Tris-glycine: 50 m*M* Tris, 2,5 m*M* EDTA, 0,4 *M* glycine	39:1	Ch 2
Tris-acetate: 6.7 m*M* Tris-HCl, pH 7.5, 3.3 m*M* sodium acetate, 1 m*M* EDTA	—	Ch 7
0.5 × Tris-borate EDTA: 45 m*M* Tris, 45 m*M* boric acid, 1 m*M* EDTA	39:1	Ch 2
1 × Tris-borate EDTA: 0.089 *M* Tris-borate, pH 8.0, 1 m*M* EDTA	37.5:1	Ch 32
1 × Tris-borate $CaCl_2$: 0.089 *M* Tris-borate, pH 7.5 to 8.0 and up to 5 m*M* $CaCl_2$	37.5:1	Ch 32
Tris-HCl: 8 m*M* Tris-HCl (recycled)	39:1	Ch 4
HEPES-NaCl: 50 m*M* NaCl, 40 m*M* HEPES-NaOH, pH 7.5 and 5% v/v glycerol. Running buffer: 50 m*M* NaCl, 20 m*M* HEPES-NaOH, pH 7.5	30:0,8 (=37.5:1)	Ch 37

Appendix II

DNA-Modification/Cleavage Reagents

Reagent	MW	DNA Duplex Attack	Comments	Uses
1,10-Phenanthroline-Copper	284	Ribose via minor groove. TAT>CG>TA	Does not attack Z-form DNA and more weakly (35%) attacks A form DNA as compared to B form.	Footprinting.
DEPC	162	G + A ¨ C, unstacked bases	Out of plane attack on bases	DNA conformation, base destacking, bending.
Dimethyl Sulphate (DMS)	126	N7-G via major groove N3-A via minor groove		Footprinting Interference
DNase I	40k	Independent attack on phosphodiester bonds from minor groove face of DNA	Sequence specific DNA cleavage	Footprinting.
Ethyl-nitroso-urea	117	60–65% Non-esterified oxygen of phosphodiester (> T-O2 (minor groove) =G-O6 (major groove) > T-O4 (major groove) >> C-O2 (minor groove))	Low sequence specificity	Interference
Exonuclease III	28k	3 terminal attack	Extremities of DNA-protein complexes	"Footprinting"

Continued...

Hydroxy-radical	17	Sugar backbone	Little sequence preference	Footprinting Interference
Osmium tetroxide	254	Unstacked T	Very specific for certain, as yet poorly defined DNA conformations including overwinding	DNA conformation
Permanganate	119	T » G, C, A	Out of plane attack on 5,6 double bond of T	DNA conformation, melting, base destacking, bending.
Diffusible singlet oxygen	16	Unstaked/unwound DNA bases	Little sequence preference in duplex DNA (though as free bases, G is by far the prefered target)	DNA conformation
uranyl(VI) ion (UO_2^{2+})	268	Phosphates-photolysis	Little sequence dependence	Footprinting of both macromolecules and drugs.
UV irradiation	_	TT >> CT, TC, or CC	Products: i) cyclobutane pyrimidine dimer (CBD). ii) pyrimidine (6-4) pyrimidone photoproduct (6-4PP)	Footprinting. Note, protein does not prevent access but changes DNA reactivity locally.

Index

MIX
Papier aus verantwortungsvollen Quellen
Paper from responsible sources
FSC® C105338

If you have any concerns about our products,
you can contact us on
ProductSafety@springernature.com

In case Publisher is established outside the EU,
the EU authorized representative is:
Springer Nature Customer Service Center GmbH
Europaplatz 3, 69115 Heidelberg, Germany

Printed by Libri Plureos GmbH
in Hamburg, Germany